Microsoft® SQL Server® 2019

A Beginner's Guide

Seventh Edition

Dušan Petković

New York   Chicago   San Francisco
Athens   London   Madrid   Mexico City
Milan   New Delhi   Singapore   Sydney   Toronto

Cataloging-in-Publication Data is on file with the Library of Congress

McGraw-Hill Education books are available at special quantity discounts to use as premiums and sales promotions, or for use in corporate training programs. To contact a representative, please visit the Contact Us pages at www.mhprofessional.com.

Microsoft® SQL Server® 2019: A Beginner's Guide, Seventh Edition

1 2 3 4 5 6 7 8 9   LCR   23 22 21 20 19

ISBN   978-1-260-45887-9
MHID   1-260-45887-3

Sponsoring Editor Lisa McClain	**Technical Editor** Todd Meister	**Production Supervisor** James Kussow
Editorial Supervisor Janet Walden	**Copy Editor** William McManus	**Composition** MPS Limited
Project Manager Ishan Chaudhary, MPS Limited	**Proofreader** Claire Splan	**Illustration** MPS Limited
Acquisitions Coordinator Emily Walters	**Indexer** Ted Laux	**Art Director, Cover** Jeff Weeks

About the Author

Dušan Petković is a computer science professor at the Polytechnic in Rosenheim, Germany. He is the bestselling author of the six previous editions of this book and has written numerous articles for *SQL Server Magazine*.

About the Technical Editor

Todd Meister has been working in the IT industry for over 20 years. He's been a technical editor on over 75 titles ranging from SQL Server to the .NET Framework. Besides technical editing books, he is the Assistant Vice President/Chief Enterprise Architect at Ball State University in Muncie, Indiana. He lives in central Indiana with his wife, Kimberly, and their five phenomenal children.

Acknowledgments

I would like to acknowledge the important contributions of my editor, Lisa McClain, my technical editor, Todd Meister, and my copy editor, Bill McManus.

Contents at a Glance

Contents

Introduction

Microsoft SQL Server is a database system that comprises many components, including the Database Engine, Analysis Services, Reporting Services, SQL Server Graph Databases, SQL Server Machine Learning Services, and several other components. The following are a few of the reasons SQL Server is the best choice for a broad spectrum of end users and database programmers building business applications:

- SQL Server is certainly the best system for Windows operating systems, because of its tight integration (and low pricing). Because the number of installed Windows systems is enormous and still increasing, SQL Server is a widely used database system.

- As of SQL Server 2017, Microsoft supports other operating systems as well: several Linux distributions, such as Ubuntu and Red Hat Linux, as well as macOS.

- The Database Engine, as the relational database system component, is the easiest database system to use. In addition to the Database Engine's well-known Windows-style user interface, Microsoft offers several different tools to help you create database objects, tune your database applications, and manage system administration tasks.

Generally, SQL Server isn't only a relational database system. It is a platform that not only manages structured, semistructured, unstructured, and graph data but also offers comprehensive, integrated operational and analysis software that enables organizations to reliably manage mission-critical information.

Goals of the Book

Microsoft SQL Server 2019: A Beginner's Guide follows six previous editions that covered SQL Server 7, 2000, 2005, 2008, 2012, and 2016. SQL Server has evolved significantly since I wrote the first edition over 20 years ago, but my overarching goal has remained the same: to provide a comprehensive introduction to SQL Server that is friendly to beginners.

Generally, all SQL Server users who want to get a good understanding of this database system and to work successfully with it will find this book very helpful. If you are a new SQL Server user but understand SQL, read the section "Differences Between SQL and Transact-SQL Syntax" later in this introduction.

This book addresses users of different components of SQL Server. For this reason, it is divided into several parts: The first three parts are most useful to users who want to learn more about Microsoft's relational database component called the Database Engine. The fourth part of the book is dedicated to Business Intelligence (BI) users who use Analysis Services and/

or Reporting Services. The fifth part of the book provides insight for users who want to learn about SQL Server features beyond those specific to relational data, such as integration of the JSON format with the relational engine, support for temporal and spatial data, and support for graph databases as an integral part of SQL Server. The last part of the book describes SQL Server Machine Learning Services, which provides the integration of the R system and the Python programming language with SQL Server

SQL Server 2019 New Features Described in the Book

SQL Server 2019 has a lot of new features, and almost all of them are discussed in this book. For each feature, at least one running example is provided to enable you to understand that feature better. The following table lists and summarizes the most important new features and identifies the chapter in which each is covered.

Feature	Summary
SQL Server runs on Windows, Linux, and Docker containers	Prior to SQL Server 2017, the only platform on which SQL Server could be installed was Windows. Now, Microsoft also supports the installation of SQL Server on Linux distributions and macOS. The installation on different Linux distributions, such as Ubuntu and Red Hat Linux, can be either direct or by using Docker containers. The installation of SQL Server on macOS can be done only via Docker containers. For purposes of demonstration, Chapter 2 shows you how to install SQL Server directly on Ubuntu.
Azure Data Studio	Azure Data Studio has been developed to support DBAs and users who do not use Windows as a platform for the Database Engine. This tool enables you to manage the instances of your database system as well as retrieve data from existing databases. Chapter 3 describes Azure Data Studio in detail.
SQL Data Discovery and Classification	Beginning with Version 17.5, SQL Server Management Studio supports a new application called Data Discovery and Classification, which is a set of services that is used to discover, classify, and report sensitive data in the database. Sensitive data includes business, financial, and healthcare information. The Data Discovery and Classification tool enables you to protect your sensitive data by meeting the data privacy standards and by controlling access to the data. This tool is described in detail in Chapter 3.
Resumable online indices	If you have a limited maintenance window, or large indices that take a long time to rebuild, you can now pause your online index rebuild operations and restart them later to complete them. With resumable online index rebuilds, DBAs will also be able to restart a failed online index rebuild operation. This feature is described in Chapter 10.
New utility tool for Linux distributions and Windows	**mssql-cli** is a new interactive command-line query tool for the Database Engine. It is an open source, cross-platform tool that can be used as an alternative to the **sqlcmd** utility. This tool is described in Chapter 15.

Automatic plan tuning	As of the release of SQL Server 2017, the Database Engine can retrieve the information captured in Query Store and make decisions automatically, forcing the last good execution plan. This feature, called automatic plan tuning, is described in Chapter 20.
Integration of SQL Server Data Tools with Visual Studio 2019	With Visual Studio 2019, the required functionality to enable Analysis Services and Reporting Services projects has been moved from SQL Server Data Tools (SSDT) into Visual Studio (VS). The support for these project types is provided through two extensions in Visual Studio: Microsoft Analysis Services Projects and Reporting Services Projects. Therefore, stand-alone SSDT installation no longer is required. Chapter 23 describes the support of VS 2019 for Microsoft Analysis Services Projects, while Chapter 25 explains how Microsoft Reporting Services Projects can be created with VS 2019.
Approximate Query Processing	The APPROX_COUNT_DISTINCT function belongs to a group of functions in relation to Approximate Query Processing and is a counterpart to the existing COUNT DISTINCT function, which can be very exhaustive, especially for larger tables with millions of rows. This function is described in Chapters 24 and 28.
Adaptive Query Processing	Adaptive Query Processing describes a group of techniques that you can use to improve performance of queries in the Database Engine. All features belonging to Adaptive Query Processing allow the Query Optimizer to generate more accurate query plans with better cardinality. There are altogether three features for adapting to application workload characteristics: Memory Grant Feedback, Adaptive Join, and Interleaved Execution. Adaptive Query Processing is described in Chapter 28.
Batch Mode on Rowstore	Batch Mode is an execution mode that targets analytical queries, which are characterized as scanning many rows and doing many complex aggregations and sorts. Starting with SQL Server 2017, Batch Mode has been reserved for queries that involve columnstore indices. Even if a query does not involve any table with a columnstore index, the Query Optimizer of SQL Server 2019 uses heuristics to decide whether to consider Batch mode or not. This feature is called Batch Mode on Rowstore. Chapter 28 discusses this feature in detail.
SQL Server Graph Databases	The SQL Server Graph Databases feature allows you to store graph data more efficiently than using Transact-SQL. Graph extensions are fully integrated in the Database Engine, meaning that the same storage engine is used to store relational as well as graph data. Chapter 31 describes this new feature.
SQL Server Machine Learning Services	SQL Server Machine Learning Services is a feature in SQL Server that enables you to run scripts written in different programming languages. You can use open source packages and frameworks for predictive analytics and machine learning. The scripts are executed in-database, meaning that you can use them to directly access data stored in relational form and managed by the Database Engine. Chapter 32 describes how R scripts can be implemented and used, while Chapter 33 is dedicated to the Python programming language.

Organization of the Book

The book has 33 chapters and is divided into six parts.

Part I, "Basic Concepts and Installation," provides an overview of database systems and explains how to install SQL Server 2019 and its components. It includes the following chapters:

- Chapter 1, "Relational Database Systems: An Introduction," discusses database systems in general and then narrows focus to relational database systems—the SQL Server Database Engine in particular. It presents the **sample** database used in examples throughout the book and introduces the Structured Query Language (SQL) that is used in relational database systems. It then provides an overview of two important database design concepts, normal forms and the entity-relationship model. The chapter also introduces the syntax conventions that are used in the rest of the book.

- Chapter 2, "Planning the Installation and Installing SQL Server," describes the first system administration task: the planning and installation of the overall system. Planning includes selection of the appropriate SQL Server edition and management tools, so the chapter begins with an overview of the editions and tools available. It then provides general recommendations to consider during the planning phase. Turning to the installation of SQL Server, which overall is a straightforward task, the chapter not only explains how to install SQL Server on Windows, but also describes the installation process on Ubuntu as an example of Microsoft's new support (as of SQL Server 2017) for several Linux distributions as well as macOS.

- Chapter 3, "Front-End Tools for the Database Engine," describes two components with similar functionality: SQL Server Management Studio (SSMS) and Azure Data Studio. The main difference between them is that the former runs exclusively on Windows, while the latter also can be used for Linux distributions and macOS. These components are presented early in the book in case you want to create database objects and query data but don't yet have a working knowledge of SQL.

Part II, "Transact-SQL Language," is intended for end users and application programmers of the Database Engine. It comprises the following chapters:

- Chapter 4, "SQL Components," describes the fundamentals of the most important part of a relational database system: a database language. For all such systems, there is only one language that counts: SQL. This chapter describes all components of SQL Server's own database language, Transact-SQL, and introduces the basic language concepts and data types. Finally, the chapter details the system functions and operators of Transact-SQL.

- Chapter 5, "Data Definition Language," describes all data definition language (DDL) statements of Transact-SQL. The DDL statements are presented in three groups, depending on their purpose. The first group contains all forms of the CREATE statement, which is used to create database objects. The second group contains all forms of the ALTER statement, which is used to modify the structure of some database objects. The third group contains all forms of the DROP statement, which is used to remove different database objects.

- Chapter 6, "Queries," discusses the most important Transact-SQL statement: SELECT. This chapter introduces you to database data retrieval and describes the use of simple and complex queries. Each SELECT clause is separately defined and explained with reference to the sample database.

- Chapter 7, "Modification of a Table's Contents," discusses the four Transact-SQL statements used for updating data: INSERT, UPDATE, DELETE, and MERGE. Each of these statements is explained through numerous examples.

- Chapter 8, "Stored Procedures and User-Defined Functions," introduces batches and routines. A *batch* is a sequence of Transact-SQL statements and procedural extensions. A *routine* can be either a stored procedure or a user-defined function (UDF). Routines can be used to create powerful programs that are stored on the server and can be reused. Because Transact-SQL is a complete computational language, all procedural extensions are inseparable parts of the language. Some stored procedures are written by users; others are provided by Microsoft and are referred to as system stored procedures.

- Chapter 9, "System Catalog," describes one of the most important parts of a database system: the system catalog, which consists of tables describing the structure of objects such as databases, base tables, views, and indices. The system catalog contains tables that are used to store the information concerning database objects and their relationships. The main characteristic of system tables of the Database Engine is that they cannot be accessed directly. The Database Engine supports several interfaces that you can use to query the system catalog.

- Chapter 10, "Indices," covers the first and most powerful method that database application programmers can use to tune their applications to get better system response and therefore better performance. This chapter describes the role of indices and gives you guidelines for how to create and use them. The end of the chapter introduces the special types of indices supported by the Database Engine.

- Chapter 11, "Views," explains how you create views, discusses the practical use of views (using numerous examples), and shows how views can be used with SELECT, INSERT, UPDATE, and DELETE statements.

- Chapter 12, "Security System of the Database Engine," provides answers to all your questions concerning security of data in the database. It addresses questions about data encryption, authorization (which user has been granted legitimate access to the database system), and authentication (which access privileges are valid for a particular user). It discusses three Transact-SQL statements—GRANT, DENY, and REVOKE— that provide the access privileges of database objects against unauthorized access. The end of the chapter explains how data changes can be tracked using the Database Engine.

- Chapter 13, "Concurrency Control," describes concurrency control in depth. The beginning of the chapter discusses the two different concurrency control models supported by the Database Engine, pessimistic and optimistic. Next, the chapter explains transactions, which are used by the database system to guarantee data consistency. It then discusses locking as a method to solve concurrency control problems. The end of the chapter explains two mechanisms for implementing concurrency control: isolation levels and row versioning.

- Chapter 14, "Triggers," describes the implementation of business logic using triggers. It discusses DML triggers and DDL triggers and shows their application areas. Each example in this chapter concerns a problem that you may face in your everyday life as a database application programmer.

Part III, "SQL Server: System Administration," describes all objectives of Database Engine system administration. It comprises the following chapters:

- Chapter 15, "System Environment of the Database Engine," discusses some internal issues concerning the Database Engine. It provides a detailed description of the Database Engine system databases, disk storage elements, and utilities. Additionally, several DBCC commands are explained, and the Policy-Based Management is introduced. (This component allows you to manage instances of the Database Engine.)

- Chapter 16, "Backup, Recovery, and System Availability," provides an overview of the fault-tolerance methods used to implement a backup strategy using either SQL Server Management Studio or corresponding Transact-SQL statements. The chapter first covers the reasons for data loss and specifies the different methods used to implement a backup strategy. The middle of the chapter discusses performing database backups and the restoration of databases. The chapter then describes in detail (and then compares) the following techniques (and a few others) for ensuring system availability: database mirroring, failover clustering, log shipping, and AlwaysOn. The chapter wraps up by describing how to maintain your database with the assistance of the Maintenance Plan Wizard.

- Chapter 17, "Automating System Administration Tasks," describes the Database Engine component called SQL Server Agent that enables you to automate certain system administration jobs, such as backing up data and using the scheduling and alert features to notify operators. This chapter also explains how to create jobs, operators, and alerts.

- Chapter 18, "Data Replication," provides an introduction to data replication, including concepts such as the publisher and subscriber. It introduces the different replication models and serves as a tutorial for how to configure publications and subscriptions using the existing wizards.

- Chapter 19, "Query Optimizer," describes the role and the work of the query optimizer. It explains in detail all the Database Engine tools (the SET statement, SQL Server Management Studio, and various dynamic management views) that can be used to edit the optimizer strategy. The end of the chapter provides optimizer hints.

- Chapter 20, "Performance Tuning," discusses performance issues and the tools for tuning the Database Engine that are relevant to daily administration of the system. After introductory notes concerning the measurements of performance, this chapter describes the factors that affect performance and presents tools for monitoring the performance of the Database Engine. The end of the chapter describes additional performance tools, including automatic plan tuning, a new feature that provides insights into potential query performance problems, recommends solutions, and automatically fixes identified problems.

- Chapter 21, "In-Memory OLTP," describes the Database Engine component that can be used to optimize OLTP (online transactional processing) queries. This means that a user can achieve significant performance improvements when specific database objects are stored entirely in memory instead of on disk. The chapter describes in detail the use of memory-optimized tables and compiled stored procedures, which are part of In-Memory OLTP.

Part IV, "SQL Server and Business Intelligence," discusses business intelligence (BI) and all related topics. The chapters in this part of the book introduce Microsoft Analysis Services and Microsoft Reporting Services. SQL/OLAP, existing optimization techniques concerning relational data storage, as well as Intelligent Query Processing are described in detail, too. This part includes the following chapters:

- Chapter 22, "Business Intelligence: An Introduction," introduces the concept of data warehousing. The first part of the chapter explains the differences between online transaction processing and data warehousing. The data store for a data warehousing process can be either a data warehouse or a data mart. Both types of data stores are discussed, and their differences are listed in the second part of the chapter. The end of the chapter explains different forms of data access.

- Chapter 23, "SQL Server Analysis Services," introduces the BI Semantic Model, which contains two submodels: Multidimensional and Tabular. The chapter explains the integration of both models into Visual Studio 2019 and provides an example showing the creation and deployment of each solution. It also introduces the MDX language for the Multidimensional model and the DAX language for the Tabular model. The end of the chapter presents security issues of SSAS.

- Chapter 24, "Business Intelligence and Transact-SQL," explains how you can use Transact-SQL to solve business intelligence problems. This chapter discusses the window construct, with its partitioning, ordering, and framing. It also discusses several operators, ranking functions, the TOP clause, and the PIVOT relational operator. The end of the chapter explains the new group of functions in relation to Approximate Query Processing.

- Chapter 25, "SQL Server Reporting Services," describes the Microsoft enterprise reporting solution. This component is used to design and deploy reports. The chapter discusses the development environment that you use to design and create reports and shows you different ways to deliver a deployed report. The end of the chapter describes performance and security issues related to SQL Server Reporting Services.

- Chapter 26, "Optimizing Techniques for Data Warehousing," describes three of the several specific optimization techniques that can be used especially in the area of business intelligence: data partitioning, star join optimization, and indexed views. The data partitioning technique called range partitioning is described. In relation to star join optimization, the role of bitmap filters in the optimization of joins is explained. The final part of the chapter explains the special index form called indexed views.

- Chapter 27, "Columnstore Indices," is devoted to the topic of columnar storage because of its importance. First, the chapter explains the benefits of this technique. After that, it discusses the technical details. The chapter also explains two forms of columnstore indices: clustered and nonclustered. The last part of the chapter, which is entirely new in this edition, compares performance of the columnstore index and a corresponding row store.

- Chapter 28, "Intelligent Query Processing," describes a group of sophisticated and related features concerning the Query Processor of the Database Engine. The main goal of these features is to improve performance of existing workloads while requiring only minimal implementation effort to adopt the features. The following features belong to Intelligent Query Processing: Adaptive Query Processing, Batch mode on Rowstore, Approximate Query Processing, Scalar UDF Inlining, and Table Variable Deferred Compilation.

Part V, "Beyond Relational Data," is dedicated to the following nonrelational topics: JSON, spatial and temporal data, and the newly added support for graph data. The following chapters are included in this part:

- Chapter 29, "JSON Integration in the Database Engine," discusses first JSON integration in SQL Server and Microsoft's set of data types and functions that supports storing JSON documents in the Database Engine, bridging the gap between JSON and relational data. The next part of the chapter explains how to present JSON documents as relational data, and vice versa, and how to query JSON documents. The last part, new to this edition, explains how JSON documents can be modified in the relational environment.

- Chapter 30, "Spatial and Temporal Data," discusses two special data forms: spatial data and temporal data. First, it describes the spatial data types, GEOMETRY and GEOGRAPHY, and presents several different standardized functions in relation to spatial data. The last part of the chapter explains temporal data. It introduces temporal tables, which are used to store temporal data, and explains some special properties of temporal data.

- Chapter 31, "SQL Server Graph Databases," discusses all extensions in the Database Engine to support graph data. It begins with a general introduction to graphs and graph databases. After that, it explains in detail the creation of node tables and edge tables. The bulk of the chapter then demonstrates how to query graph data, including how to do so using relational queries, and how to modify and edit data in graph databases.

Part VI, "Machine Learning," is dedicated to native support for the R and Python languages on SQL Server. R and Python integration includes base open source distributions, plus Microsoft-specific libraries for high-performance analytics. The following chapters are included in this part:

- Chapter 32, "SQL Server Machine Learning Services: R Support," covers the integration of the R language into the Database Engine. The first major section introduces the R language, explains generally how the R language is embedded in the SQL Server system, and defines the concept of a data frame in relation to R. The second major section shows you how to visualize data using data frames. The chapter wraps up with an exploration of predictive modeling with R and presents an example of solving linear regression problems with R.

- Chapter 33, "SQL Server Machine Learning Services: Python Support," discusses the integration of the Python language into the Database Engine. The first part of the chapter introduces the Python language and explains how the language is embedded in the SQL Server system. The second part shows you how to visualize data using data frames. The final part shows how predictive modeling with Python can be done, using an example of solving linear regression problems.

Almost all chapters include at their end numerous exercises that you can use to improve your knowledge concerning the chapter's content. All solutions to the given exercises can be found in the appendix of the book.

Changes from the Previous Edition

If you are familiar with the previous edition of this book, *Microsoft SQL Server 2016: A Beginner's Guide*, you should be aware that I have made significant changes in this edition. The following table gives you an outline of significant structural changes in the book (minor changes aren't listed).

Note that Chapter 30, "SQL Server Full-Text Search," from the previous edition of this book has been omitted in this edition. Also, the first part of Chapter 28 concerning XML has been omitted. First, the significance of Full-Text Search has diminished in the last few years. Second, there are two formats supported by SQL Server: XML and JSON. From my point of view, JSON is the format that is used to store and transport data. (All new features in SQL Server 2017 and 2019 that need transport of data from one source to the other are implemented to use JSON, and not XML.)

Chapter 2	The description of the installation of SQL Server on Ubuntu has been added.
Chapter 3	In addition to the previous edition's coverage of SQL Server Management Studio, the chapter now describes Azure Data Studio, which is mainly used to provide similar functionality as SSMS for SQL Server installed on Linux distributions and macOS. Additionally, the new feature of SSMS called Data Discovery and Classification is discussed in this chapter.
Chapter 4	The chapter is enhanced with the description of three new string functions: **trim**, **translate**, and **concat_ws**.
Chapter 6	The extension of the SELECT INTO statement to store a new table into a non-default filegroup is shown using an example.
Chapter 8	The chapter is enhanced with an example that explains how to use a cursor in SQL Server stored procedures.
Chapter 10	The chapter contains two new main sections: "Creation of Resumable Online Indices" and "Missing Indices." Resumable online indices allow you to pause the building of an index and then restart it later at the point it was paused. The latter section introduces four dynamic management views that can assist you in determining whether an index will be helpful to accelerate query performance.

Chapter 13	The chapter contains a new section, "Editing Information Concerning Transaction Logs."
Chapter 15	The section "tempdb Database" has been significantly enhanced, system utilities for Linux distributions are introduced, and a new section, "Editing Information Concerning Disk Storage," explains how to display information using a new dynamic management function.
Chapter 19	This chapter includes the description of the new dynamic management view called sys.dm_db_stats_histogram. Also, another example concerning filtered indices is given.
Chapter 20	This chapter has two new sections: "Extended Events" and "Automatic Tuning." The former introduces a tracing and troubleshooting framework that enables you to control the work of your system. The latter describes a new database feature that identifies potential query performance problems and *automatically* fixes them.
Chapter 23	The chapter has been rewritten, because the functionality of SQL Server Data Tools is now an integral part of Visual Studio 2019. Additionally, the chapter provides a short description of the Power BI Desktop component.
Chapter 24	Two new aggregate functions are described in this chapter: STRING_AGG and APPROX_COUNT_DISTINCT. The former concatenates the values of string expressions and places separator values between them, while the latter belongs to the group of approximate aggregate functions.
Chapter 25	The chapter has been rewritten, because the functionality of SQL Server Data Tools is now the integral part of Visual Studio 2019.
Chapter 26	The section "Editing Information Concerning Partitioning" has been added to this chapter.
Chapter 27	The major section "Columnstore Indices: Performance" has been added to this chapter.
Chapter 28	This chapter is new to this edition. It describes a group of new features in SQL Server 2019 that can be used for intelligent query processing.
Chapter 29	The major section "Updating JSON Documents" has been added to this chapter (formerly Chapter 28).
Chapter 31	This chapter is new to this edition. It describes the new component in SQL Server 2019 called SQL Server Graph Databases.
Chapter 32	This chapter is new to this edition. It describes the R support in SQL Server.
Chapter 33	This chapter is new to this edition. It describes the Python support in SQL Server.

Differences Between SQL and Transact-SQL Syntax

Transact-SQL, SQL Server's relational database language, has several nonstandardized properties that generally are not known to people who are familiar with SQL only:

- Whereas the semicolon (;) is used in SQL to separate two SQL statements in a statement group (and you will generally get an error message if you do not include the semicolon), in Transact-SQL, use of semicolons is strongly recommended but can be omitted.
- Transact-SQL uses the GO statement. This nonstandardized statement is generally used to separate statement groups from each other, whereas some Transact-SQL

statements (such as CREATE TABLE, CREATE INDEX, and so on) must be the only statement in the group.

- The USE statement, which is used very often in this book, changes the database context to the specified database. For example, the statement USE **sample** means that the statements that follow will be executed in the context of the **sample** database.

Downloading and Working with the Sample Databases

An introductory book like this requires a sample database that can be easily understood by each reader. For this reason, I used a very simple concept for my own sample database (fittingly named **sample**): it has only four tables with several rows each. On the other hand, its logic is complex enough to demonstrate the hundreds of examples included in the text of the book. The **sample** database that you will use in this book represents a company with departments and employees. Each employee belongs to exactly one department, which itself has one or more employees. Jobs of employees center on projects: each employee works at the same time for one or more projects, and each project engages one or more employees.

The tables of the **sample** database are presented in Chapter 1.

You can download the **sample** database either from my own website at Research Gate or from the website of this book's publisher. (The file to be downloaded additionally contains all examples from the book.) To download the file from Research Gate, go to https://www .researchgate.net. If you are not a member, click Join for Free, then select Not a Researcher at the end of the list. Click the Start Browsing button on the right side. Choose Publications and type **All Files SQL Server 2019 ABG**. Start the search process. When the title appears, click the title and download the zip file.

To download from McGraw-Hill's site, go to www.mhprofessional.com and, in the Search field, type the ISBN of the book (**1260458873**). Select the link for the book, and click the Downloads & Resources tab for a link to the Chapter Examples download.

Although the **sample** database can be used for many of the examples in this book, for some examples, tables with a lot of rows are necessary (to show optimization features, for instance). For this reason, in this book I additionally use several Microsoft-provided sample databases. The first one is **AdventureWorks**, which I use mainly in the third part of the book to show examples for OLTP application. The other one, **AdventureWorksDW**, is implemented especially for data warehousing and therefore is used in Part IV, "SQL Server and Business Intelligence." You can download both databases from the Microsoft Download center at https:// www.microsoft.com/en-us/download/details.aspx?id=49502.

In addition to those two databases, I briefly introduce two other sample databases provided by Microsoft: **AdventureWorksDW2016_EXT** and **WideWorldImporters**. Their aim is significantly different. The **AdventureWorksDW2016_EXT** database is very similar to the **AdventureWorksDW** database, but contains significantly more data. That way, you can use this sample database when you need to experiment with a huge data set. Examples 27.6 and 27.7 in Chapter 27 use this sample database, which you can download from https://github.com/ Microsoft/sql-server-samples/releases/tag/adventureworks. (In Chapter 27, the database has been renamed in **AdventureWorksDW_EXT**.)

The **WideWorldImporters** database is a new sample database from Microsoft that, in the long term, should replace all other Microsoft sample databases. Example 20.23 uses this database that you can download either from the same URL as AdventureWorksDW2016_EXT or from https://msdn.microsoft.com/library/mt748083(v=sql.1).aspx.

1 Relational Database Systems: An Introduction

In This Chapter

- Database Systems: An Overview
- Relational Database Systems
- Database Design
- Syntax Conventions

This chapter describes database systems in general. First, it discusses what a database system is, and which components it contains. Each component is described briefly, with a reference to the chapter in which it is described in detail. The second major section of the chapter is dedicated to relational database systems. It discusses the properties of relational database systems and the corresponding language used in such systems—Structured Query Language (SQL).

Generally, before you implement a database, you have to design it, with all its objects. The third major section of the chapter explains how you can use normal forms to enhance the design of your database, and also introduces the entity-relationship model, which you can use to conceptualize all entities and their relationships. The final section presents the syntax conventions used throughout the book.

Database Systems: An Overview

A database system is an overall collection of different database software components and databases containing the following parts:

- Database application programs
- Client components
- Database server(s)
- Databases

A database application program is special-purpose software that is designed and implemented by users or by third-party software companies. In contrast, client components are general-purpose database software designed and implemented by a database company. By using client components, users can access data stored on the same computer or a remote computer.

The task of a database server is to manage data stored in a database. Each client communicates with a database server by sending user queries to it. The server processes each query and sends the result back to the client.

In general, a database can be viewed from two perspectives, the users' and the database system's. Users view a database as a collection of data that logically belong together. For a database system, a database is simply a series of bytes, usually stored on a disk. Although these two views of a database are totally different, they do have something in common: the database system needs to provide not only interfaces that enable users to create databases and retrieve or modify data, but also system components to manage the stored data. Hence, a database system must provide the following features:

- Variety of user interfaces
- Physical data independence
- Logical data independence
- Query optimization
- Data integrity
- Concurrency control
- Backup and recovery
- Database security

The following sections briefly describe these features.

Variety of User Interfaces

Most databases are designed and implemented for use by many different types of users with varied levels of knowledge. For this reason, a database system should offer many distinct user interfaces. A user interface can be either graphical or textual. Graphical user interfaces (GUIs) accept user's input via the keyboard or mouse and create graphical output on the monitor. A form of textual interface, which is often used by database systems, is the command-line interface, where the user provides the input by typing a command with the keyboard and the system provides output by printing text on the computer monitor.

Physical Data Independence

Physical data independence means that the database application programs do not depend on the physical structure of the stored data in a database. This important feature enables you to make changes to the stored data without having to make any changes to database application programs. For example, if the stored data is previously ordered using one criterion, and this order is changed using another criterion, the modification of the physical data should not affect the existing database applications or the existing database *schema* (a description of a database generated by the data definition language of the database system).

Logical Data Independence

In file processing (using traditional programming languages), the declaration of a file is done in application programs, so any change to the structure of that file usually requires the modification of all programs using it. Database systems provide logical data independence—in other words, it is possible to make changes to the logical structure of the database without having to make any changes to the database application programs. For example, if the structure of an object named PERSON exists in the database system and you want to add an attribute to PERSON (say the address), you have to modify only the logical structure of the database, and not the existing application programs. (The application would have to be modified to utilize the newly added column.)

Query Optimization

Most database systems contain a subcomponent called an *optimizer* that considers a variety of possible execution strategies for querying the data and then selects the most efficient one. The selected strategy is called the *execution plan* of the query. The optimizer makes its decisions using considerations such as how big the tables are that are involved in the query, what indices exist, and what Boolean operator (AND, OR, or NOT) is used in the WHERE clause. (This topic is discussed in detail in Chapters 19 and 26.)

Data Integrity

One of the tasks of a database system is to identify logically inconsistent data and reject its storage in a database. (The date February 30 and the time 5:77:00 P.M. are two examples of inconsistent data.) Additionally, most real-life problems that are implemented using database systems have *integrity constraints* that must hold true for the data. (One example of an integrity constraint might be the company's employee number, which must be a five-digit integer.) The task of maintaining integrity can be handled by the user in application programs or by the database management system (DBMS). As much as possible, this task should be handled by the DBMS. (Data integrity is discussed in two chapters of this book: declarative integrity in Chapter 5 and procedural integrity in Chapter 14.)

Concurrency Control

A database system is a multiuser software system, meaning that many user applications access a database at the same time. Therefore, each database system must have some kind of control mechanism to ensure that several applications that are trying to update the same data do so in some controlled way. The following is an example of a problem that can arise if a database system does not contain such control mechanisms:

1. The owners of bank account 4711 at bank X have an account balance of $2000.

2. The two joint owners of this bank account, Mrs. A and Mr. B, go to two different bank tellers, and each withdraws $1000 *at the same time*.

3. After these transactions, the amount of money in bank account 4711 should be $0 and not $1000.

All database systems have the necessary mechanisms to handle cases like this example. Concurrency control is discussed in detail in Chapter 13.

Backup and Recovery

A database system must have a subsystem that is responsible for recovery from hardware or software errors. For example, if a failure occurs while a database application updates 100 rows of a table, the recovery subsystem must roll back all previously executed updates to ensure that the corresponding data is consistent after the error occurs. (See Chapter 16 for further discussion on backup and recovery.)

Database Security

The most important database security concepts are authentication and authorization. *Authentication* is the process of validating user credentials to prevent unauthorized users from using a system. Authentication is most commonly enforced by requiring the user to enter a (user) name and a password. This information is evaluated by the system to determine whether the user is allowed to access the system. This process can be strengthened by using encryption.

Authorization is the process that is applied after the identity of a user is authenticated. During this process, the system determines what resources the particular user can use. In other words, information about a particular entity is available only to principals that have permission to access that entity. (Chapter 12 discusses these concepts in detail.)

Relational Database Systems

The component of Microsoft SQL Server called the Database Engine is a relational database system. In contrast to earlier database systems (network and hierarchical), *relational database systems* are based upon the relational data model, which has a strong mathematical background.

NOTE A data model is a collection of concepts, their relationships, and their constraints that are used to represent data of a real-world problem.

The central concept of the relational data model is a relation—that is, a table. Therefore, from the user's point of view, a relational database contains tables and nothing but tables. In a table, there are one or more columns and zero or more rows. At every row and column position in a table there is always exactly one data value.

Working with the Book's sample Database

The **sample** database used in this book represents a company with departments and employees. Each employee in the example belongs to exactly one department, and each department has one or more employees. Jobs of employees center on projects: each employee works at the same time on one or more projects, and each project engages one or more employees.

The data of the **sample** database can be represented using four tables:

- department
- employee
- project
- works_on

Tables 1-1 through 1-4 show all the tables of the **sample** database.

dept_no	dept_name	location
d1	Research	Dallas
d2	Accounting	Seattle
d3	Marketing	Dallas

Table 1-1   The department Table

emp_no	emp_fname	emp_lname	dept_no
25348	Matthew	Smith	d3
10102	Ann	Jones	d3
18316	John	Barrimore	d1
29346	James	James	d2
2581	Elke	Hansel	d2
9031	Elsa	Bertoni	d2
28559	Sybill	Moser	d1

Table 1-2   The employee Table

project_no	project_name	budget
p1	Apollo	120000
p2	Gemini	95000
p3	Mercury	186500

Table 1-3   The project Table

emp_no	project_no	job	enter_date
10102	P1	Analyst	2016.10.1
10102	P3	Manager	2018.1.1
25348	P2	Clerk	2017.2.15
18316	P2	NULL	2017.6.1
29346	P2	NULL	2016.12.15
2581	P3	Analyst	2017.10.15
9031	P1	Manager	2017.4.15
28559	P1	NULL	2017.8.1
28559	P2	Clerk	2018.2.1
9031	P3	Clerk	2016.11.15
29346	P1	Clerk	2017.1.4

Table 1-4   The works_on Table

The **department** table represents all departments of the company. Each department has the following attributes:

department (dept_no, dept_name, location)

dept_no represents the unique number of each department, **dept_name** is the name of each department, and **location** is the location of the corresponding department.

The **employee** table represents all employees working for the company. Each employee has the following attributes:

employee (emp_no, emp_fname, emp_lname, dept_no)

emp_no represents the unique number of each employee. **emp_fname** and **emp_lname** are the first name and last name of each employee, respectively. Finally, **dept_no** is the number of the department to which the employee belongs.

Each project of the company is represented in the **project** table. This table has the following columns:

project (project_no, project_name, budget)

project_no represents the unique number of each project. **project_name** and **budget** specify the name and the budget of each project, respectively.

The **works_on** table specifies the relationship between employees and projects. It has the following columns:

works_on (emp_no, project_no, job, enter_date)

emp_no specifies the employee number and **project_no** specifies the number of the project on which the employee works. The combination of data values belonging to these two columns is always unique. **job** and **enter_date** specify the task and the starting date of an employee in the corresponding project, respectively.

Using the **sample** database, it is possible to describe some general properties of relational database systems:

- Rows in a table do not have any particular order.
- Columns in a table do not have any particular order.
- Every column must have a unique name within a table. On the other hand, columns from different tables may have the same name. (For example, the **sample** database has a **dept_no** column in the **department** table and a column with the same name in the **employee** table.)
- Every single data item in the table must be single valued. This means that in every row and column position of a table there is never a set of multiple data values.
- For every table, there is usually one column with the property that no two rows have the same combination of data values for all table columns. In the relational data model, such an identifier is called a *candidate key*. If there is more than one candidate key within a table, the database designer designates one of them as the *primary key* of the

table. For example, the column **dept_no** is the primary key of the **department** table; the columns **emp_no** and **project_no** are the primary keys of the tables **employee** and **project**, respectively. Finally, the primary key for the **works_on** table is the combination of the columns **emp_no** and **project_no**.

- In a table, there are never two identical rows. (This property is only theoretical; the Database Engine and all other relational database systems generally allow the existence of identical rows within a table.)

SQL: A Relational Database Language

The SQL Server relational database language is called Transact-SQL (T-SQL). It is a dialect of the most important database language today: Structured Query Language (SQL). In contrast to traditional languages like C, C++, and Java, SQL is a set-oriented language. (The former are also called record-oriented languages.) This means that SQL can query many rows from one or more tables using just one statement. This feature is one of the most important advantages of SQL, allowing the use of this language at a logically higher level than the level at which traditional languages can be used.

Another important property of SQL is its nonprocedurality. Every program written in a procedural language (C, C++, Java) describes *how* a task is accomplished, step by step. In contrast to this, SQL, as any other nonprocedural language, describes *what* it is that the user wants. Thus, the system is responsible for finding the appropriate way to solve users' requests.

SQL contains two sublanguages: a data definition language (DDL) and a data manipulation language (DML). DDL statements are used to describe the schema of database tables. The DDL contains three generic SQL statements: CREATE **object**, ALTER **object**, and DROP **object**. These statements create, alter, and remove database objects, such as databases, tables, columns, and indexes. (These statements are discussed in detail in Chapter 5.)

In contrast to the DDL, the DML encompasses all operations that manipulate the data. There are always four generic operations for manipulating the database: retrieval, insertion, deletion, and modification. The retrieval statement SELECT is described in Chapter 6, while the INSERT, DELETE, and UPDATE statements are discussed in detail in Chapter 7.

Database Design

Designing a database is a very important phase in the database life cycle, which precedes all other phases except the requirements collection and the analysis. If the database design is created merely intuitively and without any plan, the resulting database will most likely not meet the user requirements concerning performance. Another consequence of a bad database design is superfluous data redundancy, which in itself has two disadvantages: the existence of data anomalies and the use of an unnecessary amount of disk space.

Normalization of data is a process during which the existing tables of a database are tested to find certain dependencies between the columns of a table. If such dependencies exist, the table is restructured into multiple (usually two) tables, which eliminates any column dependencies. If one of these generated tables still contains data dependencies, the process of normalization must be repeated until all dependencies are resolved.

The process of eliminating data redundancy in a table is based upon the theory of functional dependencies. A *functional dependency* means that by using the known value of one column, the corresponding value of another column can always be uniquely determined.

(The same is true for column groups.) The functional dependencies between columns A and B is denoted by A ⇒ B, specifying that a value of column A can always be used to determine the corresponding value of column B. ("B is functionally dependent on A.")

Example 1.1 shows the functional dependency between two attributes of the table **employee** in the **sample** database.

Example 1.1
emp_no ⇒ emp_lname

By having a unique value for the employee number, the corresponding last name of the employee (and all other corresponding attributes) can be determined. This kind of functional dependency, where a column is dependent upon the primary key of a table, is called *trivial* functional dependency.

Normal Forms

Normal forms are used for the process of normalization of data and therefore for the database design. In theory, there are at least five different normal forms, of which the first three are the most important for practical use. The third normal form for a table can be achieved by testing the first and second normal forms at the intermediate states, and as such, the goal of good database design can usually be fulfilled if all tables of a database are in the third normal form.

First Normal Form

First normal form (1NF) means that a table has no multivalued attributes or composite attributes. (A composite attribute contains other attributes and can therefore be divided into smaller parts.) All relational tables are by definition in 1NF, because the value of any column in a row must be *atomic*—that is, single valued.

Table 1-5 demonstrates 1NF using part of the **works_on** table from the **sample** database. The rows of the **works_on** table could be grouped together, using the employee number. The resulting Table 1-6 is not in 1NF because the column **project_no** contains a set of values (p1, p3).

emp_no	project_no	....
10102	p1	....
10102	p3	....
....	....	....

Table 1-5   Part of the works_on Table

emp_no	project_no	....
10102	(p1, p3)	....
....	....	....

Table 1-6   This "Table" Is Not in 1NF

Second Normal Form

A table is in second normal form (2NF) if it is in 1NF and there is no nonkey column dependent on a partial primary key of that table. This means if (A,B) is a combination of two table columns building the key, then there is no column of the table depending either on only A or only B.

For example, Table 1-7 shows the **works_on1** table, which is identical to the **works_on** table except for the additional column, **dept_no**. The primary key of this table is the combination of columns **emp_no** and **project_no**. The column **dept_no** is dependent on the partial key **emp_no** (and is independent of **project_no**), so this table is not in 2NF. (The original table, **works_on**, is in 2NF.)

Third Normal Form

A table is in third normal form (3NF) if it is in 2NF and there are no functional dependencies between nonkey columns. For example, the **employee1** table (see Table 1-8), which is identical to the **employee** table except for the additional column, **dept_name**, is not in 3NF, because for every known value of the column **dept_no** the corresponding value of the column **dept_name** can be uniquely determined. (The original table, **employee**, as well as all other tables of the **sample** database are in 3NF.)

Entity-Relationship Model

The data in a database could easily be designed using only one table that contains all data. The main disadvantage of such a database design is its high redundancy of data. For example, if your database contains data concerning employees and their projects (assuming each employee works at the same time on one or more projects, and each project engages one or

emp_no	project_no	job	enter_date	dept_no
10102	p1	Analyst	2016.10.1	d3
10102	p3	Manager	2018.1.1	d3
25348	p2	Clerk	2017.2.15	d3
18316	p2	NULL	2017.6.1	d1
....	....	....	....	....

Note: Every table with a one-column primary key is always in 2NF.

Table 1-7   The works_on1 Table

emp_no	emp_fname	emp_lname	dept_no	dept_name
25348	Matthew	Smith	d3	Marketing
10102	Ann	Jones	d3	Marketing
18316	John	Barrimore	d1	Research
29346	James	James	d2	Accounting
....	....	....	....	....

Table 1-8   The employee1 Table

more employees), the data stored in a single table contains many columns and rows. The main disadvantage of such a table is that data is difficult to keep consistent because of its redundancy.

The *entity-relationship (ER) model* is used to design relational databases by removing all existing redundancy in the data. The basic object of the ER model is an *entity*—that is, a real-world object. Each entity has several *attributes*, which are properties of the entity and therefore describe it. Based on its type, an attribute can be

- **Atomic (or single valued)**   An atomic attribute is always represented by a single value for a particular entity. For example, a person's marital status is always an atomic attribute. Most attributes are atomic attributes.

- **Multivalued**   A multivalued attribute may have one or more values for a particular entity. For example, **Location** as the attribute of an entity called ENTERPRISE is multivalued, because each enterprise can have one or more locations.

- **Composite**   Composite attributes are not atomic because they are assembled using some other atomic attributes. A typical example of a composite attribute is a person's address, which is composed of atomic attributes, such as **City**, **Zip**, and **Street**.

The entity PERSON in Example 1.2 has several atomic attributes, one composite attribute, **Address**, and a multivalued attribute, **College_degree**.

Example 1.2
PERSON (Personal_no, F_name, L_name, Address (City,Zip,Street),{College_degree})

Each entity has one or more key attributes that are attributes (or a combination of two or more attributes) whose values are unique for each particular entity. In Example 1.2, the attribute **Personal_no** is the key attribute of the entity PERSON.

Besides entity and attribute, *relationship* is another basic concept of the ER model. A relationship exists when an entity refers to one or more other entities. The number of participating entities defines the degree of a relationship. For example, the relationship **works_on** between entities EMPLOYEE and PROJECT has degree two.

Every existing relationship between two entities must be one of the following three types: 1:1, 1:N, or M:N. (This property of a relationship is also called *cardinality ratio*.) For example, the relationship between the entities DEPARTMENT and EMPLOYEE is 1:N, because each employee belongs to exactly one department, which itself has one or more employees. Also, the relationship between the entities PROJECT and EMPLOYEE is M:N, because each project engages one or more employees and each employee works at the same time on one or more projects.

A relationship can also have its own attributes. Figure 1-1 shows an example of an ER diagram. (The ER diagram is the graphical notation used to describe the ER model.) Using this notation, entities are modeled using rectangular boxes, with the entity name written inside the box. Attributes are shown in ovals, and each attribute is attached to a particular entity (or relationship) using a straight line. Finally, relationships are modeled using diamonds, and entities participating in the relationship are attached to it using straight lines. The cardinality ratio of each entity is written on the corresponding line.

Syntax Conventions

This book uses the conventions shown in Table 1-9 for the syntax of the Transact-SQL statements and for the indication of the text.

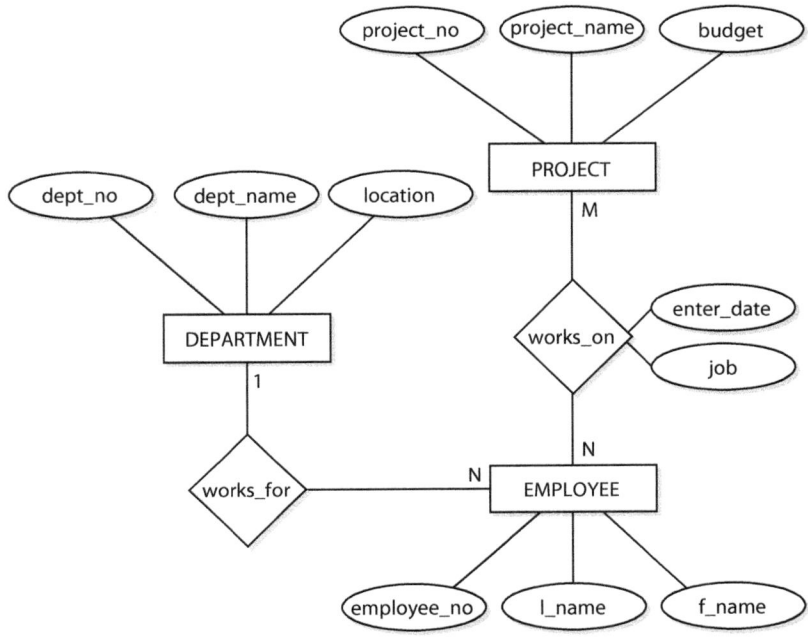

Figure 1-1   Example of an ER diagram

Convention	Indication
Italics	New terms or items of emphasis.
UPPERCASE	Transact-SQL keywords—for example, CREATE TABLE. Additional information about the keywords of the Transact-SQL language can be found in Chapter 5.
Lowercase	Variables in Transact-SQL statements—for example, CREATE TABLE *tablename*. (The user must replace *tablename* with the actual name of the table.)
Var1 | var2	Alternative use of the items var1 and var2. (You may choose only one of the items separated by the vertical bar.)
{ }	Alternative use of more items. Example: { expression | USER | NULL }
[ ]	Optional item(s). Example: [FOR LOAD]
{ } ...	Item(s) that can be repeated any number of times. Example: {, @param1 typ1} ...
bold	Name of database object (database itself, tables, columns) in the text.
Default	The default value is always underlined. Example: ALL | DISTINCT

Note: In contrast to brackets and braces, which belong to syntax conventions, parentheses, ( ), belong to the syntax of a statement and must always be typed!

Table 1-9   Syntax Conventions

Summary

All database systems provide the following features:

- Variety of user interfaces
- Physical data independence
- Logical data independence
- Query optimization
- Data integrity
- Concurrency control
- Backup and recovery
- Database security

The next chapter shows you how to install SQL Server 2019.

Exercises

E.1.1   What does "data independence" mean and which two forms of data independence exist?

E.1.2   Which is the main concept of the relational model?

E.1.3   What does the **employee** table represent in the real world? And what does the row in this table with the data for Ann Jones represent?

E.1.4   What does the **works_on** table represent in the real world (and in relation to the other tables of the **sample** database)?

E.1.5   Let **book** be a table with two columns: **isbn** and **title**. Assuming that **isbn** is unique and there are no identical titles, answer the following questions:

- a. Is **title** a key of the table?
- b. Does **isbn** functionally depend on **title**?
- c. Is the **book** table in 3NF?

E.1.6   Let **order** be a table with the following columns: **order_no**, **customer_no**, **discount**. If the column **customer_no** is functionally dependent on **order_no** and the column **discount** is functionally dependent on **customer_no**, answer the following questions and explain in detail your answers:

- a. Is **order_no** a key of the table?
- b. Is **customer_no** a key of the table?

E.1.7 Let **company** be a table with the following columns: **company_no, location**. Each company has one or more locations. In which normal form is the **company** table?

E.1.8 Let **supplier** be a table with the following columns: **supplier_no, article, city**. The key of the table is the combination of the first two columns. Each supplier delivers several articles, and each article is delivered by several suppliers. There is only one supplier in each city. Answer the following questions:

    a. In which normal form is the **supplier** table?

    b. How can you resolve the existing functional dependencies?

E.1.9 Let R(A, B, C) be a relation with the functional dependency B $\Rightarrow$ C. (The underlined attributes A and B build the composite key, and the attribute C is functionally dependent on B.) In which normal form is the relation R?

E.1.10 Let R(A, B, C) be a relation with the functional dependency C $\Rightarrow$ B. (The underlined attributes A and B build the composite key, and the attribute B is functionally dependent on C.) In which normal form is the relation R?

2

Planning the Installation and Installing SQL Server

In This Chapter

- SQL Server Editions and Management Tools
- Planning Phase: General Recommendations
- Installation of SQL Server on Windows or Ubuntu

This chapter begins by introducing the various SQL Server editions and management tools, so that you can identify which edition is appropriate for your environment and which main management tools exist for the system.

Before you proceed to the actual installation of this database system, you need to develop an installation plan. Therefore, the next part of this chapter is dedicated to general recommendations during the planning phase.

The final part of the chapter describes installation processes for two different operating systems, starting with the installation of SQL Server on Microsoft Windows. For this process, you use the component called SQL Server Installation Center to install the system on your computer. The chapter then explains the process for installing SQL Server on Ubuntu.

NOTE This chapter describes the *basic* installation of SQL Server.

SQL Server Editions and Management Tools

As you plan your installation, you need to know which SQL Server editions exist so that you can choose the most appropriate one. This section describes your options and then introduces the various management tools that you can use in all SQL Server editions.

SQL Server Editions

Microsoft supports several editions of SQL Server, but to help you find the most likely candidate for your environment, I will explain only the most important ones:

- **Express Edition**   The lightweight version of SQL Server designed for use by application developers. For this reason, the product supports the Common Language Runtime (CLR), which allows you to develop different database objects using C# and Visual Basic. Also, you can download SQL Server Management Studio (SSMS), which enables you to easily manage a database. SQL Server Express is available as a free download.

- **Standard Edition**   Designed for small and medium-sized businesses. It supports up to four processors and includes the full range of BI functionality, including Analysis Services, Reporting Services, and Integration Services. This edition does not include some enterprise-based features from Enterprise Edition.

- **Enterprise Edition**   The special form of the SQL Server system that is intended for time-critical applications with a huge number of users. In contrast to Standard Edition, this edition contains additional features that can be useful for very high-end installations with symmetrical multiprocessors or clusters. The most important additional features of Enterprise Edition are data partitioning and online database maintenance.

- **Developer Edition**   Designed for developers to build and test any type of application with SQL Server. It includes all the functionality of Enterprise Edition, but is licensed only for use in development, testing, and demonstration. Each license of Developer Edition entitles one developer to use the software on as many systems as necessary; additional developers can use the software by purchasing additional licenses. For rapid deployment into production, the database system of Developer Edition can easily be upgraded to Enterprise Edition.

Management Tools

The following management tools, among others, are available for use in all SQL Server editions:

- **SQL Server Management Studio (SSMS)**   The administrator's primary tool for interacting with the SQL Server system on Windows is SQL Server Management Studio. Both administrators and end users can use this tool. Chapter 3 describes SSMS in detail.

- **Azure Data Studio**   The functionality of Azure Data Studio is similar to the functionality of SQL Server Management Studio. In contrast to SSMS, which is implemented only for Windows, Azure Data Studio runs on Linux, Windows, and macOS. The main difference between the tools is that SSMS is a mature tool, whereas Azure Data Studio is a new tool that currently supports only a subset of features already implemented in SSMS. Chapter 3 describes Azure Data Studio, too.

- **SQL Server Configuration Manager**   SQL Server Configuration Manager is a tool to manage the services associated with SQL Server, to configure the network protocols used by SQL Server, and to manage the network connectivity configuration from SQL Server client computers. SQL Server Configuration Manager is a Microsoft Management Console snap-in that is available from the Start menu. This tool will be used and explained in Chapter 5.

- **SQL Server Profiler**   SQL Server Profiler is a graphical tool that lets system administrators monitor and record database and server activities. The tool is described in detail in Chapter 20.
- **Database Engine Tuning Advisor**   The Database Engine Tuning Advisor allows you to automate the physical design of your databases. The tool is described together with SQL Server Profiler in Chapter 20.
- **SQL Server Data Tools**   SQL Server Data Tools (SSDT) transforms database development by introducing a model that spans all the phases of database development. In SQL Server 2019, SSDT is tightly integrated with Visual Studio 2019 and uses its capabilities to build, debug, and maintain databases.

Planning Phase: General Recommendations

During the installation process, you have to make many choices. As a general guideline, it is best to familiarize yourself with the effects of each option before installing your system. At the beginning, you should answer the following questions:

- Which operating system will be used?
- Which SQL Server components should be installed?
- Where will the root directory be stored?
- Should multiple instances of the Database Engine be used?
- Which authentication mode for the Database Engine should be used?

The following subsections discuss these topics.

Which Operating System Will Be Used?

In the past, SQL Server has run only on Windows. Since SQL Server 2017, the instances of the system can be installed and used additionally on Linux and in Docker containers, which opens the possibility of running SQL Server on macOS. With extension tools now available as native Linux and macOS applications, users can choose one of several operating systems available for running SQL Server. (This book describes the installation and usage of SQL Server on Windows and on the Linux derivative called Ubuntu.)

Which SQL Server Components Should Be Installed?

Before you start the installation process, you should know exactly which SQL Server components you want to install. Figure 2-1 shows a partial list of all the components. This is a preview of the Feature Selection page of the installation wizard, which you will see again when you install SQL Server later in this chapter.

There are two groups of features on the Feature Selection page: instance features and shared features. *Instance features* are the components that are installed once for each instance, meaning you have multiple copies of them (one for each instance). *Shared features* are features that are common across all instances on a given machine. Each of these shared features is

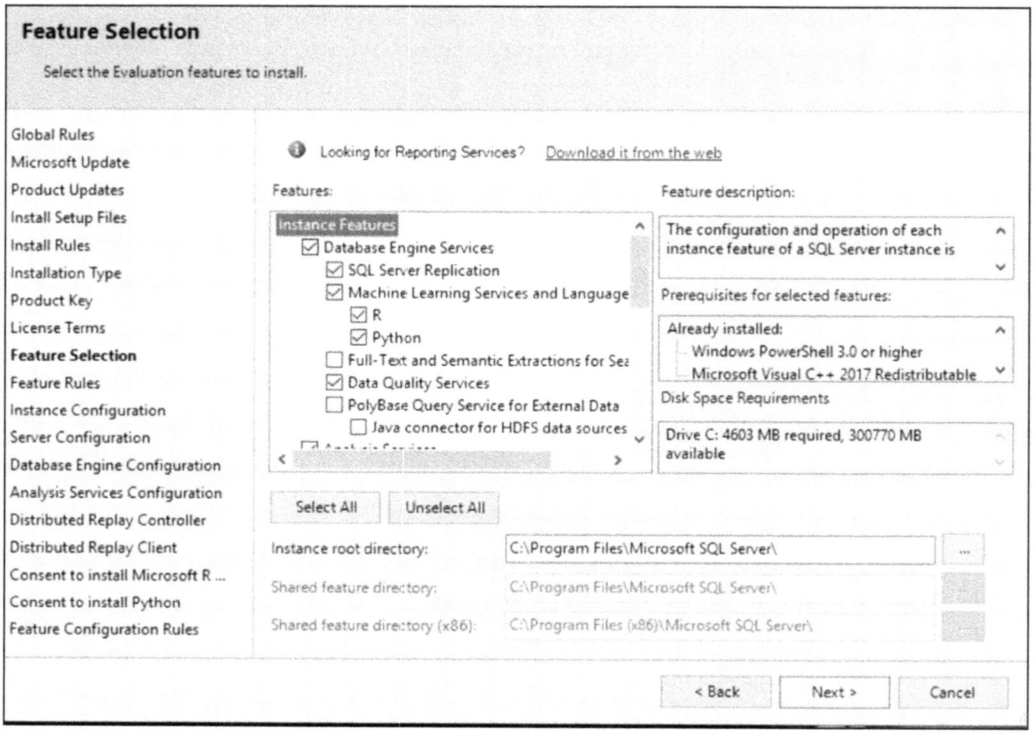

Figure 2-1   Preview of the Feature Selection page of the installation wizard

designed to be backward compatible with supported SQL Server versions that can be installed side by side. This section introduces only the instance features. For a description of the shared features, see Microsoft Docs.

The first item in the list of the instance features is Database Engine Services. The Database Engine is the relational database system of SQL Server. Parts II and III of this book describe different aspects of the Database Engine. The first instance feature under Database Engine Services, SQL Server Replication (see Figure 2-1), allows you to replicate data from one system to another. In other words, using data replication, you can achieve a distributed data environment. Detailed information on data replication can be found in Chapter 18.

The next instance feature is Machine Learning Services and Language Extensions. This feature lets you deploy R or Python packages to SQL Server, execute solutions in the context of SQL Server, and easily work with SQL Server data in R and/or Python. Detailed descriptions of R and Python integration in SQL Server can be found in Chapters 32 and 33, respectively.

The third instance feature under Database Engine Services is Full-Text and Semantic Extractions for Search (aka Full-Text Search). The Database Engine allows you to store structured data in columns of relational tables. By contrast, the unstructured data is primarily stored as text in file systems. For this reason, you will need different methods to retrieve information from unstructured data. Full-Text Search is a component of SQL Server that allows you to store and query unstructured data. (This component is described in Microsoft Docs.)

The next instance feature is Data Quality Services. These services are related to Data Quality Client. After that, the PolyBase Query Service is listed. PolyBase is a component of SQL Server that builds a gateway from SQL to Hadoop. This functionality has been available for several years in the Microsoft Analytics Platform System (APS) and is now an integrated part of SQL Server. (PolyBase will not be described in this book.)

The next category of instance features under Database Engine Services is Analysis Services (not shown in Figure 2-1), which is a component related to business intelligence (BI). Analysis Services is a group of services that is used to manage and query data that is stored in a data warehouse. (A data warehouse is a database that includes all corporate data that can be uniformly accessed by users.) Part IV of this book describes SQL Server and business intelligence in general, while Chapter 23 discusses Analysis Services in particular.

NOTE SQL Server comprises another business intelligence component called Reporting Services (SSRS), which allows you to create and manage reports. Since SQL Server 2017, Reporting Services has a separate, stand-alone installation process, which is described in Chapter 25.

Where Will the Root Directory Be Stored?

The root directory is where the Setup program stores all program files and those files that do not change as you use the SQL Server system. By default, the installation process stores all program files in the subdirectory Microsoft SQL Server, although you can change this setting during the installation process. Using the default name is recommended because it uniquely determines the version of the system.

Should Multiple Instances of the Database Engine Be Used?

With the Database Engine, you can install and use several different instances. An *instance* is a database server that does not share its system and user databases with other instances (servers) running on the same computer.

There are two instance types:

- Default
- Named

An instance is either the default (unnamed) instance, or it is a named instance. When SQL Server is installed in the default instance, it does not require a client to specify the name of the instance to make a connection. The client only has to know the server name.

Any instance of the database server other than the default instance is called a *named instance*. To identify a named instance, you have to specify its name as well as the name of the computer on which the instance is running. On one computer, there can be several named instances (in addition to the default instance). Additionally, you can configure named instances on a computer that does not have the default instance.

Although all instances running on a computer do not share most system resources (SQL Server and SQL Server Agent services, system and user databases, and registry keys), there are some components that are shared among them:

- SQL Server program group
- Development libraries

The existence of only one SQL Server program group on a computer also means that only one copy of each utility exists, which is represented by an icon in the program group. Therefore, each utility works with all instances configured on a computer.

You should consider using multiple instances if both of the following are true:

- You have different types of databases on your computer.
- Your computer is powerful enough to manage multiple instances.

The main purpose of multiple instances is to divide databases that exist in your organization into different groups. For instance, if the system manages databases that are used by different users (production databases, test databases, and sample databases), you should divide them to run under different instances. That way you can encapsulate your production databases from databases that are used by casual or inexperienced users.

A single-processor machine will not be the right hardware platform to run multiple instances of the Database Engine, because of limited resources. For this reason, you should consider the use of multiple instances only with multiprocessor computers.

Which Authentication Mode for the Database Engine Should Be Used?

In relation to the Database Engine, there are two different authentication modes:

- **Windows mode**   Specifies security exclusively at the operating system level—that is, it specifies the way in which users connect to the Windows operating system using their user accounts and group memberships.
- **Mixed mode**   Allows users to connect to the Database Engine using Windows authentication or SQL Server authentication. This means that some user accounts can be set up to use the Windows security subsystem, while others can use the SQL Server security subsystem in addition to the Windows security subsystem.

Microsoft recommends the use of Windows mode. (For details, see Chapter 12.)

Installation of SQL Server on Windows or Ubuntu

The following sections describe installation of SQL Server on two different operating systems: Windows and Ubuntu.

Planning the Installation on Windows

To start the SQL Server installation, mount the ISO image of the SQL Server software. After that, click the Setup.exe file. SQL Server contains a tool called Installation Center

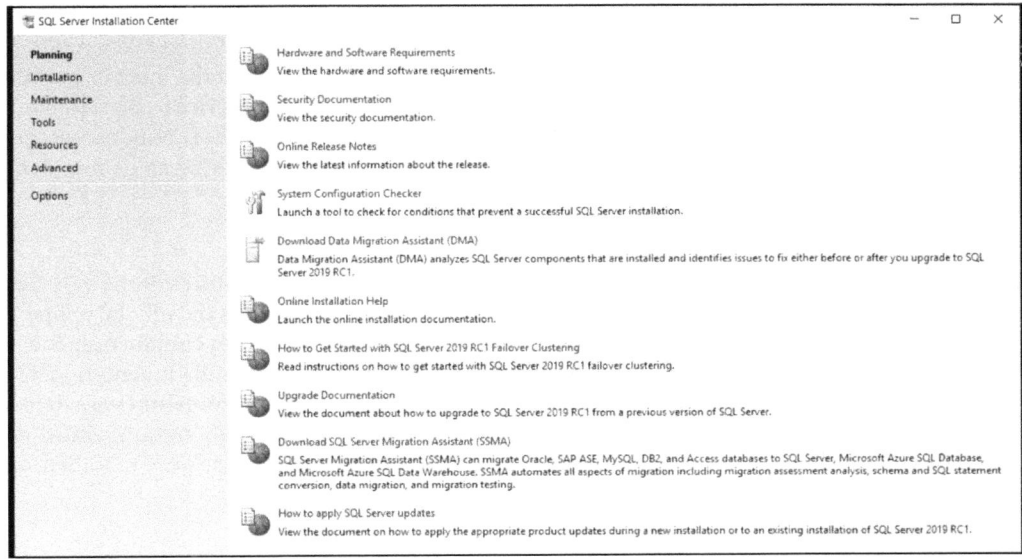

Figure 2-2   SQL Server Installation Center

(see Figure 2-2). This tool supports you during the planning, installation, and maintenance phases of your database system.

The first phase of Installation Center leads you through the process of planning the installation. As shown in Figure 2-2, when you click Planning, the following tasks, among others, can be executed:

- Hardware and Software Requirements
- Security Documentation
- Online Release Notes
- System Configuration Checker

The following subsections describe these tasks.

Hardware and Software Requirements

Microsoft recommends running SQL Server on computers with the NTFS or ReFS file system. Installing SQL Server on a computer with the FAT32 file system is supported by Microsoft but is not recommended, as it is less secure than the other two file systems.

Officially, the minimum requirement for main memory is 1MB. (The Express Edition needs only 512KB.) However, such a minimal configuration will not perform very well, and as a general guideline, main memory of your computer should be at least 2GB or more.

SQL Server requires a minimum of 6GB of available hard-disk space. Disk space requirements will vary depending on the SQL Server components that you install.

Security Documentation

When you click Security Documentation on the Planning page, the system takes you to the Microsoft page that discusses general security considerations. One of the most important security measures is to isolate services from each other. To isolate services, run separate SQL Server services under separate Windows accounts. (Chapter 12 discusses Windows accounts and other security aspects.) Information about all other security aspects can be found in Microsoft Docs.

Online Release Notes

There are two main sources to get information concerning all the features of the SQL Server system: Microsoft Docs and Online Release Notes. Microsoft Docs is the official online documentation delivered by Microsoft, whereas Online Release Notes contain only the newest information, which is not necessarily provided in the Microsoft Docs documentation. (The reason is that bugs and specific behavior issues affecting the system sometimes are detected after the official documentation is written and published.) It is strongly recommended that you read the Online Release Notes carefully to get a picture of features that were modified shortly before the delivery of the final release.

System Configuration Checker

One of the most important planning tasks is to check whether all conditions are fulfilled for a successful installation of the database system. When you click System Configuration Checker, the component called Setup Global Rules is automatically started. (The same tool is launched at the beginning of the installation phase, described next.) Setup Global Rules identifies problems that might occur when you install SQL Server support files. After finishing this task, the system shows you how many operations were checked and how many of them failed. All failures have to be corrected before the installation can continue.

Installing SQL Server on Windows

If you have ever installed a complex software product, you probably recognize that feeling of uncertainty that accompanies starting the installation. This feeling comes from the complexity of the product to be installed and the diversity of questions to be answered during the installation process. Because you may not completely understand the product, you (or the person who installs the software) may be less than confident that you can give accurate answers for all the questions that the Setup program asks to complete its tasks. This section will help you to find your way through the Windows SQL Server installation by giving you answers to most of the questions that you are likely to encounter.

NOTE   The installation process of SQL Server 2019 is slightly different, depending on whether you do a fresh installation or an installation on top of an existing one. For instance, the navigation menu on the left side of the Installation Center will not be the same for both installation processes. The figures in this chapter show the installation of SQL Server 2019 on top of an existing installation, but I also describe the differences between the two installation processes.

As its name suggests, besides planning, Installation Center supports the installation of the software on Windows, too. Installation Center shows you several options related to the installation of the database system and its components. After clicking Installation, choose

New SQL Server Stand-Alone Installation or Add Features to an Existing Installation (refer to Figure 2-2), which launches a wizard to install SQL Server.

What appears for the next step depends on whether you install SQL Server 2019 as a first installation or on top of an existing one. In the former case, the Product Key and License Terms pages appear at the top of the navigation pane, before the Global Rules page. In the latter case, the Global Rules page appears at the top, and the Product Key and License Terms pages appear later in the wizard. If you are doing a first installation, complete the Product Key and License Terms steps at this point.

The Global Rules page is identical to what you see when you start System Configuration Checker in the planning phase (refer to the previous section), so you can skip to the next page, Product Updates. On this page, you can decide whether to use the Product Updates service of Microsoft to keep the installation up to date. (I strongly recommend that you activate Microsoft's Update service.) When you click Next, the Install Setup Files page appears and the installation program installs all setup files.

Click Next, and the Install Rules page appears (see Figure 2-3). This page shows the status of some installation rules that were run. The status of rules marked "Passed" are acceptable. You can accept warnings in this step, because they are merely information for the user who installs the product. Click Next.

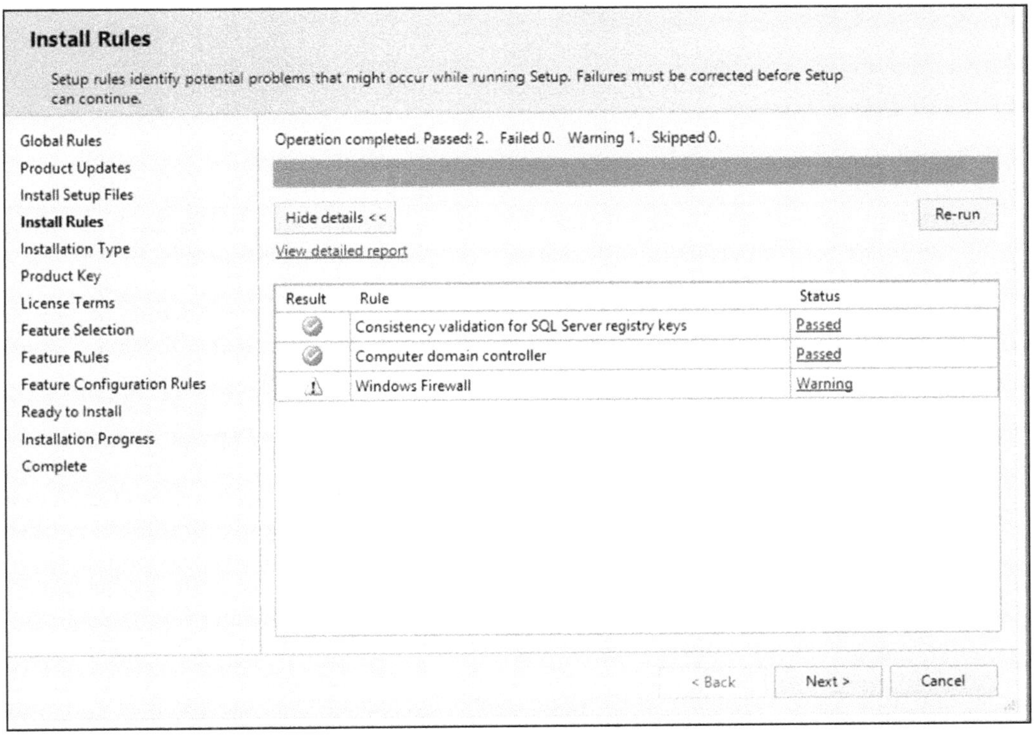

Figure 2-3   SQL Server Installation Center: Install Rules page

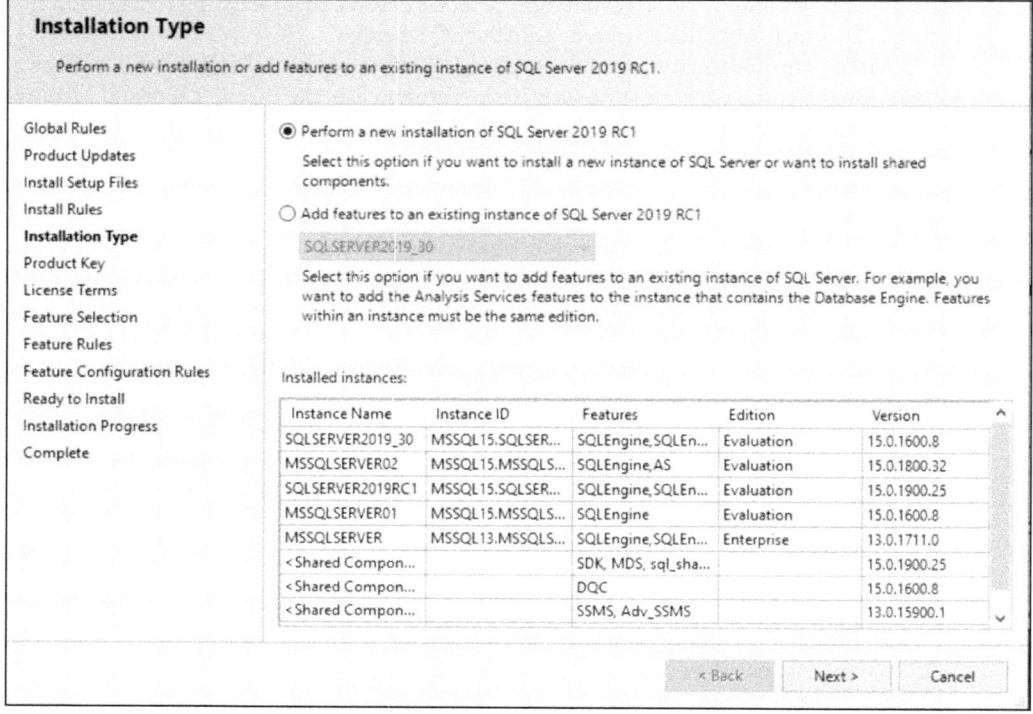

Figure 2-4   SQL Server Installation Center: Installation Type page

---

NOTE If you install SQL Server 2019 from scratch, the following page, Installation Type, will not appear as an installation step. In that case, skip the following paragraph.

On the next page, Installation Type (see Figure 2-4), you can choose either to perform a new installation of SQL Server or to add features to an existing instance of SQL Server. Choose the option to perform a new installation of SQL Server and click Next.

On the Product Key page, enter the 25-character key from the product packaging. (The alternative is to specify a free edition of the software, SQL Server Express, for instance.) Click Next to continue. On the License Terms page, click I Accept the License Terms. Click Next.

On the Feature Selection page (see Figure 2-5), select the components to install by checking the corresponding check boxes. (Recall that the instance features were introduced earlier in this chapter, in the "Which SQL Server Components Should Be Installed?" section.) Also, toward the bottom of the page, you can specify the directory in which to store the shared components. (You should not check the box for Analysis Services; save its installation for Chapter 23.) After that, click Next to continue.

CAUTION The installation of the PolyBase component requires Oracle Java SE Runtime Environment (JRE) Version 7.51 or higher. In other words, if you have selected this component on the Feature Selection page and the Java runtime system is not installed, the installation process *will fail*.

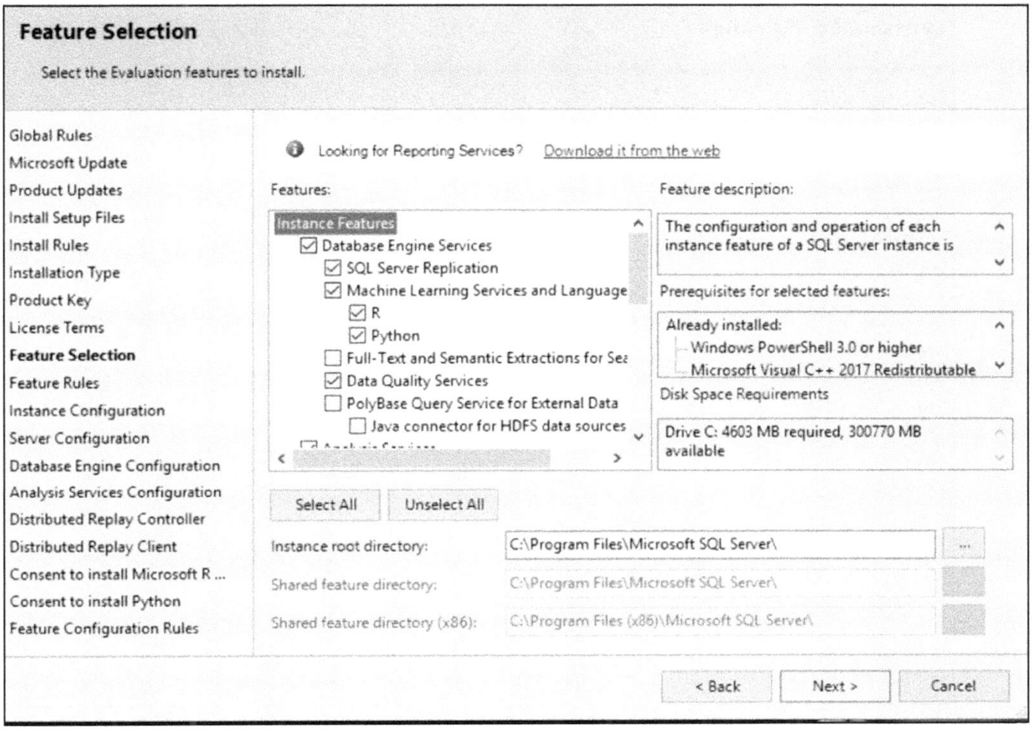

Feature Selection

Select the Evaluation features to install.

Global Rules
Microsoft Update
Product Updates
Install Setup Files
Install Rules
Installation Type
Product Key
License Terms
Feature Selection
Feature Rules
Instance Configuration
Server Configuration
Database Engine Configuration
Analysis Services Configuration
Distributed Replay Controller
Distributed Replay Client
Consent to install Microsoft R ...
Consent to install Python
Feature Configuration Rules

(i) Looking for Reporting Services?   Download it from the web

Features:

Feature description:

Instance Features
☑ Database Engine Services
   ☑ SQL Server Replication
   ☑ Machine Learning Services and Language
      ☑ R
      ☑ Python
   ☐ Full-Text and Semantic Extractions for Sea
   ☑ Data Quality Services
   ☐ PolyBase Query Service for External Data
      ☐ Java connector for HDFS data sources

The configuration and operation of each instance feature of a SQL Server instance is

Prerequisites for selected features:

Already installed:
   Windows PowerShell 3.0 or higher
   Microsoft Visual C++ 2017 Redistributable

Disk Space Requirements

Drive C: 4603 MB required, 300770 MB available

Select All   Unselect All

Instance root directory: C:\Program Files\Microsoft SQL Server\

Shared feature directory: C:\Program Files\Microsoft SQL Server\

Shared feature directory (x86):   C:\Program Files (x86)\Microsoft SQL Server\

< Back   Next >   Cancel

Figure 2-5   SQL Server Installation Center: Feature Selection page

At the next page, Feature Rules, the Setup program runs rules to determine if the installation process will be blocked. If all checks are passed (or marked "Not applicable"), click Next to continue.

On the Instance Configuration page (see Figure 2-6), you can choose between the installation of a default instance or a named instance. (See "Should Multiple Instances of the Database Engine Be Used?" earlier in this chapter for a refresher on these instance types.) To install the default instance, click Default Instance. If a default instance is already installed and you select Default Instance, the Setup program upgrades it and gives you the option to install additional components. Therefore, you have another opportunity to install components that you skipped in the previous installation processes.

To install a new named instance, click Named Instance and type a new name in the text box. In the lower part of the page, you can see the list of instances already installed on your system. (MSSQLSERVER is the name of the default instance for the Database Engine.) Click Next to continue.

The next page, Server Configuration, contains two tabs. The first one, Service Accounts (see Figure 2-7), allows you to specify usernames and corresponding passwords for services of all components that will be installed during the installation process. (You can apply one account for all services, but this is not recommended, for security reasons.)

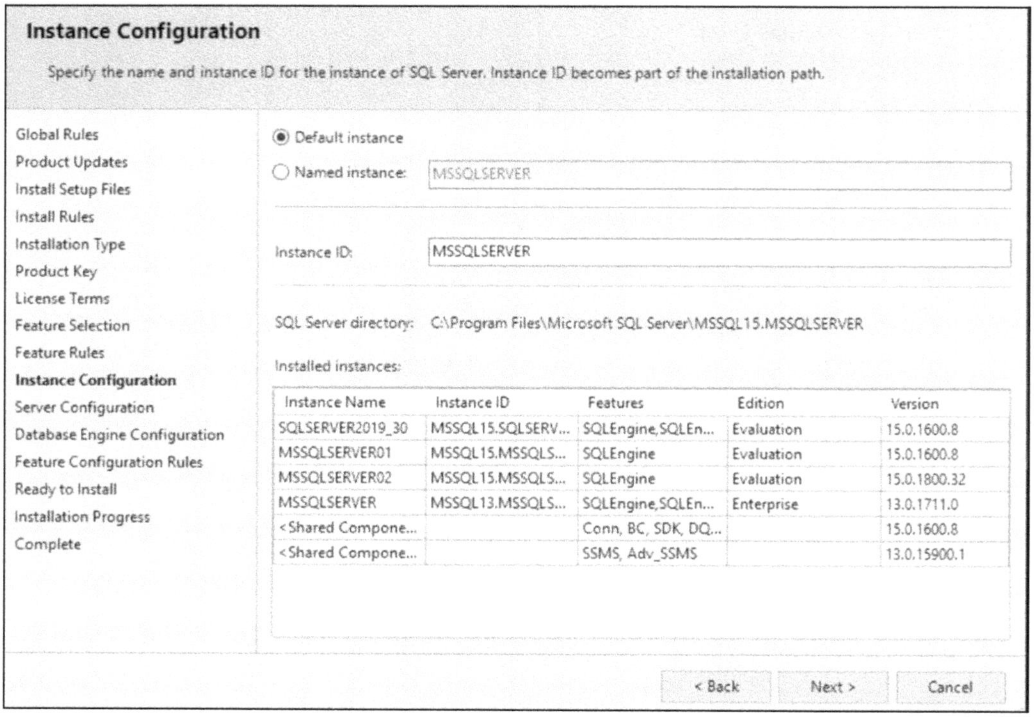

Figure 2-6   Server Installation Center: Instance Configuration page

To choose the collation of your instance, click the Collation tab. (Collation defines the sorting behavior for your instance.) You can either choose the default collations for the components that will be installed, or click Customize to select some other collations that are supported by the system. Click Next to continue.

The Database Engine Configuration page has several different tabs (see Figure 2-8). The first tab, Server Configuration, allows you to choose the authentication mode for your Database Engine system. As you already know, the Database Engine supports Windows authentication mode and Mixed mode. If you select the Windows Authentication Mode radio button, SQL Server validates the account name and password using the Windows principal token in the operating system. This means that the user identity is confirmed by Windows. SQL Server does not ask for the password, and does not perform the identity validation.

If you choose the Mixed Mode radio button, you must enter and confirm the system administrator login. Click Add Current User if you want to add one or more users that will have unrestricted access to the instance of the Database Engine. (As you can see from Figure 2-8, I added my own account for unrestricted access.)

The second tab of the Database Engine Configuration page, Data Directories, allows you to specify the locations for the directories in which Database Engine–related files are stored. In other words, the installation process lists default directories where the installation process will store different SQL Server components and which you can modify.

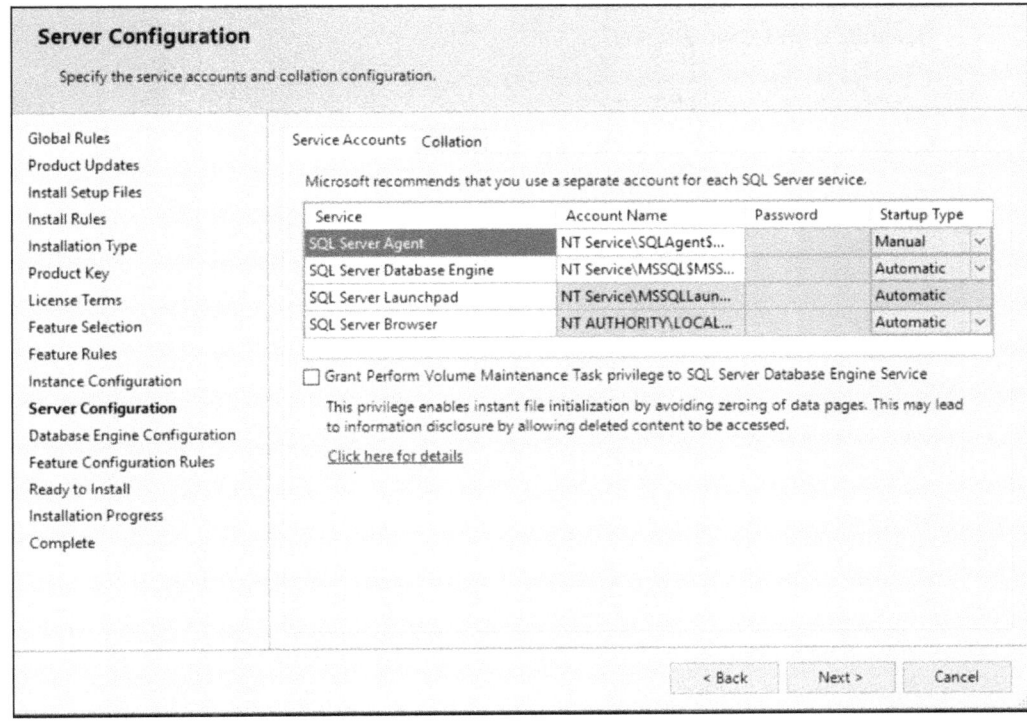

Figure 2-7   Server Configuration (Service Accounts tab)

As its name indicates, the third tab relates to the **tempdb** system database. (This database provides storage space for temporary objects that are needed.) On this tab you can configure the number of data files assigned to **tempdb** (see Figure 2-9). By default, the installation process sets the default to the number of cores on your machine. If your machine has more than eight cores, this initial number of **tempdb** data files will be set to 8.

Additionally, this tab allows you to set the initial size and autogrowth settings for the **tempdb** data and log files. Click Next.

NOTE Use the default values for MaxDOP, Memory, and FILESTRAM. These values do not have the same significance as the parameters of the **tempdb** database, and can be easily modified later.

NOTE The **tempdb** system database is often the cause of serious performance problems. For this reason, the possibility to assign a variable number of files to this database during the installation process is a good idea of the SQL Server team. (Chapter 15 provides a detailed description of the **tempdb** database.)

What appears for the next step depends on whether or not you chose to install Analysis Services. (A Configuration page appears for each SQL Server component that you chose

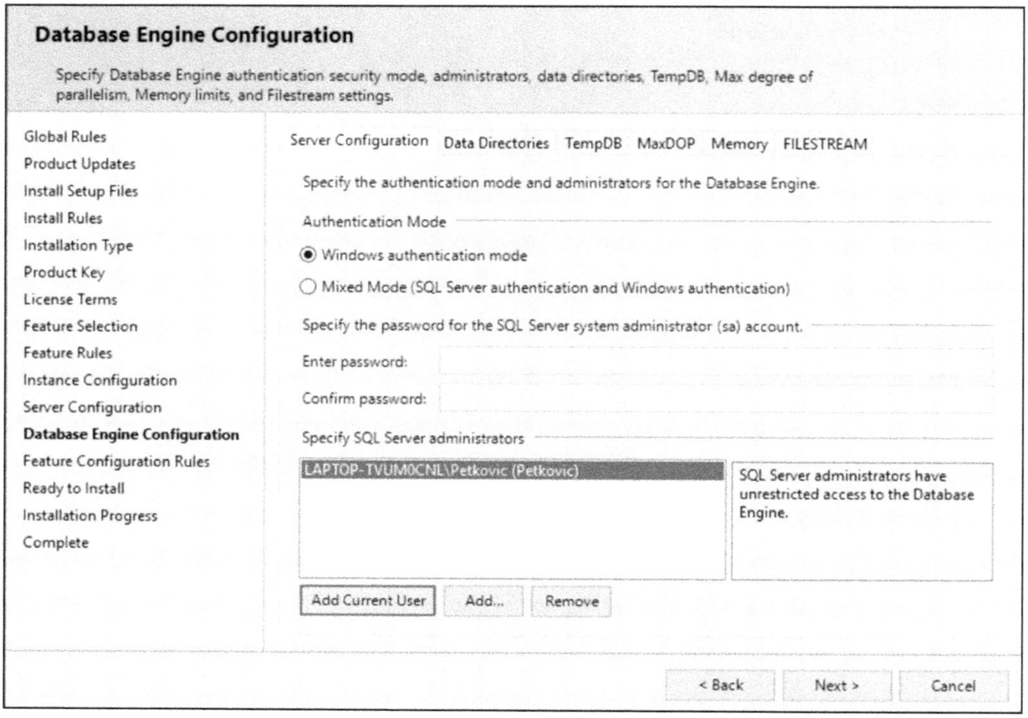

Database Engine Configuration

Specify Database Engine authentication security mode, administrators, data directories, TempDB, Max degree of parallelism, Memory limits, and Filestream settings.

Global Rules
Product Updates
Install Setup Files
Install Rules
Installation Type
Product Key
License Terms
Feature Selection
Feature Rules
Instance Configuration
Server Configuration
Database Engine Configuration
Feature Configuration Rules
Ready to Install
Installation Progress
Complete

Server Configuration   Data Directories   TempDB   MaxDOP   Memory   FILESTREAM

Specify the authentication mode and administrators for the Database Engine.

Authentication Mode

◉ Windows authentication mode

○ Mixed Mode (SQL Server authentication and Windows authentication)

Specify the password for the SQL Server system administrator (sa) account.

Enter password:

Confirm password:

Specify SQL Server administrators

LAPTOP-TVUM0CNL\Petkovic (Petkovic)

SQL Server administrators have unrestricted access to the Database Engine.

Add Current User   Add...   Remove

< Back   Next >   Cancel

Figure 2-8   Database Engine Configuration (Server Configuration tab)

to install.) If you did choose to install it, a page will appear separately for Analysis Services. (You can later easily install this component. For this reason, its installation is discussed in Chapter 23.)

The next page, Feature Configuration Rules, is similar to the earlier page called Feature Rules. During the Feature Rules step, the setup is running rules to determine if the installation process will be blocked by any of the selected features in the Feature Selection step. On the other hand, during the Feature Configuration Rules step, the setup is testing whether the process will be blocked by any rules in relation to the configuration of your system. At the end of this step, the summary of all configuration rules will be displayed. Click Next.

The last page, before the installation process actually starts, is the Ready to Install page. This page allows you to review the summary of all SQL Server components that will be installed. To start the installation process, click Install. Setup shows you the progress of your installation process. At the end, the Complete page appears, with the location of the file in which the summary log is stored. Click Close to complete the installation process.

Installing SQL Server Directly on Ubuntu

You can install and configure SQL Server 2019 on Linux derivatives either directly or by using a Docker container. This section shows you how to install SQL Server directly on Ubuntu. You

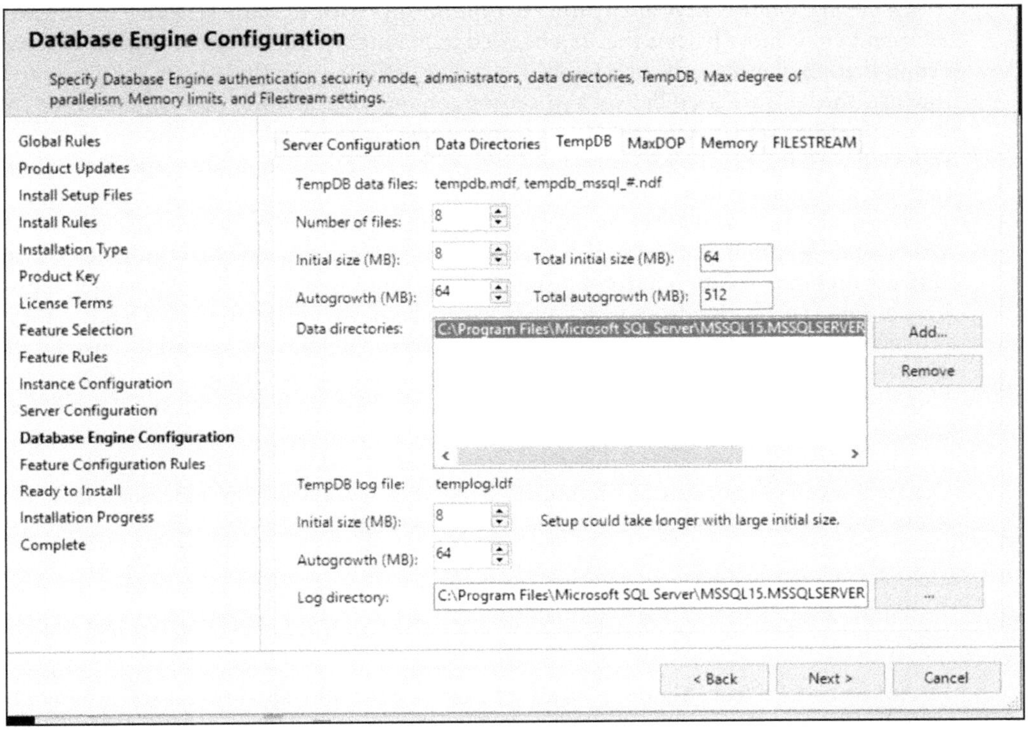

Figure 2-9   Database Engine Configuration (TempDB tab)

can find the instructions for installing SQL Server 2019 on Ubuntu using a Docker container in Microsoft Docs.

NOTE The configuration of a virtual machine and installation of Ubuntu on it are not discussed in this book. For the description of both steps, see, for example, https://www.sqlshack.com/sql-server-2019-on-linux-with-ubuntu/.

Before SQL Server can be installed directly on Ubuntu, you have to perform two introductory steps:

1. Import the public repository GNU Privacy Guard (abbreviated either GPG or GnuPG) keys.

2. Register the SQL Server Ubuntu repository for SQL Server 2019 preview.

To import GnuPG keys, execute this statement:

```
wget -qO- https://packages.microsoft.com/keys/microsoft.asc|sudo
apt-key add -
```

Generally, GnuPG keys are hybrid encryption keys used to share sensitive information between two or more parties that can be used to maintain a link. "Hybrid" means that they are a combination of symmetric and public keys. (For definitions of these keys, see Chapter 12.) In our case, the keys are used between Microsoft, on one side, and the user, on the other. In other words, the Microsoft public GPG keys, stored under https://packages.microsoft.com/keys, are imported with the first (**wget**) command and added with the second one (**apt-key**) to maintain a private information link.

The next step is to register the SQL Server Ubuntu repository. Before I discuss the corresponding command, if you're not familiar with Linux, you should be aware that installing software on Linux is different from installing it on Windows in that it is done using repositories. A *repository* is a location from which your system retrieves applications. Each repository is a collection of software hosted on a remote server and intended to be used for installing and updating software packages on Linux systems. Each such repository is managed by package managers. (You can think of package managers as an equivalent of an advanced version of Add/Remove Programs in Control Panel in Windows.)

The following command registers the SQL Server Ubuntu repository for SQL Server 2019. Change the version number of Ubuntu (16.04), if it is necessary. (The sudo command allows you to run programs as another user; by default, the root user.)

```
sudo add-apt-repository "$(wget -qO-
    https://packages.microsoft.com/config/ubuntu/16.04/mssql-server-preview.
list)"
```

As you can see from Figure 2-10, after I executed the **add-apt-repository** command, several software packages were found and added for the installation process. Some of them belong to the archive of the Ubuntu system, and some belong to the packages delivered by Microsoft.

Now, execute the following two commands to install SQL Server:

```
sudo apt-get update
```

```
sudo apt-get install -y mssql-server
```

```
petkovic@ubuntu18_sql:~$ sudo add-apt-repository "$(wget -qO- https://packages.microsoft.com/config/
ubuntu/16.04/mssql-server-preview.list)"
Hit:1 http://archive.ubuntu.com/ubuntu bionic InRelease
Get:2 http://archive.ubuntu.com/ubuntu bionic-updates InRelease [88.7 kB]
Get:3 https://packages.microsoft.com/ubuntu/16.04/mssql-server-preview xenial InRelease [3,224 B]
Get:4 http://archive.ubuntu.com/ubuntu bionic-backports InRelease [74.6 kB]
Get:5 http://archive.ubuntu.com/ubuntu bionic-security InRelease [88.7 kB]
Get:6 https://packages.microsoft.com/ubuntu/16.04/mssql-server-preview xenial/main amd64 Packages [1
4.8 kB]
Fetched 270 kB in 2s (153 kB/s)
Reading package lists... Done
petkovic@ubuntu18_sql:~$
```

Figure 2-10   The output of the add-apt-repository command

```
petkovic@ubuntu18_sql:~$
petkovic@ubuntu18_sql:~$ sudo apt-get install -y mssql-server
Reading package lists... Done
Building dependency tree
Reading state information... Done
The following additional packages will be installed:
    gcc-8-base gdb gdbserver libbabeltrace1 libc++1 libc++abi1 libc6-dbg libcc1-0 libdw1 libelf1
    libgcc1 libpython-stdlib libpython2.7-minimal libpython2.7-stdlib libsasl2-modules-gssapi-mit
    libsss-nss-idmap0 libstdc++6 python python-minimal python2.7 python2.7-minimal
Suggested packages:
    gdb-doc clang python-doc python-tk python2.7-doc binutils binfmt-support
The following NEW packages will be installed:
    gdb gdbserver libbabeltrace1 libc++1 libc++abi1 libc6-dbg libcc1-0 libdw1 libpython-stdlib
    libpython2.7-minimal libpython2.7-stdlib libsasl2-modules-gssapi-mit libsss-nss-idmap0
    mssql-server python python-minimal python2.7 python2.7-minimal
The following packages will be upgraded:
    gcc-8-base libelf1 libgcc1 libstdc++6
```

Figure 2-11   The output of the apt-get install command

The first command, **apt-get update**, updates the package lists for upgrades of software packages and inserts the names of new software packages. The second command installs the system.

Figure 2-11 shows the beginning part of the output of the **apt-get install** command. You can see that first the dependency tree of all software packages was built and that the dependent packages, together with upgrades, will be installed.

When the package installation finishes, you get the image of the SQL Server instance on the Ubuntu operating system. To use this instance, you need to configure it.

NOTE   In a Windows environment, when you install SQL Server using Installation Center, you need to specify, among other things, the software edition, the instance name, accept the license agreement, and the system administrator password. In an environment such as Ubuntu, the installation process first installs the base version and, in a second step, allows you to configure that instance.

Start the configuration process with the following command:

```
sudo /opt/mssql/bin/mssql-conf setup
```

As you can see from Figure 2-12, in the first step you have to choose the SQL Server edition you will use by entering the corresponding number. After you choose the edition, you have to accept the license terms, as shown in the middle of Figure 2-13. Provide the input as **Yes** (or No if you don't want to proceed with the installation).

In the third step you need to specify the system administrator password. As you can see from Figure 2-13, the password I provided during the first try did not meet SQL Server password policy requirements, and I had to specify a stronger one.

```
petkovic@ubuntu18_sql:~$
petkovic@ubuntu18_sql:~$
petkovic@ubuntu18_sql:~$ sudo /opt/mssql/bin/mssql-conf  setup
usermod: no changes
Choose an edition of SQL Server:
   1) Evaluation (free, no production use rights, 180-day limit)
   2) Developer (free, no production use rights)
   3) Express (free)
   4) Web (PAID)
   5) Standard (PAID)
   6) Enterprise (PAID) - CPU Core utilization restricted to 20 physical/40 hyperthreaded
   7) Enterprise Core (PAID) - CPU Core utilization up to Operating System Maximum
   8) I bought a license through a retail sales channel and have a product key to enter.

Details about editions can be found at
https://go.microsoft.com/fwlink/?LinkId=852748&clcid=0x409

Use of PAID editions of this software requires separate licensing through a
Microsoft Volume Licensing program.
By choosing a PAID edition, you are verifying that you have the appropriate
number of licenses in place to install and run this software.

Enter your edition(1-8): _
```

Figure 2-12   The beginning part of the output of the configuration process

```
Enter your edition(1-8): 1
The license terms for this product can be found in
/usr/share/doc/mssql-server or downloaded from:
https://go.microsoft.com/fwlink/?LinkId=855864&clcid=0x409

The privacy statement can be viewed at:
https://go.microsoft.com/fwlink/?LinkId=853010&clcid=0x409

Do you accept the license terms? [Yes/No]:Yes

Enter the SQL Server system administrator password:
The specified password does not meet SQL Server password policy requirements because it is too short
. The password must be at least 8 characters
Enter the SQL Server system administrator password:
Confirm the SQL Server system administrator password:
Configuring SQL Server...

This is an evaluation version.  There are [155] days left in the evaluation period.
The licensing PID was successfully processed. The new edition is [Enterprise Evaluation Edition].
ForceFlush is enabled for this instance.
ForceFlush feature is enabled for log durability.
Created symlink /etc/systemd/system/multi-user.target.wants/mssql-server.service → /lib/systemd/syst
em/mssql-server.service.
Setup has completed successfully. SQL Server is now starting.
petkovic@ubuntu18_sql:~$ _
```

Figure 2-13   The second part of the output of the configuration process

After that, the configuration starts the SQL Server instance with all specified parameters, and the successful configuration is shown with the message at the end: "Setup has completed successfully. SQL Server is now starting."

The installation described installs an instance of the database server—a SQL Server "back end." You still do not know whether you can use the system or not because a front-end tool,

```
petkovic@ubuntu18_sql:~$
petkovic@ubuntu18_sql:~$ systemctl status mssql-server --no-pager
● mssql-server.service - Microsoft SQL Server Database Engine
   Loaded: loaded (/lib/systemd/system/mssql-server.service; enabled; vendor preset: enabled)
   Active: active (running) since Sat 2019-07-20 17:43:38 UTC; 2min 24s ago
     Docs: https://docs.microsoft.com/en-us/sql/linux
 Main PID: 3133 (sqlservr)
    Tasks: 112
   CGroup: /system.slice/mssql-server.service
           ├─3133 /opt/mssql/bin/sqlservr
           └─3172 /opt/mssql/bin/sqlservr

Jul 20 17:43:59 ubuntu18_sql sqlservr[3133]: [387B blob data]
Jul 20 17:43:59 ubuntu18_sql sqlservr[3133]: [79B blob data]
Jul 20 17:43:59 ubuntu18_sql sqlservr[3133]: [96B blob data]
Jul 20 17:43:59 ubuntu18_sql sqlservr[3133]: [66B blob data]
Jul 20 17:44:00 ubuntu18_sql sqlservr[3133]: [75B blob data]
Jul 20 17:44:00 ubuntu18_sql sqlservr[3133]: [96B blob data]
Jul 20 17:44:00 ubuntu18_sql sqlservr[3133]: [100B blob data]
Jul 20 17:44:00 ubuntu18_sql sqlservr[3133]: [71B blob data]
Jul 20 17:44:00 ubuntu18_sql sqlservr[3133]: [124B blob data]
Jul 20 17:44:00 ubuntu18_sql sqlservr[3133]: [315B blob data]
petkovic@ubuntu18_sql:~$
petkovic@ubuntu18_sql:~$
```

Figure 2-14   The output of the systemctl command

such us Azure Data Studio, is still not installed. Regardless, you can check the SQL Service status from the command line by using the **systemctl** command as follows:

```
systemctl status mssql-server --no-pager
```

Figure 2-14 shows the output of the **systemctl** command. The most important part of the output is the line with the information that the system is active and running.

Additionally, the **systemctl** command starts two processes (see Figure 2-14). The parent process (#3133) handles basic configuration activities and then forks the child process (#3172). The parent process becomes a lightweight monitor and the child process runs the sqlservr.exe process.

Summary

SQL Server 2019 can be installed on Windows as well as on Linux and in Docker containers. In case of Windows, the Server Installation Center component is used to plan the installation and to accomplish it. The most important step in the planning phase is the invocation of System Configuration Checker. This component identifies problems that might occur when you install SQL Server files. The installation of SQL Server on Windows is straightforward. The most important decision that you have to make during this phase is which components to install, a decision you prepared for during the planning phase.

The installation on Linux derivatives, such as Ubuntu and Red Hat Linux, can be either direct or by using Docker containers. The installation of SQL Server on macOS can be done only via Docker containers.

The next chapter describes front-end tools for the Database Engine: SQL Server Management Studio and Azure Data Studio. These components are used by users and database administrators to interact with the system.

3

Front-End Tools for the Database Engine

In This Chapter

- SQL Server Management Studio
- Using SQL Server Management Studio with the Database Engine
- Authoring Activities Using SQL Server Management Studio
- Azure Data Studio
- SQL Server Management Studio vs. Azure Data Studio

This chapter introduces two front-end tools for the Database Engine: SQL Server Management Studio (SSMS) and Azure Data Studio. First, this chapter describes two components of SSMS, Registered Servers and Object Explorer, as well as the various user interface panes of SSMS. After that, it discusses the SSMS functions related to the Database Engine. Also, Query Editor and Solution Explorer are explained in relation to authoring activities in SSMS.

Azure Data Studio has been developed to support DBAs and users who do not use Windows as a platform for the Database Engine. Azure Data Studio enables you to manage the instances of your database system as well as retrieve data from existing databases. This component does not offer as many components as SSMS does. A comparison of the features of both front-end tools is given at the end of the chapter.

SQL Server Management Studio

The Database Engine provides various tools that are used for different purposes, such as system installation, configuration, auditing, and performance tuning. (All these tools will be discussed in subsequent chapters of this book.) The administrator's primary tool for interacting with the system on Windows platforms is SQL Server Management Studio. Both administrators and end users can use SSMS to administer multiple servers, develop databases, and replicate data, among other things.

NOTE Starting with SQL Server 2017, SQL Server Management Studio is an independent component. This means that SSMS no longer is a component of the Database Engine and therefore requires its own install. Making SSMS a completely stand-alone product, not tied to any specific version or edition of SQL Server, allows Microsoft to achieve significantly shorter release cycles for the tool. The advantage over the old, integrated model is that users sometimes had to wait months or even years for any new SSMS features. In this book we will discuss (and use) SSMS Version 18.1.

The installation of SQL Server Management Studio is straightforward. Go to www.microsoft.com, click the magnifying-glass icon in the upper-right corner, type **SSMS** in the Search box, press ENTER, and click the Download SQL Server Management Studio (SSMS) link in the search results. On the page that opens, confirm that your machine is running a supported version of Windows and then click the Download SQL Server Management Studio link for the current version of SSMS (18.1 at the time of writing). After it downloads, execute the corresponding .exe file, click Install, and wait for the installation to complete.

To open SQL Server Management Studio, just click the corresponding icon, or click Start and type **Management Studio** in Windows Desktop Search. SQL Server Management Studio comprises several different components that are used for authoring, administration, and management of the overall system. The following are the main components used for these tasks:

- Registered Servers
- Object Explorer
- Query Editor
- Solution Explorer
- Data Discovery and Classification

The first two components in the list are discussed in this section. The latter three components are explained later in this chapter in "Authoring Activities Using SQL Server Management Studio."

To get to the main SQL Server Management Studio interface, you first must connect to a server, as described next.

Connecting to a Server

When you open SQL Server Management Studio, it displays the Connect to Server dialog box (see Figure 3-1), which allows you to specify the necessary parameters to connect to a server:

- **Server Type**   For purposes of this chapter, choose Database Engine.
- **Server Name**   Select or type the name of the server that you want to use. (Generally, you can connect SQL Server Management Studio to any of the installed products on a particular server.)

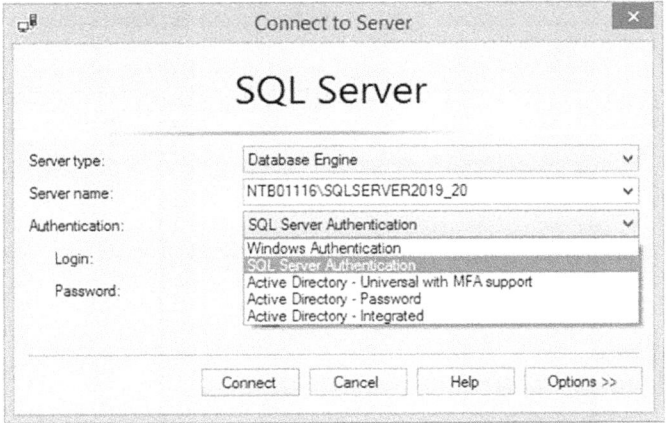

Figure 3-1   The Connect to Server dialog box

- **Authentication**   Starting with SQL Server Management Studio 17.2, the Connect to Server dialog box supports an extra three authentication methods that are used to connect to Azure SQL Database and Data Warehouse instances. The supported authentication methods are listed here:
  - **Windows Authentication**   You connect to the Database Engine using your Windows account. This option is recommended by Microsoft.
  - **SQL Server Authentication**   The Database Engine uses its own authentication.
  - **Active Directory – Universal with MFA support**   This is an interactive authentication method that supports Azure Multi-Factor Authentication, which provides strong authentication with a range of easy verification options you can choose from.
  - **Active Directory – Password and Active Directory – Integrated**   These are non-interactive authentication methods supported by Azure Active Directory Authentication and can be used in many applications such as the ODBC and JDBC.

Note that the last three types of authentication will not be discussed further in the book because it doesn't cover Microsoft Azure.

NOTE For more information concerning SQL Server Authentication, see Chapter 12.

When you click Connect, the Database Engine connects to the specified server. After connecting to the database server, the default SQL Server Management Studio window appears. The default appearance is similar to Visual Studio, so users can leverage their experience of developing in Visual Studio to use SSMS more easily. Figure 3-2 shows the SQL Server Management Studio window with several panes.

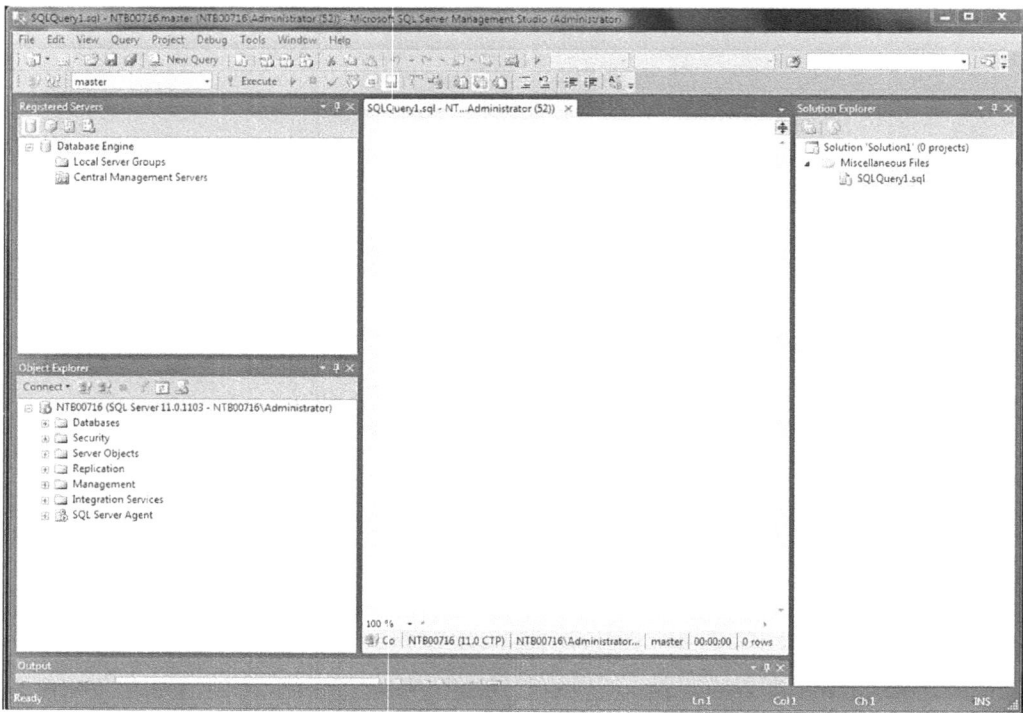

Figure 3-2   SQL Server Management Studio

NOTE SQL Server Management Studio gives you a unique interface to manage servers and create queries across all SQL Server components. This means that SSMS offers one interface for the Database Engine as well as for Analysis Services, Integration Services, and Reporting Services. This chapter demonstrates the use of SQL Server Management Studio only with the Database Engine.

Registered Servers

Registered Servers is represented as a pane that allows you to maintain connections to already used servers (see Figure 3-2). (If the Registered Servers pane isn't visible, select its name from the View menu.) You can use these connections to check a server's status or to manage its objects. Each user has a separate list of registered servers, which is stored locally.

You can add new servers to the list of all servers, or remove one or more existing servers from the list. You also can group existing servers into server groups. Each group should contain the servers that belong together logically.

Object Explorer

The Object Explorer pane contains a tree view of all the database objects in a server. (If the Object Explorer pane isn't visible, select View | Object Explorer.) The tree view shows you a hierarchy of the objects on a server instance. Hence, if you expand a tree, the logical structure of a corresponding server will be shown.

Object Explorer allows you to connect to multiple servers in the same pane. The server can be any of the existing servers. This feature is user friendly, because it allows you to manage all servers of the same type or different types from one place.

NOTE Object Explorer has several other features, explained later in this chapter.

Organizing and Navigating SQL Server Management Studio's Panes

You can dock or hide each of the panes of SQL Server Management Studio. By right-clicking the title bar at the top of the corresponding pane, you can choose between the following presentation possibilities:

- **Float** The pane becomes a separate floating pane on top of the rest of the SSMS panes. The pane can be moved anywhere around the screen.
- **Dock** Enables you to move and dock the pane in different positions. To move the pane to a different docking position, click and drag its title bar and drop it in the new position.
- **Dock as Tabbed Document** You can create a tabbed grouping using the Designer window. When this is done, the pane's state changes from dockable to tabbed document.
- **Hide** Closes the pane. (Alternatively, you can click the × in the upper-right corner of the pane.) To display a closed pane, select its name from the View menu.
- **Auto Hide** Minimizes the pane and stores it on the left side of the screen. To reopen (maximize) the pane, move your mouse over the tabs on the left side of the screen and click the push pin to pin the pane in the open position.

NOTE The difference between the Hide and Auto Hide options is that the former option removes the pane from view in SQL Server Management Studio, while the latter collapses the pane to the side panel.

To restore the default configuration of SSMS, choose Window | Reset Window Layout. The Object Explorer pane appears on the left, while the Object Explorer Details tab appears on the right. (The Object Explorer Details tab displays information about the currently selected node of Object Explorer.)

NOTE You will find that SSMS often offers several ways of accomplishing the same task. This chapter will indicate more than one way to do things, whereas only a single method will be given in subsequent chapters. Different people prefer different methods (some like to double-click, some like to click the +/− signs, some like to right-click, others like to use the pull-down menus, and others like to use the keyboard shortcuts as much as possible). Experiment with the different ways to navigate, and use the methods that feel most natural to you.

Within the Object Explorer and Registered Servers panes, a subobject appears only if you click the plus (+) sign of its direct predecessor in the tree hierarchy. To see the properties of an object, right-click the object and choose Properties. A minus (–) sign to the left of an object's name indicates that the object is currently expanded. To compress all subobjects of an object, click its minus sign. (Another possibility would be to double-click the folder, or press the LEFT ARROW key while the folder is selected.)

Using SQL Server Management Studio with the Database Engine

SQL Server Management Studio has two main purposes:

- Administration of the database servers
- Management of database objects

The following sections describe these functions of SQL Server Management Studio.

Administering Database Servers

The administrative tasks that you can perform by using SQL Server Management Studio are, among others, the following:

- Register servers
- Connect to a server
- Create new server groups
- Manage multiple servers
- Start and stop servers

The following subsections describe these administrative tasks.

Registering Servers

SQL Server Management Studio separates the activities of registering servers and exploring databases and their objects. (Both of these activities can be done using Object Explorer.) Every server (local or remote) must be registered before you can use its databases and objects. A server can be registered during the first execution of SQL Server Management Studio or later. To register a database server, right-click the icon of your database server in Object Explorer and choose Register. (If the Object Explorer pane doesn't appear on your screen, select View | Object Explorer.) The New Server Registration dialog box appears. Choose the name of the server that you want to register and the authentication mode. Click Save.

Connecting to a Server

SQL Server Management Studio also separates the tasks of registering a server and connecting to a server. This means that registering a server does not automatically connect you to the server at SSMS startup. To connect to a server from the Object Explorer window, right-click the server name and choose Connect.

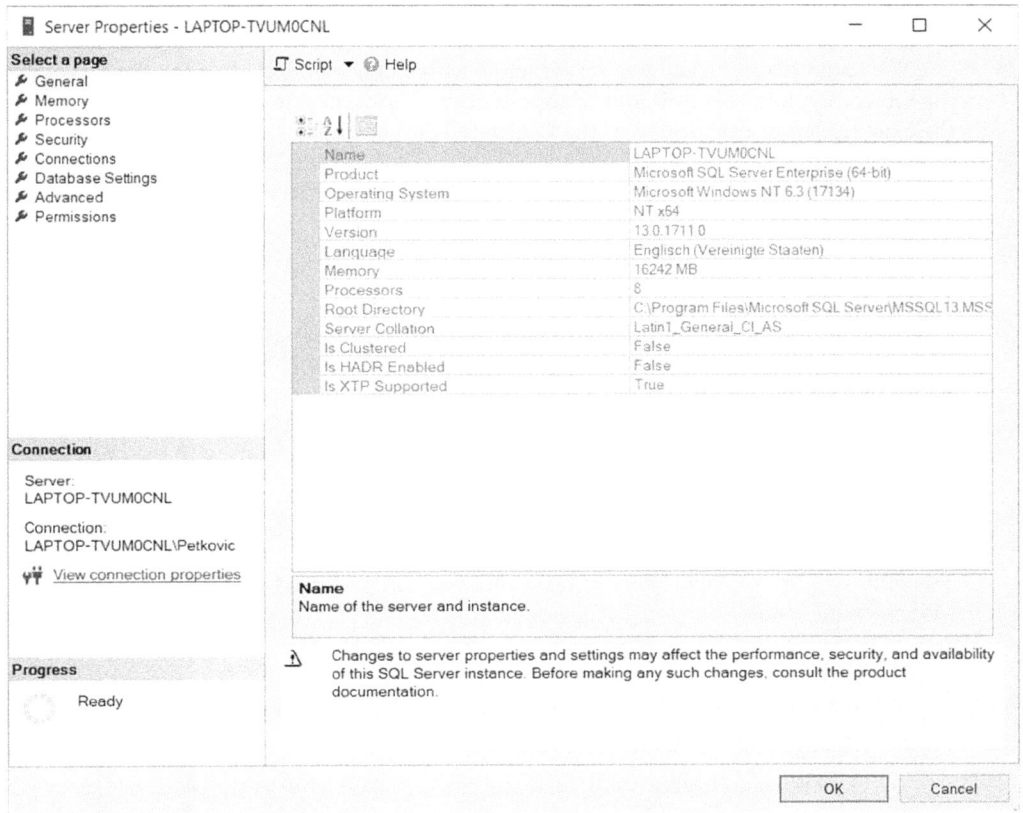

Figure 3-3   The Server Properties dialog box: the General page

Creating a New Server Group

To create a new server group in the Registered Servers pane, right-click Local Server Groups and choose New Server Group. In the New Server Group Properties dialog box, enter a (unique) group name and optionally describe the new group.

Managing Multiple Servers

SQL Server Management Studio allows you to administer multiple database servers (called instances) on one computer by using Object Explorer. Each instance of the Database Engine has its own set of database objects (system and user databases) that are not shared between different instances.

To manage a server and its configuration, right-click the server name in Object Explorer and choose Properties. The Server Properties dialog box that opens contains several different pages, such as General, Security, and Permissions.

The General page (see Figure 3-3) shows general properties of the server. The Security page contains the information concerning the authentication mode of the server and the login auditing mode. The Permissions page shows all logins and roles that can access the server. The lower part of the page shows all permissions that can be granted to the logins and roles.

You can replace the existing server name with a new name. Right-click the server name in the Object Explorer window and choose Register. Now you can rename the server and modify the existing server description in the Registered Server frame.

NOTE Do not rename servers after this point, because changing names can affect other servers that reference them.

Starting and Stopping Servers

A Database Engine server starts automatically by default each time the Windows operating system starts. To start the server using SQL Server Management Studio, right-click the selected server in the Object Explorer pane and click Start in the context menu. The menu also contains Stop and Pause functions that you can use to stop or pause the activated server, respectively. Similarly, you can resume and restart the server, using the functions with the corresponding names.

Managing Databases Using Object Explorer

The following are the management tasks that you can perform by using SQL Server Management Studio:

- Create databases without using Transact-SQL
- Modify databases without using Transact-SQL
- Manage tables without using Transact-SQL
- Generate and execute SQL statements (described later, in the section "Query Editor")

Creating Databases Without Using Transact-SQL

You can create a new database by using Object Explorer or the Transact-SQL language. (Database creation using Transact-SQL is discussed in Chapter 5.) As the name suggests, you also use Object Explorer to explore the objects within a server. From the Object Explorer pane, you can inspect all the objects within a server and manage your server and databases. The existing tree contains, among other folders, the Databases folder. This folder has several subfolders, including one for the system databases and one for each new database that is created by a user. (System and user databases are discussed in detail in Chapter 15.)

To create a database using Object Explorer, right-click Databases and select New Database. In the New Database dialog box (see Figure 3-4), type the name of the new database in the Database Name field and then click OK. Each database has several different properties, such as file type, initial size, and so on. Database properties can be selected from the left pane of the New Database dialog box. There are several different pages (property groups):

- General
- Files
- Filegroups
- Options

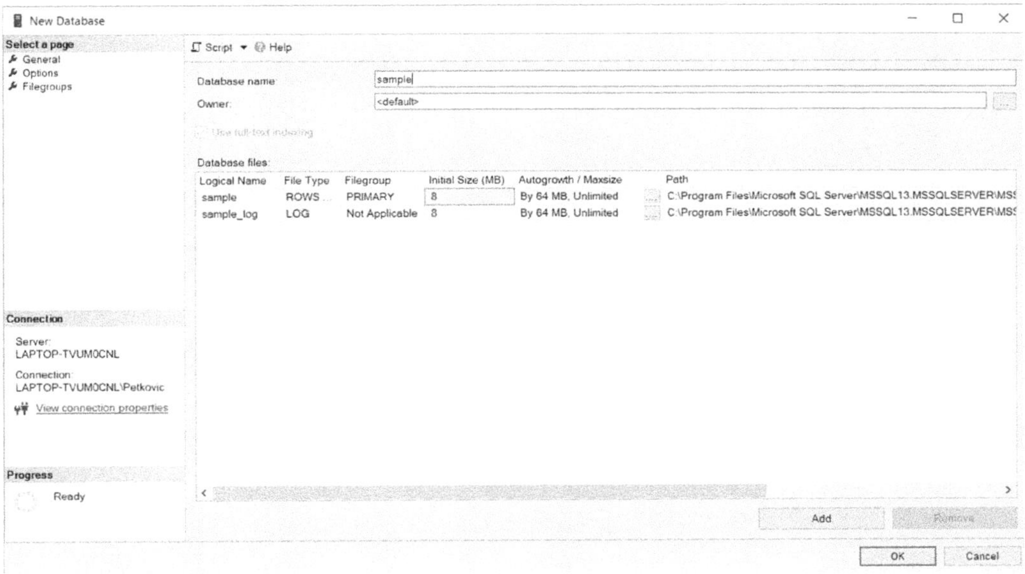

Figure 3-4   The New Database dialog box

- Change Tracking
- Permissions
- Extended Properties
- Mirroring
- Transaction Log Shipping
- Query Store

NOTE For an existing database, the system displays all property groups in the preceding list. For a new database, as shown in Figure 3-4, there are only three groups: General, Options, and Filegroups.

The General page of the Database Properties dialog box (see Figure 3-5) displays, among other things, the database name, the owner of the database, and its collation. The properties of the data files that belong to a particular database are listed in the Files page and comprise the name and initial size of the file, where the database will be stored, and the type of the file (PRIMARY, for instance). A database can be stored in multiple files.

NOTE The Database Engine has dynamic disk space management. This means that databases can be set up to automatically expand and shrink as needed. If you want to change the **Autogrowth** property in the Files page, click the ellipses (…) in the Autogrowth/Maxsize column and make your changes in the Change Autogrowth dialog box. The Enable Autogrowth check box should

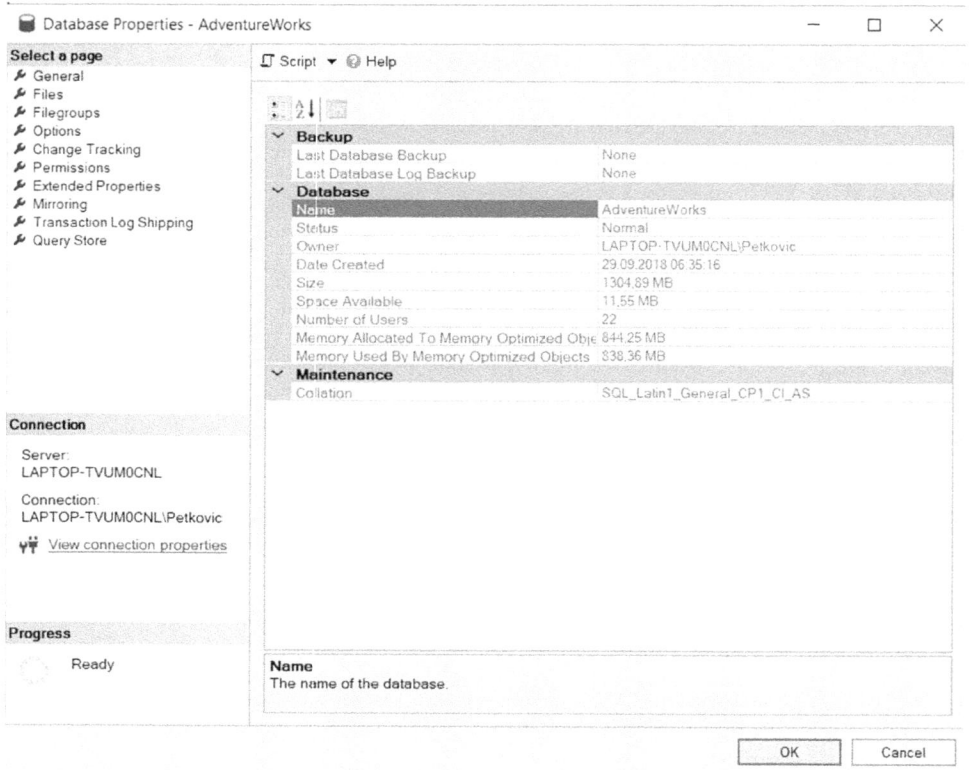

Figure 3-5   Database Properties dialog box: General page

be checked to allow the database to autogrow. Each time there is insufficient space within the file when data is added to the database, the server will request the additional space from the operating system. The amount (in megabytes) of the additional space is set by the number in the File Growth frame of the same dialog box. You can also decide whether the file can grow without any restrictions (the default value) or not. If you restrict the file growth, you have to specify the maximum file size.

The Filegroups page of the Database Properties dialog box displays the name(s) of the filegroup(s) to which the database file belongs, the art of the filegroup (default or nondefault), and the allowed operation on the filegroup (read/write or read-only).

The Options page of the Database Properties dialog box enables you to display and modify all database-level options. There are, among others, the following groups of options: Automatic, Miscellaneous, Recovery and State. For instance, the following four options exist for State:

- **Database Read-Only**   Allows read-only access to the database. This prohibits users from modifying any data. (The default value is False.)
- **Database State**   Describes the state of the database. (The default value is Normal.)

- **Restrict Access**   Restricts the use of the database to one user at a time. (The default value is MULTI_USER.)
- **Encryption Enabled**   Controls the database encryption state. (The default value is False.)

If you choose the Permissions page, the system opens the corresponding dialog box and displays all users and roles along with their permissions. (For the discussion of permissions, see Chapter 12.)

The Extended Properties page displays additional properties of the current database. Existing properties can be deleted and new properties can be added from this dialog box.

The rest of the pages (Change Tracking, Mirroring, Transaction Log Shipping, and Query Store) describe the features related to data availability and are explained in Chapter 16.

Modifying Databases Without Using Transact-SQL

Object Explorer can also be used to modify an existing database. Using this component, you can modify files and filegroups that belong to the database. To add new data files, right-click the database name and choose Properties. In the Database Properties dialog box, select Files, click Add, and type the name of the new file. You can also add a (secondary) filegroup for the database by selecting Filegroups and clicking Add Filegroup.

NOTE Only the system administrator or the database owner can modify the database properties just mentioned.

To delete a database using Object Explorer, right-click the database name and choose Delete.

Managing Tables Without Using Transact-SQL

After you create a database, your next task is to create all tables belonging to it. As with database creation, you can create tables by using either Object Explorer or Transact-SQL. Again, only Object Explorer is discussed here. (The creation of a table and all other database objects using the Transact-SQL language is discussed in detail in Chapter 5.)

To create a table using Object Explorer, expand the Databases folder, expand the database, right-click the Tables subfolder, and then click New and after that, Table.

To demonstrate the creation of a table using Object Explorer, the **department** table of the **sample** database will be used as an example. Enter the names of all columns with their properties. Enter the column names, their data types, and the NULL property of each column in the two-dimensional matrix, as shown in the top-right pane of Figure 3-6.

All data types supported by the system can be displayed (and one of them selected) by clicking the arrow sign in the Data Type column (the arrow appears after the cell has been selected). Subsequently, you can type entries in the **Length**, **Precision**, and **Scale** rows for the chosen data type on the Column Properties tab (see the bottom-right pane of Figure 3-6). Some data types, such as CHARACTER (CHAR), require a value for the **Length** row, and some, such as DECIMAL, require a value in the **Precision** and **Scale** rows. On the other hand, data types such as INTEGER do not need any of these entries to be specified. (The valid entries for a specified data type are highlighted in the list of all possible column properties.)

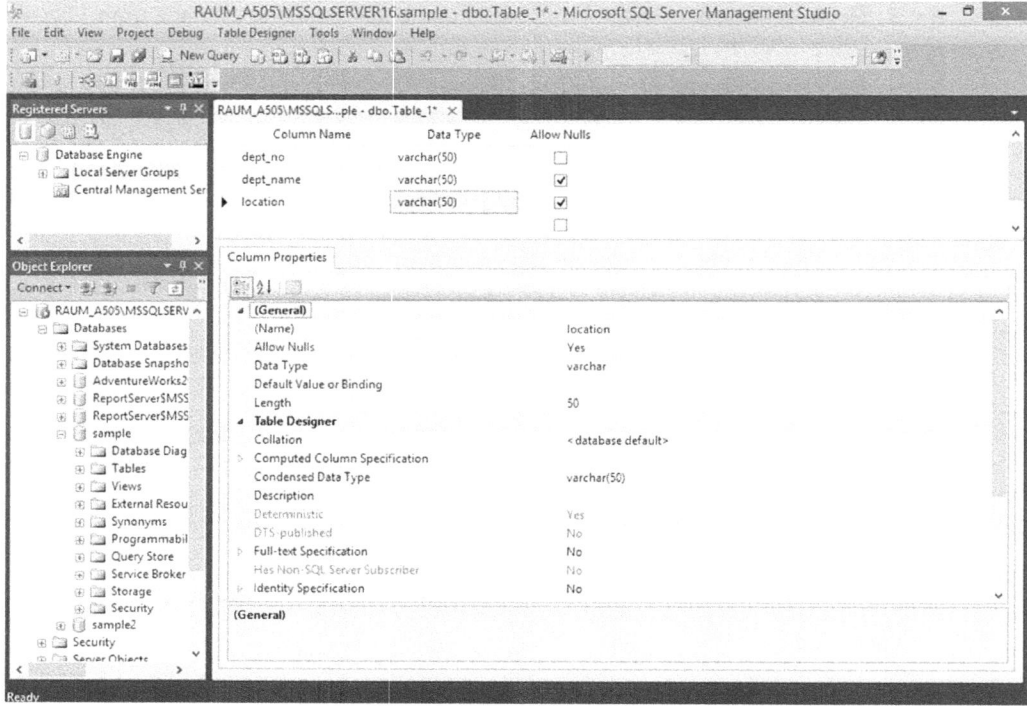

Figure 3-6   Creating the department table using Object Explorer

The check box in the Allow Nulls column must be checked if you want a table column to permit NULL values to be inserted into that column. Similarly, if there is a default value, it should be entered in the **Default Value or Binding** row of the Column Properties tab. (A default value is a value that will be inserted in a table column when there is no explicit value entered for it.)

The column **dept_no** is the primary key of the **department** table. (For the discussion of primary keys of the **sample** database, see Chapter 1.) To specify a column as the primary key of a table, you must right-click the column and choose Set Primary Key. Finally, click the × in the right pane with the information concerning the new table. After that, the system will display the Choose Name dialog box, where you can type the table name.

To view the properties of an existing table, double-click the folder of the database to which the table belongs, double-click Tables, and then right-click the name of the table and choose Properties. To rename a table, right-click the name of the table in the Tables folder and choose Rename. To remove a table, right-click the name of the table in the Tables folder in the database to which the table belongs and select Delete.

NOTE   You should now create the other three tables of the **sample** database: **employee**, **project**, and **works_on**. The structure of these tables can be found in the Introduction of the book.

Authoring Activities Using SQL Server Management Studio

SQL Server Management Studio gives you a complete authoring environment for all types of queries. You can create, save, load, and edit queries. SSMS allows you to work on queries without being connected to a particular server. This tool also gives you the option of developing your queries with different projects.

The authoring capabilities are associated with Query Editor as well as Solution Explorer, both of which are described in this section. This section also introduces the new SSMS application called Data Discovery and Classification and describes how it can be used to classify sensitive data.

Query Editor

To launch the Query Editor pane, click the New Query button in the toolbar of SQL Server Management Studio. By default, you get a new Database Engine query, but other queries are possible.

Once you open Query Editor, the status bar at the bottom of the pane tells you whether your query is in a connected or disconnected state. If you are not connected automatically to the server, the Connect to SQL Server dialog box appears, where you can type the name of the database server to which you want to connect and select the authentication mode.

NOTE Disconnected editing has more flexibility than connected editing. You can edit queries without having to choose a server, and you can disconnect a given Query Editor window from one server and connect it to another without having to open another window. (You can use disconnected editing by clicking the Cancel button in the Connect to SQL Server dialog box.)

Query Editor can be used by end users for the following tasks:

- Generating and executing Transact-SQL statements
- Storing the generated Transact-SQL statements in a file
- Generating and analyzing execution plans for generated queries
- Graphically illustrating the execution plan for a selected query

Query Editor contains an internal text editor and a selection of buttons in its toolbar. The main window is divided into a query pane (upper) and a results pane (lower). Users enter the Transact-SQL statements (queries) that they want to execute into the query pane, and after the system has processed the queries, the output is displayed in the results pane.

The example shown in Figure 3-7 demonstrates a query entered into Query Editor and the output returned. The first statement in the query pane, USE (not shown in Figure 3-7), specifies the **sample** database as the current database. The second statement, SELECT, retrieves all the rows of the **works_on** table. Select Execute or press F5 to return the results of these statements in the results pane of Query Editor.

NOTE You can open several different windows—that is, several different connections to one or more Database Engine instances. You create new connections by clicking the New Query button in the toolbar.

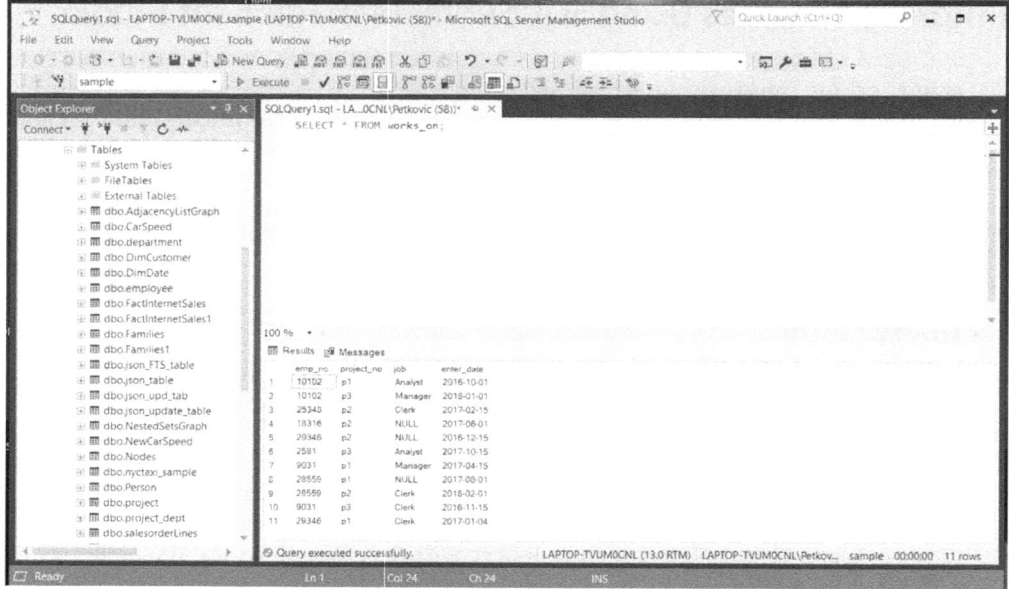

Figure 3-7 Query Editor with a query and its results

The following additional information concerning the execution of the statement(s) is displayed in the status bar at the bottom of the Query Editor window:

- The status of the current operation (for example, "Query executed successfully")
- Database server name
- Current username and server process ID
- Current database name
- Elapsed time for the execution of the last query
- The number of retrieved rows

One of the main features of SQL Server Management Studio is that it's easy to use, and that also applies to the Query Editor component. Query Editor supports a lot of features that make coding of Transact-SQL statements easier. First, Query Editor uses syntax highlighting to improve the readability of Transact-SQL statements. For example, it displays all reserved words in blue, all variables in black, strings in red, and comments in green. (For a discussion of reserved words, see the next chapter.)

There is also the context-sensitive help function called Dynamic Help that enables you to get help on a particular statement. If you do not know the syntax of a statement, just select that statement in the editor and choose Help | Dynamic Help. You can also select options of different Transact-SQL statements to get the corresponding text from Microsoft Docs.

NOTE The Database Engine supports the SQL Intellisense tool. Intellisense is a form of automated autocompletion. In other words, this add-in allows you to access descriptions of frequently used elements of Transact-SQL statements without using the keyboard.

Object Explorer can also help you edit queries. For instance, if you want to see what the corresponding Transact-SQL statement for creation of the **employee** table looks like, drill down to this database object, right-click the table name, select Script Table As, and choose CREATE to New Query Editor Window.

Object Explorer is very useful if you want to display the graphical execution plan for a particular query. (The *execution plan* is the plan selected by the optimizer to execute a given query.) If you select Query | Display Estimated Execution Plan, the system will display the graphical plan instead of the result set for the given query. This topic is discussed in detail in Chapter 19.

Solution Explorer

Query editing in SQL Server Management Studio is solution-based. If you start a blank query using the New Query button, it will still be based on a blank solution. You can see this by choosing View | Solution Explorer right after you open your blank query.

A solution can have zero, one, or more projects associated with it. A blank solution does not contain any project. If you want to associate a project with the solution, close your blank solution, Solution Explorer, and the Query Editor window, and start a new project by choosing File | New | Project. In the New Project window, type the name of your project and its location. A project is a method of organizing files in a selected location. You can choose a name for the project and select its location on disk. When you create a new project, by default you start a new solution. You can add a project to an existing solution using Solution Explorer.

Once the new project and solution are created, Solution Explorer shows nodes in each project for Connections, Queries, and Miscellaneous. To open a new Query Editor window, right-click the Queries node and choose New Query.

Data Discovery and Classification

Beginning with Version 17.5, SQL Server Management Studio supports a new application called Data Discovery and Classification, which is a set of services that is used to discover, classify, and report sensitive data in the database. Sensitive data includes business, financial, and healthcare information.

The Data Discovery and Classification tool enables you to protect your sensitive data by meeting the data privacy standards and by controlling access to the data. The following tasks are supported by the tool:

- Applying your own classification policy
- Applying recommended classification
- Summarizing the classification using a report

The following subsections describe these tasks.

Applying Your Own Classification Policy
We will use the **sample** database to demonstrate how to apply your own classification policy to the sensitive data of a database. For this example, we will classify values of the **budget** column of the **project** table as confidential.

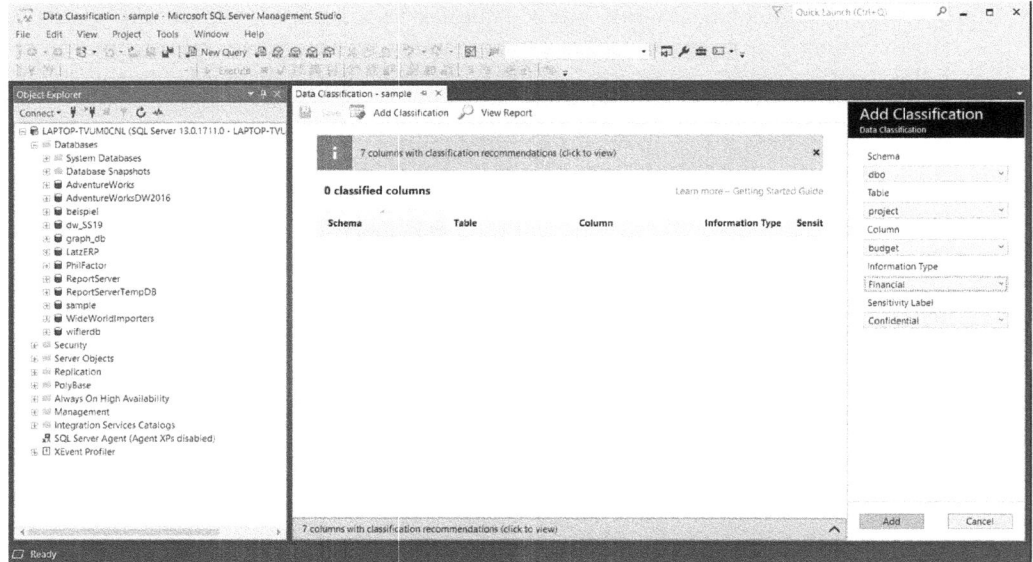

Figure 3-8 The Add Classification pane (right) for the budget column

To perform the task, right-click the **sample** database and select Tasks | Data Discovery and Classification | Classify Data. In the Data Classification window, click the Add Classification button in the middle of the pane. The Add Classification pane on the right side of the subsequent window (see Figure 3-8) allows you to specify the column that should be classified. Therefore, open the Table drop-down list and select **project** from the list of all tables. Next, open the Column drop-down list and select **budget** from the list of all columns of the **project** table.

The next drop-down list, Information Type, allows you to give a more specific description of the data you want to classify. (For the **budget** column, I selected the **Financial** option.) Finally, the Sensitivity Label field specifies the classification level of the selected data. Choose Confidential, as shown in Figure 3-8.

Generally, you can choose from among the following classification levels in the Sensitivity Label drop-down list:

- **Public** Data whose unauthorized disclosure, alteration, or destruction would result in little or no risk to the company that owns it. Examples of public data include press releases and research publications.

- **General** General personal data that can be identified by reference to an identifier, such as a name or an identification number.

- **Confidential** Data whose access is restricted according to the data classification scheme defined by the particular organization.

- **Confidential – GDPR** Confidential data that additionally falls within the scope of the European Union's General Data Protection Regulation (GDPR) rules. The GDPR is a set

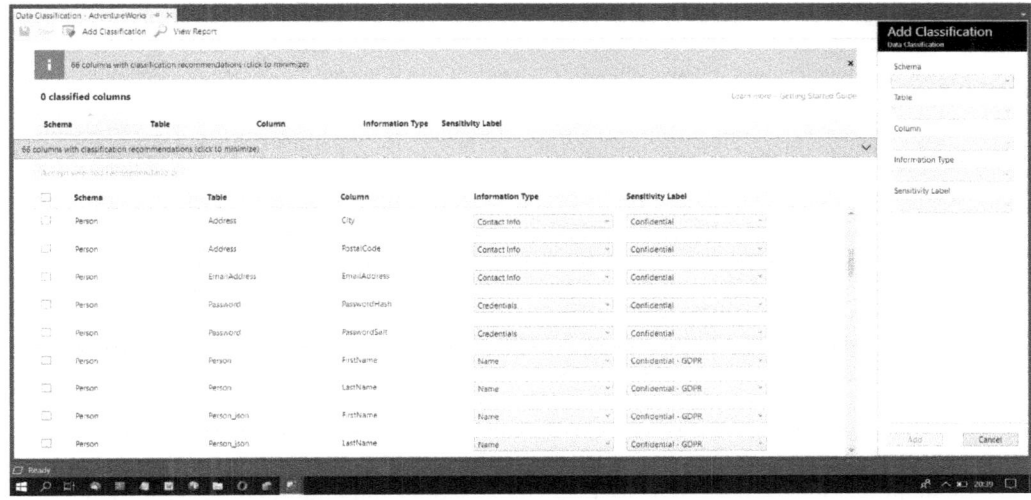

Figure 3-9   Recommendations for the AdventureWorks database

of rules issued by the EU that defines personal data as any information that can identify a natural person, directly or indirectly, by reference to an identifier. Any personal data that is collected from individuals in EU countries is subject to GDPR.

- **Highly Confidential**   Data whose access in any form, including paper or electronic, is restricted to authorized individuals only. Transmitting or storing this data without encryption is prohibited.
- **Highly Confidential – GDPR**   Highly confidential data that additionally falls within the scope of the GDPR rules.

Applying and Summarizing the Recommended Classification

Another feature of the Data Discovery and Classification tool is the capability to scan and discover the sensitive data in your database. While the scanning process of the original **sample** database does not lead to discovery of any sensitive data, I will use the **AdventureWorks** database to describe this process. (**AdventureWorks** 2016 is one of Microsoft's sample databases and can be downloaded from the MS site.)

To perform the task, you use the same sequence of steps as in the previous subsection. The only difference is the selection of the **AdventureWorks** database instead of **sample**. As you can see from Figure 3-9, the tool recommends 66 columns from the **AdventureWorks** database to be classified.

After reviewing the recommended classifications, you can change the Information Type and the Sensitivity Label values provided by the tool with any value listed previously according to your company policies and/or standards using the drop-down boxes to the right of each column. To apply the classification you have chosen, select all recommendations or a set of recommendations on specific columns by checking the check box to the left of each column. Next, click the blue Accept Selected Recommendations button.

After you accept the recommended classification on the selected columns, a notification message informs you that the changes will not be updated until you save all changes. Therefore, click the Save button in the upper-left corner of the window.

Additionally, the Discovery and Classification tool allows you to generate a report that summarizes the classification state of the database. To do this, click the View Report button in the top menu of the window, after performing the previous classifications. The displayed SQL Data Classification Report summarizes the classification state of the selected database.

Azure Data Studio

Prior to SQL Server 2017, the Database Engine was supported by only one platform—Windows. In SQL Server 2017, Microsoft added support for other operating systems, namely Linux platforms. Therefore, since SQL Server 2017, the database system can also be used with the Linux operating system and its derivatives.

Support for Linux requires another front-end tool, because SSMS runs exclusively on Windows. This tool, Azure Data Studio, is a cross-platform, desktop front-end tool for data professionals that can be used as a data platform on Linux and macOS. (Windows is supported, too.)

NOTE   Microsoft releases new versions of Azure Data Studio approximately every six weeks. For this reason, the features described in this chapter may be modified by the time you read this book.

The functionality and layout of Azure Data Studio is almost identical on Windows and Linux platforms. For this reason, only the installation of this front-end tool is presented in separate subsections for Windows and Ubuntu. Configuration of the tool and introduction of its features will be discussed without distinction between different platforms. Therefore, we will discuss the following topics:

- Installation
- Configuration
- Object Explorer
- Code Editor
- Database dashboards and customization

Installation of Azure Data Studio

Installation of Azure Data Studio on Windows is described first, followed by the installation on Linux in general, and on Ubuntu in particular.

Windows Installation

The installation of Azure Data Studio on Windows is easy to manage. The Microsoft download page for Azure Data Studio offers three options for Windows:

- User Installer
- System Installer
- .zip

The difference between User Installer and System Installer is that the former gives you different options during the installation process than the latter. (For this reason, Microsoft generally recommends to choose User Installer.) Using the .zip option, you download the .zip file with all necessary files, including the .exe file.

After downloading the files from the chosen option, you have to execute it. After that, the whole installation is straightforward.

The functionality of Azure Data Studio on Windows is similar to the functionality of SQL Server Management Studio. The main difference is that SSMS is a mature tool, while Azure Data Studio is a new tool, which, at this moment, supports only a subset of features already implemented for SSMS. (You can expect that over time the tool will be extended with several other features.)

Linux Installation

As you already know, Azure Data Studio is a "front-end" tool for SQL Server that runs on Linux, Windows, and macOS. This section describes the installation of the tool for the Ubuntu operating system.

NOTE The installation of Azure Data Studio on Red Hat Linux and SuSe Linux is similar and is explained in Microsoft Docs (https://docs.microsoft.com).

Open the web browser on your Ubuntu system and go to www.microsoft.com. Click the magnifying-glass icon in the upper-right corner, type **Download and Install Azure Data Studio** in the Search box, and press ENTER. In the search results, click the top link to open the page with the same name. For Ubuntu, you will see three different types of installers:

- **.deb**   Debian software repository
- **.rpm**   Package in .rpm format
- **.tar.gz**   Files in a compressed format

We will perform the Debian installation. Therefore, click the .deb installer. After clicking, choose Save File and click OK. The .deb file will be downloaded under the Files/Download directory of your operating system. (You can also use the default method and open the .deb file immediately using the Software installer.) When the download process is finished, double-click the file to launch the installation. The screen shown in Figure 3-10 appears. Click the Install button. To proceed with the installation process, you have to perform authentication. Enter the administrator user ID and password of your Ubuntu operating system. After that, the installation process begins.

After the installation process has completed, go to the Ubuntu applications to verify the Azure Data Studio icon is present. Double-click it to launch the Azure Data Studio initial screen.

NOTE The preceding description explains how to perform the graphical installation of Azure Data Studio. You can choose to do it manually, using the command-line interface. The description of the manual installation can be found in Microsoft Docs.

Figure 3-10   Graphical interface to start the installation of Azure Data Studio on Ubuntu

The main Azure Data Studio user interface is shown in Figure 3-11. The interface shows the following icons in the left vertical pane:

- Servers (the list of all existing servers)
- Task History (the list of all administrative and management tasks executed in the past)
- Explorer (starts Object Explorer; described later in the "Object Explorer" section)
- Search (starts the search for the specified object)
- Source Control (a version control system designed to track changes in source code)
- Extensions (described later in the "Database Dashboards and Customization" section)
- Azure (the component to explore Azure Data Studio resources)

Configuration

Each user can change the default settings of Azure Data Studio by using the File | Preferences | Settings menu options. Preferences are customized by a JSON file called settings.json. Figure 3-12 shows the configuration interface of Azure Data Studio.

The pane on the left side of Figure 3-12 shows all default user settings. The pane on the right can be used as a workspace, where you can move one or more default settings from the left pane to change them.

To demonstrate how you can change settings, I will change the default font family. Azure Data Studio provides a pencil icon that you can use to copy the default settings into the right pane and make changes to them. Go to the "Controls the font family" part of the JSON document and click Edit and Copy Settings. (Both appear subsequently when you move your cursor to the pencil icon, that appears to the left of the text.) Now, the selected settings show up

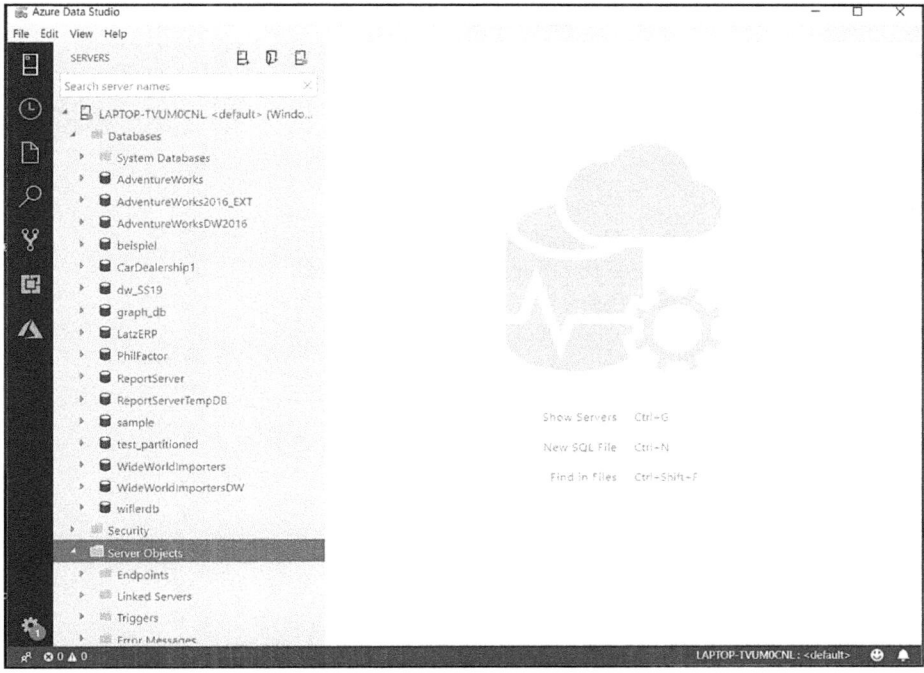

Figure 3-11   The main Azure Data Studio user interface

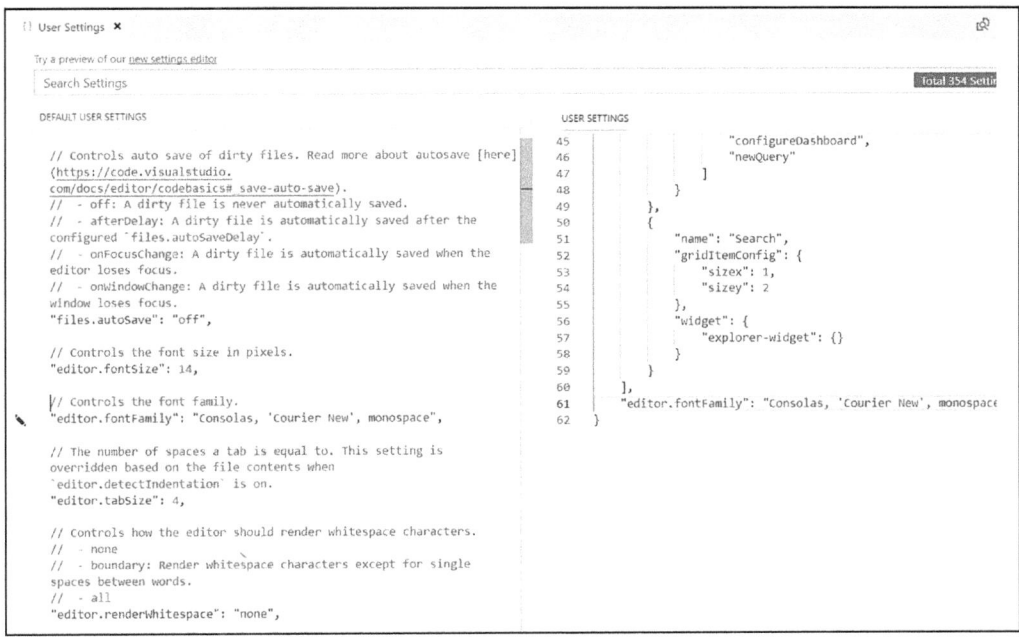

Figure 3-12   Azure Data Studio: User Settings configuration interface

in the right pane (see Figure 3-12) and you can modify them. To save updates, close the User Settings pane and click Save in the window, which appears next. The settings.json file will be modified and the new settings will be activated when you start Azure Data Studio again.

Object Explorer

Object Explorer in Azure Data Studio is similar to the SSMS component with the same name. In other words, Azure Data Studio's Object Explorer allows users to view database objects associated with SQL Server. These objects can be traversed by expanding (and collapsing) a tree view of them.

> **NOTE** Object Explorer in Azure Data Studio does not have the complete functionality available in SSMS's Object Explorer, but Microsoft continues to enhance the functionality of the former with each new version of Azure Data Studio.

Besides the navigation functionality, you can right-click several objects to display the extended functionality of them. For instance, when you right-click the name of a database, the following extended functionality appears (see Figure 3-13):

- **Manage** Opens dashboards for different tasks, such as database backup and restore
- **New Query** Opens the new Code Editor window
- **Refresh** Updates the database and its objects

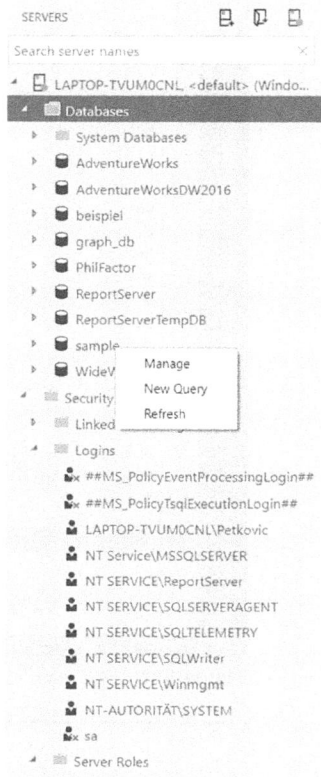

Figure 3-13 Azure Data Studio: Object Explorer interface

Code Editor

To connect to Azure Data Studio, type the name of your instance of the Database Engine in the Server dialog field. Do not forget to write the fully qualified name in the form *server_name\instance_name*. Figure 3-14 shows the server dashboard after a successful connection.

To demonstrate how Code Editor works, we will modify the first name of James James in the **employee** table of the **sample** database. On the dashboard, right-click dbo.employee and select Edit. The content of the table appears. Change the first name of Mr. James to **Andrew** and click Run. When you close the window pane, the primary form of Azure Data Studio (see Figure 3-14) appears.

Database Dashboards and Customization

Azure Data Studio has two dashboards: Server and Database. These dashboards are populated by objects called widgets. When you connect to the database system, the Server dashboard is shown (see Figure 3-14). This dashboard contains several Azure Data Studio's built-in sample widgets, which are dedicated to the following tasks:

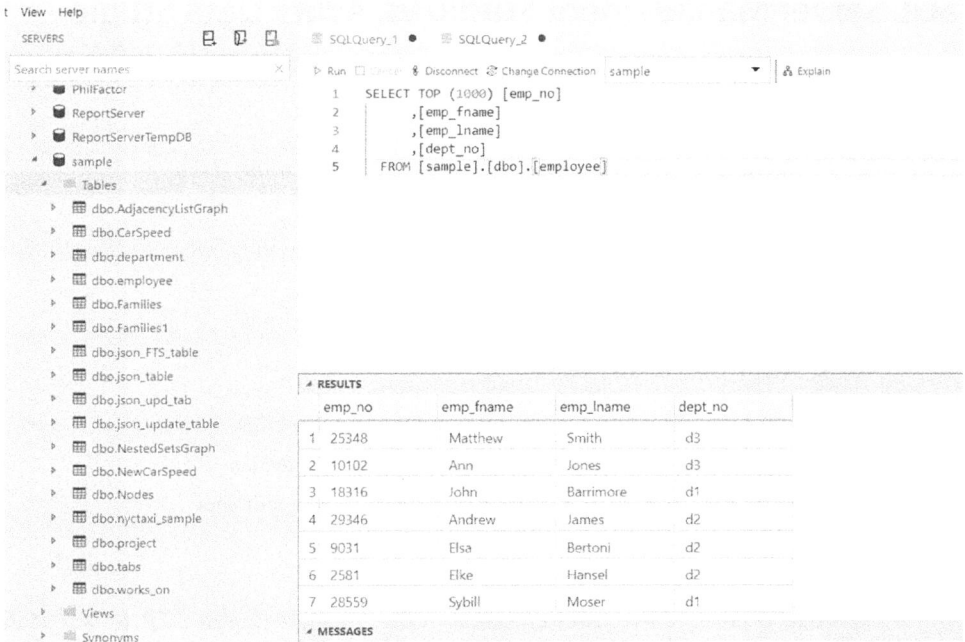

Figure 3-14   The Server dashboard

- Find database(s)
- Perform database backups
- Perform general tasks

The Server dashboard can be viewed by right-clicking the server name and selecting Manage. Similarly, you can view the Database dashboard and a particular database by right-clicking a database in the Object Explorer tree.

The customization of functionality in Azure Data Studio can be done in two different ways. First, you can create your own widgets, which are displayed with the Server dashboard or Database dashboard.

> **NOTE** At the time of writing, the creation of user-defined widgets is not very user-friendly. You have to generate several JSON documents and manually merge different parts of them to create such a widget. For this reason, we will not dive deeper into this topic.

Another way to customize Azure Data Studio is by using extensions. Extensions are (usually) built by the SQL Server community. The installed extensions can be displayed by selecting the Extensions icon, which is the second-to-last icon in the far-left pane in Azure Data Studio. (The other way to select extensions is by pressing CTRL-SHIFT-X.)

SQL Server Management Studio vs. Azure Data Studio

The following two lists give general suggestions to help you decide whether to use SSMS or Azure Data Studio.

Use SQL Server Management Studio if you:

- Spend most of your time on database administration tasks
- Perform sophisticated administrative tasks
- Perform security management, including user management, vulnerability assessment, and configuration of security features
- Need to make use of performance tuning advisors

Use Azure Data Studio if you:

- Need to run your front-end tool on Linux or macOS
- Spend most of your time editing or executing queries
- Need the ability to quickly chart and visualize result sets

Summary

This chapter covered two front-end tools of the Database Engine: SQL Server Management Studio and Azure Data Studio. SSMS is very useful for end users and administrators alike, enabling them to perform many administrative functions. SSMS contains, among others, the following components:

- **Registered Servers**   Allows you to register instances of the Database Engine and connect to them.
- **Object Explorer**   Contains a tree view of all the database objects in a server.
- **Query Editor**   Allows end users to generate, execute, and store Transact-SQL statements. Additionally, it provides the ability to analyze queries by displaying the execution plan.
- **Solution Explorer**   Allows you to create solutions. A solution can have zero or more projects associated with it.

Microsoft introduced Azure Data Studio to provide a front-end tool when database applications run on Linux or macOS. The components of this tool are

- Configuration of user settings
- Object Explorer
- Code Editor
- Database dashboards and customization

The next chapter introduces the Transact-SQL language and describes its main components. After introducing the basic concepts and existing data types, the chapter also describes system functions that Transact-SQL supports.

Exercises

E.3.1   Using SQL Server Management Studio, create a database called **test**. Store the database in a file named **testdate_a** in the directory C:\tmp and allocate 10MB of space to it. Configure the file in which the database is located to grow in increments of 2MB, not to exceed a total of 20MB.

E.3.2   Using SQL Server Management Studio, change the transaction log for the **test** database. Give the file an initial size of 3MB, with growth of 20 percent. Allow the file for the transaction log to autogrow.

E.3.3   Using SQL Server Management Studio, allow only the database owner and system administrator to use the **test** database. Is it possible that both users could use the database at the same time?

E.3.4   Using SQL Server Management Studio, create all four tables of the **sample** database (see Chapter 1) with all their columns.

E.3.5   Using SQL Server Management Studio, view which tables the **AdventureWorks** database contains. After that, choose the **Person.Address** table and view its properties.

E.3.6   Using Query Editor, how can you make the **test** database the current database?

E.3.7   Using Query Editor, make the **AdventureWorks** database the current database and execute the following Transact-SQL statement:

```
SELECT * FROM Sales.Customer
```

How can you stop the execution of the statement?

E.3.8   Store the Transact-SQL statement in E.3.7 in the file C:\tmp\createdb.sql.

E.3.9   Using Query Editor, change the output of the SELECT statement (E.3.7) so that the results appear as the text (and not as the grid).

PART

II

Transact-SQL Language

4

SQL Components

In This Chapter

- SQL's Basic Objects
- Data Types
- Transact-SQL Functions
- Scalar Operators
- NULL Values

This chapter introduces the elementary objects and basic operators supported by the Transact-SQL language. First, the basic language elements, including constants, identifiers, and delimiters, are described. Then, because every elementary object has a corresponding data type, data types are discussed in detail. Additionally, all existing operators and functions are explained. At the end of the chapter, NULL values are introduced.

SQL's Basic Objects

The language of the Database Engine, Transact-SQL, has the same basic features as other common programming languages:

- Literal values (also called constants)
- Identifiers
- Delimiters
- Comments
- Reserved keywords

The following sections describe these features.

Literal Values

A *literal* value is, for example, an alphanumerical, hexadecimal, or numeric constant. A string constant contains one or more characters of the character set enclosed in two single straight quotation marks (' ') or double straight quotation marks (" "). (Single quotation marks are preferred due to the multiple uses of double quotation marks, as discussed in a moment.) If you want to include a single quotation mark within a string delimited by single quotation marks, use two consecutive single quotation marks within the string. Hexadecimal constants are used to represent nonprintable characters and other binary data. Each hexadecimal constant begins with the characters '0x' followed by an even number of characters or numbers. Examples 4.1 and 4.2 illustrate some valid and invalid string constants and hexadecimal constants.

Example 4.1
Some valid string constants and hexadecimal constants follow:

'Philadelphia'
"Berkeley, CA 94710"
'9876'
'Apostrophe is displayed like this: can''t' (note the two consecutive single quotation marks)
0x53514C0D

Example 4.2
The following are *not* string constants:

'AB'C' (odd number of single quotation marks)
'New York" (same type of quotation mark—single or double—must be used at each end of the string)

   The numeric constants include all integer, fixed-point, and floating-point values with and without signs (see Example 4.3).

Example 4.3
The following are numeric constants:

130
−130.00
−0.357E5 (scientific notation—*nEm* means *n* multiplied by 10^m)
22.3E−3

   A constant always has a data type and a length, and both depend on the format of the constant. Additionally, every numeric constant has a precision and a scale factor. (The data types of the different kinds of literal values are explained later in this chapter.)

Identifiers

In Transact-SQL, identifiers are used to identify database objects such as databases, tables, and indices. They are represented by character strings that may include up to 128 characters and may contain letters, numerals, or the following characters: _, @, #, and $. Each name must begin with a letter or one of the following characters: _, @, or #. The character # at the beginning of a table or stored procedure name denotes a temporary object, while @ at the beginning of a name denotes a variable. As explained in the next section, these rules don't apply to delimited identifiers (also known as quoted identifiers), which can contain, or begin with, any character (other than the delimiters themselves).

Delimiters

In Transact-SQL, double quotation marks have two meanings. In addition to enclosing strings, double quotation marks can also be used as delimiters for so-called *delimited identifiers*. Delimited identifiers are a special kind of identifier usually used to allow the use of reserved keywords as identifiers and also to allow spaces in the names of database objects.

NOTE Two key differences between regular and delimited identifiers are that delimited identifiers are enclosed in double quotation marks and are case sensitive. (Transact-SQL also supports the use of square brackets instead of double quotation marks.) Double quotation marks are used only for delimiting strings. Generally, delimited identifiers were introduced to allow the specification of identifiers, which are otherwise identical to reserved keywords. Also, delimited identifiers may contain characters that are normally illegal within identifier names, such as blanks.

In Transact-SQL, the use of double quotation marks is defined using the QUOTED_ IDENTIFIER option of the SET statement. If this option is set to ON, which is the default value, an identifier in double quotation marks will be defined as a delimited identifier. In this case, double quotation marks cannot be used for delimiting strings.

Comments

There are two different ways to specify a comment in a Transact-SQL statement. Using the pair of characters /* and */ marks the enclosed text as a comment. In this case, the comment may extend over several lines. Furthermore, the characters -- (two hyphens) indicate that the remainder of the current line is a comment. (The two -- comply with the ANSI SQL standard, while /* and */ are the extensions of Transact-SQL.)

Reserved Keywords

Each programming language has a set of names with reserved meanings, which must be written and used in the defined format. Names of this kind are called *reserved keywords*. Transact-SQL uses a variety of such names, which, as in many other programming languages, cannot be used as object names, unless the objects are specified as delimited identifiers.

NOTE In Transact-SQL, the names of all data types and system functions, such as CHARACTER and INTEGER, are not reserved keywords. Therefore, they can be used to denote objects. (Do not use data types and system functions as object names! Such use makes Transact-SQL statements difficult to read and understand.)

Data Types

All the data values of a column must be of the same data type. (The only exception specifies the values of the SQL_VARIANT data type.) Transact-SQL uses different data types, which can be categorized as follows:

- Numeric data types
- Character data types
- Temporal (date and/or time) data types
- Miscellaneous data types

The following sections describe all these categories.

Numeric Data Types

Numeric data types are used to represent numbers. The following table shows the list of all numeric data types:

Data Type	Explanation
INTEGER	Represents integer values that can be stored in 4 bytes. The range of values is −2,147,483,648 to 2,147,483,647. INT is the short form for INTEGER.
SMALLINT	Represents integer values that can be stored in 2 bytes. The range of values is −32768 to 32767.
TINYINT	Represents nonnegative integer values that can be stored in 1 byte. The range of values is 0 to 255.
BIGINT	Represents integer values that can be stored in 8 bytes. The range of values is -2^{63} to $2^{63} - 1$.
DECIMAL(p,[s])	Describes fixed-point values. The argument **p** (precision) specifies the total number of digits with assumed decimal point **s** (scale) digits from the right. DECIMAL values are stored, depending on the value of **p**, in 5 to 17 bytes. DEC is the short form for DECIMAL.
NUMERIC(p,[s])	Synonym for DECIMAL.
REAL	Used for floating-point values. The range of positive values is approximately −3.40E + 38 to −1.18E − 38, and the range of negative values is approximately 1.18E − 38 to 3.40E + 38 (the value zero can also be stored).
FLOAT[(n)]	Represents floating-point values, like REAL. **n** is the number of bits that are used to store the mantissa, with **n** < 25 as single precision (stored in 4 bytes) and **n** >= 25 as double precision (stored in 8 bytes).
MONEY	Used for representing monetary values. MONEY values correspond to 8-byte DECIMAL values and are rounded to four digits after the decimal point.
SMALLMONEY	Corresponds to the MONEY data type but is stored in 4 bytes.

Character Data Types

There are two general forms of character data types. They can be strings of single-byte characters or strings of Unicode characters. (Unicode uses 2 bytes to specify one character.) Further, strings can have fixed or variable length. The following character data types are used:

Data Type	Explanation
CHAR[(n)]	Represents a fixed-length string of single-byte characters, where **n** is the number of characters inside the string. The maximum value of **n** is 8000. CHARACTER(n) is an additional equivalent form for CHAR(n). If **n** is omitted, the length of the string is assumed to be 1.
VARCHAR[(n)]	Describes a variable-length string of single-byte characters (0 < **n** ≤ 8000). In contrast to the CHAR data type, the values for the VARCHAR data type are stored in their actual length. This data type has two synonyms: CHAR VARYING and CHARACTER VARYING.
NCHAR[(n)]	Stores fixed-length strings of Unicode characters. The main difference between the CHAR and NCHAR data types is that each character of the NCHAR data type is stored in 2 bytes, while each character of the CHAR data type uses 1 byte of storage space. Therefore, the maximum number of characters in a column of the NCHAR data type is 4000.
NVARCHAR[(n)]	Stores variable-length strings of Unicode characters. The main difference between the VARCHAR and the NVARCHAR data types is that each NVARCHAR character is stored in 2 bytes, while each VARCHAR character uses 1 byte of storage space. The maximum number of characters in a column of the NVARCHAR data type is 4000.

NOTE The VARCHAR data type is identical to the CHAR data type except for one difference: if the content of a CHAR(n) string is shorter than **n** characters, the rest of the string is padded with blanks. (A value of the VARCHAR data type is always stored in its actual length.)

Temporal Data Types

Transact-SQL supports the following temporal data types:

- DATETIME
- SMALLDATETIME
- DATE
- TIME
- DATETIME2
- DATETIMEOFFSET

The DATETIME and SMALLDATETIME data types specify a date and time, with each value being stored as an integer value in 4 bytes or 2 bytes, respectively. Values of DATETIME and SMALLDATETIME are stored internally as two separate numeric values. The date value of DATETIME is stored in the range 01/01/1753 to 12/31/9999. The analog value of SMALLDATETIME is stored in the range 01/01/1900 to 06/06/2079. The time component is stored in the second 4-byte (or 2-byte for SMALLDATETIME) field as the number of three-hundredths of a second (DATETIME) or minutes (SMALLDATETIME) that have passed since midnight.

The use of DATETIME and SMALLDATETIME is rather inconvenient if you want to store only the date part or time part. For this reason, the Database Engine introduced the data types DATE and TIME, which store just the DATE or TIME component of a DATETIME, respectively. The DATE data type is stored in 3 bytes and has the range 01/01/0001 to 12/31/9999. The TIME data type is stored in 3–5 bytes and has an accuracy of 100 nanoseconds (ns).

The DATETIME2 data type stores high-precision date and time data. The data type can be defined for variable lengths depending on the requirement. (The storage size is 6–8 bytes). The accuracy of the time part is 100 ns. This data type isn't aware of Daylight Saving Time.

All the temporal data types described thus far don't have support for the time zone. The data type called DATETIMEOFFSET has the time zone offset portion. For this reason, it is stored in 6–8 bytes. (All other properties of this data type are analogous to the corresponding properties of DATETIME2.)

The date value in Transact-SQL is by default specified as a string in a format like 'mmm dd yyyy' (e.g., 'Jan 10 1993') inside two single quotation marks or double quotation marks. (Note that the relative order of month, day, and year can be controlled by the SET DATEFORMAT statement. Additionally, the system recognizes numeric month values with delimiters of / or –.) Similarly, the time value is specified in the format 'hh:mm' and the Database Engine uses 24-hour time ('23:24', for instance).

NOTE Transact-SQL supports a variety of input formats for datetime values. As you already know, both objects are identified separately; thus, date and time values can be specified in any order or alone. If one of the values is omitted, the system uses the default values. (The default value for time is 12:00 AM.)

Examples 4.4 and 4.5 show the different ways date and time values can be written using the different formats.

Example 4.4
The following date descriptions can be used:

'28/5/1959' (with SET DATEFORMAT dmy)
'May 28, 1959'
'1959 MAY 28'

Example 4.5
The following time expressions can be used:

'8:45 AM'
'4 pm'

Miscellaneous Data Types

Transact-SQL supports several data types that do not belong to any of the data type groups described previously:

- Binary data types
- BIT
- Large object data types
- CURSOR (discussed in Chapter 8)
- UNIQUEIDENTIFIER
- SQL_VARIANT
- TABLE (discussed in Chapters 5 and 8)
- XML (discussed in the previous edition of this book)
- Spatial (e.g., GEOGRAPHY and GEOMETRY) data types (discussed in Chapter 30)
- HIERARCHYID
- TIMESTAMP
- User-defined data types (discussed in Chapter 5)

The following sections describe each of these data types (other than those designated as being discussed in another chapter).

Binary and BIT Data Types

BINARY and VARBINARY are the two binary data types. They describe data objects being represented in the internal format of the system. They are used to store bit strings. For this reason, the values are entered using hexadecimal numbers.

The values of the BIT data type are stored in a single bit. Therefore, values of up to eight 1-bit columns are stored in 1 byte. The following table summarizes the properties of these data types:

Data Type	Explanation
BINARY[(n)]	Specifies a bit string of fixed length with exactly **n** bytes $(0 < \mathbf{n} \leq 8000)$.
VARBINARY[(n)]	Specifies a bit string of variable length with up to **n** bytes $(0 < \mathbf{n} \leq 8000)$.
BIT	Used for specifying the Boolean data type with three possible values: FALSE, TRUE, and NULL.

Large Object Data Types

Large objects (LOBs) are data objects with the maximum length of 2GB. These objects are generally used to store large text data and to load modules and audio/video files. Transact-SQL supports the following LOB data types:

- VARCHAR(max)
- NVARCHAR(max)
- VARBINARY(max)

You can use the same programming model to access values of standard data types and LOBs. In other words, you can use convenient system functions and string operators to work with LOBs.

The Database Engine uses the max parameter with the data types VARCHAR, NVARCHAR, and VARBINARY to define variable-length columns. When you use max by default (instead of an explicit value), the system analyzes the length of the particular string and decides whether to store the string as a convenient value or as a LOB. The max parameter indicates that the size of column values can reach the maximum LOB size of the current system.

Although the Database Engine decides how a LOB will be stored, you can override this default specification using the **sp_tableoption** system procedure with the LARGE_VALUE_ TYPES_OUT_OF_ROW option. If the option's value is set to 1, the data in columns declared using the max parameter will be stored separately from all other data. If this option is set to 0, the Database Engine stores all values for the row size < 8000 bytes as regular row data.

You can apply the FILESTREAM attribute to a VARBINARY(max) column to store large binary data directly in the NTFS file system. The main advantage of this attribute is that the size of the corresponding LOB is limited only by the file system volume size. (This storage attribute will be described in the upcoming "Storage Options" section.)

UNIQUEIDENTIFIER Data Type

As its name implies, a value of the UNIQUEIDENTIFIER data type is a unique identification number stored as a 16-byte binary string. This data type is closely related to the globally unique identifier (GUID), which guarantees uniqueness worldwide. Hence, using this data type, you can uniquely identify data and objects in distributed systems.

The initialization of a column or a variable of the UNIQUEIDENTIFIER type can be provided using the functions NEWID and NEWSEQUENTIALID, as well as with a string constant written in a special form using hexadecimal digits and hyphens. (The functions NEWID and NEWSEQUENTIALID are described in the section "System Functions" later in this chapter.)

A column of the UNIQUEIDENTIFIER data type can be referenced using the keyword ROWGUIDCOL in a query to specify that the column contains ID values. (This keyword does not generate any values.) A table can have several columns of the UNIQUEIDENTIFIER type, but only one of them can have the ROWGUIDCOL column attribute.

SQL_VARIANT Data Type

The SQL_VARIANT data type can be used to store values of various data types at the same time, such as numeric values, strings, and date values. (The only types of values that cannot be stored are TIMESTAMP values.) Each value of an SQL_VARIANT column has two parts: the data value and the information that describes the value. (This information contains all properties of the actual data type of the value, such as length, scale, and precision.)

Transact-SQL supports the SQL_VARIANT_PROPERTY function, which displays the attached information for each value of an SQL_VARIANT column. For the use of the SQL_ VARIANT data type, see Example 5.5 in Chapter 5.

NOTE Declare a column of a table using the SQL_VARIANT data type only if it is really necessary. A column should have this data type if its values may be of different types or if determining the type of a column during the database design process is not possible.

HIERARCHYID Data Type

The HIERARCHYID data type is used to store an entire hierarchy. (For instance, you can use this data type to store a hierarchy of all employees or a hierarchy of all folder lists.) It is implemented as a Common Language Runtime (CLR) user-defined type that comprises several system functions for creating and operating on hierarchy nodes. The following functions, among others, belong to the methods of this data type: GetLevel(), GetAncestor(), GetDescendant(), Read(), and Write(). (The detailed description of this data type is outside the scope of this book.)

TIMESTAMP Data Type

The TIMESTAMP data type specifies a column being defined as VARBINARY(8) or BINARY(8), depending on nullability of the column. The system maintains a current value (not a date or time) for each database, which it increments whenever any row with a TIMESTAMP column is inserted or updated and sets the TIMESTAMP column to that value. Thus, TIMESTAMP columns can be used to determine the relative time at which rows were last changed. (ROWVERSION is a synonym for TIMESTAMP.)

NOTE The value stored in a TIMESTAMP column isn't important by itself. This column is usually used to detect whether a specific row has been changed since the last time it was accessed.

Storage Options

There are two different storage options, each of which allows you to store LOBs and to save storage space:

- FILESTREAM
- Sparse columns

The following subsections describe these options.

FILESTREAM Storage

The Database Engine supports the storage of LOBs using the VARBINARY(max) data type. The property of this data type is that binary large objects (BLOBs) are stored inside the database. This solution can cause performance problems if the stored files are very large, as in the case of video or audio files. In that case, it is common to store such files outside the database, in external files.

The FILESTREAM storage option supports the management of LOBs, which are stored in the NTFS file system. The main advantage of this type of storage is that the Database Engine is able to manage FILESTREAM data even though it is stored outside the database. Therefore, this storage type has the following properties:

- You use the CREATE TABLE statement to store FILESTREAM data and use the DML statements (SELECT, INSERT, UPDATE, and DELETE) to query and update such data.

- The Database Engine assures the same level of security for FILESTREAM data as for relational data stored inside the database.

The creation of FILESTREAM data will be described in detail in Chapter 5.

Sparse Columns

The aim of sparse columns as a storage option is quite different from the FILESTREAM storage support. Whereas FILESTREAM is Microsoft's solution for the storage of LOBs outside the database, sparse columns help to minimize data storage space. These columns provide an optimized way to store column values, which are predominantly NULL. (NULL values are described at the end of this chapter.) If you use sparse columns, NULL values require no disk space, but on the other side, non-NULL data needs an additional 2 to 4 bytes, depending on the data type of the non-NULL values. For this reason, Microsoft recommends using sparse columns only when the overall storage space savings will be at least 20 percent.

You specify and access sparse columns in the same way as you specify and access all other columns of a table. This means that the statements SELECT, INSERT, UPDATE, and DELETE can be used to access sparse columns in the same way as you use them for usual columns. (These four SQL statements are described in detail in Chapters 6 and 7.) The only difference is in relation to creation of a sparse column: you use the SPARSE option (after the column name) to specify that a particular column is a sparse column: col_name data_type SPARSE.

If a table has several sparse columns, you can group them in a column set. Therefore, a column set is an alternative way to store and access all sparse columns in a table. For more information concerning column sets, see Microsoft Docs.

Transact-SQL Functions

Transact-SQL functions can be either aggregate functions or scalar functions. The following sections describe these function types.

Aggregate Functions

Aggregate functions are applied to a group of data values from a column. Aggregate functions always return a single value. Transact-SQL supports several groups of aggregate functions:

- Convenient aggregate functions
- Statistical aggregate functions
- User-defined aggregate functions
- Analytic aggregate functions

Statistical and analytic aggregate functions are discussed in Chapter 24. User-defined aggregates are beyond the scope of this book. That leaves the convenient aggregate functions, described next:

- **AVG**   Calculates the arithmetic mean (average) of the data values contained within a column. The column must contain numeric values.
- **MAX** and **MIN**   Calculate the maximum and minimum data value of the column, respectively. The column can contain numeric, string, and date/time values. The maximum (minimum) value of a string depends on the collation that your system uses.
- **SUM**   Calculates the total of all data values in a column. The column must contain numeric values.
- **COUNT**   Calculates the number of (non-NULL) data values in a column. The only aggregate function that is not applied to columns is COUNT(*). This function returns the number of rows (whether or not particular columns have NULL values).
- **COUNT_BIG**   Analogous to COUNT, the only difference being that COUNT_BIG returns a value of the BIGINT data type.

The use of convenient aggregate functions with the SELECT statement can be found in Chapter 6.

Scalar Functions

In addition to aggregate functions, Transact-SQL provides several scalar functions that are used in the construction of scalar expressions. (A scalar function operates on a single value or list of values, whereas an aggregate function operates on the data from multiple rows.) Scalar functions can be categorized as follows:

- Numeric functions
- Date functions
- String functions
- System functions
- Metadata functions

The following sections describe these function types.

Numeric Functions

Numeric functions within Transact-SQL are mathematical functions for modifying numeric values. The following numeric functions are available:

Function	Explanation
ABS(n)	Returns the absolute value (i.e., negative values are returned as positive) of the numeric expression **n**. Examples: SELECT ABS(−5.767) = 5.767 SELECT ABS(6.384) = 6.384

Function	Explanation
ACOS(n)	Calculates arc cosine of **n**. **n** and the resulting value belong to the FLOAT data type.
ASIN(n)	Calculates the arc sine of **n**. **n** and the resulting value belong to the FLOAT data type.
ATAN(n)	Calculates the arc tangent of **n**. **n** and the resulting value belong to the FLOAT data type.
ATN2(n,m)	Calculates the arc tangent of **n/m**. **n**, **m**, and the resulting value belong to the FLOAT data type.
CEILING(n)	Returns the smallest integer value greater than or equal to the specified parameter. Examples: SELECT CEILING(4.88) = 5 SELECT CEILING(−4.88) = −4
COS(n)	Calculates the cosine of **n**. **n** and the resulting value belong to the FLOAT data type.
COT(n)	Calculates the cotangent of **n**. **n** and the resulting value belong to the FLOAT data type.
DEGREES(n)	Converts radians to degrees. Examples: SELECT DEGREES(PI()/2) = 90 SELECT DEGREES(0.75) = 43
EXP(n)	Calculates the value e^n. Example: SELECT EXP(1) = 2.7183
FLOOR(n)	Calculates the largest integer value less than or equal to the specified value **n**. Example: SELECT FLOOR(4.88) = 4
LOG(n)	Calculates the natural (i.e., base e) logarithm of **n**. Examples: SELECT LOG(4.67) = 1.54 SELECT LOG(0.12) = −2.12
LOG10(n)	Calculates the logarithm (base 10) for **n**. Examples: SELECT LOG10(4.67) = 0.67 SELECT LOG10(0.12) = −0.92
PI()	Returns the value of the number pi (3.14).
POWER(x,y)	Calculates the value x^y. Examples: SELECT POWER(3.12,5) = 295.65 SELECT POWER(81,0.5) = 9
RADIANS(n)	Converts degrees to radians. Examples: SELECT RADIANS(90.0) = 1.57 SELECT RADIANS(42.97) = 0.75
RAND()	Returns a random number between 0 and 1 with a FLOAT data type.

Function	Explanation
ROUND(n,p,[t])	Rounds the value of the number **n** by using the precision **p**. Use positive values of **p** to round on the right side of the decimal point and use negative values to round on the left side. An optional parameter **t** causes **n** to be truncated. Examples: SELECT ROUND(5.4567,3) = 5.4570 SELECT ROUND(345.4567,−1) = 350.0000 SELECT ROUND(345.4567,−1,1) = 340.0000
ROWCOUNT_BIG	Returns the number of rows that have been affected by the last Transact-SQL statement executed by the system. The return value of this function has the BIGINT data type.
SIGN(n)	Returns the sign of the value **n** as a number (+1 for positive, −1 for negative, and 0 for zero). Example: SELECT SIGN(0.88) = 1.00
SIN(n)	Calculates the sine of **n**. **n** and the resulting value belong to the FLOAT data type.
SQRT(n)	Calculates the square root of **n**. Example: SELECT SQRT(9) = 3
SQUARE(n)	Returns the square of the given expression. Example: SELECT SQUARE(9) = 81
TAN(n)	Calculates the tangent of **n**. **n** and the resulting value belong to the FLOAT data type.

Date Functions

Date functions calculate the respective date or time portion of an expression or return the value from a time interval. Transact-SQL supports the following date functions:

Function	Explanation
GETDATE()	Returns the current system date and time. Example: SELECT GETDATE() = 2019-01-01 13:03:31.390
DATEPART (item,date)	Returns the specified part **item** of a date **date** as an integer. Examples: SELECT DATEPART(month, '01.01.2019') = 1 (1 = January) SELECT DATEPART(weekday, '01.01.2019') = 7 (3 = Tuesday)
DATENAME (item,date)	Returns the specified part **item** of the date **date** as a character string. Example: SELECT DATENAME(weekday, '01.01.2019') = Tuesday
DATEDIFF (item,dat1,dat2)	Calculates the difference between the two date parts **dat1** and **dat2** and returns the result as an integer in units specified by the value **item**. Example (returns the age of each employee): SELECT DATEDIFF(year, BirthDate, GETDATE()) AS age FROM employee_ext
DATEADD(i,n,d)	Adds the number **n** of units specified by the value **i** to the given date **d**. (**n** could be negative, too.) Example (adds three days to the start date of employment of every employee; see the **sample** database): SELECT DATEADD(DAY,3,HireDate) AS age FROM employee_ext

String Functions

String functions are used to manipulate data values in a column, usually of a character data type. Transact-SQL supports the following string functions:

Function	Explanation
ASCII(character)	Converts the specified character to the equivalent integer (ASCII) code. Returns an integer. Example: SELECT ASCII('A') = 65
CHAR(integer)	Converts the ASCII code to the equivalent character. Example: SELECT CHAR(65) = 'A'
CHARINDEX(z1,z2)	Returns the starting position where the partial string **z1** first occurs in the string **z2**. Returns 0 if **z1** does not occur in **z2**. Example: SELECT CHARINDEX('bl','table') = 3
CONCAT_WS (separator,string1, string2,...)	Joins two or more strings with a separator. The separator specified in the first argument is added between two strings. Example: SELECT CONCAT_WS ('!' , '1st str', '2nd str') = '1st str!2nd str'
DIFFERENCE(z1,z2)	Returns an integer, 0 through 4, that is the difference of SOUNDEX values of two strings **z1** and **z2**. (SOUNDEX returns a number that specifies the sound of a string. With this method, strings with similar sounds can be determined.) Example: SELECT DIFFERENCE('spelling','telling') = 2 (sounds a little bit similar, 0 = doesn't sound similar)
LEFT(z,length)	Returns the first **length** characters from the string **z**.
LEN(z)	Returns the number of characters, instead of the number of bytes, of the specified string expression, excluding trailing blanks.
LOWER(z1)	Converts all uppercase letters of the string **z1** to lowercase letters. Lowercase letters, digits and other characters do not change. Example: SELECT LOWER('BiG') = 'big'
LTRIM(z)	Removes leading blanks in the string **z**. Example: SELECT LTRIM(' String') = 'String'
NCHAR(i)	Returns the Unicode character with the specified integer code, as defined by the Unicode standard.
PATINDEX(%p%,expr)	Returns the starting position of the first occurrence of a pattern **p** in a specified expression **expr**, or 0 if the pattern is not found. Examples (the second query returns all family names from the **customers** column): SELECT PATINDEX('%gs%','longstring') = 4 SELECT RIGHT(ContactName, LEN(ContactName)-PATINDEX('% %', ContactName)) AS First_name FROM Customers
QUOTENAME (char_string)	Returns a Unicode string with the delimiters added to make the input string a valid delimited identifier.
REPLACE(str1,str2,str3)	Replaces all occurrences of the **str2** in the **str1** with the **str3**. Example: SELECT REPLACE('shave','s','be') = behave

Function	Explanation
REPLICATE(z,i)	Repeats string **z i** times. Example: SELECT REPLICATE('a',10) = 'aaaaaaaaaa'
REVERSE(z)	Displays the string **z** in the reverse order. Example: SELECT REVERSE('calculate') = 'etaluclac'
RIGHT(z,length)	Returns the last **length** characters from the string **z**. Example: SELECT RIGHT('Notebook',4) = 'book'
RTRIM(z)	Removes trailing blanks of the string **z**. Example: SELECT RTRIM('Notebook ') = 'Notebook'
SOUNDEX(a)	Returns a four-character SOUNDEX code to determine the similarity between two strings. Example: SELECT SOUNDEX('spelling') = S145
SPACE(length)	Returns a string with spaces of length specified by **length**. Example: SELECT SPACE(4) = ' '
STR(f,[len [,d]])	Converts the specified numeric data **f** into a string. **len** is the length of the string including decimal point, sign, digits, and spaces (10 by default), and **d** is the number of digits to the right of the decimal point to be returned. Example: SELECT STR(3.45678,4,2) = '3.46'
STUFF(z1,a,length,z2)	Replaces the partial string **z1** with the partial string **z2** starting at position **a**, replacing **length** characters of **z1**. Examples: SELECT STUFF('Notebook',5,0, ' in a ') = 'Note in a book' SELECT STUFF('Notebook',1,4, 'Hand') = 'Handbook'
SUBSTRING(z,a,length)	Creates a partial string from string **z** starting at the position **a** with a length of **length**. Example: SELECT SUBSTRING('wardrobe',1,4) = 'ward'
TRANSLATE(expr, from_str,to_str)	Returns **expr** with all occurrences of each character in **from_string** replaced by its *corresponding* character in **to_string**. Characters in **expr** that are not in **from_string** are not replaced. (This function is similar to the REPLACE function, but lets you perform several one-to-one replacements inside one function.) Example: SELECT TRANSLATE ('SQL*Server 2019*ABG','*','_') = 'SQL_Server 2019_ABG'
TRIM(z)	Removes the spaces from the beginning and end of the string **z**. Example: SELECT TRIM(' Notebook ') = 'Notebook'
UNICODE	Returns the integer value, as defined by the Unicode standard, for the first character of the input expression.
UPPER(z)	Converts all lowercase letters of string **z** to uppercase letters. Uppercase letters and digits do not change. Example: SELECT UPPER('loWer') = 'LOWER'

System Functions

System functions of Transact-SQL provide extensive information about database objects. Most system functions use an internal numeric identifier (ID), which is assigned to each database object by the system at its creation. Using this identifier, the system can uniquely identify each database object. System functions provide information about the Database Engine. The following table describes several system functions. (For the complete list of all system functions, please see Microsoft Docs.)

Function	Explanation
CAST(a AS type [(length)]	Converts an expression **a** into the specified data type **type** (if possible). **a** could be any valid expression. Example: SELECT CAST(3000000000 AS BIGINT) = 3000000000
COALESCE(a1,a2,...)	Returns for a given list of expressions **a1**, **a2**,... the value of the first expression that is not NULL.
COL_LENGTH(obj,col)	Returns the length of the column **col** belonging to the database object (table or view) **obj**. Example: SELECT COL_LENGTH('customers', 'cust_ID') = 10
CONVERT(type[(length)],a)	Equivalent to CAST, but the arguments are specified differently. CONVERT can be used with any data type.
CURRENT_TIMESTAMP	Returns the current date and time. Example: SELECT CURRENT_TIMESTAMP ='2019-01-01 17:22:55.670'
CURRENT_USER	Returns the name of the current user.
DATALENGTH(z)	Calculates the length (in bytes) of the result of the expression **z**. Example (returns the length of each field): SELECT DATALENGTH(ProductName) FROM products
GETANSINULL('dbname')	Returns 1 if the use of NULL values in the database **dbname** complies with the ANSI SQL standard. (See also the explanation of NULL values at the end of this chapter.) Example: SELECT GETANSINULL('AdventureWorks') = 1
ISNULL(expr, value)	Returns the value of **expr** if that value is not NULL; otherwise, it returns **value**.
ISNUMERIC(expression)	Determines whether an expression is a valid numeric type.
NEWID()	Creates a unique ID number that consists of a 16-byte binary string intended to store values of the UNIQUEIDENTIFIER data type.
NEWSEQUENTIALID()	Creates a GUID that is greater than any GUID previously generated by this function on a specified computer. (This function can be used only as a default value for a column.)
NULLIF(expr1,expr2)	Returns the NULL value if the expressions **expr1** and **expr2** are equal. Example (returns NULL for the project with the project_ no = 'p1'): SELECT NULLIF(project_no, 'p1') FROM project

Function	Explanation
SERVERPROPERTY (propertyname)	Returns the property information about the database server.
SYSTEM_USER	Returns the login ID of the current user. Example: SELECT SYSTEM_USER = LTB13942\dusan
TRY_CONVERT (data_type [(length)], expression [, style])	Returns a value cast to the specified data type if the cast succeeds; otherwise, returns NULL. Examples: SET DATEFORMAT mdy; SELECT TRY_CONVERT(datetime2, '12/31/2019');
USER_ID([user_name])	Returns the identifier of the user **user_name**. If no name is specified, the identifier of the current user is retrieved. Example: SELECT USER_ID('guest') = 2
USER_NAME([id])	Returns the name of the user with the identifier **id**. If no name is specified, the name of the current user is retrieved. Example: SELECT USER_NAME(1) = 'dbo'

All functions can be nested in any order; for example, REVERSE(CURRENT_USER).

Metadata Functions

Generally, metadata functions return information about the specified database and database objects. The following table describes several metadata functions. (For the complete list of all metadata functions, please see Microsoft Docs.)

Function	Explanation
COL_NAME(tab_id, col_id)	Returns the name of a column belonging to the table with the ID **tab_id** and column ID **col_id**. Example: SELECT COL_NAME(OBJECT_ID('employee') , 3) = 'emp_lname'
COLUMNPROPERTY (id, col, property)	Returns the information about the specified column. Example: SELECT COLUMNPROPERTY(object_id('project'), 'project_no', 'PRECISION') = 4
DATABASEPROPERTYEX (database, property)	Returns the named database property value for the specified database and property. Example (specifies whether the database follows the rules of the SQL-92 standard for allowing NULL values): SELECT DATABASEPROPERTYEX('sample', 'IsAnsiNullDefault') = 0
DB_ID([db_name])	Returns the identifier of the database **db_name**. If no name is specified, the identifier of the current database is returned. Example: SELECT DB_ID('AdventureWorks') = 6

Function	Explanation
DB_NAME([db_id])	Returns the name of the database with the identifier **db_id**. If no identifier is specified, the name of the current database is displayed. Example: SELECT DB_NAME(6) = 'AdventureWorks'
INDEX_COL(table, i, no)	Returns the name of the indexed column in the table **table**, defined by the index identifier **i** and the position **no** of the column in the index.
INDEXPROPERTY(obj_id, index_name, property)	Returns the named index or statistics property value of a specified table identification number, index or statistics name, and property name.
OBJECT_NAME(obj_id)	Returns the name of the database object with the identifier **obj_id**. Example: SELECT OBJECT_NAME(453576654) = 'products'
OBJECT_ID(obj_name)	Returns the identifier of the database object **obj_name**. Example: SELECT OBJECT_ID('products') = 453576654
OBJECTPROPERTY(obj_id,property)	Returns the information about the objects from the current database.

Scalar Operators

Scalar operators are used for operations with scalar values. Transact-SQL supports numeric and Boolean operators as well as concatenation.

There are unary and binary arithmetic operators. Unary operators are + and – (as signs). Binary arithmetic operators are +, –, *, /, and %. (The first four binary operators have their respective mathematical meanings, whereas % is the modulo operator.)

Boolean operators have two different notations depending on whether they are applied to bit strings or to other data types. The operators NOT, AND, and OR are applied to all data types (except BIT). They are described in detail in Chapter 6.

The bitwise operators for manipulating bit strings are listed here, and Example 4.6 shows how they are used:

Symbol	Operator
~	Complement (i.e., NOT)
&	Conjunction of bit strings (i.e., AND)
|	Disjunction of bit strings (i.e., OR)
∧	Exclusive disjunction (i.e., XOR or Exclusive OR)

Example 4.6

~(1001001) = (0110110)
(11001001) | (10101101) = (11101101)

(11001001) & (10101101) = (10001001)
(11001001) ^ (10101101) = (01100100)

The concatenation operator + can be used to concatenate two character strings or bit strings.

Global Variables

Global variables are special system variables that can be used as if they were scalar constants. Transact-SQL supports many global variables, which have to be preceded by the prefix @@. The following table describes several global variables. (For the complete list of all global variables, see Microsoft Docs.)

Variable	Explanation
@@connections	Returns the number of login attempts since starting the system.
@@cpu_busy	Returns the total CPU time (in units of milliseconds) used since starting the system.
@@error	Returns the information about the return value of the last executed Transact-SQL statement.
@@identity	Returns the last inserted value for the column with the IDENTITY property (see Chapter 6).
@@langid	Returns the identifier of the language that is currently used by the database system.
@@language	Returns the name of the language that is currently used by the database system.
@@max_connections	Returns the maximum number of actual connections to the system.
@@procid	Returns the identifier for the stored procedure currently being executed.
@@rowcount	Returns the number of rows that have been affected by the last Transact-SQL statement executed by the system.
@@servername	Retrieves information about the local database server. This information contains the name of the server.
@@spid	Returns the identifier of the server process.
@@version	Returns the current version of the database system software.

NULL Values

A NULL value is a special value that may be assigned to a column. This value normally is used when information in a column is unknown or not applicable. For example, in the case of an unknown home telephone number for a company's employee, it is recommended that the NULL value be assigned to the **home_telephone** column.

Any arithmetic expression results in a NULL if any operand of that expression is itself a NULL value. Therefore, in unary arithmetic expressions (if A is an expression with a NULL value), both +A and −A return NULL. In binary expressions, if one (or both) of the operands A or B has (have) the NULL value, A + B, A − B, A * B, A / B, and A % B also result in a NULL. (The operands A and B have to be numerical expressions.)

If an expression contains a relational operation and one (or both) of the operands has (have) the NULL value, the result of this operation will be NULL. Hence, each of the expressions A = B, A < > B, A < B, and A > B also returns NULL.

In the Boolean AND, OR, and NOT, the behavior of the NULL values is specified by the following truth tables, where T stands for true, U for unknown (NULL), and F for false. In these tables, follow the row and column represented by the values of the Boolean expressions that the operator works on, and the value where they intersect represents the resulting value.

AND	T	U	F		OR	T	U	F		NOT	
T	T	U	F		T	T	T	T		T	F
U	U	U	F		U	T	U	U		U	U
F	F	F	F		F	T	U	F		F	T

Any NULL value in the argument of aggregate functions AVG, SUM, MAX, MIN, and COUNT is eliminated before the respective function is calculated (except for the function COUNT(*)). If a column contains only NULL values, the function returns NULL. The aggregate function COUNT(*) handles all NULL values the same as non-NULL values. If the column contains only NULL values, the result of the function COUNT(DISTINCT column_name) is 0.

A NULL value has to be different from all other values. For numeric data types, there is a distinction between the value zero and NULL. The same is true for the empty string and NULL for character data types.

A column of a table allows NULL values if its definition explicitly contains NULL. On the other hand, NULL values are not permitted if the definition of a column explicitly contains NOT NULL. If the user does not specify NULL or NOT NULL for a column with a data type, the assigned value depends on whether the ANSI_NULL_DFLT_ON option is set to ON or OFF. (The columns of the TIMESTAMP data type can be declared only as NOT NULL.)

Summary

The basic features of Transact-SQL consist of data types, predicates, and functions. Data types comply with data types of the ANSI SQL standard. Transact-SQL supports a variety of useful system functions.

The next chapter introduces you to Transact-SQL statements in relation to SQL's data definition language. This part of Transact-SQL comprises all the statements needed for creating, altering, and removing database objects.

Exercises

E.4.1   What is the difference between the numeric data types INT, SMALLINT, and TINYINT?

E.4.2   What is the difference between the data types CHAR and VARCHAR? When should you use the latter (instead of the former) and vice versa?

E.4.3   How can you set the format of a column with the DATE data type so that its values can be entered in the form 'yyyy/mm/dd'?

In the following two exercises, use the SELECT statement in the Query Editor component window of SQL Server Management Studio to display the result of all system functions and global variables. (For instance, SELECT host_id() displays the ID number of the current host.)

E.4.4   Using system functions, find the ID number of the **test** database (Exercise 3.1).

E.4.5   Using the system variables, display the current version of the database system software and the language that is used by this software.

E.4.6   Using the bitwise operators &, |, and ^, calculate the following operations with the bit strings:

(11100101) & (01010111)
(10011011) | (11001001)
(10110111) ^ (10110001)

E.4.7   What is the result of the following expressions? (A is a numerical expression and B is a logical expression.)

A + NULL
NULL = NULL
B OR NULL
B AND NULL

E.4.8   When can you use both single and double quotation marks to define string and temporal constants?

E.4.9   What is a delimited identifier and when do you need it?

CHAPTER
5

Data Definition Language

In This Chapter

- Creating Database Objects
- Modifying Database Objects
- Removing Database Objects

This chapter describes all the Transact-SQL statements concerning data definition language (DDL). The DDL statements are divided into three groups, which are discussed in turn. The first group includes statements that create objects, the second group includes statements that modify the structure of objects, and the third group includes statements that remove objects.

Creating Database Objects

The organization of a database involves many different objects. All objects of a database are either physical or logical. The physical objects are related to the organization of the data on the physical device (disk). The Database Engine's physical objects are files and filegroups. Logical objects represent a user's view of a database. Databases, tables, columns, and views (virtual tables) are examples of logical objects.

The first database object that has to be created is a database itself. The Database Engine manages both system and user databases. An authorized user can create user databases, while system databases are generated during the installation of the database system.

This chapter describes the creation, alteration, and removal of user databases, while Chapter 15 covers all system databases in detail.

Creation of a Database

Two basic methods are used to create a database. The first method involves using Object Explorer in SQL Server Management Studio (see Chapter 3). The second method involves using the Transact-SQL statement CREATE DATABASE. This statement has the following general form, the details of which are discussed next:

```
CREATE DATABASE db_name
     [ON [PRIMARY] { file_spec1} ,...]
     [LOG ON {file_spec2} ,...]
       [COLLATE collation_name]
  [FOR {ATTACH | ATTACH_REBUILD_LOG } ]
```

NOTE For the syntax of the Transact-SQL statements, the conventions used are those described in the section "Syntax Conventions" in Chapter 1. According to the conventions, optional items appear in brackets, [ ]. Items written in braces, { }, and followed by "…" are items that can be repeated any number of times.

---

db_name is the name of the database. The maximum size of a database name is 128 characters. (The rules for identifiers described in Chapter 4 apply to database names.) The maximum number of databases managed by a single system is 32,767.

All databases are stored in files. These files can be explicitly specified by the system administrator or implicitly provided by the system. If the ON option exists in the CREATE DATABASE statement, all files containing the data of a database are explicitly specified.

NOTE The Database Engine uses disk files to store data. Each disk file contains data of a single database. Files themselves can be organized into filegroups. Filegroups provide the ability to distribute data over different disk drives and to back up and restore subsets of the database (useful for very large databases).

---

file_spec1 represents a file specification, which includes further options such as the logical name of the file, the physical name, and the size. The PRIMARY option specifies the first (and most important) file that contains system tables and other important internal information concerning the database. If the PRIMARY option is omitted, the first file listed in the specification is used as the primary file.

A login account of the Database Engine that is used to create a database is called a database owner. A database can have one owner, who always corresponds to a login account name. The login account, which is the database owner, has the special name **dbo**. This name is always used in relation to a database it owns.

dbo uses the LOG ON option to define one or more files as the physical destination of the transaction log of the database. If the LOG ON option is not specified, the transaction log of the database will still be created because every database must have at least one transaction log file. (The Database Engine keeps a record of each change it makes to the database. The system keeps all those records, in particular before and after values, in one or more files called the transaction log. Each database of the system has its own transaction log. Transaction logs are discussed in detail in Chapter 13.)

With the COLLATE option, you can specify the default collation for the database. If the COLLATE option is not specified, the database is assigned the default collation of the **model** database, which is the same as the default collation of the database system.

The FOR ATTACH option specifies that the database is created by attaching an existing set of files. If this option is used, you have to explicitly specify the first primary file. The FOR ATTACH_REBUILD_LOG option specifies that the database is created by attaching an existing set of operating system files. (Attaching and detaching a database is described later in this chapter.)

During the creation of a new database, the Database Engine uses the **model** database as a template. The properties of the **model** database can be changed to suit the personal conception of the system administrator.

NOTE If you have a database object that should exist in each user database, you should create that object in the **model** database first.

Example 5.1 creates a simple database without any further specifications. To execute this statement, type it in the Query Editor window of SQL Server Management Studio and press F5.

Example 5.1
```
USE master;
CREATE DATABASE sample;
```

Example 5.1 creates a database named **sample**. This concise form of the CREATE DATABASE statement is possible because almost all options of that statement have default values. The system creates, by default, two files. The logical name of the data file is **sample** and its original size is 8MB. Similarly, the logical name of the transaction log is **sample_log** and its original size is 8MB. (Both size values, as well as other properties of the new database, depend on corresponding specifications in the **model** database.)

Example 5.2 creates a database with explicit specifications for database and transaction log files. (You must create the C:\temp directory before you start the following example.)

Example 5.2
```
USE master;
CREATE DATABASE projects
  ON (NAME=projects_dat,
      FILENAME = 'C:\temp\projects.mdf',
      SIZE = 10,
      MAXSIZE = 100,
      FILEGROWTH = 5)
LOG ON
  (NAME=projects_log,
      FILENAME = 'C:\temp\projects.ldf',
      SIZE = 40,
      MAXSIZE = 100,
      FILEGROWTH = 10);
```

Example 5.2 creates a database called **projects**. Because the PRIMARY option is omitted, the first file is assumed to be the primary file. This file has the logical name **projects_dat** and is stored in the file **projects.mdf**. The original size of this file is 10MB. Additional portions of 5MB of disk storage are allocated by the system, if needed. If the MAXSIZE option is not specified or is set to UNLIMITED, the file will grow until the disk is full. (The KB and MB suffixes can be used to specify kilobytes or megabytes, respectively—the default is MB.)

There is also a single transaction log file with the logical name **projects_log** and the physical name **projects.ldf**. All options of the file specification for the transaction log have the same name and meaning as the corresponding options of the file specification for the data file.

Using the Transact-SQL language, you can apply the USE statement to change the database context to the specified database. (The alternative way is to select the database name in the Database drop-down menu in the toolbar of SQL Server Management Studio.)

The system administrator can assign a default database to a user by using the CREATE LOGIN statement or the ALTER LOGIN statement (see also Chapter 12). In this case, the users do not need to execute the USE statement if they want to use their default database.

Creation of a Database Snapshot

The CREATE DATABASE statement can also be used to create a database snapshot of an existing database (source database). A database snapshot is transactionally consistent with the source database as it existed at the time of the snapshot's creation.

The syntax for the creation of a snapshot is

```
CREATE DATABASE database_snapshot_name
    ON  (NAME = logical_file_name,
            FILENAME = 'os_file_name') [ ,...n ]
            AS SNAPSHOT OF source_database_name
```

As you can see, if you want to create a database snapshot, you have to add the AS SNAPSHOT OF clause in the CREATE DATABASE statement. Example 5.3 creates a snapshot of the **AdventureWorks** database and stores it in the C:\temp data directory. (You have to download and create the **AdventureWorks** database, if this database does not exist on your system.) The AdventureWorks database is a sample database of SQL Server and can be downloaded from Microsoft Docs (https://docs.microsoft.com).

Example 5.3
```
USE master;
CREATE DATABASE AdventureWorks_snapshot
    ON (NAME = 'AdventureWorks_Data' ,
            FILENAME = 'C:\temp\snapshot_DB.mdf')
            AS SNAPSHOT OF AdventureWorks;
```

An existing database snapshot is a read-only copy of the corresponding database that reflects the point in time when the database is copied. (For this reason, you can have multiple snapshots for an existing database.) The snapshot file (in Example 5.3, 'C:\temp\snapshot_DB.mdf') contains only the modified data that has changed from the source database. Therefore, the process of creating a database snapshot must include the logical name of each data file from the source database as well as new corresponding physical names.

While the snapshot contains only modified data, the disk space needed for each snapshot is just a small part of the overall space required for the corresponding source database.

NOTE To create snapshots of a database, you need NTFS disk volumes, because only such volumes support the sparse file technology that is used for storing snapshots.

Database snapshots are usually used as a mechanism to protect data against user errors.

Attaching and Detaching Databases

All data of a database can be detached and then attached to the same or another database server. Detaching and attaching a database should be done if you want to move the database.

You can detach a database from a database server by using the **sp_detach_db** system procedure. (The detached database must be in the single-user mode.)

To attach a database, use the CREATE DATABASE statement with the FOR ATTACH clause. When you attach a database, all data files must be available. If any data file has a different path from when the database was first created, you must specify the file's current path.

CREATE TABLE: A Basic Form

NOTE Besides traditional tables, the Database Engine supports memory-optimized tables, too. The content of these tables is always stored in memory. Hence, the performance of read and write operations on such tables can be significantly improved. Chapter 21 describes in detail how you can use the CREATE TABLE statement to create memory-optimized tables.

The CREATE TABLE statement creates a new base table with all corresponding columns and their data types. The basic form of the CREATE TABLE statement is

```
CREATE TABLE table_name
    (col_name1 type1 [NOT NULL| NULL]
    [{, col_name2 type2 [NOT NULL| NULL]} ...])
```

NOTE Besides base tables, there are also some special kinds of tables, such as temporary tables and views (see Chapters 6 and 11, respectively).

table_name is the name of the created base table. The maximum number of tables per database is limited by the number of objects in the database (there can be more than 2 billion objects in a database, including tables, views, stored procedures, triggers, and constraints). **col_name1**, **col_name2**,... are the names of the table columns. **type1**, **type2**,... are data types of corresponding columns (see Chapter 4).

NOTE The name of a database object can generally contain four parts, in the form:
```
[server_name. [db_name. [schema_name.]]] object_name
```
object_name is the name of the database object. **schema_name** is the name of the schema to which the object belongs. **server_name** and **db_name** are the names of the server and database to which the database object belongs. Table names, combined with the schema name, must be unique within the database. Similarly, column names must be unique within the table.

The first constraint that will be discussed in this book is the existence and nonexistence of NULL values within a column. If NOT NULL is specified, the assignment of NULL values for the column is not allowed. In that case, the column may not contain NULLs, and if there is a NULL value to be inserted, the system returns an error message. (NULL values are explained in detail at the end of Chapter 4.)

As already stated, a database object (in this case, a table) is always created within a schema of a database. A user can create a table only in a schema for which she has ALTER permissions. Any user in the **sysadmin**, **db_ddladmin**, or **db_owner** role can create a table in any schema. (The ALTER permissions, as well as database and server roles, are discussed in detail in Chapter 12.)

The creator of a table must not be its owner. This means that you can create a table that belongs to someone else. Similarly, a table created with the CREATE TABLE statement must not belong to the current database if some other (existing) database name, together with the schema name, is specified as the prefix of the table name.

The schema to which a table belongs has two possible default names. If a table is specified without the explicit schema name, the system checks for a table name in the corresponding default schema. If the object name cannot be found in the default schema, the system searches in the **dbo** schema.

NOTE You should always specify the table name together with the corresponding schema name. That way you can eliminate possible ambiguities.

Temporary tables are a special kind of base table. They are stored in the **tempdb** database and are automatically dropped at the end of the session. The properties of temporary tables and examples concerning them are given in Chapter 6.

Example 5.4 shows the creation of all tables of the **sample** database.

Example 5.4
```
USE sample;
CREATE TABLE employee   (emp_no INTEGER NOT NULL,
                         emp_fname CHAR(20) NOT NULL,
                         emp_lname CHAR(20) NOT NULL,
                         dept_no CHAR NULL);
CREATE TABLE department (dept_no CHAR NOT NULL,
                         dept_name CHAR(25) NOT NULL,
                         location CHAR(30) NULL);
CREATE TABLE project    (project_no CHAR NOT NULL,
                         project_name CHAR(15) NOT NULL,
                         budget FLOAT NULL);
CREATE TABLE works_on   (emp_no INTEGER NOT NULL,
                         project_no CHAR NOT NULL,
                         job CHAR (15) NULL,
                         enter_date DATE NULL);
```

Besides the data type and the nullability, the column specification can contain the following options:

- DEFAULT clause
- IDENTITY property

The DEFAULT clause in the column definition specifies the default value of the column—that is, whenever a new row is inserted into the table, the default value for the particular column will be used if there is no value specified for it. A constant value, such as the system functions USER, CURRENT_USER, SESSION_USER, SYSTEM_USER, CURRENT_TIMESTAMP, and NULL, among others, can be used as the default values.

A column with the IDENTITY property allows only integer values, which are usually implicitly assigned by the system. Each value, which should be inserted in the column, is calculated by incrementing the last inserted value of the column. Therefore, the definition of a column with the IDENTITY property contains (implicitly or explicitly) an initial value and an increment. This property will be discussed in detail in the next chapter (see Example 6.38).

To close this section, Example 5.5 shows the creation of a table with a column of the SQL_VARIANT type.

Example 5.5

```
USE sample;
CREATE TABLE Item_Attributes  (
        item_id INT NOT NULL,
        attribute NVARCHAR(30) NOT NULL,
        value SQL_VARIANT NOT NULL,
        PRIMARY KEY (item_id, attribute));
```

In Example 5.5, the table contains the **value** column, which is of type SQL_VARIANT. As you already know from Chapter 4, the SQL_VARIANT data type can be used to store values of various data types at the same time, such as numeric values, strings, and date values. Note that in Example 5.5 the SQL_VARIANT data type is used for the column values, because different attribute values may be of different data types. For example, the size attribute stores an integer attribute value, and the name attribute stores a character string attribute value.

CREATE TABLE and Declarative Integrity Constraints

One of the most important features that a DBMS must provide is a way of maintaining the integrity of data. The constraints, which are used to check the modification or insertion of data, are called *integrity constraints*. The task of maintaining integrity constraints can be handled by the user in application programs or by the DBMS. The most important benefits of handling integrity constraints by the DBMS are the following:

- Increased reliability of data
- Reduced programming time
- Simple maintenance

Using the DBMS to define integrity constraints increases the reliability of data because there is no possibility that the integrity constraints can be forgotten by a programmer. If an integrity constraint is handled by application programs, *all* programs concerning the constraint must include the corresponding code. If the code is omitted in one application program, the consistency of data is compromised.

An integrity constraint not handled by the DBMS must be defined in every application program that uses the data involved in the constraint. In contrast, the same integrity constraint must be defined only once if it is to be handled by the DBMS. Additionally, application-enforced constraints are usually more complex to code than are database-enforced constraints.

If an integrity constraint is handled by the DBMS, the modification of the structure of the constraint must be handled only once, in the DBMS. The modification of a structure in application programs requires the modification of every program that involves the corresponding code.

There are two groups of integrity constraints handled by a DBMS:

- Declarative integrity constraints
- Procedural integrity constraints that are handled by triggers (these constraints will be discussed in Chapter 14)

The declarative constraints are defined using the DDL statements CREATE TABLE and ALTER TABLE. They can be column-level constraints or table-level constraints. Column-level constraints, together with the data type and other column properties, are placed within the declaration of the column, while table-level constraints are always defined at the end of the CREATE TABLE or ALTER TABLE statement, after the definition of all columns.

NOTE There is only one difference between column-level constraints and table-level constraints: a column-level constraint can be applied only upon one column, while a table-level constraint can cover one or more columns of a table.

Each declarative constraint has a name. The name of the constraint can be explicitly assigned using the CONSTRAINT option in the CREATE TABLE statement or the ALTER TABLE statement. If the CONSTRAINT option is omitted, the Database Engine assigns an implicit name for the constraint.

NOTE Using explicit constraint names is strongly recommended. The search for an integrity constraint can be enhanced if an explicit name for a constraint is used.

All declarative constraints can be categorized into several groups:

- DEFAULT clause
- UNIQUE clause
- PRIMARY KEY clause
- CHECK clause
- FOREIGN KEY clause and referential integrity

The definition of the default value using the DEFAULT clause was shown earlier in this chapter (see also Example 5.6). All other constraints are described in the following sections.

The UNIQUE Clause

Sometimes more than one column or group of columns of the table have unique values and therefore can be used as the primary key. All columns or groups of columns that qualify to be primary keys are called *candidate keys*. Each candidate key is defined using the UNIQUE clause in the CREATE TABLE or the ALTER TABLE statement.

The UNIQUE clause has the following form:

```
[CONSTRAINT c_name]
     UNIQUE [CLUSTERED | NONCLUSTERED] ({ col_name1} ,...)
```

The CONSTRAINT option in the UNIQUE clause assigns an explicit name to the candidate key. The option CLUSTERED or NONCLUSTERED relates to the fact that the Database Engine always generates an index for each candidate key of a table. The index can be clustered—that is, the physical order of rows is specified using the indexed order of the column values. If the order is not specified, the index is nonclustered (see also Chapter 10). The default value is NONCLUSTERED. **col_name1** is a column name that builds the candidate key. (The maximum number of columns per candidate key is 16.)

Example 5.6 shows the use of the UNIQUE clause. (You have to drop the **projects** table, via DROP TABLE **projects**, before you execute the following example.)

Example 5.6

```
USE sample;
CREATE TABLE projects  (project_no CHAR  DEFAULT 'p1',
                        project_name CHAR(15) NOT NULL,
                        budget FLOAT NULL
                        CONSTRAINT unique_no UNIQUE (project_no));
```

Each value of the **project_no** column of the **projects** table is unique, including the NULL value. (Just as with any other value with a UNIQUE constraint, if NULL values are allowed on a corresponding column, there can be at most one row with the NULL value for that particular column.) If an existing value should be inserted into the column **project_no**, the system rejects it. The explicit name of the constraint that is defined in Example 5.6 is **unique_no**.

The PRIMARY KEY Clause

The *primary key* of a table is a column or group of columns whose value is different in every row. Each primary key is defined using the PRIMARY KEY clause in the CREATE TABLE or the ALTER TABLE statement.

The PRIMARY KEY clause has the following form:

```
[CONSTRAINT c_name]
 PRIMARY KEY [CLUSTERED | NONCLUSTERED] ({col_name1} ,...)
```

All options of the PRIMARY KEY clause have the same meaning as the corresponding options with the same name in the UNIQUE clause. In contrast to UNIQUE, the PRIMARY KEY column must be NOT NULL, and its default value is CLUSTERED.

Example 5.7 shows the specification of the primary key for the **employee** table of the **sample** database.

NOTE You have to drop the **employee** table (DROP TABLE **employee**) before you execute the following example.

Example 5.7
```
USE sample;
CREATE TABLE employee   (emp_no INTEGER NOT NULL,
                    emp_fname CHAR(20) NOT NULL,
                    emp_lname CHAR(20) NOT NULL,
                    dept_no CHAR NULL,
                    CONSTRAINT prim_empl PRIMARY KEY (emp_no));
```

The **employee** table is re-created and its primary key is defined in Example 5.7. The primary key of the table is specified using the declarative integrity constraint named **prim_empl**. This integrity constraint is a table-level constraint, because it is specified after the definition of all columns of the **employee** table.

Example 5.8 is equivalent to Example 5.7, except for the specification of the primary key of the **employee** table as a column-level constraint.

NOTE Again, you have to drop the **employee** table (DROP TABLE **employee**) before you execute the following example.

Example 5.8
```
USE sample;
CREATE TABLE employee
     (emp_no INTEGER NOT NULL CONSTRAINT prim_empl PRIMARY KEY,
      emp_fname CHAR(20) NOT NULL,
      emp_lname CHAR(20) NOT NULL,
      dept_no CHAR NULL);
```

In Example 5.8, the PRIMARY KEY clause belongs to the declaration of the corresponding column, together with its data type and nullability. For this reason, it is called a column-level constraint.

The CHECK Clause
The *check constraint* specifies conditions for the data inserted into a column. Each row inserted into a table or each value updating the value of the column must meet these conditions. The CHECK clause is used to specify check constraints. This clause can be defined in the CREATE TABLE or ALTER TABLE statement. The syntax of the CHECK clause is

```
[CONSTRAINT c_name]
    CHECK [NOT FOR REPLICATION] expression
```

expression must evaluate to a Boolean value (*true* or *false*) and can reference any columns in the current table (or just the current column if specified as a column-level constraint), but no other tables. Example 5.9 shows how the CHECK clause can be used.

Example 5.9
```
USE sample;
CREATE TABLE customer
     (cust_no INTEGER NOT NULL,
      cust_group CHAR NULL,
      CHECK (cust_group IN ('c1', 'c2', 'c10')));
```

The **customer** table that is created in Example 5.9 contains the **cust_group** column with the corresponding check constraint. The database system returns an error if the **cust_group** column, after a modification of its existing values or after the insertion of a new row, would contain a value different from the values in the set ('c1', 'c2', 'c10').

The FOREIGN KEY Clause

A *foreign key* is a column or group of columns in one table that contains values that match the primary key values in the same or another table. Each foreign key is defined using the FOREIGN KEY clause combined with the REFERENCES clause.

The FOREIGN KEY clause has the following form:

```
[CONSTRAINT c_name]
      [[FOREIGN KEY] ({col_name1} ,...)]
         REFERENCES table_name ({col_name2},...)
            [ON DELETE {NO ACTION| CASCADE | SET NULL | SET DEFAULT}]
            [ON UPDATE {NO ACTION | CASCADE | SET NULL | SET DEFAULT}]
```

The FOREIGN KEY clause defines all columns explicitly that belong to the foreign key. The REFERENCES clause specifies the table name with all columns that build the corresponding primary key. The number and the data types of the columns in the FOREIGN KEY clause must match the number and the corresponding data types of columns in the REFERENCES clause (and, of course, both of these must match the number and data types of the columns in the primary key of the referenced table).

The table that contains the foreign key is called the *referencing table*, and the table that contains the corresponding primary key is called the *parent table* or *referenced table*. Example 5.10 shows the specification of the foreign key in the **works_on** table of the **sample** database.

NOTE You have to drop the **works_on** table before you execute the following example.

Example 5.10

```
USE sample;
CREATE TABLE works_on  (emp_no INTEGER NOT NULL,
            project_no CHAR NOT NULL,
              job CHAR (15) NULL,
              enter_date DATE NULL,
              CONSTRAINT prim_works PRIMARY KEY(emp_no, project_no),
              CONSTRAINT foreign_works FOREIGN KEY(emp_no)
                    REFERENCES employee (emp_no));
```

The **works_on** table in Example 5.10 is specified with two declarative integrity constraints: **prim_works** and **foreign_works**. Both constraints are table-level constraints, where the former specifies the primary key and the latter the foreign key of the **works_on** table. Further, the constraint **foreign_works** specifies the **employee** table as the parent table and its **emp_no** column as the corresponding primary key of the column with the same name in the **works_on** table.

The FOREIGN KEY clause can be omitted if the foreign key is defined as a column-level constraint, because the column being constrained is the implicit column "list" of the foreign key, and the keyword REFERENCES is sufficient to indicate what kind of constraint this is.

A definition of the foreign keys in tables of a database imposes the specification of another important integrity constraint: the referential integrity, described next.

Referential Integrity

A *referential integrity* enforces insert and update rules for the tables with the foreign key and the corresponding primary key constraint. Examples 5.7 and 5.10 specify two such constraints: **prim_empl** and **foreign_works**. The REFERENCES clause in Example 5.10 determines the **employee** table as the parent table.

If the referential integrity for two tables is specified, the modification of values in the primary key and the corresponding foreign key is not always possible. The following subsection discusses when modification is possible and when it is not.

Possible Problems with Referential Integrity

There are four cases in which the modification of the values in the foreign key or in the primary key can cause problems. All of these cases will be shown using the **sample** database. The first two cases affect modifications of the referencing table, while the last two concern modifications of the parent table.

Case 1   Insert a new row into the **works_on** table with the employee number 11111.

The insertion of the new row in the referencing table **works_on** introduces a new employee number for which there is no matching employee in the parent table (**employee**). If the referential integrity for both tables is specified as is done in Examples 5.7 and 5.10, the Database Engine rejects the insertion of a new row. For readers who are familiar with the SQL language, the corresponding Transact-SQL statement is

```
USE sample;
INSERT INTO works_on (emp_no, ...)
    VALUES (11111, ...);
```

Case 2   Modify the employee number 10102 in all rows of the **works_on** table. The new number is 11111.

In Case 2, the existing value of the foreign key in the **works_on** table should be replaced using the new value, for which there is no matching value in the parent table **employee**. If the referential integrity for both tables is specified as is done in Examples 5.7 and 5.10, the database system rejects the modification of the rows in the **works_on** table. The corresponding Transact-SQL statement is

```
USE sample;
UPDATE works_on
    SET emp_no = 11111 WHERE emp_no = 10102;
```

Case 3   Modify the employee number 10102 in the corresponding row of the **employee** table. The new number is 22222.

In Case 3, the existing value of the primary key in the parent table and the foreign key of the referencing table is modified only in the parent table. The values in the referencing table are unchanged. Therefore, the system rejects the modification of the row with the employee number 10102 in the **employee** table. Referential integrity requires that no rows in the referencing table (the one with the FOREIGN KEY clause) can exist unless a corresponding row in the parent table (the one with the PRIMARY KEY clause) also exists. Otherwise, the rows in the referencing table would be "orphaned." If the modification described were permitted, then rows in the **works_on** table having the employee number 10102 would be orphaned, and the system would reject it. The corresponding Transact-SQL statement is

```
USE sample;
UPDATE employee
    SET emp_no = 22222 WHERE emp_no = 10102;
```

Case 4   Delete the row of the **employee** table with the employee number 10102.

Case 4 is similar to Case 3. The deletion would remove the employee for which matching rows exist in the referencing table. Example 5.11 shows the definition of tables of the **sample** database with all existing primary key and foreign key constraints. (If the **employee**, **department**, **project**, and **works_on** tables already exist, drop them first using the DROP TABLE **table_name** statement.)

Example 5.11

```
USE sample;
CREATE TABLE department(dept_no CHAR NOT NULL,
                        dept_name CHAR(25) NOT NULL,
                        location CHAR(30) NULL,
                        CONSTRAINT prim_dept PRIMARY KEY (dept_no));
CREATE TABLE employee  (emp_no INTEGER NOT NULL,
            emp_fname CHAR(20) NOT NULL,
            emp_lname CHAR(20) NOT NULL,
            dept_no CHAR NULL,
            CONSTRAINT prim_emp PRIMARY KEY (emp_no),
            CONSTRAINT foreign_emp FOREIGN KEY(dept_no) REFERENCES
department(dept_no));
CREATE TABLE project   (project_no CHAR NOT NULL,
            project_name CHAR(15) NOT NULL,
            budget FLOAT NULL,
            CONSTRAINT prim_proj PRIMARY KEY (project_no));
CREATE TABLE works_on  (emp_no INTEGER NOT NULL,
  project_no CHAR NOT NULL,
  job CHAR (15) NULL,
  enter_date DATE NULL,
  CONSTRAINT prim_works PRIMARY KEY(emp_no, project_no),
  CONSTRAINT foreign1_works FOREIGN KEY(emp_no) REFERENCES
employee(emp_no),
  CONSTRAINT foreign2_works FOREIGN KEY(project_no) REFERENCES
project(project_no));
```

The ON DELETE and ON UPDATE Options

The Database Engine can react differently if the values of the primary key of a table should be modified or deleted. If you try to update values of a foreign key, and those modifications result in inconsistencies in the corresponding primary key (see Case 1 and Case 2 in the previous section), the database system will always reject the modification and will display a message similar to the following:

> Server: Msg 547, Level 16, State 1, Line 1 UPDATE statement conflicted with COLUMN FOREIGN KEY constraint 'FKemployee'. The conflict occurred in database 'sample', table 'employee', column 'dept_no'. The statement has been terminated.

But if you try to modify the values of a primary key, and these modifications result in inconsistencies in the corresponding foreign key (see Case 3 and Case 4 in the previous section), a database system could react very flexibly. Generally, there are four options for how a database system can react:

- **NO ACTION**   Allows you to modify (update or delete) only those values of the parent table that do not have any corresponding values in the foreign key of the referencing table.

- **CASCADE**   Allows you to modify (update or delete) all values of the parent table. If this option is specified, a row in the referencing table is modified if the corresponding value in the primary key of the parent table has been updated. The same is true for deletion; that is, if a row in the parent table is deleted, the corresponding row in the child table is deleted, too.

- **SET NULL**   Allows you to modify (update or delete) all values of the parent table. If you want to update a value of the parent table and this modification would lead to data inconsistencies in the referencing table, the database system sets all corresponding values in the foreign key of the referencing table to NULL. Similarly, if you want to delete the row in the parent table and the deletion of the value in the primary key would lead to data inconsistencies, the database system sets all corresponding values in the foreign key to NULL. That way, all data inconsistencies are omitted.

- **SET DEFAULT**   Analogous to the SET NULL option, with one exception: all corresponding values in the foreign key are set to a default value. (Obviously, the default value must still exist in the primary key of the parent table after modification.)

NOTE The Transact-SQL language supports all four directives.

Example 5.12 shows the use of the ON DELETE and ON UPDATE options.

Example 5.12

```
USE sample;
CREATE TABLE works_on1
(emp_no INTEGER NOT NULL,
  project_no CHAR NOT NULL,
  job CHAR (15) NULL,
  enter_date DATE NULL,
  CONSTRAINT prim_works1 PRIMARY KEY(emp_no, project_no),
  CONSTRAINT foreign1_works1 FOREIGN KEY(emp_no)
      REFERENCES employee(emp_no) ON DELETE CASCADE,
  CONSTRAINT foreign2_works1 FOREIGN KEY(project_no)
      REFERENCES project(project_no) ON UPDATE CASCADE);
```

Example 5.12 creates the **works_on1** table that uses the ON DELETE CASCADE and ON UPDATE CASCADE options. If you load the **works_on1** table with the content shown in Table 1-4 (see Chapter 1), each deletion of a row in the **employee** table will cause the additional deletion of all rows in the **works_on1** table that have the corresponding value in the **emp_no** column. Similarly, each update of a value in the **project_no** column of the **project** table will cause the same modification on all corresponding values in the **project_no** column of the **works_on1** table.

Creating Other Database Objects

A relational database contains not only base tables that exist in their own right but also *views*, which are virtual tables. The data of a base table exists physically—that is, it is stored on a disk—while a view is derived from one or more base tables. The CREATE VIEW statement creates a new view from one or more existing tables (or views) using a SELECT statement, which is an inseparable part of the CREATE VIEW statement. Since the creation of a view always contains a query, the CREATE VIEW statement belongs to the data manipulation language (DML) rather than to the data definition language (DDL). For this reason, the creation and removal of views is discussed in Chapter 11, after the presentation of all Transact-SQL statements for data modification.

The CREATE INDEX statement creates a new *index* on a specified table. The indices are primarily used to allow efficient access to the data stored on a disk. The existence of an index can greatly improve the access to data. Indices, together with the CREATE INDEX statement, are discussed in detail in Chapter 10.

A *stored procedure* is an additional database object that can be created using the corresponding CREATE PROCEDURE statement. (A stored procedure is a special kind of sequence of statements written in Transact-SQL, using the SQL language and procedural extensions. Chapter 8 describes stored procedures in detail.)

A *trigger* is a database object that specifies an action as a result of an operation. This means that when a particular data-modifying action (modification, insertion, or deletion) occurs on a particular table, the Database Engine automatically invokes one or more additional actions. The CREATE TRIGGER statement creates a new trigger. Triggers are described in detail in Chapter 14.

A *synonym* is a local database object that provides a link between itself and another object managed by the same or a linked database server. Using the CREATE SYNONYM statement, you can create a new synonym for the given object. Example 5.13 shows the use of this statement.

Example 5.13
```
USE AdventureWorks;
CREATE SYNONYM  prod
            FOR AdventureWorks.Production.Product;
```

Example 5.13 creates a synonym for the **Product** table in the **Production** schema of the **AdventureWorks** database. This synonym can be used in DML statements, such as SELECT, INSERT, UPDATE, and DELETE.

NOTE The main reason to use synonyms is to omit the use of lengthy names in DML statements. As you already know, the name of a database object can generally contain four parts. Introducing a (single-part) synonym for an object that has three or four parts can save you time when typing its name.

A *schema* is a database object that includes statements for creation of tables, views, and user privileges. (You can think of a schema as a construct that collects together several tables, corresponding views, and user privileges.)

NOTE The Database Engine treats the notion of schema the same way it is treated in the ANSI SQL standard. In the SQL standard, a schema is defined as a collection of database objects that is owned by a single principal and forms a single namespace. A namespace is a set of objects that cannot have duplicate names. For example, two tables can have the same name only if they are in separate schemas. (Schema is a very important concept in the security model of the Database Engine. For this reason, you can find a detailed description of schema in Chapter 12.)

Integrity Constraints and Domains

A *domain* is the set of all possible legitimate values that columns of a table may contain. Almost all DBMSs use base data types such as INTEGER, CHARACTER, and DATE to define the set of possible values for a column. This method of enforcing "domain integrity" is incomplete, as can be seen from the following example.

The person table has a column, zip, that specifies the ZIP code of the city in which the person lives. This column can be defined using the SMALLINT or CHAR data type. The definition with the SMALLINT data type is inaccurate, because the SMALLINT data type contains all positive and negative values between $-2^{15} - 1$ and 2^{15}. The definition using CHAR is even more inaccurate, because all characters and special signs can also be used in such a case. Therefore, an accurate definition of ZIP codes requires defining an interval of positive integers between 00601 and 99950 and assigning it to the zip column.

CHECK clauses (specified in the CREATE TABLE or ALTER TABLE statement) can enforce more precise domain integrity because their expressions are flexible, and they are always enforced when the column is inserted or modified.

The Transact-SQL language provides support for domains by creating alias data types using the CREATE TYPE statement. The following two sections describe alias and Common Language Runtime (CLR) data types.

Alias Data Types

An alias data type is a special kind of data type that is defined by users using the existing base data types. Such a data type can be used with the CREATE TABLE statement to define one or more columns in a database.

The CREATE TYPE statement is generally used to create an alias data type. The syntax of this statement to specify an alias data type is as follows:

```
CREATE TYPE [ type_schema_name. ] type_name
{ [ FROM base_type [ ( precision [ , scale ] ) ] [ NULL | NOT NULL ] ]
    | [ EXTERNAL NAME assembly_name [ .class_name ] ]}
```

Example 5.14 shows the creation of an alias data type using the CREATE TYPE statement.

Example 5.14
```
USE sample;
CREATE TYPE zip
    FROM SMALLINT NOT NULL;
```

Example 5.14 creates an alias type, **zip**, based on the standard data type SMALLINT. This user-defined data type can now be used as a data type of a table column, as shown in Example 5.15.

NOTE You have to drop the **customer** table (DROP TABLE **customer**) before you execute the following example.

Example 5.15

```
USE sample;
CREATE TABLE customer
    (cust_no INT NOT NULL,
     cust_name CHAR(20) NOT NULL,
     city CHAR(20),
     zip_code ZIP,
     CHECK (zip_code BETWEEN 601 AND 99950));
```

Example 5.15 uses the new **zip** data type to specify a column of the **customer** table. The values of this column have to be constrained to the region between 601 and 99950. As can be seen from Example 5.15, this can be done using the CHECK clause.

NOTE Generally, the Database Engine implicitly converts between compatible columns of different data types. This is valid for the alias data types, too.

The Database Engine supports the creation of user-defined table types. Example 5.16 shows how you can use the CREATE TYPE statement to create such a table type.

Example 5.16

```
USE sample;
CREATE TYPE person_table_t AS TABLE
    ( name VARCHAR(30), salary DECIMAL(8,2));
```

The user-defined table type called **person_table_t** has two columns: **name** and **salary**. The main syntactical difference in relation to alias data types is the existence of the AS TABLE clause, as can be seen in Example 5.16. User-defined table types are usually used in relation to table-valued parameters (see Chapter 8).

CLR Data Types

The CREATE TYPE statement can also be applied to create a user-defined data type using .NET. In this case, the implementation of a user-defined data type is defined in a class of an assembly in the Common Language Runtime (CLR). This means that you can use one of the .NET languages like C# or Visual Basic to implement the new data type. Further description of the user-defined data types is outside the scope of this book.

Modifying Database Objects

The Transact-SQL language supports changing the structure of the following database objects, among others:

- Database
- Table

- Stored procedure
- View
- Schema
- Trigger

The following two sections describe, in turn, how you can alter a database and a table. The modification of the structure of each of the last four database objects is described in Chapters 8, 11, 12, and 14, respectively.

Altering a Database

The ALTER DATABASE statement changes the physical structure of a database. The Transact-SQL language allows you to change the following properties of a database:

- Add or remove one or more database files, log files, or filegroups
- Modify file or filegroup properties
- Set database options
- Change the name of the database using the **sp_rename** stored procedure (discussed a bit later, in the section "Altering a Table")

The following subsections describe these different types of database alterations. In this section, we will also use the ALTER DATABASE statement to show how FILESTREAM data can be stored in files and filegroups and to explain the notion of contained databases.

Adding or Removing Database Files, Log Files, or Filegroups

The ALTER DATABASE statement allows the addition and removal of database files. The clauses ADD FILE and REMOVE FILE specify the addition of a new file and the deletion of an existing file, respectively. (Additionally, a new file can be assigned to an existing filegroup using the TO FILEGROUP option.)

Example 5.17 shows how a new database file can be added to the **projects** database.

Example 5.17

```
USE master;
GO
ALTER DATABASE projects
ADD FILE (NAME=projects_dat1,
     FILENAME = 'C:\temp\projects1.mdf',   SIZE = 10,
     MAXSIZE = 100,   FILEGROWTH = 5);
```

The ALTER DATABASE statement in Example 5.17 adds a new file with the logical name **projects_dat1**. Its initial size is 10MB, and this file will grow using units of 5MB until it reaches the upper limit of 100MB. (Log files are added in the same way as database files. The only difference is that you use the ADD LOG FILE clause instead of ADD FILE.)

The REMOVE FILE clause removes one or more files that belong to an existing database. The file can be a data file or a log file. The file cannot be removed unless it is empty.

The CREATE FILEGROUP clause creates a new filegroup, while DELETE FILEGROUP removes an existing filegroup from the system. Again, you cannot remove a filegroup unless it is empty.

Modifying File or Filegroup Properties

You can use the MODIFY FILE clause to change the following file properties:

- Change the logical name of a file using the NEWNAME option of the MODIFY FILE clause
- Increase the value of the SIZE property
- Change the FILENAME, MAXSIZE, or FILEGROWTH property
- Mark the file as OFFLINE

Similarly, you can use the MODIFY FILEGROUP clause to change the following filegroup properties:

- Change the name of a filegroup using the NAME option of the MODIFY FILEGROUP clause
- Mark the filegroup as the default filegroup using the DEFAULT option
- Mark the filegroup as read-only or read-write using the READ_ONLY or READ_WRITE option, respectively

Setting Database Options

The SET clause of the ALTER DATABASE statement is used to set different database options. Some options must be set to ON or OFF, but most of them have a list of possible values. Each database option has a default value, which is set in the **model** database. Therefore, you can alter the **model** database to change the default values of specific options.

All options that you can set are divided into several groups. The most important groups are

- State options
- Auto options
- SQL options

The state options control the following:

- User access to the database (options are SINGLE_USER, RESTRICTED_USER, and MULTI_USER)
- The status of the database (options are ONLINE, OFFLINE, and EMERGENCY)
- The read/write modus (options are READ_ONLY and READ_WRITE)

The auto options control, among other things, the type of the database shutdown (the option AUTO_CLOSE) and how index statistics are built (the options AUTO_CREATE_STATISTICS and AUTO_UPDATE_STATISTICS).

The SQL options control the ANSI compliance of the database and its objects. All SQL options can be edited using the DATABASEPROPERTYEX function and modified using the ALTER DATABASE statement.

Storing FILESTREAM Data

Chapter 4 explained what FILESTREAM data is and the reason for using it. This section discusses how FILESTREAM data can be stored as a part of a database. Before you can store FILESTREAM data, you have to enable the system for this task. The following subsection explains how to enable the operating system and the instance of your database system.

Enabling FILESTREAM Storage   FILESTREAM storage has to be enabled at two levels:

- For the Windows operating system
- For the particular database server instance

You use SQL Server Configuration Manager to enable FILESTREAM storage at the OS level. To open SQL Server Configuration Manager, type **SQLServerManager15.msc** in the Search field for SQL Server 2019 or type **SQLServerManager14.msc** for SQL Server 2017. In the list of services, right-click SQL Server Services, and then click Open. In the SQL Server Configuration Manager snap-in, choose your instance of the Database Engine, right-click the instance, and then click Properties. In the SQL Server Properties dialog box, click the FILESTREAM tab. Check the Enable FILESTREAM for Transact-SQL Access check box. If you want to read and write FILESTREAM data from Windows, check the Enable FILESTREAM for File I/O Access check box. Enter the name of the Windows share in the Windows Share Name box and click Apply to apply the changes. SQL Server Configuration Manager creates a new share with the specified name for the instance.

NOTE   You need to be Windows Administrator on a local system and have administrator (**sysadmin**) rights to enable FILESTREAM storage. You also need to restart the instance for the changes to take effect.

The next step is to enable FILESTREAM storage for a particular database server instance. SQL Server Management Studio will be used to show this task. Right-click the instance in Object Explorer, click Properties, select Advanced in the left pane, and set Filestream Access Level to one of the following levels:

- **Disabled**   FILESTREAM storage is not allowed.
- **Transact-SQL Access Enabled**   FILESTREAM data can be accessed using T-SQL statements.
- **Full Access Enabled**   FILESTREAM data can be accessed using T-SQL as well as from the OS.

The alternative way is to use the **sp_configure** system procedure with the **filestream access level** option:

```
EXEC sp_configure filestream_access_level, 2
RECONFIGURE
```

The last parameter has three different values, which correspond to the three levels mentioned previously. Therefore, the value 2 in this code means "Full Access Enabled."

Adding a File to the Filegroup   After you enable FILESTREAM storage for your instance, you can use the ALTER DATABASE statement first to create a filegroup for FILESTREAM data and then to add a file to that filegroup, as shown in Example 5.18. (Of course, you can also use the CREATE DATABASE statement to accomplish this task.)

NOTE Before you execute the statement in Example 5.18, change the name of the file in the FILENAME clause.

Example 5.18
```
USE master;
ALTER DATABASE sample
   ADD FILEGROUP Employee_FSGroup CONTAINS FILESTREAM;
GO
ALTER DATABASE sample
   ADD FILE (NAME= employee_FS,
   FILENAME = 'C:\temp\emp_FS')
   TO FILEGROUP Employee_FSGroup;
```

The first ALTER DATABASE statement in Example 5.18 adds a new filegroup called **Employee_FSGroup** to the **sample** database. The CONTAINS FILESTREAM option tells the system that this filegroup will contain only FILESTREAM data. The second ALTER DATABASE statement adds a new file to the existing filegroup.

Now you can create a table with one or more FILESTREAM columns. Example 5.19 shows the creation of a table with a FILESTREAM column.

Example 5.19
```
USE sample;
CREATE TABLE employee_info
   (id UNIQUEIDENTIFIER ROWGUIDCOL NOT NULL UNIQUE,
    filestream_data VARBINARY(MAX) FILESTREAM NULL)
```

The **employee_info** table in Example 5.19 contains the **filestream_data** columns, which must be of the VARBINARY(max) data type. Such a column includes the FILESTREAM attribute, indicating that a column should store data in the FILESTREAM filegroup. All tables that store FILESTREAM data require the existence of a UNIQUE ROWGUILDCOL. For this reason, the **employee_info** table has the **id** column, defined using these two attributes.

To insert data into a FILESTREAM column, you use the standard INSERT statement, which is described in Chapter 7. Also, to read data from a FILESTREAM column, you can use the standard SELECT statement, which is described in Chapter 6. The detailed description of read and write operations on FILESTREAM data is outside the scope of this book.

Contained Databases
One of the significant problems with SQL Server databases is that they cannot be exported (or imported) easily. As you already know from this chapter, you can attach and detach a database, but many important parts and properties of the attached database will be missing. (The main problem in such a case is database security in general and existing logins in particular, which are usually incomplete or wrong after the move.)

Microsoft solved such problems by introducing contained databases. A contained database comprises all database settings and data required to specify the database and is isolated from the instance of the Database Engine on which it is installed. In other words, this form of databases has no configuration dependencies on the instance and can easily be moved from one instance of SQL Server to another.

Generally, there are three forms of databases in relation to containment:

- Fully contained databases
- Partially contained databases
- Noncontained databases

Fully contained databases are those where database objects cannot cross the application boundary. (An application boundary defines the scope of an application. For instance, user-defined functions are within the application boundary, while functions related to server instances are outside it.)

Partially contained databases allow database objects to cross the application boundary, while noncontained databases do not support the notion of an application boundary at all.

Let's take a look at how you can create a partially contained database. If a database called **my_sample** already exists, and it is created as a noncontained database (using the CREATE DATABASE statement, for instance), you can use the ALTER DATABASE statement to alter it to partial containment, as shown in Example 5.20.

Example 5.20

```
USE master;
EXEC sp_configure 'show advanced options' , 1;
RECONFIGURE WITH OVERRIDE;
EXEC sp_configure 'contained database authentication' , 1;
RECONFIGURE WITH OVERRIDE;
ALTER DATABASE my_sample SET CONTAINMENT = PARTIAL;
EXEC sp_configure 'show advanced options' , 0;
RECONFIGURE WITH OVERRIDE;
```

The ALTER DATABASE statement modifies the containment of the **my_sample** database from noncontained to partially contained. This means that the database system allows you to create both contained and noncontained database objects for the **my_sample** database. (All other statements in Example 5.20 just set the scene for the ALTER DATABASE statement.)

NOTE **sp_configure** is a system procedure that can be used to, among other things, change advanced configuration options, such as **'contained database authentication'**. To make changes to advanced configuration options, you first have to set the value of **'show advanced options'** to 1 and reconfigure the system. At the end of Example 5.20, this option has been set again to its default value (0). The **sp_configure** system procedure is discussed in detail in the section "System Procedures" in Chapter 9.

For the **my_sample** database, you can now create a user that is not tied to a login. This will be described in detail in the "Managing Authorization and Authentication of Contained Databases" section of Chapter 12.

Altering a Table

The ALTER TABLE statement modifies the schema of a table. The Transact-SQL language allows the following types of alteration:

- Add or drop one or more new columns
- Modify column properties
- Add or remove integrity constraints
- Enable or disable constraints
- Rename tables and other database objects

The following sections describe these types of changes.

Adding or Dropping a Column

You can use the ADD clause of the ALTER TABLE statement to add a new column to the existing table. Only one column can be added for each ALTER TABLE statement. Example 5.21 shows the use of the ADD clause.

Example 5.21

```
USE sample;
ALTER TABLE employee
     ADD telephone_no CHAR(12) NULL;
```

The ALTER TABLE statement in Example 5.21 adds the column **telephone_no** to the **employee** table. The Database Engine populates the new column either with NULL or IDENTITY values or with the specified default. For this reason, the new column must either be nullable or have a default constraint.

NOTE There is no way to insert a new column in a particular position in the table. The column, which is added using the ADD clause, is always inserted at the end of the table.

The DROP COLUMN clause provides the ability to drop an existing column of the table, as shown in Example 5.22.

Example 5.22

```
USE sample;
ALTER TABLE employee
    DROP COLUMN telephone_no;
```

The ALTER TABLE statement in Example 5.22 removes the **telephone_no** column, which was added to the **employee** table with the ALTER TABLE statement in Example 5.21.

Modifying Column Properties

The Transact-SQL language supports the ALTER COLUMN clause of ALTER TABLE to modify properties of an existing column. The following column properties can be modified:

- Data type
- Nullability

Example 5.23 shows the use of the ALTER COLUMN clause.

Example 5.23
```
USE sample;
ALTER TABLE department
     ALTER COLUMN location CHAR(25) NOT NULL;
```

The ALTER TABLE statement in Example 5.23 changes the previous properties (CHAR(30), nullable) of the **location** column of the **department** table to new properties (CHAR(25), not nullable).

Adding or Removing Integrity Constraints

A new integrity constraint can be added to a table using the ALTER TABLE statement and its option called ADD CONSTRAINT. Example 5.24 shows how you can use the ADD CONSTRAINT clause in relation to a check constraint.

Example 5.24
```
USE sample;
CREATE TABLE sales
     (order_no INTEGER NOT NULL,
      order_date DATE NOT NULL,
      ship_date DATE NOT NULL);
GO
ALTER TABLE sales
     ADD CONSTRAINT order_check CHECK(order_date <= ship_date);
```

The CREATE TABLE statement in Example 5.24 creates the **sales** table with two columns of the DATE data type: **order_date** and **ship_date**. The subsequent ALTER TABLE statement defines an integrity constraint named **order_check**, which compares both of the values and prevents the INSERT statement from successfully completing if the shipping date is earlier than the order date.

Example 5.25 shows how you can use the ALTER TABLE statement to additionally define the primary key of a table.

Example 5.25
```
USE sample;
ALTER TABLE sales
     ADD CONSTRAINT primaryk_sales PRIMARY KEY(order_no);
```

The ALTER TABLE statement in Example 5.25 declares the primary key for the **sales** table.

Each integrity constraint can be removed using the DROP CONSTRAINT clause of the ALTER TABLE statement, as shown in Example 5.26.

Example 5.26
```
USE sample;
ALTER TABLE sales
DROP CONSTRAINT order_check;
```

The ALTER TABLE statement in Example 5.26 removes the CHECK clause called **order_check**, specified in Example 5.24.

NOTE You cannot use the ALTER TABLE statement to modify a definition of an integrity constraint. In this case, the constraint must be re-created—that is, dropped and then added with the new definition.

Enabling or Disabling Constraints

As previously stated, an integrity constraint always has a name that can be explicitly declared using the CONSTRAINT option or implicitly declared by the system. The name of all (implicitly or explicitly) declared constraints for a table can be viewed using the system procedure **sp_helpconstraint**.

A constraint is enforced by default during future insert and update operations. Additionally, the existing values in the column(s) are checked against the constraint. Otherwise, a constraint that is created with the WITH NOCHECK option is disabled in the second case. In other words, if you use the WITH NOCHECK option, the constraint will be applied only to future insert and update operations. (Both options, WITH CHECK and WITH NOCHECK, can be applied only with the CHECK and FOREIGN KEY clauses.)

Example 5.27 shows how you can disable all existing constraints for a table.

Example 5.27

```
USE sample;
ALTER TABLE sales
     NOCHECK CONSTRAINT ALL;
```

In Example 5.27, the keyword ALL is used to disable all the constraints on the **sales** table.

NOTE Use of the NOCHECK option is not recommended. Any constraint violations that are suppressed may cause future updates to fail.

Renaming Tables and Other Database Objects

The **sp_rename** system procedure modifies the name of an existing table (and any other existing database objects, such as databases, views, or stored procedures). Examples 5.28 and 5.29 show the use of this system procedure.

Example 5.28

```
USE sample;
EXEC sp_rename @objname = department, @newname = subdivision
```

Example 5.28 renames the **department** table to **subdivision**.

Example 5.29

```
USE sample;
EXEC sp_rename @objname = 'sales.order_no' , @newname = ordernumber
```

Example 5.29 renames the **order_no** column in the **sales** table. If the object to be renamed is a column in a table, the specification must be in the form **table_name.column_name**.

NOTE Do not use the **sp_rename** system procedure, because changing object names can influence other database objects that reference them. Drop the object and re-create it with the new name.

Removing Database Objects

All Transact-SQL statements that are used to remove a database object have the following general form:

```
DROP object_type object_name
```

Each CREATE **object** statement has the corresponding DROP **object** statement. The statement

```
DROP DATABASE database1 {,  ...}
```

removes one or more databases. This means that all traces of the database are removed from your database system.

One or more tables can be removed from a database with the following statement:

```
DROP TABLE table_name1 {,  ...}
```

All data, indices, and triggers belonging to the removed table are also dropped. (In contrast, all views that are defined using the dropped table are not removed.) Only the user with the corresponding privileges can remove a table.

In addition to DATABASE and TABLE, **objects** in the DROP statement can be, among others, the following:

- TYPE
- SYNONYM
- PROCEDURE
- INDEX
- VIEW
- TRIGGER
- SCHEMA

The statements DROP TYPE and DROP SYNONYM drop a type and a synonym, respectively. (Dropping the synonym does not have any influence on the underlying object.) The rest of the statements are described in different chapters: DROP PROCEDURE in Chapter 8, DROP INDEX in Chapter 10, DROP VIEW in Chapter 11, DROP SCHEMA in Chapter 12, and DROP TRIGGER in Chapter 14.

Summary

The Transact-SQL language supports many data definition statements that create, alter, and remove database objects. The following database objects, among others, can be created and removed using the CREATE **object** and the DROP **object** statement, respectively:

- Database
- Table
- Schema
- View
- Trigger
- Stored procedure
- Index

A structure of all database objects in the preceding list can be altered using the ALTER **object** statement. Note that the ALTER TABLE statement is the only standardized statement from this list. All other ALTER **object** statements are Transact-SQL extensions to the SQL standard.

The next chapter addresses the data manipulation statement called SELECT.

Exercises

E.5.1   Using the CREATE DATABASE statement, create a new database named **test_db** with explicit specifications for database and transaction log files. The database file with the logical name **test_db_dat** is stored in the file C:\tmp\test_db.mdf and the initial size is 5MB, the maximum size is unlimited, and the file growth is 8 percent. The log file called **test_db_log** is stored in the file C:\tmp\test_db_log.ldf and the initial size is 2MB, the maximum size is 10MB, and the file growth is 500KB.

E.5.2   Using the ALTER DATABASE statement, add a new log file to the **test_db** database. The log file is stored in the file C:\tmp\emp_log.ldf and the initial size of the file is 2MB, with growth of 2MB and an unlimited maximum size.

E.5.3   Using the ALTER DATABASE statement, change the file size of the **test_db** database to 10MB.

E.5.4   In Example 5.4, there are some columns of the four created tables defined with the NOT NULL specification. For which column is this specification required and for which is it not required?

E.5.5   Why are the columns **dept_no** and **project_no** in Example 5.4 defined as CHAR values (and not as numerical values)?

E.5.6   Create the tables **customers** and **orders** with the following columns. (Do not declare the corresponding primary and foreign keys.)

customers	orders
customerid char(5) not null	orderid integer not null
companyname varchar(40) not null	customerid char(5) not null
contactname char(30) null	orderdate date null
address varchar(60) null	shippeddate date null
city char(15) null	freight money null
phone char(24) null	shipname varchar(40) null
fax char(24) null	shipaddress varchar(60) null
	quantity integer null

E.5.7   Using the ALTER TABLE statement, add a new column named **shipregion** to the **orders** table. The fields should be nullable and contain integers.

E.5.8 Using the ALTER TABLE statement, change the data type of the column **shipregion** from INTEGER to CHARACTER with length 8. The fields may contain NULL values.

E.5.9 Delete the formerly created column **shipregion**.

E.5.10 Describe exactly what happens if a table is deleted with the DROP TABLE statement.

E.5.11 Re-create the tables **customers** and **orders**, enhancing their definition with all primary and foreign keys constraints.

E.5.12 Using SQL Server Management Studio, try to insert a new row into the **orders** table with the following values:

(10, 'ord01', getdate(), getdate(), 100.0, 'Windstar', 'Ocean', 1)

Why isn't that working?

E.5.13 Using the ALTER TABLE statement, add the current system date and time as the default value to the **orderdate** column of the **orders** table.

E.5.14 Using the ALTER TABLE statement, create an integrity constraint that limits the possible values of the **quantity** column in the **orders** table to values between 1 and 30.

E.5.15 Display all integrity constraints for the **orders** table.

E.5.16 Delete the primary key of the **customers** table. Why isn't that working?

E.5.17 Delete the integrity constraint called **prim_empl** defined in Example 5.7.

E.5.18 Rename the **city** column of the **customers** table. The new name is **town**.

CHAPTER

6

Queries

In This Chapter

- SELECT Statement: Its Clauses and Functions
- Subqueries
- Temporary Tables
- Join Operator
- Correlated Subqueries
- Table Expressions

In this chapter you will learn how to use the SELECT statement to perform retrievals. This chapter describes several clauses in the SELECT statement and gives numerous examples using the **sample** and **AdventureWorks** databases to demonstrate the practical use of each clause. After that, the chapter introduces aggregate functions and the set operators, as well as computed columns and temporary tables. The second part of the chapter tells you more about complex queries. It introduces the join operator, which is the most important operator for relational database systems, and looks at all its forms. Correlated subqueries and the EXISTS function are then introduced. The end of the chapter describes common table expressions, together with the APPLY operator.

SELECT Statement: Its Clauses and Functions

The Transact-SQL language has one basic statement for retrieving information from a database: the SELECT statement. With this statement, it is possible to query information from one or more tables of a database (or even from multiple databases). The result of a SELECT statement is another table, also known as a *result set*.

The simplest form of the SELECT statement contains a SELECT list with the FROM clause. (All other clauses are optional.) This form of the SELECT statement has the following syntax:

```
SELECT [ ALL |DISTINCT] column_list
   FROM {table1 [tab_alias1] } ,...;
```

table1 is the name of the table from which information is retrieved. **tab_alias1** provides an alias for the name of the corresponding table. An *alias* is another name for the corresponding table and can be used as a shorthand way to refer to the table or as a way to refer to two logical instances of the same physical table. Don't worry; this will become clearer as examples are presented.

column_list contains one or more of the following specifications:

- The asterisk symbol (*), which specifies all columns of the named tables in the FROM clause (or from a single table when qualified, as in **table2.***)
- The explicit specification of column names to be retrieved
- The specification **column_name** [AS] **column_heading**, which is a way to replace the name of a column or to assign a new name to an expression
- An expression
- A system or an aggregate function

NOTE In addition to the preceding specifications, there are other options that will be presented later in this chapter.

A SELECT statement can retrieve either columns or rows from a table. The first operation is called *SELECT list* (or *projection*), and the second one is called *selection*. The combination of both operations is also possible in a SELECT statement.

NOTE Before you start to execute queries in this chapter, re-create the entire **sample** database.

Example 6.1 shows the simplest retrieval form with the SELECT statement.

Example 6.1

Get full details of all departments:

```
USE sample;
SELECT dept_no, dept_name, location
 FROM department;
```

The result is

dept_no	dept_name	location
d1	Research	Dallas
d2	Accounting	Seattle
d3	Marketing	Dallas

The SELECT statement in Example 6.1 retrieves all rows and all columns from the **department** table. If you include all columns of a table in a SELECT list (as in Example 6.1), you can use * as shorthand. The column names serve as column headings of the resulting output.

The simplest form of the SELECT statement just described is not very useful for queries. In practice, there are always several more clauses in a SELECT statement than in the statement shown in Example 6.1. The following is the syntax of a SELECT statement that references a table, with (almost) all possible clauses included:

```
SELECT select_list
    [INTO new_table_]
    FROM table
    [WHERE search_condition]
    [GROUP BY group_by_expression]
    [HAVING search_condition]
    [ORDER BY order_expression [ASC | DESC] ];
```

NOTE The clauses in the SELECT statement must be written in the syntactical order given in the preceding syntax—for example, the GROUP BY clause must come after the WHERE clause and before the HAVING clause. However, because the INTO clause is not as significant as the other clauses, it will be discussed later in the chapter, after the other clauses have been discussed.

The following subsections describe the clauses that can be used in a query, WHERE, GROUP BY, HAVING, and ORDER BY, as well as aggregate functions, the IDENTITY property, the sequences feature, set operators, and the CASE expression.

WHERE Clause

Often, it is necessary to define one or more conditions that limit the selected rows. The WHERE clause specifies a Boolean expression (an expression that returns a value of TRUE or FALSE) that is tested for each row to be returned (potentially). If the expression is true, then the row is returned; if it is false, it is discarded.

Example 6.2 shows the use of the WHERE clause.

Example 6.2

Get the names and numbers of all departments located in Dallas:

```
USE sample;
SELECT dept_name, dept_no
    FROM department
    WHERE location = 'Dallas';
```

The result is

dept_name	dept_no
Research	d1
Marketing	d3

In addition to the equal sign, the WHERE clause can contain other comparison operators, including the following:

< > (or !=)	not equal
<	less than
>	greater than
> =	greater than or equal
< =	less than or equal
! >	not greater than
! <	not less than

Example 6.3 shows the use of a comparison operator in the WHERE clause.

Example 6.3
Get the last and first names of all employees with employee numbers greater than or equal to 15000:

```
USE sample;
SELECT emp_lname, emp_fname
    FROM employee
    WHERE emp_no >= 15000;
```

The result is

emp_lname	emp_fname
Smith	Matthew
Barrimore	John
James	James
Moser	Sybill

An expression can also be a part of the condition in the WHERE clause, as Example 6.4 shows.

Example 6.4
Get the project names for all projects with a budget > 60000 £. The current rate of exchange is 0.51 £ per $1.

```
USE sample;
SELECT project_name
    FROM project
    WHERE budget*0.51 > 60000;
```

The result is

project_name
Apollo
Mercury

Comparisons of strings (that is, values of data types CHAR, VARCHAR, NCHAR, or NVARCHAR) are executed in accordance with the collating sequence in effect (the "sort order" specified when the Database Engine was installed). If two strings are compared using ASCII code (or any other code), each of the corresponding (first, second, third, and so on) characters will be compared. One character is lower in priority than the other if it appears in the code table before the other one. Two strings of different lengths are compared after the shorter one is padded at the right with blanks, so that the length of both strings is equal. Numbers compare algebraically. Values of temporal data types (such as DATE, TIME, and DATETIME) compare in chronological order.

Boolean Operators

WHERE clause conditions can either be simple or contain multiple conditions. Multiple conditions can be built using the Boolean operators AND, OR, and NOT. The behavior of these operators was described in Chapter 4 using truth tables.

If two conditions are connected by the AND operator, rows are retrieved for which both conditions are true. If two conditions are connected by the OR operator, all rows of a table are retrieved in which either the first or the second condition (or both) is true, as shown in Example 6.5.

Example 6.5

Get the employee numbers for all employees who work for either project p1 or project p2 (or both):

```
USE sample;
SELECT project_no, emp_no
    FROM works_on
    WHERE project_no = 'p1'
    OR project_no = 'p2';
```

The result is

project_no	emp_no
p1	10102
p2	25348
p2	18316
p2	29346
p1	9031
p1	28559
p2	28559
p1	29346

The result of Example 6.5 contains some duplicate values of the **emp_no** column. To eliminate this redundant information, use the DISTINCT option, as shown here:

```
USE sample;
SELECT DISTINCT emp_no
    FROM works_on
    WHERE project_no = 'p1'
    OR project_no = 'p2';
```

In this case, the result is

emp_no
9031
10102
18316
25348
28559
29346

Note that the DISTINCT option can be used only once in a SELECT list, and it must precede all column names in that list. Therefore, Example 6.6 is *wrong*.

Example 6.6 (Example of an Illegal Statement)

```
USE sample;
SELECT emp_fname, DISTINCT emp_no
        FROM employee
        WHERE emp_lname = 'Moser';
```

The result is

```
Server: Msg 156, Level 15, State 1, Line 1
Incorrect syntax near the keyword 'DISTINCT'.
```

NOTE When there is more than one column in the SELECT list, the DISTINCT clause displays all rows where the combination of columns is distinct.

The WHERE clause may include any number of the same or different Boolean operations. You should be aware that the three Boolean operations have different priorities for evaluation: the NOT operation has the highest priority, AND is evaluated next, and the OR operation has the lowest priority. If you do not pay attention to these different priorities for Boolean operations, you will get unexpected results, as Example 6.7 shows.

Example 6.7

```
USE sample;
SELECT emp_no, emp_fname, emp_lname
    FROM employee
    WHERE emp_no = 25348 AND emp_lname = 'Smith'
    OR emp_fname = 'Matthew' AND dept_no = 'd1';
```

```
SELECT emp_no, emp_fname, emp_lname
    FROM employee
    WHERE ((emp_no = 25348 AND emp_lname = 'Smith')
    OR emp_fname ='Matthew') AND dept_no = 'd1';
```

The result is

emp_no	emp_fname	emp_lname
25348	Matthew	Smith
emp_no	**emp_fname**	**emp_lname**

As the results of Example 6.7 show, the two SELECT statements display two different result sets. In the first SELECT statement, the system evaluates both AND operators first (from the left to the right), and then evaluates the OR operator. In the second SELECT statement, the use of parentheses changes the operation execution, with all expressions within parentheses being executed first, in sequence from left to right. As you can see, the first statement returned one row, while the second statement returned zero rows.

The existence of several Boolean operations in a WHERE clause complicates the corresponding SELECT statement and makes it error prone. In such cases, the use of parentheses is highly recommended, even if they are not necessary. The readability of such SELECT statements will be greatly improved, and possible errors can be avoided. Here is the first SELECT statement from Example 6.7, modified using the recommended form:

```
USE sample;
SELECT emp_no, emp_fname, emp_lname
    FROM employee
    WHERE (emp_no = 25348 AND emp_lname = 'Smith')
    OR (emp_fname = 'Matthew' AND dept_no = 'd1');
```

The third Boolean operator, NOT, changes the logical value of the corresponding condition. The truth table for NOT in Chapter 4 shows that the negation of the TRUE value is FALSE and vice versa; the negation of the NULL value is also NULL.

Example 6.8 shows the use of the NOT operator.

Example 6.8

Get the employee numbers and last names of all employees who do not belong to the department d2:

```
USE sample
SELECT emp_no, emp_lname
    FROM employee
    WHERE NOT dept_no = 'd2';
```

The result is

emp_no	emp_lname
25348	Smith
10102	Jones
18316	Barrimore
28559	Moser

In this case, the NOT operator can be replaced by the comparison operator <> (not equal).

NOTE This book uses the operator <> (instead of !=) to remain consistent with the ANSI SQL standard.

IN and BETWEEN Operators

An IN operator allows the specification of two or more expressions to be used for a query search. The result of the condition returns TRUE if the value of the corresponding column equals one of the expressions specified by the IN predicate.

Example 6.9 shows the use of the IN operator.

Example 6.9

Get the first and last names and the corresponding employee number for every employee whose employee number equals 29346, 28559, or 25348:

```
USE sample;
SELECT emp_no, emp_fname, emp_lname
   FROM employee
   WHERE emp_no IN (29346, 28559, 25348);
```

The result is

emp_no	emp_fname	emp_lname
25348	Matthew	Smith
29346	James	James
28559	Sybill	Moser

An IN operator is equivalent to a series of conditions, connected with one or more OR operators. (The number of OR operators is equal to the number of expressions following the IN operator minus one.)

The IN operator can be used together with the Boolean operator NOT, as shown in Example 6.10. In this case, the query retrieves rows that do not include any of the listed values in the corresponding columns.

Example 6.10

Get all columns for every employee whose employee number is neither 10102 nor 9031:

```
USE sample;
SELECT emp_no, emp_fname, emp_lname, dept_no
   FROM employee
   WHERE emp_no NOT IN (10102, 9031);
```

The result is

emp_no	emp_fname	emp_lname	dept_no
25348	Matthew	Smith	d3
18316	John	Barrimore	d1
29346	James	James	d2
2581	Elke	Hansel	d2
28559	Sybill	Moser	d1

In contrast to the IN operator, which specifies each individual value, the BETWEEN operator specifies a range, which determines the lower and upper bounds of qualifying values. Example 6.11 provides an example.

Example 6.11

Get the names and budgets for all projects with a budget between $95,000 and $120,000, inclusive:

```
USE sample;
SELECT project_name, budget
   FROM project
   WHERE budget BETWEEN 95000 AND 120000;
```

The result is

project_name	budget
Apollo	120000
Gemini	95000

The BETWEEN operator searches for all values in the range inclusively; that is, qualifying values can be between *or equal to* the lower and upper boundary values.

The BETWEEN operator is logically equal to two individual comparisons, which are connected with the Boolean operator AND. Example 6.11 is equivalent to Example 6.12.

Example 6.12

```
USE sample;
SELECT project_name, budget
   FROM project
   WHERE budget >= 95000 AND budget <= 120000;
```

Like the BETWEEN operator, the NOT BETWEEN operator can be used to search for column values that do not fall within the specified range. The BETWEEN operator can also be applied to columns with character and date values.

The two SELECT statements in Example 6.13 show a query that can be written in two different, but equivalent, ways.

Example 6.13

Get the names of all projects with a budget less than $100,000 and greater than $150,000:

```
USE sample;
SELECT project_name
  FROM project
   WHERE budget NOT BETWEEN 100000 AND 150000;
```

The result is

project_name
Gemini
Mercury

Using comparison operators, the query looks different:

```
USE sample;
SELECT project_name
  FROM project
   WHERE budget < 100000 OR budget > 150000;
```

NOTE Although the English phrasing of the requirements, "Get the names of all projects with budgets that are less than $100,000 and greater than $150,000," suggests the use of the AND operator in the second SELECT statement presented in Example 6.13, the logical meaning of the query demands the use of the OR operator, because if you use AND instead of OR, you will get no results at all. (The reason is that there cannot be a budget that is at the same time less than $100,000 and greater than $150,000.) Therefore, the second query in the example shows a possible problem that can appear between English phrasing of an exercise and its logical meaning.

Queries Involving NULL Values

A NULL in the CREATE TABLE statement specifies that a special value called NULL (which usually represents unknown or not applicable values) is allowed in the column. These values differ from all other values in a database. The WHERE clause of a SELECT statement generally returns rows for which the comparison evaluates to TRUE. The concern, then, regarding queries is, how will comparisons involving NULL values be evaluated in the WHERE clause?

All comparisons with NULL values will return FALSE (even when preceded by NOT). To retrieve the rows with NULL values in the column, Transact-SQL includes the operator feature IS NULL. This specification in a WHERE clause of a SELECT statement has the following general form:

```
column IS [NOT] NULL
```

Example 6.14 shows the use of the IS NULL operator.

Example 6.14

Get employee numbers and corresponding project numbers for employees with unknown jobs who work on project p2:

```
USE sample;
SELECT emp_no, project_no
   FROM works_on
   WHERE project_no = 'p2'
   AND job IS NULL;
```

The result is

emp_no	project_no
18316	p2
29346	p2

Because all comparisons with NULL values return FALSE, Example 6.15 shows syntactically correct, but logically incorrect, usage of NULL.

Example 6.15

```
USE sample;
SELECT project_no, job
   FROM works_on
   WHERE  job <> NULL;
```

The result is

project_no	job

The condition "column IS NOT NULL" is equivalent to the condition "NOT (column IS NULL)."

The system function ISNULL allows a display of the specified value as substitution for NULL (see Example 6.16).

Example 6.16

```
USE sample;
SELECT emp_no, ISNULL(job, 'Job unknown') AS task
    FROM works_on
    WHERE project_no = 'p1';
```

The result is

emp_no	task
10102	Analyst
9031	Manager
28559	Job unknown
29346	Clerk

Example 6.16 uses a column heading called **task** for the **job** column.

LIKE Operator

LIKE is an operator that is used for pattern matching; that is, it compares column values with a specified pattern. The data type of the column can be any character or date. The general form of the LIKE operator is

```
column [NOT] LIKE 'pattern'
```

pattern may be a string or expression (including columns of tables) and must be compatible with the data type of the corresponding column. For the specified column, the comparison between the value in a row and the pattern evaluates to TRUE if the column value matches the pattern expression.

Certain characters within the pattern—called *wildcard characters*—have a specific interpretation. Two of them are

- **%** The percent sign specifies any sequence of zero or more characters.
- **_** The underscore specifies any single character.

Example 6.17 shows the use of the wildcard characters % and _.

Example 6.17

Get the first and last names and numbers of all employees whose first name contains the letter *a* as the second character:

```
USE sample;
SELECT emp_fname, emp_lname, emp_no
    FROM employee
    WHERE emp_fname LIKE '_a%';
```

The result is

emp_fname	emp_lname	emp_no
Matthew	Smith	25348
James	James	29346

In addition to the percent sign and the underscore, Transact-SQL supports other characters that have a special meaning when used with the LIKE operator. These characters ([,], and ^) are demonstrated in Examples 6.18 and 6.19.

Example 6.18

Get full details of all departments whose location begins with a character in the range *C* through *F*:

```
USE sample;
SELECT dept_no, dept_name, location
   FROM department
   WHERE location LIKE '[C-F]%';
```

The result is

dept_no	dept_name	location
d1	Research	Dallas
d3	Marketing	Dallas

As shown in Example 6.18, the square brackets, [ ], delimit a range or list of characters. The order in which characters appear in a range is defined by the collating sequence, which is determined during the system installation.

The character ^ specifies the negation of a range or a list of characters. This character has this meaning only within a pair of square brackets, as shown in Example 6.19.

Example 6.19

Get the numbers and first and last names of all employees whose last name does not begin with the letter *J*, *K*, *L*, *M*, *N*, or *O* and whose first name does not begin with the letter *E* or *Z*:

```
USE sample;
SELECT emp_no, emp_fname, emp_lname
   FROM employee
   WHERE emp_lname LIKE '[^J-O]%'
   AND emp_fname LIKE '[^EZ]%';
```

The result is

emp_no	emp_fname	emp_lname
25348	Matthew	Smith
18316	John	Barrimore

The condition "column NOT LIKE 'pattern'" is equivalent to the condition "NOT (column LIKE 'pattern')."

Example 6.20 shows the use of the LIKE operator (together with NOT).

Example 6.20

Get full details of all employees whose first name does not end with the character *n*:

```
USE sample;
SELECT emp_no, emp_fname, emp_lname
    FROM employee
    WHERE emp_fname NOT LIKE '%n';
```

The result is

emp_no	emp_fname	emp_lname
25348	Matthew	Smith
29346	James	James
2581	Elke	Hansel
9031	Elsa	Bertoni
28559	Sybill	Moser

Any of the wildcard characters (%, _, [,], or ^) enclosed in square brackets stands for itself. An equivalent feature is available through the ESCAPE option. Therefore, both SELECT statements in Example 6.21 have the same meaning.

Example 6.21

```
USE sample;
SELECT project_no, project_name
    FROM project
    WHERE project_name LIKE '%[_]%';
SELECT project_no, project_name
        FROM project
        WHERE project_name LIKE '%!_%' ESCAPE '!';
```

The result is

project_no	project_name
project_no	project_name

Both SELECT statements search for the underscore as an actual character in the column **project_name**. In the first SELECT statement, this search is established by enclosing the sign _ in square brackets. The second SELECT statement uses a character (in Example 6.21, the character !) as an escape character. The escape character overrides the meaning of the underscore as the wildcard character and leaves it to be interpreted as an ordinary character. (The result contains no rows because there are no project names that include the underscore character.)

> **NOTE** The SQL standard supports the use of only %, _, and the ESCAPE operator. For this reason, if any wildcard character must stand for itself, using the ESCAPE operator instead of a pair of square brackets is recommended.

GROUP BY Clause

The GROUP BY clause defines one or more columns as a group such that all rows within any group have the same values for those columns. Example 6.22 shows the simple use of the GROUP BY clause.

Example 6.22

Get all jobs of the employees:

```
USE sample;
SELECT job
    FROM works_on
    GROUP BY job;
```

The result is

job
NULL
Analyst
Clerk
Manager

In Example 6.22, the GROUP BY clause builds different groups for all possible values (NULL, too) appearing in the **job** column.

> **NOTE** There is a restriction regarding the use of columns in the GROUP BY clause. Each column appearing in the SELECT list of the query must also appear in the GROUP BY clause. This restriction does not apply to constants and to columns that are part of an aggregate function. (Aggregate functions are explained in the next subsection.) This makes sense, because only columns in the GROUP BY clause are guaranteed to have a single value for each group.

A table can be grouped by any combination of its columns. Example 6.23 shows the grouping of rows of the **works_on** table using two columns.

Example 6.23

Group all employees using their project numbers and jobs:

```
USE sample;
SELECT project_no, job
   FROM works_on
   GROUP BY project_no, job;
```

The result is

project_no	job
p1	Analyst
p1	Clerk
p1	Manager
p1	NULL
p2	NULL
p2	Clerk
p3	Analyst
p3	Clerk
p3	Manager

The result of Example 6.23 shows that there are nine groups with different combinations of project numbers and jobs. The only two groups that contain more than one row are

p2	Clerk	25348, 28559
p2	NULL	18316, 29346

The sequence of the column names in the GROUP BY clause does not need to correspond to the sequence of the names in the SELECT list.

Aggregate Functions

Aggregate functions are functions that are used to get summary values. All aggregate functions can be divided into several groups:

- Convenient aggregate functions
- Statistical aggregate functions
- User-defined aggregate functions
- Analytic aggregate functions

The first three types are described in the following sections, while analytic aggregate functions are explained in detail in Chapter 24.

Convenient Aggregate Functions

The Transact-SQL language supports six aggregate functions:

- MIN
- MAX
- SUM
- AVG
- COUNT
- COUNT_BIG

All aggregate functions operate on a single argument, which can be either a column or an expression. The only exception is the second form of the COUNT and COUNT_BIG functions, COUNT(*) and COUNT_BIG(*). The result of each aggregate function is a constant value, which is displayed in a separate column of the result.

The aggregate functions appear in the SELECT list, which can include a GROUP BY clause. If there is no GROUP BY clause in the SELECT statement, and the SELECT list includes at least one aggregate function, then no simple columns can be included in the SELECT list (other than as arguments of an aggregate function). Therefore, Example 6.24 is *wrong*.

Example 6.24 (Example of an Illegal Statement)

```
USE sample;
SELECT emp_lname, MIN(emp_no)
    FROM employee;
```

The **emp_lname** column of the **employee** table must not appear in the SELECT list of Example 6.24 because it is not the argument of an aggregate function. On the other hand, all column names that are not arguments of an aggregate function may appear in the SELECT list if they are used for grouping.

The argument of an aggregate function can be preceded by one of two keywords:

- **ALL**   Indicates that all values of a column are to be considered (ALL is the default value)
- **DISTINCT**   Eliminates duplicate values of a column before the aggregate function is applied

MIN and MAX Aggregate Functions   The aggregate functions MIN and MAX compute the lowest and highest values in the column, respectively. If there is a WHERE clause, the MIN and MAX functions return the lowest or highest of values from selected rows. Example 6.25 shows the use of the aggregate function MIN.

Example 6.25

Get the lowest employee number:

```
USE sample;
SELECT MIN(emp_no) AS min_employee_no
    FROM employee;
```

The result is

min_employee_no
2581

The result of Example 6.25 is not user friendly. For instance, the name of the employee with the lowest number is not known. As already shown, the explicit specification of the **emp_name** column in the SELECT list is not allowed. To retrieve the name of the employee with the lowest employee number, use a subquery, as shown in Example 6.26, where the inner query contains the SELECT statement of the previous example.

Example 6.26
Get the number and the last name of the employee with the lowest employee number:

```
USE sample;
SELECT emp_no, emp_lname
  FROM employee
   WHERE emp_no =
   (SELECT MIN(emp_no)
       FROM employee);
```

The result is

emp_no	emp_lname
2581	Hansel

Example 6.27 shows the use of the aggregate function MAX.

Example 6.27
Get the employee number of the manager who was entered last in the **works_on** table:

```
USE sample;
SELECT emp_no
 FROM works_on
   WHERE enter_date =
   (SELECT MAX(enter_date)
       FROM works_on
        WHERE job = 'Manager');
```

The result is

emp_no
10102

The argument of the functions MIN and MAX can also be a string value or a date. If the argument has a string value, the comparison between all values will be provided using the actual collating sequence. For all arguments of temporal data types, the earliest date specifies the lowest value in the column and the latest date specifies the highest value in the column.

The DISTINCT option cannot be used with the aggregate functions MIN and MAX. All NULL values in the column that are the argument of the aggregate function MIN or MAX are always eliminated before MIN or MAX is applied.

SUM Aggregate Function   The aggregate function SUM calculates the sum of the values in the column. The argument of the function SUM must be numeric. Example 6.28 shows the use of the SUM function.

Example 6.28
Calculate the sum of all budgets of all projects:

```
USE sample;
SELECT SUM(budget) sum_of_budgets
   FROM project;
```

The result is

sum_of_budgets
401500

The aggregate function in Example 6.28 groups all values of the projects' budgets and determines their total sum. For this reason, the query in Example 6.28 (as does each analog query) implicitly contains the grouping function. The grouping function from Example 6.28 can be written explicitly in the query, as shown in Example 6.29.

Example 6.29
```
USE sample;
SELECT SUM(budget) sum_of_budgets
 FROM project
 GROUP BY();
```

The use of this syntax for the GROUP BY clause is recommended because it defines a grouping explicitly. (Chapter 24 describes several other GROUP BY features.)

The use of the DISTINCT option eliminates all duplicate values in the column before the function SUM is applied. Similarly, all NULL values are always eliminated before SUM is applied.

AVG Aggregate Function   The aggregate function AVG calculates the average of the values in the column. The argument of the function AVG must be numeric. All NULL values are eliminated before the function AVG is applied. Example 6.30 shows the use of the AVG aggregate function.

Example 6.30
Calculate the average of all budgets with an amount greater than $100,000:

```
USE sample;
SELECT AVG(budget) avg_budget
    FROM project
    WHERE budget > 100000;
```

The result is

avg_budget
153250

COUNT and COUNT_BIG Aggregate Functions The aggregate function COUNT has two different forms:

```
COUNT([DISTINCT] col_name)
COUNT(*)
```

The first form calculates the number of values in the **col_name** column. When the DISTINCT keyword is used, all duplicate values are eliminated before COUNT is applied. This form of COUNT does not count NULL values for the column.

Example 6.31 shows the use of the first form of the aggregate function COUNT.

Example 6.31

Count all different jobs in each project:

```
USE sample;
SELECT project_no, COUNT(DISTINCT job) job_count
    FROM works_on
    GROUP BY project_no;
```

The result is

project_no	job_count
p1	3
p2	1
p3	3

As can be seen from the result of Example 6.31, all NULL values are eliminated before the function COUNT(DISTINCT job) is applied.

The second form of the function COUNT, COUNT(*), counts the number of rows in the table. Or, if there is a WHERE clause in the SELECT statement, COUNT(*) returns the number of rows for which the WHERE condition is true. In contrast to the first form of the function COUNT, the second form does not eliminate NULL values, because this function operates on rows and not on columns. Example 6.32 shows the use of COUNT(*).

Example 6.32

Get the number of each job in all projects:

```
USE sample;
SELECT job, COUNT(*) job_count
   FROM works_on
  GROUP BY job;
```

The result is

job	job_count
NULL	3
Analyst	2
Clerk	4
Manager	2

The COUNT_BIG function is analogous to the COUNT function. The only difference between them is in relation to their return values: COUNT_BIG always returns a value of the BIGINT data type, while the COUNT function always returns a value of the INTEGER data type.

Statistical Aggregate Functions

The following aggregate functions belong to the group of statistical aggregate functions:

- **VAR**   Computes the variance of all the values listed in a column or expression
- **VARP**   Computes the variance for the population of all the values listed in a column or expression
- **STDEV**   Computes the standard deviation (which is computed as the square root of the corresponding variance) of all the values listed in a column or expression
- **STDEVP**   Computes the standard deviation for the population of all the values listed in a column or expression

Examples showing statistical aggregate functions will be provided in Chapter 24.

User-Defined Aggregate Functions

The Database Engine also supports the implementation of user-defined aggregate functions. Using these functions, you can implement and deploy aggregate functions that do not belong to aggregate functions included with the system. These functions are a special case of user-defined functions, which will be described in detail in Chapter 8.

HAVING Clause

The HAVING clause defines the condition that is then applied to groups of rows. Hence, this clause has the same meaning to groups of rows that the WHERE clause has to the content of the corresponding table. The syntax of the HAVING clause is

```
HAVING condition
```

where **condition** contains aggregate functions or constants.

Example 6.33 shows the use of the HAVING clause with the aggregate function COUNT(*).

Example 6.33

Get project numbers for all projects employing fewer than four persons:

```
USE sample;
SELECT project_no
    FROM works_on
    GROUP BY project_no
    HAVING COUNT(*) < 4;
```

The result is

project_no
p3

In Example 6.33, the system uses the GROUP BY clause to group all rows according to existing values in the **project_no** column. After that, it counts the number of rows in each group and selects those groups with three or fewer rows.

The HAVING clause can also be used without aggregate functions, as shown in Example 6.34.

Example 6.34

Group rows of the **works_on** table by job and eliminate those jobs that do not begin with the letter *M*:

```
USE sample;
SELECT job
    FROM works_on
    GROUP BY job
    HAVING job LIKE 'M%';
```

The result is

job
Manager

The HAVING clause can also be used without the GROUP BY clause, although doing so is uncommon in practice. In such a case, all rows of the entire table belong to a single group.

ORDER BY Clause

The ORDER BY clause defines the particular order of the rows in the result of a query. This clause has the following syntax:

```
ORDER BY {[col_name | col_number [ASC | DESC]]} , ...
```

The **col_name** column defines the order. **col_number** is an alternative specification that identifies the column by its ordinal position in the sequence of all columns in the SELECT list (1 for the first column, 2 for the second one, and so on). ASC indicates ascending order and DESC indicates descending order, with ASC as the default value.

NOTE   The columns in the ORDER BY clause need not appear in the SELECT list. However, the ORDER BY columns must appear in the SELECT list if SELECT DISTINCT is specified. Also, this clause cannot reference columns from tables that are not listed in the FROM clause.

As the syntax of the ORDER BY clause shows, the order criterion may contain more than one column, as shown in Example 6.35.

Example 6.35

Get department numbers and employee names for employees with employee numbers < 20000, in ascending order of last and first names:

```
USE sample;
SELECT emp_fname, emp_lname, dept_no
    FROM employee
    WHERE emp_no < 20000
    ORDER BY emp_lname, emp_fname;
```

The result is

emp_fname	emp_lname	dept_no
John	Barrimore	d1
Elsa	Bertoni	d2
Elke	Hansel	d2
Ann	Jones	d3

It is also possible to identify the columns in the ORDER BY clause by the ordinal position of the column in the SELECT list. The ORDER BY clause in Example 6.35 could be written in the following form:

```
ORDER BY 2,1
```

The use of column numbers instead of column names is an alternative solution if the order criterion contains any aggregate function. (The other way is to use column headings, which then appear in the ORDER BY clause.) However, using column names rather than numbers in the ORDER BY clause is recommended, to reduce the difficulty of maintaining the query if any columns need to be added or deleted from the SELECT list. Example 6.36 shows the use of column numbers.

Example 6.36

For each project number, get the project number and the number of all employees, in descending order of the employee number:

```
USE sample;
SELECT project_no, COUNT(*) emp_quantity
    FROM works_on
    GROUP BY project_no
    ORDER BY 2 DESC
```

The result is

project_no	emp_quantity
p1	4
p2	4
p3	3

The Transact-SQL language orders NULL values at the beginning of all values if the order is ascending and orders them at the end of all values if the order is descending.

Using ORDER BY to Support Paging

If you want to display rows on the current page, you can either implement that in your application or instruct the database server to do it. In the former case, all rows from the database are sent to the application, and the application's task is to retrieve the rows needed for printing and to display them. In the latter case, only those rows needed for the current page are selected from the server side and displayed. As you might guess, server-side paging generally provides better performance, because only the rows needed for printing are sent to the client.

The Database Engine supports two clauses in relation to server-side paging: OFFSET and FETCH. Example 6.37 shows the use of these two clauses.

Example 6.37

Get the business entity ID, job title, and birthday for all female employees from the **AdventureWorks** database in ascending order of job title. Display the third page. (Ten rows are displayed per page.)

```
USE AdventureWorks;
SELECT BusinessEntityID, JobTitle, BirthDate
    FROM HumanResources.Employee
    WHERE Gender = 'F'
    ORDER BY JobTitle
    OFFSET 20 ROWS
    FETCH NEXT 10 ROWS ONLY;
```

NOTE You can find further examples concerning the OFFSET clause in Chapter 24 (see Examples 24.24 and 24.25).

The OFFSET clause specifies the number of rows to skip before starting to return the rows. This is evaluated after the ORDER BY clause is evaluated and the rows are sorted. The FETCH NEXT clause specifies the number of rows to retrieve. The parameter of this clause can be a constant, an expression, or a result of a query. FETCH NEXT is analogous to FETCH FIRST.

The main purpose of server-side paging is to implement generic page forms, using variables. This can be done using a batch. The corresponding example can be found in Chapter 8 (see Example 8.5).

SELECT Statement and IDENTITY Property

The IDENTITY property allows you to specify a counter of values for a specific column of a table. Columns with numeric data types, such as TINYINT, SMALLINT, INT, and BIGINT, can have this property. The Database Engine generates values for such columns sequentially, starting with an initial value. Therefore, you can use the IDENTITY property to let the system generate unique numeric values for the table column of your choice. (The default value for the initial value and increment is 1.)

Each table can have only one column with the IDENTITY property. The table owner can specify the starting number and the increment value, as shown in Example 6.38.

Example 6.38

```
USE sample;
CREATE TABLE product
    (product_no INTEGER IDENTITY(10000,1) NOT NULL,
     product_name CHAR(30) NOT NULL,
     price MONEY);
SELECT $identity
        FROM product
        WHERE product_name = 'Soap';
```

The result could be

product_no
10005

The **product** table is created first in Example 6.38. This table has the **product_no** column with the IDENTITY property. The values of the **product_no** column are automatically generated by the system, beginning with 10000 and incrementing by 1 for every subsequent value: 10000, 10001, 10002, and so on.

Some system functions and variables are related to the IDENTITY property. Example 6.38 uses the **$identity** variable. As the result set of Example 6.38 shows, this variable automatically refers to the column with the IDENTITY property.

To find out the starting value and the increment of the column with the IDENTITY property, you can use the IDENT_SEED and IDENT_INCR functions, respectively, in the following way:

```
USE sample;
SELECT IDENT_SEED('product'), IDENT_INCR('product');
```

As you already know, the system automatically sets identity values. If you want to supply your own values for particular rows, you must set the IDENTITY_INSERT option to ON before the explicit value will be inserted:

```
SET IDENTITY_INSERT table_name ON
```

NOTE Because the IDENTITY_INSERT option can be used to specify any values for a column with the IDENTITY property, IDENTITY does not generally enforce uniqueness. Use the UNIQUE or PRIMARY KEY constraint for this task.

If you insert values after the IDENTITY_INSERT option is set to ON, the system presumes that the next value is the incremented value of the highest value that exists in the table at that moment.

CREATE SEQUENCE Statement

Using the IDENTITY property has several significant disadvantages, the most important of which are the following:

- You can use it only with the specified table.
- You cannot obtain the new value before using it.
- You can specify the IDENTITY property only when the column is created.

For these reasons, the Database Engine offers another solution called sequences. A sequence has the same semantics as the IDENTITY property but doesn't have the limitations from the preceding list. Therefore, a sequence is an independent database feature that enables you to specify a counter of values for different database objects, such as columns and variables.

Sequences are created using the CREATE SEQUENCE statement. The CREATE SEQUENCE statement is specified in the SQL standard and is implemented in other relational database systems, such as IBM Db2 and Oracle.

Example 6.39 shows how sequences can be specified.

Example 6.39

```
USE sample;
CREATE SEQUENCE dbo.Sequence1
  AS INT
  START WITH 1 INCREMENT BY 5
  MINVALUE 1 MAXVALUE 256
  CYCLE;
```

The values of the sequence called **Sequence1** in Example 6.39 are automatically generated by the system, beginning with 1 and incrementing by 5 for every subsequent value. Therefore, the START clause specifies the initial value, while the INCREMENT clause defines the incremental value. (The incremental value can be positive or negative.)

The following two optional clauses, MINVALUE and MAXVALUE, are directives, which specify a minimal and maximum value for a sequence object. (Note that MINVALUE must be less than or equal to the start value, while MAXVALUE cannot be greater than the upper boundary for the values of the data type used for the specification of the sequence.) The CYCLE clause specifies that the object should restart from the minimum value (or maximum value, for descending sequence objects) when its minimum (or maximum) value is exceeded. The default value for this clause is NO CYCLE, which means that an exception will be thrown if its minimum or maximum value is exceeded.

The main property of a sequence is that it is table-independent; that is, it can be used with any database object, such as a table's column or variable. (This property positively affects storage and, therefore, performance. You do not need storage for a specified sequence; only the last-used value is stored.)

To generate new sequence values, you can use the NEXT VALUE FOR expression. Example 6.40 shows the use of this expression.

Example 6.40

```
USE sample;
SELECT NEXT VALUE FOR dbo.sequence1;
SELECT NEXT VALUE FOR dbo.sequence1;
```

The result is

1
6

You can use the NEXT VALUE FOR expression to assign the results of a sequence to a variable or to a column. Example 6.41 shows how you can use this expression to assign the results to a table's column.

Example 6.41

```
USE sample;
CREATE TABLE dbo.table1
      (column1 INT NOT NULL PRIMARY KEY,
       column2 CHAR(10));
       INSERT INTO dbo.table1 VALUES (NEXT VALUE FOR dbo.sequence1, 'A');
       INSERT INTO dbo.table1 VALUES (NEXT VALUE FOR dbo.sequence1, 'B');
```

Example 6.41 first creates a table called **table1** with two columns. The following two INSERT statements insert two rows in this table. (For the syntax of the INSERT statement, see Chapter 7.) The first column has values 11 and 16, respectively. (These two values are subsequent values, following the generated values in Example 6.40.)

Example 6.42 shows how you can use the catalog view called **sys.sequences** to check the current value of the sequence, without using it. (Catalog views are described in detail in Chapter 9.)

Example 6.42

```
USE sample;
SELECT current_value
   FROM sys.sequences
   WHERE name = 'sequence1';
```

NOTE Generally, you use the NEXT VALUE FOR expression in the INSERT statement (see Chapter 7) to let the system insert generated values. You can also use the NEXT VALUE FOR expression as part of a multirow query by using the OVER clause (see Example 24.8 in Chapter 24).

The ALTER SEQUENCE statement modifies the properties of an existing sequence. One of the most important uses of this statement is in relation to the RESTART WITH clause, which "reseeds" a given sequence. Example 6.43 shows the use of the ALTER SEQUENCE statement to reinitialize (almost) all properties of the existing sequence called **sequence1**.

Example 6.43

```
USE sample;
ALTER SEQUENCE dbo.sequence1
    RESTART WITH 100
    INCREMENT BY 50
    MINVALUE 50
    MAXVALUE 10000
    NO CYCLE;
```

To drop a sequence, use the DROP SEQUENCE statement.

Set Operators

In addition to the operators described earlier in the chapter, three set operators are supported in the Transact-SQL language:

- UNION
- INTERSECT
- EXCEPT

NOTE The three set operators discussed in this section have different priorities for evaluation: the INTERSECT operator has the highest priority, EXCEPT is evaluated next, and the UNION operator has the lowest priority. If you do not pay attention to these different priorities, you will get unexpected results when you use several set operators together.

UNION Operator

The result of the union of two sets is the set of all elements appearing in either or both of the sets. Accordingly, the union of two tables is a new table consisting of all rows appearing in either one or both of the tables.

The general form of the UNION operator is

```
select_1 UNION [ALL] select_2 {[UNION [ALL] select_3]}...
```

select_1, select_2,... are SELECT statements that build the union. If the ALL option is used, all resulting rows, including duplicates, are displayed. The ALL option has the same meaning with the UNION operator as it has in the SELECT list, with one difference: the ALL option is the default in the SELECT list, but it must be specified with the UNION operator to display all resulting rows, including duplicates.

The **sample** database in its original form is not suitable for a demonstration of the UNION operator. For this reason, this section introduces a new table, **employee_enh**, which is identical to the existing **employee** table, up to the additional **domicile** column. The **domicile** column contains the place of residence of every employee.

The new **employee_enh** table has the following form:

emp_no	emp_fname	emp_lname	dept_no	domicile
25348	Matthew	Smith	d3	San Antonio
10102	Ann	Jones	d3	Houston
18316	John	Barrimore	d1	San Antonio
29346	James	James	d2	Seattle
2581	Elke	Hansel	d2	Portland
9031	Elsa	Bertoni	d2	Tacoma
28559	Sybill	Moser	d1	Houston

Creation of the **employee_enh** table provides an opportunity to show the use of the INTO clause of the SELECT statement. SELECT INTO has two different parts. First, it creates the new table with the columns corresponding to the columns listed in the SELECT list. Second, it inserts the existing rows of the original table into the new table. (The name of the new table appears with the INTO clause, and the name of the source table appears in the FROM clause of the SELECT statement.)

Example 6.44 shows the creation of the **employee_enh** table.

Example 6.44

```
USE sample;
SELECT emp_no, emp_fname, emp_lname, dept_no
    INTO employee_enh
    FROM employee;
ALTER TABLE employee_enh
        ADD domicile CHAR(25) NULL;
```

In Example 6.44, SELECT INTO generates the **employee_enh** table and inserts all rows from the initial table (**employee**) into the new one. Finally, the ALTER TABLE statement appends the **domicile** column to the **employee_enh** table.

NOTE In SQL Server 2016 and earlier, SELECT INTO creates a new table and stores it *always* into the default filegroup. Since SQL Server 2017, you can use the additional ON keyword of SELECT INTO to load a table into a non-default filegroup. For instance, If you want to load the **employee_enh** table in the **Employee_FSGroup** filegroup (see Example 5.18), the SELECT statement in Example 6.44 will be this:

```
SELECT emp_no, emp_fname, emp_lname, dept_no
   INTO employee_enh ON Employee_FSGroup
   FROM employee;
```

After the execution of Example 6.44, the **domicile** column contains no values. The values can be added using SQL Server Management Studio (see Chapter 3) or the following UPDATE statements:

```
USE sample;
UPDATE employee_enh SET domicile = 'San Antonio'
   WHERE emp_no = 25348;
UPDATE employee_enh SET domicile = 'Houston'
   WHERE emp_no = 10102;
UPDATE employee_enh SET domicile = 'San Antonio'
   WHERE emp_no = 18316;
UPDATE employee_enh SET domicile = 'Seattle'
   WHERE emp_no = 29346;
UPDATE employee_enh SET domicile = 'Portland'
   WHERE emp_no = 2581;
UPDATE employee_enh SET domicile = 'Tacoma'
   WHERE emp_no = 9031;
UPDATE employee_enh SET domicile = 'Houston'
   WHERE emp_no = 28559;
```

Example 6.45 shows the union of the tables **employee_enh** and **department**.

Example 6.45

```
USE sample;
SELECT domicile
 FROM employee_enh
UNION
SELECT location
   FROM department;
```

The result is

domicile
San Antonio
Houston
Portland
Tacoma
Seattle
Dallas

Two tables can be connected with the UNION operator if they are compatible with each other. This means that both the SELECT lists must have the same number of columns, and the corresponding columns must have compatible data types. (For example, INT and SMALLINT are compatible data types.)

The ordering of the result of the union can be done only if the ORDER BY clause is used with the last SELECT statement, as shown in Example 6.46. The GROUP BY and HAVING clauses can be used with the particular SELECT statements, but not with the union itself.

Example 6.46

Get the employee number for employees who either belong to department d1 or entered their project before 1/1/2017, in ascending order of employee number:

```
USE sample;
SELECT emp_no
    FROM employee
    WHERE dept_no = 'd1'
UNION
SELECT emp_no
    FROM works_on
    WHERE enter_date < '01.01.2017'
ORDER BY 1;
```

The result is

emp_no
9031
10102
18316
28559
29346

NOTE The UNION operator supports the ALL option. When UNION is used with ALL, duplicates are not removed from the result set.

The OR operator can be used instead of the UNION operator if all SELECT statements connected by one or more UNION operators reference the same table. In this case, the set of the SELECT statements is replaced through one SELECT statement with the set of OR operators.

INTERSECT and EXCEPT Operators

The two other set operators are INTERSECT, which specifies the intersection, and EXCEPT, which defines the difference operator. The intersection of two tables is the set of rows belonging to both tables. The difference of two tables is the set of all rows, where the resulting rows belong to the first table but not to the second one. Example 6.47 shows the use of the INTERSECT operator.

Example 6.47

```
USE sample;
SELECT emp_no
    FROM employee
    WHERE dept_no = 'd1'
INTERSECT
SELECT emp_no
    FROM works_on
    WHERE enter_date < '01.01.2018';
```

The result is

emp_no
18316
28559

NOTE Transact-SQL does not support the INTERSECT operator with the ALL option. (The same is true for the EXCEPT operator.)

Example 6.48 shows the use of the EXCEPT set operator.

Example 6.48

```
USE sample;
SELECT emp_no
    FROM employee
  WHERE dept_no = 'd3'
```

```
EXCEPT
SELECT emp_no
    FROM works_on
    WHERE enter_date > '01.01.2018';
```

The result is

emp_no
10102
25348

CASE Expressions

In database application programming, it is sometimes necessary to modify the representation of data. For instance, a person's status can be coded using the values 1, 2, and 3 (for female, male, and child, respectively). Such a programming technique can reduce the time for the implementation of a program. The CASE expression in the Transact-SQL language makes this type of encoding easy to implement.

NOTE CASE does not represent a statement (as in most programming languages) but an expression. Therefore, the CASE expression can be used (almost) everywhere where the Transact-SQL language allows the use of an expression.

The CASE expression has two different forms:

- Simple CASE expression
- Searched CASE expression

The syntax of the simple CASE expression is

```
CASE expression_1
    {WHEN expression_2 THEN result_1} ...
    [ELSE result_n]
END
```

A Transact-SQL statement with the simple CASE expression looks for the first expression in the list of all WHEN clauses that match **expression_1** and evaluates the corresponding THEN clause. If there is no match, the ELSE clause is evaluated. Example 6.49 shows the use of the simple CASE expression.

Example 6.49

```
USE AdventureWorks;
GO
SELECT ProductNumber, Category =
```

```
      CASE ProductLine
         WHEN 'R' THEN 'Road'
         WHEN 'M' THEN 'Mountain'
         WHEN 'T' THEN 'Touring'
         WHEN 'S' THEN 'Other sale items'
         ELSE 'Not for sale'
      END,
   Name
FROM Production.Product;
```

Example 6.49 uses the **product** table from the **production** schema of the **AdventureWorks** database. Depending on the abbreviated value stored in the **ProductLine** column of this table, the query in Example 6.49 displays the full name of each column value under the **Category** heading. (The result of this example is too large to be displayed.)

The syntax of the searched CASE expression is

```
CASE
   {WHEN condition_1 THEN result_1} ...
      [ELSE result_n]
END
```

A Transact-SQL statement with the searched CASE expression looks for the first expression that evaluates to TRUE. If none of the WHEN conditions evaluates to TRUE, the value of the ELSE expression is returned. Example 6.50 shows the use of the searched CASE expression.

Example 6.50

```
USE sample;
SELECT project_name,
    CASE
       WHEN budget > 0 AND budget < 100000  THEN 1
       WHEN budget >= 100000 AND budget < 200000  THEN 2
       WHEN budget >= 200000 AND budget < 300000 THEN 3
       ELSE 4
    END budget_weight
 FROM project;
```

The result is

project_name	budget_weight
Apollo	2
Gemini	1
Mercury	2

In Example 6.50, budgets of all projects are weighted, and the calculated weights (together with the name of the corresponding project) are displayed.

Subqueries

All previous examples in this chapter contain comparisons of column values with an expression, constant, or set of constants. Additionally, the Transact-SQL language offers the ability to compare column values with the result of another SELECT statement. Such a construct, where one or more SELECT statements are nested in the WHERE clause of another SELECT statement, is called a *subquery*. The first SELECT statement of a subquery is called the *outer query*—in contrast to the *inner query*, which denotes the SELECT statement(s) used in a comparison. The inner query will be evaluated first, and the outer query receives the values of the inner query.

NOTE An inner query can also be nested in an INSERT, UPDATE, or DELETE statement, which will be discussed later in this book.

There are two types of subqueries:

- Self-contained
- Correlated

In a self-contained subquery, the inner query is logically evaluated exactly once. A correlated subquery differs from a self-contained one in that its value depends upon a variable from the outer query. Therefore, the inner query of a correlated subquery is logically evaluated each time the system retrieves a new row from the outer query. This section shows examples of self-contained subqueries. The correlated subquery will be discussed later in the chapter, together with the join operation.

A self-contained subquery can be used with the following operators:

- Comparison operators
- IN operator
- ANY or ALL operator

Subqueries and Comparison Operators

Example 6.51 shows the self-contained subquery that is used with the operator =.

Example 6.51

Get the first and last names of employees who work in the Research department:

```
USE sample;
SELECT emp_fname, emp_lname
   FROM employee
   WHERE dept_no =
    (SELECT dept_no
       FROM department
      WHERE dept_name = 'Research');
```

The result is

emp_fname	emp_lname
John	Barrimore
Sybill	Moser

The inner query of Example 6.51 is logically evaluated first. That query returns the number of the research department (d1). Thus, after the evaluation of the inner query, the subquery in Example 6.51 can be represented with the following equivalent query:

```
USE sample
SELECT emp_fname, emp_lname
    FROM employee
    WHERE dept_no = 'd1';
```

A subquery can be used with other comparison operators, too. Any comparison operator can be used, provided that the inner query returns exactly one row. This is obvious because comparison between particular column values of the outer query and a set of values (as a result of the inner query) is not possible. The following section shows how you can handle the case in which the result of an inner query contains a set of values.

Subqueries and the IN Operator

The IN operator allows the specification of a set of expressions (or constants) that are subsequently used for the query search. This operator can be applied to a subquery for the same reason—that is, when the result of an inner query contains a set of values.

Example 6.52 shows the use of the IN operator in a subquery.

Example 6.52

Get full details of all employees whose department is located in Dallas:

```
USE sample;
SELECT *
    FROM employee
    WHERE dept_no IN
    (SELECT dept_no
        FROM department
        WHERE location = 'Dallas');
```

The result is

emp_no	emp_fname	emp_lname	dept_no
25348	Matthew	Smith	d3
10102	Ann	Jones	d3
18316	John	Barrimore	d1
28559	Sybill	Moser	d1

Each inner query may contain further queries. This type of subquery is called a subquery with multiple levels of nesting. The maximum number of inner queries in a subquery depends on the amount of memory the Database Engine has for each SELECT statement. In the case of subqueries with multiple levels of nesting, the system first evaluates the innermost query and returns the result to the query on the next nesting level, and so on. Finally, the outermost query evaluates the final outcome.

Example 6.53 shows the query with multiple levels of nesting.

Example 6.53

Get the last names of all employees who work on the project Apollo:

```
USE sample;
SELECT emp_lname
    FROM employee
    WHERE emp_no IN
    (SELECT emp_no
        FROM works_on
        WHERE project_no IN
        (SELECT project_no
                FROM project
                WHERE project_name = 'Apollo'));
```

The result is

emp_lname
Jones
James
Bertoni
Moser

The innermost query in Example 6.53 evaluates to the **project_no** value p1. The middle inner query compares this value with all values of the **project_no** column in the **works_on** table. The result of this query is the set of employee numbers: (10102, 29346, 9031, 28559). Finally, the outermost query displays the corresponding last names for the selected employee numbers.

Subqueries and ANY and ALL Operators

The operators ANY and ALL are always used in combination with one of the comparison operators. The general syntax of both operators is

```
column_name operator [ANY | ALL] query
```

where **operator** stands for a comparison operator and **query** is an inner query.

The ANY operator evaluates to TRUE if the result of the corresponding inner query contains at least one row that satisfies the comparison. The keyword SOME is the synonym for ANY. Example 6.54 shows the use of the ANY operator.

Example 6.54

Get the employee numbers, project numbers, and job names for employees who have not spent the most time on one of the projects:

```
USE sample;
SELECT DISTINCT emp_no, project_no, job
    FROM works_on
    WHERE enter_date > ANY
    (SELECT  enter_date
        FROM works_on);
```

The result is

emp_no	project_no	job
2581	p3	Analyst
9031	p1	Manager
9031	p3	Clerk
10102	p3	Manager
18316	p2	NULL
25348	p2	Clerk
28559	p1	NULL
28559	p2	Clerk
29346	p1	Clerk
29346	p2	NULL

Each value of the **enter_date** column in Example 6.54 is compared with all values of this column. For all dates of the column, except the oldest one, the comparison is evaluated to TRUE at least once. The row with the oldest date does not belong to the result because the comparison does not evaluate to TRUE in any case. In other words, the expression "enter_date > ANY (SELECT enter_date FROM works_on)" is true if there are *any* (one or more) rows in the **works_on** table with a value of the **enter_date** column less than the value of **enter_date** for the current row. This will be true for all but the earliest value of the **enter_date** column.

The ALL operator evaluates to TRUE if the evaluation of the table column in the inner query returns all values of that column.

NOTE Do not use ANY and ALL operators! Every query using ANY or ALL can be better formulated with the EXISTS function, which is explained later in this chapter (see the section "Subqueries and the EXISTS Function"). Additionally, the semantic meaning of the ANY operator can be easily confused with the semantic meaning of the ALL operator, and vice versa.

Temporary Tables

A temporary table is a database object that is temporarily stored and managed by the database system. Temporary tables can be local or global. Local temporary tables have physical representation—that is, they are stored in the **tempdb** system database. They are specified with the prefix # (for example, **#table_name**).

A local temporary table is owned by the session that created it and is visible only to that session. Such a table is thus automatically dropped when the creating session terminates. (If you define a local temporary table inside a stored procedure, it will be destroyed when the corresponding procedure terminates.)

Global temporary tables are visible to any user and any connection after they are created, and are deleted when all users that are referencing the table disconnect from the database server. In contrast to local temporary tables, global ones are specified with the prefix ##.

Examples 6.55 and 6.56 show how the temporary table **project_temp** can be created using two different Transact-SQL statements.

Example 6.55

```
USE sample;
CREATE TABLE #project_temp
    (project_no CHAR(4) NOT NULL,
     project_name CHAR(25) NOT NULL);
```

Example 6.56

```
USE sample;
SELECT project_no, project_name
    INTO #project_temp1
    FROM project;
```

Examples 6.55 and 6.56 are similar. They use two different Transact-SQL statements to create the local temporary table, **#project_temp** and **#project_temp1**, respectively. However, Example 6.55 leaves it empty, while Example 6.56 populates the temporary table with the data from the **project** table.

Join Operator

The previous sections of this chapter demonstrated the use of the SELECT statement to query rows from one table of a database. If the Transact-SQL language supported only such simple SELECT statements, the attachment of two or more tables to retrieve data would not be possible. Consequently, all data of a database would have to be stored in one table. Although the storage of all the data of a database inside one table is possible, it has one main disadvantage— the stored data are highly redundant.

Transact-SQL provides the join operator, which retrieves data from more than one table. This operator is probably the most important operator for relational database systems, because it allows data to be spread over many tables and thus achieves a vital property of database systems—nonredundant data.

NOTE The UNION operator also attaches two or more tables. However, the UNION operator always attaches two or more SELECT statements, while the join operator "joins" two or more tables using just one SELECT. Further, the UNION operator attaches rows of tables, while, as you will see later in this section, the join operator "joins" columns of tables.

The join operator is applied to base tables and views. This chapter discusses joins between base tables, while Chapter 11 discusses joins concerning views.

There are several different forms of the join operator. This section discusses the following fundamental types:

- Natural join
- Cartesian product (cross join)
- Outer join
- Theta join, self-join, and semi-join

Before explaining different join forms, this section describes the different syntax forms of the join operator.

Two Syntax Forms to Implement Joins

To join tables, you can use two different forms:

- Explicit join syntax (ANSI SQL:1992 join syntax)
- Implicit join syntax (old-style join syntax)

The ANSI SQL:1992 join syntax was introduced in the SQL92 standard and defines join operations explicitly—that is, using the corresponding name for each type of join operation. The keywords concerning the explicit definition of join are

- CROSS JOIN
- [INNER] JOIN
- LEFT [OUTER] JOIN
- RIGHT [OUTER] JOIN
- FULL [OUTER] JOIN

CROSS JOIN specifies the Cartesian product of two tables. INNER JOIN defines the natural join of two tables, while LEFT OUTER JOIN and RIGHT OUTER JOIN characterize the join operations of the same names, respectively. Finally, FULL OUTER JOIN specifies the union of the right and left outer joins. (All these different join operations are explained in the following sections.)

The implicit join syntax is "old-style" syntax, where each join operation is defined implicitly via the WHERE clause, using the so-called join columns (see the second statement in Example 6.57).

NOTE Use of the explicit join syntax is recommended. This syntax enhances the readability of queries. For this reason, all examples in this chapter concerning the join operation are solved using the explicit syntax forms. In a few introductory examples, you will see the old-style syntax, too.

Natural Join

Natural join is best explained through the use of an example, so check out Example 6.57.

NOTE The phrases "natural join" and "equi-join" are often used as synonyms, but there is a slight difference between them. The equi-join operation always has one or more pairs of columns that have identical values in every row. The operation that eliminates such columns from the equi-join is called a natural join.

Example 6.57

Get full details of each employee; that is, besides the employee's number, first and last names, and corresponding department number, also get the name of his or her department and its location, with duplicate columns displayed.

The following is the explicit join syntax:

```
USE sample;
SELECT employee.*, department.*
       FROM employee INNER JOIN department
       ON employee.dept_no = department.dept_no;
```

The SELECT list in Example 6.57 includes all columns of the **employee** and **department** tables. The FROM clause in the SELECT statement specifies the tables that are joined as well as the explicit name of the join form (INNER JOIN). The ON clause is also part of the FROM clause; it specifies the join columns from both tables. The condition employee.dept_no = department.dept_no specifies a *join condition*, and both columns are said to be *join columns*.

The equivalent solution with the old-style, implicit join syntax is as follows:

```
USE sample;
SELECT employee.*, department.*
   FROM employee, department
   WHERE employee.dept_no = department.dept_no;
```

This syntax has two significant differences from the explicit join syntax: the FROM clause of the query contains the list of tables that are joined, and the corresponding join condition is specified in the WHERE clause using join columns.

The result is

emp_no	emp_fname	emp_lname	dept_no	dept_no	dept_name	location
25348	Matthew	Smith	d3	d3	Marketing	Dallas
10102	Ann	Jones	d3	d3	Marketing	Dallas
18316	John	Barrimore	d1	d1	Research	Dallas
29346	James	James	d2	d2	Accounting	Seattle
2581	Elke	Hansel	d2	d2	Accounting	Seattle
9031	Elsa	Bertoni	d2	d2	Accounting	Seattle
28559	Sybill	Moser	d1	d1	Research	Dallas

NOTE It is strongly recommended that you use * in a SELECT list only when you are using interactive SQL, and avoid its use in an application program.

Example 6.57 can be used to show how a join operation works. Note that this is just an illustration of how you can think about the join process; the Database Engine actually has several strategies from which it chooses to implement the join operator. Imagine each row of the **employee** table combined with each row of the **department** table. The result of this combination is a table with 7 columns (4 from the table **employee** and 3 from the table **department**) and 21 rows (see Table 6-1).

In the second step, all rows from Table 6-1 that do not satisfy the join condition employee. dept_no = department.dept_no are removed. These rows are prefixed in Table 6-1 with the * sign. The rest of the rows represent the result of Example 6.57.

The semantics of the corresponding join columns must be identical. This means both columns must have the same logical meaning. It is not required that the corresponding join columns have the same name (or even an identical type), although this will often be the case.

NOTE It is not possible for a database system to check the logical meaning of a column. (For instance, project number and employee number have nothing in common, although both columns are defined as integers.) Therefore, database systems can check only the data type and the length of string data types. The Database Engine requires that the corresponding join columns have compatible data types, such as INT and SMALLINT.

---

emp_no	emp_fname	emp_lname	dept_no	dept_no	dept_name	location
*25348	Matthew	Smith	d3	d1	Research	Dallas
*10102	Ann	Jones	d3	d1	Research	Dallas
18316	John	Barrimore	d1	d1	Research	Dallas
*29346	James	James	d2	d1	Research	Dallas
*2581	Elke	Hansel	d2	d1	Research	Dallas
*9031	Elsa	Bertoni	d2	d1	Research	Dallas
28559	Sybill	Moser	d1	d1	Research	Dallas
*25348	Matthew	Smith	d3	d2	Accounting	Seattle
*10102	Ann	Jones	d3	d2	Accounting	Seattle
*18316	John	Barrimore	d1	d2	Accounting	Seattle
29346	James	James	d2	d2	Accounting	Seattle
2581	Elke	Hansel	d2	d2	Accounting	Seattle
9031	Elsa	Bertoni	d2	d2	Accounting	Seattle
*28559	Sybill	Moser	d1	d2	Accounting	Seattle
25348	Matthew	Smith	d3	d3	Marketing	Dallas
10102	Ann	Jones	d3	d3	Marketing	Dallas
*18316	John	Barrimore	d1	d3	Marketing	Dallas
*29346	James	James	d2	d3	Marketing	Dallas
*2581	Elke	Hansel	d2	d3	Marketing	Dallas
*9031	Elsa	Bertoni	d2	d3	Marketing	Dallas
*28559	Sybill	Moser	d1	d3	Marketing	Dallas

Table 6-1 Result of the Cartesian Product Between the employee and department Tables

The **sample** database contains three pairs of columns in which each column of the pair has the same logical meaning (and they have the same names as well). The **employee** and **department** tables can be joined using the columns **employee.dept_no** and **department .dept_no**. The join columns of the **employee** and **works_on** tables are the columns **employee .emp_no** and **works_on.emp_no**. Finally, the **project** and **works_on** tables can be joined using the join columns **project.project_no** and **works_on.project_no**.

The names of columns in a SELECT statement can be qualified. "Qualifying" a column name means that, to avoid any possible ambiguity about which table the column belongs to, the column name is preceded by its table name (or the alias of the table), separated by a period: **table_name.column_name**.

In most SELECT statements a column name does not need any qualification, although the use of qualified names is generally recommended for readability. If column names within a SELECT statement are ambiguous (like the columns **employee.dept_no** and **department .dept_no** in Example 6.57), the qualified names for the columns *must* be used.

In a SELECT statement with a join, the WHERE clause can include other conditions in addition to the join condition, as shown in Example 6.58.

Example 6.58
Get full details of all employees who work on the project Gemini.

Explicit join syntax:

```
USE sample;
SELECT emp_no, project.project_no, job, enter_date, project_name, budget
    FROM works_on JOIN project
    ON project.project_no = works_on.project_no
    WHERE project_name = 'Gemini';
```

Old-style join syntax:

```
USE sample;
SELECT emp_no, project.project_no, job, enter_date, project_name, budget
    FROM works_on, project
    WHERE project.project_no = works_on.project_no
    AND project_name = 'Gemini';
```

NOTE The qualification of the columns **emp_no**, **project_name**, **job**, and **budget** in Example 6.58 is not necessary, because there is no ambiguity regarding these names.

The result is

emp_no	project_no	job	enter_date	project_name	budget
25348	p2	Clerk	2017-02-15	Gemini	95000.0
18316	p2	NULL	2017-06-01	Gemini	95000.0
29346	p2	NULL	2016-12-15	Gemini	95000.0
28559	p2	Clerk	2018-02-01	Gemini	95000.0

From this point forward, all examples will be implemented using the explicit join syntax only. Example 6.59 shows another use of the inner join.

Example 6.59

Get the department number for all employees who entered their projects on October 15, 2017:

```
USE sample;
SELECT dept_no
     FROM employee JOIN works_on
     ON employee.emp_no = works_on.emp_no
     WHERE enter_date = '10.15.2017';
```

The result is

dept_no
d2

Joining More Than Two Tables

Theoretically, there is no upper limit on the number of tables that can be joined using a SELECT statement. (One join condition always combines two tables!) However, the Database Engine has an implementation restriction: the maximum number of tables that can be joined in a SELECT statement is 64.

Example 6.60 joins three tables of the **sample** database.

Example 6.60

Get the first and last names of all analysts whose department is located in Seattle:

```
USE sample;
SELECT emp_fname, emp_lname
       FROM works_on JOIN employee ON works_on.emp_no=employee.emp_no
                      JOIN department ON employee.dept_no=department.dept_no
       AND location = 'Seattle'
       AND job = 'analyst';
```

The result is

emp_fname	emp_lname
Elke	Hansel

The result in Example 6.60 can be obtained only if you join at least three tables: **works_on**, **employee**, and **department**. These tables can be joined using two pairs of join columns:

```
(works_on.emp_no, employee.emp_no)
(employee.dept_no, department.dept_no)
```

Example 6.61 uses all four tables from the **sample** database to obtain the result set.

Example 6.61
Get the names of projects (with redundant duplicates eliminated) being worked on by employees in the Accounting department:

```
USE sample;
SELECT DISTINCT project_name
      FROM project JOIN works_on
      ON project.project_no = works_on.project_no
                  JOIN employee ON works_on.emp_no = employee.emp_no
                  JOIN department ON employee.dept_no = department.dept_no
      WHERE dept_name = 'Accounting';
```

The result is

project_name
Apollo
Gemini
Mercury

Notice that when joining three tables, you use two join conditions (linking two tables each) to achieve a natural join. When you join four tables, you use three such join conditions. In general, if you join *n* tables, you need *n* – 1 join conditions to avoid a Cartesian product. Of course, using more than *n* – 1 join conditions, as well as other conditions, is certainly permissible to further reduce the result set.

Cartesian Product

The previous section illustrated a possible method of producing a natural join. In the first step of this process, each row of the **employee** table is combined with each row of the **department** table. This intermediate result was made by the operation called *Cartesian product*. Example 6.62 shows the Cartesian product of the tables **employee** and **department**.

Example 6.62
```
USE sample;
SELECT employee.*, department.*
    FROM employee CROSS JOIN department;
```

The result of Example 6.62 is shown in Table 6-1. A Cartesian product combines each row of the first table with each row of the second table. In general, the Cartesian product of two tables such that the first table has *n* rows and the second table has *m* rows will produce a result with *n* times *m* rows (or *n*m*). Thus, the result set in Example 6.62 contains 7*3 = 21 rows.

In practice, the use of a Cartesian product is highly unusual. Sometimes users generate the Cartesian product of two tables when they forget to include the join condition in the WHERE clause of the old-style join syntax. In this case, the output does not correspond to the expected result because it contains too many rows. (The existence of many and unexpected rows in the result is a hint that a Cartesian product of two tables, rather than the intended natural join, has been produced.)

Outer Join

In the previous examples of natural join, the result set included only rows from one table that have corresponding rows in the other table. Sometimes it is necessary to retrieve, in addition to the matching rows, the unmatched rows from one or both of the tables. Such an operation is called an *outer join*.

Examples 6.63 and 6.64 show the difference between a natural join and the corresponding outer join. (All examples in this section use the **employee_enh** table.)

Example 6.63

Get full details of all employees, including the location of their department, who live and work in the same city:

```
USE sample;
SELECT employee_enh.*, department.location
    FROM employee_enh JOIN department
            ON domicile = location;
```

The result is

emp_no	emp_fname	emp_lname	dept_no	domicile	location
29346	James	James	d2	Seattle	Seattle

Example 6.63 uses a natural join to display the result set of rows. If you would like to know all other existing living places of employees, you have to use the (left) outer join. This is called a *left* outer join because all rows from the table on the *left* side of the operator are returned, whether or not they have a matching row in the table on the right. In other words, if there are no matching rows in the table on the right side, the outer join will still return a row from the table on the left side, with NULL in each column of the other table (see Example 6.64). The Database Engine uses the operator LEFT OUTER JOIN to specify the left outer join.

A *right* outer join is similar, but it returns all rows of the table on the *right* of the symbol. The Database Engine uses the operator RIGHT OUTER JOIN to specify the right outer join.

Example 6.64

Get full details of all employees, including the location of their department, for all cities that are either the living place only or both the living and working place of employees:

```
USE sample;
SELECT employee_enh.*, department.location
    FROM employee_enh LEFT OUTER JOIN department
            ON domicile = location;
```

The result is

emp_no	emp_fname	emp_lname	dept_no	domicile	location
25348	Matthew	Smith	d3	San Antonio	NULL
10102	Ann	Jones	d3	Houston	NULL
18316	John	Barrimore	d1	San Antonio	NULL
29346	James	James	d2	Seattle	Seattle
2581	Elke	Hansel	d2	Portland	NULL
9031	Elsa	Bertoni	d2	Tacoma	NULL
28559	Sybill	Moser	d1	Houston	NULL

As you can see, when there is no matching row in the table on the right side (**department**, in this case), the left outer join still returns the rows from the table on the left side (**employee_enh**), and the columns of the other table are populated by NULL values. Example 6.65 shows the use of the right outer join operation.

Example 6.65

Get full details of all departments, as well as all living places of their employees, for all cities that are either the location of a department or the living and working place of an employee:

```
USE sample;
SELECT employee_enh.domicile, department.*
     FROM employee_enh RIGHT OUTER JOIN department
          ON domicile =location;
```

The result is

domicile	dept_no	dept_name	location
Seattle	d2	Accounting	Seattle
NULL	d1	Research	Dallas
NULL	d3	Marketing	Dallas

In addition to the left and right outer joins, there is also the full outer join, which is defined as the union of the left and right outer joins. In other words, all rows from both tables are represented in the result set. If there is no corresponding row in one of the tables, its columns are returned with NULL values. This operation is specified using the FULL OUTER JOIN operator.

Every outer join operation can be simulated using the UNION operator plus the NOT EXISTS function. Example 6.66 is equivalent to the example with the left outer join (Example 6.64).

Example 6.66

Get full details of all employees, including the location of their department, for all cities that are either the living place only or both the living and working place of employees:

```
USE sample;
SELECT employee_enh.*, department.location
   FROM employee_enh JOIN department
   ON domicile = location
UNION
SELECT employee_enh.*, 'NULL'
   FROM employee_enh
   WHERE NOT EXISTS
   (SELECT *
      FROM department
     WHERE location = domicile);
```

The first SELECT statement in the union specifies the natural join of the tables **employee_enh** and **department** with the join columns **domicile** and **location**. This SELECT statement retrieves all cities that are at the same time the living places and working places of each employee. The second SELECT statement in the union retrieves, additionally, all rows from the **employee_enh** table that do not match the condition in the natural join.

Further Forms of Join Operations

The preceding sections discussed the most important join forms. This section shows you three other forms:

- Theta join
- Self-join
- Semi-join

The following subsections describe these forms.

Theta Join

Join columns need not be compared using the equality sign. A join operation using a general join condition is called a theta join. Example 6.67, which uses the **employee_enh** table, shows the theta join operation.

Example 6.67

Get all the combinations of employee information and department information where the domicile of an employee alphabetically precedes any location of departments.

```
USE sample;
SELECT emp_fname, emp_lname, domicile, location
   FROM employee_enh JOIN department
   ON domicile < location;
```

The result is

emp_fname	emp_lname	domicile	location
Matthew	Smith	San Antonio	Seattle
Ann	Jones	Houston	Seattle
John	Barrimore	San Antonio	Seattle
Elsa	Bertoni	Tacoma	Seattle
Sybill	Moser	Houston	Seattle

In Example 6.67, the corresponding values of columns **domicile** and **location** are compared. In every resulting row, the value of the **domicile** column is ordered alphabetically before the corresponding value of the **location** column.

Self-Join, or Joining a Table with Itself

In addition to joining two or more different tables, a natural join operation can also be applied to a single table. In this case, the table is joined with itself, whereby a single column of the table is compared with itself. The comparison of a column with itself means that the table name appears twice in the FROM clause of a SELECT statement. Therefore, you need to be able to reference the name of the same table twice. This can be accomplished using at least one alias name. The same is true for the column names in the join condition of a SELECT statement. In order to distinguish both column names, you use the qualified names. Example 6.68 joins the **department** table with itself.

Example 6.68

Get full details of all departments located at the same location as at least one other department:

```
USE sample;
SELECT t1.dept_no, t1.dept_name, t1.location
    FROM department t1 JOIN department t2
        ON  t1.location = t2.location
    WHERE t1.dept_no <> t2.dept_no;
```

The result is

dept_no	dept_name	location
d3	Marketing	Dallas
d1	Research	Dallas

The FROM clause in Example 6.68 contains two aliases for the **department** table: **t1** and **t2**. The first condition in the WHERE clause specifies the join columns, while the second condition eliminates unnecessary duplicates by making certain that each department is compared with *different* departments.

Semi-Join

The semi-join is similar to the natural join, but the result of the semi-join is only the set of all rows from one table where one or more matches are found in the second table. Example 6.69 shows the semi-join operation.

Example 6.69

```
USE sample;
SELECT emp_no, emp_lname, e.dept_no
 FROM employee e JOIN department d
 ON e.dept_no = d.dept_no
 WHERE location = 'Dallas';
```

The result is

emp_no	emp_lname	dept_no
25348	Smith	d3
10102	Jones	d3
18316	Barrimore	d1
28559	Moser	d1

As can be seen from Example 6.69, the SELECT list of the semi-join contains only columns from the **employee** table. This is exactly what characterizes the semi-join operation. This operation is usually used in distributed query processing to minimize data transfer. The Database Engine uses the semi-join operation to implement the feature called star join (see Chapter 26).

Correlated Subqueries

A subquery is said to be a *correlated subquery* if the inner query depends on the outer query for any of its values. Example 6.70 shows a correlated subquery.

Example 6.70

Get the last names of all employees who work on project p3:

```
USE sample;
SELECT emp_lname
   FROM employee
   WHERE 'p3' IN
   (SELECT project_no
       FROM works_on
       WHERE works_on.emp_no = employee.emp_no);
```

The result is

emp_lname
Jones
Bertoni
Hansel

The inner query in Example 6.70 must be logically evaluated many times because it contains the **emp_no** column, which belongs to the **employee** table in the outer query, and the value of the **emp_no** column changes every time the Database Engine examines a different row of the **employee** table in the outer query.

Let's walk through how the system might process the query in Example 6.70. First, the system retrieves the first row of the **employee** table (for the outer query) and compares the employee number of that column (25348) with values of the **works_on.emp_no** column in the inner query. Since the only **project_no** for this employee is p2, the inner query returns the value p2. The single value in the set is not equal to the constant value p3 in the outer query, so the outer query's condition (WHERE 'p3' IN …) is not met and no rows are returned by the outer query for this employee. Then, the system retrieves the next row of the **employee** table and repeats the comparison of employee numbers in both tables. The second employee has two rows in the **works_on** table with **project_no** values of p1 and p3, so the result set of the inner query is (p1,p3). One of the elements in the result set is equal to the constant value p3, so the condition is evaluated to TRUE and the corresponding value of the **emp_lname** column in the second row (Jones) is displayed. The same process is applied to all rows of the **employee** table, and the final result set with three rows is retrieved.

More examples of correlated subqueries are shown in the next section.

Subqueries and the EXISTS Function

The EXISTS function takes an inner query as an argument and returns TRUE if the inner query returns one or more rows, and returns FALSE if it returns zero rows. This function will be explained using examples, starting with Example 6.71.

Example 6.71

Get the last names of all employees who work on project p1:

```
USE sample;
SELECT emp_lname
   FROM employee
   WHERE EXISTS
   (SELECT *
      FROM works_on
    WHERE employee.emp_no = works_on.emp_no
   AND project_no = 'p1');
```

The result is

emp_lname
Jones
James
Bertoni
Moser

The inner query of the EXISTS function almost always depends on a variable from an outer query. Therefore, the EXISTS function usually specifies a correlated subquery.

Let's walk through how the Database Engine might process the query in Example 6.71. First, the outer query considers the first row of the **employee** table (Smith). Next, the EXISTS function is evaluated to determine whether there are any rows in the **works_on** table whose employee number matches the one from the current row in the outer query, and whose **project_no** is p1. Because Mr. Smith does not work on the project p1, the result of the inner query is an empty set and the EXISTS function is evaluated to FALSE. Therefore, the employee named Smith does not belong to the final result set. Using this process, all rows of the **employee** table are tested, and the result set is displayed.

Example 6.72 shows the use of the NOT EXISTS function.

Example 6.72

Get the last names of all employees who work for departments not located in Seattle:

```
USE sample;
SELECT emp_lname
    FROM employee
    WHERE NOT EXISTS
    (SELECT *
        FROM department
        WHERE employee.dept_no = department.dept_no
        AND location = 'Seattle');
```

The result is

emp_lname
Smith
Jones
Barrimore
Moser

The SELECT list of an outer query involving the EXISTS function is not required to be of the form SELECT * as in the previous examples. The form SELECT column_list, where **column_list** is one or more columns of the table, is an alternate form. Both forms are equivalent, because the EXISTS function tests only the existence (i.e., nonexistence) of rows in the result set. For this reason, the use of SELECT * in this case is safe.

Should You Use Joins or Subqueries?

Almost all SELECT statements that join tables and use the join operator can be rewritten as subqueries, and vice versa. Writing the SELECT statement using the join operator is often easier to read and understand and can also help the Database Engine to find a more efficient strategy for retrieving the appropriate data. However, there are a few problems that can be easier solved using subqueries, and there are others that can be easier solved using joins.

Subquery Advantages

Subqueries are advantageous over joins when you have to calculate an aggregate value on-the-fly and use it in the outer query for comparison. Example 6.73 shows this.

Example 6.73

Get the employee numbers and enter dates of all employees with enter dates equal to the earliest date:

```
USE sample;
SELECT emp_no, enter_date
    FROM works_on
    WHERE enter_date = (SELECT min(enter_date)
                        FROM works_on);
```

This problem cannot be solved easily with a join, because you would have to write the aggregate function in the WHERE clause, which is not allowed. (You can solve the problem using two separate queries in relation to the **works_on** table.)

Join Advantages

Joins are advantageous over subqueries if the SELECT list in a query contains columns from more than one table. Example 6.74 shows this.

Example 6.74

Get the employee numbers, last names, and jobs for all employees who entered their projects on October 15, 2017:

```
USE sample;
SELECT employee.emp_no, emp_lname, job
    FROM employee, works_on
    WHERE employee.emp_no = works_on.emp_no
    AND enter_date = '10.15.2017';
```

The SELECT list of the query in Example 6.74 contains columns **emp_no** and **emp_lname** from the **employee** table and the **job** column from the **works_on** table. For this reason, the equivalent solution with the subquery would display an error, because subqueries can display information only from the outer table.

Table Expressions

Table expressions are subqueries that are used where a table is expected. There are two types of table expressions:

- Derived tables
- Common table expressions

The following subsections describe these two forms of table expressions.

Derived Tables

A derived table is a table expression that appears in the FROM clause of a query. You can apply derived tables when the use of column aliases is not possible because another clause is processed by the SQL translator before the alias name is known. Example 6.75 shows an attempt to use a column alias where another clause is processed before the alias name is known.

Example 6.75 (Example of an Illegal Statement)

Get all existing groups of months from the **enter_date** column of the **works_on** table:

```
USE sample;
SELECT MONTH(enter_date) as enter_month
FROM works_on
GROUP BY enter_month;
```

The result is

```
Message 207: Level 16, State 1, Line 4
          The invalid column 'enter_month'
```

The reason for the error message is that the GROUP BY clause is processed before the corresponding SELECT list, and the alias name **enter_month** is not known at the time the grouping is processed.

By using a derived table that contains the preceding query (without the GROUP BY clause), you can solve this problem, because the FROM clause is executed before the GROUP BY clause, as shown in Example 6.76.

Example 6.76

```
USE sample;
SELECT enter_month
 FROM (SELECT MONTH(enter_date) as enter_month
          FROM works_on) AS m
GROUP BY enter_month;
```

The result is

enter_month
1
2
4
6
8
10
11
12

Generally, it is possible to write a table expression any place in a SELECT statement where a table can appear. (The result of a table expression is always a table or, in a special case, an expression.) Example 6.77 shows the use of a table expression in a SELECT list.

Example 6.77

```
USE sample;
SELECT w.job, (SELECT e.emp_lname
                 FROM employee e WHERE e.emp_no = w.emp_no) AS name
    FROM works_on w
    WHERE w.job IN('Manager', 'Analyst');
```

The result is

job	name
Analyst	Jones
Manager	Jones
Analyst	Hansel
Manager	Bertoni

Common Table Expressions

A common table expression (CTE) is a named table expression that is supported by Transact-SQL. There are two types of queries that use CTEs:

- Nonrecursive queries
- Recursive queries

The following sections describe both query types.

NOTE Common table expressions are also used by the APPLY operator, which allows you to invoke a table-valued function for each row returned by an outer table expression of a query. This operator is discussed in Chapter 8.

CTEs and Nonrecursive Queries

The nonrecursive form of a CTE can be used as an alternative to derived tables and views. Generally, a CTE is defined using the WITH statement and an additional query that refers to the name used in WITH.

NOTE The WITH keyword is ambiguous in the Transact-SQL language. To avoid ambiguity, you have to use a semicolon (;) to terminate the statement preceding the WITH statement.

Examples 6.78 and 6.79 use the **AdventureWorks** database to show how CTEs can be used in nonrecursive queries. Example 6.78 uses the "convenient" features, while Example 6.79 solves the same problem using a nonrecursive query.

Example 6.78

```
USE AdventureWorks;
SELECT SalesOrderID
 FROM Sales.SalesOrderHeader
  WHERE TotalDue > (SELECT AVG(TotalDue)
                      FROM Sales.SalesOrderHeader
                      WHERE YEAR(OrderDate) = '2014')
    AND Freight > (SELECT AVG(TotalDue)
                      FROM Sales.SalesOrderHeader
                      WHERE YEAR(OrderDate) = '2014')/2.5;
```

The query in Example 6.78 finds total dues whose values are greater than the average of all dues and whose freights are greater than 40 percent of the average of all dues. The main property of this query is that it is space-consuming, because an inner query has to be written twice. One way to shorten the syntax of the query is to create a view containing the inner query, but that is rather complicated because you would have to create the view and then drop it when you are done with the query. A better way is to write a CTE. Example 6.79 shows the use of the nonrecursive CTE, which shortens the definition of the query in Example 6.78.

Example 6.79

```
USE AdventureWorks;
WITH price_calc(year_2014) AS
      (SELECT AVG(TotalDue)
            FROM Sales.SalesOrderHeader
            WHERE YEAR(OrderDate) = '2014')
SELECT SalesOrderID
      FROM Sales.SalesOrderHeader
      WHERE TotalDue > (SELECT year_2014 FROM price_calc)
AND Freight > (SELECT year_2014 FROM price_calc)/2.5;
```

The syntax for the WITH clause in nonrecursive queries is

```
WITH cte_name (column_list) AS
   ( inner_query)
outer_query
```

cte_name is the name of the CTE that specifies a resulting table. The list of columns that belong to the table expression is written in brackets. (The CTE in Example 6.79 is called **price_calc** and has one column, **year_2014**.) **inner_query** in the CTE syntax defines the SELECT statement, which specifies the result set of the corresponding table expression. After that, you can use the defined table expression in an outer query. (The outer query in Example 6.79 uses the CTE called **price_calc** and its column **year_2014** to simplify the inner query, which appears twice.)

CTEs and Recursive Queries

NOTE The material in this subsection is complex. Therefore, you might want to skip it on the first reading of the book and make a note to yourself to return to it.

You can use CTEs to implement recursion because CTEs can contain references to themselves. The basic syntax for a CTE for recursive queries is

```
WITH cte_name (column_list) AS
    (anchor_member
     UNION ALL
     recursive_member)
outer_query
```

cte_name and **column_list** have the same meaning as in CTEs for nonrecursive queries. The body of the WITH clause comprises two queries that are connected with the UNION ALL operator. The first query will be invoked only once, and it starts to accumulate the result of the recursion. The first operand of UNION ALL does not reference the CTE (see Example 6.80). This query is called the *anchor query* or *seed*.

The second query contains a reference to the CTE and represents the recursive portion of it. For this reason it is called the *recursive member*. In the first invocation of the recursive part, the reference to the CTE represents the result of the anchor query. The recursive member uses the query result of the first invocation. After that, the system repeatedly invokes the recursive part. The invocation of the recursive member ends when the result of the previous invocation is an empty set.

The UNION ALL operator joins the rows accumulated so far, as well as the additional rows that are added in the current invocation. (Inclusion of UNION ALL means that no duplicate rows will be eliminated from the result.)

Finally, **outer query** defines a query specification that uses the CTE to retrieve all invocations of the union of both members.

The table definition in Example 6.80 will be used to demonstrate the recursive form of CTEs.

Example 6.80

```
USE sample;
CREATE TABLE airplane
    (containing_assembly VARCHAR(10),
     contained_assembly VARCHAR(10),
     quantity_contained INT,
     unit_cost DECIMAL (6,2));
insert into airplane values ( 'Airplane', 'Fuselage',1, 10);
insert into airplane values ( 'Airplane', 'Wings', 1, 11);
insert into airplane values ( 'Airplane', 'Tail',1, 12);
insert into airplane values ( 'Fuselage', 'Cockpit', 1, 13);
insert into airplane values ( 'Fuselage', 'Cabin', 1, 14);
insert into airplane values ( 'Fuselage', 'Nose',1, 15);
insert into airplane values ( 'Cockpit', NULL, 1,13);
insert into airplane values ( 'Cabin', NULL, 1, 14);
insert into airplane values ( 'Nose', NULL, 1, 15);
insert into airplane values ( 'Wings', NULL,2, 11);
insert into airplane values ( 'Tail', NULL, 1, 12);
```

The **airplane** table contains four columns. The column **containing_assembly** specifies an assembly, while **contained_assembly** comprises the parts (one by one) that build the corresponding assembly. From Table 6-2 you can see that an airplane has three parts (fuselage, wings, and tail), and that its fuselage has three subparts (cockpit, cabin, and nose).

Suppose that the **airplane** table contains 11 rows, which are shown in Table 6-2. (The INSERT statements in Example 6.80 insert these rows in the **airplane** table.)

containing_assembly	contained_assembly	quantity_contained	unit_cost
Airplane	Fuselage	1	10
Airplane	Wings	1	11
Airplane	Tail	1	12
Fuselage	Cockpit	1	13
Fuselage	Cabin	1	14
Fuselage	Nose	1	15
Cockpit	NULL	1	13
Cabin	NULL	1	14
Nose	NULL	1	15
Wings	NULL	2	11
Tail	NULL	1	12

Table 6-2 Content of the airplane Table

Example 6.81 shows the use of the WITH clause to define a query that calculates the total costs of each assembly.

Example 6.81
```
USE sample;
WITH list_of_parts(assembly1, quantity, cost) AS
  (SELECT containing_assembly, quantity_contained, unit_cost
     FROM airplane
     WHERE contained_assembly IS NULL
  UNION ALL
  SELECT a.containing_assembly, a.quantity_contained,
         CAST(l.quantity*l.cost AS DECIMAL(6,2))
         FROM list_of_parts l,airplane a
         WHERE l.assembly1 = a.contained_assembly)
SELECT * FROM list_of_parts;
```

The WITH clause defines the CTE called **list_of_parts**, which contains three columns: **assembly1**, **quantity**, and **cost**. The first SELECT statement in Example 6.81 will be invoked only once, to accumulate the results of the first step in the recursion process.

The SELECT statement in the last row of Example 6.81 displays the following result:

assembly1	quantity	cost
Cockpit	1	13.00
Cabin	1	14.00
Nose	1	15.00
Wings	2	11.00
Tail	1	12.00
Airplane	1	12.00
Airplane	1	22.00
Fuselage	1	15.00
Airplane	1	15.00
Fuselage	1	14.00
Airplane	1	14.00
Fuselage	1	13.00
Airplane	1	13.00

The first five rows in the preceding output show the result set of the first invocation of the anchor member of the query in Example 6.81. All other rows are the result of the recursive member (second part) of the query in the same example. The recursive member of the query will be invoked twice: the first time for the fuselage assembly and the second time for the airplane itself.

The query in Example 6.82 is used to get the costs for each assembly with all its subparts.

Example 6.82

```
USE sample;
WITH list_of_parts(assembly, quantity, cost) AS
  (SELECT containing_assembly, quantity_contained, unit_cost
     FROM airplane
     WHERE contained_assembly IS NULL
   UNION ALL
   SELECT a.containing_assembly, a.quantity_contained,
          CAST(l.quantity*l.cost AS DECIMAL(6,2))
          FROM list_of_parts l,airplane a
          WHERE l.assembly = a.contained_assembly )
SELECT assembly, SUM(quantity) parts, SUM(cost) sum_cost
    FROM list_of_parts
    GROUP BY assembly;
```

The output of the query in Example 6.82 is as follows:

assembly	parts	sum_cost
Airplane	5	76.00
Cabin	1	14.00
Cockpit	1	13.00
Fuselage	3	42.00
Nose	1	15.00
Tail	1	12.00
Wings	2	11.00

There are several restrictions for a CTE in a recursive query:

- The CTE definition must contain at least two SELECT statements (an anchor member and one recursive member) combined by the UNION ALL operator.
- The number of columns in the anchor and recursive members must be the same. (This is the direct consequence of using the UNION ALL operator.)
- The data type of a column in the recursive member must be the same as the data type of the corresponding column in the anchor member.
- The FROM clause of the recursive member must refer only once to the name of the CTE.
- The following options are not allowed in the definition part of a recursive member: SELECT DISTINCT, GROUP BY, HAVING, aggregation functions, TOP, and subqueries. (Also, the only join operation that is allowed in the query definition is an inner join.)

Summary

This chapter covered all the features of the SELECT statement regarding data retrieval from one or more tables. Every SELECT statement that retrieves data from a table must contain at least a SELECT list and the FROM clause. The FROM clause specifies the table(s) from which the data is retrieved. The most important optional clause is the WHERE clause, containing one or more conditions that can be combined using the Boolean operators AND, OR, and NOT. Hence, the conditions in the WHERE clause place the restriction on the selected row.

Exercises

E.6.1   Get all rows of the **works_on** table.

E.6.2   Get the employee numbers for all clerks.

E.6.3   Get the employee numbers for employees working on project p2 and having employee numbers lower than 10000. Solve this problem with two different but equivalent SELECT statements.

E.6.4   Get the employee numbers for employees who didn't enter their project in 2017.

E.6.5   Get the employee numbers for all employees who have a leading job (i.e., Analyst or Manager) in project p1.

E.6.6   Get the enter dates for all employees in project p2 whose jobs have not been determined yet.

E.6.7   Get the employee numbers and last names of all employees whose first names contain two letter t's.

E.6.8   Get the employee numbers and first names of all employees whose last names have a letter o or a as the second character and end with the letters es.

E.6.9   Find the employee numbers of all employees whose department is located in Seattle.

E.6.10   Find the last and first names of all employees who entered their projects on 04.01.2017.

E.6.11   Group all departments using their locations.

E.6.12   What is a difference between the DISTINCT and GROUP BY clauses?

E.6.13   How does the GROUP BY clause manage the NULL values? Does it correspond to the general treatment of these values?

E.6.14   What is the difference between COUNT(*) and COUNT(column)?

E.6.15   Find the highest employee number.

E.6.16   Get the jobs that are done by more than two employees.

E.6.17   Find the employee numbers of all employees who are clerks or work for department d3.

E.6.18   Why is the following statement wrong?

```
SELECT project_name
    FROM project
    WHERE project_no =
        (SELECT project_no FROM works_on WHERE Job = 'Clerk')
```

Write the correct syntax form for the statement.

E.6.19   What is a practical use of temporary tables?

E.6.20   What is a difference between global and local temporary tables?

NOTE Write all solutions for the following exercises that use a join operation using the explicit join syntax.

E.6.21   For the **project** and **works_on** tables, create the following:

a.   Natural join
b.   Cartesian product

E.6.22   If you intend to join several tables in a query (say *n* tables), how many join conditions are needed?

E.6.23   Get the employee numbers and job titles of all employees working on project Gemini.

E.6.24   Get the first and last names of all employees who work for department Research or Accounting.

E.6.25   Get the enter dates of all clerks who belong to the department d1.

E.6.26   Get the names of projects on which two or more clerks are working.

E.6.27   Get the first and last names of the employees who are managers and work on project Mercury.

E.6.28   Get the first and last names of all employees who entered the project at the same time as at least one other employee.

E.6.29   Get the employee numbers of the employees living in the same location and belonging to the same department as one another. (Hint: Use the extended **sample** database.)

E.6.30   Get the employee numbers of all employees belonging to the Marketing department. Find two equivalent solutions using the following:

a.   The join operator
b.   The correlated subquery

Modification of a Table's Contents

In This Chapter

- INSERT Statement
- UPDATE Statement
- DELETE Statement
- Other T-SQL Modification Statements and Clauses

In addition to the SELECT statement, which was introduced in Chapter 6, there are three other DML statements: INSERT, UPDATE, and DELETE. Like the SELECT statement, these three modification statements operate either on tables or on views. This chapter discusses these statements in relation to tables and gives examples of their use. Additionally, it explains two other statements: TRUNCATE TABLE and MERGE. Whereas the TRUNCATE TABLE statement is a Transact-SQL extension to the SQL standard, MERGE is a standardized feature. The chapter wraps up with coverage of the OUTPUT clause, which allows you to display explicitly the inserted (or updated) rows.

INSERT Statement

The INSERT statement inserts rows (or parts of them) into a table. It has two different forms:

```
INSERT [INTO] tab_name  [(col_list)]
  DEFAULT VALUES | VALUES ({ DEFAULT | NULL | expression } [ ,...n] );

INSERT INTO tab_name | view_name [(col_list)]
    {select_statement | execute_statement};
```

Using the first form, exactly one row (or part of it) is inserted into the corresponding table. The second form of the INSERT statement inserts the result set from the SELECT statement or from the stored procedure, which is executed using the EXECUTE statement. (The stored procedure must return data, which is then inserted into the table. The SELECT statement can select values from a different table or from the same table as the target of the INSERT statement, as long as the types of the columns are compatible.)

With both forms, every inserted value must have a data type that is compatible with the data type of the corresponding column of the table. To ensure compatibility, all character-based values and data and time data must be enclosed in apostrophes, while all numeric values need no such enclosing.

Inserting a Single Row

In both forms of the INSERT statement, the explicit specification of the column list is optional. This means that omitting the list of columns is equivalent to specifying a list of all columns in the table.

The option DEFAULT VALUES inserts default values for all the columns. If a column is of the data type TIMESTAMP or has the IDENTITY property, the value, which is automatically created by the system, will be inserted. For other data types, the column is set to the appropriate non-null default value if a default exists, or NULL if it doesn't. If the column is not nullable and has no DEFAULT value, then the INSERT statement fails and an error will be indicated.

Examples 7.1 through 7.4 insert rows into the four tables of the **sample** database. This action shows the use of the INSERT statement to load a small amount of data into a database.

Example 7.1

Load data into the **employee** table:

```
USE sample;
INSERT INTO employee VALUES (25348, 'Matthew', 'Smith','d3');
INSERT INTO employee VALUES (10102, 'Ann', 'Jones','d3');
INSERT INTO employee VALUES (18316, 'John', 'Barrimore', 'd1');
INSERT INTO employee VALUES (29346, 'James', 'James', 'd2');
INSERT INTO employee VALUES (9031, 'Elsa', 'Bertoni', 'd2');
INSERT INTO employee VALUES (2581, 'Elke', 'Hansel', 'd2');
INSERT INTO employee VALUES (28559, 'Sybill', 'Moser', 'd1');
```

Example 7.2

Load data into the **department** table:

```
USE sample;
INSERT INTO department VALUES ('d1', 'Research', 'Dallas');
INSERT INTO department VALUES ('d2', 'Accounting', 'Seattle');
INSERT INTO department VALUES ('d3', 'Marketing', 'Dallas');
```

Example 7.3

Load data into the **project** table:

```
USE sample;
INSERT INTO project VALUES ('p1', 'Apollo', 120000.00);
INSERT INTO project VALUES ('p2', 'Gemini', 95000.00);
INSERT INTO project VALUES ('p3', 'Mercury', 186500.00);
```

Example 7.4

Load data into the **works_on** table:

```
USE sample;
INSERT INTO works_on VALUES (10102,'p1', 'Analyst', '2016.10.1');
INSERT INTO works_on VALUES (10102, 'p3', 'Manager', '2018.1.1');
INSERT INTO works_on VALUES (25348, 'p2', 'Clerk', '2017.2.15');
INSERT INTO works_on VALUES (18316, 'p2', NULL, '2017.6.1');
INSERT INTO works_on VALUES (29346, 'p2', NULL, '2016.12.15');
INSERT INTO works_on VALUES (2581, 'p3', 'Analyst', '2017.10.15');
INSERT INTO works_on VALUES (9031, 'p1', 'Manager', '2017.4.15');
INSERT INTO works_on VALUES (28559, 'p1', NULL, '2017.8.1');
INSERT INTO works_on VALUES (28559, 'p2', 'Clerk', '2018.2.1');
INSERT INTO works_on VALUES (9031, 'p3', 'Clerk', '2016.11.15');
INSERT INTO works_on VALUES (29346, 'p1','Clerk', '2017.1.4');
```

There are a few different ways to insert values into a new row. Examples 7.5 through 7.7 show these possibilities.

Example 7.5

```
USE sample;
INSERT INTO employee VALUES (15201, 'Dave', 'Davis', NULL);
```

The INSERT statement in Example 7.5 corresponds to the INSERT statements in Examples 7.1 through 7.4. The explicit use of the keyword NULL inserts the null value into the corresponding column.

The insertion of values into some (but not all) of a table's columns usually requires the explicit specification of the corresponding columns. The omitted columns must either be nullable or have a DEFAULT value.

Example 7.6

```
USE sample;
INSERT INTO employee (emp_no, emp_fname, emp_lname)
            VALUES (15201, 'Dave', 'Davis');
```

Examples 7.5 and 7.6 are equivalent. The **dept_no** column is the only nullable column in the **employee** table because all other columns in the **employee** table were declared with the NOT NULL clause in the CREATE TABLE statement.

The order of column names in the VALUE clause of the INSERT statement can be different from the original order of those columns, which is determined in the CREATE TABLE statement. In this case, it is absolutely necessary to list the columns in the new order.

Example 7.7

```
USE sample;
INSERT INTO employee (emp_lname, emp_fname, dept_no, emp_no)
            VALUES ('Davis', 'Dave', 'd1', 15201);
```

Inserting Multiple Rows

The second form of the INSERT statement inserts one or more rows selected with a subquery. Example 7.8 shows how a set of rows can be inserted using the second form of the INSERT statement.

Example 7.8

Get all the numbers and names for departments located in Dallas, and load the selected data into a new table:

```
USE sample;
CREATE TABLE dallas_dept
        (dept_no CHAR(4) NOT NULL,
         dept_name CHAR(20) NOT NULL);

INSERT INTO dallas_dept (dept_no, dept_name)
    SELECT dept_no, dept_name
        FROM department
        WHERE location = 'Dallas';
```

The new table created in Example 7.8, **dallas_dept**, has the same columns as the **department** table except for the **location** column. The subquery in the INSERT statement selects all rows with the value 'Dallas' in the **location** column. The selected rows will be subsequently inserted in the new table.

The content of the **dallas_dept** table can be selected with the following SELECT statement:

```
SELECT * FROM dallas_dept;
```

The result is

dept_no	dept_name
d1	Research
d3	Marketing

Example 7.9 is another example that shows how multiple rows can be inserted using the second form of the INSERT statement.

Example 7.9

Get all employee numbers, project numbers, and project enter dates for all clerks who work in project p2, and load the selected data into a new table:

```
USE sample;
CREATE TABLE clerk_t
     (emp_no INT NOT NULL,
      project_no CHAR(4),
      enter_date DATE);

INSERT INTO clerk_t (emp_no, project_no, enter_date)
   SELECT emp_no, project_no, enter_date
     FROM works_on
     WHERE job = 'Clerk'
     AND project_no = 'p2';
```

The new table, **clerk_t**, contains the following rows:

emp_no	project_no	enter_date
25348	p2	2017-02-15
28559	p2	2018-02-01

The tables **dallas_dept** and **clerk_t** (Examples 7.8 and 7.9) were empty before the INSERT statement inserted the rows. If, however, the table already exists and there are rows in it, the new rows will be appended.

NOTE You can replace both statements (CREATE TABLE and INSERT) in Example 7.9 with the SELECT statement with the INTO clause (see Example 6.44 in Chapter 6).

Table Value Constructors and INSERT

A *table* (or *row*) *value constructor* allows you to assign several tuples (rows) with a DML statement such as INSERT or UPDATE. Example 7.10 shows how you can assign several rows using such a constructor with an INSERT statement.

Example 7.10

```
USE sample;
 INSERT INTO department VALUES
              ('d4', 'Human Resources', 'Chicago'),
              ('d5', 'Distribution', 'New Orleans'),
              ('d6', 'Sales', 'Chicago');
```

The INSERT statement in Example 7.10 inserts three rows at the same time in the **department** table using the table value constructor. As you can see from the example, the syntax of the constructor is rather simple. To use a table value constructor, list the values of each row inside the pair of parentheses and separate each list from the others by using a comma.

UPDATE Statement

The UPDATE statement modifies values of table rows. This statement has the following general form:

```
UPDATE tab_name
    { SET column_1 = {expression | DEFAULT | NULL} [,...n]
    [FROM tab_name1 [,...n]];
    [WHERE condition]
```

Rows in the **tab_name** table are modified in accordance with the WHERE clause. For each row to be modified, the UPDATE statement changes the values of the columns in the SET clause, assigning a constant (or generally an expression) to the associated column. If the WHERE clause is omitted, the UPDATE statement modifies all rows of the table. This means that if you have a table with 1 million rows, all rows will be updated if the WHERE clause is missing. (The FROM clause will be discussed later in this section.)

NOTE An UPDATE statement can modify data of a single table only.

The UPDATE statement in Example 7.11 modifies exactly one row of the **works_on** table, because the combination of the columns **emp_no** and **project_no** builds the primary key of that table and is therefore unique. This example modifies the task of the employee, which was previously unknown or set to NULL.

Example 7.11

Set the task of employee number 18316, who works on project p2, to be 'Manager':

```
USE sample;
UPDATE works_on
    SET job = 'Manager'
    WHERE emp_no = 18316
    AND project_no = 'p2';
```

Example 7.12 modifies rows of a table with an expression.

Example 7.12

Change the budgets of all projects to be represented in English pounds. The current rate of exchange is 0.51£ for $1.

```
USE sample;
UPDATE project
    SET budget = budget*0.51;
```

In the example, all rows of the **project** table will be modified because of the omitted WHERE clause. The modified rows of the **project** table can be displayed with the following Transact-SQL statement:

```
SELECT * FROM project;
```

The result is

project_no	project_name	budget
p1	Apollo	61200
p2	Gemini	48450
p3	Mercury	95115

Example 7.13 uses an inner query in the WHERE clause of the UPDATE statement. Because of the use of the IN operator, more than one row can result from this query.

Example 7.13
Due to her illness, set all tasks on all projects for Mrs. Jones to NULL:

```
USE sample;
UPDATE works_on
      SET job = NULL
      WHERE emp_no IN
      (SELECT emp_no
          FROM employee
          WHERE emp_lname = 'Jones');
```

Example 7.13 can also be solved using the FROM clause of the UPDATE statement. The FROM clause contains the names of tables that are involved in the UPDATE statement. All these tables must be subsequently joined. Example 7.14 shows the use of the FROM clause. This example is identical to the previous one.

NOTE The FROM clause is a Transact-SQL extension to the ANSI SQL standard.

Example 7.14
```
USE sample;
UPDATE works_on
  SET job = NULL
  FROM works_on, employee
  WHERE emp_lname = 'Jones'
  AND works_on.emp_no = employee.emp_no;
```

Example 7.15 illustrates the use of the CASE expression in the UPDATE statement. (For a detailed discussion of this expression, refer to Chapter 6 and Example 6.50.)

Example 7.15

The budget of each project should be increased by a percentage (20, 10, or 5) depending on its previous amount of money. Those projects with a lower budget will be increased by the higher percentages.

```
USE sample;
UPDATE project
  SET budget = CASE
           WHEN budget >0 and budget < 100000  THEN budget*1.2
           WHEN budget >= 100000 and budget < 200000  THEN budget*1.1
           ELSE budget*1.05
           END
```

DELETE Statement

The DELETE statement deletes rows from a table. This statement has two different forms:

```
DELETE FROM table_name
  [WHERE predicate];

DELETE table_name
      FROM table_name [,...n]
      [WHERE condition];
```

All rows that satisfy the condition in the WHERE clause will be deleted. Explicitly naming columns within the DELETE statement is not necessary (or allowed), because the DELETE statement operates on rows and not on columns.

NOTE The TRUNCATE TABLE statement, which is semantically equivalent to UPDATE, will be explained in the next section.

Example 7.16 shows an example of the first form of the DELETE statement.

Example 7.16

Delete all managers in the **works_on** table:

```
USE sample;
DELETE FROM works_on
    WHERE job = 'Manager';
```

The WHERE clause in the DELETE statement can contain an inner query, as shown in Example 7.17.

Example 7.17

Mrs. Moser is on leave. Delete all rows in the database concerning her:

```
USE sample;
DELETE FROM works_on
    WHERE emp_no IN
    (SELECT emp_no
        FROM employee
        WHERE emp_lname = 'Moser');

DELETE FROM employee
    WHERE emp_lname = 'Moser';
```

Example 7.17 can also be performed using the FROM clause, as Example 7.18 shows. This clause has the same semantics as the FROM clause in the UPDATE statement.

Example 7.18

```
USE sample;
DELETE works_on
    FROM works_on, employee
    WHERE works_on.emp_no = employee.emp_no
    AND emp_lname = 'Moser';

DELETE FROM employee
    WHERE emp_lname = 'Moser';
```

The use of the WHERE clause in the DELETE statement is optional. If the WHERE clause is omitted, all rows of a table will be deleted, as shown in Example 7.19.

Example 7.19

```
USE sample;
DELETE FROM works_on;
```

NOTE There is a significant difference between the DELETE and the DROP TABLE statements. The DELETE statement deletes (partially or totally) the contents of a table, whereas the DROP TABLE statement deletes both the contents and the schema of a table. Thus, after a DELETE statement, the table still exists in the database (although possibly with zero rows), but after a DROP TABLE statement, the table no longer exists.

Other T-SQL Modification Statements and Clauses

The Database Engine supports two additional modification statements:

- TRUNCATE TABLE
- MERGE

and the OUTPUT clause.

Both statements, together with the OUTPUT clause, will be explained in turn in the following subsections.

TRUNCATE TABLE Statement

The Transact-SQL language also supports the TRUNCATE TABLE statement. This statement normally provides a "faster executing" version of the DELETE statement without the WHERE clause. The TRUNCATE TABLE statement deletes all rows from a table more quickly than does the DELETE statement because it drops the contents of the table page by page, while DELETE drops the contents row by row.

NOTE The TRUNCATE TABLE statement is a Transact-SQL extension to the SQL standard.

The TRUNCATE TABLE statement has the following form:

```
TRUNCATE TABLE table_name
```

TIP If you want to delete all rows from a table, use the TRUNCATE TABLE statement. This statement is significantly faster than DELETE because it is minimally logged and there are just a few entries in the log during its execution. (Logging is discussed in detail in Chapter 13.)

MERGE Statement

The MERGE statement combines the sequence of conditional INSERT, UPDATE, and DELETE statements in a single atomic statement, depending on the existence of a record. In other words, you can sync two different tables so that the content of the target table is modified based on differences found in the source table.

The main application area for MERGE is a data warehouse environment (see Chapter 24), where tables need to be refreshed periodically with new data arriving from online transaction processing (OLTP) systems. This new data may contain changes to existing rows in tables and/ or new rows that need to be inserted. If a row in the new data corresponds to an item that already exists in the table, an UPDATE or a DELETE statement is performed. Otherwise, an INSERT statement is performed.

The alternative way, which you can use instead of applying the MERGE statement, is to write a sequence of INSERT, UPDATE, and DELETE statements, where, for each row, the decision is made whether to insert, delete, or update the data. This old approach has significant performance disadvantages: it requires multiple data scans and operates on a record-by-record basis.

Examples 7.20 and 7.21 show the use of the MERGE statement.

Example 7.20

```
USE sample;
CREATE TABLE bonus
          (pr_no CHAR(4),
            bonus SMALLINT DEFAULT 100);
INSERT INTO bonus (pr_no) VALUES ('p1');
```

Example 7.20 creates the **bonus** table, which contains one row, (p1, 100). This table will be used for merging.

Example 7.21

```
USE sample;
MERGE INTO bonus B
    USING (SELECT project_no, budget
                    FROM project) E
        ON (B.pr_no = E.project_no)
        WHEN MATCHED THEN
                    UPDATE SET B.bonus = E.budget * 0.1
        WHEN NOT MATCHED THEN
                    INSERT (pr_no, bonus)
                        VALUES (E.project_no, E.budget * 0.05);
```

The MERGE statement in Example 7.21 modifies the data in the **bonus** table depending on the existing values in the **pr_no** column. If a value from the **project_no** column of the **project** table appears in the **pr_no** column of the **bonus** table, the MATCHED branch will be executed and the existing value will be updated. Otherwise, the NOT MATCHED branch will be executed and the corresponding INSERT statement will insert new rows in the **bonus** table.

The content of the **bonus** table after the execution of the MERGE statement is as follows:

pr_no	bonus
p1	12000
p2	4750
p3	9325

From the result set, you can see that a value of the **bonus** column represents 10 percent of the original value in the case of the first row, which is updated, and 5 percent in the case of the second and third rows, which are inserted.

The OUTPUT Clause

The result of the execution of an INSERT, UPDATE, or DELETE statement contains by default only the text concerning the number of modified rows ("3 rows deleted," for instance). If the content of such a result doesn't fit your needs, you can use the OUTPUT clause, which displays explicitly the rows that are inserted or updated in the table or deleted from it.

NOTE The OUTPUT clause is also part of the MERGE statement. It returns an output for each modified row in the target table (as will be demonstrated in Examples 7.25 and 7.26).

The OUTPUT clause uses the **inserted** and **deleted** tables (explained in Chapter 14) to display the corresponding result. Also, the OUTPUT clause must be used with an INTO expression to fill a table. For this reason, you use a table variable to store the result.

Example 7.22 shows how the OUTPUT statement works with a DELETE statement.

Example 7.22

```
USE sample;
DECLARE @del_table TABLE (emp_no INT, emp_lname CHAR(20));
DELETE employee
OUTPUT DELETED.emp_no, DELETED.emp_lname INTO @del_table
WHERE emp_no > 15000;
SELECT * FROM @del_table;
```

If the content of the **employee** table is in the initial state, the execution of the statements in Example 7.22 produces the following result:

emp_no	emp_lname
25348	Smith
18316	Barrimore
29346	James
28559	Moser

First, Example 7.22 declares the table variable **@del_table** with two columns: **emp_no** and **emp_lname**. (Variables are explained in detail in Chapter 8.) This table will be used to store the deleted rows. The syntax of the DELETE statement is enhanced with the OUTPUT option:

```
OUTPUT DELETED.emp_no, DELETED.emp_lname INTO @del_table
```

Using this option, the system stores the deleted rows in the **deleted** table, which is then copied in the **@del** table variable.

Example 7.23 shows the use of the OUTPUT option in an UPDATE statement.

Example 7.23

```
USE sample;
DECLARE @update_table TABLE
  (emp_no INT, project_no CHAR(20),old_job CHAR(20),new_job CHAR(20));
UPDATE works_on
SET job = NULL
OUTPUT DELETED.emp_no, DELETED.project_no,
       DELETED.job, INSERTED.job INTO @update_table
WHERE job = 'Clerk';
SELECT * FROM @update_table;
```

The result is

emp_no	project_no	old_job	new_job
25348	p2	Clerk	NULL
28559	p2	Clerk	NULL
9031	p3	Clerk	NULL
29346	p1	Clerk	NULL

The following examples show the use of the OUTPUT clause within the MERGE statement.

NOTE The use of the OUTPUT clause within the MERGE statement is complex. Therefore, you might want to skip the rest of this section on the first reading of the book and make a note to return to it later.

Suppose that your marketing department decides to give customers a price reduction of 20 percent for all bikes that cost more than $500. The SELECT statement in Example 7.24 selects all products that cost more than $500 and inserts them in the **temp_PriceList** temporary table. The consecutive UPDATE statement searches for all bikes and reduces their price. (The UPDATE statement uses three subqueries to get the necessary information from three tables: **Production.Product**, **Production.ProductSubcategory**, and **Production .ProductCategory**.)

Example 7.24

```
USE AdventureWorks;
SELECT ProductID, Product.Name as ProductName, ListPrice
INTO temp_PriceList
FROM Production.Product
WHERE ListPrice > 500;
UPDATE temp_PriceList
    SET ListPrice = ListPrice * 0.8
      WHERE ProductID IN (SELECT ProductID
      FROM Production.Product
          WHERE ProductSubcategoryID IN ( SELECT ProductCategoryID
          FROM Production.ProductSubcategory
              WHERE ProductCategoryID IN ( SELECT ProductCategoryID
              FROM Production.ProductCategory
                  WHERE Name = 'Bikes')));
```

The CREATE TABLE statement in Example 7.25 creates a new table, **temp_Difference**, that will be used to store the result set of the MERGE statement. After that, the MERGE statement compares the complete list of the products with the new list (given in the **temp_priceList** table) and inserts the modified prices for all bicycles by using the UPDATE SET clause. (Besides the insertion of the new prices for all bicycles, the statement also changes the **ModifiedDate** column for all products and sets it to the current date.) The OUTPUT clause in Example 7.25 writes the old and new prices in the temporary table called **temp_Difference**. That way, you can later calculate the aggregate differences, if needed.

Example 7.25

```
USE AdventureWorks;
CREATE TABLE temp_Difference
     (old DEC (10,2), new DEC(10,2));
GO
MERGE INTO Production. Product
USING temp_PriceList ON Product.ProductID = temp_PriceList.ProductID
WHEN MATCHED AND Product.ListPrice <> temp_PriceList.ListPrice THEN
UPDATE SET ListPrice = temp_PriceList.ListPrice, ModifiedDate = GETDATE()
WHEN NOT MATCHED BY SOURCE THEN
UPDATE SET ModifiedDate = GETDATE()
OUTPUT DELETED.ListPrice, INSERTED.ListPrice  INTO temp_Difference;
```

Example 7.26 shows the computation of the overall difference, the result of the preceding modifications.

Example 7.26

```
USE AdventureWorks;
SELECT SUM(old) -  SUM(new) AS diff
  FROM dbo.temp_Difference;
```

The result is

diff
10773.60

Summary

Generally, only three SQL statements can be used to modify a table: INSERT, UPDATE, and DELETE. These statements are generic insofar as for all types of row insertion, you use only INSERT; for all types of column modification, you use only UPDATE; and for all types of row deletion, you use only DELETE.

The nonstandard statement TRUNCATE TABLE is just another form of the DELETE statement, but the deletion of rows is executed faster with TRUNCATE TABLE than with DELETE. The MERGE statement is basically an "UPSERT" statement: it combines the UPDATE and the INSERT statements in one statement.

Chapters 5 through 7 have introduced all SQL statements that belong to DDL and DML. Most of these statements can be grouped together to build a sequence of Transact-SQL statements. Such a sequence is the basis for *stored procedures*, which will be covered in the next chapter.

Exercises

E.7.1   Insert the data of a new employee called Julia Long, whose employee number is 11111. Her department number is not known yet.

E.7.2   Create a new table called **emp_d1_d2** with all employees who work for department d1 or d2, and load the corresponding rows from the **employee** table. Find two different, but equivalent, solutions.

E.7.3   Create a new table of all employees who entered their projects in 2017 and load it with the corresponding rows from the **employee** table.

E.7.4   Modify the job of all employees in project p1 who are managers. They have to work as clerks from now on.

E.7.5   The budgets of all projects are no longer determined. Assign all budgets the NULL value.

E.7.6   Modify the jobs of the employee with the employee number 28559. From now on she will be the manager for all her projects.

E.7.7   Increase the budget of the project where the manager has the employee number 10102. The increase is 10 percent.

E.7.8   Change the name of the department for which the employee named James works. The new department name is Sales.

E.7.9   Change the enter date for the projects for those employees who work in project p1 and belong to department Sales. The new date is 12.12.2017.

E.7.10   Delete all departments that are located in Seattle.

E.7.11   The project p3 has been finished. Delete all information concerning this project in the **sample** database.

E.7.12   Delete the information in the **works_on** table for all employees who work for the departments located in Dallas.

8

Stored Procedures and User-Defined Functions

In This Chapter

- Procedural Extensions
- Stored Procedures
- User-Defined Functions

This chapter introduces batches and routines. A *batch* is a sequence of Transact-SQL statements and procedural extensions. A *routine* can be either a stored procedure or a user-defined function (UDF). The beginning of the chapter introduces all procedural extensions supported by the Database Engine. After that, procedural extensions are used, together with Transact-SQL statements, to show how batches can be implemented. A batch can be stored as a database object, as either a stored procedure or a UDF. Some stored procedures are written by users, and others are provided by Microsoft and are referred to as *system stored procedures*. In contrast to user-defined stored procedures, UDFs return a value to a caller. All routines can be written either in Transact-SQL or in another programming language such as C# or Visual Basic.

Procedural Extensions

The preceding chapters introduced Transact-SQL statements that belong to the data definition language and the data manipulation language. Most of these statements can be grouped together to build a batch. As previously mentioned, a batch is a sequence of Transact-SQL statements and procedural extensions that are sent to the Database Engine for execution together. The number of statements in a batch is limited by the size of the compiled batch object. The main advantage of a batch over a group of singleton statements is that executing all statements at once brings significant performance benefits.

There are a number of restrictions concerning the appearance of different Transact-SQL statements inside a batch. The most important is that the data definition statements CREATE VIEW, CREATE PROCEDURE, and CREATE TRIGGER must each be the only statement in a batch.

NOTE To separate DDL statements from one another, use the GO statement.

The following sections describe each procedural extension of the Transact-SQL language separately.

Block of Statements

A block allows the building of units with one or more Transact-SQL statements. Every block begins with the BEGIN statement and terminates with the END statement, as shown in the following example:

```
BEGIN
statement_1
statement_2
...
END
```

A block can be used inside the IF statement to allow the execution of more than one statement, depending on a certain condition (see Example 8.1).

IF Statement

The Transact-SQL statement IF corresponds to the statement with the same name that is supported by almost all programming languages. IF executes one Transact-SQL statement (or more, enclosed in a block) *if* a Boolean expression, which follows the keyword IF, evaluates to TRUE. If the IF statement contains an ELSE statement, a second group of statements can be executed if the Boolean expression evaluates to FALSE.

NOTE Before you start to execute batches, stored procedures, and UDFs in this chapter, re-create the entire **sample** database.

Example 8.1

```
USE sample;
IF (SELECT COUNT(*)
                FROM works_on
                WHERE project_no = 'p1'
                GROUP BY project_no ) > 3
            PRINT 'The number of employees in the project p1 is 4 or more'
            ELSE BEGIN
                PRINT 'The following employees work for the project p1'
                SELECT emp_fname, emp_lname
```

```
                    FROM employee, works_on
                    WHERE employee.emp_no = works_on.emp_no
                    AND project_no = 'p1'
          END
```

Example 8.1 shows the use of a block inside the IF statement. The Boolean expression in the IF statement,

```
(SELECT COUNT(*)
                    FROM works_on
                    WHERE project_no = 'p1'
                     GROUP BY project_no) > 3
```

is evaluated to TRUE for the **sample** database. Therefore, the single PRINT statement in the IF part is executed. Notice that this example uses a subquery to return the number of rows (using the COUNT aggregate function) that satisfy the WHERE condition (project_no='p1'). The result of Example 8.1 is

```
The number of employees in the project p1 is 4 or more
```

NOTE The ELSE part of the IF statement in Example 8.1 contains two statements: PRINT and
        SELECT. Therefore, the block with the BEGIN and END statements is required to enclose the two
        statements. (The PRINT statement is another statement that belongs to procedural extensions; it
        returns a user-defined message.)

WHILE Statement

The WHILE statement repeatedly executes one Transact-SQL statement (or more, enclosed in a block) *while* the Boolean expression evaluates to TRUE. In other words, if the expression is true, the statement (or block) is executed, and then the expression is evaluated again to determine if the statement (or block) should be executed again. This process repeats until the expression evaluates to FALSE.

A block within the WHILE statement can optionally contain one of two statements used to control the execution of the statements within the block: BREAK or CONTINUE. The BREAK statement stops the execution of the statements inside the block and starts the execution of the statement immediately following this block. The CONTINUE statement stops the current execution of the statements in the block and starts the execution of the block from its beginning.

Example 8.2 shows the use of the WHILE statement.

Example 8.2

```
USE sample;
WHILE (SELECT SUM(budget)
                    FROM project) < 500000
```

```
BEGIN
   UPDATE project SET budget = budget*1.1
   IF (SELECT MAX(budget)
             FROM project) > 240000
     BREAK
   ELSE CONTINUE
END
```

In Example 8.2, the budget of all projects will be increased by 10 percent until the sum of budgets is greater than $500,000. However, the repeated execution will be stopped if the budget of one of the projects is greater than $240,000. The execution of Example 8.2 gives the following output:

```
(3 rows affected)
(3 rows affected)
(3 rows affected)
```

NOTE If you want to suppress the output, such as that in Example 8.2 (indicating the number of affected rows in SQL statements), use the SET NOCOUNT ON statement.

Local Variables

Local variables are an important procedural extension to the Transact-SQL language. They are used to store values (of any type) within a batch or a routine. They are "local" because they can be referenced only within the same batch in which they were declared. (The Database Engine also supports global variables, which are described in Chapter 4.)

Every local variable in a batch must be defined using the DECLARE statement. (For the syntax of the DECLARE statement, see Example 8.3.) The definition of each variable contains its name and the corresponding data type. Variables are always referenced in a batch using the prefix @. The assignment of a value to a local variable is done:

- Using the special form of the SELECT statement
- Using the SET statement
- Directly in the DECLARE statement using the = sign (for instance, @extra_budget MONEY = 1500)

The usage of the first two assignment statements for a value assignment is demonstrated in Example 8.3. The batch in this example calculates the average of all project budgets and compares this value with the budget of all projects stored in the **project** table. If the latter value is smaller than the calculated value, the budget of project p1 will be increased by the value of the local variable **@extra_budget**. We will implement this batch in two different ways. Example 8.3a uses Transact-SQL statements only.

Example 8.3a (using T-SQL only)

```
-- set-oriented way to retrieve values
USE sample;
DECLARE @avg_budget MONEY, @extra_budget MONEY
DECLARE @pr_nr CHAR(4)
          SET @extra_budget = 15000
          SELECT @avg_budget = AVG(budget) FROM project
          IF (SELECT budget
                    FROM project
                    WHERE project_no=@pr_nr ) < @avg_budget
        BEGIN
          UPDATE project
                  SET budget = budget + @extra_budget
                   WHERE project_no=@pr_nr
             PRINT 'Budget for @pr_nr increased by @extra_budget'
        END
```

In Example 8.3a only T-SQL statements have been used. The sole use of these statements guarantees that all retrieved rows, independent of their number, will be sent to the system to process them at the same time. (This is called *set-oriented processing*.)

Example 8.3b applies a concept called *cursors*.

Example 8.3b (using cursor)

```
-- record-oriented way to retrieve values
USE sample;
DECLARE @avg_budget MONEY;
DECLARE @extra_budget MONEY;
DECLARE @budget MONEY;
DECLARE @pr_nr CHAR(4)
DECLARE @P_cursor as CURSOR;
SET @extra_Budget = 15000;
SELECT @avg_budget = AVG(budget) FROM project;
SET @budget = 0;
 SET @P_cursor = CURSOR FOR
               SELECT project_no, budget FROM project;
OPEN @P_cursor;
FETCH NEXT FROM @P_cursor INTO @pr_nr, @budget
WHILE @@FETCH_STATUS = 0
BEGIN
          PRINT @pr_nr
          PRINT @budget
        IF (SELECT budget FROM project
                    WHERE project_no=@pr_nr) >= @avg_budget
```

```
                          BEGIN
                          GOTO L1
                          END
              ELSE                UPDATE project
                          SET budget = budget + @extra_budget
                          WHERE project_no =@pr_nr
                    PRINT 'Budget for @pr_nr increased'
      L1:
         FETCH NEXT FROM @P_cursor INTO @pr_nr, @budget
END
CLOSE @P_cursor;
DEALLOCATE @P_cursor;
```

In Example 8.3b, the cursor feature is used to solve the same problem as in Example 8.3a. The main difference between these two solutions is that 8.3b retrieves each row from the result separately; i.e., in this case the system processes one record (row) at a time. (For this reason, the type of processing is called *record-oriented processing*.)

Before discussing the differences between set-oriented processing and record-oriented processing using cursors, I will explain the cursor features in Example 8.3b.

Generally, there are several steps in creating and using cursors in batches and stored procedures. First, you declare your cursor by using the DECLARE statement and assigning the CURSOR data type. After that, you use the SET statement to assign the set of rows, which will be retrieved (one by one) with the cursor. This is followed by opening the cursor using the OPEN statement. Immediately after the OPEN statement is executed, the cursor points before the first row of the selected set of rows.

Now the data processing starts. To move the cursor to the first row in the result set, you use the FETCH NEXT statement:

```
FETCH NEXT FROM @P_cursor INTO @pr_nr, @budget
```

This statement fetches a record from the result set and assigns values retrieved with the SELECT statement to the variables @**pr_nr** and @**budget**, respectively. The THEN part of the IF statement uses the fetched values to calculate the average of all project budgets and compare the average value with the budget of the particular project.

The WHILE statement uses the system function called @@FETCH_STATUS to create a loop, which will be terminated when all records from the result set are processed. (In other words, this system function returns the status of the last cursor FETCH statement issued against any cursor currently opened by the connection. The return value 0 means that the FETCH statement was successful, while –1 specifies that the record (row) was beyond the result set.)

Once all the data has been processed, you use the CLOSE statement to close the cursor. Finally, the DEALLOCATE statement deallocates the particular cursor. The difference between CLOSE and DEALLOCATE is that after the CLOSE statement is executed, you can still reopen the cursor, whereas after the execution of the DEALLOCATE statement, the link between the cursor and the result set is abandoned. (The explicit use of the DEALLOCATE statement is highly recommended, because that way you release all of the internal resources.)

NOTE Do not use the implementation with CURSOR unless absolutely necessary. The record-oriented processing of rows is significantly slower than the set-oriented processing. The more rows that have to be processed, the better performance you will achieve with set-oriented processing.

Miscellaneous Procedural Statements

The procedural extensions of the Transact-SQL language also contain the following statements:

- RETURN
- GOTO
- RAISEERROR()
- WAITFOR

The RETURN statement has the same functionality inside a batch as the BREAK statement inside WHILE. This means that the RETURN statement causes the execution of the batch to terminate and the first statement following the end of the batch to begin executing.

The GOTO statement branches to a label, which stands in front of a Transact-SQL statement within a batch. The RAISERROR() statement generates a user-defined error message and sets a system error flag. A user-defined error number must be greater than 50000. (All error numbers <= 50000 are system defined and are reserved by the Database Engine.) The error values are stored in the global variable **@@error**. (Example 17.3 shows the use of the RAISERROR() statement.)

The WAITFOR statement defines either the time interval (if the DELAY option is used) or a specified time (if the TIME option is used) that the system has to wait before executing the next statement in the batch. The syntax of this statement is

```
WAITFOR {DELAY 'time' | TIME 'time' | TIMEOUT 'timeout' }
```

The DELAY option tells the database system to wait until the specified amount of time has passed. TIME specifies a time in one of the acceptable formats for temporal data. TIMEOUT specifies the amount of time, in milliseconds, to wait for a message to arrive in the queue. (Example 13.8 shows the use of the WAITFOR statement.)

Exception Handling with TRY, CATCH, and THROW

The Database Engine supports two statements, TRY and CATCH, to capture and handle exceptions. An *exception* is a problem (usually an error) that prevents the continuation of a program. In other words, an unhandled exception prevents the application from continuing. In the case of a handled exception, the existing problem will be relegated to another part of the program, which will handle the exception.

The role of the TRY statement is to capture the exception. (Because this process usually comprises several statements, the term "TRY block" typically is used instead of "TRY statement.") If an exception occurs within the TRY block, the part of the system called the *exception handler* delivers the exception to the other part of the program, which will handle the exception. This program part is denoted by the keyword CATCH and is therefore called the CATCH block.

> **NOTE** You can handle errors using the **@@error** global variable (see Example 13.1), but exception handling using the TRY and CATCH statements is the common way modern programming languages like C# and Java treat errors.

Exception handling with the TRY and CATCH blocks gives a programmer a lot of benefits, such as:

- Exceptions provide a clean way to check for errors without cluttering code.
- Exceptions provide a mechanism to signal errors directly rather than using some side effects.
- Exceptions can be seen by the programmer and checked during the compilation process.

The third statement in relation to handling errors is THROW. This statement allows you to throw an exception caught in the exception handling block. Simply stated, the THROW statement is another return mechanism, which behaves similarly to the already described RAISERROR() statement.

Example 8.4 shows how exception handling with the TRY/CATCH/THROW works. It shows how you can use exception handling to insert all statements in a batch or to roll back the entire statement group if an error occurs. The example is based on the referential integrity between the **department** and **employee** tables. For this reason, you have to create both tables using the PRIMARY KEY and FOREIGN KEY clauses, as done in Example 5.11.

Example 8.4

```
USE sample;
BEGIN TRY
    BEGIN TRANSACTION
    insert into employee values(11111, 'Ann', 'Smith','d2');
    insert into employee values(22222, 'Matthew', 'Jones','d4'); --
referential integrity error
    insert into employee values(33333, 'John', 'Barrimore', 'd2');
    COMMIT TRANSACTION
    PRINT 'Transaction committed'
END TRY
BEGIN CATCH
    ROLLBACK
    PRINT 'Transaction rolled back';
    THROW
END CATCH
```

After the execution of the batch in Example 8.4, all three statements in the batch won't be executed at all, and the output of this example is

```
Transaction rolled back

Msg 547, Level 16, State 0, Line 4
The INSERT statement conflicted with the FOREIGN KEY constraint
"foreign_emp". The conflict occurred in database "sample", table
"dbo.department", column 'dept_no'.
```

The execution of Example 8.4 works as follows. The first INSERT statement is executed successfully. Then, the second statement causes the referential integrity error. Because all three statements are written inside the TRY block, the exception is "thrown" and the exception handler starts the CATCH block. CATCH rolls back all statements and prints the corresponding message. After that the THROW statement returns the execution of the batch to the caller. For this reason, the content of the **employee** table won't change.

NOTE The statements BEGIN TRANSACTION, COMMIT TRANSACTION, and ROLLBACK are Transact-SQL statements concerning transactions. These statements start, commit, and roll back transactions, respectively. See Chapter 13 for the discussion of these statements and transactions generally.

Example 8.5 shows the batch that supports server-side paging (for the description of server-side paging, see Chapter 6).

Example 8.5

```
USE AdventureWorks;
DECLARE
    @PageSize    TINYINT = 20,
    @CurrentPage INT    = 4;
SELECT BusinessEntityID, JobTitle, BirthDate
   FROM HumanResources.Employee
   WHERE Gender = 'F'
ORDER BY JobTitle
OFFSET (@PageSize * (@CurrentPage - 1)) ROWS
    FETCH NEXT @PageSize ROWS ONLY;
```

The batch in Example 8.5 uses the **AdventureWorks** database and its **Employee** table (from the **HumanResources** schema) to show how *generic* server-side paging can be implemented. The @**Pagesize** variable is used with the FETCH NEXT statement to specify the number of rows per page (20, in this case). The other variable, @**CurrentPage**, specifies which particular page should be displayed. In this example, the content of the third page will be displayed. (The result is not shown because it is too lengthy.)

Stored Procedures

A *stored procedure* is a special kind of batch written in Transact-SQL, using the SQL language and its procedural extensions. The main difference between a batch and a stored procedure is that the latter is stored as a database object. In other words, stored procedures are saved on the server side to improve the performance and consistency of repetitive tasks.

The Database Engine supports stored procedures and system procedures. Stored procedures are created in the same way as all other database objects—that is, by using the DDL. System procedures are provided with the Database Engine and can be used to access and modify the information in the system catalog. This section describes (user-defined) stored procedures, while system procedures are explained in Chapter 9.

When a stored procedure is created, an optional list of parameters can be defined. The procedure accepts the corresponding arguments each time it is invoked. Stored procedures can optionally return a value that displays the user-defined information or, in the case of an error, the corresponding error message.

A stored procedure is precompiled before it is stored as an object in the database. The precompiled form is stored in the database and used whenever the stored procedure is executed. This property of stored procedures offers an important benefit: the repeated compilation of a procedure is (almost always) eliminated, and the execution performance is therefore increased. This property of stored procedures offers another benefit concerning the volume of data that must be sent to and from the database system. It might take less than 50 bytes to call a stored procedure containing several thousand bytes of statements. The accumulated effect of this savings when multiple users are performing repetitive tasks can be quite significant.

> **NOTE** Stored procedures can be *natively compiled*, meaning that the particular procedure is compiled when it is created, rather than when it is executed. This special form of stored procedures is described in Chapter 21 (see Example 21.7).

Creation and Execution of Stored Procedures

Stored procedures are created with the CREATE PROCEDURE statement, which has the following syntax:

```
CREATE PROC[EDURE] [schema_name.]proc_name
[({@param1} type1 [ VARYING] [= default1] [OUTPUT])] {, …}
[WITH {RECOMPILE | ENCRYPTION | EXECUTE AS 'user_name'}]
[FOR REPLICATION]
AS batch | EXTERNAL NAME method_name
```

schema_name is the name of the schema to which the ownership of the created stored procedure is assigned. **proc_name** is the name of the new stored procedure. **@param1** is a parameter, while **type1** specifies its data type. The parameter in a stored procedure has the same logical meaning as the local variable for a batch. Parameters are values passed from the caller of the stored procedure and are used within the stored procedure. **default1** specifies the optional default value of the corresponding parameter. (Default can also be NULL.)

The OUTPUT option indicates that the parameter is a return parameter and can be returned to the calling procedure or to the system (demonstrated a bit later in Example 8.9).

As you already know, the precompiled form of a procedure is stored in the database and used whenever the stored procedure is executed. If you want to generate the compiled form each time the procedure is executed, use the WITH RECOMPILE option.

NOTE The use of the WITH RECOMPILE option destroys one of the most important benefits of the stored procedures: the performance advantage gained by a single precompilation. For this reason, the WITH RECOMPILE option should be used only when database objects used by the stored procedure are modified frequently or when the parameters used by the stored procedure are volatile.

The EXECUTE AS clause specifies the security context under which to execute the stored procedure after it is accessed. By specifying the context in which the procedure is executed, you can control which user account the Database Engine uses to validate permissions on objects referenced by the procedure.

By default, only the members of the **sysadmin** fixed server role, and the **db_owner** and **db_ddladmin** fixed database roles, can use the CREATE PROCEDURE statement. However, the members of these roles may assign this privilege to other users by using the GRANT CREATE PROCEDURE statement. (For the discussion of user permissions, fixed server roles, and fixed database roles, see Chapter 12.)

Example 8.6 shows the creation of the simple stored procedure for the **project** table.

Example 8.6

```
USE sample;
GO
CREATE PROCEDURE increase_budget (@percent INT=5)
          AS UPDATE project
                  SET budget = budget + budget*@percent/100;
```

NOTE The GO statement is used to separate two batches. (The CREATE PROCEDURE statement must be the first statement in the batch.)

The stored procedure **increase_budget** increases the budgets of all projects for a certain percentage value that is defined using the parameter **@percent**. The procedure also defines the default value (5), which is used if there is no argument at the execution time of the procedure.

NOTE It is possible to create stored procedures that reference nonexistent tables. This feature allows you to debug procedure code without creating the underlying tables first, or even connecting to the target server.

In contrast to "base" stored procedures that are placed in the current database, it is possible to create temporary stored procedures that are always placed in the temporary system database called **tempdb**. You might create a temporary stored procedure to avoid executing a particular group of statements repeatedly within a connection. Analogous to local and global temporary tables, you can create *local* or *global* temporary procedures by preceding the procedure name with a single pound sign (#**proc_name**) for local temporary procedures and a double pound sign (##**proc_name**) for global temporary procedures. A local temporary stored procedure can be executed only by the user who created it, and only during the same connection. A global temporary procedure can be executed by all users, but only until the last connection executing it (usually the creator's) ends.

The life cycle of a stored procedure has two phases: its creation and its execution. Each procedure is created once and executed many times. The EXECUTE statement executes an existing procedure. The execution of a stored procedure is allowed for each user who either is the owner of or has the EXECUTE privilege for the procedure (see Chapter 12). The EXECUTE statement has the following syntax:

```
[[EXEC[UTE]] [@return_status =] {proc_name
        | @proc_name_var}
        {[[@parameter1 =] value | [@parameter1=] @variable [OUTPUT]] |
DEFAULT}..
        [WITH RECOMPILE]
```

All options in the EXECUTE statement, other than **return_status**, have the equivalent logical meaning as the options with the same names in the CREATE PROCEDURE statement. **return_status** is an optional integer variable that stores the return status of a procedure. The value of a parameter can be assigned using either a value (**value**) or a local variable (**@variable**). The order of parameter values is not relevant if they are named, but if they are not named, parameter values must be supplied in the order defined in the CREATE PROCEDURE statement.

The DEFAULT clause supplies the default value of the parameter as defined in the procedure. When the procedure expects a value for a parameter that does not have a defined default and either a parameter is missing or the DEFAULT keyword is specified, an error occurs.

NOTE   When the EXECUTE statement is the first statement in a batch, the word "EXECUTE" can be omitted from the statement. Despite this, it would be safer to include this word in every batch you write.

Example 8.7 shows the use of the EXECUTE statement.

Example 8.7

```
USE sample;
EXECUTE increase_budget 10;
```

The EXECUTE statement in Example 8.7 executes the stored procedure **increase_budget** (Example 8.6) and increases the budgets of all projects by 10 percent each.

Example 8.8 shows the creation of a procedure that references the tables **employee** and **works_on**.

Example 8.8

```
USE sample;
GO
CREATE PROCEDURE modify_empno (@old_no INTEGER, @new_no INTEGER)
        AS UPDATE employee
                SET emp_no = @new_no
                WHERE emp_no = @old_no
```

```
UPDATE works_on
        SET emp_no = @new_no
        WHERE emp_no = @old_no
```

The procedure **modify_empno** in Example 8.8 demonstrates the use of stored procedures as part of the maintenance of the referential integrity (in this case, between the **employee** and **works_on** tables). Such a stored procedure can be used inside the definition of a trigger, which actually maintains the referential integrity (see Example 14.3). To execute this procedure, use the EXECUTE statement and assign values to both input parameters, **@old_no** and **@new_no**.

Example 8.9 shows the use of the OUTPUT clause.

Example 8.9

```
USE sample;
GO
CREATE PROCEDURE delete_emp @employee_no INT, @counter INT OUTPUT
        AS SELECT @counter = COUNT(*)
                FROM works_on
                  WHERE emp_no = @employee_no;
        DELETE FROM employee
                WHERE emp_no = @employee_no;
          DELETE FROM works_on
                WHERE emp_no = @employee_no;
```

This stored procedure can be executed using the following batch:

```
DECLARE @quantity INT;
EXECUTE delete_emp @employee_no=28559, @counter=@quantity OUTPUT;
```

The preceding example contains the creation of the **delete_emp** procedure as well as its execution. This procedure calculates the number of projects on which the employee (with the employee number **@employee_no**) works. The calculated value is then assigned to the **@counter** parameter. After the deletion of all rows with the assigned employee number from the **employee** and **works_on** tables, the calculated value will be assigned to the **@quantity** variable.

NOTE The value of the parameter will be returned to the calling procedure if the OUTPUT option is used. In Example 8.9, the **delete_emp** procedure passes the @**counter** parameter to the calling statement, so the procedure returns the value to the system. Therefore, the @**counter** parameter must be declared with the OUTPUT option in the procedure as well as in the EXECUTE statement.

The EXECUTE Statement with RESULT SETS Clause

Using the WITH RESULT SETS clause for the EXECUTE statement, you can change conditionally the form of the result set of a stored procedure.

The following two examples help to explain this clause. Example 8.10 is an introductory example that shows how the output looks when the WITH RESULT SETS clause is omitted.

Example 8.10

```
USE sample;
GO
CREATE PROCEDURE employees_in_dept (@dept CHAR(4))
 AS SELECT emp_no, emp_lname
    FROM employee
    WHERE dept_no IN (SELECT @dept FROM department
                            GROUP BY dept_no)
```

employees_in_dept is a simple stored procedure that displays the numbers and family names of all employees working for a particular department. (The department number is a parameter of the procedure and must be specified when the procedure is invoked.) The result of this procedure is a table with two columns, named according to the names of the corresponding columns (**emp_no** and **emp_lname**). To change these names (and their data types, too), use the WITH RESULTS SETS clause (see Example 8.11).

Example 8.11

```
USE sample;
EXEC employees_in_dept 'd1'
  WITH RESULT SETS
  ( ([EMPLOYEE NUMBER] INT NOT NULL,
     [NAME OF EMPLOYEE] CHAR(20) NOT NULL));
```

The output is

Employee Number	Name Of Employee
18316	Barrimore
28559	Moser

As you can see, the WITH RESULT SETS clause in Example 8.11 allows you to change the name and data types of columns displayed in the result set. Therefore, this functionality gives you the flexibility to execute stored procedures and display the output result sets in another form.

NOTE The Database Engine provides capabilities to create temporary stored procedures, too. The semantics of temporary stored procedures is the same as the semantics of the temporary tables. In other words, a scope and lifetime of temporary stored procedures is for the duration of a session, and you use the symbol # as a prefix to specify temporary stored procedures.

Changing the Structure of Stored Procedures

The Database Engine also supports the ALTER PROCEDURE statement, which modifies the structure of a stored procedure. The ALTER PROCEDURE statement is usually used to modify Transact-SQL statements inside a procedure. All options of the ALTER PROCEDURE statement correspond to the options with the same name in the CREATE PROCEDURE statement. The main purpose of this statement is to avoid reassignment of existing privileges for the stored procedure.

A stored procedure is removed using the DROP PROCEDURE statement. Only the owner of the stored procedure and the members of the **db_owner** and **sysadmin** fixed roles can remove the procedure.

User-Defined Functions

In programming languages, there are generally two types of routines:

- Stored procedures
- User-defined functions (UDFs)

As discussed in the previous major section of this chapter, stored procedures are made up of several statements that have zero or more input parameters but usually do not return any output parameters. In contrast, functions always have one return value. This section describes the types of UDFs and the creation and use of UDFs.

Types of User-Defined Functions

User-defined functions can be

- Scalar
- Table-valued

The return value of scalar functions is always a single value. Scalar functions are generally used within a query. Also, scalar functions can be called using the EXECUTE statement, the same way as stored procedures. (An example of a scalar UDF is given in the upcoming "Scalar UDF Inlining" section.)

A table-valued function is a user-defined function that returns data of a table type. The return type of a table-valued function is a table. For this reason, you can use the table-valued function just like you would use a table. (A detailed description of these functions is given in the upcoming "Table-Valued Functions" section.)

Creation and Execution of User-Defined Functions

UDFs are created with the CREATE FUNCTION statement, which has the following syntax:

```
CREATE FUNCTION [schema_name.]function_name
        [({@param } type [= default])  {,...}
        RETURNS {scalar_type | [@variable] TABLE}
        [WITH {ENCRYPTION | SCHEMABINDING}
        [AS] {block | RETURN (select_statement)}
```

schema_name is the name of the schema to which the ownership of the created UDF is assigned. **function_name** is the name of the new function. **@param** is an input parameter, while **type** specifies its data type. Parameters are values passed from the caller of the UDF and are used within the function. **default** specifies the optional default value of the corresponding parameter. (Default can also be NULL.)

The RETURNS clause defines a data type of the value returned by the UDF. This data type can be any of the standard data types supported by the database system, including the TABLE data type. (The only standard data type that you cannot use is TIMESTAMP.)

UDFs are either scalar-valued or table-valued. A scalar-valued function returns an atomic (scalar) value. This means that in the RETURNS clause of a scalar-valued function, you specify one of the standard data types. Functions are table-valued if the RETURNS clause returns a set of rows (see the next subsection).

The WITH ENCRYPTION option encrypts the information in the system catalog that contains the text of the CREATE FUNCTION statement. In that case, you cannot view the text used to create the function. (Use this option to enhance the security of your database system.)

The alternative clause, WITH SCHEMABINDING, binds the UDF to the database objects that it references. Any attempt to modify the structure of the database object that the function references fails. (The binding of the function to the database objects it references is removed only when the function is altered, so the SCHEMABINDING option is no longer specified.)

Database objects that are referenced by a function must fulfill the following conditions if you want to use the SCHEMABINDING clause during the creation of that function:

- All views and UDFs referenced by the function must be schema-bound.
- All database objects (tables, views, or UDFs) must be in the same database as the function.

block is the BEGIN/END block that contains the implementation of the function. The final statement of the block must be a RETURN statement with an argument. (The value of the argument is the value returned by the function.) In the body of a BEGIN/END block, only the following statements are allowed:

- Assignment statements such as SET
- Control-of-flow statements such as WHILE and IF
- DECLARE statements defining local data variables
- SELECT statements containing SELECT lists with expressions that assign to variables that are local to the function
- INSERT, UPDATE, and DELETE statements modifying variables of the TABLE data type that are local to the function

By default, only the members of the **sysadmin** fixed server role and the **db_owner** and **db_ddladmin** fixed database roles can use the CREATE FUNCTION statement. However, the members of these roles may assign this privilege to other users by using the GRANT CREATE FUNCTION statement (see Chapter 12).

Example 8.12 shows the creation of the scalar user-defined function called **compute_costs**.

Example 8.12

```
-- This function computes additional total costs that arise
-- if budgets of projects increase
USE sample;
GO
CREATE FUNCTION compute_costs (@percent INT =10)
      RETURNS DECIMAL(16,2)
  BEGIN
   DECLARE @additional_costs DEC (14,2), @sum_budget dec(16,2)
   SELECT @sum_budget = SUM (budget) FROM project
   SET @additional_costs = @sum_budget * @percent/100
   RETURN @additional_costs
  END
```

The **compute_costs** function computes additional costs that arise when all budgets of projects increase. The single input variable, **@percent**, specifies the percentage of increase of budgets. The BEGIN/END block first declares two local variables: **@additional_costs** and **@sum_budget**. The function then assigns to **@sum_budget** the sum of all budgets, using the SELECT statement. After that, the function computes total additional costs and returns this value using the RETURN statement.

Invoking User-Defined Functions

Each UDF can be invoked in Transact-SQL statements, such as SELECT, INSERT, UPDATE, or DELETE. To invoke a function, specify the name of it, followed by parentheses. Within the parentheses, you can specify one or more arguments. (*Arguments* are values or expressions that are passed to the input parameters that are defined immediately after the function name.) When you invoke a function, and all parameters have no default values, you must supply argument values for all of the parameters and you must specify the argument values in the same sequence in which the parameters are defined in the CREATE FUNCTION statement. (As you already know, the scalar UDFs can be invoked using the EXEC statement, too.)

Example 8.13 shows the use of the **compute_costs** function (Example 8.12) in a SELECT statement.

Example 8.13

```
USE sample;
SELECT project_no, project_name
     FROM project
   WHERE budget < dbo.compute_costs(25)
```

The result is

project_no	project_name
p2	Gemini

The SELECT statement in Example 8.13 displays names and numbers of all projects where the budget is lower than the total additional costs of all projects for a given percentage.

NOTE Each function used in a Transact-SQL statement must be specified using its two-part name—that is, **schema_name.function_name**.

Scalar UDF Inlining

In all versions prior to SQL Server 2019, scalar user-defined functions are generally a performance issue because they are generally processed in a row-oriented way, meaning that they run for every returned row. That way, they do not allow parallel execution of rows.

SQL Server 2019 supports a new feature called *scalar UDF inlining*, the goal of which is to improve performance of queries that invoke scalar UDFs. In other words, the feature allows certain scalar UDFs to have their definitions placed directly into the query so that the query does not call the UDF when executing each row.

Scalar UDF inlining will be discussed in detail in Chapter 28 (see Example 28.14).

Table-Valued Functions

As you already know, functions are table-valued if the RETURNS clause returns a set of rows. Depending on how the body of the function is defined, table-valued functions can be classified as inline or multistatement functions. If the RETURNS clause specifies TABLE with no accompanying list of columns, the function is an inline function. Inline functions return the result set of a SELECT statement as a table variable; that is, a variable of the TABLE data type. In other words, an inline function can contain only one root SELECT statement that is used to describe its result. This is good from an optimization prospective, because the Query Optimizer may inline a function's text into a query and thus optimize a query as a whole.

A multistatement table-valued function includes a name followed by TABLE. (The name defines an internal variable of the type TABLE.) You can use this variable to insert rows into it and then return the variable as the return value of the function. Multistatement table-valued functions are difficult to optimize, and are not considered by the Query Optimizer.

Example 8.14 shows an inline function that returns a table variable.

Example 8.14

```
USE sample;
GO
CREATE FUNCTION employees_in_project (@pr_number CHAR(4))
    RETURNS TABLE
    AS RETURN (SELECT emp_fname, emp_lname
                    FROM works_on, employee
                  WHERE employee.emp_no = works_on.emp_no
               AND project_no = @pr_number)
```

The **employees_in_project** function is used to display names of all employees that belong to a particular project. The input parameter **@pr_number** specifies a project number. While the function generally returns a set of rows, the RETURNS clause contains the TABLE data type. (Note that the BEGIN/END block in Example 8.14 must be omitted, while the RETURN clause contains a SELECT statement.)

Example 8.15 shows the use of the **employees_in_project** function.

Example 8.15

```
USE sample;
SELECT *
     FROM employees_in_project('p3')
```

The result is

emp_fname	emp_lname
Ann	Jones
Elsa	Bertoni
Elke	Hansel

Table-Valued Functions and APPLY

The APPLY operator is a relational operator that allows you to invoke a table-valued function for each row of a table expression. This operator is specified in the FROM clause of the corresponding SELECT statement in the same way as the join operator is applied. There are two forms of the APPLY operator:

- CROSS APPLY
- OUTER APPLY

The CROSS APPLY operator returns those rows from the inner (left) table expression that match rows in the outer (right) table expression. Therefore, the CROSS APPLY operator is logically the same as the INNER JOIN operator.

NOTE CROSS APPLY is Microsoft's extension to the SQL standard. You can rewrite most queries with the CROSS APPLY operator using INNER JOIN, but the advantage of the former is that it can yield a better execution plan and better performance, since it can limit the set being joined, before the join occurs.

The OUTER APPLY operator returns all the rows from the inner (left) table expression. (For the rows for which there are no corresponding matches in the outer table expression, it contains NULL values in columns of the outer table expression.) OUTER APPLY is logically equivalent to LEFT OUTER JOIN.

Examples 8.16 and 8.17 show how you can use APPLY.

Example 8.16

```
-- generate function
CREATE FUNCTION dbo.fn_getjob(@empid AS INT)
     RETURNS TABLE AS
```

```
RETURN
        SELECT job
            FROM works_on
            WHERE emp_no = @empid
        AND job IS NOT NULL  AND project_no = 'p1';
```

The **fn_getjob()** function in Example 8.16 returns the set of rows from the **works_on** table. This result set is "joined" in Example 8.17 with the content of the **employee** table.

Example 8.17

```
-- use CROSS APPLY
SELECT E.emp_no, emp_fname, emp_lname, job
 FROM employee as E
    CROSS APPLY dbo.fn_getjob(E.emp_no) AS A
-- use OUTER APPLY
SELECT E.emp_no, emp_fname, emp_lname, job
 FROM employee as E
    OUTER APPLY dbo.fn_getjob(E.emp_no) AS A
```

The result is

emp_no	emp_fname	emp_lname	job
10102	Ann	Jones	Analyst
29346	James	James	Clerk
9031	Elsa	Bertoni	Manager
28559	Sybill	Moser	NULL

emp_no	emp_fname	emp_lname	job
25348	Matthew	Smith	NULL
10102	Ann	Jones	Analyst
18316	John	Barrimore	NULL
29346	James	James	Clerk
9031	Elsa	Bertoni	Manager
2581	Elke	Hansel	NULL
28559	Sybill	Moser	NULL

In the first query of Example 8.17, the result set of the table-valued function **fn_getjob()** is "joined" with the content of the **employee** table using the CROSS APPLY operator. **fn_getjob()** acts as the right input, and the **employee** table acts as the left input. The right input is evaluated for each row from the left input, and the rows produced are combined for the final output.

The second query is similar to the first one, but uses OUTER APPLY, which corresponds to the outer join operation of two tables.

Table-Valued Parameters

Sometimes it is necessary to send many parameters to a routine. One way to do it is to use a temporary table, insert the values into it, and then call the routine. The better way is to use table-valued parameters (see Example 8.18).

Example 8.18

```
USE sample;
GO
CREATE TYPE departmentType AS TABLE
   (dept_no CHAR(4),dept_name CHAR(25),location CHAR(30));
GO
CREATE TABLE #dallasTable
   (dept_no CHAR(4),dept_name CHAR(25),location CHAR(30));
GO
CREATE PROCEDURE insertProc
  @Dallas departmentType READONLY
  AS SET NOCOUNT ON
  INSERT INTO #dallasTable (dept_no, dept_name, location)
   SELECT * FROM @Dallas
GO
DECLARE @Dallas AS departmentType;
INSERT INTO @Dallas( dept_no, dept_name, location)
SELECT * FROM department
WHERE location = 'Dallas'
EXEC insertProc @Dallas;
```

Example 8.18 first defines the type called **departmentType** as a table. This means that its type is the TABLE data type, so rows can be inserted in it. The second statement creates the temporary table called **#dallasTable**. In the **insertProc** procedure, the **@Dallas** variable, which is of the **departmentType** type, is specified. (The READONLY clause specifies that the content of the table variable cannot be modified.) In the subsequent batch, data is added to the table variable, and after that the procedure is executed. The procedure, when executed, inserts rows from the table variable into the temporary table **#dallasTable**. The content of the temporary table is as follows:

dept_no	dept_name	location
d1	Research	Dallas
d3	Marketing	Dallas

The use of table-valued parameters gives you the following benefits:

- It simplifies the programming model in relation to routines.
- It reduces the round trips to the server.
- The resulting table can have different numbers of rows.

Changing the Structure of UDFs

The Transact-SQL language also supports the ALTER FUNCTION statement, which modifies the structure of a UDF. This statement is usually used to remove the schema binding. All options of the ALTER FUNCTION statement correspond to the options with the same name in the CREATE FUNCTION statement.

A UDF is removed using the DROP FUNCTION statement. Only the owner of the function (or the members of the **db_owner** and **sysadmin** fixed database roles) can remove the function.

Summary

A stored procedure is a special kind of batch, written either in the Transact-SQL language or using the Common Language Runtime (CLR). (This version of the book describes only how T-SQL can be used to implement stored procedures.) Stored procedures are used for the following purposes:

- To improve the performance of repetitive tasks
- To control access authorization
- To create an audit trail of activities in database tables
- To enforce consistency and business rules with respect to data modification

User-defined functions have a lot in common with stored procedures. The main difference is that UDFs return a *single* data value, which can also be a table.

The next chapter discusses the system catalog of the Database Engine.

Exercises

E.8.1   Create a batch that inserts 3000 rows in the **employee** table. The values of the **emp_no** column should be unique and between 1 and 3000. All values of the columns **emp_lname**, **emp_fname**, and **dept_no** should be set to 'Jane', 'Smith', and 'd1', respectively.

E.8.2   Modify the batch from E.8.1 so that the values of the **emp_no** column should be generated randomly using the RAND function. (Hint: Use the temporal system functions DATEPART and GETDATE to generate the random values.)

9 System Catalog

In This Chapter

- Introduction to the System Catalog
- General Interfaces
- Proprietary Interfaces

This chapter discusses the system catalog of the Database Engine. The introduction is followed by a description of the structure of several catalog views, each of which allows you to retrieve metadata. The use of dynamic management views and dynamic management functions is also covered in the first part of the chapter. Three property interfaces for retrieving metadata information are discussed in the second part: system stored procedures, system functions, and property functions.

Introduction to the System Catalog

The system catalog consists of tables describing the structure of objects such as databases, base tables, views, and indices. (These tables are called *system base tables*.) The Database Engine frequently accesses the system catalog for information that is essential for the system to function properly.

The Database Engine distinguishes the system base tables of the **master** database from those of a particular user-defined database. System tables of the **master** database belong to the system catalog, while system tables of a particular database form the database catalog. Therefore, system base tables occur only once in the entire system (if they belong exclusively to the **master** database), while others occur once in each database, including the **master** database. (The detailed description of the **master** database can be found in Chapter 15.)

In all relational database systems, system base tables have the same logical structure as base tables. As a result, the same Transact-SQL statements used to retrieve information in the base tables can also be used to retrieve information in system base tables.

NOTE The system base tables cannot be accessed directly: you have to use existing interfaces to query the information from the system catalog.

There are several different interfaces that you can use to access the information in the system base tables:

- **Catalog views**   Present the primary interface to the metadata stored in system base tables. (Metadata is data that describes the attributes of objects in a database system.)
- **Dynamic management views (DMVs) and functions (DMFs)**   Generally used to observe active processes and the contents of the memory.
- **Information schema**   A standardized solution for the access of metadata that gives you a general interface not only for the Database Engine, but for all existing relational database systems (assuming that the system supports the information schema).
- **System and property functions**   Allow you to retrieve system information. The difference between these two function types is mainly in their structure. Also, property functions can return more information than system functions.
- **System stored procedures**   Some system stored procedures can be used to access and modify the content of the system base tables.

Figure 9-1 shows a simplified form of the Database Engine's system information and different interfaces that you can use to access it.

NOTE This chapter shows you just an overview of the system catalog and the ways in which you can access metadata. Particular catalog views, as well as all other interfaces, that are specific for different topics (such as indices, security, etc.) are discussed in the corresponding chapters.

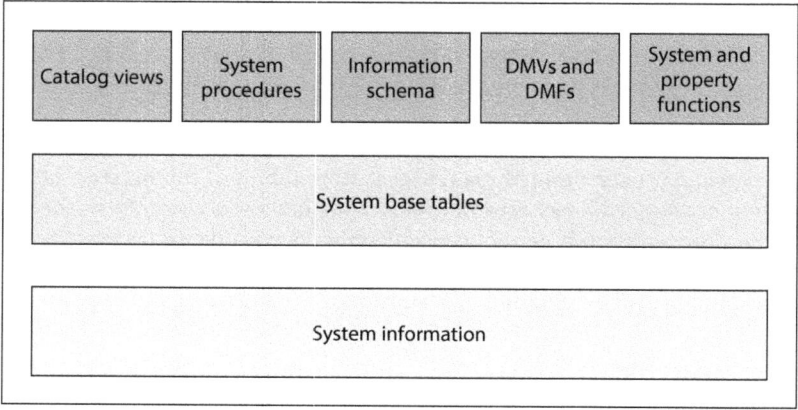

Figure 9-1   Graphical presentation of different interfaces for the system catalog

These interfaces can be grouped in two groups: *general* interfaces (catalog views, DMVs and DMFs, and the information schema), and *proprietary* interfaces in relation to the Database Engine (system stored procedures and system and property functions).

NOTE "General" means that all relational database systems support such interfaces, but use different terminology. For instance, in Oracle's terminology, catalog views and DMVs are called "data dictionary views" and "V$ views," respectively.

The following section describes general interfaces. Proprietary interfaces are discussed later in the chapter.

General Interfaces

As already stated, the following interfaces are general interfaces:

- Catalog views
- DMVs and DMFs
- Information schema

Catalog Views

Catalog views are the most general interface to the metadata and provide the most efficient way to obtain customized forms of this information (see Examples 9.1 through 9.3).

Catalog views belong to the **sys** schema, so you have to use the schema name when you access one of the objects. This section describes the three most important catalog views:

- sys.objects
- sys.columns
- sys.database_principals

NOTE You can find the description of other views either in different chapters of this book or in Microsoft Docs.

The **sys.objects** catalog view contains a row for each user-defined object in relation to the user's schema. There are two other catalog views that show similar information: **sys.system_objects** and **sys.all_objects**. The former contains a row for each system object, while the latter shows the union of all schema-scoped user-defined objects and system objects. (All three catalog views have the same structure.) Table 9-1 lists and describes the most important columns of the **sys.objects** catalog view.

The **sys.columns** catalog view contains a row for each column of an object that has columns, such as tables and views. Table 9-2 lists and describes the most important columns of the **sys.columns** catalog view.

The **sys.database_principals** catalog view contains a row for each security principal (that is, user, group, or role in a database). Table 9-3 lists and describes the most important columns of the **sys.database_principals** catalog view. (For a detailed discussion of principal types, see Chapter 12.)

Column Name	Description
name	The object name
object_id	The object identification number, unique within a database
schema_id	The ID of the schema in which the object is contained
type	The object type, which could be a table, view, or any other database object that is supported through the Database Engine

Table 9-1   Selected Columns of the sys.objects Catalog View

Column Name	Description
object_id	The ID of the object to which this column belongs
name	The column name
column_id	The ID of the column (unique within the object)

Table 9-2   Selected Columns of the sys.columns Catalog View

Column Name	Description
name	The name of principal
principal_id	The ID of the principal (unique within the database)
type	The principal type

Table 9-3   Selected Columns of the sys.database_principals Catalog View

Querying Catalog Views

As already stated in this chapter, all system tables have the same structure as base tables. Because system tables cannot be referenced directly, you have to query catalog views, which correspond to particular system tables. (Views can be queried in the same way as user tables.) Examples 9.1 through 9.3 use existing catalog views to demonstrate how information concerning database objects can be queried.

Example 9.1

Get the table ID, user ID, and table type of the **employee** table:

```
USE sample;
SELECT object_id, principal_id, type
    FROM sys.objects
    WHERE name = 'employee';
```

The result is

object_id	principal_id	type
530100929	NULL	U

The **object_id** column of the **sys.objects** catalog view displays the unique ID number for the corresponding database object. The NULL value in the **principal_id** column indicates that the object's owner is the same as the owner of the schema. *U* in the **type** column stands for the user (table).

Example 9.2

Get the names of all tables of the **sample** database that contain the **project_no** column:

```
USE sample;
SELECT sys.objects.name
  FROM sys.objects INNER JOIN sys.columns
  ON sys.objects.object_id = sys.columns.object_id
   WHERE sys.objects.type = 'U'
   AND sys.columns.name = 'project_no';
```

The result is

Name
project
works_on

Example 9.3

Who is the owner of the **employee** table?

```
SELECT sys.database_principals.name
FROM sys.database_principals INNER JOIN sys.objects
ON sys.database_principals.principal_id = sys.objects.schema_id
WHERE sys.objects.name = 'employee'
AND sys.objects.type = 'U';
```

The result is

name
dbo

Dynamic Management Views and Functions

Dynamic management views (DMVs) and functions (DMFs) return server state information that can be used to observe active processes and therefore to tune system performance or to monitor the actual system state. In contrast to catalog views, the DMVs and DMFs are based on internal structures of the system.

NOTE The main difference between catalog views and DMVs is in their application: catalog views display the static information about metadata, while DMVs (and DMFs) are used to access dynamic properties of the system. In other words, you use DMVs to get insightful information about the database, individual queries, or an individual user.

DMVs and DMFs belong to the **sys** schema and their names start with the prefix **dm_**, followed by a text string that indicates the category to which the particular DMV or DMF belongs.

The following list identifies and describes some of these categories:

- **sys.dm_db_*** Contains information about databases and their objects
- **sys.dm_tran_*** Contains information in relation to transactions
- **sys.dm_io_*** Contains information about I/O activities
- **sys.dm_exec_*** Contains information related to the execution of user code

NOTE The functionality and application areas of DMVs and DMFs are identical. (Only the syntax is slightly different.) For this reason, I will use "DMV" as a common name for both of them.

Information Schema

The information schema consists of read-only views that provide information about all tables, views, and columns of the Database Engine to which you have access. In contrast to the system catalog that manages the metadata applied to the system as a whole, the information schema primarily manages the environment of a database.

NOTE The information schema was originally introduced in the SQL92 standard. The Database Engine provides information schema views so that applications developed on other database systems can obtain its system catalog without having to use it directly. These standard views use different terminology, so when you interpret the column names, be aware that *catalog* is a synonym for *database* and that *domain* is a synonym for *user-defined data type*.

The following sections provide a description of the most important information schema views.

information_schema.tables

The **information_schema.tables** view contains one row for each table in the current database to which the user has access. The view retrieves the information from the system catalog using the **sys.objects** catalog view. Table 9-4 lists and describes the four columns of this view.

Column	Description
TABLE_CATALOG	The name of the catalog (database) to which the view belongs
TABLE_SCHEMA	The name of the schema to which the view belongs
TABLE_NAME	The table name
TABLE_TYPE	The type of the table (can be BASE TABLE or VIEW)

Table 9-4 The information_schema.tables View

Column	Description
TABLE_CATALOG	The name of the catalog (database) to which the column belongs
TABLE_SCHEMA	The name of the schema to which the column belongs
TABLE_NAME	The name of the table to which the column belongs
COLUMN_NAME	The column name
ORDINAL_POSITION	The ordinal position of the column
DATA_TYPE	The data type of the column

Table 9-5   The information_schema.columns View

information_schema.columns

The **information_schema.columns** view contains one row for each column in the current database accessible by the user. The view retrieves the information from the **sys.columns** and **sys .objects** catalog views. Table 9-5 lists and describes the six most important columns of this view.

Example 9.4 shows the use of the **information_schema.tables** and **information_schema .columns** views. The query is equivalent to the query from Example 9.2.

Example 9.4

Get the names of all tables of the **sample** database that contain the **project_no** column:

```
USE sample;
SELECT   t.table_name
    FROM information_schema.tables t INNER JOIN
         information_schema.columns c
    ON t.table_name = c.table_name
    AND c.column_name = 'project_no';
```

Proprietary Interfaces

The previous section describes the use of the general interfaces for accessing system base tables. (As a reminder, "general" means that all relational database systems support such interfaces.) You can also retrieve system information using one of the following proprietary mechanisms of the Database Engine:

- System stored procedures
- System functions
- Property functions

The following sections describe these interfaces.

System Stored Procedures

System stored procedures are used to provide many administrative and end-user tasks, such as renaming database objects, identifying users, and monitoring authorization and resources.

Almost all existing system stored procedures access system base tables to retrieve and modify system information.

NOTE The most important property of system stored procedures is that they can be used for easy and reliable modification of system base tables.

This section describes two system stored procedures: **sp_help** and **sp_configure**. Depending on the subject matter of the chapters, certain system stored procedures were discussed in previous chapters, and additional procedures will be discussed in later chapters of the book.

The **sp_help** system stored procedure displays information about one or more database objects. The name of any database object or data type can be used as a parameter of this procedure. If **sp_help** is executed without any parameter, information on all database objects of the current database will be displayed.

The **sp_configure** system stored procedure displays or changes global configuration settings for the current server.

Example 9.5 shows the use of the **sp_configure** system stored procedure.

Example 9.5

```
USE sample;
EXEC sp_configure 'show advanced options' , 1;
RECONFIGURE WITH OVERRIDE;
EXEC sp_configure 'fill factor', 100;
RECONFIGURE WITH OVERRIDE;
```

Generally, you do not have access to advanced configuration options of the Database Engine. For this reason, the first EXECUTE statement in Example 9.5 tells the system to allow changes of advanced options. With the next statement, RECONFIGURE WITH OVERRIDE, these changes will be installed. Now it is possible to change any of the existing advanced options. Example 9.5 changes the fill factor to 100 and installs this change. (Fill factor specifies the storage percentage for index pages and will be described in detail in Chapter 10.)

System Functions

System functions are described in Chapter 5. Some of them can be used to access system base tables. Example 9.6 shows two SELECT statements that retrieve the same information using different interfaces.

Example 9.6

```
USE sample;
SELECT object_id
  FROM sys.objects
  WHERE name = 'employee';
SELECT object_id('employee');
```

The second SELECT statement in Example 9.6 uses the system function **object_id** to retrieve the ID of the **employee** table. (This information can be stored in a variable and used when calling a command, or a system stored procedure, with the object's ID as a parameter.)

The following system functions, among others, access system base tables. The names of these functions are self-explanatory.

- OBJECT_ID(object_name)
- OBJECT_NAME(object_id)
- USER_ID([user_name])
- USER_NAME([user_id])
- DB_ID([db_name])
- DB_NAME([db_id])

Property Functions

Property functions return properties of database objects, data types, or files. Generally, property functions can return more information than system functions can return, because property functions support dozens of properties (as parameters), which you can specify explicitly.

Almost all property functions return one of the following three values: 0, 1, or NULL. If the value is 0, the object does not have the specified property. If the value is 1, the object has the specified property. Similarly, the value NULL specifies that the existence of the specified property for the object is unknown to the system.

The Database Engine supports, among others, the following property functions:

- OBJECTPROPERTY(id, property)
- COLUMNPROPERTY(id, column, property)
- FILEPROPERTY(filename, property)
- TYPEPROPERTY(type, property)

The OBJECTPROPERTY function returns information about objects in the current database (see Exercise E.9.2). The COLUMNPROPERTY function returns information about a column or procedure parameter. The FILEPROPERTY function returns the specified filename and property value for a given filename and property name. The TYPEPROPERTY function returns information about a data type. (The description of existing properties for each property function can be found in Microsoft Docs.)

Summary

The system catalog is a collection of system base tables belonging to the **master** database and existing user databases. Generally, system base tables cannot be queried directly by a user. The Database Engine supports several different interfaces that you can use to access the information from the system catalog. Catalog views are the most general interface that you can apply to obtain system information. Dynamic management views (DMVs) and functions (DMFs) are similar to catalog views, but you use them to access dynamic properties of the system. System stored procedures provide easy and reliable read and write access to system base tables. It is strongly recommended to exclusively use system stored procedures for modification of system information.

The information schema is a collection of views defined on system base tables that provides unified access to the system catalog for all database applications developed on other database systems. The use of the information schema is recommended if you intend to port your databases from one database system to another.

The next chapter introduces you to database indices.

Exercises

E.9.1   Using catalog views, find the operating system path and filename of the **sample** database.

E.9.2   Using catalog views, find how many integrity constraints are defined for the **employee** table of the **sample** database.

E.9.3   Using catalog views, find out if there is any integrity constraint defined for the **dept_no** column of the **employee** table.

E.9.4   Using the information schema, display all user tables that belong to the **AdventureWorks** database.

E.9.5   Using the information schema, find all columns of the **employee** table with their ordinal positions and the corresponding data types.

Indices

In This Chapter

- Introduction to Indices
- Transact-SQL and Indices
- Guidelines for Creating and Using Indices
- Special Types of Indices

This chapter describes indices and their role in optimizing the response time of queries. The first part of the chapter discusses how indices are stored and describes the existing forms of them. The main part of the chapter explains three Transact-SQL statements pertaining to indices: CREATE INDEX, ALTER INDEX, and DROP INDEX. After that, index fragmentation and its impact on the performance of the system will be explained. The next part of the chapter gives you several general recommendations for how and when indices should be created. The final part of the chapter describes several special types of indices.

Introduction to Indices

Database systems generally use indices to provide fast access to relational data. An index is a separate physical data structure that enables queries to access one or more data rows fast. Proper tuning of indices is therefore a key for good query performance.

An index is in many ways analogous to a book index. When you are looking for a topic in a book, you use its index to find the page(s) where that topic is described. Similarly, when you search for a row of a table, the Database Engine uses an index to find the row's physical location. However, there are two main differences between a book index and a database index:

- As a book reader, you can decide whether or not to use the book's index. This possibility generally does not exist if you use a database system: the system component called the query optimizer decides whether or not to use an existing index. (A user can manipulate the use of indices by using index hints, but their use is recommended only in a few special cases. Index hints are described in Chapter 19.)

• A particular book's index is edited together with the book and does not change at all. This means that you can find a topic exactly on the page where it is determined in the index. By contrast, a database index can change each time the corresponding data is changed.

If a table does not have an appropriate index, the database system uses the table scan method to retrieve rows. *Table scan* means that each row is retrieved and examined in sequence (from first to last) and returned in the result set if the search condition in the WHERE clause evaluates to TRUE. Therefore, all rows are fetched according to their physical memory location. This method is less efficient than an index access, as explained next.

Indices are stored in additional data structures called *index pages*. (The structure of index pages is very similar to the structure of data pages, as you will see in Chapter 15.) For each indexed row there is an *index entry*, which is stored in an index page. Each index entry consists of the index key plus a pointer. For this reason, each index entry is significantly shorter than the corresponding row. Therefore, the number of index entries per (index) page is significantly higher than the number of rows per (data) page. This index property plays a very important role, because the number of I/O operations required to traverse the index pages is significantly lower than the number of I/O operations required to traverse the corresponding data pages. In other words, a table scan would probably result in many more I/O operations than a corresponding index access would.

The Database Engine's indices are constructed using the B+-tree data structure. As its name suggests, a B+-tree has a treelike structure in which all of the bottommost nodes (leaf nodes) are the same number of levels away from the top (root node) of the tree. This property is maintained even when new data is added to or deleted from the indexed column.

Figure 10-1 illustrates the structure of the B+-tree and the direct access to the row of the **employee** table with the value 25348 in its **emp_no** column. (It is assumed that the **employee** table has an index on the **emp_no** column.) You can also see that each B+-tree consists of a root node, leaf nodes, and zero or more intermediate nodes.

Searching for the data value 25348 can be executed as follows: Starting from the root of the B+-tree, a search proceeds for a lowest key value greater than or equal to the value to be retrieved. Therefore, the value 29346 is retrieved from the root node; then the value 28559 is fetched from the intermediate level, and the searched value, 25348, is retrieved at the leaf level. With the help of the respective pointers, the appropriate row is retrieved. (An alternative, but equivalent, search method would be to search for smaller or equal values.)

Index access is generally the preferred and obviously advantageous method for accessing tables with many rows. With index access, it takes only a few I/O operations to find any row of a table in a very short time, whereas sequential access (i.e., table scan) requires much more time to find a row physically stored at the end of the table.

The two existing index types, clustered and nonclustered indices, are described next, after which you will find out how to create an index.

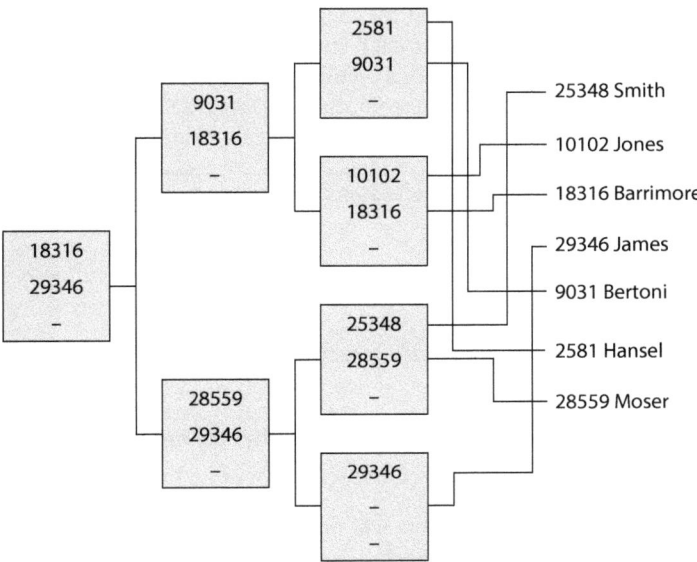

Figure 10-1   B⁺-tree for the emp_no column of the employee table

Clustered Indices

A clustered index determines the physical order of the data in a table. The Database Engine allows the creation of a single clustered index per table, because the rows of the table cannot be physically ordered more than one way. When using a clustered index, the system navigates down from the root of the B⁺-tree structure to the leaf nodes, which are linked together in a doubly linked list called a *page chain*. The important property of a clustered index is that its leaf pages contain data pages. (All other levels of a clustered index structure are composed of index pages.) If a clustered index is (implicitly or explicitly) defined for a table, the table is called a *clustered table*. Figure 10-2 shows the B⁺-tree structure of a clustered index.

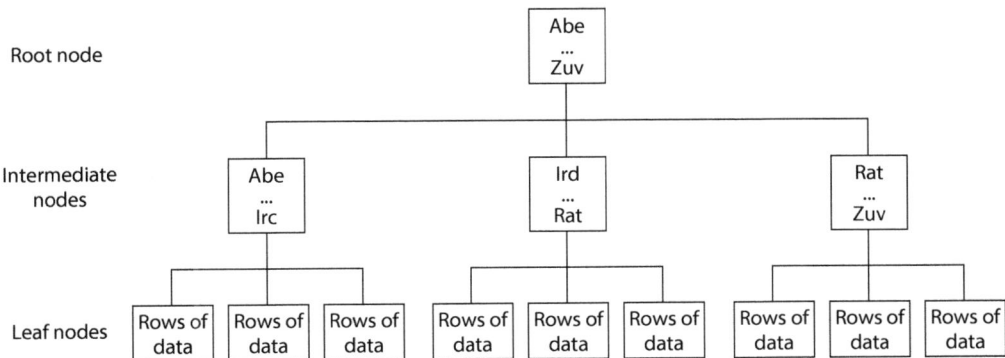

Figure 10-2   Physical structure of a clustered index

A clustered index is built by default for each table for which you define the primary key using the primary key constraint. Also, each clustered index is unique by default—that is, each data value can appear only once in a column for which the clustered index is defined. If a clustered index is built on a nonunique column, the Database Engine will force uniqueness by adding a 4-byte identifier to the rows that have duplicate values.

NOTE Clustered indices allow very fast access in cases where a query searches for a range of values (see Chapter 19).

Nonclustered Indices

A nonclustered index has the same index structure as a clustered index, with two important differences:

- A nonclustered index does not change the physical order of the rows in the table.
- The leaf pages of a nonclustered index consist of an index key plus a bookmark.

The physical order of rows in a table will not be changed if one or more nonclustered indices are defined for that table. For each nonclustered index, the Database Engine creates an additional index structure that is stored in index pages.

A bookmark of a nonclustered index shows where to find the row corresponding to the index key. The bookmark part of the index key can have two forms, depending on the form of the table—that is, the table can be a clustered table or a heap. (In Microsoft terminology, a *heap* is a table without a clustered index.) If a clustered index exists, the bookmark of the nonclustered index shows the B+-tree structure of the table's clustered index. If the table has no clustered index, the bookmark is identical to the row identifier (RID), which contains three parts: the address of the file to which the corresponding table belongs, the address of the physical block (page) in which the row is stored, and the offset, which is the position of the row inside the page.

As the preceding discussion indicates, searching for data using a nonclustered index could proceed in either of two different ways, depending on the form of the table:

- **Heap**   Traversal of the nonclustered index structure is followed by the retrieval of the row using the RID.
- **Clustered table**   Traversal of the nonclustered index structure is followed by traversal of the corresponding clustered index.

In both cases, the number of I/O operations is quite high, so you should design a nonclustered index with care and only when you are sure that there will be significant performance gains by using it. Figure 10-3 shows the B+-tree structure of a nonclustered index.

Transact-SQL and Indices

Now that you are familiar with the physical structure of indices, this section describes how you can create, alter, and drop indices; obtain index fragmentation information; and edit index

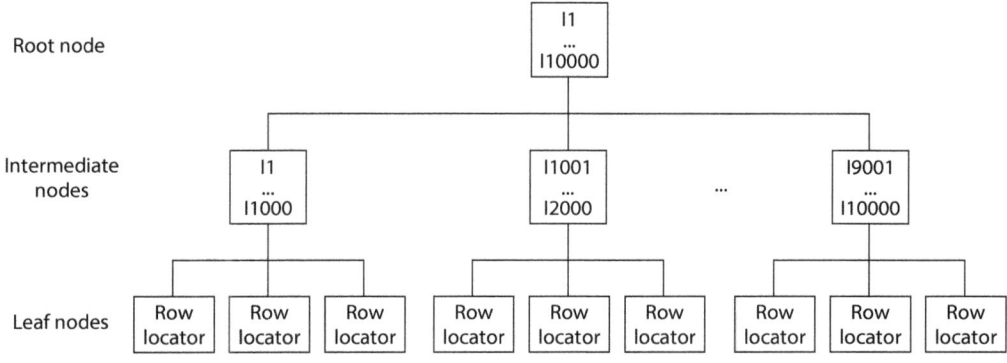

Row locator = RID or pointer to clustered index log

Figure 10-3   Structure of a nonclustered index

information; all of which will prepare you for the subsequent discussion of how you can use indices to improve performance of the system.

Creating Indices

The CREATE INDEX statement creates an index for the particular table. The general form of this statement is

```
CREATE [UNIQUE] [CLUSTERED |NONCLUSTERED] INDEX index_name
    ON table_name (column1 [ASC | DESC] ,...)
           [ INCLUDE ( column_name [ ,... ] ) ]
 [WITH
       [FILLFACTOR=n]
       [[, ] PAD_INDEX = {ON | OFF}]
       [[, ] DROP_EXISTING = {ON | OFF}]
      [[, ] SORT_IN_TEMPDB = {ON | OFF}]
      [[, ] IGNORE_DUP_KEY = {ON | OFF}]
              [[, ] ALLOW_ROW_LOCKS = {ON | OFF}]
              [[, ] ALLOW_PAGE_LOCKS = {ON | OFF}]
      [[, ] STATISTICS_NORECOMPUTE = {ON | OFF}]
              [[, ] ONLINE = {ON | OFF}]]
              [ON file_group | "default"]
```

index_name identifies the name of the created index. An index can be established for one or more columns of a single table (**table_name**). **column1** is the name of the column for which the index is created. (As you can see from the syntax of the CREATE INDEX statement, you can specify an index for several columns of a table.) The Database Engine supports indices on views too. Such views, called *indexed views*, are discussed in Chapter 26.

NOTE Each column of a table can be indexed. This means that columns with VARBINARY(max), BIGINT, and SQL_VARIANT data types can be indexed too.

An index can be either single or composite. A single index has one column, whereas a composite index is built on more than one column. Each composite index has certain restrictions concerning its length and number of columns. The maximum size of an index is 900 bytes, while the index can contain up to 16 columns.

The UNIQUE option specifies that each data value can appear only once in an indexed column. For a unique composite index, the combination of data values of all columns in each row must be unique. If UNIQUE is not specified, duplicate values in the indexed column(s) are allowed.

The CLUSTERED option specifies a clustered index. The NONCLUSTERED option (the default) specifies that the index does not change the order of the rows in the table. The Database Engine allows a maximum of 249 nonclustered indices per table.

The Database Engine also supports indices with descending order on column values. The ASC option after the column name specifies that the index is created on the ascending order of the column's values, while DESC specifies the descending order. This gives you more flexibility for using an index. Descending indices should be used when you create a composite index on columns that have opposite sorting directions.

The INCLUDE option allows you to specify the nonkey columns, which are added to the leaf pages of the nonclustered index. Column names cannot be repeated in the INCLUDE list and cannot be used simultaneously as both key and nonkey columns. To understand the benefit of the INCLUDE option, you have to know what a *covering index* is. Significant performance gains can be achieved when *all* columns in a query are included in the index, because the query optimizer can locate all the column values within the index pages without having to access pages with table data. This feature is called a *covering index* or *covered query*. So, if you include additional nonkey columns in the leaf pages of the nonclustered index, more queries will be covered and their performance will be significantly better. (Further discussion of this topic, as well as an example of how the query optimizer handles a covering index, can be found later in this chapter in the section "Covering Index.")

FILLFACTOR=n defines the storage percentage for each index page at the time the index is created. You can set the value of FILLFACTOR from 1 to 100. If the value of **n** is set to 100, each index page will be 100 percent filled—that is, the existing index leaf pages as well as nonleaf pages will have no space for the insertion of new rows. Therefore, this value is recommended only for static tables. (The default value, 0, also indicates that the leaf index pages are filled and the intermediate nonleaf pages contain one free entry each.)

If you set the FILLFACTOR option to a value between 1 and 99, the new index structure will be created with leaf pages that are not completely full. The bigger the value of FILLFACTOR, the smaller the space that is left free on an index page. For instance, setting FILLFACTOR to 60 means that 40 percent of each leaf index page is left free for future insertion of index rows. (Index rows will be inserted when you execute either the INSERT statement or the UPDATE statement.) For this reason, the value 60 could be a reasonable value for tables with rather frequent data modification. For all values of the FILLFACTOR option between 1 and 99, the intermediate nonleaf pages contain one free entry each.

NOTE   The FILLFACTOR value is not maintained—that is, it specifies only how much storage space is reserved with the existing data at the time the storage percentage is defined. If you want to reestablish the original value of the FILLFACTOR option, you need to use the ALTER INDEX statement, which is described later in this chapter.

The PAD_INDEX option is tightly connected to the FILLFACTOR option. The FILLFACTOR option mainly specifies the percentage of space that is left free on leaf index pages. The PAD_INDEX option specifies that the FILLFACTOR setting should be applied to the index pages as well as to the data pages in the index.

The DROP_EXISTING option allows you to enhance performance when re-creating a clustered index on a table that also has a nonclustered index. See the section "Rebuilding an Index" later in the chapter for more details.

The SORT_IN_TEMPDB option is used to place into the **tempdb** system database the data from intermediate sort operations used while creating the index. This can result in a performance benefit if the **tempdb** database is placed on another disk drive from the data itself.

The IGNORE_DUP_KEY option causes the system to ignore the attempt to insert duplicate values in the indexed column(s). This option should be used only to avoid the termination of a long transaction in cases where the INSERT statement inserts duplicate data in the indexed column(s). If this option is activated and an INSERT statement attempts to insert rows that would violate the uniqueness of the index, the database system returns a warning rather than causing the entire statement to fail. The Database Engine does *not* insert the rows that would add duplicate key values; it merely ignores those rows and adds the rest. (If this option is not set, the statement as a whole will be aborted.)

The ALLOW_ROW_LOCKS option specifies that the system uses row locks when this option is activated (set to ON). Similarly, the ALLOW_PAGE_LOCKS option specifies that the system uses page locks when this option is set to ON. (For the description of page and row locks, see Chapter 13.)

The STATISTICS_NORECOMPUTE option specifies that statistics of the specified index should not be automatically recomputed. (Statistics will be explained in Chapter 19.) The ON option creates the specified index either on the default file group (**"default"**) or on the specified file group (**file_group**).

If you activate the ONLINE option, you can create, rebuild, or drop an index online. This option allows concurrent modifications to the underlying table or clustered index data and any associated indices during index execution. For example, while a clustered index is being rebuilt, you can continue to make updates to the underlying data and perform queries against the data.

NOTE Before you start to execute queries in this chapter, re-create the entire **sample** database.

Example 10.1 shows the creation of a nonclustered index.

Example 10.1
Create an index for the **emp_no** column of the **employee** table:

```
USE sample;
CREATE INDEX i_empno ON employee (emp_no);
```

Example 10.2 shows the creation of a unique composite index.

Example 10.2
Create a composite index for the columns **emp_no** and **project_no** on the **works_on** table. The compound values in both columns must be unique. Eighty percent of each index leaf page should be filled.

```
USE sample;
CREATE UNIQUE INDEX i_empno_prno
        ON works_on (emp_no, project_no)
        WITH FILLFACTOR=80;
```

The creation of a unique index for a column is not possible if the column already contains duplicate values. The creation of such an index is possible if each existing data value (including the NULL value) occurs only once. Also, any attempt to insert an existing data value into, or modify an existing data value within, a column with an existing unique index will be rejected by the system.

Editing Information Concerning Indices

You can use the following system features to edit information concerning indices:

- **sys.indexes** catalog view
- **sys.index_columns** catalog view
- **sp_helpindex** system procedure
- OBJECTPROPERTY property function
- SQL Server Management Studio
- **sys.dm_db_index_usage_stats** DMV
- **sys.dm_db_missing_index_details** DMV (explained later in the chapter in the section "Missing Indices")

The **sys.indexes** catalog view contains a row for each index and a row for each table without a clustered index. The most important columns of this view are **object_id**, **name**, and **index_id**. **object_id** is the name of the database object to which the index belongs, while **name** and **index_id** are the name and the ID of that index, respectively.

The **sys.index_columns** catalog view contains a row per column that is part of an index or a heap. This information can be used together with the information from the **sys.indexes** catalog view to obtain further properties of a specific index.

The **sp_helpindex** system procedure displays all indices on a table as well as column statistics. The syntax of this procedure is

```
sp_helpindex [@db_object = ] 'name'
```

where **db_object** is the name of a table.

The OBJECTPROPERTY property function has two properties in relation to indices: **IsIndexed** and **IsIndexable**. The former informs you whether a table or view has an index, while the latter specifies whether a table or view can be indexed.

To edit information about an existing index using SQL Server Management Studio, choose the database in the Databases folder and expand the Tables folder. Expand the Indexes folder. The list of all existing indices for that table is shown. After you double-click one of the indices, the system shows you the Index Properties dialog box with all properties of that index. (You can also use Management Studio to create a new index or drop an existing one.)

The **sys.dm_db_index_usage_stats** dynamic management view (DMV) returns counts of different types of index operations and the time each type of operation was last performed. Every individual seek, lookup, or update on the specified index by one query execution is counted as a use of that index and increments the corresponding counter in this DMV. That way you can get general information about how often an index is used to determine which indices are used more heavily than the others.

Editing Information Concerning Index Fragmentation

During the life cycle of an index, it can become *fragmented*, meaning the storage of data in its pages is done inefficiently. There are two forms of index fragmentation: internal and external. Internal fragmentation occurs when there is empty space in the index page, which can happen due to write operations. External fragmentation occurs when the logical order of the pages is wrong.

To get information concerning internal index fragmentation, you use the DMV called **sys .dm_db_index_physical_stats**. This DMV returns size and fragmentation information for the data and indices of the specified table. For each index, one row is returned for each level of the B+-tree. Using this DMV, you can obtain information about the degree of fragmentation of rows on data pages. You can use this information to decide whether reorganization of the data is necessary. (The next section, "Altering Indices," explains how you can reorganize a fragmented index.)

Example 10.3 shows how you can use the **sys.dm_db_index_physical_stats** view. The view displays the average amount of fragmentation of all indices of the **employee** table.

Example 10.3

```
DECLARE @db_id INT;
DECLARE @tab_id INT;
DECLARE @ind_id INT;
SET @db_id = DB_ID('sample');
SET @tab_id = OBJECT_ID('employee');
SELECT avg_fragmentation_in_percent,
 FROM sys.dm_db_index_physical_stats
(@db_id, @tab_id, NULL, NULL, NULL)
```

As you can see from Example 10.3, the **sys.dm_db_index_physical_stats** view has five parameters. The first three specify the IDs of the current database, table, and index, respectively. The fourth specifies the partition ID (see Chapter 26), and the last one specifies the scan level that is used to obtain statistics. (You can always use NULL to specify the default value of the particular parameter.)

The **sys.dm_db_index_physical_stats** view has several columns, of which **avg_ fragmentation_in_percent** and **avg_page_space_used_in_percent** are the most important. The former specifies the average fragmentation in percent, while the latter defines the percentage of the used space.

Altering Indices

The Database Engine is one of a few database systems that support the ALTER INDEX statement. This statement can be used for index maintenance activities. The syntax of the ALTER INDEX statement is very similar to the syntax of the CREATE INDEX statement. In other words, this statement allows you to change the settings for the options ALLOW_ROW_ LOCKS, ALLOW_PAGE_LOCKS, IGNORE_DUP_KEY, and STATISTICS_NORECOMPUTE, previously described in relation to the CREATE INDEX statement.

In addition to the preceding options, the ALTER INDEX statement supports three other activities:

- Reorganizing leaf index pages using the REORGANIZE option
- Disabling an index using the DISABLE option
- Rebuilding an index using the REBUILD option

Rebuilding an index can also be accomplished without the ALTER INDEX statement, by using the DROP_EXISTING option of the CREATE INDEX statement. Both rebuilding methods are described and compared in the upcoming "Rebuilding an Index" subsection.

Reorganizing Leaf Index Pages

The REORGANIZE option of the ALTER INDEX statement specifies that the leaf pages of the corresponding index structure will be reorganized so that the physical order of the pages matches the left-to-right logical order of the leaf nodes. Therefore, this option removes some of the fragmentation from an index, thus improving performance.

Disabling an Index

The DISABLE option of the ALTER INDEX statement disables an existing index. Each disabled index is unavailable for use until you enable it again. Note that a disabled index won't be maintained as changes to the corresponding data are made. For this reason, indices must be completely rebuilt if you want to use them again. To enable a disabled index, use the REBUILD option of the ALTER TABLE statement.

NOTE If you disable the clustered index of a table, the data won't be available, because all data pages are stored in the leaf level of the clustered index.

Rebuilding an Index

When you perform any data modifications using an INSERT, UPDATE, or DELETE statement, data fragmentation can occur. If the data is indexed, index fragmentation can occur as well, and the information in the index can get scattered on different physical pages. Fragmented index data can cause the Database Engine to perform additional data reads, which decreases the overall performance of the system. In such a case, you have to rebuild all fragmented indices.

There are two ways in which you can rebuild an index:

- Use the REBUILD option of the ALTER INDEX statement
- Use the DROP_EXISTING option of the CREATE INDEX statement

With the REBUILD option, you can rebuild an index. If you specify ALL instead of an index name, all indices of the table will be rebuilt. (By allowing indices to be rebuilt dynamically, you don't have to drop and re-create them.)

The DROP_EXISTING option of the CREATE INDEX statement allows you to enhance performance when re-creating a clustered index on a table that also has nonclustered indices. It specifies that the existing clustered or nonclustered index should be dropped and the specified index rebuilt. As you already know, each nonclustered index in a clustered table contains in its leaf nodes the corresponding values of the table's clustered index. For this reason, all nonclustered indices must be rebuilt when a table's clustered index is dropped. Using the DROP_EXISTING option, you can prevent the nonclustered indices from being rebuilt twice.

NOTE The DROP_EXISTING option is more powerful than REBUILD, because it is more flexible and offers several options, such as changing the columns that make up the index and changing a nonclustered index to a clustered one.

Creation of Resumable Online Indices

The creation of an index for a big table, especially if the index is a composite one, can significantly impact performance of the whole database system. Additionally, the longer the creation of the index takes, the longer users are affected. In versions prior to SQL Server 2019, the only solution to this problem is to kill the index creation process. In such a case, the system rolls back the process and you have to start it again, from the beginning.

SQL Server 2019 supports the creation of resumable online indices. This feature allows you to pause the build and then restart it later at the point it was paused. The new option of the CREATE INDEX called RESUMABLE allows you to create a resumable online index.

Example 10.4 creates a table that will be used to show the creation of a resumable online index.

Example 10.4

```
USE AdventureWorks;
GO
SELECT    PersonID,
          StoreID,
          TerritoryID,
          AccountNumber,
          ModifiedDate
     INTO  sample.dbo.My_Customers
     FROM AdventureWorks.Sales.Customer;
GO
USE sample;
INSERT INTO My_Customers
SELECT
              PersonID,
              StoreID,
              TerritoryID,
              AccountNumber,
              ModifiedDate
FROM AdventureWorks.Sales.Customer
GO 80
```

In the first part of Example 10.4, the SELECT ... INTO statement uses the **Customers** table of the **AdventureWorks** database to create and load data into the **My_Customers** table of the **sample** database. After the execution of the SELECT ... INTO statement, the structure and the content of both tables is identical. (For a detailed explanation of the SELECT ... INTO statement, see Example 6.44.)

In the second part of the example, the new table is loaded again with the content of the **Customers** table of the **AdventureWorks** database, but this time using the INSERT statement. The final statement, GO 80, executes the INSERT statement 80 times. That way, a table with 1,605,420 (81 × 19,820) rows is created.

Example 10.5 comprises two statements: CREATE INDEX and ALTER INDEX. These two statements have to be executed in the given order and in two separate windows.

Example 10.5

```
USE sample;
GO
  CREATE INDEX I_MyCustomers
  ON My_Customers
      (PersonID, StoreID, TerritoryID, AccountNumber,ModifiedDate)
        WITH (RESUMABLE = ON, ONLINE = ON);
USE sample;
GO
ALTER INDEX I_MyCustomers ON My_Customers
  PAUSE;
GO
```

The first statement of Example 10.5, CREATE INDEX, creates a composite nonclustered index for the **My_Customers** table. The index comprises five columns of the table. The RESUMABLE option of the WITH clause is set to ON. That way the **I_MyCustomers** index is specified as resumable.

NOTE The RESUMABLE option cannot be set to ON when the ONLINE option is set to OFF. In other words, a resumable index is only supported with online operations. (For discussion of the ONLINE option, see the earlier section "Creating Indices.")

Now, the second statement of Example 10.5 (ALTER INDEX) must be started as fast as possible in another window. This statement uses the PAUSE clause to temporarily stop the creation of the resumable online index. If the PAUSE option is executed successfully, the process of creating the resumable online index in the first window will be stopped with the following error messages:

```
Msg 1219, Level 16, State 1, Line 3
Your session has been disconnected because of a high priority DDL operation.
Msg 1219, Level 16, State 1, Line 3
Your session has been disconnected because of a high priority DDL operation.
Msg 596, Level 21, State 1, Line 2
```

```
Cannot continue the execution because the session is in the kill state.
Msg 0, Level 20, State 0, Line 2
A severe error occurred on the current command.  The results, if any, should be
discarded.
```

Besides the RESUMABLE option, the CREATE INDEX statement supports an additional clause called MAX_DURATION. You use this option to specify the maximum time interval for building a resumable online index. Once the specified time is up, the index build automatically gets paused if it has not completed.

You can check the percentage of index creation using the **sys.index_resumable_operations** system view. This view monitors and checks the current execution status for a resumable online index. Example 10.6 shows the selection of three columns (among others) of this system view.

Example 10.6
```
SELECT name, percent_complete, state
    FROM sys.index_resumable_operations;
```

The result is

name	percent_complete	state
I_MyCustomers	50	1

The **name** column displays the name of the resumable index, **percent_complete** displays the percentage of index creation that has completed, and **state** displays the operational state for the resumable index (0 means "running" and 1 means "paused"). As you can see from the result of Example 10.6, the creation of the **I_MyCustomers** index has been paused and the index creation operation is 50 percent complete.

Now, the ALTER INDEX statement can be applied once more to resume the creation of the index, as shown in Example 10.7.

Example 10.7
```
ALTER INDEX I_MyCustomers ON My_Customers  RESUME;
```

Besides the PAUSE and RESUME options, the ALTER INDEX statement supports an additional clause called ABORT. Using this option you can permanently cancel the already started creation of an online index.

Removing and Renaming Indices

The DROP INDEX statement removes one or more existing indices from the current database. Note that removing the clustered index of a table can be a very resource-intensive operation, because all nonclustered indices will have to be rebuilt. (All the nonclustered indices use the index key of the clustered index as a pointer in their leaf index pages.) Example 10.8 shows how the **i_empno** index can be dropped.

Example 10.8
Remove the index created in Example 10.1:

```
USE sample;
DROP INDEX i_empno ON employee;
```

The DROP INDEX statement has an additional option, MOVE TO, which is analogous to the ON option of CREATE INDEX. In other words, you can use this option to specify a location to which to move the data rows that are currently in the leaf pages of the clustered index. The data is moved to the new location in the form of a heap. You can specify either a default or named file group as the new location.

NOTE The DROP INDEX statement cannot be used to remove indices that are implicitly generated by the system for integrity constraints, such as PRIMARY KEY or UNIQUE. To remove such indices, you must drop the constraint.

The **sp_rename** system procedure, which is discussed in Chapter 5, can be used to rename indices.

NOTE You can create, alter, and drop indices using SQL Server Management Studio, too. To manage indices inside Management Studio, you can use Object Explorer. The simplest way is to use the Indexes node of a particular table. Index management with Management Studio is analogous to table management with the same tool. (For details, see Chapter 3.)

Guidelines for Creating and Using Indices

Although the Database Engine does not have any practical limitations concerning the number of indices, it is advisable to limit them, for a couple of reasons. First, each index uses a certain amount of disk space, so it is possible that the total number of index pages could exceed the number of data pages within a database. Second, in contrast to the benefits of using an index for retrievals, inserts and updates have a direct impact on the maintenance of the index. The more indices on the tables, the more index reorganizations that are necessary. The rule of thumb is to choose indices wisely for frequent queries and evaluate index usage afterward.

This section gives some recommendations for creating and using indices.

NOTE The following recommendations are general rules of thumb. The choice of whether to incorporate any particular recommendation ultimately depends on how your database will be used in production and which queries are used most frequently. An index on a column that is never used will be counterproductive.

Indices and Conditions in the WHERE Clause

If the WHERE clause in a SELECT statement contains a search condition with a single column, you should create an index on this column. The use of an index is especially recommended if the selectivity of the condition is high. The *selectivity* of a condition is defined as the ratio of the number of rows satisfying the condition to the total number of rows in the table.

(High selectivity corresponds to a small ratio.) The most successful processing of a retrieval with the indexed column will be achieved if the selectivity of a condition is 5 percent or less.

The column should not be indexed if the selectivity of the condition is constantly 80 percent or more. In such a case, additional I/O operations will be needed for the existing index pages, which would eliminate any time savings gained by index access. In this particular case, a table scan would be faster, and the query optimizer would usually choose to use a table scan, rendering the index useless.

If a search condition in a frequently used query contains one or more AND operators, it is best to create a composite index that includes all the columns of the table specified in the WHERE clause of the SELECT statement. Example 10.9 shows the creation of a composite index that includes all the columns specified in the WHERE clause of the SELECT statement.

Example 10.9

```
USE sample;
CREATE INDEX i_works ON works_on(emp_no, enter_date);
SELECT emp_no, project_no, enter_date
   FROM works_on
   WHERE emp_no = 29346 AND enter_date='1.4.2016';
```

The AND operator in this query contains two conditions. As such, both of the columns appearing in each condition should be indexed using a composite nonclustered index.

Indices and the Join Operator

In the case of a join operation, it is recommended that you index each join column. Join columns often represent the primary key of one table and the corresponding foreign key of the other or the same table. If you specify the PRIMARY KEY and FOREIGN KEY integrity constraints for the corresponding join columns, only a nonclustered index for the column with the foreign key should be created, because the system will implicitly create the clustered index for the PRIMARY KEY column.

Example 10.10 shows the creation of indices, which should be used if you have a query with a join operation and an additional filter.

Example 10.10

```
USE sample;
SELECT emp_lname, emp_fname
   FROM employee, works_on
   WHERE employee.emp_no = works_on.emp_no
   AND enter_date = '10.15.2017';
```

For Example 10.10, the creation of two separate indices for the **emp_no** column in both the **employee** and **works_on** tables is recommended. Also, an additional index should be created for the **enter_date** column.

Covering Index

As you already know, significant performance gains can be achieved when *all* columns in the query are included in the index. Example 10.11 shows a covering index.

Example 10.11
```
USE AdventureWorks;
GO
DROP INDEX Person.Address.IX_Address_StateProvinceID;
GO
CREATE INDEX i_address_zip
  ON Person.Address (PostalCode)
  INCLUDE (City, StateProvinceID);
GO
SELECT City, StateProvinceID
    FROM Person.Address
    WHERE PostalCode = '84407';
```

Example 10.11 first drops the **IX_Address_StateProvinceID** index of the **Address** table. In the second step, it creates a new index, which additionally includes two other columns, on the **PostalCode** column. Finally, the SELECT statement at the end of the example shows a query covered by the index. For this query, the system does not have to search for data in data pages, because the optimizer can find all the column values in the leaf pages of the nonclustered index.

NOTE The use of covering indices is recommended because index pages generally contain many more entries than the corresponding data pages contain. Also, to use this method, the filtered columns must be the first key columns in the index.

Missing Indices

The Database Engine provides several ways to assist in determining whether an index will be helpful to accelerate query performance. One of these ways is to use a group of several dynamic management views, which are listed here:

- sys.dm_db_missing_index_details
- sys.dm_db_missing_index_group_stats
- sys.dm_db_missing_index_groups
- sys.dm_db_missing_index_columns

We will discuss only the first one. The **sys.dm_db_missing_index_details** DMV returns detailed information about missing indices, excluding spatial indices. Example 10.12 shows how you can use this view.

Example 10.12
```
-- 1
USE AdventureWorks;
GO
sp_helpindex [Sales.Store]
GO
--2
```

```
SELECT Name, BusinessEntityID, ModifiedDate FROM Sales.Store
    WHERE (Name='Sharp Bikes' and BusinessEntityID <> ''
         AND ModifiedDate > '2013-10-01');
-- 3
SELECT equality_columns, inequality_columns, statement
         FROM sys.dm_db_missing_index_details
```

The **sp_helpindex** system stored procedure in the code displays all existing indices of the **Store** table from the **Sales** schema of the **AdventureWorks** database. The result of this procedure is

index_name	index_description	index_key
AK_Store_rowguid	nonclustered, located on PRIMARY	rowguid
IX_Store_SalesPersonID	nonclustered, located on PRIMARY	SalesPersonID
PK_Store_BusinessEntityID	clustered, primary key	BusinessEntityID

The table contains three indices: two nonclustered indices and one clustered index on the column of the table with the primary key constraint. Now, we execute the query, and use the information concerning missing indices. The last statement in Example 10.12 allows us to display this information using the **sys.dm_db_missing_index_details** DMV. The DMV displays the following information:

```
equality_columns      inequality_columns            statement
-----------------------------------------------------------------------------

[Name]                [BusinessEntityID],[ModifiedDate]  [AdventureWorks].
                                                         [Sales].[Store]
```

The most important columns in **sys.dm_db_missing_index_details** are: **equality_columns**, **inequality_columns**, and **statement**. The first, **equality_columns**, displays all columns that have been used in the WHERE clause of the query and with an equal (=) operator. Similarly, **inequality_columns** displays a comma-separated list of columns that contribute to inequality predicates, e.g., predicates of the form: *table.column > constant_value*. (Any comparison operator other than "=" expresses inequality.) Finally, the **statement** column identifies the table against which the query was executed.

Next, you should convert the information returned by the **sys.dm_db_missing_index_details** DMV into a corresponding CREATE INDEX statement. Generally, equality columns should be put before the inequality ones, and together they should make the key of the index. Therefore, the missing index of Example 10.12 in the **AdventureWorks.Sales.Store** should be

```
CREATE INDEX i1 ON AdventureWorks.Sales.Store (Name,BusinessEntityID,ModifiedDate)
```

NOTE There are also other ways to view and find indices that might improve performance. One of them is to use the Performance Dashboard report in SQL Server Management Studio.

Special Types of Indices

The Database Engine allows you to create the following special types of indices:

- Indexed views
- Filtered indices
- Indices on computed columns
- Partitioned indices
- Column store indices

Indexed views are discussed in Chapter 26. Filtered indices are explained in Chapter 19. Partitioned indices are used with partitioned tables and are described in Chapter 26. Column store indices will be explained in detail in Chapter 27.

Summary

Indices are used to access data more efficiently. They can affect not only SELECT statements but also performance of INSERT, UPDATE, and DELETE statements. An index can be clustered or nonclustered, unique or nonunique, and single or composite. A clustered index physically sorts the rows of the table in the order of the specified column(s). A unique index specifies that each value can appear only once in that column of the table. A composite index is composed of more than one column.

A great feature in relation to indices is the Database Engine Tuning Advisor (DTA), which will, among other things, analyze a sample of your actual workload (supplied via either a script file from you or a captured trace file from SQL Server Profiler) and recommend indices for you to add or delete based on that workload. Use of DTA is highly recommended. For more information on SQL Server Profiler and DTA, see Chapter 20.

The next chapter discusses the notion of a view.

Exercises

E.10.1 Create a nonclustered index for the **enter_date** column of the **works_on** table. Sixty percent of each index leaf page should be filled.

E.10.2 Create a unique composite index for the **l_name** and **f_name** columns of the **employee** table. Is there any difference if you change the order of the columns in the composite index?

E.10.3 How can you drop the index that is implicitly created for the primary key of a table?

E.10.4 Discuss the benefits and disadvantages of an index.

In the following four exercises, create indices that will improve performance of the queries. (Assume that all tables of the **sample** database that are used in the following exercises have a very large number of rows.)

E.10.5

```
USE sample;
SELECT emp_no, emp_fname, emp_lname
     FROM employee
     WHERE emp_lname = 'Smith'
```

E.10.6

```
USE sample;
SELECT emp_no, emp_fname, emp_lname
     FROM employee
     WHERE emp_lname = 'Hansel'
     AND emp_fname = 'Elke'
```

E.10.7

```
USE sample;
SELECT job
   FROM works_on, employee
   WHERE employee.emp_no = works_on.emp_no
```

E.10.8

```
USE sample;
SELECT emp_lname, emp_fname
     FROM employee, department
     WHERE employee.dept_no = department.dept_no
     AND dept_name = 'Research'
```

11 Views

In This Chapter

- DDL Statements and Views
- DML Statements and Views

This chapter is dedicated exclusively to the database object called a *view*. The structure of this chapter corresponds to the structure of Chapters 5 to 7, in which the DDL and DML statements for base tables were described. The first section of this chapter covers the DDL statements concerning views: CREATE VIEW, ALTER VIEW, and DROP VIEW. The second part of the chapter describes the DML statements SELECT, INSERT, UPDATE, and DELETE with views. The SELECT statement will be looked at separately from the other three statements. In contrast to base tables, views cannot be used for modification operations without certain limitations. These limitations are described at the end of each corresponding section.

The alternative form of a view, called an indexed view, is described in Chapter 26. This type of index materializes the corresponding query and allows you to achieve significant performance gains in relation to queries with aggregated data.

DDL Statements and Views

In the previous chapters, base tables were used to describe DDL and DML statements. A base table contains data stored on the disk. By contrast, views, by default, do not exist physically— that is, their content is not stored on the disk. (This is not true for indexed views, which are discussed in Chapter 26.) Views are database objects that are always derived from one or more base tables (or views) using metadata information. This information (including the name of the view and the way the rows from the base tables are to be retrieved) is the only information concerning views that is physically stored. Thus, views are also called *virtual tables*.

Creating a View

A view is created using the CREATE VIEW statement. The general form of this statement is

```
CREATE VIEW view_name [(column_list)]
  [WITH {ENCRYPTION | SCHEMABINDING | VIEW_METADATA}]
  AS select_statement
  [WITH CHECK OPTION]
```

NOTE The CREATE VIEW statement must be the only statement in a batch. (This means that you have to use the GO statement to separate this statement from other statements in a statement group.)

---

view_name is the name of the defined view. **column_list** is the list of names to be used for columns in a view. If this optional specification is omitted, column names of the underlying tables are used. **select_statement** specifies the SELECT statement that retrieves rows and columns from one or more tables (or views). The WITH ENCRYPTION option encrypts the SELECT statement, thus enhancing the security of the database system.

The SCHEMABINDING clause binds the view to the schema of the underlying table. When SCHEMABINDING is specified, the base table or tables cannot be modified in a way that would affect the view definition. The view definition itself must first be modified or dropped to remove dependencies on the table that is to be modified. When SCHEMABINDING is specified, tables referenced in the SELECT statement must include the two-part names in the form **owner.tab_name**. (Besides being used with a table, SCHEMABINDING can be used with a view or a user-defined function, too.) Example 26.12 shows the use of this clause.

Any attempt to modify the structure of views or tables that are referenced in a view created with the SCHEMABINDING clause fails. You have to drop the view or change it so that it no longer has this clause if you want to apply the ALTER or DROP statement to the referenced objects. (The WITH CHECK OPTION clause is discussed in detail later in this chapter in the section "INSERT Statement and a View.")

When a view is created with the VIEW_METADATA option, all of its columns (except columns with the TIMESTAMP data type) can be updated if the view has INSERT or UPDATE INSTEAD OF triggers. (Triggers are described in Chapter 14.)

NOTE The SELECT statement in a view cannot include the ORDER BY clause or INTO option. Additionally, a temporary table cannot be referenced in the query.

Views can be used for different purposes:

- To restrict the use of particular columns and/or rows of tables. Therefore, views can be used for controlling access to a particular part of one or more tables (as discussed in Chapter 12).

- To hide the details of complicated queries. If database applications need queries that involve complicated join operations, the creation of corresponding views can simplify the use of such queries.

- To restrict inserted and updated values to certain ranges.

Example 11.1 shows the creation of a view.

Example 11.1

```
USE sample;
GO
CREATE VIEW v_clerk
  AS SELECT emp_no, project_no, job, enter_date
    FROM works_on
    WHERE job = 'Clerk';
```

The query in Example 11.1 retrieves the rows of the **works_on** table for which the condition job = 'Clerk' evaluates to TRUE. The **v_clerk** view is defined as the rows and columns returned by this query. Table 11-1 shows the **works_on** table with the rows that belong to the **v_clerk** view bolded.

Example 11.1 specifies the selection of rows—that is, it creates a horizontal subset from the base table **works_on**. It is also possible to create a view that limits the columns as well as the rows to be included in the view. Example 11.2 shows the creation of such a view.

Example 11.2

```
USE sample;
GO
CREATE VIEW v_without_budget
  AS SELECT project_no, project_name
      FROM project;
```

The **v_without_budget** view in Example 11.2 contains all columns of the **project** table except the **budget** column.

emp_no	project_no	job	enter_date
10102	p1	Analyst	2016.10.1 00:00:00
10102	p3	Manager	2018.1.1 00:00:00
25348	**p2**	**Clerk**	**2017.2.15 00:00:00**
18316	p2	NULL	2017.6.1 00:00:00
29346	p2	NULL	2016.12.15 00:00:00
2581	p3	Analyst	2017.10.15 00:00:00
9031	p1	Manager	2017.4.15 00:00:00
28559	p1	NULL	2017.8.1. 00:00:00
28559	**p2**	**Clerk**	**2018.2.1 00:00:00**
9031	**p3**	**Clerk**	**2016.11.15 00:00:00**
29346	**p1**	**Clerk**	**2017.1.4 00:00:00**

Table 11-1   The Base Table works_on

As already stated, specifying column names with a view in the general format of the CREATE VIEW statement is optional. On the other hand, there are also two cases in which the explicit specification of column names is required:

- If a column of the view is derived from an expression or an aggregate function
- If two or more columns of the view have the same name in the underlying tables

Example 11.3 shows the explicit specification of column names in relation to a view.

Example 11.3

```
USE sample;
GO
CREATE VIEW v_count(project_no, count_project)
  AS SELECT project_no, COUNT(*)
     FROM works_on
     GROUP BY project_no;
```

The column names of the **v_count** view in Example 11.3 must be explicitly specified because the SELECT statement contains the aggregate function COUNT(*), and all columns in a view must be named.

You can avoid the explicit specification of the column list in the CREATE VIEW statement if you use column headers, as in Example 11.4.

Example 11.4

```
USE sample;
GO
CREATE VIEW v_count1
  AS SELECT project_no, COUNT(*) count_project
     FROM works_on
     GROUP BY project_no;
```

A view can be derived from another existing view, as shown in Example 11.5.

Example 11.5

```
USE sample;
GO
CREATE VIEW v_project_p2
  AS SELECT emp_no
     FROM v_clerk
     WHERE project_no ='p2';
```

The **v_project_p2** view in Example 11.5 is derived from the **v_clerk** view (see Example 11.1). Every query using the **v_project_p2** view is converted into the equivalent query on the underlying base table **works_on**.

You can also create a view using Object Explorer of SQL Server Management Studio. Select the database under which you want to create the view, right-click Views, and choose New View. The corresponding editor appears. Using the editor, you can do the following:

- Select underlying tables and columns from these tables for the view.
- Name the view and define conditions in the WHERE clause of the corresponding query.

Altering and Removing Views

The Transact-SQL language supports the nonstandard ALTER VIEW statement, which is used to modify the definition of the view query. The syntax of ALTER VIEW is analogous to that of the CREATE VIEW statement.

You can use the ALTER VIEW statement to avoid reassigning existing privileges for the view. Also, altering an existing view using this statement does not affect database objects that depend upon the view. Otherwise, if you use the DROP VIEW and CREATE VIEW statements to remove and re-create a view, any database object that uses the view will not work properly, at least in the time period between removing and re-creating the view.

Example 11.6 shows the use of the ALTER VIEW statement.

Example 11.6

```
USE sample;
GO
ALTER VIEW v_without_budget
  AS SELECT project_no, project_name
     FROM project
     WHERE project_no >= 'p3';
```

The ALTER VIEW statement in Example 11.6 extends the SELECT statement of the **v_without_budget** view (see Example 11.2) with the new condition in the WHERE clause.

The DROP VIEW statement removes the definition of the specified view from the system tables. Example 11.7 shows the use of the DROP VIEW statement.

Example 11.7

```
USE sample;
GO
DROP VIEW v_count;
```

If the DROP VIEW statement removes a view, all other views derived from it will be influenced, too, as demonstrated in Example 11.8.

Example 11.8

```
USE sample;
GO
DROP VIEW v_clerk;
```

The DROP VIEW statement in Example 11.8 also implicitly influences the **v_project_p2** view (see Example 11.5). The **v_project_p2** view still exists, but cannot be used. In other words, if you query the **v_project_p2** view, you will get the binding error (error number 4413).

NOTE A view is not automatically dropped if the underlying table is removed. This means that any view from the removed table must be removed using the DROP VIEW statement. On the other hand, if a table with the same logical structure as the removed one is subsequently created, the view can be used again.

DML Statements and Views

Views are retrieved and modified with the same Transact-SQL statements that are used to retrieve and modify base tables. The following subsections discuss all four DML statements in relation to views.

View Retrieval

A view is used exactly like any base table of a database. You can think of selecting from a view as if the statement were transformed into an equivalent operation on the underlying base table(s). Example 11.9 shows this.

Example 11.9

```
USE sample;
GO
CREATE VIEW v_d2
  AS SELECT emp_no, emp_lname
     FROM employee
     WHERE dept_no ='d2';
GO
SELECT emp_lname
   FROM v_d2
   WHERE emp_lname LIKE 'J%';
```

The result is

emp_lname
James

The SELECT statement in Example 11.9 is transformed into the following equivalent form, using the underlying table of the **v_d2** view:

```
SELECT emp_lname
  FROM employee
  WHERE emp_lname LIKE 'J%'
  AND dept_no ='d2';
```

The next three sections describe the use of views with the other three DML statements: INSERT, UPDATE, and DELETE. Data modification with these statements is treated in a manner similar to a retrieval. The only difference is that there are some restrictions on a view used for insertion, modification, and deletion of data from the table that it depends on.

INSERT Statement and a View

A view can be used with the INSERT statement as if it were a base table. When a view is used to insert rows, the rows are actually inserted into the underlying base table.

The **v_dept** view, which is created in Example 11.10, contains the first two columns of the **department** table. The subsequent INSERT statement inserts the row into the underlying table using the values 'd4' and 'Development'. The **location** column, which is not referenced by the **v_dept** view, is assigned a NULL value.

Example 11.10

```
USE sample;
GO
CREATE VIEW v_dept
  AS SELECT dept_no, dept_name
     FROM department;
GO
INSERT INTO v_dept
   VALUES('d4', 'Development');
```

Using a view, it is generally possible to insert a row that does not satisfy the conditions of the view query's WHERE clause. The option WITH CHECK OPTION is used to restrict the insertion of only such rows that satisfy the conditions of the query. If this option is used, the Database Engine tests every inserted row to ensure that the conditions in the WHERE clause evaluate to TRUE. If this option is omitted, there is no check of conditions in the WHERE clause, and therefore every row is inserted into the underlying table. This could lead to the confusing situation of a row being inserted using a view but subsequently not being returned by a SELECT statement against that view, because the WHERE clause is enforced for the SELECT. WITH CHECK OPTION is also applied to the UPDATE statement.

Examples 11.11 and 11.12 show the difference of applying and not applying WITH CHECK OPTION, respectively.

Example 11.11

```
USE sample;
GO
CREATE VIEW v_2016_check
  AS SELECT emp_no, project_no, enter_date
     FROM works_on
     WHERE enter_date BETWEEN '01.01.2016' AND '12.31.2016'
     WITH CHECK OPTION;
GO
INSERT INTO v_2016_check
   VALUES (22334, 'p2', '1.15.2017');
```

In Example 11.11, the system tests whether the inserted value of the **enter_date** column evaluates to TRUE for the condition in the WHERE clause of the SELECT statement. The attempted insert fails because the condition is not met.

Example 11.12

```
USE sample;
GO
CREATE VIEW v_2016_nocheck
  AS SELECT emp_no, project_no, enter_date
     FROM works_on
     WHERE enter_date BETWEEN '01.01.2016' AND '12.31.2016';
GO
INSERT INTO v_2016_nocheck
  VALUES (22334, 'p2', '1.15.2017');
SELECT *
  FROM v_2016_nocheck;
```

The result is

emp_no	project_no	enter_date
10102	p1	2016-10-01
29346	p2	2016-12-15
9031	p3	2016-11-15

Because Example 11.12 does not use WITH CHECK OPTION, the INSERT statement is executed and the row is inserted into the underlying **works_on** table. Notice that the subsequent SELECT statement does not display the inserted row because it cannot be retrieved using the **v_2016_nocheck** view.

The insertion of rows into the underlying tables is *not* possible if the corresponding view contains any of the following features:

- The FROM clause in the view definition involves two or more tables and the column list includes columns from more than one table.
- A column of the view is derived from an aggregate function.
- The SELECT statement in the view contains the GROUP BY clause or the DISTINCT option.
- A column of the view is derived from a constant or an expression.

Example 11.13 shows a view that cannot be used to insert rows in the underlying base table.

Example 11.13

```
USE sample;
GO
CREATE VIEW v_sum(sum_of_budget)
  AS SELECT SUM(budget)
     FROM project;
GO
SELECT *
  FROM v_sum;
```

Example 11.13 creates the **v_sum** view, which contains an aggregate function in its SELECT statement. Because the view in the example represents the result of an aggregation of many rows (and not a single row of the **project** table), it does not make sense to try to insert a row into the underlying table using this view.

UPDATE Statement and a View

A view can be used with the UPDATE statement as if it were a base table. When a view is used to modify rows, the content of the underlying base table is actually modified.

Example 11.14 creates a view that is then used to modify the **works_on** table.

Example 11.14

```
USE sample;
GO
CREATE VIEW v_p1
  AS SELECT emp_no, job
     FROM works_on
     WHERE project_no = 'p1';
GO
UPDATE v_p1
   SET job = NULL
   WHERE job = 'Manager';
```

You can think of updating the view in Example 11.14 as if the UPDATE statement were transformed into the following equivalent statement:

```
UPDATE works_on
   SET job = NULL
   WHERE job = 'Manager'
   AND project_no = 'p1'
```

WITH CHECK OPTION has the same logical meaning for the UPDATE statement as it has for the INSERT statement. Example 11.15 shows the use of WITH CHECK OPTION with the UPDATE statement.

Example 11.15

```
USE sample;
GO
CREATE VIEW v_100000
  AS SELECT project_no, budget
     FROM project
     WHERE budget > 100000
     WITH CHECK OPTION;
GO
UPDATE v_100000
  SET budget = 93000
  WHERE project_no = 'p3';
```

In Example 11.15, the Database Engine tests whether the modified value of the **budget** column evaluates to TRUE for the condition in the WHERE clause of the SELECT statement. The attempted modification fails because the condition is not met—that is, the value 93000 is not greater than the value 100000.

The modification of columns in the underlying tables is *not* possible if the corresponding view contains any of the following features:

- The FROM clause in the view definition involves two or more tables and the column list includes columns from more than one table.
- A column of the view is derived from an aggregate function.
- The SELECT statement in the view contains the GROUP BY clause or the DISTINCT option.
- A column of the view is derived from a constant or an expression.

Example 11.16 shows a view that cannot be used to modify row values in the underlying base table.

Example 11.16

```
USE sample;
GO
CREATE VIEW v_uk_pound (project_number, budget_in_pounds)
  AS SELECT project_no, budget*0.65
     FROM project
     WHERE budget > 100000;
GO
SELECT *
  FROM v_uk_pound;
```

The result is

project_number	budget_in_pounds
p1	78000
p3	121225

The **v_uk_pound** view in Example 11.16 cannot be used with an UPDATE statement (nor with an INSERT statement) because the **budget_in_pounds** column is calculated using an arithmetic expression, and therefore does not represent an original column of the underlying table.

DELETE Statement and a View

A view can be used to delete rows of a table that it depends on, as shown in Example 11.17.

Example 11.17

```
USE sample;
GO
CREATE VIEW v_project_p1
  AS SELECT emp_no, job
     FROM works_on
     WHERE project_no = 'p1';
GO
DELETE FROM v_project_p1
  WHERE job = 'Clerk';
```

Example 11.17 creates a view that is then used to delete rows from the **works_on** table.

The deletion of rows in the underlying tables is *not* possible if the corresponding view contains any of the following features:

- The FROM clause in the view definition involves two or more tables and the column list includes columns from more than one table.
- A column of the view is derived from an aggregate function.
- The SELECT statement in the view contains the GROUP BY clause or the DISTINCT option.

In contrast to the INSERT and UPDATE statements, the DELETE statement allows the existence of a constant or an expression in a column of the view that is used to delete rows from the underlying table.

Example 11.18 shows a view that can be used to delete rows, but not to insert rows or modify column values.

Example 11.18

```
USE sample;
GO
CREATE VIEW v_budget (budget_reduction)
  AS SELECT budget*0.9
      FROM project;
GO
DELETE FROM v_budget;
```

The DELETE statement in Example 11.18 deletes all rows of the **project** table, which is referenced by the **v_budget** view.

Editing Information Concerning Views

sys.objects is the most important catalog view concerning views. As you already know, this catalog view contains information in relation to all objects of the current database. All rows of this view that have the value **V** for the **type** column contain information concerning views.

Another catalog view called **sys.views** displays additional information about existing views. The most important column of this view is **with_check_option**, which instructs you whether or not WITH CHECK OPTION is specified.

Using the system procedure **sp_helptext**, you can display the query belonging to a particular view.

Summary

Views are created, retrieved, and modified with the same Transact-SQL statements that are used to create, retrieve, and modify base tables. The query on a view is always transformed into the equivalent query on an underlying base table. An update operation is treated in a manner similar to a retrieval. The only difference is that there are some restrictions on a view used for insertion, modification, and deletion of data from a table that it depends on. Even so, the way in which the Database Engine handles the modification of rows and columns is more systematic than the way in which other relational database systems handle such modification.

The following chapter explains in detail the security issues of the Database Engine.

Exercises

E.11.1   Create a view that comprises the data of all employees who work for the department d1.

E.11.2   For the **project** table, create a view that can be used by employees who are allowed to view all data of this table except the **budget** column.

E.11.3   Create a view that comprises the first and last names of all employees who entered their projects in the second half of the year 2017.

E.11.4   Solve Exercise E.11.3 so that the original columns **f_name** and **l_name** have new names in the view: **first** and **last**, respectively.

E.11.5   Use the view in E.11.1 to display full details of every employee whose last name begins with the letter M.

E.11.6   Create a view that comprises full details of all projects on which the employee named Smith works.

E.11.7   Using the ALTER VIEW statement, modify the condition in the view in E.11.1. The modified view should comprise the data of all employees who work for department d1, department d2, or both.

E.11.8   Delete the view created in E.11.3. What happens with the view created in E.11.4?

E.11.9   Using the view from E.11.2, insert the details of the new project with the project number p2 and the name Moon.

E.11.10   Create a view (with the WITH CHECK OPTION clause) that comprises the first and last names of all employees whose employee number is less than 10,000. After that, use the view to insert data for a new employee named Kohn with the employee number 22123, who works for the department d3.

E.11.11   Solve Exercise E.11.10 without the WITH CHECK OPTION clause and find the differences in relation to the insertion of the data.

E.11.12   Create a view (with the WITH CHECK OPTION clause) with full details from the **works_on** table for all employees who entered their projects during the years 2017 and 2018. After that, modify the entering date of the employee with the employee number 29346. The new date is 06/01/2016.

E.11.13   Solve Exercise E.11.12 without the WITH CHECK OPTION clause and find the differences in relation to the modification of the data.

CHAPTER

12

Security System of the Database Engine

In This Chapter

- Encrypting Data
- Authentication
- Schemas
- Database Security
- Roles
- Authorization
- Change Tracking
- Data Security and Views

This chapter begins with an overview of the most important concepts of database security. It then discusses the specific features of the security system of the Database Engine.

The following are the most important database security concepts:

- Data encryption
- Authentication
- Authorization
- Change tracking

Data encryption is the process of scrambling information so that it is incomprehensible until it is decrypted by the intended recipient. The Database Engine supports two different encryption methods. The first one allows you to encrypt the whole database, while the second one supports column encryption.

Authentication validates the identity of the user. Therefore, this security concept specifies the process of validating user credentials to prevent unauthorized users from using a system. Authentication can be checked by requesting the user to provide, for example:

- Something the user knows (usually a password)
- A physical object the user possesses, such as a magnetic card or badge
- A physical characteristic of the user, such as a signature or fingerprints

Authentication is most commonly confirmed using a name and a password. This information is evaluated by the system. This process can be strengthened using encryption.

Authorization is the process that is applied after the identity of a user is verified through authentication. During this process, the system determines which resources the particular user can use.

Change tracking means that actions of users are followed and documented on your system. In other words, all insert, update, and delete operations that are applied to database objects are documented. After that, they can be viewed by the authorized users. (This process is useful to protect the system against users with elevated privileges.)

Before I describe these four concepts in the following sections, I will give you a concise definition of the security model that the Database Engine uses. The security model comprises three different categories, which interact among themselves:

- **Principals**   Subjects that have permission to access a particular entity. Typical principals are Windows user accounts and SQL Server logins. In addition to these principals, there are also Windows groups and SQL Server roles. A Windows group is a collection of Windows user accounts and groups. Similarly, a role is a collection of logins and other roles. (Principals will be discussed in detail in the "Authentication" section.)

- **Securables**   The resources to which the database authorization system regulates access. Most securables build a hierarchy, meaning that some of them can be contained within others. Most of them have a certain number of permissions that apply to them. There are three main securable scopes: server, database, and schema.

- **Permissions**   Every securable has associated *permissions* that can be granted to a principal. Permissions are discussed in the section "Authorization" later in this chapter.

Encrypting Data

Encryption is a process of obfuscating data, thereby enhancing the data's security. Generally, the concrete encryption procedure is carried out using an algorithm. The most important algorithm for encryption is called RSA. (It is an acronym for Rivers, Shamir, and Adelman, the last names of the three men who invented it.)

The Database Engine secures data with hierarchical encryption layers and a key management infrastructure. Each layer secures the layer beneath it, using a combination of certificates, asymmetric keys, and symmetric keys (see Figure 12-1).

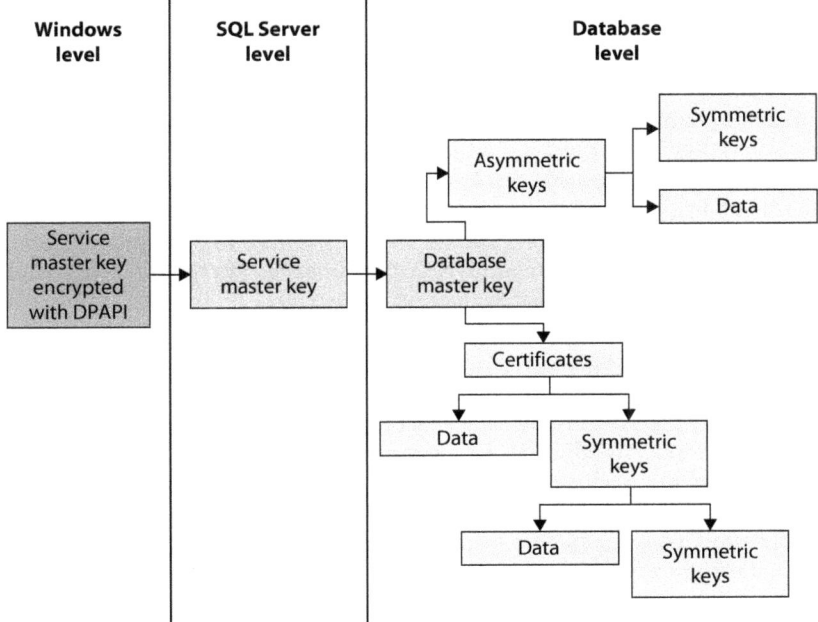

Figure 12-1   The Database Engine hierarchical encryption layers

The service master key in Figure 12-1 specifies the key that rules all other keys and certificates. The service master key is created automatically when you install the Database Engine. This key is encrypted using the Windows Data Protection API (DPAPI). The important property of the service master key is that it is managed by the system. Although the system administrator can perform several maintenance tasks, the only task he or she should perform is to back up the service master key, so that it can be restored if it becomes corrupted.

As you can see in Figure 12-1, the database master key is the root encryption object for all keys, certificates, and data at the database level. Each database has a single database master key, which is created using the CREATE MASTER KEY statement (shown in upcoming Example 12.1). Because the database master key is protected by the service master key, it is possible for the system to automatically decrypt the database master key.

Once the database master key exists, users can use it to create keys. There are three forms of user keys:

- Symmetric keys
- Asymmetric keys
- Certificates

The following subsections describe the user keys.

Symmetric Keys

An encryption system that uses symmetric keys is one in which the sender and receiver of a message share a common key. Thus, this single key is used for both encryption and decryption.

Using symmetric keys has several benefits and one disadvantage. One advantage of using symmetric keys is that they can protect a significantly greater amount of data than can the other two types of user keys. Also, using this key type is faster than using an asymmetric key.

On the other hand, in a distributed environment, using this type of key can make it almost impossible to keep encryption secure, because the same key is used to decrypt and encrypt data on both ends. So, the general recommendation is that symmetric keys should be used only with applications in which data is stored as encrypted text at one place.

The Transact-SQL language supports several statements and system functions related to symmetric keys. The CREATE SYMMETRIC KEY statement creates a new symmetric key, while the DROP SYMMETRIC KEY statement removes an existing symmetric key. Each symmetric key must be opened before you can use it to encrypt data or protect another new key. Therefore, you use the OPEN SYMMETRIC KEY statement to open a key. (You use the CLOSE SYMMETRIC KEY statement to close a key.)

After you open a symmetric key, you need to use the **EncryptByKey** system function for encryption. This function has two input parameters: the ID of the symmetric key to be used to encrypt the text, and the text to be encrypted. For decryption, you use the **DecryptByKey** function.

> **NOTE** See Microsoft Docs for detailed descriptions of all Transact-SQL statements related to symmetric keys as well as the system functions **EncryptByKey** and **DecryptByKey**.

Asymmetric Keys

If you have a distributed environment or if a symmetric key does not keep your encryption secure, use asymmetric keys. An asymmetric key consists of two parts: a private key and the corresponding public key. Each key can decrypt data encrypted by the other key. Because of the existence of a private key, asymmetric encryption provides a higher level of security than does symmetric encryption.

The Transact-SQL language supports several statements and system functions related to asymmetric keys. The CREATE ASYMMETRIC KEY statement creates a new asymmetric key, while the ALTER ASYMMETRIC KEY statement changes the properties of an asymmetric key. The DROP ASYMMETRIC KEY statement drops an existing asymmetric key.

After you create an asymmetric key, use the **EncryptByAsymKey** system function to encrypt data. This function has two input parameters: the ID of the asymmetric key to be used to encrypt the text, and the text to be encrypted. For decryption, use the **DecryptByAsymKey** function.

> **NOTE** See Microsoft Docs for detailed descriptions of all Transact-SQL statements related to asymmetric keys as well as the system functions **EncryptByAsymKey** and **DecryptByAsymKey**.

Certificates

A public key certificate, usually simply called a certificate, is a digitally signed statement that binds the value of a public key to the identity of the person, device, or service that holds the corresponding private key. Certificates are issued and signed by a certification authority (CA). The entity that receives a certificate from a CA is the subject of that certificate.

NOTE There is no significant functional difference between certificates and asymmetric keys. Both use the RSA algorithm. The main difference is that certificates can contain more information.

Certificates contain the following information:

- The subject's public key value
- The subject's identifier information
- Issuer identifier information
- The digital signature of the issuer

A primary benefit of certificates is that they relieve hosts of the need to maintain a set of passwords for individual subjects. When a host, such as a secure web server, designates an issuer as a trusted authority, the host implicitly trusts that the issuer has verified the identity of the certificate subject.

NOTE Certificates provide the highest level of encryption in the Database Engine security model. The encryption algorithms for certificates are very processor-intensive. For this reason, use certificates sparingly.

Example 12.1 shows how you can create a database master key, as well as a certificate.

Example 12.1

```
USE master;
CREATE MASTER KEY
ENCRYPTION BY PASSWORD = 'p1s4w9d16!'
GO
CREATE CERTIFICATE cert01
      WITH SUBJECT = 'Certificate for dbo';
```

If you want to create a certificate without the ENCRYPTION BY option, you first have to create the database master key. (Each CREATE CERTIFICATE statement that does not include this option is protected by the database master key.) For this reason, the first statement in Example 12.1 is the CREATE MASTER KEY statement. After that, the CREATE CERTIFICATE statement is used to create a new certificate, **cert01**.

Editing Metadata Concerning User Keys

The most important catalog views in relation to user keys are the following:

- sys.symmetric_keys
- sys.asymmetric_keys
- sys.certificates
- sys.database_principals

The first three catalog views provide information about all symmetric keys, all asymmetric keys, and all the certificates installed in the current database, respectively. The **sys.database_principals** catalog view provides information about each of the principals in the current database. Example 12.2 shows how you can use the **sys.database_principals** view to retrieve the information concerning existing certificates. You can get the analogous information concerning symmetric and asymmetric keys using the **sys.symmetric_keys** view and **sys.asymmetric_keys** view, respectively.

Example 12.2

```
select p.name, c.name, certificate_id
    from sys.database_principals p, sys.certificates c
    where p.principal_id = c.principal_id;
```

Extensible Key Management

Another step in achieving greater key security is the use of Extensible Key Management (EKM). EKM has the following main goals:

- Enhanced key security through a choice of encryption provider
- General key management across your enterprise

EKM allows third-party vendors to register their devices in the Database Engine. Once the devices are registered, logins can use the encryption keys stored on these modules as well as leverage advanced encryption features that these modules support. EKM also allows data protection from database administrators (except members of the **sysadmin** group). That way, you can protect the system against users with elevated privileges. Data can be encrypted and decrypted using Transact-SQL cryptographic statements, and the Database Engine uses the external EKM device as the key store.

Methods of Data Encryption

The Database Engine supports two methods of data encryption:

- Transparent Data Encryption
- Always Encrypted

The following two subsections describe these two methods.

Transparent Data Encryption

Transparent Data Encryption (TDE) introduces a new database option that encrypts the database files automatically, without needing to alter any applications. That way, you can prevent database access by unauthorized persons even if they obtain the database files or database backup files.

Encryption of the database file is performed at the page level. The pages in an encrypted database are encrypted before they are written to disk and are decrypted when they are read into memory.

TDE, like most other encryption methods, is based on an encryption key. It uses a symmetric key, which secures the encrypted database.

For a particular database, TDE can be implemented in four steps:

1. Create a database master key using the CREATE MASTER KEY statement (see the first statement in Example 12.1).

2. Create a certificate using the CREATE CERTIFICATE statement (see the second statement in Example 12.1).

3. Create an encryption key (see Example 12.3).

4. Configure the database to use encryption (see Example 12.4).

Example 12.3 creates an encryption key for the **sample** database. The CREATE DATABASE ENCRYPTION KEY statement associates the certificate **cert01** to the current database. The ALGORITHM option specifies the algorithm that is used for encryption. You can choose from one of the following algorithms: AES_128, AES_192, AES_256, or TRIPLE_DES_3KEY.

Example 12.3

```
USE sample;
GO
-- Associate the certificate to the sample database
CREATE DATABASE ENCRYPTION KEY
WITH ALGORITHM = AES_128
ENCRYPTION BY SERVER CERTIFICATE cert01;
```

After the database encryption key is created, Example 12.4 changes the **sample** database using the ALTER DATABASE statement to enable encryption. The encryption process runs as a background task, so the database remains available while it's being encrypted.

Example 12.4

```
-- Encrypt the database.
ALTER DATABASE sample SET ENCRYPTION ON;
```

Monitoring Transparent Data Encryption   The most important dynamic management view in relation to Transparent Data Encryption is **sys.dm_database_encryption_keys**, which can be used to display all databases that are encrypted. Example 12.5 shows the use of the **sys .dm_database_encryption_keys** view. (Note that the value 3 of the **encryption_state** column means that the corresponding database is encrypted.)

Example 12.5

```
USE master;
SELECT * FROM sys.dm_database_encryption_keys
  WHERE encryption_state = 3;
```

Always Encrypted

In contrast to Transparent Data Encryption, which encrypts a whole database, the method called Always Encrypted encrypts particular columns. Always Encrypted guarantees that neither the database nor the database server ever sees unencrypted values of sensitive columns. In other words, only the application code works with unencrypted data.

NOTE Columns belonging to the primary key, as well as all other non-sensitive columns, should not be encrypted, for performance reasons.

In the Always Encrypted method, the database driver handles the process of encrypting and decrypting of data. When a query is executed, the driver automatically looks up the master key in the Windows Certificate Store or other OS-dependent location. The master key is then used to decrypt a column-specific key, which in turn is used for encrypting and decrypting fields and parameters.

To use Always Encrypted, you have to perform the following three steps:

1. Create a column master key (CMK).
2. Create a column encryption key (CEK).
3. Create a table with one or more encrypted columns.

The column master key is used to protect a column encryption key. Additionally, the column master key definition object is created in the database. This object will store the information about the location of the column master key.

To create a column master key with Object Explorer, open the Always Encrypted Keys folder under Security for the database that will contain the table with encrypted columns. Right-click Column Master Key and select New Column Master Key. This opens the New Column Master Key dialog box, shown in Figure 12-2, which you will use to define a column master key for your database. The easiest option for developing new applications using Always Encrypted is to use a certificate, stored in your personal Certificate Store, as a column master key (see Figure 12-2). Alternatively, you can use a certificate of the local computer.

After that, enter a name for your column master key (in our example, **MyCMK**), click Generate Certificate, and click OK. This will generate a self-signed certificate, put it in your personal store, and create a definition of the column master key in the database (see Figure 12-2).

NOTE You can also create a column master key using the CREATE COLUMN MASTER KEY statement.

The next step is to create a column encryption key. This key is used to encrypt sensitive data in one or more columns of particular tables. To create such a key, right-click the Column Encryption Keys folder, select New Column Encryption Key, enter a key name, and select the already created column master key name as an encrypting column master key for your new column encryption key (in our example, **MyCEK**). Once you click OK, a new column encryption key gets created, encrypted with the certificate you configured in the first step, and the encrypted value is uploaded to the database.

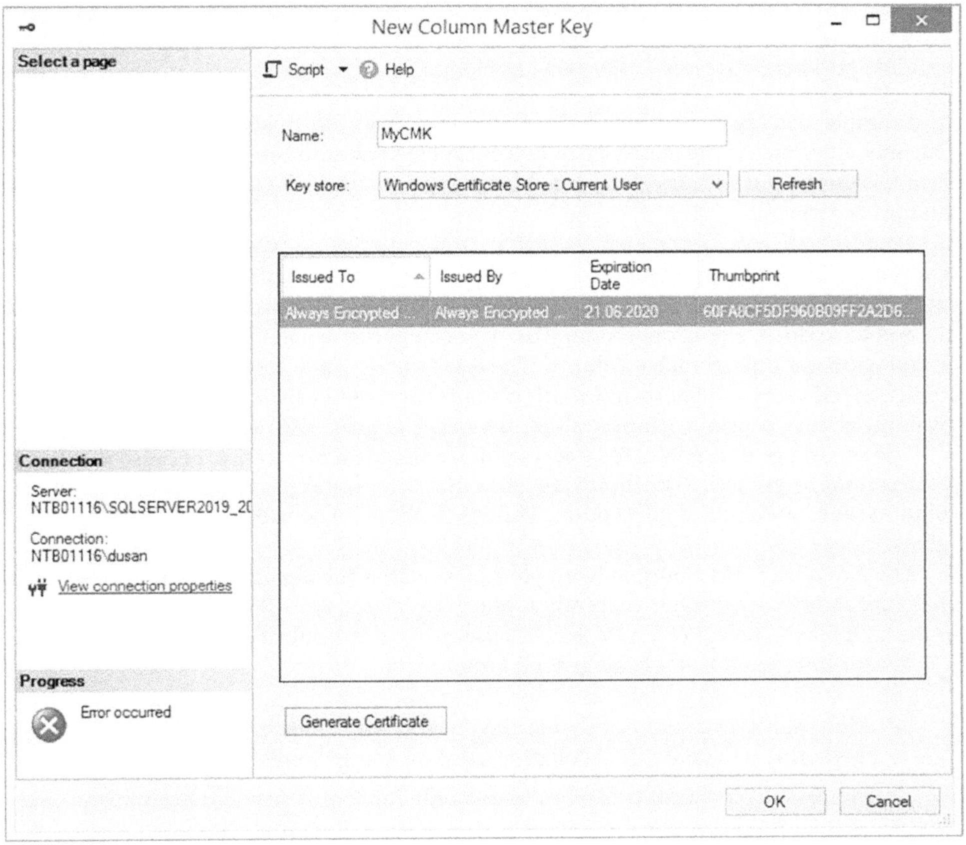

Figure 12-2   Specification of a column master key

NOTE You can also create a column encryption key using the CREATE COLUMN ENCRYPTION KEY statement.

In the third step, you use the convenient CREATE TABLE statement to create a table with encrypted columns. Example 12.6 shows this statement.

Example 12.6

```
USE sample;
CREATE TABLE   employee_encr(
 emp_no INT NOT NULL,
  emp_lname nvarchar (11) COLLATE Latin1_General_BIN2
 ENCRYPTED WITH (ENCRYPTION_TYPE = DETERMINISTIC,
 ALGORITHM = 'AEAD_AES_256_CBC_HMAC_SHA_256',
 COLUMN_ENCRYPTION_KEY = MyCEK) NOT NULL,
```

```
salary MONEY ENCRYPTED WITH (ENCRYPTION_TYPE = RANDOMIZED,
ALGORITHM = 'AEAD_AES_256_CBC_HMAC_SHA_256',
COLUMN_ENCRYPTION_KEY = MyCEK) NOT NULL);
```

Example 12.6 creates a table similar to the **employee** table used in the previous chapters. The only difference is the existence of two encrypted columns: **emp_lname** and **salary**. The former is encrypted with the deterministic encryption, while the latter is encrypted with the randomized one.

Deterministic encryption ensures that a given value always has the same (encrypted) representation, while *randomized encryption* delivers (theoretically) a different value every time. The advantage of deterministic encryption is that you can use this encryption method to perform a seek operation. Also, you can perform equality comparison, joins, and grouping operations on such a column. The disadvantage of deterministic encryption is that unauthorized users can guess information about encrypted values by examining patterns. (The fewer number of possible column values, the easier to guess the decrypted value.)

The advantage of randomized encryption is that it is more secure, because the decrypted value cannot be guessed. The disadvantage is that you cannot perform any operation on columns with randomized encryption. Therefore, they can be used only for display of column values.

The ALGORITHM clause in Example 12.6 specifies the algorithm used for encryption. Only one algorithm option is currently supported: AEAD_AES_256_CBC_HMAC_SHA_256. In the COLUMN_ENCRYPTION_KEY clause, you have to specify the encryption key name.

Always Encrypted has several general limitations:

- Alphanumerical columns that are encrypted with deterministic encryption have to use the Latin1_General_BIN2 collation, as shown in Example 12.6.

- Indices and constraints can be created only for columns with deterministic encryption.

- All range-like operations (greater than/less than, pattern matching using LIKE, etc.) are disallowed.

- Client libraries need to be updated to support encryption and decryption of columns. Not all drivers will support this functionality. (For the list of drivers, Microsoft Docs.)

- The following data types are not supported: SQL_VARIANT, XML, GEOGRAPHY, GEOMETRY, and user-defined data types (UDTs).

Transparent Data Encryption vs. Always Encrypted

There are several significant differences between TDE and Always Encrypted. The most important one is that TDE encrypts at the database level, while Always Encrypted encrypts at the column level of a table. Therefore, Always Encrypted reduces further the attack's surface area and the number of people who have access to the data in relation to TDE.

Another difference relates to the layers where the data is encrypted and decrypted. With TDE, each data page is encrypted when written to disk and is decrypted when read from disk. In other words, data is encrypted "at rest." On the other hand, Always Encrypted is encrypted *at rest and in memory*, meaning that data is encrypted on the disk, in memory, as well as across the network, and the only place where it is possible to decrypt data is in the user application.

Monitoring Always Encrypted   sys.columns_encryption_keys is the most important catalog view in relation to Always Encrypted. This view returns information about column encryption keys. Each row represents a column encryption key. Example 12.7 shows how you can use this view (together with the **sys.columns** view) to find out which columns in your database are encrypted and with which encryption keys.

Example 12.7

```
USE sample;
SELECT t.name AS table_name, c.name AS column_name,
 c.encryption_type_desc AS encr, k.name
 FROM sys.columns c JOIN sys.column_encryption_keys k
 ON  (c.column_encryption_key_id= k.column_encryption_key_id)
 JOIN sys.tables t ON (c.object_id = t.object_id);
```

The result is

table_name	column_name	encr	name
employee_encr	emp_lname	DETERMINISTIC	MyCEK
employee_encr	salary	RANDOMIZED	MyCEK

Authentication

The Database Engine's security system includes two different security subsystems:

- Windows security
- Database Engine security

Windows security specifies security at the operating system level—that is, the method by which a user connects to Windows using her *Windows user account* (called "user account" in the rest of the book). Authentication using this subsystem is also called Windows authentication.

Database Engine security specifies the additional security necessary at the system level—that is, how users who have already logged on to the operating system can subsequently connect to the database server. Database Engine security defines a SQL Server login (called "login" in the rest of the book) that is created within the system and is associated with a password. Some logins are identical to the existing user accounts. (Authentication using this subsystem is called SQL Server authentication.)

Based on these two security subsystems, the Database Engine can operate in one of the following authentication modes:

- Windows mode
- Mixed mode

Windows mode requires users to use user accounts exclusively to log in to the system. The system accepts the user account, assuming it has already been validated at the operating system level. This kind of connection to a database system is called a *trusted connection*, because the Database Engine trusts that the operating system already validated the account and the corresponding password.

Mixed mode allows users to connect to the Database Engine either using Windows authentication or SQL Server authentication. This means that some user accounts can be set up to use the Windows security subsystem, while others can be set up to use both the Database Engine security subsystem and the Windows security subsystem.

NOTE SQL Server authentication is provided for backward compatibility only. For this reason, use Windows authentication instead.

You use SQL Server Management Studio to choose one of the existing authentication modes. To set up Windows mode, right-click the server and click Properties. In the Server Properties dialog box, choose the Security page and click Windows Authentication Mode. To choose Mixed mode, the only difference is that you have to click SQL Server and Windows Authentication Mode in the Server Properties dialog box.

Setting Up the Database System Security

The security of the database system can be set up using:

- SQL Server Management Studio
- T-SQL statements

The following subsections discuss these two alternatives.

Managing Security Using SQL Server Management Studio

To create a new login using Management Studio, expand the server, expand Security, right-click Logins, and click New Login. The Login dialog box (see Figure 12-3) appears. First, you have to decide between Windows authentication and SQL Server authentication. If you choose Windows authentication, the login name must be a valid Windows name, which is written in the form **domain\user_name**. If you choose SQL Server authentication, you have to type the new login name and the corresponding password. Optionally, you may also specify the default database and language for the new login. (The default database is the database that the user is automatically connected to immediately after logging in to the Database Engine.) After that, the user can log in to the system under the new account.

Managing Security Using Transact-SQL Statements

The three Transact-SQL statements that are used to manage security of the Database Engine are CREATE LOGIN, ALTER LOGIN, and DROP LOGIN.

The CREATE LOGIN statement creates a new login. The syntax is as follows:

```
CREATE LOGIN login_name
{ WITH option_list1  |
FROM {WINDOWS [ WITH option_list2 [,...] ]
| CERTIFICATE certname  | ASYMMETRIC KEY key_name }}
```

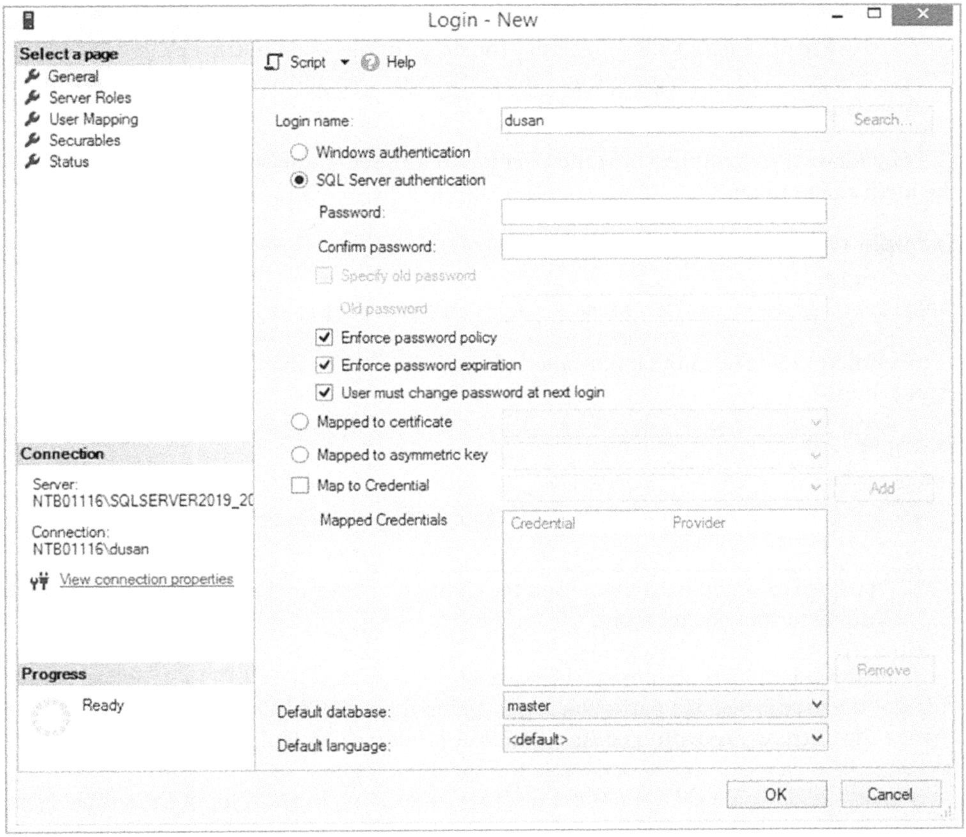

Figure 12-3   Login dialog box

login_name specifies the name of the login that is being created. As you can see from the syntax of the statement, you can use the WITH clause to specify one or more options for the login or use the FROM clause to define a certificate, asymmetric key, or user account associated with the corresponding login.

option_list1 contains several options. The most important one is the PASSWORD option, which specifies the password of the login (see Example 12.8). (The other possible options are DEFAULT_DATABASE, DEFAULT_LANGUAGE, and CHECK_EXPIRATION.)

As you can see from the syntax of the CREATE LOGIN statement, the FROM clause contains one of the following options:

- **WINDOWS**   Specifies that the login will be mapped to an existing user account (see Example 12.9). This clause can be specified with other suboptions, such as DEFAULT_ DATABASE and DEFAULT_LANGUAGE.

- **CERTIFICATE**   Specifies the name of the certificate to be associated with this login.
- **ASYMMETRIC KEY**   Specifies the name of the asymmetric key to be associated with this login. (The certificate and the asymmetric key must already exist in the **master** database.)

The following examples show the creation of different login forms. Example 12.8 specifies the login called **mary**.

Example 12.8

```
USE sample;
CREATE LOGIN mary WITH PASSWORD = 'youlknow4it9!';
```

Example 12.9 creates the login called **pete**, which will be mapped to a user account with the same name.

Example 12.9

```
USE sample;
CREATE LOGIN [NTB11901\pete] FROM WINDOWS;
```

NOTE You have to alter the username and the computer name (in the form **domain\username**) according to your environment.

The second security statement supported by Transact-SQL is ALTER LOGIN, which changes the properties of a particular login. Using the ALTER LOGIN statement, you can change the current password and its expiration properties, default database, and default language. You can also enable or disable the specified login.

Finally, the DROP LOGIN statement drops an existing login. A login cannot be dropped if it references other objects.

The sa Login

System administrator (**sa**) is a special login provided for backward compatibility. By default, it is assigned to the **sysadmin** fixed server role and cannot be changed. The **sa** login is the login to which was granted all possible permissions for system administration tasks.

NOTE Use the **sa** login only when there is not another way to log in to the database system.

Schemas

The Database Engine uses schemas in its security model to simplify the relationship between users and objects, and thus schemas have a very big impact on how you interact with the Database Engine. This section describes the role of schemas in Database Engine security. The first subsection describes the relationship between schemas and users; the second subsection discusses all three Transact-SQL statements related to schema creation and modification.

User–Schema Separation

A schema is a collection of database objects that is owned by a single person. The main purpose of a schema is to group logically related objects of a large database in different subunits. The Database Engine supports named schemas using the notion of a *principal*. A principal can be either of the following:

- An indivisible principal
- A group principal

An indivisible principal represents a single user, such as a login or user account. A group principal can be a group of users, such as a role or Windows group. Principals are ownerships of schemas, but the ownership of a schema can be transferred easily to another principal and without changing the schema name.

The separation of database users from schemas provides significant benefits, such as:

- One principal can own several schemas.
- Several indivisible principals can own a single schema via membership in roles or Windows groups.
- Dropping a database user does not require the renaming of objects contained by that user's schema.

Each database has a default schema, which is used to resolve the names of objects that are referred to without their fully qualified names. The default schema specifies the first schema that will be searched by the database server when it resolves the names of objects. The default schema can be set and changed using the DEFAULT_SCHEMA option of the CREATE USER or ALTER USER statement. If DEFAULT_SCHEMA is left undefined, the database user will have **dbo** as its default schema. (All default schemas are described in detail in the section "Default Database Schemas" later in this chapter.)

DDL Schema-Related Statements

There are three Transact-SQL schema-related statements:

- CREATE SCHEMA
- ALTER SCHEMA
- DROP SCHEMA

The following subsections describe in detail these statements.

CREATE SCHEMA

Example 12.10 shows how schemas can be created and used to control database security.

NOTE Before you start Example 12.10, you have to create database users **peter** and **mary**. For this reason, first execute the Transact-SQL statements in Example 12.12, located in the section "Managing Database Security Using Transact-SQL Statements."

Example 12.10
```
USE sample;
GO
CREATE SCHEMA my_schema AUTHORIZATION peter
GO
CREATE TABLE product
     (product_no CHAR(10) NOT NULL UNIQUE,
      product_name CHAR(20) NULL,
       price MONEY NULL);
GO
CREATE VIEW product_info
    AS SELECT product_no, product_name
          FROM product;
GO
GRANT SELECT TO mary;
DENY UPDATE TO mary;
```

Example 12.10 creates the **my_schema** schema, which comprises the **product** table and the **product_info** view. The database user called **peter** is the database-level principal that owns the schema. (You use the AUTHORIZATION option to define the principal of a schema. The principal may own other schemas and may not use the current schema as his or her default schema.)

NOTE The two other statements concerning permissions of database objects, GRANT and DENY, are discussed in detail later in this chapter. In Example 12.10, GRANT grants the SELECT permissions for all objects created in the schema, while DENY denies the UPDATE permissions for all objects of the schema.

The CREATE SCHEMA statement can create a schema; create the tables and views it contains; and grant, revoke, or deny permissions on a securable in a single statement. As you already know, securables are resources to which the system regulates access. There are three main securable scopes: server, database, and schema, which contain other securables, such as logins, database users, tables, and stored procedures.

The CREATE SCHEMA statement is atomic. In other words, if any error occurs during the execution of a CREATE SCHEMA statement, none of the Transact-SQL statements specified in the schema will be executed.

Database objects that are created in a CREATE SCHEMA statement can be specified in any order, with one exception: a view that references another view must be specified after the referenced view.

A database-level principal could be a user, role, or application role. (Roles and application roles are discussed in the "Roles" section later in the chapter.) The principal that is specified in the AUTHORIZATION clause of the CREATE SCHEMA statement is the owner of all objects created within the schema. Ownership of schema-contained objects can be transferred to any other database-level principal using the ALTER AUTHORIZATION statement.

The user needs the CREATE SCHEMA permission on the database to execute the CREATE SCHEMA statement. Also, to create the objects specified within the CREATE SCHEMA statement, the user needs the corresponding CREATE permissions.

ALTER SCHEMA

The ALTER SCHEMA statement transfers an object between different schemas of the same database. The syntax of the ALTER SCHEMA statement is as follows:

```
ALTER SCHEMA schema_name TRANSFER object_name
```

Example 12.11 shows the use of the ALTER SCHEMA statement.

Example 12.11

```
USE AdventureWorks;
ALTER SCHEMA HumanResources TRANSFER Person.ContactType;
```

Example 12.11 alters the schema called **HumanResources** of the **AdventureWorks** database by transferring into it the **ContactType** table from the **Person** schema of the same database.

The ALTER SCHEMA statement can only be used to transfer objects between different schemas in the same database. (Single objects within a schema can be altered using the ALTER TABLE statement or the ALTER VIEW statement.)

DROP SCHEMA

The DROP SCHEMA statement removes a schema from the database. You can successfully execute the DROP SCHEMA statement for a schema only if the schema does not contain any objects. If the schema contains any objects, the DROP SCHEMA statement will be rejected by the system.

As previously stated, the system allows you to change the ownership of a schema by using the ALTER AUTHORIZATION statement. This statement modifies the ownership of an entity.

NOTE The Transact-SQL language does not support the CREATE AUTHORIZATION and DROP AUTHORIZATION statements. You specify the ownership of an entity by using the CREATE SCHEMA statement.

Default Database Schemas

Each database within the system has the following default database schemas:

- guest
- dbo
- INFORMATION_SCHEMA
- sys

The Database Engine allows users without user accounts to access a database using the **guest** schema. (After creation, each database contains this schema.) You can apply permissions to the **guest** schema in the same way as you apply them to any other schema. Also, you can drop and add the **guest** schema from any database except the **master** and **tempdb** system databases.

Each database object belongs to one and only one schema, which is the default schema for that object. The default schema can be defined explicitly or implicitly. If the default schema isn't defined explicitly during the creation of an object, that object belongs to the **dbo** schema. Also, the login that is the owner of a database always has the special username **dbo** when using the database it owns.

The INFORMATION_SCHEMA schema contains all information schema views (see Chapter 11). The **sys** schema, as you may have already guessed, contains system objects, such as catalog views.

Database Security

A user account or a login allows a user to log in to the system. A user who subsequently wants to access a particular database of the system also needs a database user account to work with the database. Therefore, users must have a database user account for each database they want to use. The database user account can be mapped from the existing user accounts, groups (of which the user is a member), logins, or roles.

To manage database security, you can use

- SQL Server Management Studio
- Transact-SQL statements

The following subsections describe both ways to manage database security.

Managing Database Security Using SQL Server Management Studio

To add users to a database using Management Studio, expand the server, expand the Databases folder, expand the database, and expand Security. Right-click Users and click New User. In the Database User dialog box, enter a username and choose a corresponding login name. Optionally, you can choose a default schema for this user.

Managing Database Security Using Transact-SQL Statements

The CREATE USER statement adds a user to the current database. The syntax of this statement is

```
CREATE USER user_name
  [FOR {LOGIN login |CERTIFICATE cert_name |ASYMMETRIC KEY key_name}]
    [ WITH DEFAULT_SCHEMA = schema_name ]
```

user_name is the name that is used to identify the user inside the database. **login** specifies the login for which the user is being created. **cert_name** and **key_name** specify the corresponding certificate and asymmetric key, respectively. Finally, the WITH DEFAULT SCHEMA option specifies the first schema that will be searched by the server when it resolves the names of objects for this database user.

Example 12.12 demonstrates the use of the CREATE USER statement.

Example 12.12
```
USE sample;
CREATE USER peter FOR LOGIN [NTB11901\pete];
CREATE USER mary FOR LOGIN mary WITH DEFAULT_SCHEMA=my_schema;
```

NOTE To execute the first statement successfully, create the user account named **pete** and change the server (domain) name.

The first CREATE USER statement creates the database user called **peter** for the Windows login called **pete**. **pete** will use **dbo** as its default schema because the DEFAULT SCHEMA option is omitted. The second CREATE USER statement creates a new database user with the name **mary**. This user has **my_schema** as her default schema.

NOTE Each database has its own specific users. Therefore, the CREATE USER statement must be executed once for each database where a user account should exist. Also, a login can have only a single corresponding database user for a given database.

The ALTER USER statement modifies a database username, changes its default schema, or remaps a user to another login. Similar to the CREATE USER statement, it is possible to assign a default schema to a user before the creation of the schema.

The DROP USER statement removes a user from the current database. Users that own securables (that is, database objects) cannot be dropped from the database.

Roles

When several users need to perform similar activities in a particular database (and there is no corresponding Windows group), you can add a *database role*, which specifies a group of database users that can access the same objects of the database.

Members of a database role can be any of the following:

- Windows groups and user accounts
- Logins
- Other roles

The security architecture in the Database Engine includes several "system" roles that have special implicit permissions. There are two types of predefined roles (in addition to user-defined roles):

- Fixed server roles
- Fixed database roles

Beside these two, the following sections also describe the following types of roles:

- Application roles
- User-defined server roles
- User-defined database roles

The following sections describe in detail these five role types.

Fixed Server Roles

Fixed server roles are defined at the server level and therefore exist outside of databases belonging to the database server. Table 12-1 lists all existing fixed server roles.

Fixed Server Role	Description
sysadmin	Performs any activity in the database system
serveradmin	Configures server settings
setupadmin	Installs replication and manages extended procedures
securityadmin	Manages logins and CREATE DATABASE permissions and reads audits
processadmin	Manages system processes
dbcreator	Creates and modifies databases
bulkadmin	Allows non-administrative users to use the BULK INSERT command
diskadmin	Manages disk files

Table 12-1   Fixed Server Roles

Managing Fixed Server Roles

You can add members to and delete members from fixed server roles in two ways:

- Using SQL Server Management Studio
- Using T-SQL statements

To add a login to a fixed server role using Management Studio, expand the server, expand Security, and expand Server Roles. Right-click the role to which you want to add a login and then click Properties. On the Members page of the Server Role Properties dialog box (see Figure 12-4), click Add. Search for the login you want to add. Such login is then a member of the role and inherits all authorization properties assigned to that role.

The Transact-SQL statements CREATE SERVER ROLE and DROP SERVER ROLE are used, respectively, to add members to and delete members from a fixed server role. The ALTER SERVER ROLE statement modifies the membership of a server role. Example 12.14, later in the chapter, shows the use of the CREATE SERVER ROLE and ALTER SERVER ROLE statements.

NOTE   You cannot add, remove, or rename fixed server roles. Additionally, only the members of fixed server roles can execute the system procedures to add or remove logins to or from the role.

Fixed Database Roles

Fixed database roles are defined at the database level and therefore exist in each database belonging to the database server. Table 12-2 lists all of the fixed database roles. Members of the fixed database role can perform different activities. Use Microsoft Docs to learn which activities are allowed for each of the fixed database roles.

Besides the fixed database roles listed in Table 12-2, there is a special fixed database role called **public**, which is explained next.

public Role

The **public** role is a special fixed database role to which every legitimate user of a database belongs. It captures all default permissions for users in a database. This provides a mechanism for giving all users without appropriate permissions a set of (usually limited) permissions. The

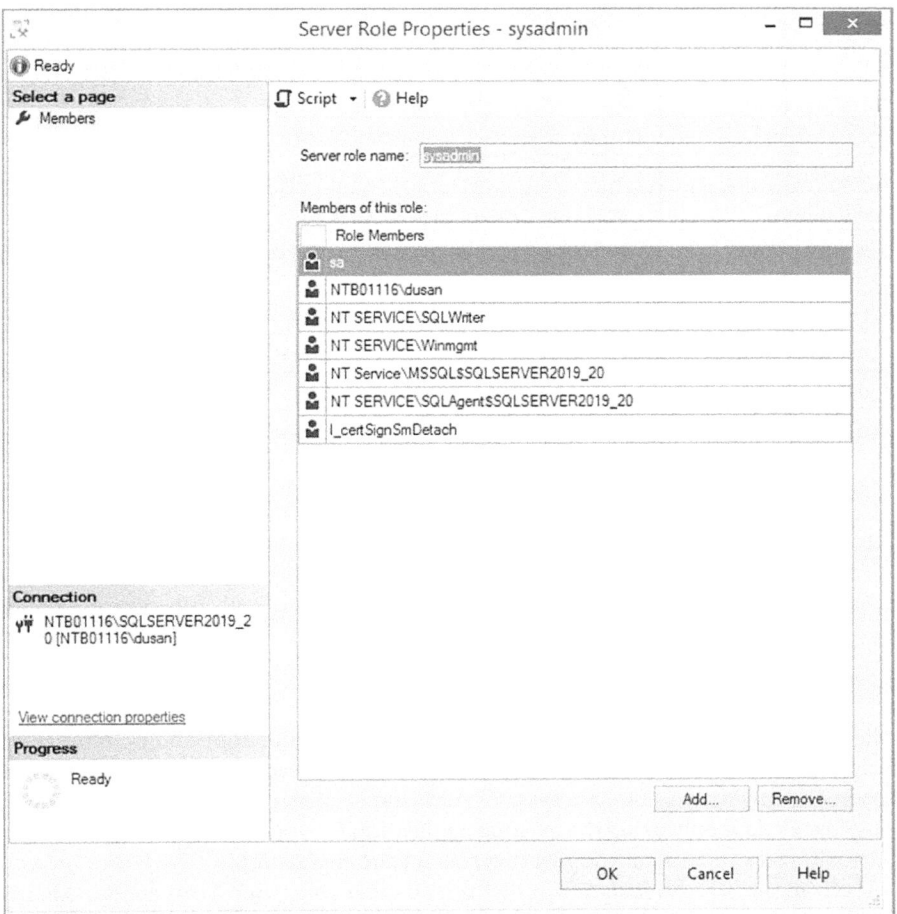

Figure 12-4   Server Role Properties dialog box

public role maintains all default permissions for users in a database and cannot be dropped. This role cannot have users, groups, or roles assigned to it because they belong to the role by default. (Example 12.24, later in the chapter, shows the use of the **public** role.)

By default, the **public** role allows users to do the following:

- View catalog views and display information from the **master** system database using certain system procedures
- Execute statements that do not require permissions—for example, PRINT

Assigning a User to a Fixed Database Role

To assign a user to a fixed database role using SQL Server Management Studio, expand the server, expand Databases, expand the database, expand Security, expand Roles, and then expand

Fixed Database Role	Description
db_owner	Users who can perform almost all activities in the database
db_accessadmin	Users who can add or remove users
db_datareader	Users who can see data from all user tables in the database
db_datawriter	Users who can add, modify, or delete data in all user tables in the database
db_ddladmin	Users who can perform all DDL operations in the database
db_securityadmin	Users who can manage all activities concerning security permissions in the database
db_backupoperator	Users who can back up the database
db_denydatareader	Users who cannot see any data in the database
db_denydatawriter	Users who cannot change any data in the database

Table 12-2 Fixed Database Roles

Database Roles. Right-click the role to which you want to add a user and then click Properties. In the Database Role Properties dialog box, click Add and browse for the user(s) you want to add. Such an account is then a member of the role and inherits all authorization properties assigned to that role.

Application Roles

Application roles allow you to enforce security for a particular application. In other words, application roles allow the application itself to accept the responsibility of user authentication, instead of relying on the database system. For instance, if clerks in your company may change an employee's data only using the existing application (and not Transact-SQL statements or any other tool), you can create an application role for the application.

Application roles differ significantly from all other role types. First, application roles have no members, because they use the application only and therefore do not need to grant permissions directly to users. Second, you need a password to activate an application role.

When an application role is activated for a session by the application, the session loses all permissions applied to the logins, user accounts and groups, or roles in all databases for the duration of the session. Because these roles are applicable only to the database in which they exist, the session can gain access to another database only by virtue of permissions granted to the **guest** user account in the other database. For this reason, if there is no **guest** user account in a database, the session cannot gain access to that database.

The next two subsections describe the management of application roles.

Managing Application Roles Using SQL Server Management Studio

To create an application role using SQL Server Management Studio, expand the server, expand Databases, and then expand the database and its Security folder. Right-click Roles, click New, and then click New Application Role. In the Application Role dialog box, enter the name of the new role. Additionally, you must enter the password and may enter the default schema for the new role.

Managing Application Roles Using T-SQL

You can create, modify, and delete application roles using the Transact-SQL statements CREATE APPLICATION ROLE, ALTER APPLICATION ROLE, and DROP APPLICATION ROLE.

The CREATE APPLICATION ROLE statement creates an application role for the current database. This statement has two options: one to specify the password and one to define the default schema—that is, the first schema that will be searched by the server when it resolves the names of objects for this role.

Example 12.13 adds a new application role called **weekly_reports** to the sample database.

Example 12.13

```
USE sample;
CREATE APPLICATION ROLE weekly_reports
WITH PASSWORD ='x1y2z3w4!', DEFAULT_SCHEMA =my_schema;
```

The ALTER APPLICATION ROLE statement changes the name, password, or default schema of an existing application role. The syntax of this statement is similar to the syntax of the CREATE APPLICATION ROLE statement. To execute the ALTER APPLICATION ROLE statement, you need the ALTER permission on the role.

The DROP APPLICATION ROLE statement removes the application role from the current database. If the application role owns any objects (securables), it cannot be dropped.

Activating Application Roles

After a connection is started, it must execute the **sp_setapprole** system procedure to activate the permissions that are associated with an application role. This procedure has the following syntax:

```
sp_setapprole [@rolename =] 'role' ,
                       [@password =] 'password'
                       [, [@encrypt =] 'encrypt_style']
```

role is the name of the application role defined in the current database, **password** specifies the corresponding password, and **encrypt_style** defines the encryption style specified for the password.

When you activate an application role using **sp_setapprole**, you need to know the following:

- After the activation of an application role, you cannot deactivate it in the current database until the session is disconnected from the system.

- An application role is always database bound—that is, its scope is the current database. If you change the current database within a session, you are allowed to perform other activities based on the permissions in that database.

NOTE The design of application roles in SQL Server is suboptimal, because it is not uniform. To create and delete application roles, you use Transact-SQL. After that, the activation of application roles is done by a system procedure.

User-Defined Server Roles

You can create and remove user-defined server roles using T-SQL statements CREATE SERVER ROLE and DROP SERVER ROLE, respectively. To add or delete members from a role, use the ALTER SERVER ROLE statement. Example 12.14 shows the use of the CREATE SERVER ROLE and ALTER SERVER ROLE statements. It creates a user-defined server role called **programadmin** and adds a new member to it.

Example 12.14

```
USE master;
GO
CREATE SERVER ROLE programadmin;
ALTER SERVER ROLE programadmin ADD MEMBER mary;
```

User-Defined Database Roles

Generally, user-defined database roles are applied when a group of database users needs to perform a common set of activities within a database and no applicable Windows group exists. These roles are created and deleted using either Management Studio or the Transact-SQL statements CREATE ROLE, ALTER ROLE, and DROP ROLE.

The following two subsections describe the management of user-defined database roles.

Managing User-Defined Database Roles Using SSMS

To create a user-defined role using Management Studio, expand the server, expand Databases, and then expand the database and its Security folder. Right-click Roles, click New, and then click New Database Role. In the Database Role dialog box (see Figure 12-5), enter the name of the new role. Click Add to add members to the new role. Choose the members (users and/or other roles) of the new role and click OK.

Managing User-Defined Database Roles Using T-SQL

The CREATE ROLE statement creates a new user-defined database role in the current database. The syntax of this statement is

```
CREATE ROLE role_name [AUTHORIZATION owner_name]
```

role_name is the name of the user-defined role to be created. **owner_name** specifies the database user or role that will own the new role. (If no user is specified, the role will be owned by the user that executes the CREATE ROLE statement.)

The ALTER ROLE statement changes the name of a user-defined database role. Similarly, the DROP ROLE statement removes a role from the database. Roles that own database objects (securables) cannot be dropped from the database. To drop such a role, you must first transfer the ownership of those objects.

Example 12.15 shows how you can create and add members to a user-defined role.

Example 12.15

```
USE sample;
CREATE ROLE marketing AUTHORIZATION peter;
GO
```

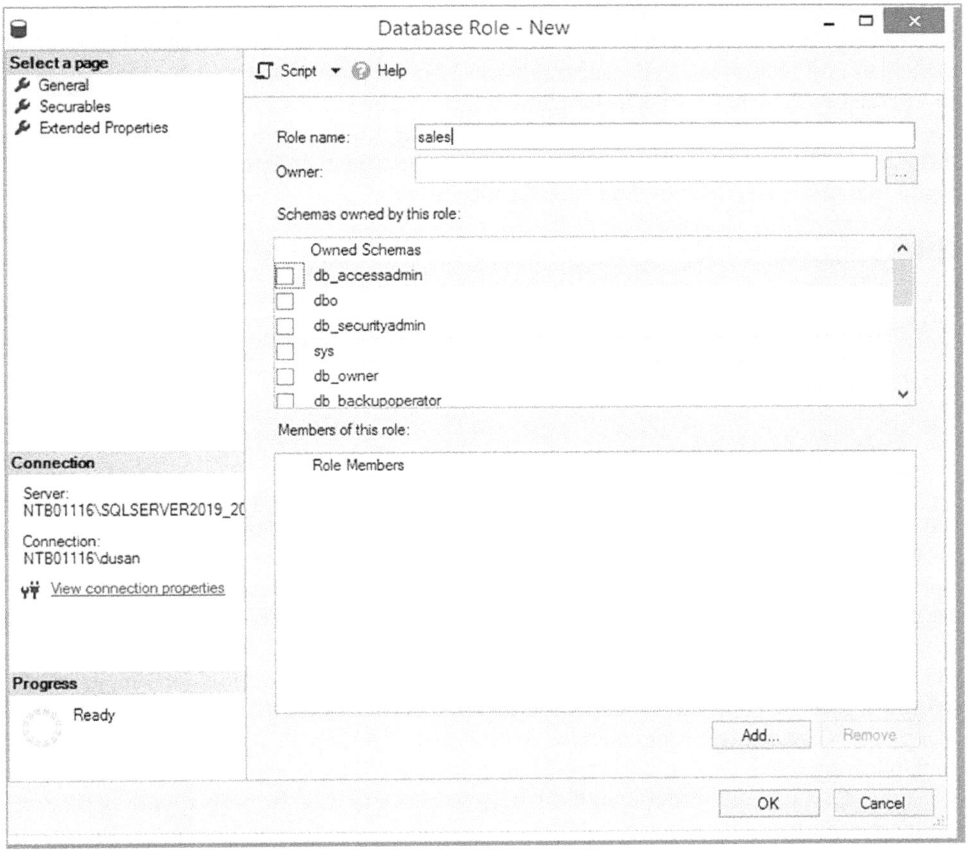

Figure 12-5   Database Role dialog box

```
ALTER ROLE marketing ADD MEMBER peter;
ALTER ROLE marketing ADD MEMBER mary;
```

Example 12.15 first creates the user-defined role called **marketing**, and then, using the ADD MEMBER clause of the ALTER ROLE statement, adds two members, **peter** and **mary**, to the role.

Authorization

Only authorized users are able to execute statements or perform operations on an entity. If an unauthorized user attempts to do either task, the execution of the Transact-SQL statement or the operation on the database object will be rejected.

There are three Transact-SQL statements related to authorization:

- GRANT
- DENY
- REVOKE

Before you read about these three statements, I will repeat the most important facts concerning the security model of the Database Engine. The model separates the world into principals and securables. Every securable has associated permissions that can be granted to a principal. Principals, such as individuals, groups, or applications, can access securables. Securables are the resources to which the authorization subsystem regulates access. There are three securable classes: server, database, and schema, which contain other securables, such as logins, database users, tables, and stored procedures.

GRANT Statement

The GRANT statement grants permissions to securables. The syntax of the GRANT statement is

```
GRANT {ALL [PRIVILEGES]} | permission_list
    [ON [class::] securable]  TO principal_list [WITH GRANT OPTION]
    [AS principal ]
```

The ALL clause is a deprecated feature and is maintained only for backward compatibility. It does not grant all possible permissions, as the name implies. (For the list of specific permissions, see Microsoft Docs.) **permission_list** specifies either statements or objects (separated by commas) for which the permissions are granted. **class** specifies either a securable class or a securable name for which permission is granted. ON **securable** specifies the securable for which permissions are granted (see Example 12.20 later in this section). **principal_list** lists all accounts (separated by commas) to which permissions are granted. **principal** and the components of **principal_list** can be a user account, a login or user account mapped to a certificate, a login mapped to an asymmetric key, a database user, a database role, or an application role.

Table 12-3 lists and describes all the permissions and lists the corresponding securables to which they apply.

NOTE Table 12-3 shows only the most important permissions. The security model of the Database Engine is hierarchical. Hence, there are many granular permissions that are not listed in the table. You can find the description of these permissions in Microsoft Docs.

The following examples demonstrate the use of the GRANT statement. To begin, Example 12.16 demonstrates the use of the CREATE permission.

Example 12.16

```
USE sample;
GRANT CREATE TABLE, CREATE PROCEDURE
                    TO peter, mary;
```

In Example 12.16, the users **peter** and **mary** can execute the Transact-SQL statements CREATE TABLE and CREATE PROCEDURE. (As you can see from this example, the GRANT statement with the CREATE permission does not include the ON option.)

Example 12.17 allows the user **mary** to create user-defined functions in the **sample** database.

Permission	Applies To	Description
SELECT	Tables + columns, synonyms, views + columns, table-valued functions	Provides the ability to select (read) rows. You can restrict this permission to one or more columns by listing them. (If the list is omitted, all columns of the table can be selected.)
INSERT	Tables + columns, synonyms, views + columns	Provides the ability to insert rows.
UPDATE	Tables + columns, synonyms, views + columns	Provides the ability to modify column values. You can restrict this permission to one or more columns by listing them. (If the list is omitted, all columns of the table can be modified.)
DELETE	Tables + columns, synonyms, views + columns	Provides the ability to delete rows.
REFERENCES	User-defined functions (SQL and CLR), tables + columns, synonyms, views + columns	Provides the ability to reference columns of the foreign key in the referenced table when the user has no SELECT permission for the referenced table.
EXECUTE	Stored procedures (SQL and CLR), user-defined functions (SQL and CLR), synonyms	Provides the ability to execute the specified stored procedure or user-defined function.
CONTROL	Stored procedures (SQL and CLR), user-defined functions (SQL and CLR), synonyms	Provides ownership-like capabilities; the grantee effectively has all defined permissions on the securable. A principal that has been granted CONTROL also has the ability to grant permissions on the securable. CONTROL at a particular scope implicitly includes CONTROL on all the securables under that scope (see Example 12.21 later in this section).
ALTER	Stored procedures (SQL and CLR), user-defined functions (SQL and CLR), tables, views	Provides the ability to alter the properties (except ownership) of a particular securable. When granted on a scope, it also bestows the ability to ALTER, CREATE, or DROP any securable contained within that scope.
TAKE OWNERSHIP	Stored procedures (SQL and CLR), user-defined functions (SQL and CLR), tables, views, synonyms	Provides the ability to take ownership of the securable on which it is granted.
VIEW DEFINITION	Stored procedures (SQL and CLR), user-defined functions (SQL and CLR), tables, views, synonyms	Controls the ability of the grantee to see the metadata of the securable (see Example 12.20).
CREATE (Server securable)	n/a	Provides the ability to create the server securable.
CREATE (DB securable)	n/a	Provides the ability to create the database securable.

Table 12-3 Permissions with Corresponding Securables

Part II

Example 12.17

```
USE sample;
GRANT CREATE FUNCTION TO mary;
```

Example 12.18 shows the use of the SELECT permission within the GRANT statement.

Example 12.18

```
USE sample;
GRANT SELECT ON employee
    TO peter, mary;
```

In Example 12.18, the users **peter** and **mary** can read rows from the **employee** table.

NOTE When a permission is granted to a user account or a login, this account (login) is the only one affected by the permission. On the other hand, if a permission is granted to a group or role, the permission affects all users belonging to the group (role).

Example 12.19 shows the use of the UPDATE permission within the GRANT statement.

Example 12.19

```
USE sample;
GRANT UPDATE ON works_on (emp_no, enter_date) TO peter;
```

After the GRANT statement in Example 12.19 is executed, the user **peter** can modify values of two columns of the **works_on** table: **emp_no** and **enter_date**.

Example 12.20 shows the use of the VIEW DEFINITION permission, which controls the ability of users to read metadata.

Example 12.20

```
USE sample;
GRANT VIEW DEFINITION ON OBJECT::employee TO peter;
GRANT VIEW DEFINITION ON SCHEMA::dbo TO peter;
```

Example 12.20 shows two GRANT statements for the VIEW DEFINITION permission. The first one allows the user **peter** to see metadata about the **employee** table of the **sample** database. (OBJECT is one of the base securables, and you can use this clause to give permissions for specific objects, such as tables, views, and stored procedures.) Because of the hierarchical structure of securables, you can use a "higher" securable to extend the VIEW DEFINITION (or any other base) permission. The second statement in Example 12.20 gives the user **peter** access to metadata of all the objects of the **dbo** schema of the **sample** database.

NOTE The VIEW DEFINITION permission allows you to grant or deny access to different pieces of your metadata and hence to decide which part of metadata is visible to other users.

Example 12.21 shows the use of the CONTROL permission.

Example 12.21

```
USE sample;
GRANT CONTROL ON DATABASE::sample TO peter;
```

In Example 12.21, the user **peter** effectively has all defined permissions on the securable (in this case, the **sample** database). A principal that has been granted CONTROL also implicitly has the ability to grant permissions on the securable; in other words, the CONTROL permission includes the WITH GRANT OPTION clause (see Example 12.22). The CONTROL permission is the highest permission in relation to several base securables. For this reason, CONTROL at a particular scope implicitly includes CONTROL on all the securables under that scope. Therefore, the CONTROL permission of user **peter** on the **sample** database implies all permissions on this database, all permissions on all assemblies in the database, all permissions on all schemas in the **sample** database, and all permissions on objects within the **sample** database.

By default, if user A grants a permission to user B, then user B can use the permission only to execute the Transact-SQL statement listed in the GRANT statement. The WITH GRANT OPTION gives user B the additional capability of granting the privilege to other users, as shown in Example 12.22.

Example 12.22

```
USE sample;
GRANT SELECT ON works_on  TO mary
   WITH GRANT OPTION;
```

In Example 12.22, the user **mary** can use the SELECT statement to retrieve rows from the **works_on** table and also may grant this privilege to other users of the **sample** database.

DENY Statement

The DENY statement prevents users from performing actions. This means that the statement removes existing permissions from user accounts or prevents users from gaining permissions through their group/role membership that might be granted in the future. This statement has the following syntax:

```
DENY {ALL [PRIVILEGES] } | permission_list
  [ON [class::] securable]  TO principal_list
[CASCADE]   [ AS principal ]
```

All options of the DENY statement have the same logical meaning as the options with the same name in the GRANT statement. DENY has an additional option, CASCADE, which specifies that permissions will be denied to user A and any other users to whom user A passed this permission. (If the CASCADE option is not specified in the DENY statement, and the corresponding object permission was granted with the WITH GRANT OPTION, an error is returned.)

The DENY statement prevents the user, group, or role from gaining access to the permission granted through their group or role membership. This means that if a user belongs to a group (or role) and the granted permission for the group is denied to the user, this user will be the only one of the group who cannot use this permission. On the other hand, if a permission is denied for a whole group, all members of the group will be denied the permission.

NOTE   You can think of the GRANT statement as a "positive" user authorization and the DENY statement as a "negative" user authorization. Usually, the DENY statement is used to deny already existing permissions for groups (roles) to a few members of the group.

Examples 12.23 and 12.24 show the use of the DENY statement.

Example 12.23

```
USE sample;
DENY CREATE TABLE, CREATE PROCEDURE
    TO peter;
```

The DENY statement in Example 12.23 denies two previously granted statement permissions to the user **peter**.

Example 12.24

```
USE sample;
GRANT SELECT ON project
    TO PUBLIC;
DENY SELECT ON project
        TO peter, mary;
```

Example 12.24 shows the negative authorization of some users of the **sample** database. First, the retrieval of all rows of the **project** table is granted to all users of the **sample** database. After that, this permission is denied to two users: **peter** and **mary**.

NOTE Permissions denied at a higher scope of the Database Engine security model override granted permissions at a lower scope. For instance, if SELECT permission is denied on the level of the **sample** database, and SELECT is granted on the **employee** table, the result is that SELECT is denied to the **employee** table as well as all other tables.

REVOKE Statement

The REVOKE statement removes one or more previously granted or denied permissions. This statement has the following syntax:

```
REVOKE [GRANT OPTION FOR]
    { [ALL [PRIVILEGES] ]  | permission_list ]}
 [ON  [class:: ] securable ]
 FROM principal_list [CASCADE]  [ AS principal ]
```

The only new option in the REVOKE statement is GRANT OPTION FOR. (All other options have the same logical meaning as the options with the same names in the GRANT or DENY statement.) GRANT OPTION FOR is used to remove the effects of the WITH GRANT OPTION in the corresponding GRANT statement. This means that the user will still have the previously granted permissions but will no longer be able to grant the permission to other users.

NOTE The REVOKE statement revokes "positive" permissions specified with the GRANT statement as well as "negative" permissions generated by the DENY statement. Therefore, its function is to neutralize the specified (positive or negative) permissions.

Example 12.25 shows the use of the REVOKE statement.

Example 12.25

```
USE sample;
REVOKE SELECT ON project FROM PUBLIC;
```

The REVOKE statement in Example 12.25 revokes the granted permission for the **public** role. At the same time, the existing "negative" permissions for the users **peter** and **mary** are not revoked (as in Example 12.24), because the explicitly granted or denied permissions are not affected by revoking roles or groups.

Managing Permissions Using SQL Server Management Studio

Database users can perform activities that are granted to them. In this case, there is a corresponding entry in the **sys.database_permissions** catalog view (that is, the value of the **state** column is set to **G** for grant). A negative entry in the table prevents a user from performing activities. The entry **D** (deny) in the **state** column overrides a permission that was granted to a user explicitly or implicitly using a role to which the user belongs. Therefore, the user cannot perform this activity in any case. In the last case (value **R**), the user has no explicit privileges but can perform an activity if a role to which the user belongs has the appropriate permission.

To manage permissions for a user or role using Management Studio, expand the server and expand Databases. Right-click the database and click Properties. Choose the Permissions page and click the Search button. In the Database Properties dialog box, shown in Figure 12-6 for the user **mary**, you can select one or more object types (users and/or roles) to which you want to grant or deny permissions. To grant a permission, check the corresponding box in the Grant column and click OK. To deny a permission, check the corresponding box in the Deny column. (The With Grant column specifies that the user has the additional capability of granting the privilege to other users.) Blanks in these columns mean no permission.

To manage permissions for a single database object using Management Studio, expand the server, expand Databases, expand the database, and then expand Tables, Views, or Synonyms, depending on the database object for which you want to manage permissions. Right-click the object, choose Properties, and select the Permissions page. (Figure 12-7 shows the Table Properties dialog box for the **department** table.) Click the Search button to open the Select Users or Roles dialog box. Click Object Types and select one or more object types (users, database roles, application roles). After that, click Browse and check all objects to which permissions should be granted. To grant a permission, check the corresponding box in the Grant column. To deny a permission, check the corresponding box in the Deny column.

Managing Authorization and Authentication of Contained Databases

As you already know from Chapter 5, contained databases have no configuration dependencies on the server instance where they are created and can therefore be easily moved from one instance of the Database Engine to another one. In this section you will learn how to authenticate users for contained databases. Each user that belongs to a contained database is not tied to a login, because such a user has no external dependencies and can be attached elsewhere.

Example 12.26 shows the creation of such a user.

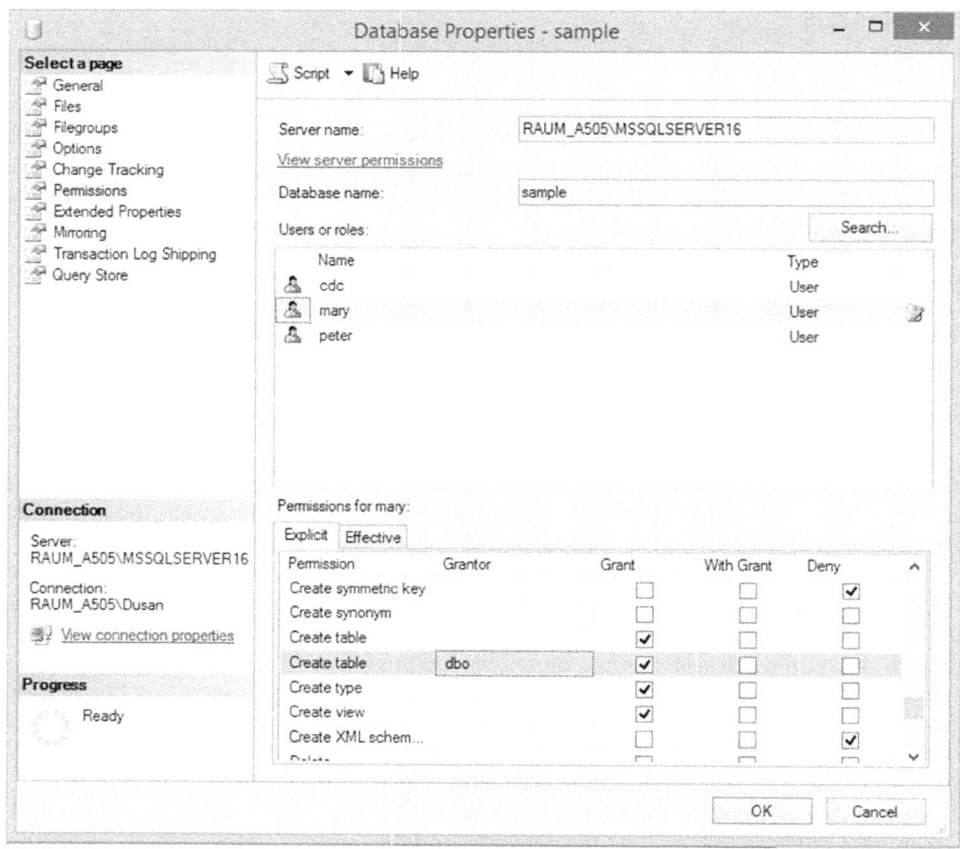

Figure 12-6 Managing statement permissions using SQL Server Management Studio

Example 12.26

```
USE my_sample;
CREATE USER my_login WITH PASSWORD = 'x1y2z3w4?';
```

Example 12.26 creates a user **my_login** that is not tied to a login. (The **my_sample** database is a contained database that was created in Example 5.20.) If you try to create such a user in a conventional database, you get the following error:

```
Msg 33233, Level 16, State 1, Line 1
You can only create a user with a password in a contained database.
```

The **sp_migrate_user_to_contained** procedure converts a database user that is mapped to a login to a contained database user with a password. **sp_migrate_user_to_contained** separates the user from the original login, so that settings such as password and default

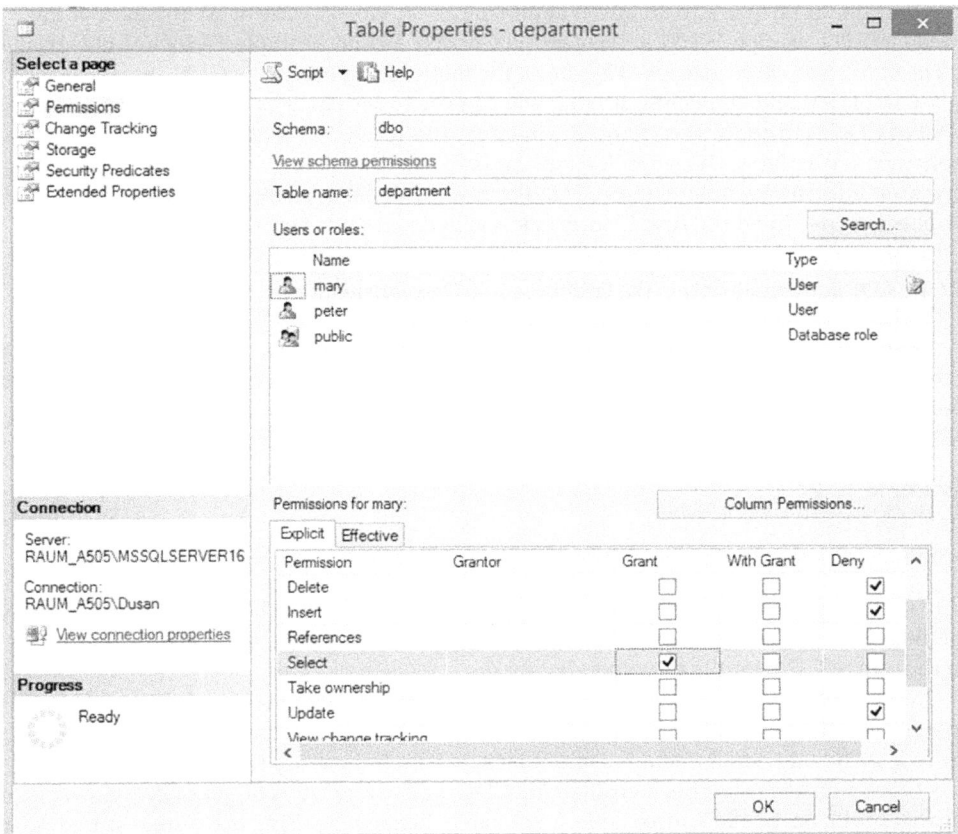

Figure 12-7   Managing object permissions for the department table

language can be administered separately for the contained database. This system stored procedure removes dependencies on the instance of the Database Engine and can be used before moving the contained database to a different server instance.

Also, you can use the dynamic management view called **sys.dm_db_uncontained_entities** to learn which parts of your database cannot be moved to a different server instance.

Change Tracking

Change tracking refers to documenting all insert, update, and delete activities that are applied to tables of the database. These changes can then be viewed to find out who accessed the data and when they accessed it. There are two ways to do it:

- Using triggers
- Using change data capture (CDC)

You can use triggers to create an audit trail of activities in one or more tables of the database. The section "AFTER Triggers" in Chapter 14 and Example 14.1 show how triggers can be used to track such changes. Therefore, the focus of this section is CDC.

CDC is a tracking mechanism that you can use to see changes as they happen. The primary goal of CDC is to audit who changed what data and when, but it can also be used to support concurrency updates. (If an application wants to modify a row, CDC can check the change tracking information to make sure that the row hasn't been changed since the last time the application modified the row. This check is called a *concurrency update*.)

NOTE CDC is available only in the Enterprise and Developer editions.

Before a capture instance can be created for individual tables, the database that contains the tables must be enabled for CDC, which you do with the system stored procedure **sys.sp_cdc_enable_db**, as shown in Example 12.27. (Only members of the **sysadmin** fixed server role can execute this procedure.)

Example 12.27

```
USE sample;
EXECUTE sys.sp_cdc_enable_db
```

To determine whether the **sample** database is enabled for CDC, you can retrieve the value of the column **is_cdc_enabled** in the **sys.databases** catalog view. The value 1 indicates the activation of CDC for the particular database.

When a database is enabled for CDC, the **cdc** schema, **cdc** user, metadata tables, and other system objects are created for the database. The **cdc** schema contains the CDC metadata tables as well as the individual tracking tables that serve as a repository for CDC.

Once a database has been enabled for CDC, you can create a target table that will capture changes for a particular source table. You enable the table by using the system stored procedure **sys.sp_cdc_enable_table**. Example 12.28 shows the use of this stored procedure.

NOTE The SQLServerAgent service must be running before you enable tables for CDC.

Example 12.28

```
USE sample;
EXECUTE sys.sp_cdc_enable_table
    @source_schema = N'dbo', @source_name = N'works_on',
    @role_name = N'cdc_admin';
```

The **sys.sp_cdc_enable_table** system procedure in Example 12.28 enables CDC for the specified source table in the current database. When a table is enabled for CDC, all DML statements are read from the transaction log and captured in the associated change table. The **@source_schema** parameter specifies the name of the schema in which the source table belongs. **@source_name** is the name of the source table on which you enable CDC. The **@role_name** parameter specifies the name of the database role used to allow access to data.

Creating a capture instance also creates a tracking table that corresponds to the source table. You can specify up to two capture instances for a source table. Example 12.29 changes the content of the source table (**works_on**).

Example 12.29

```
USE sample;
INSERT INTO works_on VALUES (10102, 'p2', 'Analyst', NULL);
INSERT INTO works_on VALUES (9031, 'p2', 'Analyst', NULL);
INSERT INTO works_on VALUES (29346, 'p3', 'Clerk', NULL);
```

By default, at least one table-valued function is created to access the data in the associated change table. This function allows you to query all changes that occur within a defined interval. The function name is the concatenation of **cdc.fn_cdc_get_all_changes_** and the value assigned to the **@capture_instance** parameter. In this case, the suffix of the parameter is **dbo_works_on**, as Example 12.30 shows.

Example 12.30

```
USE sample;
SELECT *
FROM cdc.fn_cdc_get_all_changes_dbo_works_on
     (sys.fn_cdc_get_min_lsn('dbo_works_on'), sys.fn_cdc_get_max_lsn(), 'all');
```

The following output shows part of the result of Example 12.30:

__$start_lsn	__$update_mask	emp_no	project_no	job	enter_date
0x0000001C000001EF0003	0x0F	10102	p2	Analyst	NULL
0x0000001D000000100003	0x0F	9031	p2	Analyst	NULL
0x0000001D000000110003	0x0F	29346	p3	Clerk	NULL

Example 12.30 shows all changes that happened after the execution of the three INSERT statements. If you want to track all changes in a certain time interval, you can use a batch similar to the one shown in Example 12.31.

Example 12.31

```
USE sample;
  DECLARE @from_lsn binary(10), @to_lsn binary(10);
    SELECT @from_lsn =
        sys.fn_cdc_map_time_to_lsn('smallest greater than', GETDATE() - 1);
    SELECT @to_lsn =
        sys.fn_cdc_map_time_to_lsn('largest less than or equal', GETDATE());
    SELECT * FROM
        cdc.fn_cdc_get_all_changes_dbo_works_on (@from_lsn, @to_lsn, 'all');
```

The only difference between Example 12.31 and Example 12.30 is that Example 12.31 uses two parameters (**@from_lsn** and **@to_lsn**) to define the beginning and end of the time interval. (The assignment of time boundaries is done using the **sys.fn_cdc_map_time_to_lsn()** function.)

Data Security and Views

As already stated in Chapter 11, views can be used for the following purposes:

- To restrict the use of particular columns and/or rows of tables
- To hide the details of complicated queries
- To restrict inserted and updated values to certain ranges

Restricting the use of particular columns and/or rows means that the view mechanism provides itself with the control of data access. For example, if the **employee** table also contains the salaries of each employee, then access to these salaries can be restricted using a view that accesses all columns of the table except the **salary** column. Subsequently, retrieval of data from the table can be granted to all users of the database using the view, while only a small number of (privileged) users will have the same permission for all data of the table.

Examples 12.32, 12.33, and 12.34 show the use of views to restrict the access to data.

Example 12.32

```
USE sample;
GO
CREATE VIEW v_without_budget
   AS SELECT project_no, project_name
             FROM project;
```

Using the **v_without_budget** view, as shown in Example 12.32, it is possible to divide users into two groups: the group of privileged users who can access the budget of all projects, and the group of common users who can access all rows from the **projects** table but not the data from the **budget** column.

Example 12.33

```
USE sample;
GO
ALTER TABLE employee
   ADD user_name CHAR(60) DEFAULT SYSTEM_USER;
GO
CREATE VIEW v_my_rows
        AS SELECT emp_no, emp_fname, emp_lname, dept_no
           FROM employee
           WHERE user_name = SYSTEM_USER;
```

The schema of the **employee** table is modified in Example 12.33 by adding the new column **user_name**. Every time a new row is inserted into the **employee** table, the system login is inserted into the **user_name** column. After the creation of the corresponding views, every user who uses this view can retrieve only the rows that he or she inserted into the table.

Example 12.34

```
USE sample;
GO
CREATE VIEW v_analyst
    AS SELECT employee.emp_no, emp_fname, emp_lname
            FROM employee, works_on
            WHERE employee.emp_no = works_on.emp_no
            AND job = 'Analyst';
```

The **v_analyst** view in Example 12.34 represents a horizontal subset and a vertical subset (in other words, it limits the rows and columns that can be accessed) of the **employee** table.

Summary

The following are the most important concepts of database system security:

- Encryption
- Authentication
- Authorization
- Change tracking

Data encryption is the process of scrambling information so that it is incomprehensible until it is decrypted by the intended recipient. Several different methods can be used to encrypt data. Authentication is the process of validating user credentials to prevent unauthorized users from using a system. It is most commonly enforced by requiring a username and password.

During the authorization process, the system determines which resources the particular user can use. The Database Engine supports authorization with the following Transact-SQL statements: GRANT, DENY, and REVOKE. Change tracking means that actions of unauthorized users are followed and documented on your system. This process is useful to protect the system against users with elevated privileges.

The next chapter discusses the features concerning the Database Engine as a multiuser software system and describes the notions of optimistic and pessimistic concurrency.

Exercises

E.12.1   What is a difference between Windows mode and Mixed mode?

E.12.2   What is a difference between a login and a user account?

E.12.3   Create three logins called **ann**, **burt**, and **chuck**. The corresponding passwords are **a1b2c3d4e5!**, **d4e3f2g1h0!**, and **f102gh285!**, respectively. The default database is the **sample** database. After creating the logins, check their existence using the system catalog.

E.12.4 Create three new database usernames for the logins in E.12.3. The new names are **s_ann**, **s_burt**, and **s_charles**.

E.12.5 Create a new user-defined database role called **managers** and add three members (see E.12.4) to the role. After that, display the information for this role and its members.

E.12.6 Using the GRANT statement, allow the user **s_burt** to create tables and the user **s_ann** to create stored procedures in the **sample** database.

E.12.7 Using the GRANT statement, allow the user **s_charles** to update the columns **lname** and **fname** of the **employee** table.

E.12.8 Using the GRANT statement, allow the users **s_burt** and **s_ann** to read the values from the columns **emp_lname** and **emp_fname** of the **employee** table. (Hint: Create the corresponding view first.)

E.12.9 Using the GRANT statement, allow the user-defined role **managers** to insert new rows in the **project** table.

E.12.10 Revoke the SELECT rights from the user **s_burt**.

E.12.11 Using Transact-SQL, do not allow the user **s_ann** to insert the new rows in the **project** table either directly or indirectly (using roles).

E.12.12 Discuss the difference between the use of views and Transact-SQL statements GRANT, DENY, and REVOKE in relation to security.

E.12.13 Display the existing information about the user **s_ann** in relation to the **sample** database. (Hint: Use the system procedure **sp_helpuser**.)

13 Concurrency Control

In This Chapter

- Concurrency Models
- Transactions
- Locking
- Isolation Levels
- Row Versioning

As you already know, data in a database is generally shared between many user application programs. The situation in which several user application programs read and write the same data at the same time is called *concurrency*. Thus, each DBMS must have some kind of control mechanism to solve concurrency problems.

A high level of concurrency is possible in a database system that can manage many active user applications without them interfering with each other. Conversely, a database system in which different active applications interfere with each other supports a low level of concurrency.

This chapter begins by describing the two concurrency control models that the Database Engine supports. The next section explains how concurrency problems can be solved using transactions. This discussion includes an introduction to the four properties of transactions, known as ACID properties, an overview of the Transact-SQL statements related to transactions, and an introduction to transaction logs. The third major section addresses locking and the three general lock properties: lock modes, lock resources, and lock duration. Deadlock, an important problem that can arise as a consequence of locking, is also introduced.

The behavior of transactions depends on the selected isolation level. The five isolation levels are introduced, including whether each belongs to the pessimistic or the optimistic concurrency model. The differences between existing isolation levels and their practical meaning will be explained, too.

The end of the chapter introduces row versioning, which is how the Database Engine implements the optimistic concurrency model. The two isolation levels related to this model—SNAPSHOT and READ COMMITTED SNAPSHOT—are discussed, as well as use of the **tempdb** system database as a version store.

Concurrency Models

The Database Engine supports two different concurrency models:

- Pessimistic concurrency
- Optimistic concurrency

Pessimistic concurrency uses locks to block access to data that is used by another process at the same time. In other words, a database system that uses pessimistic concurrency assumes that a conflict between two or more processes can occur at any time and therefore locks resources (row, page, table), as they are required, for the duration of a transaction. As you will see in the section "Locking," pessimistic concurrency issues shared locks on data being read so that no other process can modify that data. Also, pessimistic concurrency issues exclusive locks for data being modified so that no other processes can read or modify that data.

Optimistic concurrency works on the assumption that a transaction is unlikely to modify data that another transaction is modifying at the same time. The Database Engine supports optimistic concurrency so that older versions of data rows are saved, and any process that reads the same data uses the row version that was active when it started reading data. For that reason, a process that modifies the data can do so without any limitation, because all other processes that read the same data access the saved versions of the data. The only conflict scenario occurs when two or more write operations use the same data. In that case, the system displays an error so that the client application can handle it.

NOTE The notion of optimistic concurrency is generally defined in a broader sense. Optimistic concurrency control works on the assumption that resource conflicts between multiple users are unlikely, and allows transactions to execute without using locks. Only when a user is attempting to change data are resources checked to determine if any conflicts have occurred. If a conflict occurs, the application must be restarted.

Transactions

A transaction specifies a sequence of Transact-SQL statements that is used by database programmers to package together read and write operations, so that the database system can guarantee the consistency of data. There are two forms of transactions:

- **Implicit**   Specifies any single INSERT, UPDATE, or DELETE statement as a transaction unit
- **Explicit**   Generally, a group of Transact-SQL statements, where the beginning and the end of the group are marked using statements such as BEGIN TRANSACTION, COMMIT, and ROLLBACK

The notion of a transaction is best explained through an example. Suppose that in the **sample** database, the employee Ann Jones should be assigned a new employee number. The employee number must be modified in two different tables at the same time. The row in the **employee** table and all corresponding rows in the **works_on** table must be modified at the same time. (If only one of these tables were modified, data in the **sample** database would be inconsistent, because the values of the primary key in the **employee** table and the corresponding values of the foreign key in the **works_on** table for Ann Jones would not match.) Example 13.1 shows the implementation of this transaction using Transact-SQL statements.

Example 13.1

```
USE sample;
BEGIN TRANSACTION /* The beginning of the transaction */
UPDATE employee
    SET emp_no = 39831
    WHERE emp_no = 10102
    IF (@@error <> 0)
        ROLLBACK /* Rollback of the transaction */
UPDATE works_on
    SET emp_no = 39831
    WHERE emp_no = 10102
    IF (@@error <> 0)
        ROLLBACK
COMMIT /*The end of the transaction */
```

The consistent state of data used in Example 13.1 can be obtained only if both UPDATE statements are executed or neither of them is executed. The global variable **@@error** is used to test the execution of each Transact-SQL statement. If an error occurs, **@@error** is set to a negative value and the execution of all statements is rolled back. (The Transact-SQL statements BEGIN TRANSACTION, COMMIT, and ROLLBACK are defined in the upcoming section "Transact-SQL Statements and Transactions.")

NOTE The Transact-SQL language supports exception handling. Instead of using the global variable **@@error**, used in Example 13.1, you can use TRY and CATCH statements to implement exception handling in a transaction. The use of these statements is discussed in Chapter 8.

The next section explains the ACID properties of transactions. These properties guarantee that the data used by application programs will be consistent.

Properties of Transactions

Transactions have the following properties, which are known collectively by the acronym ACID:

- Atomicity
- Consistency
- Isolation
- Durability

The atomicity property guarantees the indivisibility of a set of statements that modifies data in a database and is part of a transaction. This means that either all data modifications in a transaction are executed or, in the case of any failure, all already executed changes are undone.

Consistency guarantees that a transaction will not allow the database to contain inconsistent data. In other words, the transactional transformations on data bring the database from one consistent state to another.

The isolation property separates concurrent transactions from each other. In other words, an active transaction can't see data modifications in a concurrent and incomplete transaction. This means that some transactions might be rolled back to guarantee isolation.

Durability guarantees one of the most important database concepts: persistence of data. This property ensures that the effects of the particular transaction persist even if a system error occurs. For this reason, if a system error occurs while a transaction is active, all statements of that transaction will be undone.

Transact-SQL Statements and Transactions

There are six Transact-SQL statements related to transactions:

- BEGIN TRANSACTION
- BEGIN DISTRIBUTED TRANSACTION
- COMMIT [WORK]
- ROLLBACK [WORK]
- SAVE TRANSACTION
- SET IMPLICIT_TRANSACTIONS

The BEGIN TRANSACTION statement starts the transaction. It has the following syntax:

```
BEGIN TRANSACTION [ {transaction_name | @trans_var }
    [WITH MARK ['description']]]
```

transaction_name is the name assigned to the transaction, which can be used only on the outermost pair of nested BEGIN TRANSACTION/COMMIT or BEGIN TRANSACTION / ROLLBACK statements. **@trans_var** is the name of a user-defined variable containing a valid transaction name. The WITH MARK option specifies that the transaction is to be marked in the log. **description** is a string that describes the mark. If WITH MARK is used, a transaction name must be specified. (The transaction log is discussed in detail later in this chapter, as well as in Chapter 16.)

The BEGIN DISTRIBUTED TRANSACTION statement specifies the start of a distributed transaction managed by the Microsoft Distributed Transaction Coordinator (DTC). A *distributed* transaction is one that involves databases on more than one server. For this reason, there is a need for a coordinator that will coordinate execution of statements on all involved servers. The server executing the BEGIN DISTRIBUTED TRANSACTION statement is the transaction coordinator and therefore controls the completion of the distributed transaction. (See Chapter 18 for a discussion of distributed transactions.)

The COMMIT WORK statement successfully ends the transaction started with the BEGIN TRANSACTION statement. This means that all modifications made by the transaction are stored on the disk. The COMMIT WORK statement is a standardized SQL statement. (The WORK clause is optional.)

NOTE The Transact-SQL language also supports the COMMIT TRANSACTION statement, which is functionally equivalent to COMMIT WORK, with the exception that COMMIT TRANSACTION accepts a user-defined transaction name. COMMIT TRANSACTION is an extension of Transact-SQL in relation to the SQL standard.

In contrast to the COMMIT WORK statement, the ROLLBACK WORK statement reports an unsuccessful end of the transaction. Programmers use this statement if they assume that the database might be in an inconsistent state. In this case, all executed modification operations within the transaction are rolled back. The ROLLBACK WORK statement is a standardized SQL statement. (The WORK clause is optional.)

NOTE Transact-SQL also supports the ROLLBACK TRANSACTION statement, which is functionally equivalent to ROLLBACK WORK, with the exception that ROLLBACK TRANSACTION accepts a user-defined transaction name.

The SAVE TRANSACTION statement sets a savepoint within a transaction. A *savepoint* marks a specified point within the transaction so that all updates that follow can be canceled without canceling the entire transaction. (To cancel an entire transaction, use the ROLLBACK statement.)

NOTE The SAVE TRANSACTION statement actually does not commit any modification operation; it only creates a target for the subsequent ROLLBACK statement with the label with the same name as the SAVE TRANSACTION statement.

Example 13.2 shows the use of the SAVE TRANSACTION statement.

Example 13.2

```
BEGIN TRANSACTION;
INSERT INTO department (dept_no, dept_name)
     VALUES ('d4', 'Sales');
SAVE TRANSACTION a;
INSERT INTO department (dept_no, dept_name)
     VALUES ('d5', 'Research');
SAVE TRANSACTION b;
INSERT INTO department (dept_no, dept_name)
    VALUES ('d6', 'Management');
ROLLBACK TRANSACTION b;
INSERT INTO department (dept_no, dept_name)
    VALUES ('d7', 'Support');
ROLLBACK TRANSACTION a;
COMMIT TRANSACTION;
```

The only statement in Example 13.2 that is executed is the first INSERT statement. The third INSERT statement is rolled back by the ROLLBACK TRANSACTION **b** statement, while the other two INSERT statements are rolled back by the ROLLBACK TRANSACTION **a** statement.

NOTE The SAVE TRANSACTION statement, in combination with the IF or WHILE statement, is a useful transaction feature for the execution of parts of an entire transaction. On the other hand, the use of this statement is contrary to the principle of operational databases that a transaction should be as short as possible, because long transactions generally reduce data availability.

As you already know, each Transact-SQL statement always belongs either implicitly or explicitly to a transaction. The Database Engine provides implicit transactions for compliance with the SQL standard. When a session operates in the implicit transaction mode, selected statements implicitly issue the BEGIN TRANSACTION statement. This means that you do nothing to start an implicit transaction. However, the end of each implicit transaction must be explicitly committed or rolled back using the COMMIT or ROLLBACK statement. (If you do not explicitly commit the transaction, the transaction and all the data changes that it contains are rolled back when the user disconnects.)

To enable an implicit transaction, you have to enable the IMPLICIT_TRANSACTIONS clause of the SET statement. This statement sets the implicit transaction mode for the current session. When a connection is in the implicit transaction mode and the connection is not currently in a transaction, executing any of the following statements starts a transaction:

ALTER TABLE	FETCH	REVOKE
CREATE TABLE	GRANT	SELECT
DELETE	INSERT	TRUNCATE TABLE
DROP TABLE	OPEN	UPDATE

In other words, if you have a sequence of statements from the preceding list, each statement will represent a single transaction.

The beginning of an explicit transaction is marked with the BEGIN TRANSACTION statement. The end of an explicit transaction is marked with the COMMIT or ROLLBACK statement. Explicit transactions can be nested. In this case, each pair of statements BEGIN TRANSACTION/COMMIT or BEGIN TRANSACTION/ROLLBACK is used inside one or more such pairs. (The nested transactions are usually used in stored procedures, which themselves contain transactions and are invoked inside another transaction.) The global variable **@@trancount** contains the number of active transactions for the current user.

BEGIN TRANSACTION, COMMIT, and ROLLBACK can be specified using a name assigned to the transaction. (The named ROLLBACK statement corresponds either to a named transaction or to the SAVE TRANSACTION statement with the same name.) You can use a named transaction only in the outermost statement pair of nested BEGIN TRANSACTION/ COMMIT or BEGIN TRANSACTION/ROLLBACK statements.

Transaction Log

Relational database systems keep a record of each change they make to the database during a transaction. This is necessary in case an error occurs during the execution of the transaction. In this situation, all previously executed statements within the transaction have to be rolled back. As soon as the system detects the error, it uses the stored records to return the database to the consistent state that existed before the transaction was started.

The Database Engine keeps all stored records, in particular the before and after values, in one or more files called the *transaction log*. Each database has its own transaction log. Thus, if it is necessary to roll back one or more modification operations executed on the tables of the current database, the Database Engine uses the entries in the transaction log to restore the values of columns that the database had before the transaction was started.

Before Images, After Images, and Write-Ahead Log

The transaction log is used to roll back or restore a transaction. If an error occurs and the transaction does not completely execute, the system uses all existing before values from the transaction log (called *before images*) to roll back all modifications since the start of the transaction. The process in which before images from the transaction log are used to roll back all modifications is called the *undo* activity.

Transaction logs also store after images. *After images* are modified values that are used to roll forward all modifications since the start of the transaction. This process is called the *redo* activity and is applied during recovery of a database. (For further details concerning transaction logs and recovery, see Chapter 16.)

Every entry written into the log is uniquely identified using the log sequence number (LSN). All log entries that are part of the particular transaction are linked together, so that all parts of a transaction can be located for undo and redo activities.

In case of a system failure, the Database Engine must be able to restore all data. If the data would be committed first, the data written to disk but not logged before failure could not be restored. For this reason, the Database Engine (and all other relational database systems) writes data changes to a log *before* the transaction is committed. This process is called *write-ahead logging*. Hence, the task of write-ahead logging is to provide high availability and consistency in case of failure.

On the other hand, heavy transaction log writes may become the bottleneck, and performance of the whole system can significantly suffer. In that case, the Database Engine supports an option to trade the (moderate) data loss for performance. This feature is called delayed durability and is discussed next.

Delayed Durability

Delayed durability enables transactions to continue running as if the data, prepared for logging, had been flushed to disk. Actually, all write operations to disk are deferred and are sent to disk together with write operations of other transactions. The system uses a 60KB chunk of log buffer and attempts to flush the log to disk when this 60KB block is full. Delayed durability can be set either at the database level or at the transaction level.

The ALTER DATABASE statement is used together with the DELAYED_DURABILITY option of the SET clause to set delayed durability at the database level. This option has three values: DISABLED, ALLOWED, and FORCED. ALLOWED means that any individual transaction can use delayed durability. FORCED means that all transactions that can use delayed durability will use it. (This can be useful in the case of an existing application where you want to use this mode throughout and also minimize the amount of code that has to be modified.)

If you want to apply delayed durability at the transaction level, use the COMMIT TRANSACTION statement in the following way:

```
COMMIT TRANSACTION WITH (DELAYED_DURABILITY = ON);
```

The main advantage of delayed durability can be achieved if your applications contain mainly short transactions and when the log disk is slow. Delayed durability should not be used if your transactions are usually long-running or if you have high throughput and high concurrency.

Editing Information Concerning Transactions and Logs

There are several dynamic management views that can be used to display different information concerning transactions and logs. This section describes two DMVs concerning transactions and two DMVs concerning transaction logs:

- **sys.dm_tran_active_transactions**
- **sys.dm_tran_database_transactions**
- **sys.dm_db_log_space_usage**
- **sys.dm_db_log_stats** (introduced in SQL Server 2017)

The **sys.dm_tran_active_transactions** DMV returns information about transactions of the particular Database Engine instance. The **transaction_id** column provides the unique ID number of each transaction, the **transaction_begin_time** column displays the starting time of each transaction, and the numerical values of the **transaction_type** column provide the type of the particular transaction. The most important values of the **transaction_type** column are 1 (for read/write transaction), 2 (for read/only transaction), and 3 (for system transaction). Finally, the **name** column specifies the transaction name.

The following example displays the unique IDs of all active transactions, their types, and the starting times.

Example 13.3

```
SELECT transaction_id ID, name,transaction_begin_time start,
transaction_type type
        FROM sys.dm_tran_active_transactions;
```

The result is

```
ID      name                 start                      type
429     worktable            2019-03-26 19:14:31.290    2
432     worktable            2019-03-26 19:14:31.290    2
436     worktable            2019-03-26 19:14:31.290    2
438     worktable            2019-03-26 19:14:31.290    2
441     user_transaction     2019-03-26 19:14:31.290    2
450     user_transaction     2019-03-26 19:14:31.290    2
```

I will only explain the values of the **name** column. You can observe that apart from user transactions (the last two rows of the result set), **sys.dm_tran_active_transactions** also lists worktable in the **name** column. Worktables are used when you need the **tempdb** system database for storing temporary result sets. (Note that the output will vary based on the state of your database.)

The **sys.dm_tran_database_transactions** DMV displays detailed information about the transactions occurring on your Database Engine instance. It provides snapshot data, so the results may vary each time the view is queried. The information from this DMV is similar to information you get using the **sys.dm_tran_active_transactions** view (see the previous example), but the **sys.dm_tran_database_transactions** view can provide a more granular level of detail about each transaction.

NOTE The name of this dynamic management view is a misnomer, because the DMV is server-scoped.

The most important columns of the **sys.dm_tran_database_transactions** DMV are **transaction_id**, **database_id**, **database_transaction_type**, and **database_transaction_state**. The first two columns display the unique ID of the transaction and the unique database ID of the database, to which the particular transaction belongs, respectively. The **database_transaction_type** column displays the type of a transaction, and the **database_transaction_state** column displays the state of a transaction.

Several properties of transaction logs have an impact on the maintenance of databases. The total size of the log and the amount of space used are the most important parameters related to the performance of your system.

There are two ways to display the information concerning logs:

- DBCC SQLPERF
- sys.dm_db_log_space_usage

The DBCC SQLPERF command displays statistics concerning transaction log space usage for *all* databases.

The **sys.dm_db_log_space_usage** DMV returns space usage information for the transaction log per database, as shown in Example 13.4.

Example 13.4

```
USE AdventureWorks;
SELECT DB_NAME(database_id) AS DBName,
   ROUND(CONVERT(FLOAT,total_log_size_in_bytes/1024)/1024,2)
 AS LogSize_in_MB,
   ROUND(CONVERT(FLOAT,used_log_space_in_bytes/1024)/1024,2)
       AS LogUsedSize_in_MB,
   ROUND(used_log_space_in_percent,2) AS LogUsed_Percent
FROM sys.dm_db_log_space_usage;
```

The result is

DBName	LogSize_in_MB	LogUsedSize_in_MB	LogUsed_Percent
AdventureWorks	529.99	516.95	97.54

The query in Example 13.4 displays the content of the **total_log_size_in_bytes** and **used_log_space_in_bytes** columns, in megabytes. For this reason, both values are converted in the FLOAT data type, and then calculated in megabytes and rounded.

The **sys.dm_db_log_stats** DMV returns summary-level attributes and information on transaction log files of databases. You can use this information for monitoring and diagnostics of transaction log health. The query in Example 13.5 displays the last log backup times for the databases in my instance of the Database Engine.

Example 13.5

```
SELECT name AS 'DBName', log_backup_time AS 'Last log backup time'
FROM sys.databases AS s
CROSS APPLY sys.dm_db_log_stats(s.database_id);
```

The result is

```
DBName        Last log backup time
----------------------------------------
master        1900-01-01 00:00:00.000
graph_db      2019-03-30 12:03:59.893
sample        2019-03-30 12:02:36.780
```

Example 13.5 "joins" the **sys.dm_db_log_stats** DMV with the **sys.databases** catalog view to get database names of the corresponding transaction logs. That way, the last log backup time can be provided for all existing databases.

As you can see from the result of Example 13.5, the backup of the transaction logs of the **graph_db** database and the **sample** database happened on March 30, 2019, while the backup of the transaction log of the **master** database has never been created on my instance.

Locking

Concurrency can lead to several negative effects, such as the reading of nonexistent data or loss of modified data. Consider this real-world example illustrating one of these negative effects, called *dirty read*: User U_1 in the personnel department gets notice of an address change for the employee Jim Smith. U_1 makes the address change, but when viewing the bank account information of Mr. Smith in the consecutive dialog step, he realizes that he modified the address of the wrong person. (The enterprise employs two persons with the name Jim Smith.) Fortunately, the application allows the user to cancel this change by clicking a button. U_1 clicks the button, knowing that he has committed no error.

At the same time, user U_2 in the technical department retrieves the data of the latter Mr. Smith to send the newest technical document to his home, because the employee seldom comes to the office. As the employee's address was wrongly changed just before U_2 retrieved the address, U_2 prints out the wrong address label and sends the document to the wrong person.

To prevent problems like these in the pessimistic concurrency model, every DBMS must have mechanisms that control the access of data by all users at the same time. The Database Engine, like all relational DBMSs, uses locks to guarantee the consistency of the database in case of multiuser access. Each application program requires locks for the data it needs, guaranteeing that no other program can modify the same data. When another application program requests the modification of the locked data, the system either stops the program with an error or makes a program wait.

Locking has several different aspects:

- Lock duration
- Lock modes
- Lock granularity

Lock duration specifies a time period during which a resource holds the particular lock. Duration of a lock depends on, among other things, the mode of the lock and the choice of the isolation level.

The next two sections describe lock modes and lock granularity.

> **NOTE** The following discussion concerns the pessimistic concurrency model. The optimistic concurrency model is handled using row versioning, and will be explained at the end of this chapter.

Lock Modes

Lock modes specify different kinds of locks. The choice of which lock mode to apply depends on the resource that needs to be locked. The following three lock types are used for row- and page-level locking:

- Shared (S)
- Exclusive (X)
- Update (U)

A *shared lock* reserves a resource (page or row) for reading only. Other processes cannot modify the locked resource while the lock remains. On the other hand, several processes can hold a shared lock for a resource at the same time—that is, several processes can read the resource locked with the shared lock.

An *exclusive lock* reserves a page or row for the exclusive use of a single transaction. It is used for DML statements (INSERT, UPDATE, and DELETE) that modify the resource. An exclusive lock cannot be set if some other process holds a shared or exclusive lock on the resource—that is, there can be only one exclusive lock for a resource. Once an exclusive lock is set for the page (or row), no other lock can be placed on the same resource.

> **NOTE** The database system automatically chooses the appropriate lock mode according to the operation type (read or write).

An *update lock* can be placed only if no other update or exclusive lock exists. On the other hand, it can be placed on objects that already have shared locks. (In this case, the update lock acquires another shared lock on the same object.) If a transaction that modifies the object is committed, the update lock is changed to an exclusive lock if there are no other locks on the object. There can be only one update lock for an object.

> **NOTE** Update locks prevent certain common types of deadlocks. (Deadlocks are described at the end of this section.)

	Shared	**Update**	**Exclusive**
Shared	Yes	Yes	No
Update	Yes	No	No
Exclusive	No	No	No

Table 13-1   Compatibility Matrix for Shared, Exclusive, and Update Locks

Table 13-1 shows the compatibility matrix for shared, exclusive, and update locks. The matrix is interpreted as follows: Suppose transaction T_1 holds a lock as specified in the first column of the matrix, and suppose some other transaction, T_2, requests a lock as specified in the corresponding column heading. In this case, "yes" indicates that a lock of T_2 is possible, whereas "no" indicates a conflict with the existing lock.

NOTE   The Database Engine also supports other lock forms, such as latches and spinlocks. The description of these lock forms can be found in Microsoft Docs.

At the table level, there are five different types of locks:

- Shared (S)
- Exclusive (X)
- Intent shared (IS)
- Intent exclusive (IX)
- Shared with intent exclusive (SIX)

Shared and exclusive locks correspond to the row-level (or page-level) locks with the same names. Generally, an *intent* lock shows an intention to lock the next-lower resource in the hierarchy of the database objects. Therefore, intent locks are placed at a level in the object hierarchy above that which the process intends to lock. This is an efficient way to tell whether such locks will be possible, and it prevents other processes from locking the higher level before the desired locks can be attained.

Table 13-2 shows the compatibility matrix for all kinds of table locks. The matrix is interpreted exactly as the matrix in Table 13-1.

	S	**X**	**IS**	**SIX**	**IX**
S	Yes	No	Yes	No	No
X	No	No	No	No	No
IS	Yes	No	Yes	Yes	Yes
SIX	No	No	Yes	No	No
IX	No	No	Yes	No	Yes

Table 13-2   Compatibility Matrix for All Kinds of Table Locks

Lock Granularity

Lock granularity specifies which resource is locked by a single lock attempt. The Database Engine can lock the following resources:

- Row
- Page
- Index key or range of index keys
- Table
- Extent
- Database itself

NOTE The system automatically chooses the appropriate lock granularity.

A row is the smallest resource that can be locked. The support of row-level locking includes both data rows and index entries. Row-level locking means that only the row that is accessed by an application will be locked. Hence, all other rows that belong to the same page are free and can be used by other applications. The Database Engine can also lock the page on which the row that has to be locked is stored.

NOTE For clustered tables, the data pages are stored at the leaf level of the (clustered) index structure and are therefore locked with index key locks instead of row locks.

Locking is also done on disk units, called *extents*, that are 64K in size (see Chapter 15 for further discussion of extents). Extent locks are set automatically when a table (or index) grows and the additional disk space is needed.

Lock granularity affects concurrency. In general, the more granular the lock, the more concurrency is reduced. This means that row-level locking maximizes concurrency because it leaves all but one row on the page unlocked. On the other hand, system overhead is increased because each locked row requires one lock. Page-level locking (and table-level locking) restricts the availability of data but decreases the system overhead.

Lock Escalation

If many locks of the same granularity are held during a transaction, the Database Engine automatically upgrades these locks into a table lock. This process of converting many page-, row-, or index-level locks into one table lock is called *lock escalation*. The escalation threshold is the boundary at which the database system applies the lock escalation. Escalation thresholds are determined dynamically by the system and require no configuration. (Currently, the threshold boundary is 5000 locks.)

The general problem with lock escalation is that the Database Engine decides when to escalate a particular lock, and this decision might be suboptimal for applications with different requirements. You can use the ALTER TABLE statement to change the lock escalation mechanism. This statement supports the TABLE option with the following syntax:

```
SET ( LOCK_ESCALATION = { TABLE | AUTO | DISABLE } )
```

The TABLE option is the default value and specifies that lock escalation will be done at table-level granularity. The AUTO option allows the Database Engine to select the lock escalation granularity that is appropriate for the table schema. Finally, the DISABLE option allows you to disable lock escalation in most cases. (There are some cases in which the Database Engine must take a table lock to protect data integrity.)

Example 13.6 disables the lock escalation for the **employee** table.

Example 13.6

```
USE sample;
ALTER TABLE employee SET (LOCK_ESCALATION = DISABLE);
```

Affecting Locks

You can use either locking hints or the LOCK_TIMEOUT option of the SET statement to affect locks. The following subsections describe these features.

Locking Hints

Locking hints specify the type of locking used by the Database Engine to lock table data. Table-level locking hints can be used when finer control of the types of locks acquired on a resource is required. (Locking hints override the current transaction isolation level for the session.)

All locking hints are written as a part of the FROM clause in the SELECT statement. You can use the following locking hints:

- **UPDLOCK**   Places update locks for each row of the table during the read operation. All update locks are held until the end of the transaction.
- **TABLOCK (TABLOCKX)**   Places a shared (or exclusive) table lock on the table. All locks are held until the end of the transaction.
- **ROWLOCK**   Replaces the existing shared table lock with shared row locks for each qualifying row of the table.
- **PAGLOCK**   Replaces a shared table lock with shared page locks for each page containing qualifying rows.
- **NOLOCK**   Synonym for READUNCOMMITTED (see the description of isolation-level hints in the "Setting and Editing Isolation Levels" section later in this chapter).
- **HOLDLOCK**   Synonym for REPEATABLEREAD (see the description of isolation-level hints later in this chapter).
- **XLOCK**   Specifies that exclusive locks are to be taken and held until the transaction completes. If XLOCK is specified with ROWLOCK, PAGLOCK, or TABLOCK, the exclusive locks apply to the appropriate level of granularity.
- **READPAST**   Specifies that the Database Engine does not read rows that are locked by other transactions.

NOTE All these options can be combined in any order if the combination makes sense. (For example, the combination of TABLOCK and PAGLOCK does not make sense, because both options are applied to different resources.)

LOCK_TIMEOUT Option

If you don't want your process to wait without any time limitations, you can use the LOCK_TIMEOUT option of the SET statement. This option specifies the number of milliseconds a transaction will wait for a lock to be released. For instance, if you want your processes to wait eight seconds, you write the following statement:

```
SET LOCK_TIMEOUT 8000
```

If the particular resource cannot be granted to your process within this time period, the statement will be aborted with the corresponding error message.

The value of –1 (the default value) indicates no time-out; in other words, the transaction won't wait at all. (The READPAST locking hint provides an alternative to the LOCK_TIMEOUT option.)

Displaying Lock Information

The most important utility to display lock information is a dynamic management view called **sys.dm_tran_locks**. This view returns information about currently active lock manager resources. Each row represents a currently active request for a lock that has been granted or is waiting to be granted. The columns of this view relate to two groups: resource and request. The resource group describes the resource on which the lock request is being made, and the request group describes the lock request. The most important columns of this view are as follows:

- **resource_type**   Represents the resource type
- **resource_database_id**   Specifies the ID of the database under which this resource is scoped
- **request_mode**   Specifies the mode of the request
- **request_status**   Specifies the current status of the request

Example 13.7 displays all the locks that are in a wait state.

Example 13.7

```
USE AdventureWorks;
SELECT resource_type, DB_NAME(resource_database_id) as db_name,
    request_session_id, request_mode, request_status
    FROM sys.dm_tran_locks
    WHERE request_status = 'WAIT';
```

Deadlock

A *deadlock* is a special concurrency problem in which two transactions block the progress of each other. The first transaction has a lock on some database object that the other transaction wants to access, and vice versa. (In general, several transactions can cause a deadlock by building a circle of dependencies.) Example 13.8 shows the deadlock situation between two transactions.

> **NOTE** The parallelism of processes cannot be achieved naturally using the small **sample** database, because every transaction in it is executed very quickly. Therefore, Example 13.8 uses the WAITFOR statement to pause both transactions for ten seconds to simulate the deadlock.

Example 13.8

```
USE sample;
BEGIN TRANSACTION
UPDATE works_on
    SET job = 'Manager'
    WHERE emp_no = 18316
    AND project_no = 'p2'
WAITFOR DELAY '00:00:10'
UPDATE employee
    SET emp_lname = 'Green'
    WHERE emp_no = 9031
COMMIT

BEGIN TRANSACTION
UPDATE employee
    SET dept_no = 'd2'
    WHERE emp_no = 9031
WAITFOR DELAY '00:00:10'
DELETE FROM works_on
  WHERE emp_no = 18316
  AND project_no = 'p2'
COMMIT
```

If both transactions in Example 13.8 are executed at the same time, the deadlock appears and the system returns the following output:

```
Server: Msg 1205, Level 13, State 45
Transaction (Process id 56) was deadlocked with another process and
has been chosen as deadlock victim. Rerun your command.
```

As the output of Example 13.8 shows, the Database Engine handles a deadlock by choosing one of the transactions as a "victim" (actually, the one that closed the loop in lock requests) and rolling it back. (The other transaction is executed after that.) A programmer can handle a deadlock by implementing the conditional statement that tests for the returned error number (1205) and then executes the rolled-back transaction again.

You can affect which transaction the system chooses as the "victim" by using the DEADLOCK_PRIORITY option of the SET statement. There are 21 different priority levels, from –10 to 10. The "victim" session is chosen according to the session's deadlock priority.

Isolation Levels

In theory, each transaction should be fully isolated from other transactions. But, in such a case, data availability is significantly reduced, because read operations in a transaction block write operations in other transactions, and vice versa. If data availability is an important issue, this property can be loosened using isolation levels. *Isolation levels* specify the degree to which data being retrieved in a transaction is protected from changes to the same data by other transactions. Before you are introduced to the existing isolation levels, the following section takes a look at scenarios that can arise if locking isn't used and, hence, there is no isolation between transactions.

Concurrency Problems

If locking isn't used, and thus no isolation exists between transactions, the following four problems may appear:

- Lost update
- Dirty reads (discussed earlier, in the "Locking" section)
- Nonrepeatable reads
- Phantoms

The *lost update* concurrency problem occurs when no isolation is provided to a transaction from other transactions. This means that several transactions can read the same data and modify it. The changes to the data by all transactions, except those by the last transaction, are lost.

The *nonrepeatable read* concurrency problem occurs when one process reads data several times and another process changes the same data between two read operations of the first process. Therefore, the values read by both read operations of the first process are different.

The *phantom* concurrency problem is similar to the nonrepeatable read concurrency problem, because two subsequent read operations can display different values, but in this case, the reason for this behavior lies in the different number of rows being read the first time and the second time. (Additional rows, called *phantoms*, are inserted by other transactions.)

The Database Engine and Isolation Levels

Using isolation levels, you can specify which of the concurrency problems discussed in the preceding section may occur and which you want to avoid. The Database Engine supports the following five isolation levels, which control how your read operations are executed:

- READ UNCOMMITTED
- READ COMMITTED
- REPEATABLE READ
- SERIALIZABLE
- SNAPSHOT

READ UNCOMMITTED, REPEATABLE READ, and SERIALIZABLE are available only in the pessimistic concurrency model, whereas SNAPSHOT is available only in the optimistic concurrency model. READ COMMITTED is available in both models. The four isolation levels available in the pessimistic concurrency model are described next. SNAPSHOT is described in the section titled "Row Versioning."

READ UNCOMMITTED

READ UNCOMMITTED provides the simplest form of isolation between transactions, because it does not isolate the read operations from other transactions at all. When a transaction retrieves a row at this isolation level, it acquires no locks and respects none of the existing locks. The data that is read by such a transaction may be inconsistent. In this case, a transaction reads data that is updated from some other active transaction. If the latter transaction rolls back later, the former transaction reads data that never really existed.

Of the four concurrency problems described in the preceding section, READ UNCOMMITTED allows dirty reads, nonrepeatable reads, and phantoms.

> **NOTE** The READ UNCOMMITTED isolation level is usually very undesirable and should be used only when the accuracy of the data read is not important or the data is seldom modified.

READ COMMITTED

As you already know, the READ COMMITTED isolation level has two forms. The first form applies to the pessimistic concurrency model, while the second form applies to the optimistic concurrency model. This section discusses the former. The second form, READ COMMITTED SNAPSHOT, is discussed in the upcoming section "Row Versioning."

A transaction that reads a row and uses the READ COMMITTED isolation level tests only whether an exclusive lock is placed on the row. If no such lock exists, the transaction fetches the row. (This is done using a shared lock.) This action prevents the transaction from reading data that is not committed and that can be subsequently rolled back. After reading the data values, the data can be changed by some other transaction.

Shared locks used by this isolation level are released immediately after the data is processed. (Generally, all locks are released at the end of the transaction.) For this reason, the access to the concurrent data is improved, but nonrepeatable reads and phantoms can still happen.

> **NOTE** The READ COMMITTED isolation level is the default isolation level of the Database Engine.

REPEATABLE READ

In contrast to the READ COMMITTED isolation level, REPEATABLE READ places shared locks on all data that is read and holds these locks until the transaction is committed or rolled back. Therefore, in this case, the execution of a query several times inside a transaction will always display the same result. The disadvantage of this isolation level is that concurrency is further reduced, because the time interval during which other transactions cannot update the same data is significantly longer than in the case of READ COMMITTED.

This isolation level does not prevent another transaction from inserting new rows, which are included in subsequent reads, so phantoms can appear.

SERIALIZABLE

SERIALIZABLE is the strongest isolation level, because it prevents all four concurrency problems already discussed. It acquires a range lock on all data that is read by the corresponding transaction. Therefore, this isolation level also prevents the insertion of new rows by another transaction until the former transaction is committed or rolled back.

> **NOTE** The SERIALIZABLE isolation level is implemented using a key-range locking method. This method locks individual rows and the ranges between them. A key-range lock acquires locks for index entries rather than locks for the particular pages or the entire table. In this case, any modification operation of another transaction cannot be executed, because the necessary changes of index entries are not possible.

As a conclusion to the previous discussion of all four isolation levels, you have to know that each isolation level in the preceding description reduces the concurrency more than the previous one. Thus, the isolation level READ UNCOMMITTED reduces concurrency the least. On the other hand, it also has the smallest isolation from concurrent transactions. SERIALIZABLE reduces concurrency the most, but guarantees *full isolation* between concurrent transactions.

Setting and Editing Isolation Levels

You can set an isolation level by using the following:

- The TRANSACTION ISOLATION LEVEL clause of the SET statement
- Isolation-level hints

The TRANSACTION ISOLATION LEVEL option of the SET statement provides five constant values, which have the same names and meanings as the standard isolation levels just described. The FROM clause of the SELECT statement supports four hints for isolation levels:

- READUNCOMMITTED
- READCOMMITTED
- REPEATABLEREAD
- SERIALIZABLE

These hints correspond to the isolation levels with the same name (but with a space in the name for the first three). The specification of isolation levels in the FROM clause of the SELECT statement overrides the current value set by the SET TRANSACTION ISOLATION LEVEL statement.

The DBCC USEROPTIONS command returns, among other things, information about the isolation level. Look at the value of the ISOLATION LEVEL option of this statement to find out the isolation level of your process.

Row Versioning

The Database Engine supports an optimistic concurrency control mechanism based on row versioning. When data is modified using row versioning, logical copies of the data are maintained for all data modifications performed in the database. Every time a row is modified, the Database Engine stores the before image of the previously committed row in **tempdb**. Each version is marked with the transaction sequence number (XSN) of the transaction that made the change. (The XSN is used to identify all operations to be managed under the corresponding transaction.) The newest version of a row is always stored in the database and chained in the linked list with old versions of the same row. An old row version in the **tempdb** database might contain pointers to other, even older versions. Each row version is kept in the **tempdb** database as long as there are operations that might require it.

Row versioning isolates transactions from the effects of modifications made by other transactions without the need for requesting shared locks on rows that have been read. This significant reduction in the total number of locks acquired by this isolation form significantly increases availability of data. However, exclusive locks are still needed.

The Database Engine supports two forms of row versioning:

- Read Committed Snapshot Isolation (RCSI)
- Snapshot Isolation (SI)

The following subsections describe Read Committed Snapshot Isolation and Snapshot Isolation.

Read Committed Snapshot Isolation

RCSI is a statement-level isolation, which means that any other transaction will read the committed values as they exist at the beginning of the statement. The most important property of RCSI is that read operations do not block updates, and updates do not block read operations.

RCSI will be applied only when activated and in the following way:

- Write operations will continue to use the pessimistic locking model, meaning that each data modification acquires exclusive lock(s). On the other hand, these locks will not affect reading operations. (The locks only affect other write operations.)
- Read operations are supplied with versioned copies of the data they need. This ensures that the data will not be changed while a process is performing read operations.

You use the SET clause of the ALTER DATABASE statement to enable the READ COMMITTED SNAPSHOT isolation level. After activation, no further changes are necessary. Any transaction specified with the READ COMMITTED SNAPSHOT isolation level will now run under RCSI.

Snapshot Isolation

Snapshot Isolation is a transaction-level isolation, which means that any other transaction will read the committed values as they exist just before the snapshot transaction starts. Also, the snapshot transaction will return the initial value until it completes, even if another transaction changed it in the meantime. Therefore, only after the snapshot transaction ends will the other transaction read a modified value.

Transactions running under SI acquire exclusive locks on data before performing the modification only to enforce constraints. Otherwise, locks are not acquired on data until the data is to be modified. When a data row meets the update criteria, the snapshot transaction verifies that the data row has not been modified by a concurrent transaction that committed after the transaction began. If the data row has been modified in a concurrent transaction, an update conflict occurs and the snapshot transaction is terminated. The update conflict is handled by the Database Engine, and no way exists to disable the update conflict detection.

Enabling SI is a two-step process. First, on the database level, enable the ALLOW_ SNAPSHOT_ISOLATION database option (using SQL Server Management Studio, for instance). Second, for each session that will use this isolation level, set the SET TRANSACTION ISOLATION LEVEL statement to SNAPSHOT. When these options are set, versions are built for all rows that are modified in the database.

RCSI vs. SI

The most important difference between the two forms of row versioning is that SI can result in update conflicts when a process sees the same data for the duration of its transaction and is not blocked. By contrast, RCSI does not use its own XSN when choosing row versions. Each time a statement is started, such a transaction reads the latest XSN issued for that instance of the database system and selects the row based on that number.

Another difference is that RCSI allows other transactions to modify the data before the snapshot transaction completes. This can lead to a conflict if another transaction modifies the data between the time the former performs a read and subsequently tries to execute the corresponding write operation. (For an application based on SI, the system detects the possible conflicts and sends the corresponding error message.)

The third difference is in relation to code changes. If you choose SI, your applications must be written to trap concurrency problems. Otherwise, you will have to rewrite your application code. On the other hand, RCSI does not require any code changes in applications that use the pessimistic model of the Database Engine.

Summary

Concurrency in multiuser database systems can lead to several negative effects, such as the reading of nonexistent data or loss of modified data. The Database Engine, like all other DBMSs, solves this problem by using transactions. A transaction is a sequence of Transact-SQL statements that logically belong together. All statements inside a transaction build an atomic unit. This means that either all statements are executed or, in the case of failure, all statements are canceled.

The locking mechanism is used to implement transactions. The effect of the lock is to prevent other transactions from changing the locked object. Locking has the following aspects: lock modes, lock granularity, and lock duration. Lock mode specifies different kinds of locks, the choice of which depends on the resource that needs to be locked. Lock duration specifies a time period during which a resource holds the particular lock.

The Database Engine provides a mechanism called a trigger that enforces, among other things, general integrity constraints. This mechanism is discussed in detail in the next chapter.

Exercises

E.13.1   What is a purpose of transactions?

E.13.2   What is the difference between a local transaction and a distributed transaction?

E.13.3   What is the difference between implicit transaction mode and explicit transaction mode?

E.13.4 What kinds of locks are compatible with an exclusive lock?

E.13.5 How can you test the successful execution of each T-SQL statement?

E.13.6 When should you use the SAVE TRANSACTION statement?

E.13.7 Discuss the difference between row-level and page-level locking.

E.13.8 Can a user explicitly influence the locking behavior of the system?

E.13.9 What is a difference between basic lock types (shared and exclusive) and an intent lock?

E.13.10 What does lock escalation mean?

E.13.11 Discuss the difference between the READ UNCOMMITTED and SERIALIZABLE isolation levels.

E.13.12 What is deadlock?

E.13.13 Which process is used as a victim in a deadlock situation? Can a user influence the decision of the system?

14 Triggers

In This Chapter

- Introduction to Triggers
- Application Areas for DML Triggers
- DDL Triggers and Their Application Areas

This chapter is dedicated to a mechanism called a *trigger*. The beginning of the chapter describes Transact-SQL statements for creating, deleting, and modifying triggers. After that, examples of different application areas for DML triggers are given. Each example is created using one of three statements, INSERT, UPDATE, or DELETE. The second part of the chapter covers DDL triggers, which are based on DDL statements such as CREATE TABLE. Again, examples of different application areas related to DDL triggers are given.

Introduction to Triggers

A trigger is a mechanism that is invoked when a particular action occurs on a particular table. Each trigger has three general parts:

- A name
- The action
- The execution

The maximum size of a trigger name is 128 characters. The action of a trigger can be either a DML statement (INSERT, UPDATE, or DELETE) or a DDL statement. Therefore, there are two trigger forms: DML triggers and DDL triggers. The execution part of a trigger usually contains a stored procedure or a batch.

> **NOTE** The Database Engine allows you to create triggers using either Transact-SQL or CLR programming languages such as C# and Visual Basic. This section describes the use of Transact-SQL to implement triggers. The implementation of triggers using CLR programming languages is not covered in this edition of the book. You can find it in the previous edition.

Creating a DML Trigger

A DML trigger is created using the CREATE TRIGGER statement, which has the following form:

```
CREATE TRIGGER [schema_name.]trigger_name
   ON {table_name | view_name}
      [WITH dml_trigger_option [,…]]
   {FOR | AFTER | INSTEAD OF} { [INSERT] [,] [UPDATE] [,] [DELETE]}
   [WITH APPEND]
   {AS  sql_statement   | EXTERNAL NAME method_name}
```

> **NOTE** The preceding syntax covers only DML triggers. DDL triggers have a slightly different syntax, which will be shown later in this chapter.

schema_name is the name of the schema to which the trigger belongs. **trigger_name** is the name of the trigger. **table_name** is the name of the table for which the trigger is specified. (Triggers on views are also supported, as indicated by the inclusion of **view_name**.)

AFTER and INSTEAD OF are two additional options that you can define for a trigger. (The FOR clause is a synonym for AFTER.) AFTER triggers fire after the triggering action occurs. INSTEAD OF triggers are executed instead of the corresponding triggering action. AFTER triggers can be created only on tables, while INSTEAD OF triggers can be created on both tables and views. Examples showing the use of these two trigger types are provided later in this chapter.

The INSERT, UPDATE, and DELETE options specify the trigger action. (The trigger action is the type of Transact-SQL statement that activates the trigger.) The DELETE statement is not allowed if the IF UPDATE option is used.

As you can see from the syntax of the CREATE TRIGGER statement, the AS **sql_statement** specification is used to determine the action(s) of the trigger. (You can also use the EXTERNAL NAME option, which is explained later in this chapter.)

> **NOTE** The Database Engine allows you to create multiple triggers for each table and for each action (INSERT, UPDATE, and DELETE). By default, there is no defined order in which multiple triggers for a given modification action are executed. (You can define the order by using the first and last triggers, as described later in this chapter.)

Only the database owner, DDL administrators, and the owner of the table on which the trigger is defined have the authority to create a trigger for the current database. (In contrast to the permissions for other CREATE statements, this permission is not transferable.)

Modifying a Trigger's Structure

Transact-SQL also supports the ALTER TRIGGER statement, which modifies the structure of a trigger. The ALTER TRIGGER statement is generally used to modify the body of the trigger. All clauses and options of the ALTER TRIGGER statement correspond to the clauses and options with the same names in the CREATE TRIGGER statement.

The DROP TRIGGER statement removes one or more existing triggers from the current database.

The following section describes deleted and inserted virtual tables, which play a significant role in a triggered action.

Using deleted and inserted Virtual Tables

When creating a triggered action, you usually must indicate whether you are referring to the value of a column before or after the triggering action changes it. For this reason, two virtual tables with special names are used to test the effect of the triggering statement:

- **deleted**   Contains copies of rows that are deleted from the triggered table
- **inserted**   Contains copies of rows that are inserted into the triggered table

The structure of these tables is equivalent to the structure of the table for which the trigger is specified.

The **deleted** virtual table is used if the DELETE or UPDATE clause is specified in the CREATE TRIGGER statement. The **inserted** virtual table is used if the INSERT or UPDATE clause is specified in the CREATE TRIGGER statement. This means that for each DELETE statement executed in the triggered action, the **deleted** virtual table is created. Similarly, for each INSERT statement executed in the triggered action, the **inserted** virtual table is created.

An UPDATE statement is treated as a DELETE, followed by an INSERT. Therefore, for each UPDATE statement executed in the triggered action, the **deleted** and **inserted** virtual tables are created (in this sequence).

The materialization of the **inserted** and **deleted** virtual tables is done using row versioning, which is discussed in detail in Chapter 13. When DML statements such as INSERT, UPDATE, and DELETE are executed on a table with corresponding triggers, all changes to the table are always versioned. When the trigger needs the information from the **deleted** virtual table, it accesses the data from the version store. In the case of the **inserted** virtual table, the trigger accesses the most recent versions of the rows.

NOTE Row versioning uses the **tempdb** database as the version store. For this reason, you must expect significant growth of this system database if your database contains many triggers that are often used.

Application Areas for DML Triggers

The first part of the chapter introduced how you can create a DML trigger and modify its structure. This trigger form can be used to solve different problems. This section describes several application areas for DML triggers (AFTER triggers and INSTEAD OF triggers).

AFTER Triggers

As you already know, AFTER triggers fire after the triggering action has been processed. You can specify an AFTER trigger by using either the AFTER or FOR reserved keyword. AFTER triggers can be created only on base tables.

AFTER triggers can be used to perform the following actions, among others:

- Create an audit trail of activities in one or more tables of the database (see Example 14.1)
- Implement business rules (see Example 14.2)
- Enforce referential integrity (see Examples 14.3 and 14.4)

Creating an Audit Trail

Chapter 12 discussed how you can capture data changes using the mechanism called CDC (change data capture). DML triggers can also be used to solve the same problem. Example 14.1 shows how triggers can create an audit trail of activities in one or more tables of the database.

Example 14.1

```
/* The audit_budget table is used as an audit trail in the project table */
USE sample;
GO
CREATE TABLE audit_budget
  (project_no CHAR(4) NULL,
   user_name CHAR(16) NULL,
   date DATETIME NULL,
   budget_old FLOAT NULL,
   budget_new FLOAT NULL);
GO
CREATE TRIGGER modify_budget
   ON project AFTER UPDATE
   AS IF UPDATE(budget)
   BEGIN
   DECLARE @budget_old FLOAT
   DECLARE @budget_new FLOAT
   DECLARE @project_number CHAR(4)
   SELECT @budget_old = (SELECT budget FROM deleted)
   SELECT @budget_new = (SELECT budget FROM inserted)
   SELECT @project_number = (SELECT project_no FROM deleted)
   INSERT INTO audit_budget VALUES
   (@project_number,USER_NAME(),GETDATE(),@budget_old, @budget_new)
   END
```

Example 14.1 shows how triggers can be used to implement an audit trail of the activity within a table. This example creates the **audit_budget** table, which stores all modifications of the **budget** column of the **project** table. Recording all the modifications of this column will be executed using the **modify_budget** trigger.

Every modification of the **budget** column using the UPDATE statement activates the trigger. In doing so, the values of the rows of the **deleted** and **inserted** virtual tables are assigned to the corresponding variables **@budget_old**, **@budget_new**, and **@project_number**. The assigned values, together with the username and the current date, will be subsequently inserted into the **audit_budget** table.

NOTE Example 14.1 assumes that only one row will be updated at a time. Therefore, it is a simplification of a general case in which a trigger handles multirow updates. The implementation of such a general (and complicated) trigger is beyond the introductory level of this book.

If the following Transact-SQL statement is executed,

```
UPDATE project
   SET budget = 200000
   WHERE project_no = 'p2';
```

the content of the **audit_budget** table is as follows:

project_no	user_name	date	budget_old	budget_new
p2	dbo	2019-01-31 14:00:05	95000	200000

Implementing Business Rules

Triggers can be used to create business rules for an application. Example 14.2 shows the creation of such a trigger.

Example 14.2

```
USE sample;
GO
CREATE TRIGGER total_budget
    ON project AFTER UPDATE
    AS IF UPDATE (budget)
      BEGIN
      DECLARE @sum_old1 FLOAT
      DECLARE @sum_old2 FLOAT
      DECLARE @sum_new FLOAT
      SELECT @sum_new = (SELECT SUM(budget) FROM inserted)
      SELECT @sum_old1 = (SELECT SUM(p.budget)
                          FROM project p WHERE p.project_no
          NOT IN (SELECT d.project_no FROM deleted d))
      SELECT @sum_old2 = (SELECT SUM(budget) FROM deleted)
      IF @sum_new > (@sum_old1 + @sum_old2)   *1.5
      BEGIN
         PRINT 'No modification of budgets'
         ROLLBACK TRANSACTION
         END
```

```
        ELSE
                PRINT 'The modification of budgets executed'
        END
```

Example 14.2 creates the rule controlling the modification of the budget for the projects. The **total_budget** trigger tests every modification of the budgets and executes only such UPDATE statements where the modification does not increase the sum of all budgets by more than 50 percent. Otherwise, the UPDATE statement is rolled back using the ROLLBACK TRANSACTION statement.

Enforcing Integrity Constraints

As previously stated in Chapter 5, the Database Engine handles two types of integrity constraints:

- Declarative integrity constraints, defined by using the CREATE TABLE and ALTER TABLE statements
- Procedural integrity constraints (handled by triggers)

Generally, you should use declarative integrity constraints, because they are supported by the system and you do not have to implement them. The use of triggers is recommended only for cases where declarative integrity constraints do not exist.

Example 14.3 shows how you can enforce the referential integrity for the **employee** and **works_on** tables using triggers.

Example 14.3

```
USE sample;
GO
CREATE TRIGGER works_integrity
    ON works_on AFTER UPDATE
    AS IF UPDATE(emp_no)
       BEGIN
       IF (SELECT employee.emp_no
           FROM employee, inserted
           WHERE employee.emp_no = inserted.emp_no) IS NULL
         BEGIN
         ROLLBACK TRANSACTION
         PRINT 'No insertion/modification of the row'
         END
       ELSE PRINT 'The row inserted/modified'
       END
```

The **works_integrity** trigger in Example 14.3 checks one form of the referential integrity for the **employee** and **works_on** tables. This means that every UPDATE of the **emp_no** column in the referenced **works_on** table is checked, and any violation of the constraint is rejected. The ROLLBACK TRANSACTION statement in the second BEGIN block rolls back the INSERT or UPDATE statement after a violation of the referential constraint.

The trigger in Example 14.3 checks Case 2 for referential integrity between the **employee** and **works_on** tables (see the section "Referential Integrity" in Chapter 5).

If the following statement is executed,

```
update works_on set emp_no = 12 where emp_no = 10102;
```

the system will display the following error message:

```
No insertion/modification of the row
Msg 3609, Level 16, State 1, Line 1
The transaction ended in the trigger. The batch has been aborted.
```

Example 14.4 introduces the trigger that checks for the violation of integrity constraints between the same tables in Case 3.

Example 14.4

```
USE sample;
GO
CREATE TRIGGER refint_workson2
   ON employee AFTER UPDATE
   AS IF UPDATE (emp_no)
   BEGIN
    IF (SELECT COUNT(*)
       FROM WORKS_ON, deleted
       WHERE works_on.emp_no = deleted.emp_no) > 0
      BEGIN
      ROLLBACK TRANSACTION
      PRINT 'No modification/deletion of the row'
      END
    ELSE PRINT 'The row is deleted/modified'
   END
```

If this statement is executed,

```
update employee set emp_no = 1 where emp_no = 2581;
```

the system will display the following error message:

```
No modification/deletion of the row
Msg 3609, Level 16, State 1, Line 4
The transaction ended in the trigger. The batch has been aborted.
```

INSTEAD OF Triggers

A trigger with the INSTEAD OF clause replaces the corresponding triggering action. It is executed after the corresponding **inserted** and **deleted** virtual tables are created, but before any integrity constraint or any other action is performed.

INSTEAD OF triggers can be created on tables as well as on views. When a Transact-SQL statement references a view that has an INSTEAD OF trigger, the database system executes the trigger instead of taking any action against any table. The trigger always uses the information in the **inserted** and **deleted** virtual tables built for the view to create any statements needed to build the requested event.

Several restrictions apply to column values that are supplied by an INSTEAD OF trigger:

- Values cannot be specified for computed columns.
- Values cannot be specified for columns with the TIMESTAMP data type.
- Values cannot be specified for columns with an IDENTITY property, unless the IDENTITY_INSERT option is set to ON.

These restrictions are valid only for INSERT and UPDATE statements that reference a base table. An INSERT statement that references a view that has an INSTEAD OF trigger must supply values for all non-nullable columns of that view. (The same is true for an UPDATE statement: an UPDATE statement that references a view that has an INSTEAD OF trigger must supply values for each view column that does not allow nulls and that is referenced in the SET clause.)

First and Last Triggers

The Database Engine allows multiple triggers to be created for each table or view and for each modification action (INSERT, UPDATE, and DELETE) on them. By default, there is no defined order in which multiple triggers for a given modification action are executed. Additionally, you can specify the order of multiple triggers defined for a given action. Using the system stored procedure **sp_settriggerorder**, you can specify that one of the AFTER triggers associated with a table be either the first AFTER trigger or the last AFTER trigger executed for each triggering action. This system procedure has a parameter called **@order** that can contain three values:

- **first** Specifies that the trigger is the first AFTER trigger fired for a modification action.
- **last** Specifies that the trigger is the last AFTER trigger fired for a triggering action.
- **none** Specifies that there is no specific order in which the trigger should be fired. (This value is generally used to reset a trigger from being either first or last.)

NOTE If you use the ALTER TRIGGER statement to modify the structure of a trigger, the order of that trigger (first or last) will be dropped.

Example 14.5 shows the use of the system stored procedure **sp_settriggerorder**.

Example 14.5
```
EXEC sp_settriggerorder @triggername = 'modify_budget',
                        @order = 'first', @stmttype='update'
```

NOTE There can be only one first and one last AFTER trigger on a table. The sequence in which all other AFTER triggers fire is undefined.

To display the order of a trigger, you can use the following:

- sp_helptrigger
- OBJECTPROPERTY function

The system procedure **sp_helptrigger** contains the **order** column, which displays the order of the specified trigger. Using the OBJECTPROPERTY function, you can specify either **ExecIsFirstTrigger** or **ExecIsLastTrigger** as the value of the second parameter of this function. The first parameter is always the identification number of the database object. The OBJECTPROPERTY function displays 1 if the particular property is TRUE.

NOTE Because an INSTEAD OF trigger is fired before data modifications are made to the underlying table, INSTEAD OF triggers cannot be specified as first or last triggers.

DDL Triggers and Their Application Areas

The first part of this chapter described DML triggers, which specify an action that is performed by the server when a modification of the table using an INSERT, UPDATE, or DELETE statement is executed. The Database Engine allows you to define triggers for DDL statements, such as CREATE DATABASE, DROP TABLE, and ALTER TABLE. The syntax for DDL triggers is

```
CREATE TRIGGER [schema_name.]trigger_name
   ON {ALL SERVER | DATABASE }
[WITH {ENCRYPTION | EXECUTE AS clause_name]
   {FOR | AFTER } { event_group | event_type | LOGON}
   AS {batch   | EXTERNAL NAME method_name}
```

As you can see from the preceding syntax, DDL triggers are created the same way DML triggers are created. (The ALTER TRIGGER and DROP TRIGGER statements are used to modify and drop DDL triggers, too.) Therefore, this section describes only those options of CREATE TRIGGER that are new in the syntax for DDL triggers.

When you define a DDL trigger, you first must decide on the scope of your trigger. The DATABASE clause specifies that the scope of a DDL trigger is the current database. The ALL SERVER clause specifies that the scope of a DDL trigger is the current server.

After specifying the trigger's scope, you have to decide whether the trigger fires to a single DDL statement or a group of statements. **event_type** specifies a DDL statement that, after execution, causes a trigger to fire. **event_group** defines a name of a predefined group of Transact-SQL language events. The DDL trigger fires after execution of any Transact-SQL language event belonging to **event_group**. You can find the list of all event groups and types in Microsoft Docs. The LOGON keyword specifies a logon trigger (see Example 14.7, later in this section).

Besides the similarities that exist between DML and DDL triggers, there are several significant differences. The main difference between these two trigger forms is that a DDL trigger can be used to define as its scope an entire database or even an entire server, not just a single object. Also, DDL triggers do not support INSTEAD OF triggers. As you might have guessed, **inserted** and **deleted** virtual tables are not necessary, because DDL triggers do not change a table's content.

The two different forms of DDL triggers, database-level and server-level, are described next.

Database-Level Triggers

Example 14.6 shows how you can implement a DDL trigger whose scope is the current database.

Example 14.6

```
USE sample;
GO
CREATE TRIGGER prevent_drop_synonyms
  ON DATABASE FOR DROP_SYNONYM
  AS PRINT 'You must disable "prevent_drop_synonyms" to drop any synonym'
```

The trigger in Example 14.6 prevents all users from deleting any synonym that belongs to the **sample** database. The DATABASE clause specifies that the **prevent_drop_synonyms** trigger is a database-level trigger. The DROP_SYNONYM keyword is a predefined event type that prevents a deletion of any synonym.

Server-Level Triggers

Server-level triggers respond to changes on the server. You use the ALL SERVER clause to implement server-level triggers. Depending on the action, there are two different flavors of server-level triggers: conventional DDL triggers and logon triggers. The triggering action of conventional DDL triggers is based on DDL statements, while the triggering action of logon triggers is a logon event.

Example 14.7 shows a server-level trigger that is at the same time a logon trigger.

Example 14.7

```
USE master;
GO
CREATE LOGIN login_test WITH PASSWORD = 'dpetkovic§$!',
    CHECK_EXPIRATION = ON;
GO
GRANT VIEW SERVER STATE TO login_test;
GO
CREATE TRIGGER connection_limit_trigger
ON ALL SERVER WITH EXECUTE AS 'login_test'
FOR LOGON AS
BEGIN
IF ORIGINAL_LOGIN()= 'login_test' AND
    (SELECT COUNT(*) FROM sys.dm_exec_sessions
            WHERE is_user_process = 1 AND
                original_login_name = 'login_test') > 1
    ROLLBACK;
END;
```

Example 14.7 first creates the login called **login_test**. This login is subsequently used in a server-level trigger. For this reason, it requires server permission VIEW SERVER STATE, which is given to it with the GRANT statement. After that, the **connection_limit_trigger** trigger is created. This trigger belongs to logon triggers, because of the LOGON keyword. The use of the **sys.dm_exec_sessions** view allows you to check if there is already a session established using the **login_test** login. In that case, the ROLLBACK statement is executed. That way, the **login_test** login can establish only one session at a time.

Summary

A trigger is a mechanism that resides in the database server and comes in two flavors: DML triggers and DDL triggers. DML triggers specify one or more actions that are automatically performed by the database server when a modification of the table using an INSERT, UPDATE, or DELETE statement is executed. (A DML trigger cannot be used with the SELECT statement.) DDL triggers are based on DDL statements. They come in two different forms, depending on the scope of the trigger. The DATABASE clause specifies that the scope of a DDL trigger is the current database. The ALL SERVER clause specifies that the scope of a DDL trigger is the current server.

This chapter is the last chapter of the second part of the book. The next chapter starts the third part and discusses the system environment of the Database Engine.

Exercises

E.14.1   Using triggers, define the referential integrity for the primary key of the **department** table, the **dept_no** column, which is the foreign key of the **works_on** table.

E.14.2   With the help of triggers, define the referential integrity for the primary key of the **project** table, the **project_no** column, which is the foreign key of the **works_on** table.

PART

III

SQL Server: System Administration

15

System Environment of the Database Engine

In This Chapter

- System Databases
- Disk Storage
- Utilities
- DBCC Commands
- Policy-Based Management

This chapter describes several features of the Database Engine that belong to the system environment. First, the chapter provides a detailed description of the system databases that are created during the installation process. It then discusses data storage by examining several types of data pages and describing how different data types are stored on the disk. Next, the chapter presents the **bcp**, **sqlcmd**, **mssql-cli**, and **sqlservr** system utilities and the DBCC system commands. The final major section of the chapter introduces Policy-Based Management, a component for managing an instance of the Database Engine.

System Databases

During the installation of the Database Engine, the following system databases are generated:

- master
- model
- tempdb
- msdb

> **NOTE** There is another, "hidden" system database, called the **resource** database, which is used to store system objects, such as system stored procedures and functions. The content of this database is generally used for system upgrades.

The following sections describe each of the system databases in turn.

master Database

The **master** database is the most important system database of the Database Engine. It comprises all system tables that are necessary for your work. For example, the **master** database contains information about all other databases managed by the Database Engine, system connections to clients, and user authorizations.

Because of the importance of this system database, you should always keep a current backup copy of it. Also, the **master** database is modified each time you perform an operation such as creating user databases or user tables. For this reason, you should back it up after the execution of such operations. (The section "Backing Up the master Database" in Chapter 16 explains when it is necessary to back up the **master** database.)

model Database

The **model** database is used as a template when user-defined databases are created. It contains the subset of all system tables of the **master** database, which every user-defined database needs. The system administrator can change the properties of the **model** database to adapt it to the specific needs of their system.

> **NOTE** Because the **model** database is used as a model each time you create a new database, you can extend it with certain database objects and/or permissions. After that, all new databases will inherit the new properties. Use the ALTER DATABASE statement to extend or modify the **model** database, the same way as you modify user databases.

tempdb Database

The **tempdb** system database is a shared database for all users of the Database Engine instance. It provides, among other things, the storage space for temporary tables and other temporary objects that are needed. Its content is destroyed every time the system is restarted.

> **NOTE** I will describe the **tempdb** system database in more detail than the other system databases because it can be a significant factor in relation both to performance and to the management of the whole Database Engine instance to which it belongs.

This section covers the following topics in relation to the **tempdb** database:

- The use of **tempdb**
- Differences between **tempdb** and other databases
- Specification of size and number of files for **tempdb**
- Editing information concerning **tempdb**

The following subsections describe these features.

Use of the tempdb Database

Three types of objects are stored in the **tempdb** system database:

- Temporary user objects, such as temporary tables and stored procedures
- Internal objects
- Version store (see the "Row Versioning" section in Chapter 13)

Local and global temporary tables as well as temporary stored procedures, all of which are created by users, are stored in the **tempdb** database. The other objects stored in this system database are table variables and table-valued functions. All user objects stored in **tempdb** are treated by the system in the same way as any other database object. This means that entries concerning a temporary object are stored in the system catalog and you can retrieve information about the object using the **sys.objects** catalog view.

Internal objects are similar to user objects, except that they are not visible using catalog views or other tools to retrieve metadata. There are three types of internal objects:

- **Work files**   Created when the system retrieves information using particular operators
- **Work tables**   Created by the system when certain operations, such as spooling and recovering databases and tables by using the DBCC commands, are executed
- **Sort units**   Created when a sort operation is executed

The third use of the **tempdb** system database is for storing row versions when using the READ COMMITTED SNAPSHOT level or SNAPSHOT ISOLATION level with user tables. (For the detailed description of these isolation levels, see Chapter 13.) Both forms of isolation level mentioned use the feature called *optimistic concurrency* to save older versions of rows as updates are made, so you can see the prior committed values. During this process, the **tempdb** database grows each time the system performs the following operations, among others:

- An INSERT, UPDATE, or DELETE statement is executed in a snapshot transaction.
- A trigger is executed.

NOTE Because of optimistic concurrency, the system makes heavy use of the **tempdb** database. For this reason, make sure that **tempdb** is large enough and monitor its space regularly. (Monitoring the space of **tempdb** is described at the end of this section.)

Differences Between tempdb and Other Databases

The most important difference between **tempdb** and other databases is that **tempdb** is re-created every time the corresponding Database Engine instance starts up. (You can use this fact to quickly determine the last time a Database Engine instance was restarted, checking the **create_date** column of the **tempdb** system database.)

Another difference is a caching behavior. To reduce the impact on structures of the **tempdb** database, the system can cache temporary objects in the same way as caching is done for persistent database objects. Instead of dropping a temporary object, the system keeps the system metadata and truncates the table data. The process of truncation reduces the storage requirement to a single data page and reduces the allocation information to a single Index Allocation Map (IAM) page. (An IAM page tracks the pages/extents allocation for a specific allocation unit of a table.) The main benefit of caching is that it avoids almost all of the allocation and metadata costs of re-creating the temporary object.

There is also a difference in the logging process that is done for **tempdb**. The **tempdb** system database works in simple recovery mode, but only the minimal amount of information concerning modification operations is stored in the transaction log. (Simple recovery mode will be described in the Chapter 16 section "Recovery Models.")

Specification of Size and Number of Files for tempdb

Two important issues concerning **tempdb** are its total size and the number of files used to store data in it. There is no simple way to specify the size of **tempdb**. The best way is an empirical one: Use your system for a couple of days or weeks and monitor how large the **tempdb** database is. Generally, you should allow **tempdb** to auto-grow. For monitoring the size of your **tempdb** database, use the **sys.dm_db_file_space_usage** dynamic management view (see the upcoming Example 15.2).

During the installation process, on the Database Engine Configuration page of the installation wizard (see Chapter 2), you can specify the number of **tempdb** files on the TempDB tab. The default value is 4. There are several important recommendations for specifying the number of files. First, you should create a data file per logical processor for your **tempdb**, with a maximum of eight data files. Second, each of the multiple data files should have the same size so the Database Engine can apply the proportional fill optimization procedure. Third, according to Microsoft, you should set the Autogrowth property for **tempdb**.

Editing Information Concerning tempdb

You can edit information of temporary objects (tables and stored procedures) in the **tempdb** database the same way you would do for any other such objects. Example 15.1 shows this.

Example 15.1

```
USE tempdb;
SELECT name, create_date FROM tempdb.sys.tables
    WHERE name LIKE '#%'
```

The result for my instance at present is

```
Name                    create_date
-----------------------------------------
#employee               2019-05-03 11:44
#project                2019-05-03 11:45
```

If you want to see how much space is being used for the **tempdb** database of your Database Engine instance, you can use the **sys.dm_db_file_space_usage** dynamic management view. Example 15.2 shows the use of this DMV.

Example 15.2

```
USE tempdb;
SELECT user_object_reserved_page_count AS user_pages,
internal_object_reserved_page_count AS internal_obj_pages,
version_store_reserved_page_count AS version_store_pages
    FROM tempdb.sys.dm_db_file_space_usage;
```

Example 15.2 displays the information concerning three types of objects in the **tempdb** system database. The **user_object_reserved_page_count** column displays the number of pages allocated for user objects. The **internal_object_reserved_page_count** column displays the number of pages allocated for internal objects. The **version_store_reserved_page_count** column specifies the number of pages allocated for the version store.

msdb Database

The **msdb** database is used by the component called SQL Server Agent to schedule alerts and jobs. This system database contains task scheduling, exception handling, alert management, and system operator information; for example, the **msdb** database holds information for all the operators, such as e-mail addresses and pager numbers, and history information about all the backups and restore operations. For more information regarding how this system database can be restored, see Chapter 16.

Disk Storage

The storage architecture of the Database Engine contains several units for storing database objects:

- Page
- Extent
- File
- Filegroup

NOTE Files and filegroups will not be discussed in this chapter. They are described in Chapter 5.

The main unit of data storage is the *page*. The size of a page is always 8KB. Each page has a 96-byte header used to store the system information. Data rows are placed on the page immediately after the header.

The Database Engine supports different page types. The most important are

- Data pages
- Index pages

NOTE Data and index pages are actually physical parts of a database where the corresponding tables and indices are stored. The content of a database is stored in one or more files, and each file is divided into page units. Therefore, each table or index page (as a database physical unit) can be uniquely identified using a database ID, database file ID, and a page number.

When you create a table or index, the system allocates a fixed amount of space to contain the data belonging to the table or index. When the space fills, the space for additional storage must be allocated. The physical unit of storage in which space is allocated to a table (index) is called an *extent*. An extent comprises eight contiguous pages, or 64KB. There are two types of extents:

- **Uniform extents**   Owned by a single table or index
- **Mixed extents**   Shared by up to eight tables or indices

The system always allocates pages from mixed extents first. After that, if the size of the table (index) is greater than eight pages, it switches to uniform extents.

Properties of Data Pages

All types of data pages have a fixed size (8KB) and consist of the following three parts:

- Page header
- Space reserved for data
- Row offset table

NOTE This chapter does not include a separate discussion of the properties of index pages because index pages are almost identical to data pages.

The following sections describe these parts.

Page Header

Each page has a 96-byte page header used to store the system information, such as page ID, the ID of the database object to which the page belongs, and the previous page and next page in a page chain. As you may have already guessed, the page header is stored at the beginning of each page. Table 15-1 shows the information stored in the page header.

Page Header Information	Description
pageId	Database file ID plus the page ID
level	For index pages, the level of the page (leaf level is level 0, first intermediate level is level 1, and so on)
flagBits	Additional information concerning the page
nextPage	Database file ID plus the page ID of the next page in the chain (if a table has a clustered index)
prevPage	Database file ID plus the page ID of the previous page in the chain (if a table has a clustered index)
objId	ID of the database object to which the page belongs
lsn	Log sequence number (see Chapter 13)
slotCnt	Total number of slots used on this page
indexId	Index ID of the page (0, if the page is a data page)
freeData	Byte offset of the first available free space on the page
pminlen	Number of bytes in fixed-length part of rows
freeCnt	Number of free bytes on page
reservedCnt	Number of bytes reserved by all transactions
xactReserved	Number of bytes reserved by the most recently started transaction
xactId	ID of the most recently started transaction
tornBits	One bit per sector for detecting torn page write

Table 15-1 Information Contained in the Page Header

Space Reserved for Data

The part of the page reserved for data has a variable length that depends on the number and length of rows stored on the page. For each row stored on the page, there is an entry in the space reserved for data and an entry in the row offset table at the end of the page. [A data row cannot span two or more pages, except for values of VARCHAR(MAX) and VARBINARY(MAX) data that are stored in their own specific pages.] Each row is stored subsequently after already-stored rows, until the page is filled. If there is not enough space for a new row of the same table, it is stored on the next page in the chain of pages.

For all tables that have only fixed-length columns, the same number of rows is stored at each page. If a table has at least one variable-length column (a VARCHAR column, for instance), the number of rows per page may differ and the system then stores as many rows per page as will fit on it.

Row Offset Table

The last part of a page is tightly connected to a space reserved for data, because each row stored on a page has a corresponding entry in the row offset table (see Figure 15-1). The row offset table contains 2-byte entries consisting of the row number and the offset byte address of the row on the page. (The entries in the row offset table are in reverse order from the sequence of the rows on the page.) Suppose that each row of a table is fixed-length, 36 bytes in length. The first table row is stored at byte offset 96 of a page (because of the page header). The corresponding entry in the row offset table is written in the last 2 bytes of a page, indicating the row number (in the first byte) and the row offset (in the second byte). The next row is stored subsequently in the next 36 bytes of the page. Therefore, the corresponding entry in the row offset table is stored in the third- and fourth-to-last bytes of the page, indicating again the row number (1) and the row offset (132).

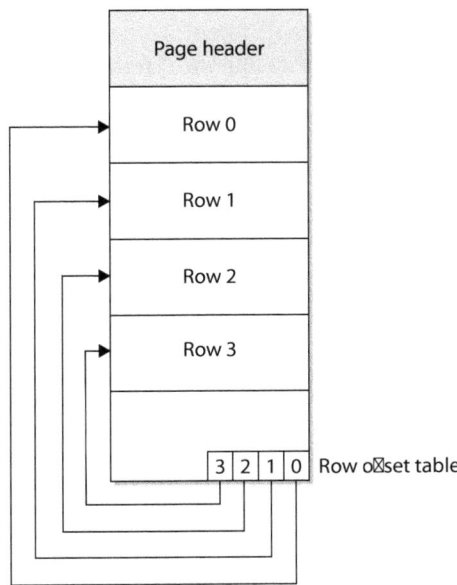

Figure 15-1 The structure of a data page

Types of Data Pages

Data pages are used to store data of a table. There are two types of data pages, each of which is used to store data in a different format:

- In-row data pages
- Row-overflow data pages

In-Row Data Pages

There is nothing special to say about in-row data pages: they are pages in which it is convenient to store data and index information. All data that doesn't belong to large objects is always stored in-row. Also, VARCHAR(MAX), NVARCHAR(MAX), VARBINARY(MAX), and XML values can be stored in-row, if the **large value types out of row** option of the **sp_tableoption** system procedure is set to 0. In this case, all such values are stored directly in the data row, up to a limit of 8000 bytes and as long as the value can fit in the record. If the value does not fit in the record, a pointer is stored in-row and the rest is stored out of row in the storage space for large objects.

Row-Overflow Data Pages

Values of the VARCHAR(MAX), NVARCHAR(MAX), and VARBINARY(MAX) columns can be stored outside of the actual data page. As you already know, 8KB is the maximum size of a row on a data page, but you can exceed this size limit if you use columns of such large data types. In this case, the system stores the values of these columns in extra pages, which are called row-overflow pages.

The storage in row-overflow pages is done only under certain circumstances. The primary factor is the length of the row: if the row needs more than 8060 bytes, some of the column's values will be stored on overflow pages. (A value of a column cannot be split between the actual data page and a row-overflow page.)

As an example of how content of a table with large values is stored, Example 15.3 creates such a table and inserts a row into it.

Example 15.3

```
USE sample;
CREATE TABLE mytable
        (col1 VARCHAR(1000),
         col2 VARCHAR(3000),
         col3 VARCHAR(3000),
          col4 VARCHAR(3000));
    INSERT INTO mytable
       SELECT REPLICATE('a', 1000), REPLICATE('b', 3000),
               REPLICATE('c', 3000), REPLICATE('d', 3000);
```

The CREATE TABLE statement in Example 15.3 creates the **mytable** table. The subsequent INSERT statement inserts a new row in the table. The length of the inserted row is 10,000 bytes. For this reason, the row doesn't fit in a page.

The query in Example 15.4 uses several catalog views to display information concerning page type description.

Example 15.4

```
USE sample;
SELECT  p.rows, a.type_desc AS page_type, a.total_pages AS pages
        FROM sys.partitions p JOIN sys.allocation_units a ON
                         p.partition_id = a.container_id
        WHERE p.object_id = object_id('mytable');
```

The result is

rows	page_type	pages
1	IN_ROW_DATA	9
1	ROW_OVERFLOW_DATA	9

As you can see from Example 15.4, the two columns in the SELECT list (**type_desc** and **total_pages**) are from the **sys.allocation_units** catalog view. (This view contains a row for each allocation unit in the database.) Different allocation units can be displayed using the **type_desc** column of the **sys.allocation_units** catalog view. (A set of pages of one particular data page type is called an *allocation unit*.)

The only reason use of the **sys.partition** view is required is to specify **mytable** as the corresponding table using the **object_id** column of the view. The **sys.partition** view contains one row for each partition of each table or index. (Nonpartitioned tables, such as **mytable**, have only one partition unit.)

The result of Example 15.4 shows that for the single row of the **mytable** table, nine data pages plus nine row-overflow pages are allocated by the system.

NOTE The performance of a system can significantly degrade if your queries access many row-overflow data pages.

Editing Information Concerning Disk Storage

In earlier versions of SQL Server, the only way to inspect data or index pages of a table has been to use two DBCC commands: DBCC PAGE and DBCC IND. The main problem with these commands is that they were (and still are) undocumented, meaning that the functionality has been added to the system but there is no official support for it.

In SQL Server 2012, Microsoft released a new dynamic management function (DMF), called **sys.dm_db_database_page_allocations**. The purpose of this DMF is to replace DBCC PAGE and DBCC IND. We will use an example to introduce this DMF.

In Example 15.5, we inspect data and index pages of the **employee** table.

Example 15.5

```
USE sample;
SELECT allocated_page_page_id AS page, index_id AS ind,
                    page_type_desc AS description
     FROM sys.dm_db_database_page_allocations
```

```
        (DB_ID('sample'), OBJECT_ID('employee'),NULL, NULL, 'DETAILED')
    WHERE is_allocated = 1;
```

The result is

page	ind	description
34	0	IAM_PAGE
264	0	DATA_PAGE
121	2	IAM_PAGE
21320	2	INDEX_PAGE

As you can see from Example 15.5, the **sys.dm_db_database_page_allocations** function has five parameters. The first four parameters—**@databaseid**, **@tableid**, **@indexid**, and **@partitionid**—detail the object whose pages you are interested in. Therefore, in Example 15.5, we analyze the **employee** table (**@tableid**) of the **sample** database (**@databaseid**) in relation to all indexes (**@indexid** = NULL) and all partitions (**@partitionid** = NULL).

@tableid, **@indexid**, and **@partitionid** can be NULL, but **@databaseid** has to be specified. If, for example, you specify the **@tableid** and **@indexid** parameters of a particular index but you pass in NULL for the **@partitionid** parameter, the function will return all pages in all partitions for that particular index. If only the **@databaseid** parameter is valued, all (table-related) pages in that database are returned.

The last parameter of this DMF, **@mode**, specifies the amount of information that is displayed and can be either 'LIMITED' or 'DETAILED'. 'LIMITED' returns less information. On the other hand, 'DETAILED' uses significantly more resources. Therefore, the value 'DETAILED' should be used only on small sets of pages.

The **is_allocated** column of the DMF in the WHERE clause of the SELECT statement restricts the result set only to the pages that are allocated.

NOTE Before the execution of the DMF in Example 15.5, I created an index for the **emp_no** column of the **employee** table to display the information of index pages, too.

There are two problems with the **sys.dm_db_database_page_allocations** DMF. First, it is undocumented. The other problem is more significant and concerns performance. The function reads the information of all the pages for the entire table or index, even when the filter in the WHERE clause is restricted to a single page. For this reason, the bigger the table (index), the worse the performance of this DMF.

Therefore, Microsoft introduced in SQL Server 2019 a new and improved dynamic management function called **sys.dm_db_page_info**. This function takes a similar set of arguments as **sys.dm_db_database_page_allocations**, still allowing you to identify an object by its page, but performs more efficiently. Example 15.6 shows the use of this DMF.

Example 15.6

```
USE sample;
DECLARE @db_id  CHAR(10) = db_id('sample')
SELECT OBJECT_NAME(object_id) table_name, page_id
```

```
FROM sys.dm_db_page_info
(@db_id,1,  21320,   N'LIMITED');
```

The result is

table_name	page_id
employee	2

The **sys.dm_db_page_info** DMF has four parameters: **@DatabaseId**, **@FileId**, **@PageId**, and **@Mode**. The first and the last parameter have the same meaning as the parameters with the same name for the **sys.dm_db_database_page_allocations** DMF. **@FileId** is the ID of the file, while **@PageId** is the ID of the corresponding page. (The page_ID = 21320, which we use as the fourth parameter of the **sys.dm_db_page_info** view, has been determined in Example 15.5.)

Parallel Processing of Tasks

The Database Engine can execute different database tasks in parallel. The following tasks can be parallelized:

- Bulk load
- Backup
- Query execution
- Indices

The Database Engine allows data to be loaded in parallel using the **bcp** utility. (For the description of the **bcp** utility, see the next section.) The table into which the data is loaded must not have any indices, and the load operation must not be logged. (Only applications using the ODBC or OLE DB–based APIs can perform parallel data loads into a single table.)

The Database Engine can back up databases or transaction logs to multiple devices using parallel striped backup. In this case, database pages are read by multiple threads one extent at a time (see also Chapter 16).

The Database Engine provides parallel queries to enhance the query execution. With this feature, the independent parts of a SELECT statement can be executed using several native threads on a computer. Each query that is planned for the parallel execution contains an exchange operator in its query execution plan. (An *exchange operator* is an operator in a query execution plan that provides process management, data redistribution, and flow control.) For such a query, the Database Engine generates a parallel query execution plan. Parallel queries significantly improve the performance of the SELECT statements that process very large amounts of data.

On computers with multiple processors, the Database Engine automatically uses more processors to perform index operations, such as creation and rebuilding of an index. The number of processors employed to execute a single index statement is determined by the configuration option **max degree of parallelism** as well as the current workload. If the Database Engine detects that the system is busy, the degree of parallelism is automatically reduced before the statement is executed.

Utilities

Utilities are components that provide different features such as data reliability, data definition, and statistics maintenance functions. The following utilities are described next:

- bcp
- sqlcmd
- mssql-cli
- sqlservr

> **NOTE** The first three utilities—**bcp**, **sqlcmd**, and **mssql-cli**—are cross-platform utilities, meaning that you can use them with Windows as well as with macOS and Linux operating systems. The Windows installation of **bcp** and **sqlcmd** is performed when you install the corresponding instance of the Database Engine. The Linux installation of **bcp** and **sqlcmd** is similar to the installation of **mssql-cli**, which is described in detail later in this chapter using the Ubuntu operating system.

bcp Utility

bcp (Bulk Copy Program) is a useful utility that copies database data to or from a data file. Therefore, **bcp** is often used to transfer a large amount of data into a Database Engine database from another relational DBMS using a text file, or vice versa.

The syntax of the **bcp** utility is

```
bcp [[db_name.]schema_name.]table_name {IN | OUT | QUERYOUT | FORMAT}
                        file_name    [{-option parameter} ...]
```

db_name is the name of the database to which the table (**table_name**) belongs. IN or OUT specifies the direction of data transfer. The IN option copies data from the **file_name** file into the **table_name** table, and the OUT option copies rows from the **table_name** table into the **file_name** file. The FORMAT option creates a format file based on the options specified. If this option is used, the option **–f** must also be used.

> **NOTE** The IN option appends the content of the file to the content of the database table, whereas the OUT option overwrites the content of the file.

Data can be copied as either specific text or standardized (ASCII) text. Copying data as specific text is referred to as working in native mode, whereas copying data as ASCII text is referred to as working in character mode. The parameter **–n** specifies native mode, and the parameter **–c** specifies character mode. Native mode is used to export and import data from one system managed by the Database Engine to another system managed by the Database Engine, and character mode is commonly used to transfer data between a Database Engine instance and other database systems.

Example 15.7 shows the use of the **bcp** utility. (You have to execute this statement from a command line of your Windows operating system.)

Example 15.7
```
bcp AdventureWorks.Person.Address out "address.txt" -T -c
```

The **bcp** command in Example 15.7 exports the data from the **address** table of the **AdventureWorks** database in the output file **address.txt**. The option **–T** specifies that the trusted connection is used. (*Trusted connection* means that the system uses integrated security instead of the SQL Server authentication.) The option **–c** specifies character mode; thus, the data is stored in the ASCII file.

NOTE Be aware that the BULK INSERT statement is an alternative to **bcp**. It supports all of the **bcp** options (although the syntax is a bit different) and offers much greater performance.

To import data from a file to a database table, you must have INSERT and SELECT permissions on the table. To export data from a table to a file, you must have SELECT permission on the table.

sqlcmd Utility

sqlcmd allows you to enter Transact-SQL statements, system procedures, and script files at the command prompt. The general form of this utility is

```
sqlcmd {option [parameter]} ...
```

where **option** is the specific option of the utility, and **parameter** specifies the value of the defined option. The **sqlcmd** utility has many options, the most important of which are described in Table 15-2.

Example 15.8 shows the use of **sqlcmd**. (You have to execute this statement from a command line of your Windows operating system.)

NOTE Before you execute Example 15.8, you have to change the server name and make sure the input file is available.

Example 15.8
```
sqlcmd -S NTB11901 -i C:\ms0510.sql -o C:\ms0510.rpt
```

In Example 15.8, a user of the database system named NTB11900 executes the batch stored in the file ms0510.sql and stores the result in the output file ms0510.rpt. Depending on the authentication mode, the system prompts for the username and password (SQL Server authentication) or just executes the statement (Windows authentication).

One of the most important options of the **sqlcmd** utility is the **–A** option. As you already know from Table 15-2, this option allows you to start a dedicated administrator connection (DAC) to an instance of the Database Engine. Usually, you make the connection to an instance of the Database Engine with SQL Server Management Studio. But, there are certain extraordinary situations in which users cannot connect to the instance. In that case, the use of the DAC can help.

Option	Description
–S server_name [\instance_name], [port]	Specifies the name of the database server and the instance to which the connection is made. If this option is omitted, the connection is made to the database server set with the environment variable SQLSERVER. If this environment variable is not set, the connection is established to the local machine.
–U login_id	Specifies the login. If this option is omitted, the value of the environment variable SQLCMDUSER is used.
–P password	Specifies a password corresponding to the login. If neither the –U option nor the –P option is specified, **sqlcmd** attempts to connect by using Windows authentication mode. Authentication is based on the account of the user who is running **sqlcmd**.
–c command_end	Specifies the batch terminator. (The default value is GO.) This option can be used to set the command terminator to a semicolon (;), which is the default terminator for almost all other database systems.
–i input_file	Specifies the name of the file that contains a batch or a stored procedure. The file must contain (at least one) command terminator. The sign < can be used instead of –**i**.
–o output_file	Specifies the name of the file that receives the result from the utility. The sign > can be used instead of –**o**.
–E	Uses a trusted connection (see Chapter 12) instead of requesting a password.
–A	Starts the dedicated administrator connection (DAC), which is described following this table.
–L	Shows a list of all database instances found on the network.
–t seconds	Specifies the number of seconds. The time interval defines how long the utility should wait before it considers the connection to the server to be a failure.
–?	Specifies a standard request for all options of the **sqlcmd** utility.
–d dbname	Specifies which database should be the current database when **sqlcmd** is started.

Table 15-2 Most Important Options of the sqlcmd Utility

The DAC is a special connection that can be used by DBAs in case of extreme server resource depletion. Even when there are not enough resources for other users to connect, the Database Engine will attempt to free resources for the DAC. That way, administrators can troubleshoot problems on an instance without having to take down that instance.

The **sqlcmd** utility supports several specific commands that can be used within the utility, in addition to Transact-SQL statements. Table 15-3 describes the most important commands of the **sqlcmd** utility. Example 15.9 shows how a query in relation to the sample database can be executed using this utility. (To execute Example 15.9, start **sqlcmd**, and at the command prompt type each line separately.)

Command	Description
:ED	Starts the text editor. This editor can be used to edit the current batch or the last executed batch. The editor is defined by the SQLCMDEDITOR environment variable. For instance, if you want to set the text editor to Microsoft WordPad, type **SET SQLCMDEDITOR=wordpad**.
:!!	Executes operating system commands. For example, **:!! dir** lists all files and directories in the current directory.
:r filename	Parses additional Transact-SQL statements and **sqlcmd** commands from the file specified by **filename** into the statement cache. It is possible to issue multiple **:r** commands. Hence, you can use this command to chain scripts with the **sqlcmd** utility.
:List	Prints the content of the statement cache.
:QUIT	Ends the session started by **sqlcmd**.
:EXIT [(statement)]	Allows you to use the result of a SELECT statement as the return value from **sqlcmd**.

Table 15-3   Most Important Commands of the sqlcmd Utility

Example 15.9

```
1>USE sample;
2>SELECT * FROM project
3>:EXIT(SELECT @@rowcount)
```

This example displays the number of rows of the **project** table (e.g., the number 3 if **project** contains three rows).

mssql-cli Utility

mssql-cli is an open source, cross-platform, interactive command-line query tool for the Database Engine. This tool can be used as an alternative to the **sqlcmd** utility.

This section demonstrates the Windows installation and the Ubuntu installation of the tool, as well as how you can use it. The final subsection compares the properties of **mssql-cli** to those of the **sqlcmd** utility.

Windows Installation of mssql-cli

The **mssql-cli** utility is implemented in Python. Python is not installed by default on Windows. Therefore, before you install **mssql-cli**, you have to install Python.

You can download the latest Python installation package from https://www.python.org/downloads/. When installing, check the Add Python to PATH option, which you can find as the last line in the installation screen. That way, the necessary information concerning Python directories and files can be found by the system in the PATH environment variable.

Once Python is installed and the PATH environment variable is set, open a command prompt and install **mssql-cli** using the following command:

```
pip install mssql-cli
```

Figure 15-2 Installation of mssql-cli on Windows

Figure 15-2 shows the installation process of the **mssql-cli** utility under Windows.

Ubuntu Installation of mssql-cli

As you already know, **mssql-cli** is a tool based on the Python programming language. You can find the source code for this tool on GitHub. To install the tool, go to https://github.com/dbcli/mssql-cli. Click Installation Guide under the Ubuntu platform. The page that appears contains different forms of installations. Scroll down to "Installation for latest preview version via Direct Downloads." Download the Debian package and save the .deb file under the desired location.

Go to the download folder and double-click the file to launch it. It opens the screen shown in Figure 15-3. Click Install. After that, you will be prompted to enter the credentials for the authentication.

NOTE As you can see from the description, the **mssql-cli** installation process is similar to the installation process of Azure Data Studio (see Chapter 3). Analogously, you do not get any message when the installation of **mssql-cli** is finished.

Using mssql-cli

The **mssql-cli** utility allows you to write T-SQL statements, system procedures, and script files at the command prompt. The general form of this utility is

```
mssql-cli {option [parameter]} ...
```

where **option** is the specific option of the utility, and **parameter** specifies the value of the particular option. The utility has many options, the most important of which are described in Table 15-4.

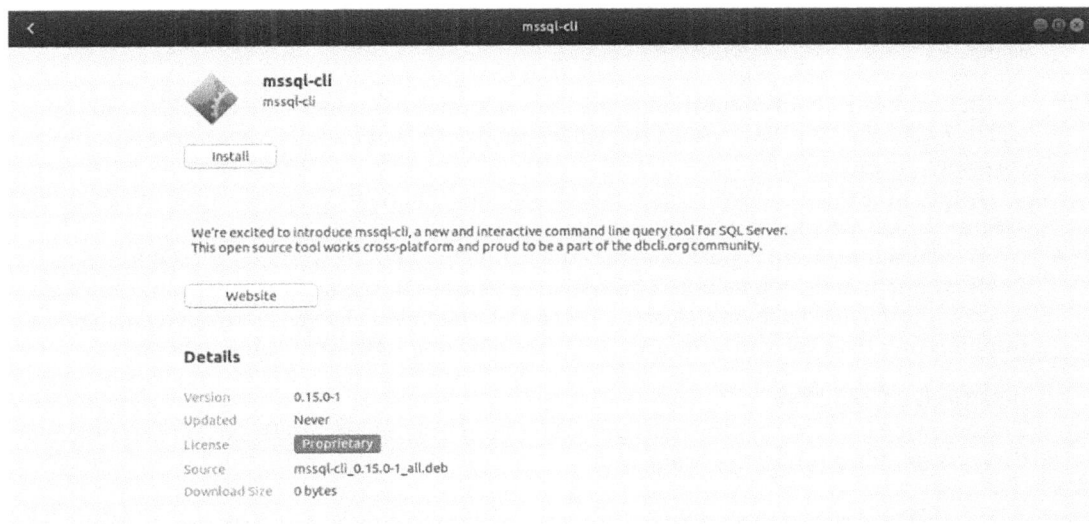

Figure 15-3   Installation screen for mssql-cli on Ubuntu

NOTE The functionality of **mssql-cli** is identical for all platforms.

Example 15.10 shows the use of the **mssql-cli** tool. (You have to execute this statement from a command line of your Windows operating system.)

NOTE Before you execute Example 15.10, you have to change the server name.

Option	Description
–h	Lists all options.
–S server_name	Specifies the SQL Server instance name or address.
–U username	Specifies the username to connect to the database.
–P password	Specifies a password corresponding to the username. (The default value is stored in the MSSQL_CLI_PASSWORD variable.)
–E	Uses a trusted connection instead of requesting a password,
–A	Starts the dedicated administrator connection. (The DAC is described in detail in the previous section.)
–l connect-timeout	Specifies the time interval for how long the utility should wait before it considers the connection to the server to be a failure.
–d database	Specifies which database should be the current database.

Table 15-4   Most Important Options of the mssql-cli utility

Example 15.10

```
mssql-cli -S LAPTOP-TVUM0CNL -U sa
```

In Example 15.10, a connection to the Database Engine instance LAPTOP-TVUM0CNL will be established for the **sa** user. (You will be prompted by the system to type the corresponding password.)

Figure 15-4 shows how you can establish a connection (with Windows Authentication) and execute T-SQL statements.

As you can see from Figure 15-4, the connection to the server instance is established using a trusted (integrated) connection. The system establishes the connection and shows you the name of the default database ("master>"). The USE statement changes the default database to **sample**. After that, you can write any T-SQL statement in relation to that database.

mssql-cli vs. sqlcmd

The following list describes the most important properties of **mssql-cli** and compares them to the corresponding properties of **sqlcmd**:

- **mssql-cli** is an open source contribution from Microsoft, with its source code available at GitHub, whereas **sqlcmd** is Microsoft proprietary software.

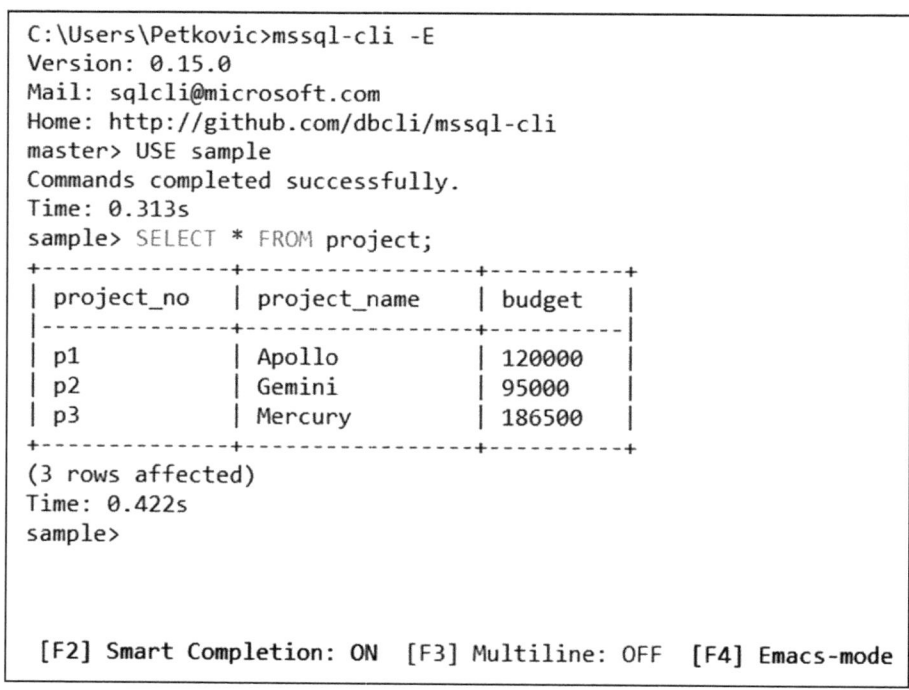

Figure 15-4 Connecting to the instance and executing statements with mssql-cli

- **mssql-cli** has user-friendly interactive features like T-SQL IntelliSense, multiline editing, syntax highlighting, and formatting of results. (You can see most of these features by looking at the output of the SELECT statement in Figure 15-4.) **sqlcmd** is a simple command-line tool with none of these interactive features.

- **mssql-cli** and **sqlcmd** are both cross-platform utilities, meaning that you can use them with Windows as well as with macOS and Linux derivatives. **mssql-cli** is a new tool and does not have all the command arguments available in **sqlcmd**. For example, at the time of writing, the tool does not have an option to specify the output file.

- Generally, the **mssql-cli** tool can be used as an enhanced alternative to **sqlcmd**. On the other hand, **sqlcmd** is a mature tool and has more options.

sqlservr Utility

The most convenient way to start an instance of the Database Engine is automatically with the boot process of the computer. However, certain circumstances might require different handling of the system. Therefore, the Database Engine offers, among others, the **sqlservr** utility for starting an instance.

NOTE You can also use Management Studio or the **net** command to start or stop an instance of the Database Engine.

The **sqlservr** utility is invoked using the following command:

```
sqlservr option_list
```

option_list contains all options that can be invoked using the application. Table 15-5 describes the most important options.

Option	Description
–f	Indicates that the instance is started with the minimal configuration.
–m	Indicates that the instance is started in single-user mode. Use this option if you have problems with the system and want to perform maintenance on it (this option must be used to restore the **master** database).
–s instance_name	Specifies the instance of the Database Engine. If no named instance is specified, **sqlservr** starts the default instance of the Database Engine.

Table 15-5   Most Important Options of the sqlservr Utility

DBCC Commands

The Transact-SQL language supports the DBCC (Database Console Commands) statements, which are commands for the Database Engine. Depending on the options used with DBCC, the DBCC commands can be divided into the following groups:

- Maintenance
- Informational
- Validation
- Performance

In contrast to utilities of the Database Engine, which have to be installed separately for all Linux derivatives, the DBCC commands are already installed during the installation process.

> **NOTE** This section discusses only the validation commands and a command in relation to performance. Other commands will be discussed in relation to their application. For instance, DBCC FREEPROCCACHE is discussed in detail in Chapter 19, while the description of DBCC USEROPTIONS can be found in Chapter 13.

Validation Commands

The validation commands do consistency checking of the database. The following commands belong to this group:

- DBCC CHECKALLOC
- DBCC CHECKTABLE
- DBCC CHECKCATALOG
- DBCC CHECKDB

The DBCC CHECKALLOC command validates whether every extent indicated by the system has been allocated, as well as that there are no allocated extents that are not indicated by the system. Therefore, this command performs cross-referencing checks for extents.

The DBCC CHECKTABLE command checks the integrity of all the pages and structures that make up the table or indexed view. All performed checks are both physical and logical. The physical checks control the integrity of the physical structure of the page. The logical checks control, among other things, whether every row in the base table has a matching row in each nonclustered index, and vice versa, and whether indices are in their correct sort order. Using the PHYSICAL_ONLY option, you can validate only the physical structure of the page. This option causes a much shorter execution time of the command and is therefore recommended for frequent use on production systems.

The DBCC CHECKCATALOG command checks for catalog consistency within the specified database. It performs many cross-referencing checks between tables in the system catalog. After the DBCC CATALOG command finishes, a message is written to the error log. If the DBCC command successfully executes, the message indicates a successful completion and the amount of time that the command ran. If the DBCC command stops because of an error, the message indicates the command was terminated, a state value, and the amount of time the command ran.

If you want to check the allocation and the structural and logical integrity of all the objects in the specified database, use DBCC CHECKDB. (As a matter of fact, this command performs all checks previously described, in the given order.)

NOTE All DBCC commands that validate the system use snapshot transactions (see Chapter 13) to provide transactional consistency. In other words, the validation operations do not interfere with the other, ongoing database operations, because they use versions of current rows for validation.

Performance Command

The DBCC MEMORYSTATUS command provides a snapshot of the current memory status of the Database Engine. The command's output is useful in troubleshooting issues that relate to the memory consumption of the Database Engine or to specific out-of-memory errors (many of which automatically print this output in the error log).

The output of this command has several parts, including the "Process/System Counts" part, which delivers important information concerning the total amount of memory (the Working Set parameter) and the actual memory used by the Database Engine (the Available Physical Memory parameter).

Policy-Based Management

The Database Engine supports Policy-Based Management, a framework for managing one or more server instances, databases, or other database objects. Before you learn how this framework works, though, you need to understand some key terms and concepts of it.

Key Terms and Concepts

The following is a list of the key terms regarding Policy-Based Management, which is followed by a description of the concepts related to these terms:

- Managed target
- Target set
- Facet
- Condition
- Policy
- Category

The system manages entities called *managed targets*, which may be server instances, databases, tables, or indices. All managed targets that belong to an instance form a hierarchy. A *target set* is the set of managed targets that results from applying filters to the target hierarchy. For instance, if your managed target is a table, a target set could comprise all indices that belong to that table.

A *facet* is a set of logical properties that models the behavior or characteristics for certain types of managed targets. The number and characteristics of the properties are built into the facet and can be added or removed only by the maker of the facet. Some facets can be applied only to certain types of managed targets.

A *condition* is a Boolean expression that specifies a set of allowed states of a managed target with regard to a facet. Again, some conditions can be applied only to certain types of managed targets.

A *policy* is a condition and its corresponding behavior. A policy can contain only one condition. Policies can be enabled or disabled. They are managed by users through the use of categories.

A policy belongs to one and only one *category*. A category is a group of policies that is introduced to give a user more flexibility in cases where third-party software is hosted. Database owners can subscribe a database to a set of categories. Only policies from the database's subscribed categories can govern that database. All databases implicitly subscribe to the default policy category.

Using Policy-Based Management

This section presents an example that shows how you can use Policy-Based Management. This example will create a policy whose condition is that the index fill factor will be 60 percent for all databases of the instance. (For a description of the FILLFACTOR option, see Chapter 10.)

The following are the three main steps to implement Policy-Based Management:

1. Create a condition based on a facet.
2. Create a policy.
3. Categorize the policy.

Generally, to create a policy, open SQL Server Management Studio, expand the server, and then expand Management | Policy Management.

The first step is to create a condition. Right-click Conditions and choose New Condition. In the Create New Condition dialog box (see Figure 15-5), type the condition name in the Name field (**SetFillFactor** in this example), and choose Server Configuration in the Facet drop-down list. (Setting a fill factor for all databases of an instance is server-bound and thus belongs to the server configuration.) In the Field column of the Expression area, choose @FillFactor from the drop-down menu and choose = as the operator. Finally, enter **60** in the Value field. Click OK.

The next step is to create a policy based on the condition. In the Policy Management folder, right-click Policies and choose New Policy. In the Name field of the Create New Policy dialog box, type the name for the new policy (**PolicyFillFactor60** in this example). In the Check Condition drop-down list, choose the condition that you have created (SetFillFactor). (This condition can be found under the node called Server Configurations.) Choose On Demand from the Evaluation Mode drop-down list.

NOTE Policy administrators can run policies on demand, or enable automated policy execution by using one of the existing execution modes.

After you create a policy, you should categorize it. To categorize a policy, click the Description page in the Create New Policy dialog box. You can place policies in the Default

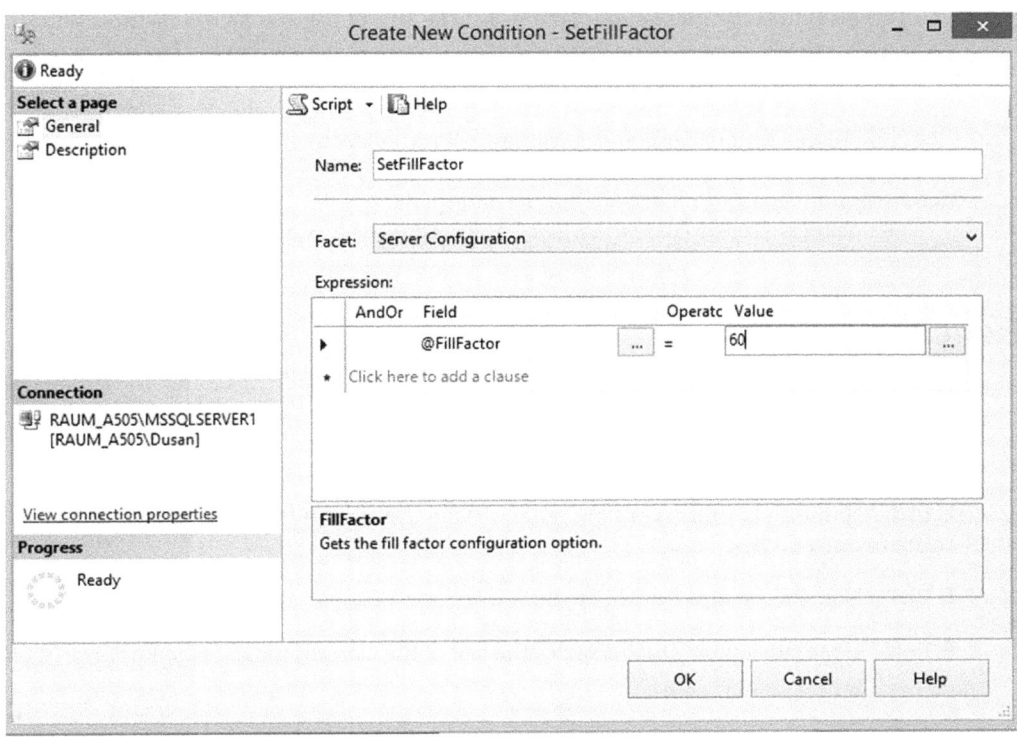

Figure 15-5   The Create New Condition dialog box

category or in a more specific category. (You can also create your own category by clicking the New button.)

The process described in this section can be applied in the same way to dozens of different policies in relation to servers, databases, and database objects.

Summary

This chapter described several features of the system environment of the Database Engine:

- System databases
- Disk storage
- Utilities
- DBCC commands
- Policy-Based Management

The system databases contain system information and high-level information about the whole database system. The most important of them is the **master** database.

The main unit of disk storage is the page. The size of pages is 8KB. The most important page type is the data page. (The form of an index page is almost identical to that of a data page.)

The Database Engine supports many utilities and commands. This chapter discussed four utilities (**sqlcmd**, **sqlservr**, **mssql-cli**, and **bcp**) and the DBCC validation commands.

Policy-Based Management is a framework for DBAs to configure and manage objects at the system level. It allows you to define and enforce policies for configuring and managing databases and database objects across the enterprise.

The next chapter discusses how you can prevent the loss of data, using backup and recovery.

Exercises

E.15.1   If you create a temporary table, where will it be stored? Also, how you can display information concerning such a table?

E.15.2   Change the properties of the **model** database so that its size is 4MB.

E.15.3   Name all key terms of Policy-Based Management and discuss their roles and how they relate to one another.

E.15.4   Name all groups for which you can specify a condition.

E.15.5   Generate a policy that disables the use of the Common Language Runtime (CLR).

16

Backup, Recovery, and System Availability

In This Chapter

- Reasons for Data Loss
- Introduction to Backup Methods
- Performing Database Backup
- Performing Database Recovery
- System Availability
- Maintenance Plan Wizard

This chapter first covers two of the most important tasks related to system administration: backup and recovery. Backup refers to the process of making copies of the database(s) and/or transaction logs to separate media that can later be used for recovery, if necessary. Recovery is the process of using the backup media to replace uncommitted, inconsistent, or lost data.

System availability refers to keeping the downtime of the database system as low as possible. This chapter describes in detail the following options available for system availability: failover clustering, database mirroring, log shipping, and AlwaysOn. It also discusses the benefits and disadvantages of each option.

At the end of the chapter, the Maintenance Plan Wizard is discussed. This wizard provides you with the set of basic tasks you need understand to maintain a database. Therefore, the wizard can be used to, among other things, back up and restore user databases.

NOTE SQL Azure is outside the scope of this introductory book. For this reason, backup and recovery in the cloud will not be described here. Please see Microsoft Docs if you want to use backup and recovery features of SQL Azure for your database.

Reasons for Data Loss

Performing backups is a precautionary measure that you have to take to prevent data loss. The reasons for data loss can be divided into the following groups:

- Program errors
- Administrator (human) errors
- Computer failures (e.g., system crash)
- Disk failures
- Catastrophes (fire, flood, earthquake) or theft

During execution of a program, conditions may arise that abnormally terminate the program. Such program errors affect only the database application and usually have no impact on the entire database system. Because these errors are based on faulty program logic, the database system cannot recover in such situations. The recovery should therefore be done by the programmer, who has to handle such exceptions using the COMMIT and ROLLBACK statements (see Chapter 13).

Another source of data loss is human error. Users with sufficient permissions (DBA, for instance) may accidentally lose or corrupt data (people have been known to drop the wrong table, update or delete data incorrectly, and so on). Of course, ideally, this would never happen, and you can establish practices that make it unlikely that production data will be compromised in this way, but you have to recognize that people make mistakes, and data can be affected. The best that you can do is to try to avoid it, and be prepared to recover when it happens.

A computer failure may occur as a result of various different hardware or software errors. A hardware crash is an example of a system failure. In this case, the contents of the computer's main memory may be lost. A disk failure occurs either when a read/write head of the disk crashes or when the I/O system discovers corrupted disk blocks during I/O operations.

In the case of catastrophes or theft, the system must have enough information available to recover from the failure. This is normally done by means of media that offer the needed recovery information on a piece of hardware that is stored separately and thus has not been damaged or lost by the catastrophe or theft.

For most of the errors just described, backups, discussed next, can provide a recovery solution.

Introduction to Backup Methods

Database backup is the process of dumping data (from a database, a transaction log, or a file) into backup devices that the system creates and maintains. A backup device can be a disk file or a tape. The Database Engine provides both static and dynamic backups. *Static backup* means that during the backup process, the only active session supported by the system is the one that creates the backup. In other words, user processes are not allowed during backup. *Dynamic backup* means that a database backup can be performed without stopping the database server, removing users, or even closing the files. (The users will not even know that the backup process is in progress.)

The Database Engine provides four different backup methods:

- Full database backup
- Differential backup
- Transaction log backup
- File (or filegroup) backup

The following sections describe these backup methods.

Full Database Backup

A full database backup captures the state of the database at the time the backup is started. During the full database backup, the system copies the data as well as the schema of all tables of the database and the corresponding file structures. If the full database backup is executed dynamically, the database system records any activity that takes place during the backup. Therefore, even all uncommitted transactions in the transaction log are written to the backup media.

Differential Backup

A differential backup creates a copy of only the parts of the database that have changed since the last full database backup. (As in a full database backup, any activity that takes place during a differential backup is backed up too.) The advantage of a differential backup is speed. It minimizes the time required to back up a database, because the amount of data to be backed up has the potential to be considerably smaller than in the case of a full database backup. (Remember that a full database backup includes a copy of all database pages.)

Transaction Log Backup

A transaction log backup considers only the changes recorded in the log. This form of backup is therefore not based on physical parts (pages) of the database, but rather on logical operations— that is, changes executed using the DML statements INSERT, UPDATE, and DELETE. Again, because the amount of data to be backed up has the potential to be considerably smaller, this process can be performed significantly quicker than a full database backup and quicker than a differential backup.

NOTE It does not make sense to perform a differential backup or to back up a transaction log unless a full database backup has been performed at least once.

There are two main reasons to perform a transaction log backup: first, to store the data that has changed since the last transaction log backup or full database backup on a secure medium; second (and more importantly), to properly close the transaction log up to the beginning of the active portion of it. (The active portion of the transaction log contains all uncommitted transactions.)

Using a full database backup and a valid chain of all closed transaction logs, it is possible to propagate a database copy on a different computer. This database copy can then be used to replace the original database in case of a failure. (The same scenario can be established using a full database backup and the last differential backup.)

The Database Engine does not allow you to store the transaction log in the same file in which the database is stored. One reason for this is that if the file is damaged, the use of the transaction log to restore all changes since the last backup will not be possible.

Using a transaction log to record changes in the database is a common feature used by nearly all existing relational DBMSs. Nevertheless, situations may arise in which it becomes helpful to switch this feature off. For example, the execution of a heavy load can last for hours. Such a program runs much faster when the logging process is switched off. On the other hand, switching off the logging process is dangerous, as it destroys the valid chain of transaction logs. To ensure successful database recovery, it is strongly recommended that you perform a full database backup after the successful end of the load.

One of the most common system failures occurs because the transaction log is filled up. Be aware that such a problem may cause a complete standstill of the system. If the storage used for the transaction log fills up to 100 percent, the system must stop all running transactions until the transaction log storage is freed again. This problem can be avoided only by making frequent backups of the transaction log: each time you close a portion of the actual transaction log and store it to a different storage media, that portion of the log becomes reusable, and the system thus regains disk space.

NOTE A differential backup and a transaction log backup both minimize the time required to back up the database. But there is one significant difference between them: the transaction log backup contains all changes of a row that has been modified several times since the last backup, whereas a differential backup contains only the last modification of that row.

Some differences between log backups and differential backups are worth noting. The benefit of differential backups is that you save time in the restore process, because to recover a database completely, you need a full database backup and only the *latest* differential backup. If you use log backups for the same scenario, you have to apply a full database backup and *all* existing log backups to bring the database to a consistent state. A disadvantage of differential backups is that you cannot use them to recover data to a specific point in time, because they do not store intermediate changes to the database.

File or Filegroup Backup

File (or filegroup) backup allows you to back up specific database files (or filegroups) instead of the entire database. In this case, the Database Engine backs up only files you specify. Individual files (or filegroups) can be restored from a database backup, allowing recovery from a failure that affects only a small subset of the database files. You can use either a database backup or a filegroup backup to restore individual files or filegroups. This means that you can use database and transaction log backups as your backup procedure and still be able to restore individual files (or filegroups) from the database backup.

NOTE File backup is also called file-level backup. This type of backup is recommended only when a database that should be backed up is very large and there is not enough time to perform a full database backup.

Performing Database Backup

You can perform backup operations using the following:

- Transact-SQL statements
- SQL Server Management Studio

Each of these backup methods is described in the following sections.

Backing Up Using Transact-SQL Statements

All types of backup operations can be executed using two Transact-SQL statements:

- BACKUP DATABASE
- BACKUP LOG

Before these two Transact-SQL statements are described, the existing types of backup devices will be explained.

Types of Backup Devices

The Database Engine allows you to back up databases, transaction logs, and files to the following backup devices:

- Disk
- Tape

NOTE There is also another form of backup device called a network share. I will not describe it separately because it is simply a special form of a disk drive that specifies a network drive to use for backups.

Disk files are the most common media used for storing backups. Disk backup devices can be located on a server's local hard disk or on a remote disk on a shared network resource. The Database Engine allows you to append a new backup to a file that already contains backups from the same or different databases. By appending a new backup set to existing media, the previous contents of the media remain intact, and the new backup is written after the end of the last backup on the media. (The *backup set* includes all stored data of the object you chose to back up.) By default, the Database Engine always appends new backups to disk files.

CAUTION Do not back up to a file on the same physical disk where the database or its transaction log is stored! If the disk with the database crashes, the backup that is stored on the same disk will also be damaged.

Tape backup devices are generally used in the same way as disk devices. However, when you back up to a tape, the tape drive must be attached locally to the system. The advantage of tape devices relative to disk devices is their simple administration and operation.

NOTE You can use third-party backup utilities, too. They provide a variety of options beyond what is delivered by the Database Engine.

BACKUP DATABASE Statement

The BACKUP DATABASE statement is used to perform a full database backup or a differential database backup. This statement has the following syntax:

```
BACKUP DATABASE {db_name | @variable}
    TO device_list
    [MIRROR TO device_list2]
    [WITH | option_list]
```

db_name is the name of the database that should be backed up. (The name of the database can also be supplied using a variable, **@variable**.) **device_list** specifies one or more device names, where the database backup will be stored. **device_list** can be a list of names of disk files or tapes. The syntax for a device is

```
{ logical_device_name | @logical_device_name_var }
    |{DISK| TAPE| URL}= {'physical_device_name'| @physical_device_name_var}
```

where the device name can be either a logical name (or a variable) or a physical name beginning with the DISK, TAPE, or URL keyword. The meaning of the first two options is obvious. The URL format is used for creating backups to the Microsoft Azure storage service. (The TAPE option will be removed in a future version of the Database Engine.)

The MIRROR TO option indicates that the accompanying set of backup devices is a mirror within a mirrored media set. The backup devices must be identical in type and number to the devices specified in the TO clause. In a mirrored media set, all the backup devices must have the same properties. (See also the description of mirrored media in the section "Database Mirroring" later in this chapter.)

option_list comprises several options that can be specified for the different backup types. The most important options are the following:

- DIFFERENTIAL
- NOSKIP/SKIP
- NOINIT/INIT
- NOFORMAT/FORMAT
- UNLOAD/NOUNLOAD
- MEDIANAME and MEDIADESCRIPTION
- BLOCKSIZE
- COMPRESSION

The first option, DIFFERENTIAL, specifies a differential database backup. All other clauses in the list concern full database backups.

The SKIP option disables the backup set expiration and name checking, which is usually performed by BACKUP DATABASE to prevent overwrites of backup sets. The NOSKIP option, which is the default, instructs the BACKUP statement to check the expiration date and name of all backup sets before allowing them to be overwritten.

The INIT option is used to overwrite any existing data on the backup media. This option does not overwrite the media header, if one exists. If there is a backup that has not yet expired, the backup operation fails. In this case, use the combination of SKIP and INIT options to overwrite the backup device. The NOINIT option, which is the default, appends a backup to existing backups on the media.

The FORMAT option is used to write a header on all of the files (or tape volumes) that are used for a backup. Therefore, use this option to initialize a backup medium. When you use the FORMAT option to back up to a tape device, the INIT option and the SKIP option are implied. Similarly, the INIT option is implied if the FORMAT option is specified for a file device. NOFORMAT, which is the default, specifies that the backup operation processes the existing media header and backup sets on the media volumes.

The UNLOAD and NOUNLOAD options are performed only if the backup medium is a tape device. The UNLOAD option, which is the default, specifies that the tape is automatically rewound and unloaded from the tape device after the backup is completed. Use the NOUNLOAD option if the database system should not rewind (and unload) the tape from the tape device automatically.

MEDIADESCRIPTION and MEDIANAME specify the description and the name of the media set, respectively. The BLOCKSIZE option specifies the physical block size, in bytes. The supported sizes are 512, 1024, 2048, 4096, 8192, 16384, 32768, and 65536 bytes. The default is 65536 bytes for tape devices and 512 bytes otherwise.

The Database Engine supports backup compression. To specify backup compression, use the COMPRESSION option of the BACKUP DATABASE statement. Example 16.1 backs up the **sample** database and compresses the backup file.

Example 16.1

```
USE master;
BACKUP DATABASE sample
   TO DISK = 'C:\temp\sample.bak'
   WITH INIT, COMPRESSION;
```

If you want to know whether the particular backup file is compressed, view the output of the RESTORE HEADERONLY statement, which is described later in this chapter.

BACKUP LOG Statement

The BACKUP LOG statement is used to perform a backup of the transaction log. This statement has the following syntax:

```
BACKUP LOG {db_name | @variable}
   TO device_list
     [MIRROR TO device_list2]
   [WITH option_list]
```

db_name, **@variable**, **device_list**, and **device_list2** have the same meanings as the parameters with the same names in the BACKUP DATABASE statement. **option_list** has the same options as the BACKUP DATABASE statement and also supports the specific log options NO_TRUNCATE, NORECOVERY, and STANDBY.

You should use the NO_TRUNCATE option if you want to back up the transaction log without truncating it—that is, this option does not clear the committed transactions in the log. After the execution of this option, the system writes all recent database activities in the transaction log. Therefore, the NO_TRUNCATE option allows you to recover data right up to the point of the database failure.

The NORECOVERY option backs up the tail of the log and leaves the database in the restoring state. NORECOVERY is useful when failing over to a secondary database or when saving the tail of the log before a restore operation. The STANDBY option backs up the tail of the log and leaves the database in a read-only and standby state. (The restore operation and the standby state are explained later in this chapter.)

Backing Up Using SQL Server Management Studio

Before you can perform a database or transaction log backup, you must specify (or create) backup devices. Management Studio allows you to create disk devices and tape devices in a similar manner. In both cases, expand the server, expand Server Objects, right-click Backup Devices, and choose New Backup Device. In the Backup Device dialog box, enter the name of either the disk device (if you clicked File) or the tape device (if you clicked Tape). In the former case, you can click the ... button on the right side of the field to display existing backup device locations. In the latter case, if Tape cannot be activated, then no tape devices exist on the local computer. Click OK.

After you specify backup devices, you can do a database backup. Expand the server, expand Databases, right-click the database, and choose Tasks | Back Up. The Back Up Database dialog box appears (see Figure 16-1). On the General page of the dialog box, choose the backup type in the Backup Type drop-down list (Full, Differential, or Transaction Log). In the same dialog box, you can choose an expiration date for the backup.

In the Destination frame, select an existing device by clicking Add. (The Remove button allows you to remove one or more backup devices from the list of devices to be used.)

On the Media Options page (see Figure 16-2), to append to an existing backup on the selected device, click the Append to the Existing Backup Set radio button. Choosing the Overwrite All Existing Backup Sets radio button in the same frame overwrites any existing backups on the selected backup device.

For verification of the database backup, check the Verify Backup When Finished check box in the Reliability frame. On the Media Options page, you can also choose to back up to a new media set by clicking the Back Up to a New Media Set, and Erase All Existing Backup Sets radio button and then entering the media set name and description.

For creation and verification of a differential database backup or transaction log backup, follow the same steps, but choose the corresponding backup type in the Backup Type field on the General page.

After you have chosen all your options, click OK. The database or the transaction log is then backed up. You can display the name, physical location, and the type of the backup devices

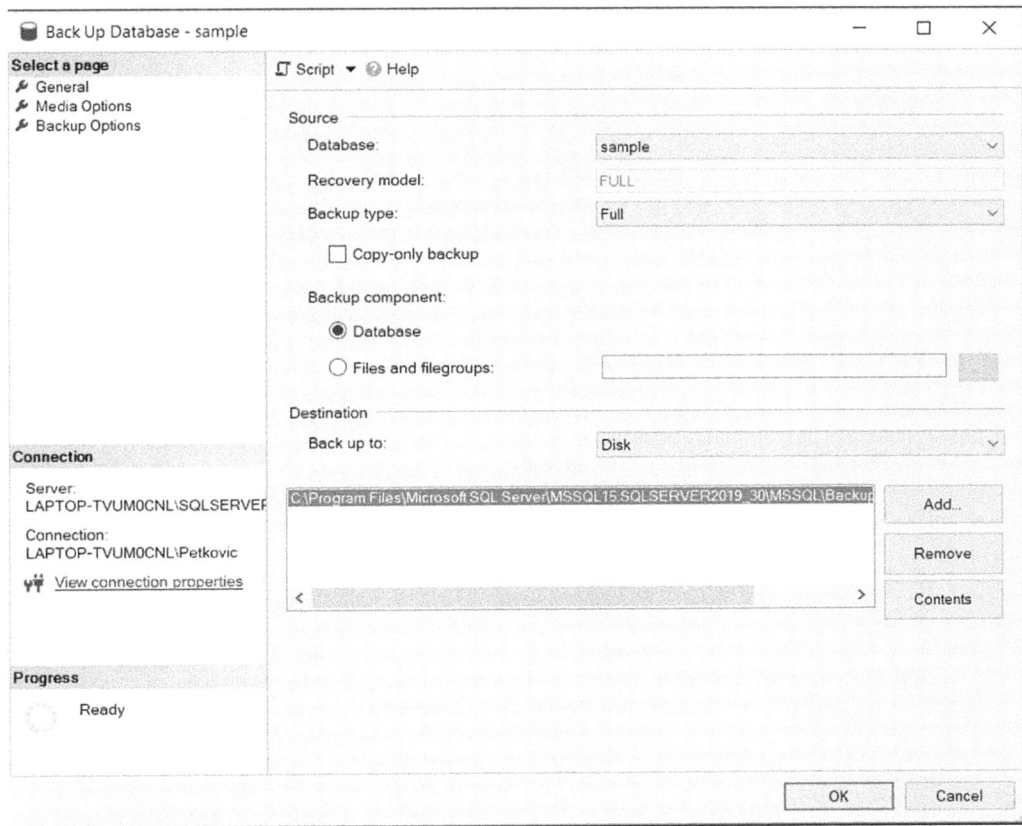

Figure 16-1   The Back Up Database dialog box, General page

by selecting the server, expanding the Server Objects folder, expanding the Backup Devices folder, and then selecting the particular file.

Scheduling Backups with SQL Server Management Studio

A well-planned timetable for the scheduling of backup operations will help you avoid system shortages when users are working. Management Studio supports this planning by offering an easy-to-use graphical interface for scheduling backups. Scheduling backups using Management Studio is explained in detail in Chapter 17.

Determining Which Databases to Back Up

The following databases should be backed up regularly:

- The **master** database
- All production databases

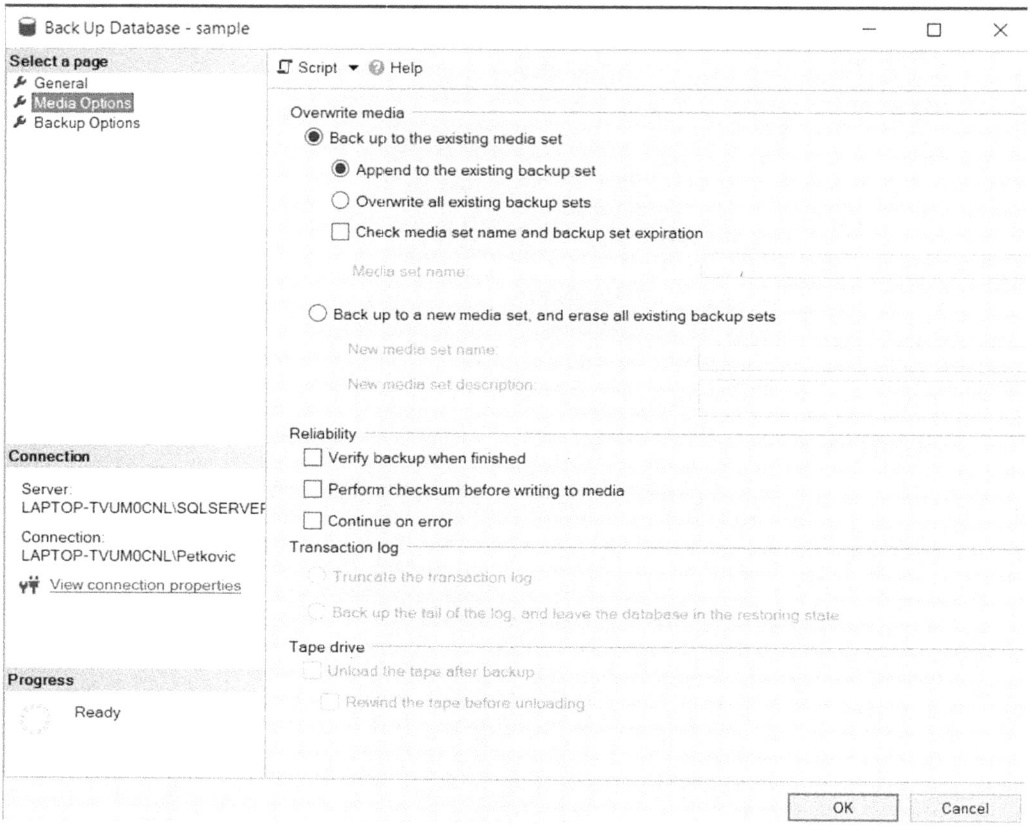

Figure 16-2 The Back Up Database dialog box, Media Options page

Backing Up the master Database

The **master** database is the most important system database because it contains information about all of the databases in the system. Therefore, you should back up the **master** database on a regular basis. Additionally, you should back up the **master** database anytime certain statements and stored procedures are executed, because the Database Engine modifies the **master** database automatically.

NOTE You can perform only full database backups of the **master** database. (The system does not support differential, transaction log, and file backups for the **master** database.)

Many activities cause the modification of the **master** database. Some of them are listed here:

- The creation, alteration, and removal of a database
- The alteration of the transaction log

NOTE Without a backup of the **master** database, you must completely rebuild all system databases, because if the **master** database is damaged, all references to the existing user-defined databases are lost.

Backing Up Production Databases

You should back up each production database on a regular basis. Additionally, you should back up any production database when the following activities are executed:

- After creating it
- After creating indices
- After clearing the transaction log
- After performing nonlogged operations

Always make a full database backup after it has been created, in case a failure occurs between the creation of the database and the first regular database backup. Remember that backups of the transaction log cannot be applied without a full database backup.

Backing up the database after creation of one or more indices saves time during the restore process, because the index structures are backed up together with the data. Backing up the transaction log after creation of indices does not save time during the restore process at all, because the transaction log records only the fact that an index was created (and does not record the modified index structure).

Backing up the database after clearing the transaction log is necessary because the transaction log no longer contains a record of database activity, which is used to recover the database. All operations that are not recorded to the transaction log are called nonlogged operations. Therefore, all changes made by these operations cannot be restored during the recovery process.

Performing Database Recovery

Whenever a transaction is submitted for execution, the Database Engine is responsible either for executing the transaction completely and recording its changes permanently in the database or for guaranteeing that the transaction has no effect at all on the database. This approach ensures that the database is consistent in case of a failure, because failures do not damage the database itself, but instead affect transactions that are in progress at the time of the failure. The Database Engine supports both automatic and manual recovery, which are discussed next in turn.

Automatic Recovery

Automatic recovery is a fault-tolerant feature that the Database Engine executes every time it is restarted after a failure or shutdown. The automatic recovery process checks to see if the restoration of databases is necessary. If it is, each database is returned to its last consistent state using the transaction log.

During automatic recovery, the Database Engine examines the transaction log from the last checkpoint to the point at which the system failed or was shut down. (A *checkpoint* is the most recent point at which all data changes are written permanently to the database from memory. Therefore, a checkpoint ensures the physical consistency of the data.) The transaction log contains committed transactions (transactions that are successfully executed, but their changes have not yet been written to the database) and uncommitted transactions (transactions that are not successfully executed before a shutdown or failure occurred). The Database Engine rolls forward all committed transactions, thus making permanent changes to the database, and undoes the part of the uncommitted transactions that occurred before the checkpoint.

The Database Engine first performs the automatic recovery of the **master** database, followed by the recovery of all other system databases. Then, all user-defined databases are recovered.

Manual Recovery

A manual recovery of a database specifies the application of the full backup of your database and subsequent application of all transaction logs in the sequence of their creation. (Alternatively, you can use the full database backup together with the last differential backup of the database.) After this, the database is in the same (consistent) state as it was at the point when the transaction log was backed up for the last time.

When you recover a database using a full database backup, the Database Engine first re-creates all database files and places them in the corresponding physical locations. After that, the system re-creates all database objects.

The Database Engine can process certain forms of recovery dynamically (in other words, while an instance of the database system is running). Dynamic recovery improves the availability of the system, because only the data being restored is unavailable. Dynamic recovery allows you to restore either an entire database file or a filegroup. (Microsoft calls dynamic recovery "online restore.")

Is My Backup Set Ready for Recovery?

After finishing the backup process, the selected device (tape or disk) contains all data of the object you chose to back up. The stored data is called a *backup set*. Before you start a recovery process, you should be sure that

- The backup set contains the data you want to restore.
- The backup set is usable.

The Database Engine supports a set of Transact-SQL statements that allows you to confirm that the backup set is usable and contains the proper data. The following four statements, among others, belong to it:

- RESTORE LABELONLY
- RESTORE HEADERONLY
- RESTORE FILELISTONLY
- RESTORE VERIFYONLY

The following subsection describes these statements.

RESTORE LABELONLY   This statement is used to display the header information of the media (disk or tape) used for a backup process. The output of the RESTORE LABELONLY statement is a single row that contains the summary of the header information (name of the media, description of the backup process, and date of a backup process).

NOTE RESTORE LABELONLY reads just the header file, so use this statement if you want to get a quick look at what your backup set contains.

RESTORE HEADERONLY   Whereas the RESTORE LABELONLY statement gives you concise information about the header file of your backup device, the RESTORE HEADERONLY statement gives you information about backups that are stored on a backup device. This statement displays a one-line summary for each backup on a backup device. In contrast to RESTORE LABELONLY, using RESTORE HEADERONLY can be time consuming if the device contains several backups.

RESTORE FILELISTONLY   The RESTORE FILELISTONLY statement returns a result set with a list of the database and log files contained in the backup set. You can display information about only one backup set at a time. For this reason, if the specified backup device contains several backups, you have to specify the position of the backup set to be processed.

You should use RESTORE FILELISTONLY if you don't know exactly either which backup sets exist or where the files of a particular backup set are stored. In both cases, you can check all or part of the devices to make a global picture of existing backups.

RESTORE VERIFYONLY   In contrast to the previous three statements, which display information about your backup set, the RESTORE VERIFYONLY statement verifies the backup without using it for the restore process. This statement checks the existence of all backup devices (tapes or files) and whether the existing information can be read.

RESTORE VERIFYONLY supports two specific options:

- **LOADHISTORY**   Causes the backup information to be added to the backup history tables
- **STATS**   Displays a message each time another percentage of the reading process completes, and is used to gauge progress (the default value is 10)

Restoring Databases and Logs Using Transact-SQL Statements

All restore operations can be executed using two Transact-SQL statements:

- RESTORE DATABASE
- RESTORE LOG

The RESTORE DATABASE statement is used to perform the restore process for a database. The general syntax of this statement is

```
RESTORE DATABASE {db_name | @variable}
    [FROM device_list]
    [WITH option_list]
```

db_name is the name of the database that will be restored. (The name of the database can be supplied using a variable, **@variable**.) **device_list** specifies one or more names of devices on which the database backup is stored. (If you do not specify the FROM clause, only the process of automatic recovery takes place, not the restore of a backup, and you must specify either the RECOVERY, NORECOVERY, or STANDBY option. This action can take place if you want to switch over to a standby server.) **device_list** can be a list of names of disk files or tapes. **option_list** comprises several options that can be specified for the different backup forms. The most important options are

- RECOVERY/NORECOVERY/STANDBY
- CHECKSUM/NO_CHECKSUM
- REPLACE
- PARTIAL
- STOPAT/STOPATMARK/STOPBEFOREMARK

The RECOVERY option instructs the Database Engine to roll forward any committed transaction and to roll back any uncommitted transaction from the corresponding transaction log. After the RECOVERY option is applied, the database is in a consistent state and is ready for use. This option is the default.

NOTE Use the RECOVERY option either with the last transaction log to be restored or to restore with a full database backup without subsequent transaction log backups.

With the NORECOVERY option, the Database Engine does not roll back uncommitted transactions because you will be applying further backups. After the NORECOVERY option is applied, the database is unavailable for use.

NOTE Use the NORECOVERY option with all but the last transaction log to be restored.

The STANDBY option is an alternative to the RECOVERY and NORECOVERY options and is used with the standby server. (The standby server is discussed later, in the section "Using a Standby Server.") In order to access data stored on the standby server, you usually recover the database after a transaction log is restored. On the other hand, if you recover the database on the standby server, you cannot apply additional logs from the production server for the restore process. In that case, you use the STANDBY option to allow users read access to the standby server. Additionally, you allow the system to restore additional transaction logs. The STANDBY option implies the existence of the undo file that is used to roll back changes when additional logs are restored.

The CHECKSUM option initiates the verification of both the backup checksums and page checksums, if present. If checksums are absent, RESTORE proceeds without verification. The NO_CHECKSUM option explicitly disables the validation of checksums by the restore operation.

The REPLACE option replaces an existing database with data from a backup of a different database. In this case, the existing database is first destroyed, and the differences regarding the names of the files in the database and the database name are ignored. (If you do not use the REPLACE option, the database system performs a safety check that guarantees an existing database is not replaced if the names of files in the database, or the database name itself, differ from the corresponding names in the backup set.)

The PARTIAL option specifies a partial restore operation. With this option you can restore a portion of a database, consisting of its primary filegroup and one or more secondary filegroups, which are specified in an additional option called FILEGROUP. (The PARTIAL option is not allowed with the RESTORE LOG statement.)

The STOPAT option allows you to restore a database to the state it was in at the exact moment before a failure occurred by specifying a point in time. The Database Engine restores all committed transactions that were recorded in the transaction log before the specified point in time. If you want to restore a database by specifying a point in time, execute the RESTORE DATABASE statement using the NORECOVERY clause. After that, execute the RESTORE LOG statement to apply each transaction log backup, specifying the name of the database, the backup device from which the transaction log backup will be restored, and the STOPAT clause. (If the backup of a log does not contain the requested time, the database will not be recovered.)

The STOPATMARK and STOPBEFOREMARK options specify to recover to a mark. This topic is described a bit later, in the section "Recovering to a Mark."

The RESTORE DATABASE statement is also used to restore a database from a differential backup. The syntax and the options for restoring a differential backup are the same as for restoring from a full database backup. During a restoration from a differential backup, the Database Engine restores only that part of the database that has changed since the last full database backup. Therefore, restore the full database backup *before* you restore a differential backup!

The RESTORE LOG statement is used to perform a restore process for a transaction log. This statement has the same syntax form and the same options as the RESTORE DATABASE statement.

Restoring Databases and Logs Using SQL Server Management Studio

To restore a database from a full database backup using Management Studio, expand the server, choose Databases, right-click the database, and choose Tasks | Restore | Database. The Restore Database dialog box appears (see Figure 16-3). On the General page, select a database to which you want to restore. Then check the backup set that you want to use for your backup process.

NOTE If you restore from the log backup, do not forget the sequence of restoring different types of backups. First restore the full database backup. Then restore all corresponding transaction logs in the sequence of their creation.

To select the appropriate restore options, choose the Options page of the Restore Database dialog box. In the upper part of the window you can choose one or more of the restore options, which are self-explanatory.

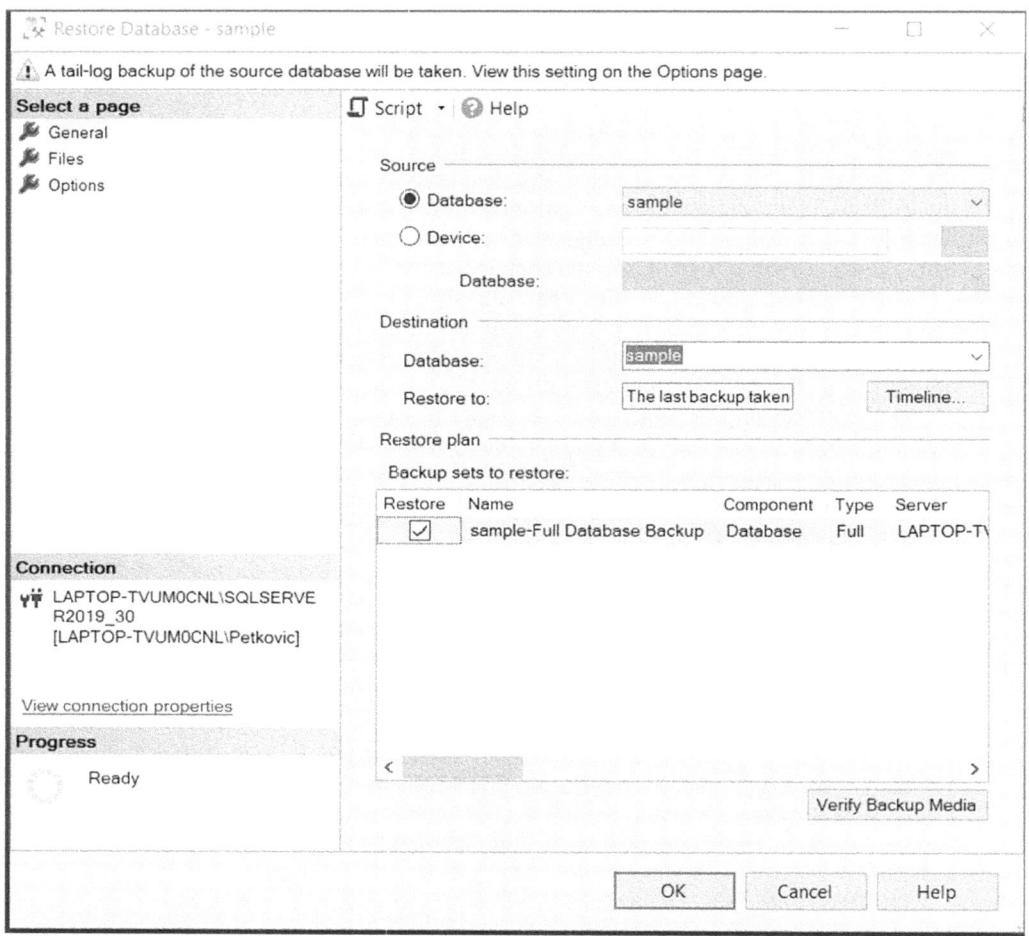

Figure 16-3 The Restore Database dialog box, General page

In the Recovery State box, you can choose between the following states:

- RESTORE WITH RECOVERY
- RESTORE WITH NORECOVERY
- RESTORE WITH STANDBY

Choosing the first option, RESTORE WITH RECOVERY (see Figure 16-4), instructs the Database Engine to roll forward any committed transaction and to roll back any uncommitted transaction. After applying this option, the database is in a consistent state and is ready for use. This option is equivalent to the RECOVERY option of the RESTORE DATABASE statement.

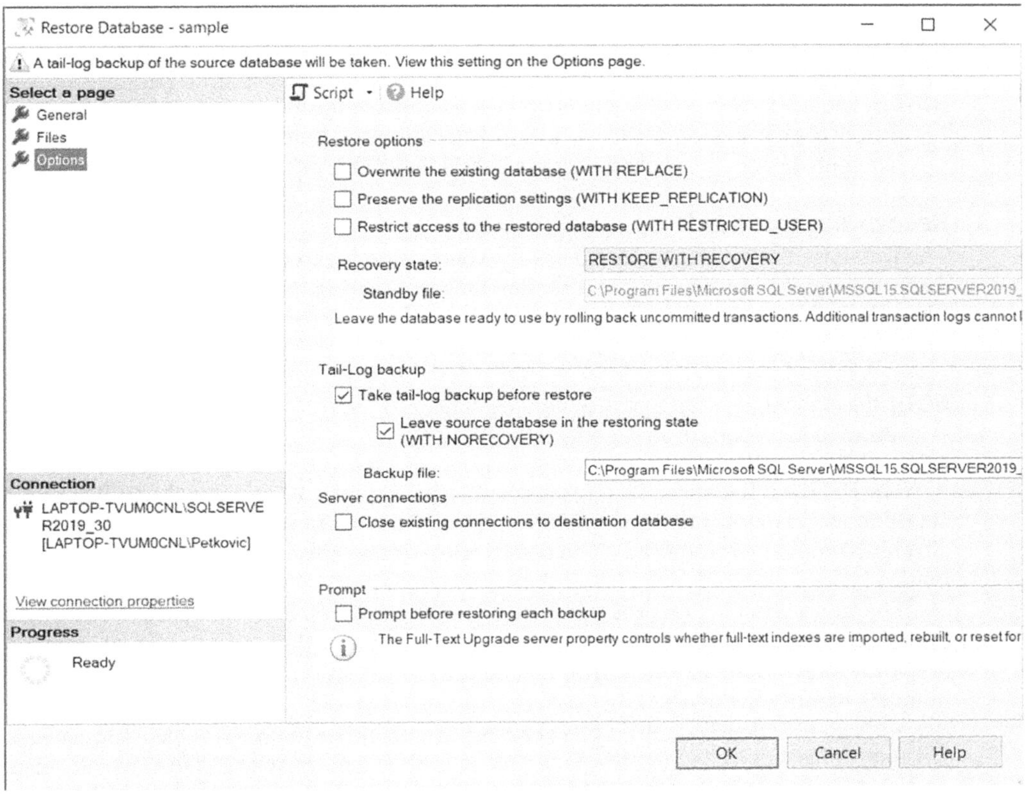

Figure 16-4 The Restore Database dialog box, Options page

NOTE Use this option only with the last transaction log to be restored or with a full database restore when no subsequent transaction logs need to be applied.

If you choose the second option, RESTORE WITH NORECOVERY, the Database Engine does not roll back uncommitted transactions because you will be applying further backups. After you apply this option, the database is unavailable for use, and additional transaction logs should be restored. This option is equivalent to the NORECOVERY option of the RESTORE DATABASE statement.

NOTE Use this option with all but the last transaction log to be restored or with a differential database restore.

Choosing the third option, RESTORE WITH STANDBY, specifies the file that is subsequently used to roll back the recovery effects. This option is equivalent to the STANDBY option in the RESTORE DATABASE statement.

The process of a database restoration from a differential database backup is equivalent to the process of a restoration from a full database backup.

> **NOTE** If you restore from a differential backup, first restore the full database backup before you restore the corresponding differential one. In contrast to transaction log backups, only the latest differential backup is applied, because it includes all changes since the full backup.

Recovering to a Mark

The Database Engine allows you to use the transaction log to recover to a specific mark. Log marks correspond to a specific transaction and are inserted only if the transaction commits. This allows the marks to be tied to a particular amount of work and provides the ability to recover to a point that includes or excludes this work.

> **NOTE** If a marked transaction spans multiple databases on the same database server, the marks are recorded in the logs of all the affected databases.

The BEGIN TRANSACTION statement (see Chapter 13) supports the WITH MARK clause to insert marks into the logs. Because the name of the mark is the same as its transaction, a transaction name is required. (The **description** option specifies a textual description of the mark.)

The transaction log records the mark name, description, database, user, date and time information, and the log sequence number (LSN). To allow their reuse, the transaction names are not required to be unique. The date and time information is used along with the name to uniquely identify the mark.

You can use the RESTORE LOG statement (with either the STOPATMARK clause or the STOPBEFOREMARK clause) to specify recovering to a mark. The STOPATMARK clause causes the recovery process to roll forward to the mark and include the transaction that contains the mark. If you specify the STOPBEFOREMARK clause, the recovery process excludes the transaction that contains the mark.

Both clauses just described support AFTER **datetime**. If this option is omitted, recovery stops at the first mark with the specified name. If the option is specified, recovery stops at the first mark with the specified name exactly at or after **datetime**.

Restoring the master Database

The corruption of the **master** system database can be devastating for the whole system because it comprises all system tables that are necessary to work with the database system. The restore process for the **master** database is quite different from the same process for user-defined databases.

A damaged **master** database makes itself known through different failures. These failures include the following:

- Inability to start the MSSQLSERVER process
- An input/output error
- Execution of the DBCC command points to such a failure

Two different ways exist to restore the **master** database. The easier way, which is available only if you can start your database system, is to restore the **master** database from the full database backup. If you can't start your system, then you must go the more difficult route and use the **sqlservr** utility. (The **sqlservr** utility is described in detail in Chapter 15.)

To restore your **master** database, start your instance in single-user mode. Of the two ways to do it, my favorite is to use the command window and execute the **sqlservr** utility (from the command prompt) with the option **–m**. Although the use of this utility is more difficult, this approach allows you to restore the **master** database in most cases. In the second step, you restore the **master** database together with all other databases using the last full database backup.

> **NOTE** If there have been any changes to the **master** database since the last full database backup, you will need to re-create those changes manually.

Restoring Other System Databases

The restore process for all system databases other than **master** is similar. Therefore, I will explain this process using the **msdb** database. The **msdb** database needs to be restored from a backup when either the **master** database has been rebuilt or the **msdb** database itself has been damaged. If the **msdb** database is damaged, restore it using the existing backups. If there have been any changes after the **msdb** database backup was created, re-create those changes manually. (You can find the description of the **msdb** system database in Chapter 15.)

> **NOTE** You cannot restore a database that is being accessed by users. Therefore, when restoring the **msdb** database, the SQL Server Agent service should be stopped. (SQL Server Agent accesses the **msdb** database.)

Recovery Models

A recovery model allows you to control *to what extent* you are ready to risk losing committed transactions if a database is damaged. It also determines the speed and size of your transaction log backups. Additionally, the choice of a recovery model has an impact on the size of the transaction log and therefore on the time period needed to back up the log. The Database Engine supports three recovery models:

- Full
- Bulk-logged
- Simple

The following sections describe these recovery models.

Full Recovery Model

During full recovery, all operations are written to the transaction log. Therefore, this model provides complete protection against media failure. This means that you can restore your database up to the last committed transaction that is stored in the log file. Additionally, you can recover data to any point in time (prior to the point of failure). To guarantee this, such operations as SELECT INTO and the execution of the **bcp** utility are fully logged too.

Besides point-in-time recovery, the full recovery model allows you also to recover to a log mark. Log marks correspond to a specific transaction and are inserted only if the transaction commits. The full recovery model also logs all operations concerning the CREATE INDEX statement, implying that the process of data recovery now includes the restoration of index creations. That way, the re-creation of the indices is faster, because you do not have to rebuild them separately.

The disadvantage of this recovery model is that the corresponding transaction log may be very voluminous and the files on the disk containing the log will be filled up very quickly. Also, for such a voluminous log, you will need significantly more time for backup.

NOTE If you use the full recovery model, the transaction log must be protected from media failure. For this reason, using RAID 1 to protect transaction logs is strongly recommended. (RAID 1 is explained in the section "Using RAID Technology" later in this chapter.)

Bulk-Logged Recovery Model

Bulk-logged recovery supports log backups by using minimal space in the transaction log for certain large-scale or bulk operations. The logging of the following operations is minimal and cannot be controlled on an operation-by-operation basis:

- SELECT INTO
- CREATE INDEX (including indexed views)
- **bcp** utility

Although bulk operations are not fully logged, you do not have to perform a full database backup after the completion of such an operation. During bulk-logged recovery, transaction log backups contain both the log and the results of a bulk operation. This simplifies the transition between full and bulk-logged recovery models.

The bulk-logged recovery model allows you to recover a database to the end of a transaction log backup (that is, up to the last committed transaction). Additionally, you can restore your database to any point in time if you haven't performed any bulk operations. The same is true for the restore operation to a named log mark.

The advantage of the bulk-logged recovery model is that bulk operations are performed much faster than under the full recovery model, because they are not fully logged. On the other side, the Database Engine backs up all the modified extents, together with the log itself. Therefore, the log backup needs a lot more space than in the case of the full recovery. (The time to restore a log backup is significantly increased, too.)

Simple Recovery Model

In the simple recovery model, the transaction log is truncated whenever a checkpoint occurs. Therefore, you can recover a damaged database only by using the full database backup or the differential backup, because they do not require log backups. Backup strategy for this model is very simple: restore the database using existing database backups and, if differential backups exist, apply the most recent one.

NOTE The simple recovery model doesn't mean that there is no logging at all. The log content won't be used for backup purposes, but it is used at the checkpoint time, where all the transactions in the log are committed or rolled back.

The advantages of the simple recovery model are that the performance of all bulk operations is very high and requirements for the log space are very small. On the other hand, in the case of failure, this model requires the most manual work because all changes since the most recent database (or differential) backup must be redone. Point-in-time and page restore are not possible with this recovery model. Also, file restore is available only for read-only secondary filegroups.

NOTE Do not use the simple recovery model for production databases.

Changing and Editing a Recovery Model

You can change the recovery model by using the RECOVERY option of the ALTER DATABASE statement. The part of the syntax of the ALTER DATABASE statement concerning recovery models is

```
SET RECOVERY [FULL | BULK_LOGGED | SIMPLE]
```

If you want to edit the current recovery model, before changing it, you can use the following two alternatives:

- Using the **databasepropertyex** property function
- Using the **sys.databases** catalog view

If you want to display the current model of your database, use the recovery value for the second parameter of the **databasepropertyex** function. Example 16.2 shows the query that displays the recovery model for the **sample** database. (The function displays one of the values FULL, BULK_LOGGED, or SIMPLE.)

Example 16.2

```
SELECT databasepropertyex('sample', 'recovery');
```

The **recovery_model_desc** column of the **sys.databases** catalog view displays the same information as the **databasepropertyex** function, as Example 16.3 shows.

Example 16.3

```
SELECT name, database_id, recovery_model_desc AS model
          FROM sys.databases
          WHERE name = 'sample';
```

The result is

name	database_id	model
sample	7	FULL

System Availability

Ensuring the availability of your instance and databases is one of the most important issues today. There are several techniques that you can use to ensure their availability, which can be divided in two groups: those that are components of the Database Engine and those that are not implemented in the database server. The following two techniques are not part of the Database Engine:

- Using a standby server
- Using RAID technology

The following proprietary techniques belong to the Database Engine:

- Database mirroring
- Failover clustering
- Log shipping
- AlwaysOn
- Replication

The following sections describe these techniques, other than replication, which is discussed in Chapter 18.

Using a Standby Server

A standby server is just what its name implies—another server that is standing by in case something happens to the production server (also called the primary server). The standby server contains files, databases (system and user-defined), and user accounts identical to those on the production server.

A standby server is implemented by initially restoring a full database backup of the database and applying transaction log backups to keep the database on the standby server synchronized with the production server. To set up a standby server, set the **read only** database option to TRUE. This option prevents users from performing any write operations in the database.

The general steps to use a copy of a production database are as follows:

- Restore the production database using the RESTORE DATABASE statement with the STANDBY clause.
- Apply each transaction log to the standby server using the RESTORE LOG statement with the STANDBY clause.
- When applying the final transaction log backup, use the RESTORE LOG statement with the RECOVERY clause. (This final statement recovers the database without creating a file with before images, making the database available for write operations, too.)

After the database and transaction logs are restored, users can work with an exact copy of the production database. Only the noncommitted transactions at the time of failure will be permanently lost.

NOTE If the production server fails, user processes are not automatically brought to the standby server. Additionally, all user processes need to restart any tasks with the uncommitted transactions due to the failure of the production server.

Using RAID Technology

RAID (redundant array of inexpensive disks) is a special disk configuration in which multiple disk drives build a single logical unit. This process allows files to span multiple disk devices. RAID technology provides improved reliability at the cost of performance decrease. Generally, there are six RAID levels, 0 through 5. Only three of these levels, levels 0, 1, and 5, are significant for database systems.

RAID can be hardware or software based. Hardware-based RAID is more costly (because you have to buy additional disk controllers), but it usually performs better. Software-based RAID can be supported usually by the operating system. Windows operating systems provide RAID levels 0, 1, and 5. RAID technology has impacts on the following features:

- Fault tolerance
- Performance

The benefits and disadvantages of each RAID level in relation to these two features are explained next.

RAID provides protection from hard disk failure and accompanying data loss with three methods: disk striping, mirroring, and parity. These three methods correspond to RAID levels 0, 1, and 5, respectively.

RAID 0 (Disk Striping)

RAID 0 specifies disk striping without parity. Using RAID 0, the data is written across several disk drives in order to allow data access more readily, and all read and write operations can be speeded up. For this reason, RAID 0 is the fastest RAID configuration. The disadvantage of disk striping is that it does not offer fault tolerance at all. This means that if one disk fails, all the data on that array become inaccessible.

RAID 1 (Mirroring)

RAID 1 uses the space on a disk drive to maintain a duplicate copy of all files. Therefore, RAID 1, which specifies disk mirroring, protects data against media failure by maintaining a copy of the database (or a part of it) on another disk. If there is a drive loss with RAID 1 in place, the files for the lost drive can be rebuilt by replacing the failed drive and rebuilding the damaged files. The hardware configurations of RAID 1 are more expensive, but they provide additional speed. (Also, hardware configurations of RAID 1 implement some caching options that provide better throughput.) The advantage of the Windows solution for RAID 1 is that it can be configured to mirror disk partitions, while the hardware solutions are usually implemented on the entire disk.

In contrast to RAID 0, RAID 1 is much slower, but the reliability is higher. Also, RAID 1 costs much more than RAID 0 because each mirrored disk drive must be doubled. It can sustain at least one failed drive and may be able to survive failure of up to half the drives in the set of mirrored disks without forcing the system administrator to shut down the server and recover from file backup. (RAID 1 is the best-performing RAID option when fault tolerance is required.)

RAID 1 also has performance impacts in relation to read and write operations. When RAID 1 is used, write operations decrease performance, because each such operation costs two disk I/O operations, one to the original and one to the mirrored disk drive. On the other hand, RAID 1 increases performance of read operations, because the system will be able to read from either disk drive, depending on which one is least busy at the time.

RAID 5 (Parity)

Parity is implemented by calculating recovery information about data written to disk and writing that parity information on the other drives that form the RAID array. If a drive fails, a new drive is inserted into the RAID array and the data on that failed drive is recovered by taking the recovery information (parity) written on the other drives and using this information to regenerate the data from the failed drive.

The advantage of parity is that you need one additional disk drive to protect any number of existing disk drives. The disadvantages of parity concern performance and fault tolerance. Due to the additional costs associated with calculating and writing parity, additional disk I/O operations are required. (Read I/O operation costs are the same for RAID 1 and parity.) Also, using parity, you can sustain only one failed drive before the array must be taken offline and recovery from backup media must be performed. Generally, RAID 5 requires four disk I/O operations, whereas RAID 0 requires only one operation and RAID 1 two operations.

Database Mirroring

As you already know, mirroring can be supported through hardware or software. The advantage of the software support for mirroring is that it can be configured to mirror disk partitions, while the hardware solutions are usually implemented on the entire disk. This section discusses the Windows solution for database mirroring and how you can set it up.

To set up database mirroring, use two servers with a database that will be mirrored from one server to the other. The former is called the *principal server*, while the latter is called the *mirrored server*. (The copy of the database on the mirrored server is called the *mirrored database*.)

Database mirroring allows continuous streaming of the transaction log from the principal server to the mirrored server. The copy of the transaction log activity is written to the log of the mirrored database, and the transactions are executed on it. If the principal server becomes unavailable, applications can reconnect to the database on the mirrored server without waiting for recovery to finish. Unlike failover clustering, the mirrored server is fully cached and ready to accept workloads because of its synchronized state. It is possible to implement up to four mirrored backup sets. (To implement mirroring, use the MIRROR TO option of either the BACKUP DATABASE statement or the BACKUP LOG statement.)

There is also a third server, called the *witness server*. It determines which server is the principal server and which is the mirrored server. This server is only needed when automatic failover is required. (To enable automatic failover, you must turn on the synchronous operating mode—that is, set the SAFETY option of the ALTER DATABASE statement to FULL.)

Another performance issue in relation to database mirroring is the possibility to automatically compress the data sent to the mirror. The Database Engine compresses the stream data if at least a 12.5 percent compression ratio can be achieved. That way, the system reduces the consumption of log data that is sent from the principal server to mirrored server(s).

Failover Clustering

Failover clustering is a process in which the operating system and database system work together to provide availability in the event of failures. A failover cluster consists of a group of redundant servers, called *nodes*, that share an external disk system. When a node within the cluster fails, the instance of the Database Engine on that machine shuts down. Microsoft Cluster Service transfers resources from a failing machine to an equally configured target node automatically. The transfer of resources from one node to the other node in a cluster occurs very quickly.

The advantage of failover clustering is that it protects your system against hardware failures, because it provides a mechanism to automatically restart the database system on another node of the cluster. On the other hand, this technology has a single point of failure in the set of disks, which cluster nodes share and cannot protect from data errors. Another disadvantage of this technology is that it does not increase performance or scalability. In other words, an application cannot scale any further on a cluster than it can on one node.

In summary, failover clustering provides server redundancy, but it doesn't provide data file redundancy. (See also the section "Comparison of High-Availability Components" later in this chapter.)

Log Shipping

Log shipping allows the transaction logs from one database to be constantly sent and used by another database. This allows you to have a warm standby server and also provides a way to offload data from the source machine to read-only destination computers. The target database is an exact copy of the primary database, because the former receives all changes from the latter. You have the ability to make the target database a new primary database if the primary server, which hosts the original database, becomes unavailable. When the primary server becomes available again, you can reverse the server roles again.

Log shipping does not support automatic failover. Therefore, if the source database server fails, you must recover the target database yourself, either manually or through custom code.

In summary, log shipping is similar to database mirroring in that it provides database redundancy. (See also the upcoming section "Comparison of High-Availability Components.")

AlwaysOn

Database mirroring as a technique to achieve high availability has several drawbacks:

- Read-only queries cannot be executed on the mirror.
- Database mirroring can be applied only on two instances of the Database Engine.
- Database mirroring mirrors only objects inside the database; objects such as logins cannot be protected using mirroring.

To overcome these drawbacks of database mirroring, the Database Engine supports a technique called AlwaysOn, which allows you to maximize availability for your databases.

AlwaysOn is based upon three concepts: availability groups, replicas, and modes, which will be discussed next.

NOTE AlwaysOn is the most sophisticated and hence the most complex technique to support high availability of Database Engine databases. For this reason, this chapter gives you just an introductory description of AlwaysOn. For all other topics concerning AlwaysOn, such as configuration of server instances and creation and configuration of availability groups, see Microsoft Docs.

Availability Groups, Replicas, and Modes

An *availability group* comprises a set of failover servers called *availability replicas*. Each availability replica has a local copy of each of the databases in the availability group. One of these replicas, called the *primary replica*, maintains the primary copy of each database. The primary replica makes these databases, called *primary databases*, available to users for read-write access. For each primary database, up to eight other availability replicas, known as *secondary replicas*, maintain a failover copy of the database, known as a *secondary database*.

An availability replica provides redundancy only at the database level, for the set of databases in one availability group. The primary replica makes the primary databases available for read-write connections from clients. Also, in a process known as *data synchronization*, which occurs at the database level, the primary replica sends transaction log records of each primary database to every secondary database.

Every secondary replica stores the transaction log records on the disk and, after that, applies them to its corresponding secondary database(s). Data synchronization occurs between the primary database and each connected secondary database, independently of the other databases. Optionally, you can configure one or more secondary replicas to support read-only access to secondary databases, and you can configure any secondary replica to permit backups on secondary databases.

Availability replicas can be hosted only by instances that reside on Windows Server Failover Clustering (WSFC) nodes. Instances can be either failover cluster instances or stand-alone instances. The instances that host availability replicas for a given availability group must reside on separate WSFC nodes. The server instance on which the primary replica is located is known as the *primary location*. An instance on which a secondary replica is located is known as a *secondary location*.

The *availability mode* is a property that is set independently for each availability replica. The availability mode of a secondary replica determines whether the primary replica waits to commit transactions on a database until the secondary replica has written the records in the corresponding transaction logs to disk. Therefore, AlwaysOn supports two availability modes:

- Asynchronous-commit mode
- Synchronous-commit mode

Asynchronous-commit mode means that the primary replica does not wait for any of the secondary replicas to store the log to disk. In other words, after writing the log record to the local log file, the primary replica sends the transaction confirmation to the client. The primary replica runs with minimum transaction latency in relation to a secondary replica that is configured for asynchronous-commit mode.

Synchronous-commit mode emphasizes high availability over performance, at the cost of increased transaction latency. Under synchronous-commit mode, transactions concerning the primary database wait to send the transaction confirmation to the client until the secondary replica has stored the log to disk. When data synchronization begins on a secondary database, the secondary replica begins applying incoming log records from the corresponding primary database. As soon as every log record has been permanently stored, the secondary database enters the SYNCHRONIZED state.

Besides these three concepts, there is also a notion of a primary role and a secondary role. The primary and secondary roles of availability replicas are interchangeable in a process called *failover*. Three forms of failover exist: automatic failover (without data loss), planned manual failover (without data loss), and forced manual failover (with possible data loss), typically called *forced failover*.

NOTE Starting with SQL Server 2016, AlwaysOn has been included with the Standard Edition of SQL Server. (Earlier, only Enterprise Edition supported this technique.) AlwaysOn in Standard Edition will have several restrictions. The most important one is that only two nodes (one primary and one secondary) can be used.

Comparison of High-Availability Components

The Database Engine supports several proprietary components that are used to enhance the availability of your instance of the Database Engine and databases:

- Failover clustering
- Log shipping
- Mirroring
- AlwaysOn

In relation to these techniques, there are three important issues:

- Server redundancy
- Database redundancy
- Data file redundancy

Server redundancy means that an application runs on two or more servers in such a way as to provide fault tolerance. (Clustering is one of the most important server redundancy technologies.) *Database redundancy* means that a fault tolerance is guaranteed for a database with all its applications. (*Data file redundancy* is defined similarly.)

Failover clustering provides server redundancy, but doesn't provide database and data file redundancy. Log shipping provides database redundancy, but doesn't provide server redundancy. The disadvantage of log shipping is that it doesn't provide automatic failover.

Database mirroring doesn't provide server redundancy, but provides database redundancy and data file redundancy. (This component belongs to deprecated features and won't be supported in one of the future versions of the Database Engine.)

AlwaysOn is similar to database mirroring, but additionally supports clustering. AlwaysOn provides server redundancy as well as database and data file redundancy. The primary goal of AlwaysOn is to support database availability while also giving you the benefits of disaster recovery.

Maintenance Plan Wizard

The Maintenance Plan Wizard provides you with the set of basic tasks needed to maintain a database. It ensures that your database performs well, is regularly backed up, and is free of inconsistencies.

NOTE To create or manage maintenance plans, you have to be a member of the **sysadmin** fixed server role.

To start the Maintenance Plan Wizard, expand the server in SQL Server Management Studio, expand Management, right-click Maintenance Plans, and choose Maintenance Plan Wizard. As you can see on the starting page of the Maintenance Plan Wizard, you can perform the following administration tasks:

- Check database integrity
- Perform index maintenance
- Update database statistics
- Perform database backups

NOTE I will show you how to use the Maintenance Plan Wizard to perform database backups. All other tasks can be performed in a similar manner.

Click Next on the starting page, and the next wizard page, Select Plan Properties (see Figure 16-5), enables you to select properties for your plan, enter the plan's name, and, optionally, describe the plan. Also, you can choose between separate schedules for each task or a single schedule for the entire plan. This example will perform the backup of the **sample** database, so name the plan **Backup-sample** and choose the Single Schedule for the Entire Plan radio button. The Schedule field allows you to create a schedule for the execution of the plan or to execute it on demand. (Chapter 17 describes in detail how you can create such a schedule. For purposes of this example, leave the Schedule field set to Not Scheduled (On Demand).)

Click Next, and the wizard enables you to choose, among other tasks, full, differential, and transaction log backups. (For the description of these options, see "Introduction to Backup Methods" at the beginning of this chapter.) Check Back Up Database (Full) and click Next, which opens the Select Maintenance Task Order page. You can then specify the order in which the tasks should be performed. (In this case, there is no order, because there is only one task to be performed.) Click Next, and the Back Up Database Full page appears.

The next page, Define Back Up Database (Full) Task, enables you to specify several different options. First, select the database(s) on which the task should be performed. Then, select a destination for the backup files. The destination includes the media type and their location. (You can also specify an expiration date for your backup set.)

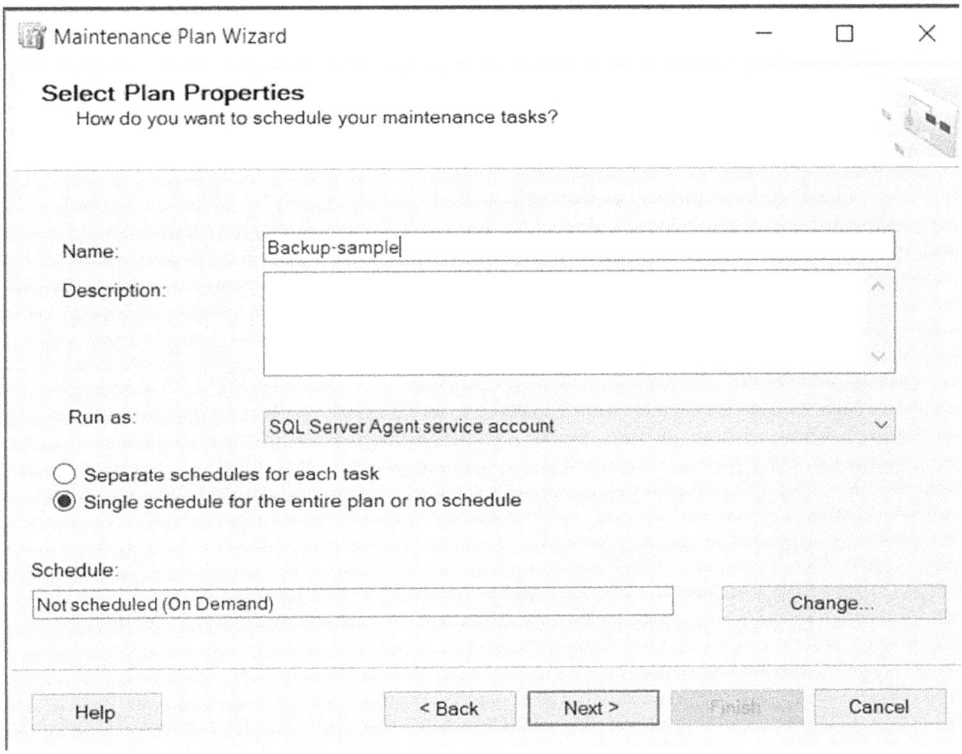

Figure 16-5   The Select Plan Properties wizard page

The next option, Create a Backup File for Every Database, allows you to create a separate file for each database you have specified in the Database(s) drop-down list box. Click this radio button, because this is the preferred way to maintain the backup of several databases. Check the last option, Verify Backup Integrity, so that the Database Engine checks the integrity of the backup files. Click Next to continue.

The Select Report Options wizard page allows you to write a report to a specific file and/or send an e-mail message. An e-mail message can be sent only to an existing operator. (Chapter 17 describes in detail how you can create an operator.)

To complete the wizard, click Next, and after that, click Finish. The wizard performs the task and creates a corresponding report.

To view the history of an existing maintenance plan, expand Management, expand Maintenance Plans, right-click the name of the plan, and choose View History. The Log File Viewer with the history of the selected plan is shown.

Summary

The system administrator or database owner should periodically make a backup copy of the database and its transaction log. The Database Engine enables you to make two kinds of backup copies of the database: full and differential. A full backup captures the state of the database at the time the statement is issued and copies it to the backup media (file or tape device). A differential backup copies the parts of the database that have changed since the last full database backup. The benefit of the differential backup is that it completes more rapidly than the full database backup for the same database. (There is also a transaction log backup, which copies transaction logs to a backup media.)

The Database Engine performs automatic recovery each time a system failure occurs that does not cause any media failure. (Automatic recovery is also performed when the system is started after each shutdown of the system.) During automatic recovery, any committed transaction found in the transaction log is written to the database, and any uncommitted transaction is rolled back. After any media failure, it may be necessary to manually recover the database from the archived copy of it and its transaction logs. To recover a database, a full database backup and only the latest differential backup must be used. If you use transaction logs to restore a database, use the full database backup first and then apply all existing transaction logs in the sequence of their creation to bring the database to the consistent state that it was in before the last transaction log backup was created.

The Database Engine supports the following high-availability components:

- Failover clustering
- Log shipping
- Database mirroring
- AlwaysOn

Failover clustering provides server redundancy, but doesn't provide database and data file redundancy. Log shipping provides database redundancy, but doesn't provide server redundancy.

Database mirroring doesn't provide server redundancy, but provides database redundancy and data file redundancy. AlwaysOn provides server redundancy as well as database and data file redundancy.

The Maintenance Plan Wizard is a general tool that you can use for a set of basic tasks needed to maintain a database. It ensures that your database is regularly backed up and, therefore, free of inconsistencies. (The wizard was described in this chapter because it is usually used in relation to backup and restore operations.)

The next chapter describes all the system features that allow you to automate system administration tasks.

Exercises

E.16.1   Discuss the differences between the differential backup and transaction log backup.

E.16.2   When should you back up your production database?

E.16.3   How can you make a differential backup of the **master** database?

E.16.4   Discuss the use of different RAID technologies with regard to fault tolerance of a database and its transaction log.

E.16.5   What are the main differences between manual and automatic recovery?

E.16.6   Which statement should you use to verify your backup, without using it for the restore process?

E.16.7   Discuss the advantages and disadvantages of the three recovery models.

E.16.8   Discuss the similarities and differences between failover clustering, database mirroring, and log shipping.

17 Automating System Administration Tasks

In This Chapter

- Starting SQL Server Agent
- Creating Jobs and Operators
- Alerts

One of the most important advantages of the Database Engine in relation to other relational DBMSs is its capability to automate administrative tasks and hence to reduce costs. The following are examples of some important tasks that are performed frequently and therefore could be automated:

- Backing up the database and transaction log
- Transferring data
- Dropping and re-creating indices
- Checking data integrity

You can automate all these tasks so that they occur on a regular schedule. For example, you can set the database backup task to occur every Friday at 8:00 p.m. and the transaction log backup task to occur daily at 10:00 p.m.

The components of the Database Engine that are used in automation processes include the following:

- SQL Server service (MSSQLSERVER)
- Windows Application log
- SQL Server Agent service

Why does the Database Engine need these three components to automate processes? In relation to automation of administration tasks, the MSSQLSERVER service is needed to write

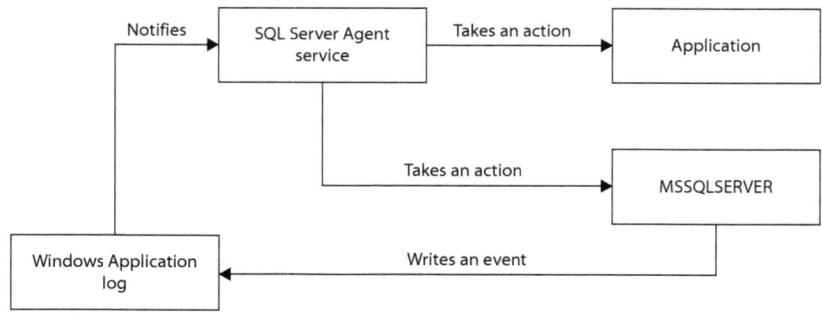

Figure 17-1 SQL Server automation components

events to the Windows Application log. Some events are written automatically, and some must be raised by the system administrator (see the detailed explanation later in this chapter).

The Windows Application log is where application events are written. The role of the Windows Application log in the automation process is to notify SQL Server Agent about existing events.

SQL Server Agent is another service that writes events to the Windows Application log and the MSSQLSERVER service. The role of SQL Server Agent in the automation process is to take an action after a notification through the Windows Application log. The action can be performed in connection with the MSSQLSERVER service or some other application. Figure 17-1 shows how these three components work together.

Starting SQL Server Agent

SQL Server Agent executes jobs and fires alerts. As you will see in the upcoming sections, jobs and alerts are defined separately and can be executed independently. Nevertheless, jobs and alerts may also be complementary processes, because a job can invoke an alert and vice versa.

Consider an example: A job is executed to inform the system administrator about an unexpected filling of the transaction log that exceeds a tolerable limit. When this event occurs, the associated alert is invoked and, as a reaction, the system administrator may be notified by e-mail or SMS text alert.

Another critical event is a failure in backing up the transaction log. When this happens, the associated alert may invoke a job that truncates the transaction log. This reaction will be appropriate if the reason for the backup failure is an overflow (filling up) of the transaction log. In other cases (for example, the target device for the backup copy is full), such a truncation will have no effect. This example shows the close connection that may exist between events that have similar symptoms.

SQL Server Agent allows you to automate different administrative tasks. Before you can do this, the process has to be started. To start SQL Server Agent under SQL Server Management Studio, right-click the SQL Server Agent folder (under your Database Engine instance) and choose Start.

As already stated, the invocation of an alert can also include the notification of one or more operators by e-mail using Database Mail. Database Mail is an enterprise solution for sending e-mail messages from the Database Engine. Using Database Mail, your applications can send e-mail messages to users. The messages may contain query results, and may also include files from any resource on your network.

Creating Jobs and Operators

Generally, there are three steps to follow if you want to create a job:

1. Create a job and its steps.
2. Create a schedule of the job execution if the job is not to be executed on demand.
3. Notify operators about the status of the job.

The following sections explain these steps using an example.

Creating a Job and Its Steps

A job may contain one or more steps. There are different ways in which a job step can be defined. The following list contains some of them.

- **Using Transact-SQL statements**   Many job steps contain Transact-SQL statements. For example, if you want to automate database or transaction log backups, you use the BACKUP DATABASE statement or BACKUP LOG statement, respectively.

- **Using the operating system (CmdExec)**   Some jobs may require the execution of a Database Engine utility, which usually will be started with the corresponding command. For example, if you want to automate the data transfer from your database server to a data file, or vice versa, you could use the **bcp** utility.

- **Invoking a program**   As another alternative, it may be necessary to execute a program that has been developed using Visual Basic or some other programming language. In this case, you should always include the path drive letter in the Command text box when you start such a program. This is necessary because SQL Server Agent has to find the executable file.

If the job contains several steps, it is important to determine which actions should be taken in case of a failure. Generally, the Database Engine starts the next job step if the previous one was successfully executed. However, if a job step fails, any job steps that follow will not be executed. Therefore, you should always specify how often each step should be retried in the case of failure. And, of course, it will be necessary to eliminate the reason for the abnormal termination of the job step. (Obviously, a repeated job execution will always lead to the same error if the cause is not repaired.)

NOTE The number of attempts depends on the type and content of the executed job step (batch, command, or application program).

You can create a job using the following:

- SQL Server Management Studio
- System stored procedures (**sp_add_job** and **sp_add_jobstep**)

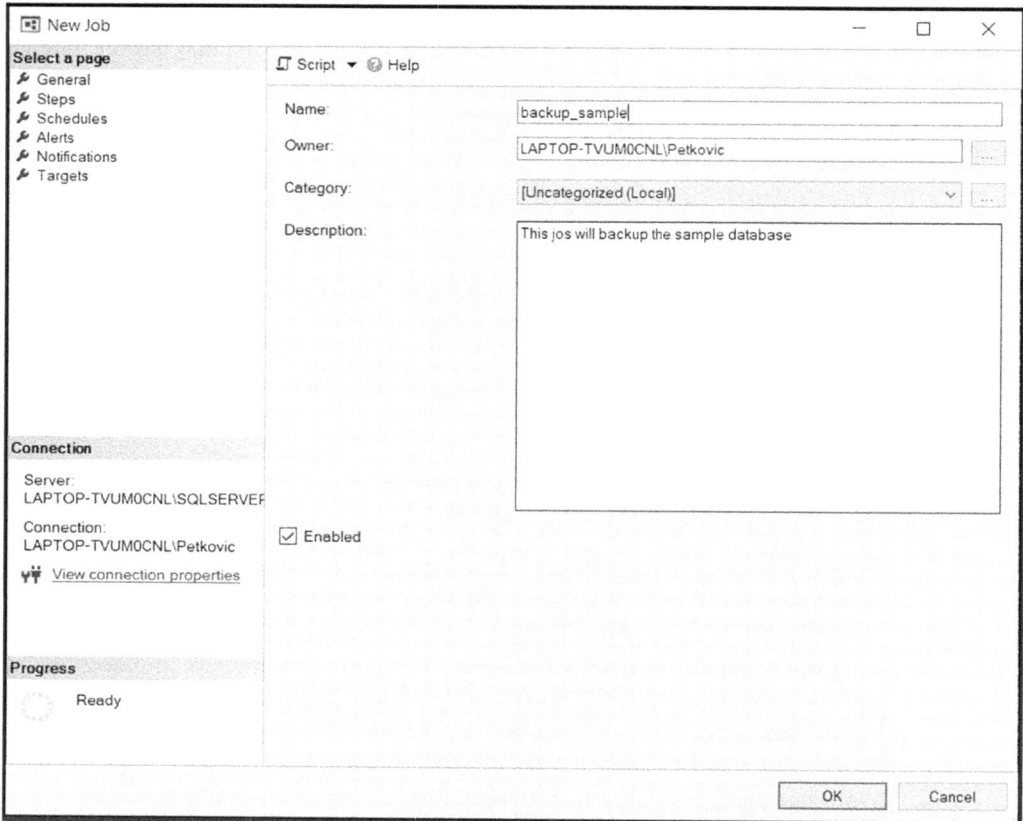

Figure 17-2   The New Job dialog box

Management Studio is used in this example, which creates a job that backs up the **sample** database. Expand SQL Server Agent, right-click Jobs, and choose New Job. (SQL Server Agent must be running.) The New Job dialog box appears (see Figure 17-2). On the General page, enter a name for the job in the Name box. (The name of the job for backing up the **sample** database will be **backup_sample**.)

For the Owner field, click the ellipsis (...) button and choose the owner responsible for performing the job. In the Category drop-down list, choose the category to which the job belongs. You can add a description of the job in the Description box, if you wish.

NOTE   If you have to manage several jobs, categorizing them is recommended. This is especially useful if your jobs are executed in a multiserver environment.

Check the Enabled check box to enable the job.

NOTE   All jobs are enabled by default. SQL Server Agent disables jobs if the job schedule is defined either at a specific time that has passed or on a recurring basis with an end date that has also passed. In both cases, you must re-enable the job manually.

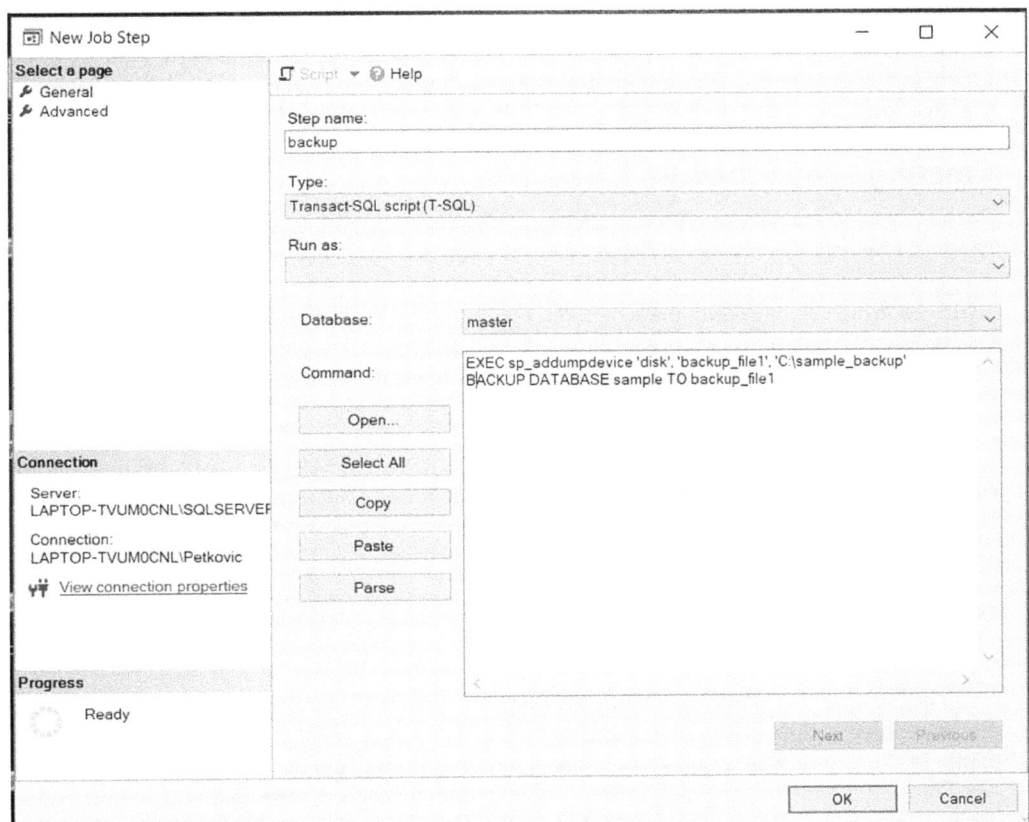

Figure 17-3   The New Job Step dialog box, General page

Each job must have one or more steps. Therefore, in addition to defining job properties, you must create at least one step before you can save the job. To define one or more steps, click the Steps page in the New Job dialog box and click New. The New Job Step dialog box appears, as shown in Figure 17-3. Enter a name for the job step. (It is called **backup** in the example.) In the Type drop-down list, choose Transact-SQL Script (T-SQL), because the backup of the sample database will be executed using the Transact-SQL statement BACKUP DATABASE.

In the Database drop-down list, choose the **master** database, because this system database must be the current database if you want to back up a database.

You can either enter the Transact-SQL statement directly in the Command box or invoke it from a file. In the former case, enter the following statements, after you change the path for the backup file:

```
EXEC sp_addumpdevice 'disk', 'backup_file1', 'C:\sample_backup'
BACKUP DATABASE sample TO backup_file1
```

As you probably guessed, the **sp_addumpdevice** system procedure adds a backup device to an instance of the Database Engine. To invoke the Transact-SQL statement from a file, click Open and select the file. The syntax of the statement(s) can be checked by clicking Parse. Click OK to continue.

Creating a Job Schedule

Each created job can be executed on demand (that is, manually by the user) or by using one or more schedules. A scheduled job can occur at a specific time or on a recurring schedule.

NOTE Each job can have multiple schedules. For example, the backup of the transaction log of a production database can be executed with two different schedules, depending on the time of day. This means that during peak business hours, you can execute the backup more frequently than during non-peak hours.

To create a schedule for an existing job using SQL Server Management Studio, select the Schedules page in the New Job dialog box and click New. (The New Job dialog box is the same dialog box as shown in Figure 17-2). If the New Job dialog box is not active, expand SQL Server Agent, expand Jobs, and click the job you want to process.

NOTE If you get the warning, "The On Access action of the last step will be changed from Get Next Step to Quit with Success," click Yes.

The New Job Schedule dialog box appears (see Figure 17-4). For the **sample** database, set the schedule for the backup to be executed every Sunday at 10:00 P.M. To do this, enter the job name in the Name dialog box and choose Recurring in the Schedule Type drop-down list. In the Frequency section, choose Weekly in the Occurs drop-down list, and check Sunday. In the Daily Frequency section, click the Occurs Once At radio button, and enter the time (**22:00:00**). In the Duration section, choose the start date in the Start Date drop-down list, and then click the End Date radio button and choose the end date in the corresponding drop-down list. (If the job should be scheduled without the end date, click No End Date.)

Notifying Operators About the Job Status

When a job completes, several methods of notification are possible. For example, you can instruct the system to write a corresponding message to the Windows Application log, hoping that the system administrator reads this log from time to time. A better choice is to explicitly notify one or more operators using e-mail name and/or pager e-mail name.

Before an operator can be assigned to a job, you have to create an entry for it. To create an operator using SQL Server Management Studio, expand SQL Server Agent, right-click Operators, and choose New Operator. The New Operator dialog box appears (see Figure 17-5). On the General page, enter the name of the operator in the Name box. Specify one or more methods of notifying the operator (via e-mail). In the Pager on Duty Schedule section, enter the working hours of the operator. Click OK.

To notify one or more operators after the job finishes (successfully or unsuccessfully), return to the New Job dialog box of the job, select the Notifications page (see Figure 17-6), and

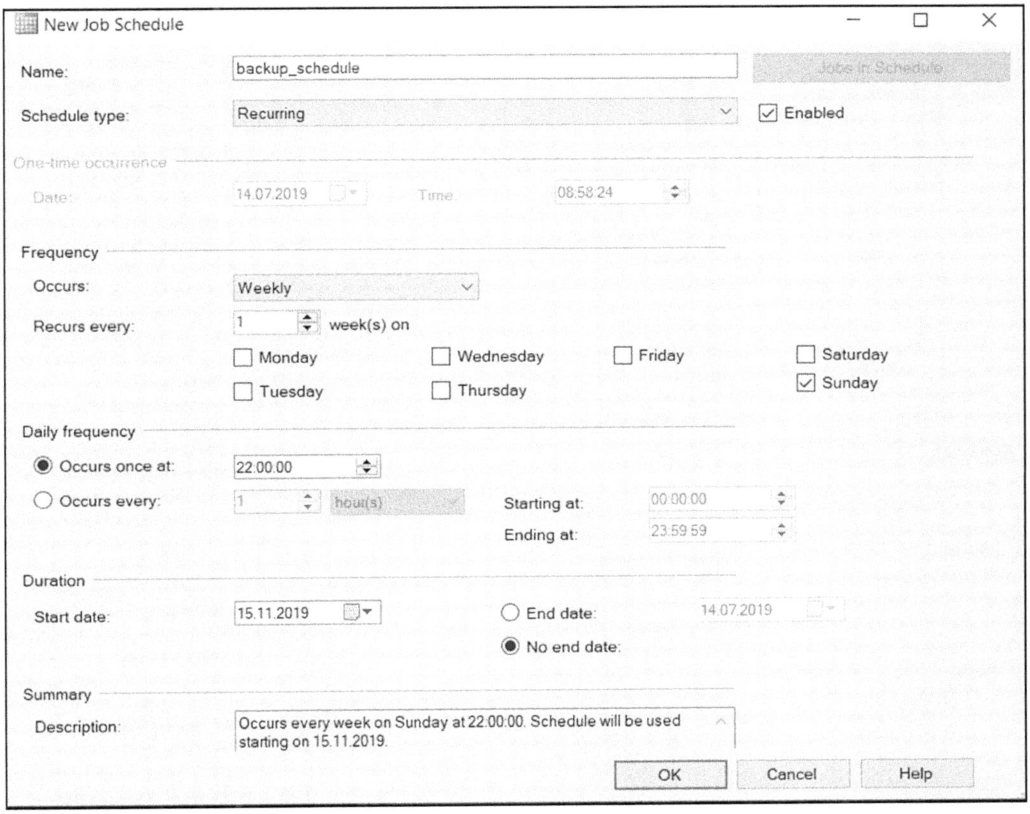

Figure 17-4   The New Job Schedule dialog box

check the corresponding boxes. (Besides e-mail, in this dialog box you also have the option of writing the message to the Windows Application log and/or deleting the job.)

Viewing the Job History Log

The Database Engine stores the information concerning all job activities in the **sysjobhistory** system table of the **msdb** system database. Therefore, this table represents the job history log of your system. You can view the information in this table using SQL Server Management Studio. To do this, expand SQL Server Agent, expand Jobs, right-click the job, and choose View History. The Log File Viewer dialog box shows the history log of the job.

Each row of the job history log is displayed in the details pane, which contains, among other information, the following:

- Date and time when the job step occurred
- Whether the job step completed successfully or unsuccessfully
- Operators who were notified

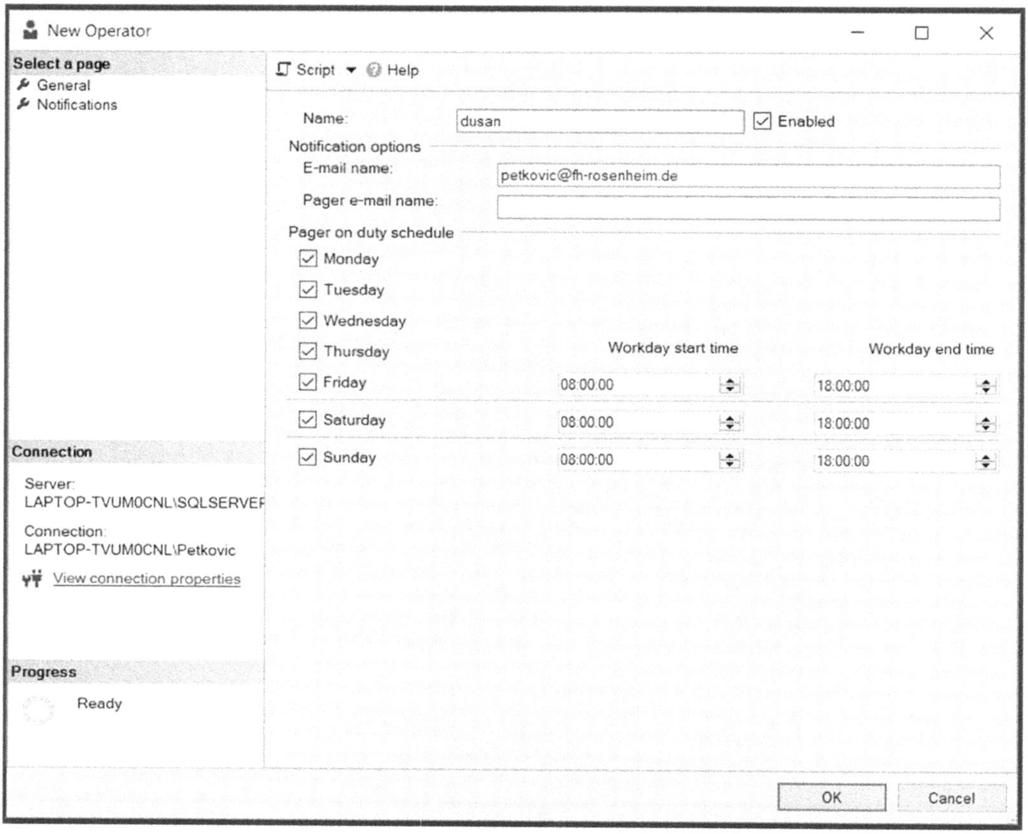

Figure 17-5   The New Operator dialog box

- Duration of the job
- Errors or messages concerning the job step

By default, the maximum size of the job history log is 1000 rows, while the number of rows for a particular job is limited to 100. (The job history log is automatically cleared when the maximum size of rows is reached.) If you want to store the information about each job, and your system has several jobs, increase the size of the job history log and/or the number of rows per job. Using Management Studio, right-click SQL Server Agent and choose Properties. In the SQL Server Agent Properties dialog box, select the History page and enter the new values for the Maximum Job History Log Size and Maximum Job History Rows per Job fields. You can also check Remove Agent History and specify a time interval after which logs should be deleted.

Alerts

The information about execution of jobs and system error messages is stored in the Windows Application log. SQL Server Agent reads this log and compares the stored messages with the

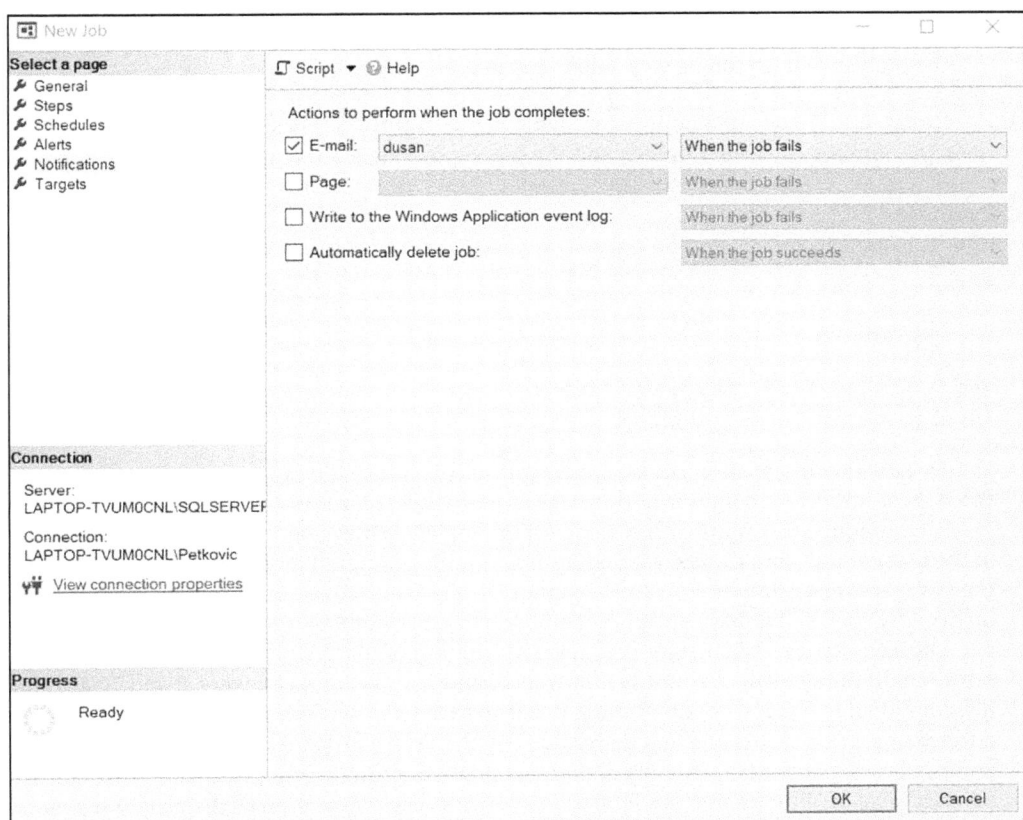

Figure 17-6   The New Job dialog box, Notifications page

alerts defined for the system. If there is a match, SQL Server Agent fires the alert. Therefore, alerts can be used to respond to potential problems (such as filling up the transaction log), different system errors, or user-defined errors. Before explaining how you create alerts, this section discusses system error messages and two logs, the SQL Server Agent error log and the Windows Application log, which are used to capture all system messages (and thus most of the errors).

Error Messages

System errors are grouped in four different groups. The Database Engine provides extensive information about each error. The information is structured and includes the following:

- A unique error message number
- An additional number between 0 and 25, which represents the error's severity level
- A line number, which identifies the line where the error occurred
- The error text

NOTE The error text not only describes the detected error but also may recommend how to resolve the problem, which can be very helpful to the user.

Example 17.1 queries a nonexistent table in the **sample** database, thus showing the error message number, the level number, and the corresponding error text.

Example 17.1
```
USE sample;
SELECT * FROM authors;
```

The result is

```
Msg 208, Level 16, State 1, Line 2
Invalid object name 'authors'.
```

To view the information concerning error messages, use the **sys.messages** catalog view. The three most important columns of this view are **message_id**, **severity**, and **text**.

Each unique error number has a corresponding error message. (The error message is stored in the **text** column, and the corresponding error number is stored in the **message_id** column of the **sys.messages** catalog view.) In Example 17.1, the message concerning the nonexistent or incorrectly spelled database object corresponds to error number –208.

The severity level of an error (the **severity** column of the **sys.messages** catalog view) is represented in the form of a number between 0 and 25. The levels between 0 and 10 are simply informational messages, where nothing needs to be fixed. All levels from 11 through 16 indicate different program errors and can be resolved by the user. The values 17 and 18 indicate software and hardware errors that generally do not terminate the running process. All errors with a severity level of 19 or greater are fatal system errors. The connection of the program generating such an error is closed, and its process will then be removed.

The messages relating to program errors (that is, the levels between 11 and 16) are shown on the screen only. All system errors (errors with a severity level of 19 or greater) will also be written to the log.

System error messages are written to the SQL Server Agent error log and to the Windows Application log. The following two sections describe these two components.

SQL Server Agent Error Log

SQL Server Agent creates an error log that records warnings and errors by default. The following warnings and errors are displayed in the log:

- Warning messages that provide information about potential problems
- Error messages that usually require intervention by a system administrator

The system maintains up to ten SQL Server Agent error logs. The current log is called **Current**, while all other logs have an extension that indicates the relative age of the log. For example, **Archive #1** indicates the newest archived error log.

The SQL Server Agent error log is an important source of information for the system administrator. With it, he or she can trace the progress of the system and determine which corrective actions to take.

To view the SQL Server Agent error logs from Management Studio, expand the instance in Object Explorer, expand SQL Server Agent, and expand Error Logs. Double-click one of the files to view the desired log. The log details appear in the details pane of the Log File Viewer dialog box.

Windows Application Log

The Database Engine also writes system messages to the Windows Application log. The Windows Application log is the location of all operating system messages for the Windows operating systems, and it is where all application messages are stored. You can view the Windows Application log using the Event Viewer.

NOTE The Windows Application log is also called the Windows Application Event log.

Viewing errors in the Windows Application log has some advantages compared to viewing them in the SQL Server Agent error log. The most important is that the Windows Application log provides an additional component for the search for desired strings.

To view information stored in the Windows Application log, choose Start | Control Panel | Administrative Tools | Event Viewer. In the Event Viewer window, you can choose between system, security, and application messages. For database system messages, click Application. Database system events are identified by the entry MSSQLSERVER in the source column.

Defining Alerts to Handle Errors

An alert can be defined to raise a response to a particular error number or to the group of errors that belongs to a specific severity code. Furthermore, the definition of an alert for a particular error is different for system errors and user-defined errors. (The creation of alerts on user-defined errors is described later in this chapter.)

The rest of this section shows how you can create alerts using Management Studio.

Creating Alerts on System Errors

Example 13.8 (see Chapter 13), in which one transaction was deadlocked by another transaction, will be used to show how to create an alert about a system error number. If a transaction is deadlocked by another transaction, the "victim" must be executed again. This can be done, among other ways, by using an alert.

To create the deadlock (or any other) alert, expand SQL Server Agent, right-click Alerts, and choose New Alert. In the New Alert dialog box (see Figure 17-7), enter the name of the alert in the Name box, choose SQL Server Event Alert in the Type drop-down list, and choose <all databases> from the Database Name drop-down list. Click the Error Number radio button, and enter **1205**. (This error number indicates a deadlock problem, where the current process was selected as the "victim.")

The second step defines the response for the alert. In the same dialog box, click the Response page. First check Execute Job, and then choose the job to execute when the alert occurs (**backup_sample**, in our case). Check Notify Operators, and then, in the Operator List

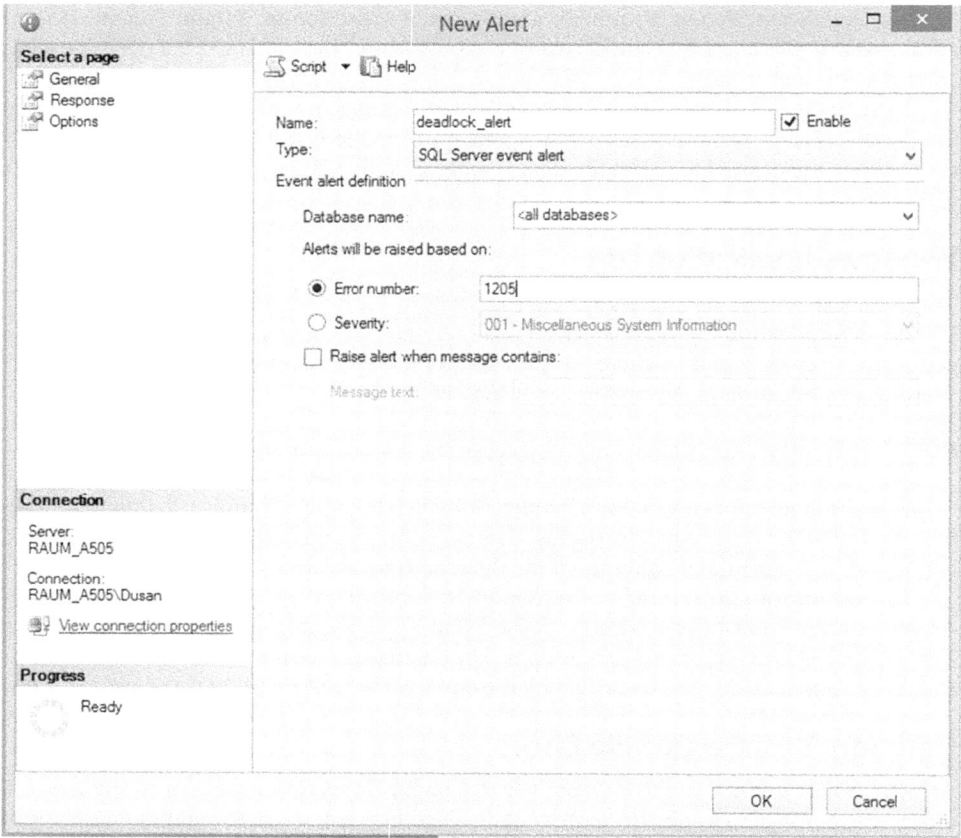

Figure 17-7   The New Alert dialog box, General page

pane, select operators and choose the methods of their notifications (e-mail name, and/or pager e-mail name).

Creating Alerts on Error Severity Levels

You can also define an alert that will raise a response on error severity levels. As you already know, each system error has a corresponding severity level that is a number between 0 and 25. The higher the severity level is, the more serious the error. Errors with severity levels 20 through 25 are fatal errors. Errors with severity levels 19 through 25 are written to the Windows Application log.

NOTE Always define an operator to be notified when a fatal error occurs.

As an example of how you can create alerts in relation to severity levels, here's how you use Management Studio to create the particular alert for severity level 25. First, expand SQL Server Agent, right-click Alerts, and choose New Alert. In the Name box, enter a name for this alert (for example, **Severity 25 errors**). In the Type drop-down list, choose SQL Server Event Alert. In the Database Name drop-down list, choose the **sample** database. Click the Severity radio button and choose 025 – Fatal Error.

On the Response page, enter one or more operators to be notified via e-mail and/or pager, when an error of severity level 25 occurs.

Creating Alerts on User-Defined Errors

In addition to creating alerts on system errors, you can create alerts on customized error messages for individual database applications. Using such messages (and alerts), you can define solutions to problems that might occur in an application.

The following steps are necessary if you want to create an alert on a user-defined message:

1. Create the error message.

2. Raise the error from a database application.

3. Define an alert on the error message.

An example is the best way to illustrate the creation of such an alert. We will use the **sales** table (see Example 5.24 in Chapter 5) and define an alert that fires when the value of the shipping date (the **ship_date** column) is earlier than the order date (the **order_date** column).

NOTE Only the first two steps are described here, because an alert on a user-defined message is defined similarly to an alert on a system error message.

Creating an Error Message   To create a user-defined error message, you can use either Management Studio or the **sp_addmessage** stored procedure. Example 17.2 creates the error message for the example using the **sp_addmessage** stored procedure.

Example 17.2

```
sp_addmessage @msgnum=50010, @severity=16,
@msgtext='The shipping date of a product is earlier than the order date',
@lang='us_english', @with_log='true'
```

The **sp_addmessage** system stored procedure in Example 17.2 creates a user-defined error message with error number 50010 (the **@msgnum** parameter) and severity level 16 (the **@severity** parameter). All user-defined error messages are stored in the **sysmessages** system table of the **master** database and can be viewed by using the **sys.messages** catalog view. The error number in Example 17.2 is 50010 because all user-defined errors must be greater than 50000. (All error message numbers less than 50000 are reserved for the system.)

For each user-defined error message, you can optionally use the **@lang** parameter to specify the language in which the message is displayed. This specification may be necessary if multiple languages are installed on your computer. (When the **@lang** parameter is omitted, the session language is the default language.)

By default, user-defined messages are not written to the Windows Application log. On the other hand, you must write the message to this log if you want to raise an alert on it. If you set the **@with_log** parameter of the **sp_addmessage** system procedure to TRUE, the message will be written to the log.

Raising an Error Using Triggers To raise an error from a database application, you invoke the RAISERROR statement. This statement returns a user-defined error message and sets a system flag in the **@@error** global variable. (You can also handle error messages using TRY/CATCH blocks.)

Example 17.3 creates the trigger **t_date_comp**, which returns a user-defined error of 50010 if the shipping date of a product is earlier than the order date.

NOTE To execute Example 17.3, the **sales** table must exist.

Example 17.3
```
USE sample;
GO
CREATE TRIGGER t_date_comp
   ON sales
   FOR INSERT AS
   DECLARE @order_date DATE
   DECLARE @shipped_date DATE
SELECT @order_date=order_date, @shipped_date=ship_date FROM INSERTED
   IF @order_date > @shipped_date
           RAISERROR (50010, 16, -1)
```

Now, if you insert the following row in the **sales** table, the shipping date of a product is earlier than the order date:

```
INSERT INTO sales VALUES (1, '01.01.2017', '01.01.2016')
```

The system will return the user-defined error message:

```
Msg 50010, Level 16, State 1, Procedure t_date_comp, Line 8
```

The shipping date of a product is earlier than the order date.

Summary

The Database Engine allows you to automate and streamline many administrator tasks, such as database backups, data transfers, and index maintenance. For the execution of such tasks, SQL Server Agent must be running.

To automate a task, you have to execute several steps:

• Create a job

• Create operators

• Create alerts

Job and *task* are synonymous, so when you create a job, you create the particular task that you want to automate. The easiest way to create a job is to use SQL Server Management Studio, which allows you to define one or more job steps and create an execution schedule.

When a job (successfully or unsuccessfully) completes, you can notify one or more persons, using operators. Again, the general way to create an operator is to use Management Studio.

Alerts are defined separately and can also be executed independently of jobs. An alert can handle individual system errors, user-defined errors, or groups of errors belonging to one of 25 severity levels.

The next chapter discusses data replication.

Exercises

E.17.1 Name several administrative tasks that could be automated.

E.17.2 You want to back up the transaction log of your database every hour during peak business hours and every four hours during nonpeak hours. What should you do?

E.17.3 You want to test performance of your production database in relation to locks and want to know whether the lock wait time is more than 30 seconds. How could you be notified automatically when this event occurs?

E.17.4 Specify all parts of a SQL Server error message.

E.17.5 Which are the most important columns of the **sys.messages** catalog view concerning errors?

CHAPTER

18

Data Replication

In This Chapter

- Distributed Data and Methods for Distributing
- SQL Server Replication: An Overview
- Managing Replication

Today, market forces require most companies to set up their computers (and the applications running on them) so that they focus on business and on customers. As a result, data used by these applications must be available ad hoc at different locations and at different times. Such a data environment is provided by several distributed databases that include multiple copies of the same information.

The traveling salesperson represents a good example of how a distributed data environment is used. During the day, the salesperson usually uses a laptop to query all necessary information from the database (prices and availability of products, for example) to inform customers on the spot. Afterwards, in the hotel room, the salesperson again uses the laptop—this time to transmit data (about the sold products) to headquarters.

From this scenario, you can see that a distributed data environment has several benefits compared to centralized computing:

- It is directly available to the people who need it, when they need it.
- It allows local users to operate autonomously.
- It reduces network traffic.
- It makes nonstop processing cheaper.

On the other hand, a distributed data environment is much more complex than the corresponding centralized model and therefore requires more planning and administration.

The introductory part of this chapter discusses distributed transactions and compares them with data replication, which is the topic of this chapter. After that, the chapter introduces replication elements and explains the existing replication types. The last part of the chapter describes how you can use SQL Server Management Studio to manage replication.

Distributed Data and Methods for Distributing

There are two general methods for distributing data on multiple database servers:

- Distributed transactions
- Data replication

A *distributed transaction* is a transaction in which all updates to all locations (where the distributed data is stored) are gathered together and executed synchronously. Distributed database systems use a method called *two-phase commit* to implement distributed transactions.

Each database involved in a distributed transaction has its own recovery technique, which is used in case of error. (Remember that all statements inside a transaction are executed in their entirety or are cancelled.) A global recovery manager (called a coordinator) coordinates the two phases of distributed processing.

In the first phase of this process, the coordinator checks whether all participating sites are ready to execute their part of the distributed transaction. The second phase consists of the actual execution of the transaction at all participating sites. During this process, any error at any site causes the coordinator to stop the transaction. In this case, it sends a message to each local recovery manager to undo the part of the transaction that is already executed at that site.

NOTE   The Microsoft Distributed Transaction Coordinator (DTC) supports distributed transactions using two-phase commit.

During the data replication process, copies of data are distributed from a source database to one or more target databases located on separate computers. Because of this, data replication differs from distributed transactions in two ways: timing and delay in time.

In contrast to the distributed transaction method, in which all data is distributed on all participating sites at the same time, *data replication* allows sites to have different data at the same time. Additionally, data replication is an asynchronous process. This means that there is a certain delay during which all copies of data are matched on all participating sites. (This delay can last from a couple of milliseconds to several days or weeks.)

Data replication is, in most cases, a better solution than distributed transactions because it is more reliable and cheaper. Experience with two-phase commit has shown that administration becomes very difficult if the number of participating sites increases. Also, the increased number of participating sites decreases the reliability, because the probability that a local part of a distributed transaction will fail increases with the increased number of nodes. (If one local part fails, the entire distributed transaction will fail, too.)

Another reason to use data replication instead of centralized data is performance: clients at the site where the data is replicated experience improved performance because they can access data locally rather than using a network to connect to a central database server.

SQL Server Replication: An Overview

Generally, replication is based on one of two different concepts:

- Using transaction logs
- Using triggers

As already stated in Chapter 16, the Database Engine keeps all values of modified rows ("before" as well as "after" values) in system files called transaction logs. If selected rows need to be replicated, the system starts a new process that reads the data from the transaction log and sends it to one or more target databases.

The other method is based on triggers. The modification of a table that contains data to be replicated fires the corresponding trigger, which in turn creates a new table with the data and starts a replication process.

Both concepts have their benefits and disadvantages. The log-based replication is characterized by improved performance, because the process that reads data from the transaction log runs asynchronously and has little effect on the performance of the overall system. On the other hand, the implementation of log-based replication is very complex for companies, because the Database Engine not only has to manage additional processes and buffers but also has to solve the concurrency problems between system and replication processes that access the transaction log.

NOTE The Database Engine uses both concepts: the transaction log method for transactional replication and triggers for merge replication. (Transactional and merge replications are described in detail later in this chapter.)

Publishers, Distributors, and Subscribers

The Database Engine replication is based on the so-called publisher–subscriber metaphor. This metaphor describes the different roles servers can play in a replication process. One or more servers publish data that other servers can subscribe to. In between there exists a distributor that stores the changes and forwards them further (to the subscribers). Hence, a node can have one (or more) different roles in a replication scenario:

- **Publisher (or publishing server)**   Maintains its source databases, makes data available for replication, and sends the modified data to the distributor
- **Distributor (or distribution server)**   Receives all changes to the replicated data from the publisher and stores and forwards them to the appropriate subscribers
- **Subscriber (or subscription server)**   Receives and maintains published data

A database server can play one or more roles in a replication process. For example, a server can act as the publisher and the distributor at the same time. This scenario is appropriate for a process with few replications and few subscribers. If there are a lot of subscribers for the publishing information, the distributor can be located on its own server. Figure 18-1 shows a simple scenario in which one instance is both the publishing and distribution server and three

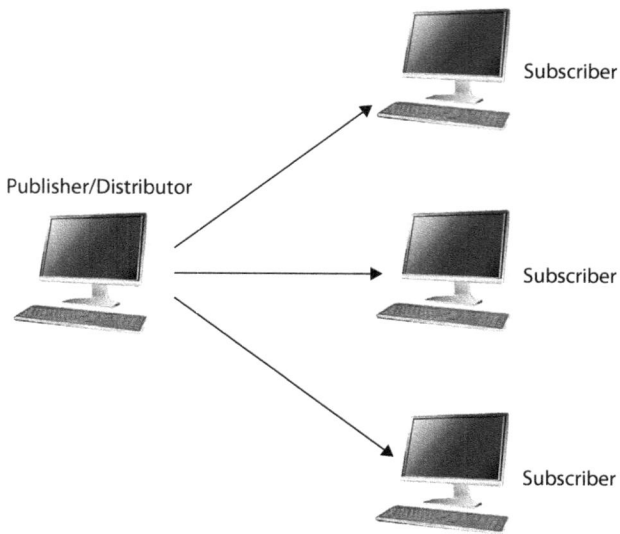

Figure 18-1   Central publisher with the distributor

other instances are subscription servers. (The section "Replication Models" later in this chapter discusses in detail possible replication scenarios.)

NOTE   You can replicate only user-defined databases.

Publications and Articles

The unit of data to be published is called a *publication*. An *article* contains data from a table and/or one or more stored procedures. A table article can be a single table or a subset of data in a table. A stored procedure article can contain one or more stored procedures that exist at the publication time in the database. A publication contains one or more articles. Each publication can contain data only from one database.

NOTE   A publication is the basis of a subscription. This means that you cannot subscribe directly to an article, because an article is always part of a publication.

A *filter* is the process that restricts information, producing a subset. Therefore, a publication contains one or more of the following items that specify types of table articles:

- Table
- Vertical filter
- Horizontal filter
- A combination of vertical and horizontal filters

A vertical filter contains a subset of the columns in a table. A horizontal filter contains a subset of rows in a table.

Publications are tightly connected to subscriptions. A subscription can be initiated in two different ways:

- Using a push subscription
- Using a pull subscription

With a *push subscription*, all the administration of setting up subscriptions is performed on the publisher during the definition of a publication. (Besides the publisher, the distributor also creates and manages push subscriptions.) Push subscriptions simplify and centralize administration, because the usual replication scenario contains one publisher and many subscribers. The benefit of a push subscription is higher security, because the initialization process is managed at one place. On the other hand, the performance of the distributor can suffer because the overall distribution of subscriptions runs at once.

With a *pull subscription*, the subscriber initiates and manages the subscription. The pull subscription is more selective than the push subscription, because the subscriber can select publications to subscribe to. In contrast to the push subscription, the pull subscription should be used for publications with a high number of subscribers.

NOTE The downloading of data using the Internet is a typical form of pull subscription.

There is a special type of pull subscription called *anonymous subscription*. Generally, information concerning subscribers is kept on the distribution server. If the workload on this server should be reduced (because of too many subscribers, for instance), it is possible to allow subscribers to initiate their own ("anonymous") subscriptions.

Agents

During the data replication process, the Database Engine uses several agents to manage different tasks. The system supports, among others, the following agents:

- Snapshot agent
- Log Reader agent
- Distribution agent
- Merge agent

These agent types are discussed following a brief introduction to the **distribution** database.

The distribution Database

The **distribution** database is a system database that is installed on the distribution server when the replication process is initiated. This database holds all replicated transactions from the publisher that need to be forwarded to the subscribers.

In many cases, a single **distribution** database is sufficient. However, if multiple publishing servers communicate with a single distribution server, you can create a **distribution** database for each publishing server. Doing so ensures that the data flowing through each **distribution** database is distinct.

Snapshot Agent

The Snapshot agent generates the schema and data of the published tables and stores them on the distribution server. The schema of a table and the corresponding data file build the synchronization set that represents the snapshot of the table at a particular time. (A *snapshot* is essentially what it sounds like: a snapshot of the data to be replicated.) The status of the synchronization of that set is recorded in the **distribution** database. Whether the Snapshot agent creates new snapshot files each time it runs depends on the type of replication and options chosen.

Log Reader Agent

If the transaction log of the system is used to replicate data, all transactions that contain the data to be replicated are marked for replication. A component called the Log Reader agent searches for marked transactions and copies them from the transaction log on the publisher to the distribution server. These transactions are stored in the **distribution** database. Each database that uses the transaction log for replication has its own Log Reader agent running on the distribution server.

Distribution Agent

After the transactions and snapshots are stored in the **distribution** database, they have to be moved to the subscribers. This task is handled by the Distribution agent, which moves transactions and snapshots to subscribers, where they are applied to the target tables in the subscription databases.

The task of the Distribution agent is different for pull and push subscriptions. For push subscriptions, the agent pushes out the changes to the subscriber. For pull subscriptions, the agent pulls the transactions from the distribution server. (All actions that change data on the publisher are applied to the subscriber in chronological order.)

Merge Agent

As you already know, the Snapshot agent prepares files containing the table schema and data and stores them at the distributor site. If both the publisher and subscribers can update replicated data, then a synchronization job is necessary that sends all changed data to the other sites. This job is performed by the Merge agent. In other words, the Merge agent can send replicated data to the subscribers and to the publisher. Before the send process is started, the Merge agent also stores the appropriate information that is used to track possible conflicts.

Replication Types

The Database Engine supports several replication types, which are discussed in the following subsections:

- Transactional
- Peer-to-peer
- Snapshot
- Merge

Transactional Replication

In transactional replication, the transaction log of the system is used to replicate data. All transactions that contain the data to be replicated are marked for replication. The Log Reader agent searches for marked transactions and copies them from the transaction log on the publisher to the **distribution** database. The Distribution agent moves transactions to subscribers, where they are applied to the target tables in the subscription databases.

NOTE All tables published using transactional replication must explicitly contain a primary key. The primary key is required to uniquely identify the rows of the published table, because a row is the transfer unit in transactional replication.

Transactional replication can replicate tables (or parts of tables) and one or more stored procedures. The use of stored procedures by transactional replication increases performance, because the amount of data to be sent over a network is usually significantly smaller. Instead of replicated data, only the stored procedure is sent to the subscribers, where it is executed. You can configure the delay of synchronization time between the publisher on one side and subscribers on the other during a transactional replication. (All these changes are propagated by the Log Reader and Distribution agents.)

NOTE Before transactional replication can begin, a copy of the entire database must be transferred to each subscriber; this is performed by executing a snapshot.

A special form of transactional replication is peer-to-peer transactional replication, discussed next.

Peer-to-Peer Transactional Replication

Peer-to-peer is another form of transactional replication, in which each server is at the same time a publisher, distributor, and subscriber for the same data. In other words, all servers contain the same data, but each server is responsible for the modification of its own partition of data. (Note that data partitions on different servers can intersect.)

Peer-to-peer transactional replication is best explained through an example. Suppose that a company has several branch offices in different cities and that each office server has the same data set as all other servers. On the other hand, the entire data set is partitioned in subsets, and each office server can update only its own subset of data. When data is modified on one of the office servers, the changes are replicated to all other servers (subscribers) in the peer-to-peer network. (Users in each office can read data without any restrictions.)

The benefits of this replication form are

- The entire system scales well.
- The entire system provides high availability.

A system that supports peer-to-peer transactional replication scales well because each server serves only local users. (Users can update only the data partition that belongs to their local server. For read operations, all data is stored locally, too.)

The high availability is based on the fact that if one or more servers go offline, all other servers can continue to operate, because all data they need for read and write operations is stored locally. When an offline server is online again, the replication process restarts and the server receives all data modifications that have happened at the other sites.

Conflict Detection in Peer-to-Peer Replication   With peer-to-peer replication, you can change data at any node. Therefore, data changes at different nodes could conflict with each other. (If a row is modified at more than one node, it can cause a conflict.)

The Database Engine supports the option to enable conflict detection across a configured topology. With this option enabled, a conflicting change is treated as a critical error that causes the failure of the Distribution agent. In the event of a conflict, the scenario remains in an inconsistent state until the conflict is resolved and the data is made consistent on all participating servers.

NOTE You can enable conflict detection using the system procedures **sp_addpublication** and **sp_configure_peerconflictdetection**.

Conflicts in peer-to-peer transactional replication are detected by the stored procedures that apply changes to each node, based on a hidden column in each published table. This hidden column stores an identifier that combines a unique ID that you specify for each node and the version of the row. The procedures are executed by the Distribution agent and they apply insert, update, and delete operations from other peers. If one of the procedures detects a conflict when it reads the hidden column value, it raises an error.

NOTE The hidden column can be accessed only by a user that is logged in through the dedicated administrator connection (DAC). For the description of DAC, see Chapter 15.

When a conflict occurs in peer-to-peer transactional replication, the "Peer-to-peer conflict detection alert" is raised. You should configure this alert so that you are notified when a conflict occurs. (The previous chapter explains how alerts can be configured and discusses the ways to notify operators.) Microsoft Docs describes several approaches for handling the conflicts that occur.

NOTE You should try to avoid conflicts in a peer-to-peer replication, even if conflict detection is enabled.

Snapshot Replication
The simplest type of replication, snapshot replication, copies the data to be published from the publisher to all subscribers. (The difference between snapshot replication and transactional replication is that the former sends all the published data to the subscribers and the latter sends only the changes of data to the subscribers.)

NOTE Transactional and snapshot replications are one-way replications, meaning the only changes to the replicated data are made at the publishing server. Therefore, the data at all subscription servers is read-only, except for the changes made by replication processes.

In contrast to transactional replication, snapshot replication requires no primary key for tables. The reason is obvious: the unit of transfer in snapshot replication is a snapshot file and not a row of a table. Another difference between these two replication types concerns a delay in time: snapshot replication is replicated periodically, which means the delay is significant because all published data (changed and unchanged) is transferred from the publisher to the subscribers.

NOTE Snapshot replication does not use the **distribution** database directly. However, the **distribution** database contains status information and other details that are used by snapshot replication.

Merge Replication

In transactional and snapshot replication, the publisher sends the data, and a subscriber receives it. (There is no possibility that a subscriber sends replicated data to the publisher.) Merge replication allows the publisher and subscribers to update data to be replicated. Because of that, conflicts can arise during a replication process.

When you use the merge replication scenario, the system makes three important changes to the schema of the publication database:

- It identifies a unique column for each replicated row.
- It adds several system tables.
- It creates triggers for tables in which data is replicated.

The Database Engine creates or identifies a unique column in the table with the replicated data. If the base table already contains a column with the UNIQUEIDENTIFIER data type and the ROWGUIDCOL property, the system uses that column to identify each replicated row. If there is no such column in the table, the system adds the column **rowguid** of the UNIQUEIDENTIFIER data type with the ROWGUIDCOL property.

NOTE UNIQUEIDENTIFIER columns may contain multiple occurrences of a value. The ROWGUIDCOL property additionally indicates that the values of the column of the UNIQUEIDENTIFIER data type uniquely identify rows in the table. Therefore, a column of the data type UNIQUEIDENTIFIER with the ROWGUIDCOL property contains unique values for each row across all networked computers in the world and thus guarantees the uniqueness of replicated rows across multiple copies of the table on the publisher and subscribers.

The addition of new system tables provides the way to detect and resolve any update conflict. The Database Engine stores all changes concerning the replicated data in the merge system tables **msmerge_contents** and **msmerge_tombstone** and joins them with the table that contains replicated data to resolve the conflict.

The Database Engine creates triggers on tables that contain replicated data on all sites to track changes to the data in each replicated row. These triggers determine the changes made to the table, and they record them in the **msmerge_contents** and **msmerge_tombstone** system tables. Conflict detection is done by the Merge agent using the column lineage of the

msmerge_contents system table when a conflict is detected. The resolution of it can be either priority based or custom based.

Priority-based resolution means that any conflict between new and old values in the replicated row is resolved automatically based on assigned priorities. (The special case of the priority-based method specifies the "first wins" method, where the timely first change of the replicated row is the winner.) The priority-based method is the default. The *custom-based* method uses customized triggers based on business rules defined by the database administrator to resolve conflicts.

Replication Models

The previous section introduced different replication types that the Database Engine uses to distribute data between different nodes. The replication types (transactional, snapshot, peer-to-peer, and merge) provide the functionality for maintaining replicated data. *Replication models* are used by a company to design its own data replication. Each replication model can be implemented using one or more existing replication types. Both the replication type and replication model are usually specified at the same time.

Depending on requirements, several replication models can be used. The basic ones are as follows:

- Central publisher with distributor
- Central publisher with a remote distributor
- Central subscriber with multiple publishers
- Multiple publishers with multiple subscribers

The following sections describe these models.

Central Publisher with Distributor

In the central publisher with distributor model, there is one publisher and usually one distributor, which are hosted on one instance of the Database Engine (see Figure 18-1 at the beginning of this chapter). The publisher creates publications that are distributed by the distributor to several subscribers. The publications designed by this model and received at a subscriber are usually read-only.

The advantage of this model is its simplicity. For this reason, the model is usually used to create a copy of a database, which is then used for interactive queries and simple report generation. (Another situation for using this model is to maintain a remote copy of a database, which could be used by remote systems in the case of communication breakdown.)

For this model, you can use the transactional replication type.

Central Publisher with a Remote Distributor

If the amount of publishing data is not very large, the publisher and distributor can reside on one server. Otherwise, if you have a heavy load, using two separate servers for publishing and distribution is recommended because of performance issues. (If there is a heavy load of data to be published, the distributor is usually the bottleneck.) Figure 18-2 shows the replication model with the central publisher and a separate distributor.

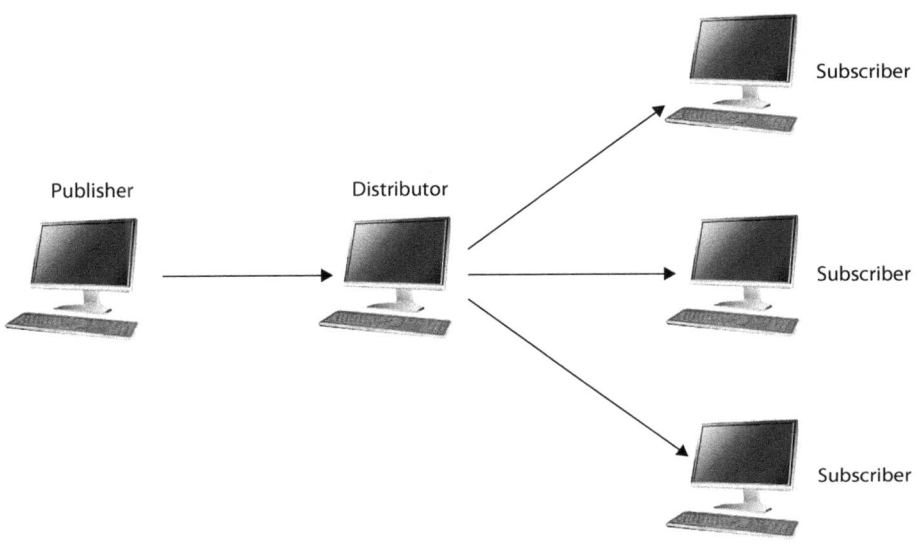

Figure 18-2   Central publisher with a remote distributor

NOTE This scenario can be used as a starting point to increase a number of publishing servers and/or subscribing servers.

Central Subscriber with Multiple Publishers

The scenario described at the beginning of this chapter of the traveling salesperson who transmits data to headquarters is a typical example of the central subscriber with multiple publishers. The data is gathered at a centralized subscriber, and several publishers send their data.

For this model, you can use either the peer-to-peer transactional replication type or the merge replication type, depending on the use of replicated data. If publishers publish (and therefore update) the same data to the subscriber, merge replication should be used. If each publisher has its own data to publish, peer-to-peer transactional replication should be used. (In this case, published tables will be filtered horizontally, and each publisher will be the owner of a particular table fragment.)

Multiple Publishers with Multiple Subscribers

The replication model in which some or all of the servers participating in data replication play the role of the publisher and the subscriber is known as multiple publishers with multiple subscribers. In most cases, this model includes several distributors that are usually placed at each publisher (see Figure 18-3).

This model can be implemented using merge replication only, because publications are modified at each publishing server. (The only other way to implement this model is to use the distributed transactions with two-phase commit.)

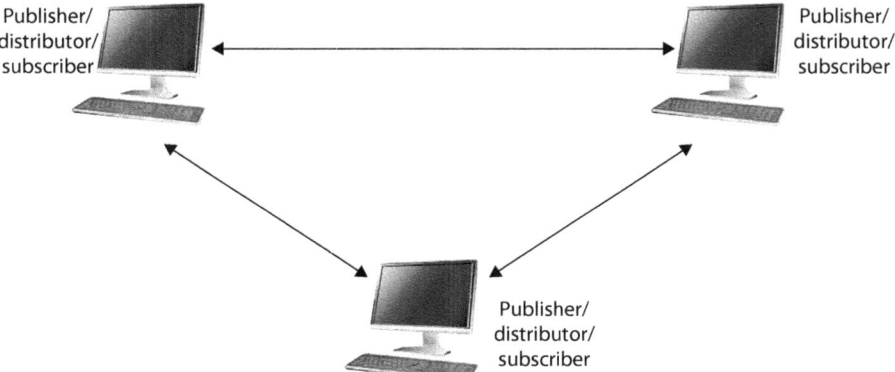

Figure 18-3 Multiple publishers with multiple subscribers

Managing Replication

All servers that participate in a replication must be registered. (Server registration is described in Chapter 3.) After registering servers, the distribution server, publishing server(s), and subscription server(s) must be set up. The following sections describe configuration of these processes using the corresponding wizards.

Configuring the Distribution and Publication Servers

Before you install publishing databases, you must install the distribution server and configure the **distribution** database. You can set up a distribution server by using the Configure Distribution Wizard. This wizard allows you to configure the distributor and the **distribution** database and to enable publisher(s). With the wizard you can

- Configure your server to be a distributor that can be used by other publishers
- Configure your server to be a publisher that acts as its own distributor
- Configure your server to be a publisher that uses another server as its distributor

This section shows a scenario for data replication of the **sample** database using the following servers: **LAPTOP-TVUM0CNL** and **NTB01112**. The former will be used as a publisher and distributor, while the latter will be the subscriber. The first step is to use the Configure Distribution Wizard to set up the **LAPTOP-TVUM0CNL** server to be a publisher that acts as its own distributor. (Additionally, the wizard will create the **distribution** database.)

NOTE You can also use the system procedures **sp_adddistributor** and **sp_adddistributiondb** to set up the distribution server and the **distribution** database. **sp_adddistributor** sets up the distribution server by creating a new row in the **sysservers** system table. **sp_adddistributiondb** creates a new **distribution** database and installs the distribution schema.

To start the wizard, start SQL Server Management Studio, expand the instance, right-click Replication, and select Configure Distribution. The Configure Distribution Wizard appears. On the Distributor page, choose the **LAPTOP-TVUM0CNL** server as the distribution server and click Next. After that, select the folder in which snapshots from publisher(s) that use the distribution server will be stored and click Next. On the Distribution Database page, select the name of the **distribution** database and log files and click Next. On the Publishers page, enable the publisher(s) (the **LAPTOP-TVUM0CNL** server in this example), choose whether to finish the configuration process immediately or generate the script file to start the distribution configuration later, and then click Next. Figure 18-4 shows the summary of all steps that you have made to configure the **LAPTOP-TVUM0CNL** server as the distributor and publisher.

NOTE The existing publishing and distribution on a server can be disabled using the Disable Publishing and Distribution Wizard. To start the wizard, right-click Replication and choose Disable Publishing and Distribution.

After you configure the distribution and publishing servers, you must set up the publishing process. This is done with the New Publication Wizard, explained in the following section.

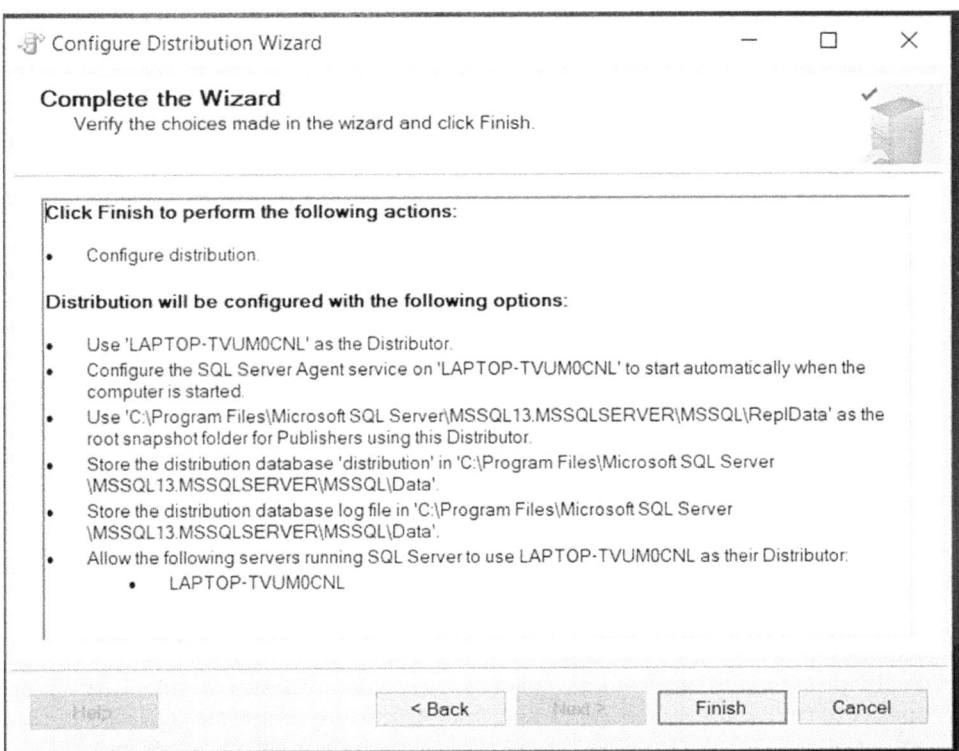

Figure 18-4   Complete the Wizard page for the distributor and publisher(s)

Setting Up Publications

You can use the New Publication Wizard to:

- Select the data and database objects you want to replicate
- Filter the published data so the subscribers receive only the data they need

Assume that you want to publish data of the **employee** table from the **LAPTOP-TVUM0CNL** server using the snapshot replication type. In this case, the entire **employee** table is the publication unit.

To create a publication, expand the server node of the publishing server (**LAPTOP-TVUM0CNL**), expand the Replication folder, right-click the Local Publications folder, and choose New Publication. The New Publication Wizard appears. On the first two pages, choose the database to publish (**sample**) and the publication type (in this case, the snapshot publication) and click Next. Then select at least one object for publication and click Next (in this example, select the entire **employee** table). The New Publication Wizard also allows you to filter (horizontally or vertically) the data that you want to publish. The next page, Snapshot Agent, allows you to create the snapshot of the selected data immediately and/or to run periodically. (For our example, we will create the snapshot immediately.)

On the Agent Security page, specify the security settings for the Snapshot agent. To do this, click the Security Settings button and type the Windows user account under which the Snapshot agent process will run. (The user account must be entered in the form *domain_name\ account_name*.) Click OK on the Security Setting dialog box and click Next. In the Wizard Actions page, you can decide to finish the configuration process immediately or generate the script file to start the publication creation later. Figure 18-5 shows the summary of all steps made to set up the **employee** table as a publication unit.

The last step is to configure the subscription servers, discussed in the following section.

Configuring Subscription Servers

A task that concerns subscribers but has to be performed at the publisher is enabling the publisher to subscribe. Use Management Studio to enable a subscriber at the publishing server. First expand the publishing server, expand Replication, right-click Local Subscriptions, and choose New Subscriptions. The New Subscription Wizard appears. You can use the wizard to:

- Create one or more subscriptions to a publication
- Specify where and when to run agents that synchronize the subscription

On the Publication page, choose the publication for which you want to create one or more subscriptions and then click Next. (In this example, choose the publication already generated with the New Publication Wizard.)

On the Distribution Agent Location page, you must choose between the push and pull subscriptions. A push subscription means that the synchronization of subscriptions is administered centrally. For this replication, check Run All Agents at the Distributor. To specify the pull subscription, check Run Each Agent at Its Subscriber. Click Next. (Because our subscription is pushed from the central publisher, we choose the first option.)

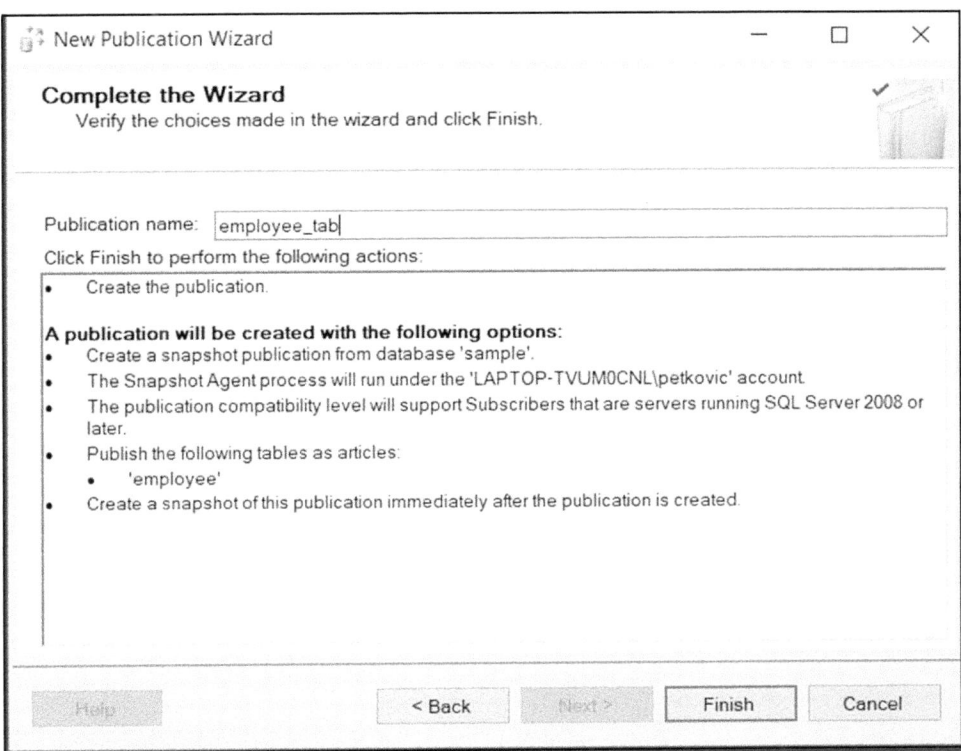

Figure 18-5   Complete the Wizard page for the publication unit

On the Subscribers page, you must specify all subscription servers. If the subscription servers have not been added, click Add Subscriber, and after selecting all servers to which data will be replicated, click Next. Before you finish the process, the wizard shows you the summary of the subscription configuration.

Summary

Data replication is the preferred method for data distribution because it is simpler than using distributed transactions. The Database Engine allows you to choose one of four possible replication types (snapshot, transactional, merge, and peer-to-peer replication), depending on the physical model you use. Theoretically, any replication model can use any of the replication types, although each (basic) model has a corresponding type that is used in most cases.

A publication is the smallest unit of replication. A single database can have many publications with different replication types. (Otherwise, each publication corresponds to only one database.)

To configure the replication process, you must first set up the distribution server and the distribution system database and configure the publishing server(s). In the next step, you have to define one or more publications. Finally, you have to configure the subscription server(s). The Database Engine supports these steps with three different wizards: the Configure Distribution Wizard, New Publication Wizard, and New Subscription Wizard.

The next three chapters discuss the overall performance of the system. Chapter 19 explains how the query optimizer of the Database Engine works, while Chapter 20 discusses performance tuning. Chapter 21 explains a relatively new and powerful technique, In-Memory OLTP, that allows you to optimize OLTP queries. These chapters close the third part of the book.

Exercises

E.18.1   Why do you need a primary key for data replication? Which replication type requires a primary key?

E.18.2   How can you limit network traffic and/or database size?

E.18.3   Update conflicts are not recommended. How can you minimize them?

E.18.4   When does the system use the Log Reader agent, the Merge agent, and the Snapshot agent, respectively?

19 Query Optimizer

In This Chapter

- Phases of Query Processing
- How Query Optimization Works
- Tools for Editing the Optimizer Strategy
- Optimizer Hints

The question that generally arises when the Database Engine (or any other relational database system) executes a query is how the data that is necessary for the query can be accessed and processed in the most efficient manner. The component of a database system that is responsible for the processing is called the query optimizer.

The task of the *query optimizer* (or just *optimizer*) is to consider a variety of possible execution strategies for querying the data in relation to a given query and to select the most efficient strategy. The selected strategy is called the *execution plan* of the query. The optimizer makes its decisions using considerations such as how big the tables are that are involved in the query, what indices exist, and what Boolean operator(s) (AND, OR, NOT) are used in the WHERE clause. Generally, these considerations are called *statistics*.

The beginning of the chapter introduces the phases of query processing and then explains in depth how the third phase, query optimization, works. This lays the foundation for the practical examples presented in the subsequent sections. Following that, you are introduced to the different tools that you can use to edit how the query optimizer does its work. The end of the chapter presents optimizer hints that you can give to the optimizer in special situations where it cannot find the optimal solution.

Phases of Query Processing

The task of the optimizer is to work out the most efficient execution plan for a given query. This task is done using the following four phases (see Figure 19-1).

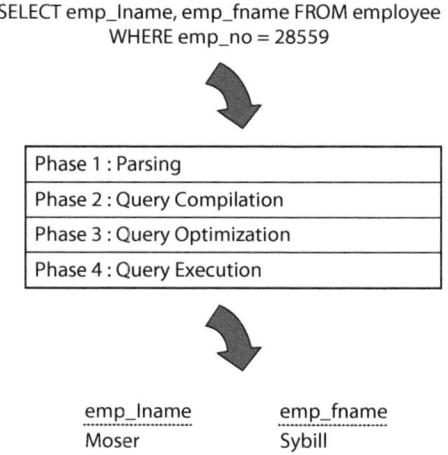

SELECT emp_lname, emp_fname FROM employee
WHERE emp_no = 28559

| Phase 1 : Parsing |
| Phase 2 : Query Compilation |
| Phase 3 : Query Optimization |
| Phase 4 : Query Execution |

emp_lname   emp_fname
Moser   Sybill

Figure 19-1 Phases in processing a query

NOTE This chapter refers to using the query optimizer for queries in SELECT statements. The query optimizer is also used for INSERT, UPDATE, and DELETE statements. The INSERT statement can contain a subquery, while the UPDATE and DELETE statements often have a WHERE clause, which might contain subqueries, too.

1. **Parsing** The query's syntax is validated and the query is transformed in a tree. After that, the validation of all database objects referenced by the query is checked. (For instance, the existence of all columns referenced in the query is checked and their IDs are determined.) After the validation process, the final query tree is formed.

2. **Query compilation** The query tree is compiled by the query optimizer.

3. **Query optimization** The query optimizer takes as input the compiled query tree generated in the previous step and investigates several access strategies before it decides how to process the given query. To find the most efficient execution plan, the query optimizer first makes the query analysis, during which it searches for search arguments and join operations. The optimizer then selects which indices to use. Finally, if join operations exist, the optimizer selects the join order and chooses one of the join processing techniques. (These optimization tasks are discussed in detail in the following section.)

4. **Query execution** After the execution plan is generated, it is permanently stored and executed.

NOTE For some statements, parsing and optimization can be avoided if the Database Engine knows that there is only one viable plan. (This process is called *trivial plan optimization*.) An example of a statement for which a trivial plan optimization can be used is the simple form of the INSERT statement.

How Query Optimization Works

As you already know from the previous section, the query optimization phase can be divided into the following phases:

- Query analysis
- Index selection
- Join order selection
- Join processing techniques

The following sections describe these phases. Also, at the end of this section, plan caching will be introduced.

Query Analysis

During the query analysis, the optimizer examines the query for search arguments, the use of the OR operator, and the existence of join criteria, in that order. Because the use of the OR operator and the existence of join criteria are self-explanatory, only search arguments are discussed.

A *search argument* is the part of a query that restricts the intermediate result set of the query. The main purpose of search arguments is to allow the use of existing indices in relation to the given expression. The following are examples of search arguments:

- emp_fname = 'Moser'
- salary >= 50000
- emp_fname = 'Moser' AND salary >= 50000

There are several expression forms that cannot be used by the optimizer as search arguments. To the first group belongs all expressions with the NOT (<>) operator. Also, if you use the expression on the left side of the operator, the existing expression cannot be used as a search argument.

The following are examples of expressions that are not search arguments:

- NOT IN ('d1', 'd2')
- emp_no <> 9031
- budget * 0.59 > 55000

The main disadvantage of expressions that cannot be used as search arguments is that the optimizer cannot use existing indices in relation to the expression to speed up the performance of the corresponding query. In other words, the only access the optimizer uses in this case is the table scan.

Index Selection

The identification of search arguments allows the optimizer to decide whether one or more existing indices will be used. In this phase, the optimizer checks each search argument to see if there are indices in relation to the corresponding expression. If an index exists, the optimizer decides whether or not to use it. This decision depends mainly on the selectivity of the corresponding expression. The *selectivity* of an expression is defined as the ratio of the number of rows satisfying the condition to the total number of rows in the table. The smaller the number of rows satisfying the condition, the higher the selectivity.

The optimizer checks the selectivity of an expression with the indexed column by using statistics that are created in relation to the distribution of values in a column. The query optimizer uses this information to determine the optimal query execution plan by estimating the cost of using an index to execute the query.

The following sections discuss in detail selectivity of an expression with the indexed column and statistics. (Because statistics exist in relation to both indices and columns, they are discussed separately in two sections.)

NOTE The Database Engine automatically creates (index and column) statistics if the database option called AUTO_CREATE_STATISTICS is activated. (This option is described later in this chapter.) Also, you can explicitly create statistics using the CREATE STATISTICS statement (as demonstrated in Example 19.12 later in this chapter).

Selectivity of an Expression with the Indexed Column

As you already know, the optimizer uses indices to improve query execution time. When you query a table that doesn't have indices, or if the optimizer decides not to use an existing index, the system performs a table scan. During the table scan, the Database Engine sequentially reads the table's data pages to find the rows that belong to the result set. *Index access* is an access method in which the database system reads and writes data pages using an existing index. Because index access significantly reduces the number of I/O read operations, it often outperforms a table scan.

The Database Engine uses a nonclustered index to search for data in one of two ways. If you have a heap (a table without a clustered index), the system first traverses the nonclustered index structure and then retrieves a row using the row identifier (RID). If you have a clustered table, however, the traversal of the nonclustered index structure is followed by the traversal of the index structure of the table's clustered index. On the other hand, the use of a clustered index to search for data is always unique: the Database Engine starts from the root of the corresponding B+-tree and usually after three or four read operations reaches the leaf nodes, where the data is stored. For this reason, the traversing of the index structure of a clustered index is almost always significantly faster than the traversing of the index structure of the corresponding nonclustered index.

From the preceding discussion, you can see that the answer to which access method (index scan or table scan) is faster isn't straightforward and depends on the selectivity and the index type.

Tests that I performed showed that a table scan often starts to perform better than a nonclustered index access when at least 10 percent of the rows are selected. In this case, the optimizer's decision of when to switch from nonclustered index access to a table scan must not be correct. (If you think that the optimizer forces a table scan prematurely, you can use the INDEX query hint to change its decision, as discussed later in this chapter.)

For several reasons, the clustered index usually performs better than the nonclustered index. When the system scans a clustered index, it doesn't need to leave the B+-tree structure to scan data pages, because the pages already exist at the leaf level of the tree. Also, a nonclustered index requires more I/O operations than a corresponding clustered index. The nonclustered index either needs to read data pages after traversing the B$^+$-tree or, if a clustered index for another table's column(s) exists, needs to read the clustered index's B$^+$-tree structure.

Therefore, you can expect a clustered index to perform significantly better than a table scan even when selectivity is poor (that is, the percentage of returned rows is high, because the query returns many rows). The tests that I performed showed that when the selectivity of an expression is 75 percent or less, the clustered index access is generally faster than the table scan.

Index Statistics

Index statistics are generally created when an index for the particular column(s) is created. The creation of index statistics for an index means that the Database Engine creates a *histogram* based on up to 200 values of the column. (Therefore, up to 199 intervals are built.) The histogram specifies, among other things, how many rows exactly match each interval, the average number of rows per distinct value inside the interval, and the density of values.

NOTE Index statistics are always created for one column. If your index is a composite (multicolumn) index, the system generates statistics for the first column in the index.

If you want to create index statistics explicitly, you can use the following tools:

- **sp_createstats** system procedure
- SQL Server Management Studio

The **sp_createstats** system procedure creates single-column statistics for all columns of all user tables in the current database. The new statistic has the same name as the column where it is created.

To use Management Studio for index statistics creation, expand the server, expand the Databases folder, expand the database, expand the Tables folder, expand the table, right-click Statistics, and choose New Statistics. The New Statistics on Table dialog box appears. In the dialog box, specify first the name for the new statistics. After that, click the Add button, select column(s) of the table to which to add the statistics, and click OK. Finally, click OK in the New Statistics on Table dialog box.

As the data in a column changes, index statistics become out of date. The out-of-date statistics can significantly influence the performance of the query. The Database Engine can automatically update index statistics if the database option AUTO_UPDATE_STATISTICS is activated (set to ON). In that case, any out-of-date statistics required by a query for optimization are automatically updated during query optimization.

Column Statistics

As you already know from the previous section, the Database Engine creates statistics for every existing index and the corresponding column. The system can create statistics for nonindexed columns too. These statistics are called *column statistics*. Together with index statistics, column statistics are used to optimize execution plans. The Database Engine creates statistics for a nonindexed column that is a part of the condition in the WHERE clause.

There are several situations in which the existence of column statistics can help the optimizer to make the right decision. One of them is when you have a composite index on two or more columns. For such an index, the system generates statistics only for the first column in the index. The existence of column statistics for the second column (and all other columns) of the composite index can help the optimizer to choose the optimal execution plan.

The Database Engine supports two catalog views in relation to column statistics (these views can be used to edit the information concerning index statistics, too):

- sys.stats
- sys.stats_columns

The **sys.stats** view contains a row for each statistic of a table or a view. Besides the **name** column, which specifies the name of the statistics, this catalog view has, among others, two other columns:

- **auto_created**   Statistics created by the optimizer
- **user_created**   Statistics explicitly created by the user

The **sys.stats_columns** view contains additional information concerning columns that are part of the **sys.stats** view. (To ascertain this additional information, you have to join both views using the **object_id** column in both tables as join columns.)

The information provided by the **sys.stats** and **sys.stats_columns** catalog views is similar to the information provided by the **sys.dm_exec_query_stats** dynamic management view (as shown in Example 19.10 later in the chapter).

Join Order Selection

Generally, the order in which two or more joined tables are written in the FROM clause of a SELECT statement doesn't influence the decision made by the optimizer in relation to their processing order.

As you will see in the next section, many different factors influence the decision of the optimizer regarding which table will be accessed first. On the other hand, you can influence the join order selection by using the FORCE ORDER hint (discussed in detail later in the chapter).

Join Processing Techniques

The join operation is the most time-consuming operation in query processing. The Database Engine supports the following three different join processing techniques, so the optimizer can choose one of them depending on the statistics for both tables:

- Nested loop
- Merge join
- Hash join

The following subsections describe these techniques.

Nested Loop

The nested loop processing technique works by "brute force." In other words, for each row of the outer table, each row from the inner table is retrieved and compared. The pseudo-code in Algorithm 19.1 demonstrates the nested loop processing technique for two tables.

Algorithm 19.1

```
(A and B are two tables.)
for each row in the outer table A do:
     read the row
     for each row in the inner table B do:
        read the row
           if A.join_column = B.join_column then
                accept the row and add it to the resulting set
           end if
     end for
end for
```

In Algorithm 19.1, every row selected from the outer table (table A) causes the access of all rows of the inner table (table B). After that, the comparison of the values in the join columns is performed and the row is added to the result set if the values in both columns are equal.

The nested loop technique is very slow if there is no index for the join column of the *inner table*. Without such an index, the Database Engine would have to scan the outer table once and the inner table n times, where n is the number of rows of the outer table. Therefore, the query optimizer usually chooses this method if the join column of the *inner* table is indexed, so the inner table does not have to be scanned for each row in the outer table.

Merge Join

The merge join technique provides a cost-effective alternative to the nested loop technique. The rows of the joined tables must be physically sorted using the values of the join column. Both tables are then scanned in order of the join columns, matching the rows with the same value for the join columns. The pseudo-code in Algorithm 19.2 demonstrates the merge join processing technique for two tables.

Algorithm 19.2

```
a. Sort the outer table A in ascending order using the join column
b. Sort the inner table B in ascending order using the join column
for each row in the outer table A do:
     read the row
     for each row from the inner table B with a value less than or equal to
the join column do:
        read the row
        if A.join_column = B.join_column then
            accept the row and add it to the resulting set
        end if
     end for
end for
```

The merge join processing technique has a high overhead if the rows from both tables are unsorted. However, this method is preferable when the values of both join columns are sorted in advance. (This is always the case when both join columns are primary keys of corresponding tables, because the Database Engine creates by default the clustered index for the primary key of a table.)

Hash Join

The hash join technique is usually used when there are no indices for join columns. In the case of the hash join technique, both tables that have to be joined are considered as two inputs: the build input and the probe input. (The smaller table usually represents the build input.) The process works as follows:

1. The value of the join column of a row from the build input is stored in a hash bucket depending on the number returned by the hashing algorithm.

2. Once all rows from the build input are processed, the processing of the rows from the probe input starts.

3. Each value of the join column of a row from the probe input is processed using the same hashing algorithm.

4. The corresponding rows in each bucket are retrieved and used to build the result set.

NOTE The hash join technique requires no index. Therefore, this method is highly applicable for ad hoc queries, where indices cannot be expected. Also, if the optimizer uses this processing technique, it could be a hint that you should create additional indices for one or both join columns.

Plan Caching

The Database Engine uses a set of memory caches as a storage space for data and for execution plans of queries. The first time such a query is executed, the compiled version of the query is stored in the *plan cache* (the part of memory used to store compiled query plans). When the same query is executed for the second time, the Database Engine checks whether an existing plan is stored in the plan cache. If so, the plan is used, and the recompilation of the query does not take place.

NOTE The process of plan caching for stored procedures is analogous.

Influencing Plan Caching

There are several ways in which you can influence plan caching. This section describes two of them:

- The **optimize for ad hoc workloads** option
- The DBCC FREEPROCCACHE command

optimize for ad hoc workloads is an advanced configuration option that prevents the system from placing an execution plan in cache on the *first execution* of the corresponding statement. In other words, the Database Engine places only a stub of the execution plan instead of the entire plan. This stub contains the minimum information necessary for the system to find matches with future queries. The idea behind this is to reduce the uncontrolled growth of the plan cache by storing execution plans only for such queries that are executed more than one time. (Keep in mind that the execution plan for the simple query with a couple of indexed columns in the SELECT list needs about 20KB of memory. The plan cache for complicated queries can be significantly larger.)

DBCC FREEPROCCACHE removes all plans from the plan cache. This command can be useful for testing purposes. In other words, if you want to determine which plans are cached, you can clear the cache using this command. You can also use the command to remove a specific plan from the plan cache by specifying the unique value for its plan handle. (Each query plan stored in the cache is identified by a unique identifier called a *plan handle*.)

Displaying Plan Cache Information

To display information concerning the plan cache, you can use the following dynamic management views (functions):

- sys.dm_exec_cached_plans
- sys.dm_exec_query_stats
- sys.dm_exec_sql_text

All these views and functions will be described in the section "Dynamic Management Views and Query Optimizer" later in the chapter.

Tools for Editing the Optimizer Strategy

The Database Engine supports several tools that enable you to edit what the query optimizer is doing. You can use the following tools, among others:

- SET statement (to display textual or XML execution plans)
- SQL Server Management Studio (to display graphical execution plans)
- Dynamic management views and functions
- SQL Server Profiler (discussed in detail in Chapter 20)

The following sections describe the first three tools.

NOTE Almost all examples in this chapter use the **AdventureWorks** database. If your system doesn't contain this database, the introductory part of the book describes how you can download it.

SET Statement

To understand the different options of the SET statement, you have to know that there are three different forms for how the execution plan of a query can be displayed:

- Textual form
- Using XML
- Graphical form

The first two forms use the SET statement, so these two forms are discussed in the following subsections. (The graphical form of execution plans is discussed a bit later, in the section "SQL Server Management Studio and Graphical Execution Plans.")

Textual Execution Plan

The phrase "textual execution plan" means that the execution plan of a query is displayed in text form. Therefore, the output of a textual execution plan is returned in the form of rows. The Database Engine uses vertical bars to show the dependencies between the operations taking place. Textual execution plans can be displayed using the following options of the SET statement:

- SHOWPLAN_TEXT
- SHOWPLAN_ALL

Users running a query can display the textual execution plan for the query by activating (setting the option to ON) either SHOWPLAN_TEXT or SHOWPLAN_ALL, before they enter the corresponding SELECT statement. The SHOWPLAN_ALL option displays the same detailed information about the selected execution plan for the query as SHOWPLAN_TEXT with the addition of an estimate of the resource requirements for that statement.

Example 19.1 shows the use of the SET SHOWPLAN_TEXT option.

> **NOTE** Once you activate the SHOWPLAN_TEXT option, all consecutive Transact-SQL statements will not be executed until you deactivate this option with SET SHOWPLAN_TEXT OFF.

Example 19.1

```
SET SHOWPLAN_TEXT ON;
GO
USE AdventureWorks;
SELECT * FROM HumanResources.Employee e JOIN Person.Address a
  ON e.BusinessEntityID = a.AddressID AND e.BusinessEntityID = 10;
GO
SET SHOWPLAN_TEXT OFF;
```

The following textual plan shows the output of Example 19.1:

```
|--Nested Loops(Inner Join)
     |--Clustered Index Seek
(OBJECT:([AdventureWorks].[Person].[Address].[PK_Address_AddressID] AS
[a]), SEEK:([a].[AddressID]=(10)) ORDERED FORWARD)
        |--Compute Scalar
(DEFINE:([e].[OrganizationLevel]=[AdventureWorks].[HumanResources]
.[Employee].[OrganizationLevel] as [e].[OrganizationLevel]))
           |--Compute Scalar
(DEFINE:([e].[OrganizationLevel]=[AdventureWorks].[HumanResources]
.[Employee].[OrganizationNode] as [e].[OrganizationNode].GetLevel()))
              |--Clustered Index Seek
(OBJECT:([AdventureWorks].[HumanResources].[Employee].[PK_Employee_
BusinessEntityID] AS [e]), SEEK:([e].[BusinessEntityID]=(10)) ORDERED
FORWARD)
```

Before we discuss the execution plan of Example 19.1, you need to understand how to read its textual form. The indentation of operators determines the operator execution: the operator that is indented the furthest is executed first. If two or more operators have the same indentation, they are processed from the top downward. Now, if you take a look at the output of Example 19.1, you will see that there are three operators: Nested Loops, Clustered Index Seek (appears twice), and Compute Scalar (appears twice). (All operators are preceded by a bar, |.)

NOTE The Compute Scalar operator evaluates an expression to produce a computed scalar value. This may then be referenced elsewhere in the query, as in this example.

Further, the Clustered Index Seek operator for the **Employee** table is executed first. After that, the Compute Scalar operator is applied twice to the **Employee** table and the Clustered Index Seek operator is applied to the **Address** table. At the end, both tables (**Employee** and **Address**) are joined using the nested loop technique.

NOTE Generally, index access has two forms: index scan and index seek. Index scan processes the entire leaf level of an index tree, while index seek returns index values (or rows) from one or more ranges of an index. (Index seek is described later in the chapter.)

XML Execution Plan

The phrase "XML execution plan" means that the execution plan of a query is displayed as an XML document. (For more information about XML, see Microsoft Docs.) The most important advantage of using XML execution plans is that such plans can be ported from one system to another, allowing you to use them in another environment. (How to save execution plans in a file is explained a bit later.)

The SET statement has two options in relation to XML:

- SHOWPLAN_XML
- STATISTICS XML

The SHOWPLAN_XML option is very similar to the SHOWPLAN_TEXT option, but returns information as a set of XML documents. In other words, if you activate this option, the Database Engine returns detailed information about how the statements are going to be executed in the form of a well-formed XML document, without executing them. Each statement is reflected in the output by a single document. Each document contains the text of the statement, followed by the details of the execution steps.

NOTE Once you activate the SHOWPLAN_XML option, all consecutive Transact-SQL statements will not be executed until you deactivate this option with SET SHOWPLAN_XML OFF.

The main difference between the SHOWPLAN_XML and STATISTICS XML options is that the output of the latter is generated at run time. For this reason, STATISTICS XML includes the result of the SHOWPLAN_XML option as well as additional run-time information.

To save an XML execution plan in a file, in the Results pane, right-click the SQL Server XML Showplan that contains the query plan and choose Save Results As. In the Save <Grid or Text> Results dialog box, in the Save As Type box, click All Files (*.*). In the File Name box, provide a name with the **.sqlplan** suffix, and then click Save. (You can open a saved XML execution plan by double-clicking it.)

Other Options of the SET Statement

The SET statement has many other options, which are used in relation to locking, transaction, and date/time statements. Concerning statistics, the Database Engine supports the following three options of the SET statement:

- STATISTICS IO
- STATISTICS TIME
- STATISTICS PROFILE

The STATISTICS IO option causes the system to display statistical information concerning the amount of disk activity generated by the query—for example, the number of read and write I/O operations processed with the query. The STATISTICS TIME option causes the system to display the processing, optimization, and execution time of the query.

When the STATISTICS PROFILE option is activated, each executed query returns its regular result set, followed by an additional result set that shows the profile of the query execution.

SQL Server Management Studio and Graphical Execution Plans

A graphical execution plan is the best way to display the execution plan of a query if you are a beginner or want to take a look at different plans in a short time. This form of display uses icons to represent operators in the query plan.

As an example of how graphical execution plans can be initiated and what they look like, Figure 19-2 shows the graphical execution plan for the query in Example 19.1. To display an execution plan in the graphical form, write the query in Query Editor of Management Studio and click the Display Estimated Execution Plan button in the toolbar of SQL Server Management Studio. (The alternative way is to choose Query | Display Estimated Execution Plan.)

If you take a look at Figure 19-2, you will see that there is one icon for each operator of the execution plan. If you move the mouse over one of the icons, its detailed information appears, including the estimated I/O and CPU costs, estimated number of rows and their size, and the cost of the operator. The arrows between the icons represent the data flow. (You can also click an arrow, in which case the related information, such as estimated number of rows and estimated row size, will be displayed.)

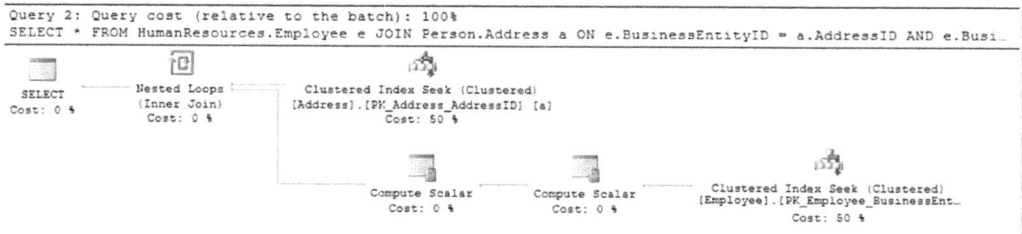

Figure 19-2   Graphical execution plan for the query in Example 19.1

To "read" the graphical execution plan of a query, you have to follow the flow of information *from right to left and from top to bottom*. (It is analogous to the flow of information in a textual execution plan.)

NOTE The query optimizer of the Database Engine uses operators to build a query execution plan. Each operator has its own unique icon, which is used in the graphical display of execution plans. As you can see from Figure 19-2, there are several operators, such as Clustered Index Seek and Compute Scalar, each represented with the corresponding icon.

As its name suggests, clicking the Display Estimated Execution Plan button displays the estimated plan of the query, without executing it. There is another button, Include Actual Execution Plan, that executes the query and additionally displays its execution plan. The actual execution plan contains additional information in relation to the estimated one, such as the actual number of processed rows and the actual number of executions for each operator.

Examples of Execution Plans

This section presents several queries related to the **AdventureWorks** database, together with their execution plans. These examples demonstrate the topics already discussed, enabling you to see how the query optimizer works in practice.

Example 19.2 introduces a new table (**new_addresses**) in the **sample** database.

Example 19.2

```
USE sample;
SELECT * into new_addresses
 FROM AdventureWorks.Person.Address;
GO
CREATE INDEX i_stateprov on new_addresses(StateProvinceID)
```

Example 19.2 copies the content of the **Address** table from the **Person** schema of the **AdventureWorks** database into the **new_addresses** table of the **sample** database. This is necessary because the former table contains several indices, which hinders the direct use of the **Address** table of the **AdventureWorks** database to show specific properties of the query optimizer. (Besides that, the example creates an index on the **StateProvinceID** column of that table.)

Example 19.3 shows a query with high selectivity and shows the textual plan that the optimizer chooses in such a case.

Example 19.3

```
-- high selectivity
USE sample;
SELECT * FROM new_addresses a
    WHERE a.StateProvinceID = 32;
```

The graphical output of Example 19.3 is shown in Figure 19-3.

The filter in Example 19.3 selects only one row from the **new_addresses** table. (The total number of rows in this table is 19614.) For this reason, the selectivity of the expression in the

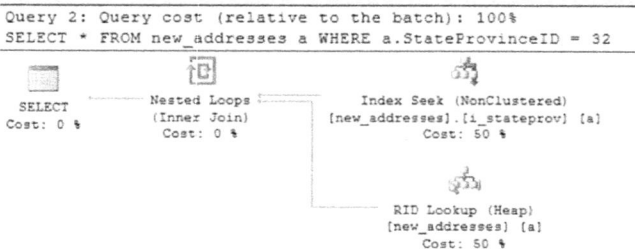

```
Query 2: Query cost (relative to the batch): 100%
SELECT * FROM new_addresses a WHERE a.StateProvinceID = 32
```

SELECT
Cost: 0 %

Nested Loops
(Inner Join)
Cost: 0 %

Index Seek (NonClustered)
[new_addresses].[i_stateprov] [a]
Cost: 50 %

RID Lookup (Heap)
[new_addresses] [a]
Cost: 50 %

Figure 19-3   Graphical execution plan for the query in Example 19.3

```
Query 2: Query cost (relative to the batch): 100%
SELECT * FROM new_addresses a WHERE a.StateProvinceID = 9
Missing Index (Impact 56.417): CREATE NONCLUSTERED INDEX [<Name of Missing Index, sysname,>] ON [dbo].[new_add_
```

SELECT
Cost: 0 %

Table Scan
[new_addresses] [a]
Cost: 100 %

Figure 19-4   Graphical execution plan for the query in Example 19.4

WHERE clause is very high (1/19614). In such a case, as you can see from Figure 19-3, the existing index on the **StateProvinceID** column is used by the optimizer.

Example 19.4 shows the same query as in Example 19.3, but with another filter.

Example 19.4

```
-- low selectivity
USE sample;
SELECT * FROM new_addresses a
    WHERE a.StateProvinceID = 9;
```

The graphical plan of Example 19.4 is given in Figure 19-4.

Although the query in Example 19.4 differs from the query in Example 19.3 only by a value on the right side of the condition in the WHERE clause, the execution plan that the optimizer chooses differs significantly. In this case, the existing index won't be used, because the selectivity of the filter is low. (The ratio of the number of rows satisfying the condition to the total number of rows in the table is 4564/19614 = 0.23, or 23 percent.)

Example 19.5 shows the use of the clustered index.

Example 19.5

```
USE AdventureWorks;
SELECT * FROM HumanResources.Employee
    WHERE HumanResources.Employee.BusinessEntityID = 10;
```

The graphical output of Example 19.5 is given in Figure 19-5.

The query in Example 19.5 uses the **PK_Employee_BusinessEntityID** clustered index. This clustered index is created implicitly by the system because the **BusinessEntityID** column is the primary key of the **Employee** table.

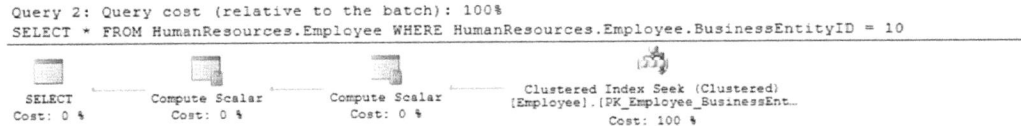

Figure 19-5   Graphical execution plan for the query in Example 19.5

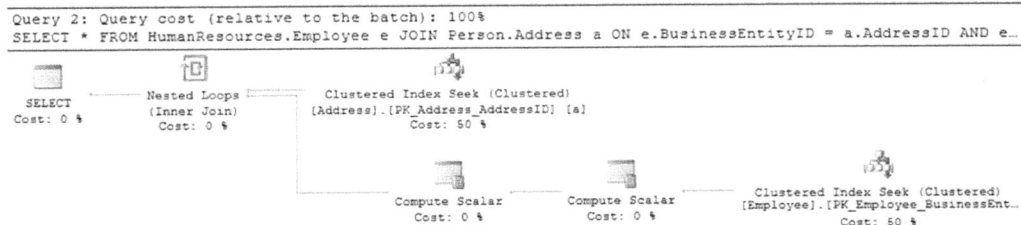

Figure 19-6   Graphical execution plan for the query in Example 19.6

Example 19.6 shows the use of the nested loop technique.

Example 19.6

```
USE AdventureWorks;
SELECT * FROM HumanResources.Employee e JOIN
                Person.Address a
                     ON e.BusinessEntityID = a.AddressID
                     AND e.BusinessEntityID = 10;
```

The graphical output of Example 19.6 is given in Figure 19-6.

The query in Example 19.6 uses the nested loop technique even though the join columns of the tables are at the same time their primary keys. For this reason, one could expect that the merge join technique would be used. The query optimizer decides to use the nested loop technique because there is an additional filter (e.EmployeeID = 10) that reduces the result set of the query to a single row.

Example 19.7 shows the use of the hash join technique.

Example 19.7

```
USE AdventureWorks;
SELECT * FROM Person.Address a JOIN Person.StateProvince s
   ON a.StateProvinceID = s.StateProvinceID;
```

The graphical output of the query in Example 19.7 is given in Figure 19-7.

Although both join columns in the ON clause are primary keys of the particular tables (**Address** and **StateProvince**), the query optimizer doesn't choose the merge join method. The reason is that *all* (19,614) rows of the **Address** table belong to the result set. In such a case, the use of the hash join method is more beneficial than the other two join processing techniques.

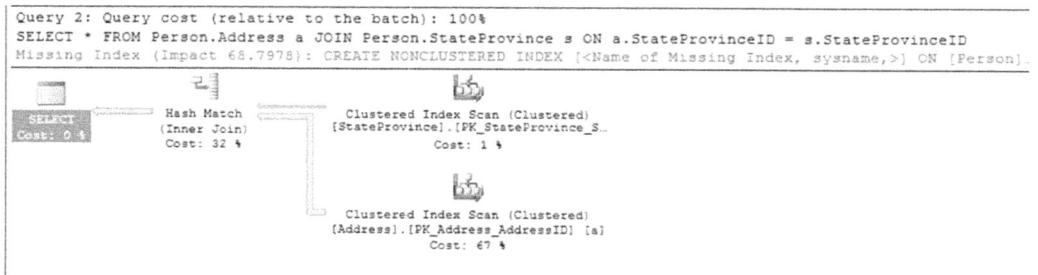

Figure 19-7 Graphical execution plan for the query in Example 19.7

Examples of Execution Plans with Filtered Indices

Filtered indices are exactly what the name suggests: indices specified with a filter in the WHERE clause of the SELECT statement. This feature provides benefits when you frequently retrieve a specific subset of tables' rows. The first benefit is that by creating a filtered (instead of a regular) index, you create only a subset of index rows of the underlying table, hence saving a significant amount of disk space. Second, rebuilding or reorganizing such indices will be faster.

Example 19.8 shows the creation of a filtered index and its advantage in relation to the corresponding regular index. The corresponding execution plan is displayed in Figure 19-8.

Example 19.8

```
USE AdventureWorks;
CREATE INDEX i_unitprice
ON Sales.SalesOrderDetail(UnitPrice)
WHERE UnitPrice > 1000;
SELECT SalesOrderDetailID, UnitPrice
FROM Sales.SalesOrderDetail
    WHERE UnitPrice > 2000;
```

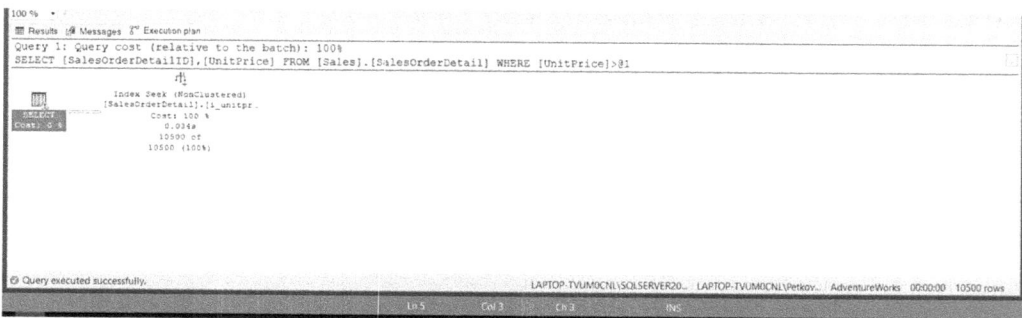

Figure 19-8 Graphical execution plan for the query in Example 19.8

Example 19.8 creates a nonclustered index called **i_unitprice** on the **UnitPrice** column of the **SalesOrderDetail** table. This index is a filtered index, because the definition of that index contains the filter (WHERE UnitPrice > 1000). If you take a closer look on the Index Seek operator in Figure 19-8, you will see that the query used the **i_unitprice** index. The query will benefit from the **i_unitprice** index, because the index itself restricts the search of the subsequent query (WHERE UnitPrice > 1000 is the condition in the filtered index and WHERE UnitPrice > 2000 is the corresponding condition in the query).

Dynamic Management Views and Query Optimizer

There are many dynamic management views (and functions) that are directly related to query optimization. In this section, the following DMVs are discussed:

- sys.dm_exec_query_optimizer_info
- sys.dm_exec_query_plan
- sys.dm_exec_cached_plans
- sys.dm_exec_query_stats
- sys.dm_exec_sql_text
- sys.dm_exec_text_query_plan
- sys.dm_db_stats_histogram

sys.dm_exec_query_optimizer_info

The **sys.dm_exec_query_optimizer_info** view is probably the most important DMV in relation to the work of the query optimizer because it returns detailed statistics about its operation. You can use this view when tuning a workload to identify query optimization problems or improvements.

The **sys.dm_exec_query_optimizer_info** view contains exactly three columns: **counter**, **occurrence**, and **value**. The **counter** column specifies the name of the optimizer event, while the **occurrence** column displays the cumulative number of occurrences of these events. The value of the **value** column contains additional information concerning events. (Not all events deliver a **value** value.)

Using this view, you can, for example, display the total number of optimizations, the elapsed time value, and the final cost value to compare the query optimizations of the current workload and any changes observed during the tuning process.

Example 19.9 shows the use of the **sys.dm_exec_query_optimizer_info** view.

Example 19.9

```
USE sample;
SELECT counter, occurrence, value
            FROM sys.dm_exec_query_optimizer_info
            WHERE value IS NOT NULL
            AND counter LIKE 'search 1%';
```

The result is

counter	occurrence	value
search 1	117	1
search 1 time	95	0.0120736842105263
search 1 tasks	117	513.982905982906

The **counter** column displays the phases of the optimization process. Therefore, Example 19.9 investigates how many times optimization Phase 1 is executed.

NOTE Because of its complexity, the optimization process is broken into three phases. The first phase (Phase 0) considers only nonparallel execution plans. If the cost of Phase 0 isn't optimal, Phase 1 will be executed, in which both nonparallel plans and parallel plans are considered. Phase 2 takes into account only parallel plans.

---

sys.dm_exec_query_plan

As you already know, execution plans for batches and Transact-SQL statements are placed in the cache. That way, they can be used anytime by the optimizer. You can examine the cache using several DMVs. One of these is the **sys.dm_exec_query_plan** view, which returns all execution plans that are stored in the cache of your system. (The execution plans are displayed in XML format.)

Each query plan stored in the cache is identified by a unique identifier called a *plan handle*. The **sys.dm_exec_query_plan** view requires a plan handle to retrieve the execution plan for a particular Transact-SQL query or batch.

sys.dm_exec_cached_plans

This view returns a row for each query plan that is cached by the Database Engine for faster query execution. You can use this view to find cached query plans, query text of the cached plans, the amount of memory taken by cached plans, and the reuse count of the cached plans.

The most important columns of this view are **cacheobjtype**, which specifies the type of object in the cache, and **usecount**, which determines the number of times this cache object has been used since its inception.

sys.dm_exec_query_stats

The **sys.dm_exec_query_stats** view returns aggregate performance statistics for cached query plans. The view contains one row per query statement within the cached plan, and the lifetime of the rows is tied to the plan itself.

Example 19.10 shows the use of the **sys.dm_exec_query_stats** and **sys.dm_exec_cached_plans** views.

Example 19.10

```
SELECT ecp.objtype AS  Object_Type ,
(SELECT t.text  FROM sys.dm_exec_sql_text(qs.sql_handle) AS t) AS
 Adhoc_Batch ,qs.execution_count  AS  Counts ,
qs.total_worker_time  AS  Total_Worker_Time ,
```

```
(qs.total_physical_reads  / qs.execution_count ) AS  Avg_Physical_Reads ,
(qs.total_logical_writes  / qs.execution_count ) AS  Avg_Logical_Writes ,
(qs.total_logical_reads  / qs.execution_count ) AS  Avg_Logical_Reads ,
qs.total_elapsed_time  AS  Total_Elapsed_Time,
(qs.total_elapsed_time  / qs.execution_count ) AS  Avg_Elapsed_Time ,
qs.last_execution_time  AS  Last_Exec_Time,
qs.creation_time  AS  Creation_Time
        FROM sys.dm_exec_query_stats AS qs
    JOIN sys.dm_exec_cached_plans ecp ON qs.plan_handle =
ecp.plan_handle
        ORDER BY  Counts  DESC;
```

Example 19.10 joins the **sys.dm_exec_query_stats** and **sys.dm_exec_cached_plans** views to return information for execution plans for all cached plans, which are ordered by the count of their execution times. (The displayed information from the different columns is self-explanatory.)

sys.dm_exec_sql_text and sys.dm_exec_text_query_plan

The previous view, **sys.dm_exec_query_stats**, can be used with several other DMVs to display different properties of queries. In other words, each DMV that needs the plan handle to identify the query will be "joined" with the **sys.dm_exec_query_stats** view to display the required information. One such view is **sys.dm_exec_sql_text**. This view returns the text of the SQL batch that is identified by the specified handle. Example 19.11 shows the use of the **sys.dm_exec_sql_text** DMV.

Example 19.11

```
SELECT TOP 5 total_worker_time/execution_count AS [Avg CPU Time],
    SUBSTRING(st.text, (qs.statement_start_offset/2)+1,
        ((CASE qs.statement_end_offset
          WHEN -1 THEN DATALENGTH(st.text)
          ELSE qs.statement_end_offset
          END - qs.statement_start_offset)/2) + 1) AS statement_text
FROM sys.dm_exec_query_stats AS qs
CROSS APPLY sys.dm_exec_sql_text(qs.sql_handle) AS st
ORDER BY total_worker_time/execution_count DESC;
```

The first column in the SELECT list of Example 19.11 displays the average time of the top five queries. The second column displays the selected execution plan of the query in XML format.

In contrast to the **sys.dm_exec_sql_text** view, **sys.dm_exec_text_query_plan** returns the execution plan of the batch in XML format. Similar to the previous views, this one is specified by the plan handle. (The plan specified by the plan handle can either be cached or currently executing.)

sys.dm_db_stats_histogram

The **sys.dm_db_stats_histogram** view returns the statistics histogram for the specified database object (table or indexed view) in the current database. Example 19.12 creates statistics for the **State_Name** column of the **State** table.

Example 19.12

```
USE sample;
CREATE TABLE State
(State_ID int IDENTITY PRIMARY KEY,
State_name varchar(120) NOT NULL);
INSERT State (State_name)
    VALUES ('Idaho'), ('Iowa'), ('Indiana'), ('Texas');
GO
CREATE STATISTICS State_Stats
    ON State (State_Name) ;
GO
SELECT object_id,stats_id,range_high_key,range_rows,equal_rows
FROM sys.dm_db_stats_histogram(OBJECT_ID('State'), 2);
```

The result is

object_id	stats_id	range_high_key	range_rows	equal_rows
1394104007	2	Idaho	0	1
1394104007	2	Indiana	0	1
1394104007	2	Iowa	0	1
1394104007	2	Texas	0	1

Example 19.12 creates a **State** table and inserts a row into it. The next statement, CREATE STATISTICS, creates query optimization statistics on one or more columns of a table, an indexed view, or an external table. Therefore, Example 19.12 creates statistics for the **State_Name** column of the **State** table.

NOTE For most queries, the query optimizer already generates the necessary statistics for a execution plan; in a few cases, you need to create additional statistics with CREATE STATISTICS or modify the query design to improve query performance. In Example 19.12 the execution of the CREATE STATISTICS statement is required because the **State** table just has been created and populated.

After that, the DMV is executed using the value 2 for the **stats_id** column. The primary key occupies **stats_id** number 1. For this reason, we call the DMV for **stats_id** number 2 to get the histogram for the **State** table.

Optimizer Hints

In most cases, the query optimizer chooses the fastest execution plan. However, there are some special situations in which the optimizer, for some particular reasons, cannot find the optimal solution. In such cases, you should use optimizer hints to force it to use a particular execution plan that could perform better.

Optimizer hints are optional parts in a SELECT statement that instruct the query optimizer to execute one specific behavior. In other words, by using optimizer hints, you do not allow the query optimizer to search and find the way to execute a query because you tell it exactly what to do.

Why Use Optimizer Hints

You should use optimizer hints only temporarily and for testing. In other words, avoid using them as a permanent part of a query. There are two reasons for this statement. First, if you force the optimizer to use a particular index and later define an index that results in better performance of the query, the query and the application to which it belongs cannot benefit from the new index. Second, Microsoft continuously strives to make the query optimizer better. If you bind a query to a specific execution plan, the query cannot benefit from new and improved features in the subsequent versions of the database system.

There are two reasons why the optimizer sometimes does not choose the fastest execution plan:

- The query optimizer is not perfect.
- The system does not provide the optimizer with the appropriate information.

NOTE Optimizer hints can help you only if the execution plan chosen by the optimizer is not optimal. If the system does not provide the optimizer with the appropriate information, use the AUTO_CREATE_STATISTICS and AUTO_UPDATE_STATISTICS database options to create or modify existing statistics.

Types of Optimizer Hints

The Database Engine supports the following types of optimizer hints:

- Table hints
- Join hints
- Query hints
- Plan guides

The following sections describe these hints.

NOTE The examples that follow demonstrate the use of optimizer hints, but they don't give you any recommendations about using them in any particular query. (In most cases shown in these examples, the use of hints would be counterproductive.)

Table Hints

You can apply table hints to a single table. The following table hints are supported:

- INDEX
- NOEXPAND
- FORCESEEK

The INDEX hint is used to specify one or more indices that are then used in a query. This hint is specified in the FROM clause of the query. You can use this hint to force index access if the optimizer for some reason chooses to perform a table scan for a given query. (Also, the INDEX hint can be used to prevent the optimizer from using a particular index.)

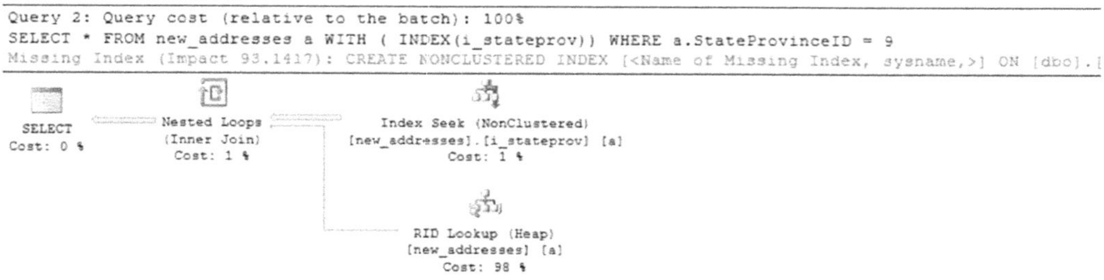

```
Query 2: Query cost (relative to the batch): 100%
SELECT * FROM new_addresses a WITH ( INDEX(i_stateprov)) WHERE a.StateProvinceID = 9
Missing Index (Impact 93.1417): CREATE NONCLUSTERED INDEX [<Name of Missing Index, sysname,>] ON [dbo].[
```

Figure 19-9   Graphical execution plan for the query in Example 19.13

Examples 19.13 and 19.14 show the use of the INDEX hint.

Example 19.13

```
USE sample;
SELECT * FROM new_addresses a WITH ( INDEX(i_stateprov))
WHERE a.StateProvinceID = 9;
```

The graphical output of Example 19.13 is shown in Figure 19-9.

Example 19.13 is identical to Example 19.4, but contains the additional INDEX hint. This hint forces the query optimizer to use the **i_stateprov** index. Without this hint, the optimizer chooses the table scan, as you can see from the output of Example 19.4.

The other form of the INDEX query hint, INDEX(0), forces the optimizer to not use any of the existing nonclustered indices. Example 19.14 shows the use of this hint.

Example 19.14

```
SET SHOWPLAN_TEXT ON;
GO
USE AdventureWorks;
SELECT * FROM Person.Address a
        WITH(INDEX(0))
        WHERE a.StateProvinceID = 32;
GO
SET SHOWPLAN_TEXT OFF;
```

The textual output of Example 19.14 is

```
|--Clustered Index Scan
(OBJECT:([AdventureWorks].[Person].[Address].[PK_Address_AddressID]
AS [a]),
 WHERE:([AdventureWorks].[Person].[Address].[StateProvinceID] as
 [a].[StateProvinceID]=(32)))
```

NOTE If a clustered index exists, INDEX(0) forces a clustered index scan and INDEX(1) forces a clustered index scan or seek. If no clustered index exists, INDEX(0) forces a table scan and INDEX(1) is interpreted as an error.

The execution plan of the query in Example 19.14 shows that the optimizer uses the clustered index scan, because of the INDEX(0) hint. Without this hint, the query optimizer itself would choose the nonclustered index scan.

The NOEXPAND hint specifies that any indexed view isn't expanded to access underlying tables when the query optimizer processes the query. The query optimizer treats the view like a table with the clustered index. (For a discussion of indexed views, see Chapter 26.)

The FORCESEEK table hint forces the optimizer to use only an index seek operation as the access path to the data in the table (or view) referenced in the query. You can use this table hint to override the default plan chosen by the query optimizer, to avoid performance issues caused by an inefficient query plan. For example, if a plan contains table or index scan operators, and the corresponding tables cause a high number of reads during the execution of the query, forcing an index seek operation may yield better query performance. This is especially true when inaccurate cardinality or cost estimations cause the optimizer to favor scan operations at plan compilation time.

NOTE The FORCESEEK hint can be applied to both clustered and nonclustered indices.

Join Hints

Join hints instruct the query optimizer how join operations in a query should be performed. They force the optimizer either to join tables in the order in which they are specified in the FROM clause of the SELECT statement or to use the join processing techniques explicitly specified in the statement. The Database Engine supports the following join hints:

- FORCE ORDER
- LOOP
- HASH
- MERGE

The FORCE ORDER hint forces the optimizer to join tables in the order in which they are specified in a query. Example 19.15 shows the use of this join hint.

Example 19.15

```
USE AdventureWorks;
SELECT e.BusinessEntityID, e.LoginID, d.DepartmentID
    FROM HumanResources.Employee e, HumanResources.Department d,
        HumanResources.EmployeeDepartmentHistory h
    WHERE d.DepartmentID = h.DepartmentID
    AND h.BusinessEntityID = e.BusinessEntityID
    AND h.EndDate IS NOT NULL
    OPTION(FORCE ORDER);
```

The graphical output of Example 19.15 is shown in Figure 19-10.

As you can see from the graphical output of Example 19.15, the optimizer performs the join operation in the order in which the tables appear in the query. This means that the **EmployeeDepartmentHistory** table will be processed first, then the **Department** table,

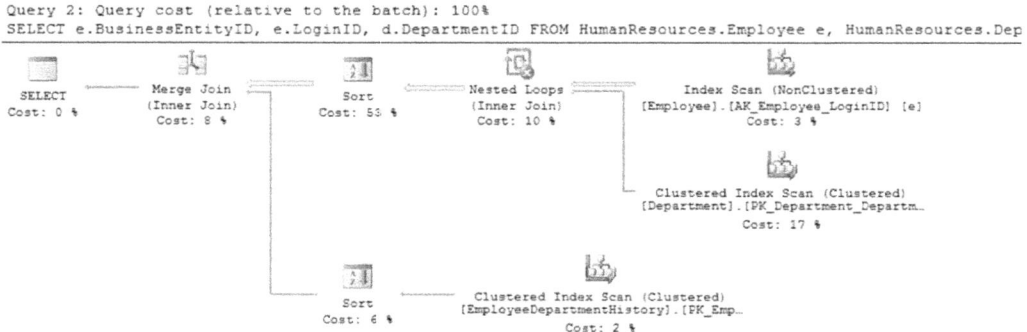

```
Query 2: Query cost (relative to the batch): 100%
SELECT e.BusinessEntityID, e.LoginID, d.DepartmentID FROM HumanResources.Employee e, HumanResources.Dep
```

Figure 19-10 Graphical execution plan for the query in Example 19.15

and finally the **Employee** table. (If you execute the query without the FORCE ORDER hint, the query optimizer will process the tables in the opposite order: first **Employee**, then **Department**, and then **EmployeeDepartmentHistory**.)

> **NOTE** Keep in mind that this does not necessarily mean that the new execution plan performs better than that chosen by the optimizer.

The query hints LOOP, MERGE, and HASH force the optimizer to use the nested loop technique, merge join technique, and hash join technique, respectively. These three join hints can be used only when the join operation conforms to the SQL standard—that is, when the join is explicitly indicated with the JOIN keyword in the FROM clause of a SELECT statement.

Example 19.16 shows a query that uses the merge join technique because the hint with the same name is explicitly defined in the SELECT statement. (You can apply the other two hints, HASH and LOOP, in the same way.)

Example 19.16

```
USE AdventureWorks;
SELECT * FROM Person.Address a JOIN Person.StateProvince s
  ON a.StateProvinceID = s.StateProvinceID
    OPTION (MERGE JOIN);
```

The graphical output of Example 19.16 is given in Figure 19-11.

As you can see from the output of Example 19.16, the query optimizer is forced to use the merge join processing technique. (If the hint is removed, the query optimizer chooses the hash join technique.)

The specific join hint can be written either in the FROM clause of a query or using the OPTION clause at the end of it. The use of the OPTION clause is recommended if you want to

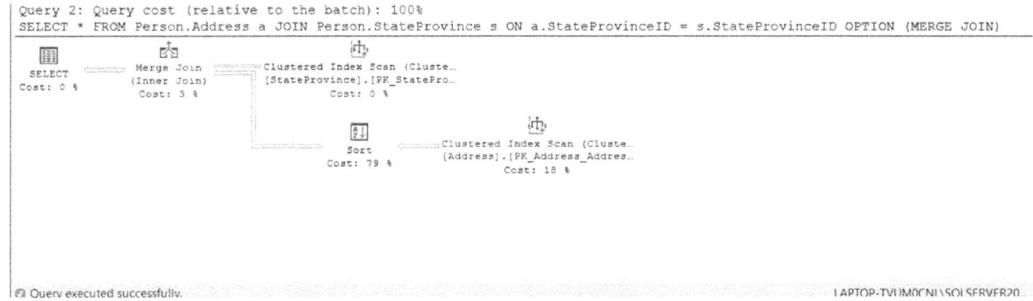

Figure 19-11   Graphical execution plan for the query in Example 19.16

write several *different* hints together. Example 19.17 is identical to Example 19.16, but specifies the join hint in the FROM clause of the query. (Note that in this case the INNER keyword is required.)

Example 19.17

```
USE AdventureWorks;
SELECT * FROM Person.Address a INNER MERGE JOIN Person.StateProvince s
  ON a.StateProvinceID = s.StateProvinceID;
```

Query Hints

There are several query hints, which are used for different purposes. This section discusses the following query hints:

- FAST *n*
- OPTIMIZE FOR
- OPTIMIZE FOR UNKNOWN
- USE PLAN

The FAST *n* hint specifies that the query is optimized for fast retrieval of the first *n* rows. After the first *n* rows are returned, the query continues execution and produces its full result set.

NOTE This hint can be very helpful if you have a complex query with many result rows, requiring a lot of time for processing. Generally, a query is processed completely and then the system displays the result. This query hint forces the system to display the first *n* rows immediately after their processing.

The OPTIMIZE FOR hint forces the query optimizer to use a particular value for a local variable when the query is compiled and optimized. The value is used only during query optimization, and not during query execution. This query hint can be used when you create plan guides, which are discussed in the next section.

Example 19.18 shows the use of the OPTIMIZE FOR query hint.

Example 19.18
```
USE AdventureWorks;
DECLARE @city_name nvarchar(30)
SET @city_name = 'Newark'
SELECT * FROM Person.Address
       WHERE City = @city_name
             OPTION ( OPTIMIZE FOR (@city_name = 'Seattle') );
```

Although the value of the **@city_name** variable is set to Newark, the OPTIMIZE FOR hint forces the optimizer to use the value Seattle for the variable when optimizing the query.

The OPTIMIZE FOR UNKNOWN hint instructs the query optimizer to use statistical data instead of the initial values for all local variables when the query is compiled and optimized, including parameters created with forced parameterization. (*Forced parameterization* means that any literal value that appears in a SELECT, INSERT, UPDATE, or DELETE statement, submitted in any form, is converted to a parameter during query compilation. That way, all queries with a different parameter value will be able to reuse the cached query plan instead of having to recompile the statement each time when the parameter value is different.)

The USE PLAN hint takes a plan stored as an XML document as the parameter and advises the Database Engine to use the specified execution plan for the query. (For the storage of an execution plan as an XML document, see the description of the SHOWPLAN_XML option of the SET statement earlier in this chapter, in the section "XML Execution Plan.")

Plan Guides

As you know from the previous section, hints are explicitly specified in the SELECT statement to influence the work of the query optimizer. Sometimes you cannot or do not want to change the text of the SELECT statement directly. In that case, it is still possible to influence the execution of queries by using plan guides. In other words, plan guides allow you to use a particular optimizer hint without changing the syntax of the SELECT statement.

NOTE The main purpose of plan guides is to avoid hard-coding of hints in cases where it is not recommended or not possible (for third-party application code, for instance).

Plan guides are created using the **sp_create_plan_guide** system procedure. This procedure creates a plan guide for associating query hints or actual query plans with queries in a database. Another system procedure, **sp_control_plan_guide**, enables, disables, or drops an existing plan guide. (You can also use SQL Server Management Studio to drop a plan guide. Examples for creation and deletion of plan guides are given in Microsoft Docs.)

NOTE There are no Transact-SQL DDL statements for creation and deletion of plan guides. A subsequent SQL Server version will hopefully support such statements.

The Database Engine supports three types of plan guides:

- **SQL**　Matches queries that execute in the context of stand-alone Transact-SQL statements and batches that are not part of a database object
- **OBJECT**　Matches queries that execute in the context of routines and DML triggers
- **TEMPLATE**　Matches stand-alone queries that are parameterized to a specified form

Example 19.19 shows how you can create the optimizer hint from Example 19.16, without the modification of the corresponding SELECT statement.

Example 19.19

```
sp_create_plan_guide  @name = N'Example_19_16',
@stmt = N'SELECT * FROM Person.Address a JOIN Person.StateProvince s
  ON a.StateProvinceID = s.StateProvinceID',
@type = N'SQL',
@module_or_batch = NULL,
@params = NULL,
@hints = N'OPTION (HASH JOIN)'
```

As you can see from Example 19.19, the **sp_create_plan_guide** system procedure has several parameters. The **@name** parameter specifies the name of the new plan guide. The **@stmt** parameter comprises the T-SQL statement, while the **@type** parameter specifies the type of the plan guide (SQL, OBJECT, or TEMPLATE). The optimizer hint is specified in the **@hints** parameter. (You can also use Management Studio to create plan guides.)

To edit information related to plan guides, use the **sys.plan_guides** catalog view. This view contains a row for each plan guide in the current database. The most important columns are **plan_guide_id**, **name**, and **query_text**. The **plan_guide_id** column specifies the unique identifier of the plan guide, while the **name** column defines its name. The **query_text** column specifies the text of the query on which the plan guide is created.

Summary

The query optimizer is the part of the Database Engine that decides how to best perform a query. It generates several execution plans for the given query and selects the plan with the lowest cost.

The query optimization phase can be divided into the following phases: query analysis, index selection, and join order selection. During the query analysis phase, the optimizer examines the query for search arguments, the use of the OR operator, and the existence of join criteria, in that order. The identification of search arguments allows the optimizer to decide whether one or more existing indices will be used.

The order in which two or more joined tables are written in the FROM clause of a SELECT statement doesn't influence the optimizer's decision regarding their processing order. The Database Engine supports three different join processing techniques that can be used by the optimizer. Which technique the optimizer chooses depends on existing statistics for joined tables.

The Database Engine supports many tools that can be used to edit execution plans. The most important are textual and graphical display of the plan and dynamic management views.

You can influence the work of the optimizer by using optimizer hints. The Database Engine supports many optimizer hints, which can be grouped as follows: table hints, join hints, and query hints. Plan guides allow you to influence the optimization process of queries, without modifying a particular SELECT statement, as in the case of query hints.

The next chapter discusses performance tuning.

Exercises

E.19.1   What is the difference between the merge join technique and the nested loop technique? When does the SQL Server Optimizer use the former and when the latter technique?

E.19.2   What is the difference between the hash join technique and the nested loop technique? When does the SQL Server Optimizer use the former and when the latter technique?

20 | Performance Tuning

In This Chapter

- Factors That Affect Performance
- Monitoring Performance
- Choosing the Right Tool for Monitoring
- Other Performance Tools of the Database Engine

Improving the performance of a database system requires many decisions, such as where to store data and how to access the data. This task is different from other administrative tasks because it comprises several different steps that concern all aspects of software and hardware. If the database system is not performing optimally, the system administrator must check many factors and possibly tune software (operating system, database system, database applications) and hardware.

The performance of the Database Engine (and any other relational DBMS) is measured by two criteria:

- Response time
- Throughput

Response time measures the performance of an individual transaction or program. Response time is treated as the length of time from the moment a user enters a command or statement until the time the system indicates that the command (statement) has completed. To achieve optimum response time of an overall system, almost all existing commands and statements (90 to 95 percent of them) must not cross the specified response time limit.

Throughput measures the overall performance of the system by counting the number of transactions that can be handled by the Database Engine during the given time period. (The throughput is typically measured in transactions per second.) Therefore, there is a direct relation between response time of the system and its throughput: when the response time of a system degrades (for example, because many users concurrently use the system), the throughput of the system degrades too.

This chapter discusses performance issues and the tools for tuning the Database Engine that are relevant to daily administration of the system. The first part of the chapter describes the factors that affect performance. After that, some recommendations are given for how to choose the right tool for the administration job. At the end of the chapter, tools for monitoring the database system are presented.

Factors That Affect Performance

Factors affecting performance fall into three general categories:

- Database applications
- Database system
- System resources

These factors in turn can be affected by several other factors, as discussed in the following sections.

Database Applications and Performance

The following factors can affect the performance of database applications:

- Application-code efficiency
- Physical design

Application-Code Efficiency

Applications introduce their own load on the system software and on the Database Engine. For this reason, they can contribute to performance problems if you make poor use of system resources. Most performance problems in application programs are caused by the improper choice of Transact-SQL statements and their sequence in an application program.

The following list gives some of the ways you can improve overall performance by modifying code in an application:

- Use clustered indices
- Do not use the NOT IN predicate

Clustered indices generally improve performance. Performance of a range query is the best if you use a clustered index for the column in the filter. When you retrieve only a few rows, there is no significant difference between the use of a nonclustered index and a clustered index.

The NOT IN predicate is not optimizable; in other words, the query optimizer cannot use it as a search argument (the part of a query that restricts the intermediate result set of the query). Therefore, the expression with the NOT IN predicate always results in a table scan.

NOTE More hints on how code modification can improve overall performance are given in Chapter 19.

Physical Design

During physical database design, you choose the specific storage structures and access paths for the database files. In this design step, it is sometimes recommended that you denormalize some of the tables in the database to achieve good performance for various database applications. *Denormalizing* tables means coupling together two or more normalized tables, resulting in some redundant data.

To demonstrate the process of denormalization, consider this example: Table 20-1 shows two tables from the **sample** database, **department** and **employee**, that are normalized. (For more information on data normalization, see Chapter 1.) Data in those two tables can be specified using just one table, **dept_emp** (see Table 20-2), which shows the denormalized form of data stored in the tables **department** and **employee**. In contrast to the tables **department** and **employee**, which do not contain any data redundancies, the **dept_emp** table contains a lot of redundancies, because two columns of this table (**dept_name** and **location**) are dependent on the **dept_no** column.

Data denormalization has two benefits and two disadvantages. First the benefits: If you have a column that is dependent on another column of the table (such as the **dept_name** column in the **dept_emp** table, which is dependent on the **dept_no** column) for data often required by queries, you can avoid the use of the join operation, which would affect the performance of applications. Second, denormalized data requires fewer tables than are required to store the same amount of normalized data.

On the other hand, a denormalized table requires additional amounts of disk space, and data modification is difficult because of data redundancy.

Another option in the physical database design that contributes to good performance is the creation of indices. Chapter 10 gives several guidelines for the creation of indices, and examples are given later in this chapter.

emp_no	emp_fname	emp_lname	dept_no
25348	Matthew	Smith	d3
10102	Ann	Jones	d3
18316	John	Barrimore	d1
29346	James	James	d2
2581	Elke	Hansel	d2
28559	Sybill	Moser	d1
9031	Elsa	Bertoni	d2

dept_no	dept_name	location	
d1	Research	Dallas	
d2	Accounting	Seattle	
d3	Marketing	Dallas	

Table 20-1   The employee and department Tables

emp_no	emp_fname	emp_lname	dept_no	dept_name	location
25348	Matthew	Smith	d3	Marketing	Dallas
10102	Ann	Jones	d3	Marketing	Dallas
18316	John	Barrimore	d1	Research	Dallas
29346	James	James	d2	Accounting	Seattle
2581	Elke	Hansel	d2	Accounting	Seattle
28559	Sybill	Moser	d1	Research	Dallas
9031	Elsa	Bertoni	d2	Accounting	Seattle

Table 20-2 The dept_emp Table

The Database Engine and Performance

The Database Engine can substantially affect the performance of an entire system. The two most important components of the Database Engine that affect performance are

- Query optimizer
- Locks

Query Optimizer

The optimizer formulates several execution plans for fetching the data rows that are required to process a query and then decides which plan should be used. The decision concerning the selection of the most appropriate execution plan includes which indices should be used, how to access tables, and the order of joining tables. All of these decisions can significantly affect the performance of database applications. The optimizer is discussed in detail in Chapter 19.

Locks

The database system uses locks as the mechanism for protecting one user's work from another's. Therefore, locks are used to control the access of data by all users at the same time and to prevent possible errors that can arise from the concurrent access of the same data.

Locking affects the performance of the system through its granularity—that is, the size of the object that is being locked and the isolation level. Row-level locking provides the best system performance, because it leaves all but one row on the page unlocked and hence allows more concurrency than page- or table-level locking.

Isolation levels affect the duration of the lock for SELECT statements. By using the lower isolation levels, such as READ UNCOMMITTED and READ COMMITTED, the data availability, and hence the concurrency, of the data can be improved. (Locking and isolation levels are explained in detail in Chapter 13.)

System Resources and Performance

The Database Engine runs on an operating system, which itself uses underlying system resources. These resources have a significant impact on the performance of both the operating system and the database system. Performance of any database system depends on four main system resources:

- Central processing unit (CPU)
- Memory
- Disk I/O
- Network

The CPU, together with memory, is the key component for marking the speed of a computer. It is also the key to the performance of a system, because it manages other resources of the system and executes all applications. It executes user processes and interacts with other resources of your system. Performance problems in relation to the CPU can occur when the operating system and user programs are making too many requests on it. Generally, the more CPU power available for your computer, the better the overall system is likely to perform.

The Database Engine dynamically acquires and frees memory as needed. Performance problems concerning memory can occur only if there is not enough of it to do the required work. When this occurs, many memory pages are written to a pagefile. (The notion of a *pagefile* is explained in detail later in this chapter.) If the process of writing to a pagefile happens very often, the performance of the system can degrade. Therefore, similar to the CPU rule, the more memory available for your computer, the better the system is likely to perform.

There are two issues concerning disk I/O: disk speed and disk transfer rate. The disk speed determines how fast read and write operations to disk are executed. The disk transfer rate specifies how much data can be written to disk during a time unit (usually measured in seconds). Obviously, the faster the disk, the larger the amount of data being processed. Also, more disks are generally better than a single disk when many users are using the database system concurrently. (In this case, access to data is usually spread across many disks, thus improving the overall performance of the system.)

For a client/server configuration, a database system sometimes performs poorly if there are many client connections. In that case, the amount of data that needs to be transferred across the network possibly exceeds the network capacity. To avoid such a performance bottleneck, the following general recommendations should be taken into account:

- If a database server sends any rows to an application, only the rows needed by the application should be sent.
- If a long-lasting user application executes strictly on the client side, move it to the server side (by executing it as a stored procedure, for example).

All four of these system resources are dependent on each other. This means that performance problems in one resource can cause performance problems in the other resources. Similarly, an improvement concerning one resource can significantly increase performance of some other (or even all) resources. For example:

- If you increase the number of CPUs, each CPU can share the load evenly and therefore can remedy the disk I/O bottleneck. On the other hand, the inefficient use of the CPU is often the result of a preexisting heavy load on disk I/O and/or memory.

- If more memory is available, there is a better chance of finding in memory a page needed by the application (rather than reading the page from disk), which results in a performance gain. By contrast, reading from the disk drive instead of drawing from the immensely faster memory slows the system down considerably, especially if there are many concurrent processes.

The following sections describe in detail disk I/O and memory.

Disk I/O

One purpose of a database is to store, retrieve, and modify data. Therefore, the Database Engine, like any other database system, must perform a lot of disk activity. In contrast to other system resources, a disk subsystem has two moving parts: the disk itself and the disk head. The rotation of the disk and the movement of the disk head need a great deal of time; therefore, disk reads and writes are two of the highest-cost operations that a database system performs. (For instance, access to a disk is significantly slower than memory access.)

The Database Engine stores the data in 8KB pages. The buffer cache of RAM is also divided into 8KB pages. The system reads data in units of pages. Reads occur not only for data retrieval, but also for any modification operations such as UPDATE and DELETE because the database system must read the data before it can be modified.

If the needed page is in the buffer cache, it will be read from memory. This I/O operation is called *logical I/O* or *logical read*. If it is not in memory, the page is read from disk and put in the buffer cache. This I/O operation is called *physical I/O* or *physical read*. The buffer cache is shared because the Database Engine uses the architecture with only one memory address space. Therefore, many users can access the same page. A logical write occurs when data is modified in the buffer cache. Similarly, a physical write occurs when the page is written from the buffer cache to disk. Therefore, more logical write operations can be made on one page before it is written to disk.

The Database Engine has a few components that have great impact on performance because they significantly consume the I/O resources:

- Read ahead
- Checkpoint

Read ahead is described in the following section, while the checkpoint is explained in Chapter 16.

Read Ahead The optimal behavior of a database system would be to read data into memory and never have to wait for a disk read request. The best way to perform this task is to know the next several pages that the user will need and to read them from the disk into the buffer pool *before* they are requested by the user process. This mechanism is called *read ahead*, and it allows the system to optimize performance by processing large amounts of data effectively.

The component of the Database Engine called Read Ahead Manager manages the read-ahead processes completely internally, so a user has no way to influence this process. Instead of using the usual 8KB pages, the Database Engine uses 64KB blocks of data as the unit for read-ahead reads. That way, the throughput for I/O requests is significantly increased. The read-ahead mechanism is used by the Database Engine to perform large table scans and index range scans. Table scans are performed using the information that is stored in Index Allocation

Map (IAM) pages to build a serial list of the disk addresses that must be read. (IAM pages are allocation pages containing information about the extents that a table or index uses.) This allows the database system to optimize its I/O as large, sequential reads in disk order. Read Ahead Manager reads up to 2MB of data at a time. Each extent is read with a single operation.

> **NOTE** The Database Engine provides multiple serial read-ahead operations at once for each file involved in the table scan. This feature can take advantage of striped disk sets.

For index ranges, the Database Engine uses the information in the intermediate level of index pages immediately above the leaf level to determine which pages to read. The system scans all these pages and builds a list of the leaf pages that must be read. During this operation, the contiguous pages are recognized and read in one operation. When there are many pages to be retrieved, the Database Engine schedules a block of reads at a time.

The read-ahead mechanism can also have negative impacts on performance if too many pages for a process are read and the buffer cache is unnecessarily filled up. The only thing you can do in this case is create the indices you will actually need.

If you want to know whether the Database Engine is using read ahead to read database pages for a query, you can use the following methods:

- Turn on SET STATISTICS IO in the session where the particular query is executed (for the description of SET STATISTICS IO, see Chapter 19).

- Use an extended event to track the event **file_read completed** (Example 20.14, later in the chapter, shows how you can programmatically create such an extended event).

Memory

Memory is a crucial resource component, not only for the running applications but also for the operating system. When an application is executed, it is loaded into memory and a certain amount of memory is allocated to the application. (In Microsoft terminology, the total amount of memory available for an application is called its *address space*.)

Operating systems generally support virtual memory. This means that the total amount of memory available to applications is the amount of physical memory (or RAM) in the computer plus the size of the specific file on the disk drive called *pagefile*. (The name of the pagefile on Windows is **pagefile.sys**.) Once data is moved out of its location in RAM, it resides in the pagefile. If the system is asked to retrieve data that is not in the proper RAM location, it will load the data from the location where it is stored and produce a so-called *page fault*.

> **NOTE** The pagefile should be placed on a different drive from the drive on which files used by the Database Engine are placed, because the paging process can have a negative impact on disk I/O activities.

For an entire application, only a portion of it resides in RAM. (Recently referenced pages can usually be found in RAM.) When the information the application needs is not in RAM, the operating system must *page* (that is, read the page from the pagefile into RAM). This process is called *demand paging*. The more the system has to page, the worse the performance is.

NOTE When a page in RAM is required, the oldest page of the address space for an application is moved to the pagefile to make room for the new page. The replacement of pages is always limited to the address space of the current application. Therefore, there is no chance that pages in the address space of other running applications will be replaced.

As you already know, a page fault occurs if the application makes a request for information and the data page that contains that information is not in the proper RAM location of the computer. The information may either have been paged out to the pagefile or be located somewhere else in RAM. Therefore, there are two types of page fault:

- **Hard page fault**   The page has been paged out (to the pagefile) and has to be brought into RAM from the disk drive.
- **Soft page fault**   The page is found in another location in RAM.

Soft page faults consume only RAM resources. Therefore, they are significantly better for performance than hard page faults, which cause disk reads and writes to occur.

NOTE Page faults are normal in an operating system environment because the operating system requires pages from the running applications to satisfy the need for memory of the starting applications. However, excessive paging (especially with hard page faults) is a serious performance problem because it can cause disk bottlenecks and start to consume the additional power of the processor.

Monitoring Performance

All the factors that affect performance can be monitored using different components. These components can be grouped in the following categories:

- Performance Monitor
- Dynamic management views (DMVs) and catalog views
- DBCC commands
- System stored procedures

This section first gives an overview of Performance Monitor and then describes all components for monitoring performance in relation to the four factors: CPU, memory, disk access, and network.

Performance Monitor: An Overview

Performance Monitor is a Windows graphical tool that enables you to monitor Windows activities and database system activities. The benefit of this tool is that it is tightly integrated with the Windows operating system and therefore displays reliable values concerning different performance issues. Performance Monitor provides a lot of performance objects, and each performance object contains several counters. These counters can be monitored locally or over the network.

Performance Monitor supports three different presentation modes:

- **Graphic mode**   Displays the selected counters as colored lines, with the X axis representing time and the Y axis representing the value of the counter. (This is the default display mode.)
- **Histogram mode**   Displays the selected counters as colored horizontal bars that represent the data sampling values.
- **Report mode**   Displays the values of counters textually.

To start Performance Monitor, click Start in Windows, type **perfmon** in the Search bar, click its name in the search results, and then click Performance Monitor under Monitoring Tools. The starting window of Performance Monitor, shown in Figure 20-1, contains one default counter: **% Processor Time** (Object: Processor). This counter is very important, but you will need to display the values of several other counters.

To add a counter for monitoring, click the plus sign in the toolbar of Performance Monitor, select the performance object to which the counter belongs, choose the counter, and click Add. To remove a counter, highlight the line in the bottom area and click Delete. (The following sections describe, among other things, the most important counters in relation to the CPU, memory, I/O, and network.)

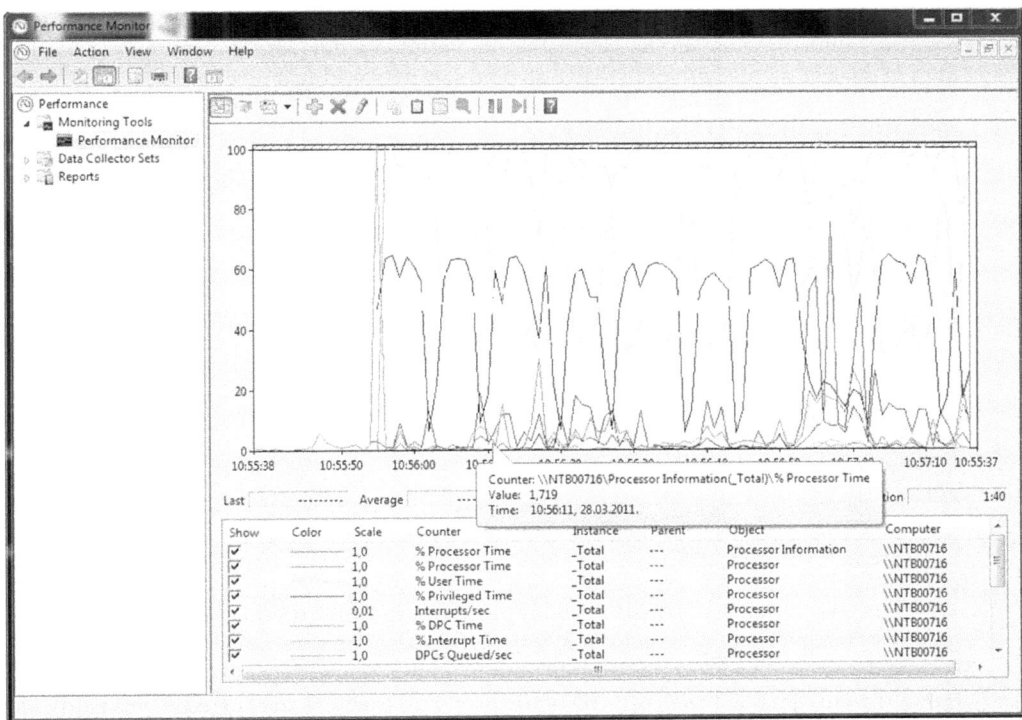

Figure 20-1   Performance Monitor

Monitoring the CPU

This section contains two subsections related to monitoring the CPU. The first subsection describes several Performance Monitor counters, while the second discusses catalog views and DMVs that you can use for the same purpose.

Monitoring the CPU Using Counters

The following counters are related to monitoring the CPU:

- % Processor Time (Object: Processor)
- % Interrupt Time (Object: Processor)
- Interrupts/sec (Object: Processor)

The **% Processor Time** counter displays system-wide CPU usage and acts as the primary indicator of processor activity. The lower the value of this counter, the better CPU usage that can be achieved. You should try to reduce CPU usage if the value of the counter is constantly greater than 90. (CPU usage of 100 percent is acceptable only if it happens for short periods of time.)

The **% Interrupt Time** counter shows you the percentage of time that the CPU spends servicing hardware interrupts. The values of this counter are important if at least one piece of hardware is trying to get processor time.

The **Interrupts/sec** counter displays the number of times per second the processor receives hardware interrupts for service requests from peripheral devices. This number generally may be high in environments with high disk utilization or networking demands.

Monitoring the CPU Using Views

The following catalog views and DMVs are used, among others, to monitor CPU usage:

- sys.sysprocesses
- sys.dm_exec_requests
- sys.dm_exec_query_stats

The **sys.sysprocesses** catalog view can be useful if you want to identify processes that use the most processor time. Example 20.1 shows the use of this catalog view.

Example 20.1

```
USE master;
SELECT spid, dbid, uid, cpu
     FROM master.dbo.sysprocesses
   order by cpu DESC;
```

The view contains information about processes that are running on an instance. These processes can be client processes or system processes. The view belongs to the **master** system database. The most important columns of the view are **spid** (session ID), **dbid** (ID of the current database), **uid** (ID of the user who executes the current command), and **cpu** (cumulative CPU time for the process).

The **sys.dm_exec_requests** dynamic management view provides the same information as the **sys.sysprocesses** catalog view, but the names of the corresponding columns are different. Example 20.2 displays, separate from the additional values for the **sp_handle** column, the same information as Example 20.1.

Example 20.2

```
USE master;
SELECT session_id, database_id, user_id, cpu_time, sql_handle
    FROM sys.dm_exec_requests
    order by cpu_time DESC;
```

The **sql_handle** column of the view points to the area in which the entire batch is stored. If you want to reduce the information to only one statement, you have to use the columns **statement_start_offset** and **statement_end_offset** to shorten the result. (All other column names are self-explanatory.)

NOTE **sys.dm_exec_requests** is especially worthwhile if you want to identify long-running queries.

Another DMV that can be used to display the information of such cached Transact-SQL statements and stored procedures using the most CPU time is **sys.dm_exec_query_stats**. (You can find the description and another example of this DMV in Chapter 19.) Example 20.3 shows the use of this view.

Example 20.3

```
USE master;
 SELECT TOP 20 SUM(total_worker_time) AS cpu_total,
        SUM(execution_count) AS exec_ct, COUNT(*) AS all_stmts, plan_handle
        FROM sys.dm_exec_query_stats
      GROUP BY plan_handle
        ORDER BY cpu_total;
```

The **total_worker_time** column of the **sys.dm_exec_query_stats** view displays the total amount of CPU time that was consumed by executions of cached SQL statements and stored procedures since it was compiled. The **execution_count** column displays the number of times that the cached plans have been executed since they were last compiled. (The TOP clause reduces the number of the displayed rows according to the parameter and the ORDER BY clause. This clause is described in detail in Chapter 24.)

Monitoring Memory

This section contains two subsections related to monitoring memory. The first subsection describes several Performance Monitor counters and the second one discusses DMVs you can use for the same purpose. (The DBCC MEMORYSTATUS command can also be used to monitor memory. This command is described in Chapter 15.)

Monitoring Memory Using Counters

The following Performance Monitor counters are used to monitor memory:

- Buffer Cache Hit Ratio (Object: SQLServer:BufferManager)
- Pages/sec (Object: Memory)
- Page Faults/sec (Object: Memory)

The **Buffer Cache Hit Ratio** counter displays the percentage of pages that did not require a read from disk. The higher this ratio, the less often the system has to go to the hard disk to fetch data, and performance overall is boosted. Note that there is no ideal value for this counter because it is application specific.

NOTE   This counter is different from most other counters, because it is not a real-time measurement, but rather an average value of all the days since the last restart of the Database Engine.

The **Pages/sec** counter displays the amount of paging (that is, the number of pages read or written to disk per second). The counter is an important indicator of the types of faults that cause performance problems. If the value of this counter is too high, you should consider adding more memory.

The **Page Faults/sec** counter displays the average number of page faults per second. This counter includes both soft page and hard page faults. As you already know, page faults occur when a system process refers to a virtual memory page that is not currently within the working set in the physical memory. If the requested page is on the standby list or a page currently shared with another process, a *soft page fault* is generated and the memory reference is resolved without physical disk access. However, if the referenced page is currently in the paging file, a *hard page fault* is generated and the data must be fetched from the disk.

Monitoring Memory Using Dynamic Management Views

The following DMVs are related to memory:

- sys.dm_os_memory_clerks
- sys.dm_os_memory_objects

The **sys.dm_os_memory_clerks** view returns the set of all memory clerks that are active in the current instance. You can use this view to find memory allocations by different memory types. Example 20.4 shows the use of this view.

Example 20.4

```
USE master;
SELECT type, SUM(pages_kb)
    FROM sys.dm_os_memory_clerks
    WHERE pages_kb != 0
    GROUP BY type
     ORDER BY 2 DESC;
```

The **type** column of the **sys.dm_os_memory_clerks** view describes the type of memory clerk. The **pages_kb** column specifies the amount of memory allocated by using the single page allocator of a memory node.

NOTE The Database Engine memory manager consists of a three-layer hierarchy. At the bottom of the hierarchy are memory nodes. The next level consists of memory clerks, memory caches, and memory pools. The last layer consists of memory objects. These objects are generally used to allocate memory.

The **sys.dm_os_memory_objects** view returns memory objects that are currently allocated by the database system. This DMV is primarily used to analyze memory usage and to identify possible memory leaks, as shown in Example 20.5.

Example 20.5

```
USE master;
SELECT type, SUM(pages_in_bytes) AS total_memory
   FROM sys.dm_os_memory_objects
   GROUP BY type
   ORDER BY total_memory DESC;
```

Example 20.5 groups all memory objects according to their type and then uses the values of the **pages_in_bytes** column to display the total memory of each group.

Monitoring the Disk System

This section contains two subsections that discuss monitoring the disk system. The first subsection describes several Performance Monitor counters, while the second describes the corresponding DMVs that you can use to monitor the disk system.

Monitoring the Disk System Using Counters

The following counters are related to monitoring the disk system:

- % Disk Time (Object: Physical Disk)
- Current Disk Queue Length (Object: Physical Disk)
- Disk Read Bytes/sec (Object: Physical Disk)
- Disk Write Bytes/sec (Object: Physical Disk)
- % Disk Time (Object: Logical Disk)
- Current Disk Queue Length (Object: Logical Disk)
- Disk Read Bytes/sec (Object: Logical Disk)
- Disk Write Bytes/sec (Object: Logical Disk)

As you can see from the preceding list, the names of the Performance Monitor counters for the Physical Disk object and the Logical Disk object are the same. (The difference between physical and logical objects is explained in Chapter 5.) These counters have the same purpose for each of the objects as well, so the following descriptions explain the counters only for the Physical Disk object.

The **% Disk Time** counter displays the amount of time that the hard disk actually has to work. It provides a good relative measure of how busy your disk system is, and it should be used over a longer period of time to indicate a potential need for more I/O capacity.

The **Current Disk Queue Length** counter tells you how many I/O operations are waiting for the disk to become available. This number should be as low as possible.

The **Disk Read Bytes/sec** counter shows the rate at which bytes were transferred from the hard disk during read operations, while **Disk Write Bytes/sec** provides the rate at which bytes were transferred to the hard disk during write operations.

Monitoring the Disk System Using DMVs

The following DMVs can be useful to display information concerning the disk system:

- sys.dm_os_wait_stats
- sys.dm_io_virtual_file_stats

The **sys.dm_os_wait_stats** view returns information about the waits encountered by threads that are in execution. Use this view to diagnose performance issues with the Database Engine and with specific queries and batches. Example 20.6 shows the use of this view.

Example 20.6

```
USE master;
SELECT wait_type, waiting_tasks_count, wait_time_ms
        FROM sys.dm_os_wait_stats
        ORDER BY wait_type;
```

The most important columns of this view are **wait_type** and **waiting_tasks_count**. The former displays the names of the wait types, while the latter displays the number of waits on the corresponding wait type.

The second view, **sys.dm_io_virtual_file_stats**, displays the file activity within a database allocation. Example 20.7 shows the use of this view.

Example 20.7

```
USE master;
SELECT database_id, file_id, num_of_reads,
      num_of_bytes_read, num_of_bytes_written
        FROM sys.dm_io_virtual_file_stats (NULL, NULL);
```

The columns of the **sys.dm_io_virtual_file_stats** view are self-explanatory. As you can see from Example 20.7, this view has two parameters. The first, **database_id**, specifies the unique ID number of the database, while the second, **file_id**, specifies the ID of the file. (When NULL is specified, all databases—i.e., all files in the instance of the Database Engine—are returned.)

Monitoring the Network Interface

This section comprises three subsections related to monitoring the network interface. The first subsection describes several Performance Monitor counters, the second discusses the corresponding DMV, and the last one describes the **sp_monitor** system procedure.

Monitoring the Network Interface Using Counters

The following Performance Monitor counters are related to monitoring the network:

- Bytes Total/sec (Object: Network Interface)
- Bytes Received/sec (Object: Network Interface)
- Bytes Sent/sec (Object: Network Interface)

The **Bytes Total/sec** counter monitors the number of bytes that are sent and received over the network per second. (This includes both the Database Engine and non–Database Engine network traffic.) Assuming your server is a dedicated database server, the vast majority of the traffic measured by this counter should be from the Database Engine. A consistently low value for this counter indicates that network problems may be interfering with your application.

To find out how much data is being sent back and forth from your server to the network, use the **Bytes Received/sec** and **Bytes Sent/sec** counters. The former displays the rate at which network data (in bytes) are received, while the latter checks the outbound rate. These counters will help you to find out how busy your actual server is over the network.

Monitoring the Network Interface Using a DMV

The **sys.dm_exec_connections** view returns information about the connections established to the instance of the Database Engine and the details of each connection. Examples 20.8 and 20.9 show the use of this view.

Example 20.8

```
USE master;
SELECT net_transport, auth_scheme
    FROM sys.dm_exec_connections
    WHERE session_id=@@SPID;
```

Example 20.8 displays the basic information about the current connection: network transport protocol and authentication mechanism. The condition in the WHERE clause reduces the output to the current session. (The @@spid global variable, which is described in Chapter 4, returns the identifier of the current server process.)

Example 20.9

```
USE master;
SELECT num_reads, num_writes
    FROM sys.dm_exec_connections;
```

Example 20.9 uses two important columns of this DMV, **num_reads** and **num_writes**. The former displays the number of packet reads that have occurred over the current connection, while the latter provides information about the number of packet writes that have occurred over this connection.

Monitoring the Network Interface Using a System Procedure

The **sp_monitor** system procedure can be very useful to monitor data concerning the network interface because it displays the information in relation to packets sent and received as a running total. This system procedure also displays statistics, such as the number of seconds the CPU has been doing system activities, the number of seconds the system has been idle, and the number of logins (or attempted logins) to the system.

Choosing the Right Tool for Monitoring

The choice of an appropriate tool depends on the performance factors to be monitored and the type of monitoring. The type of monitoring can be

- Real time
- Delayed (by saving information in the file, for example)

Real-time monitoring means that performance issues are investigated as they are happening. If you want to display the actual values of one or a few performance factors, such as the number of users or number of attempted logins, use dynamic management views because of their simplicity. In fact, DMVs can only be used for real-time monitoring. Therefore, if you want to trace performance activities during a specific time period, you have to use a tool such as SQL Server Profiler or Extended Events, both of which are described in this section.

Probably the best all-around tool for monitoring is the already described tool called Performance Monitor because of its many options. First, you can choose the performance activities you want to track and then display them simultaneously. Second, Performance Monitor allows you to set thresholds on specific counters (performance factors) to generate alerts that notify operators. This way, you can react promptly to any performance bottlenecks. Third, you can report performance activities and investigate the resulting chart log files later.

The following sections describe SQL Server Profiler, the Database Engine Tuning Advisor, and Extended Events.

SQL Server Profiler

SQL Server Profiler is a graphical tool that lets system administrators monitor and record database and server activities, such as login, user, and application information. SQL Server Profiler can display information about several server activities in real time, or it can create filters to focus on particular events of a user, types of commands, or types of Transact-SQL statements. Among others, you can monitor the following events using SQL Server Profiler:

- Login connections, attempts, failures, and disconnections
- CPU use of a batch
- Deadlock problems
- All DML statements (SELECT, INSERT, UPDATE, and DELETE)
- The start and/or end of a stored procedure

NOTE Since SQL Server 2017, SQL Server Profiler has been identified by Microsoft as a deprecated feature. This means that this tool may be removed any time. Microsoft recommends using Extended Events for tracing. (Extended Events is described in detail later in this chapter.)

The most useful feature of SQL Server Profiler is the ability to capture activities in relation to queries. These activities can be used as input for the Database Engine Tuning Advisor, which allows you to select indices and indexed views for one or more queries. For this reason, the following section discusses the features of SQL Server Profiler together with the Database Engine Tuning Advisor.

Database Engine Tuning Advisor

The Database Engine Tuning Advisor is part of the overall system and allows you to automate the physical design of your databases. As mentioned earlier, the Database Engine Tuning Advisor is tightly connected to SQL Server Profiler, which can display information about several server activities in real time, or it can create filters to focus on particular events of a user, types of commands, or Transact-SQL statements.

The specific feature of SQL Server Profiler that is used by the Database Engine Tuning Advisor is its ability to watch and record batches executed by users and to provide performance information, such as CPU use of a batch and corresponding I/O statistics, as explained next.

Providing Information for the Database Engine Tuning Advisor

The Database Engine Tuning Advisor is usually used together with SQL Server Profiler to automate tuning processes. You use SQL Server Profiler to record into a trace file information about the workload being examined. (As an alternative to a workload file, you can use any file that contains a set of Transact-SQL statements. In this case, you do not need SQL Server Profiler.) The Database Engine Tuning Advisor can then read the file and recommend several physical objects, such as indices, indexed views, and partitioning schema, that should be created for the given workload.

Example 20.10 creates two new tables, **orders** and **order_details**. These tables will be used to demonstrate the recommendation of physical objects by the Database Engine Tuning Advisor.

NOTE If your **sample** database already contains the **orders** table, you have to drop it using the DROP TABLE statement.

Example 20.10

```
USE sample;
CREATE TABLE orders
    (orderid INTEGER NOT NULL,
    orderdate DATE,
    shippeddate DATE,
    freight money);
CREATE TABLE order_details
    (productid INTEGER NOT NULL,
    orderid INTEGER NOT NULL,
    unitprice money,
    quantity INTEGER);
```

To demonstrate the use of the Database Engine Tuning Advisor, many more rows are needed in both tables. Examples 20.11 and 20.12 insert 3000 rows in the **orders** table and 30,000 rows in the **order_details** table, respectively.

Example 20.11

```
-- This batch inserts 3000 rows in the table orders
USE sample;
declare @i int, @order_id integer
          declare @orderdate datetime
          declare @shipped_date datetime
          declare @freight money
          set @i = 1
          set @orderdate = getdate()
          set @shipped_date = getdate()
          set @freight = 100.00
        while @i < 3001
        begin
        insert into orders (orderid, orderdate, shippeddate, freight)
          values(@i, @orderdate, @shipped_date, @freight)
        set @i = @i+1
        end
```

Example 20.12

```
-- This batch inserts 30000 rows in order_details and modifies some of them
USE sample;
declare @i int, @j int
          set @i = 3000
          set @j = 10
          while @j > 0
          begin
          if @i > 0
              begin
              insert into order_details (productid, orderid, quantity)
                    values (@i, @j, 5)
            set @i = @i - 1
            end
          else begin
              set @j = @j - 1
              set @i = 3000
              end
          end
          go
          update order_details set quantity = 3
                  where productid in (1511, 2678)
```

The query in Example 20.13 will be used as an input file for SQL Server Profiler. (Assume that no indices are created for the columns that appear in the SELECT statement.) Click the Windows Start icon or press the Windows key and type **SQL Server Profiler**. When the SQL Server Profiler tile appears, click it. On the File menu, choose New Trace. After connecting to the server, the Trace Properties dialog box appears. Type a name for the trace and select an output .trc file for the Profiler information (in the Save to File field). Click Run to start the capture and use SQL Server Management Studio to execute the query in Example 20.13.

Example 20.13

```
USE sample;
SELECT orders.orderid, orders.shippeddate
      FROM orders
    WHERE orders.orderid between 806 and 1600
    and not exists (SELECT order_details.orderid
         FROM order_details
 WHERE order_details.orderid = orders.orderid);
```

Finally, stop SQL Server Profiler by choosing File | Stop Trace and selecting the corresponding trace.

Working with the Database Engine Tuning Advisor

The Database Engine Tuning Advisor analyzes a workload and recommends the physical design of one or more databases. The analysis will include recommendations to add, remove, or modify the physical database structures, such as indices, indexed views, and partitions. The Database Engine Tuning Advisor will recommend a set of physical database structures that will optimize the tasks included in the workload.

To start the Database Engine Tuning Advisor, shown in Figure 20-2, begin typing its name in the Windows Search bar on the Start menu, and click its name when it appears in the results. (The alternative way is to start SQL Server Profiler and choose Tools | Database Engine Tuning Advisor.)

In the Session Name field, type the name of the session for which the Database Engine Tuning Advisor will create tuning recommendations. In the Workload frame, choose either File or Table. If you choose File, enter the name of the trace file. If you choose Table, you must enter the name of the table that is created by SQL Server Profiler. (Using SQL Server Profiler, you can capture and save data about each workload to a file or to a SQL Server table.)

NOTE Running SQL Server Profiler can place a heavy burden on a busy instance of the Database Engine.

In the Select Databases and Tables to Tune frame, choose one or more databases and/or one or more tables that you want to tune. (The Database Engine Tuning Advisor can tune a workload that involves multiple databases. This means that the tool can recommend indices, indexed views, and partitioning schema on any of the databases in the workload.) For our example, select the **sample** database and its two tables, **orders** and **order_details**.

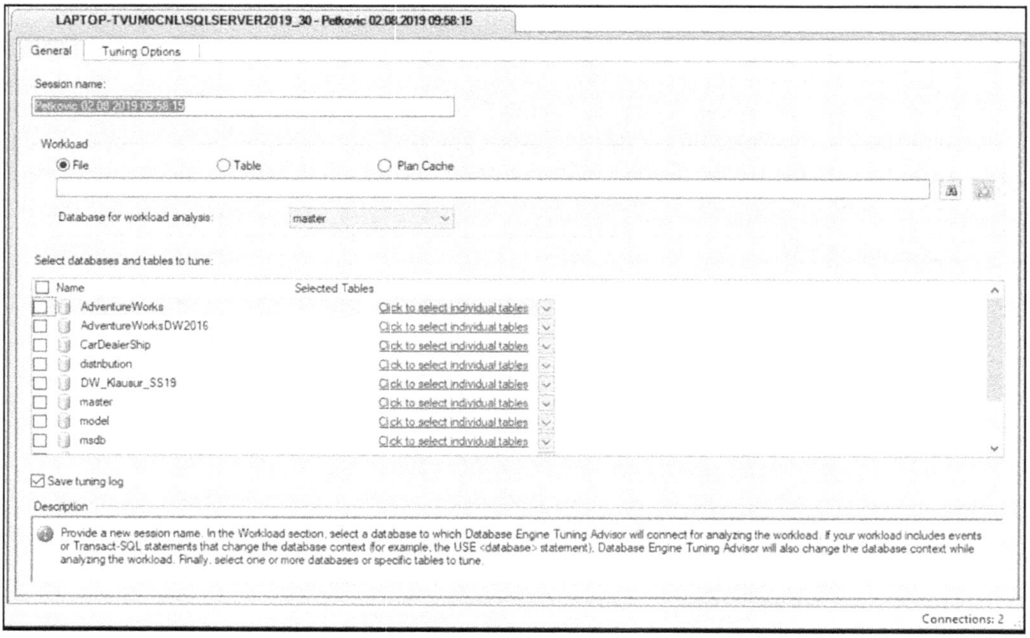

Figure 20-2   Database Engine Tuning Advisor: General tab

NOTE If you choose Plan Cache as the workload option (see Figure 20-2), the Database Engine Tuning Advisor selects the top 1000 events from the plan cache to use for analysis. For a detailed description of how to tune a database using the plan cache, see Microsoft Docs.

To choose options for tuning, click the Tuning Options tab (see Figure 20-3). Most of the options on this tab are divided into three groups:

- **Physical Design Structures (PDS) to use in database**   Allows you to choose which physical structures (indices and/or indexed views) should be recommended by the Database Engine Tuning Advisor, after tuning the existing workload. (The Evaluate Utilization of Existing PDS Only option causes the Database Engine Tuning Advisor to analyze the existing physical structures and recommend which of them should be deleted.)

- **Partitioning strategy to employ**   Allows you to choose whether or not partitioning recommendations should be made. If you opt for partitioning recommendations, you can also choose the type of partitioning, full or aligned. (Partitioning is discussed in detail in Chapter 26.)

- **Physical Design Structures (PDS) to keep in database**   Enables you to decide which, if any, existing structures should remain intact in the database after the tuning process.

For large databases, tuning physical structures usually requires a significant amount of time and resources. Instead of starting an exhaustive search for possible indices, the Database Engine

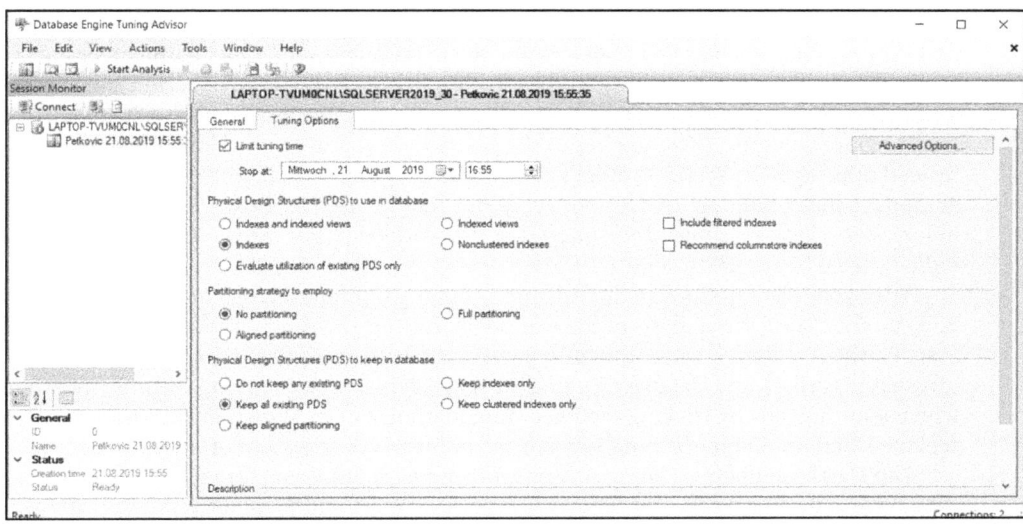

Figure 20-3   Database Engine Tuning Advisor: Tuning Options tab

Tuning Advisor offers (by default) the restrictive use of resources. This operation mode still gives very accurate results, although the number of resources tuned is significantly reduced.

During the specification of tuning options, you can define additional customization options by clicking Advanced Options, which opens the Advanced Tuning Options dialog box (see Figure 20-4). Checking the check box at the top of the dialog box enables you to define the maximum space for recommendations. Increase the maximum space to 20MB if you intend to start an exhaustive search. (For large databases, selection of physical structures usually requires a significant amount of resources. Instead of starting an exhaustive search, the Database Engine Tuning Advisor offers you the option to restrict the space used for tuning.)

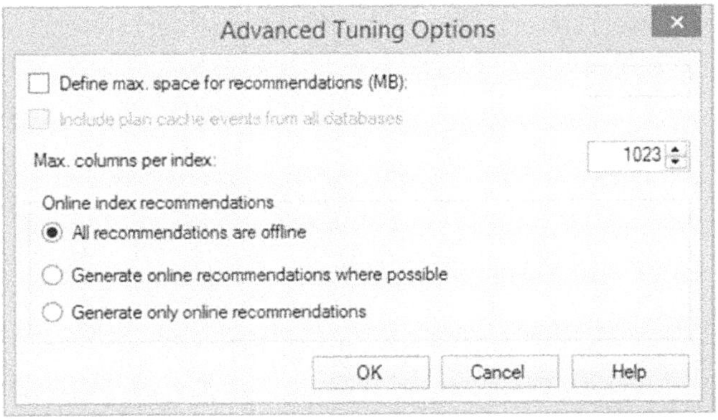

Figure 20-4   Advanced Tuning Options dialog box

Of all index tuning options, one of the most interesting is the second option in this dialog box, which enables you to determine the maximum number of columns per index. A single-column index or a composite index built on two columns can be used several times for a workload with many queries and requires less storage space than a composite index built on four or more columns. (This applies in the case where you use a workload file on your own instead of using SQL Server Profiler's trace for the specific workload.) On the other hand, a composite index built on four or more columns may be used as a covering index to enable index-only access for some of the queries in the workload. (For more information on covering indices, see Chapter 10.)

After you select options in the Advanced Tuning Options dialog box, click OK to close it. You can then start the analysis of the workload. To start the tuning process, choose Actions | Start Analysis. After you start the tuning process for the trace file of the query in Example 20.13 (presented earlier), the Database Engine Tuning Advisor creates tuning recommendations, which you can view by clicking the Recommendations tab, as shown in Figure 20-5. As you can see, the Database Engine Tuning Advisor recommends the creation of two indices.

The Database Engine Tuning Advisor recommendations concerning physical structures can be viewed using a series of reports that provide information about very interesting options.

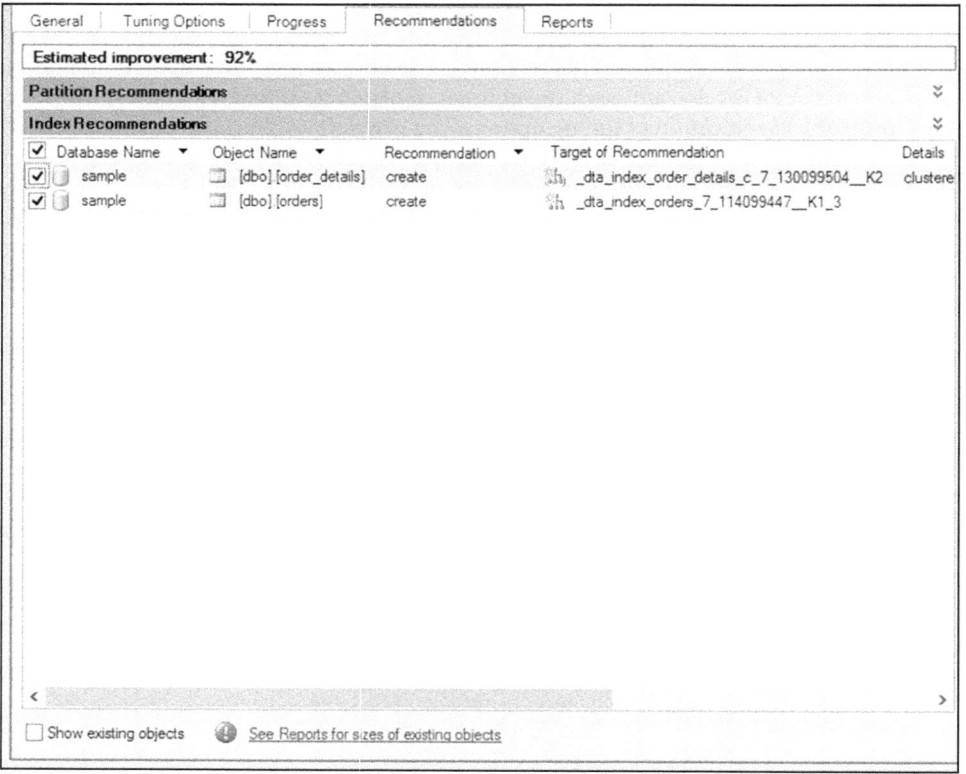

Figure 20-5 Database Engine Tuning Advisor: Recommendations tab

These reports enable you to see how the Database Engine Tuning Advisor evaluated the workload. To see these reports, click the Reports tab in the Database Engine Tuning Advisor dialog box after the tuning process is finished. You can see the following reports, among others:

- **Index Usage Report (recommended)**   Displays information about the expected usage of the recommended indices and their estimated sizes
- **Index Usage Report (current)**   Presents the same information about expected usage for the existing configuration
- **Index Detail Report (recommended)**   Displays information about the names of all recommended indices and their types
- **Index Detail Report (current)**   Presents the same information for the actual configuration, before the tuning process was started
- **Table Access Report**   Displays information about the costs of all queries in the workload (using tables in the database)
- **Workload Analysis Report**   Provides information about the relative frequencies of all data modification statements (costs are calculated relative to the most expensive statement with the current index configuration)

There are two ways in which you can apply recommendations: immediately or after saving to the file. If you choose Actions | Apply Recommendations, the recommendations will be applied immediately. Similarly, if you choose Actions | Save Recommendations, the recommendations will be saved to the file. (This alternative is useful if you generate the script with a test system and intend to use the tuning recommendation with a production system.) The third option, Actions | Evaluate Recommendations, is used to evaluate the recommendations produced by the Database Engine Tuning Advisor.

Extended Events

Extended Events is a tracing and troubleshooting framework that enables you to control information at a granular level. That way, DBAs and programmers can collect information to find any performance bottlenecks and other system properties in the Database Engine.

Extended Events: Architecture

The architecture of Extended Events is based upon objects. There are four groups of objects:

- Events
- Actions
- Targets
- Predicates

Events indicate points of interest in an execution path. You can think of events as important places in the programming code where the execution is traced.

Actions are functions, which are invoked by events. In other words, an action is triggered when an event fires. At that point in time, all the columns in the event are available.

Targets consume events, either synchronously on the thread that fires the event, or asynchronously on a system-provided thread. In other words, a target is a destination where events can be published and stored. There are two different target types: **event_file** and **ring_buffer**. The former permanently stores events to a file, while the latter saves events to memory buffers, which store volatile data.

Predicates retrieve values from event sources for use in comparison operations. They compare specific data types and return a Boolean value.

Another important concept of Extended Events is a *session*. To use Extended Events, you have to create a session. A session is used to trace the particular event(s). Extended Events sessions are stored in system tables, which belong to the **master** database and are exposed through catalog views and dynamic management views. To activate a created session, you have to start it. After starting, the session traces the corresponding event(s) and publishes information to the defined target, until it is explicitly stopped.

The creation of a session can be done either with SQL Server Management Studio or programmatically. The following sections describe both ways.

Creating a Session Using SQL Server Management Studio

To create a session in SQL Server Management Studio, select Management | Extended Events in Object Explorer. Right-click the Sessions folder and select New Session. The New Session window appears. In the window, there are four different pages (see the left pane of Figure 20-6):

- General
- Events
- Data Storage
- Advanced

The General page allows you to specify the session name and (optionally) to define the schedule of that session. In the Events page, shown in Figure 20-6, you select the events you want to capture from the event library and specify the properties to be traced.

Extended Events targets event data and stores it persistently or volatile. The Data Storage page allows you to specify file(s) where the data is stored persistently. Using the Advanced page you can modify miscellaneous values, such as maximum memory size and maximum event size.

As an example of how to enable an extended event and start it, suppose that you are interested in tracing all long-running queries with an execution time longer than one second. To start, select the General page and enter the name **Special_queries** in the Session Name field. Additionally, you can choose between two different schedule forms. One starts the event session at server startup and the other starts the event session immediately after the session is created.

Now, select Events in the left pane. On the Events page, you can select the events you want to capture from the event library. To select one or more related events, type in the Event Library search box a phrase that is common for all related events. In the case of long-running queries, you are concerned with SQL statements, so type, for instance, the phrase **sql_stat** (see Figure 20-6). The system finds three hits: sql statement completed, sql statement recompile,

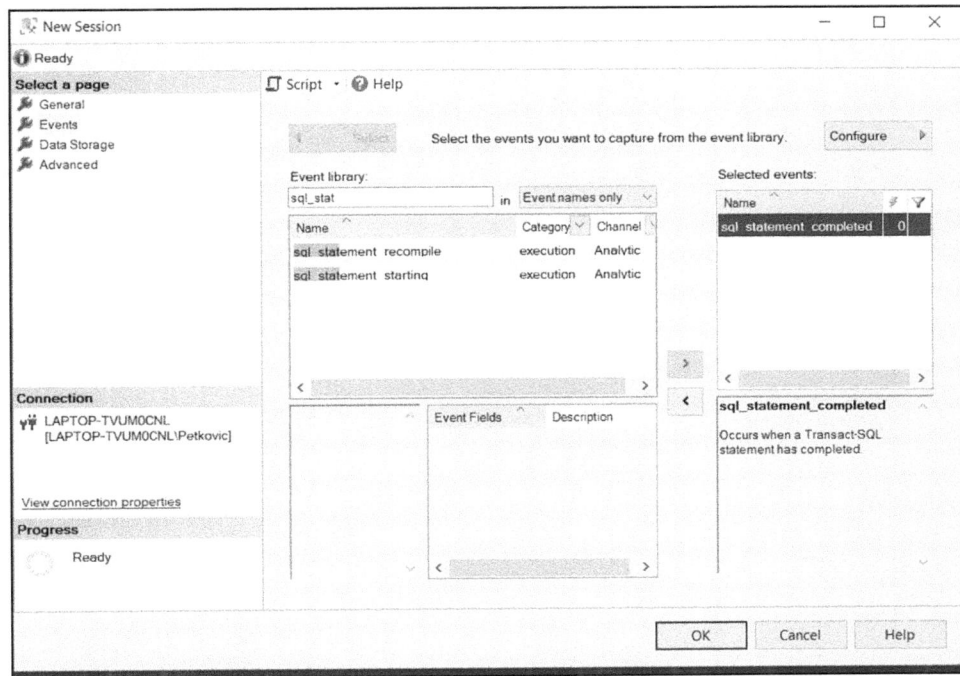

Figure 20-6   Extended Events: The New Session window

and sql statement starting. You are interested in the completed queries, so select sql statement completed and click the right-pointing arrow between the left pane and right pane to move the selected event to the Selected Events list.

The Events page is also used to configure the event and add a filter condition. To configure the event, click the Configure button in the top right of the Events page. The Event Configuration Options section appears to the right of the Selected Events section, as shown in Figure 20-7. In the Event Configuration Options section, you first specify actions. Suppose that you want to know the name of users who execute long-running queries. Therefore, choose the Global Fields (Actions) tab. The list of all possible actions appears. Check the username check box.

Next, you specify the corresponding predicate, which states that the event should fire when the particular query runs too long. Therefore, on the Events page, choose the Filter (Predicate) tab (see Figure 20-7). Click where indicated to add a filter clause. Click in the Field column and select duration. (To select the desired action, you have to scroll through the list of all actions.) Under Operator, select >=. Under Value, enter **1000**. That way, you restrict the created session to capture only those queries that run longer than 1000 milliseconds (1 second).

Move to the next page of the New Session window, Data Storage, where you specify how event data should be stored. You want to store the gathered information in a file. Therefore, in the Targets section, under Type, select event_file. In the Properties section, specify in the File Name on Server field the location and the name of your event file. (As you already know, the

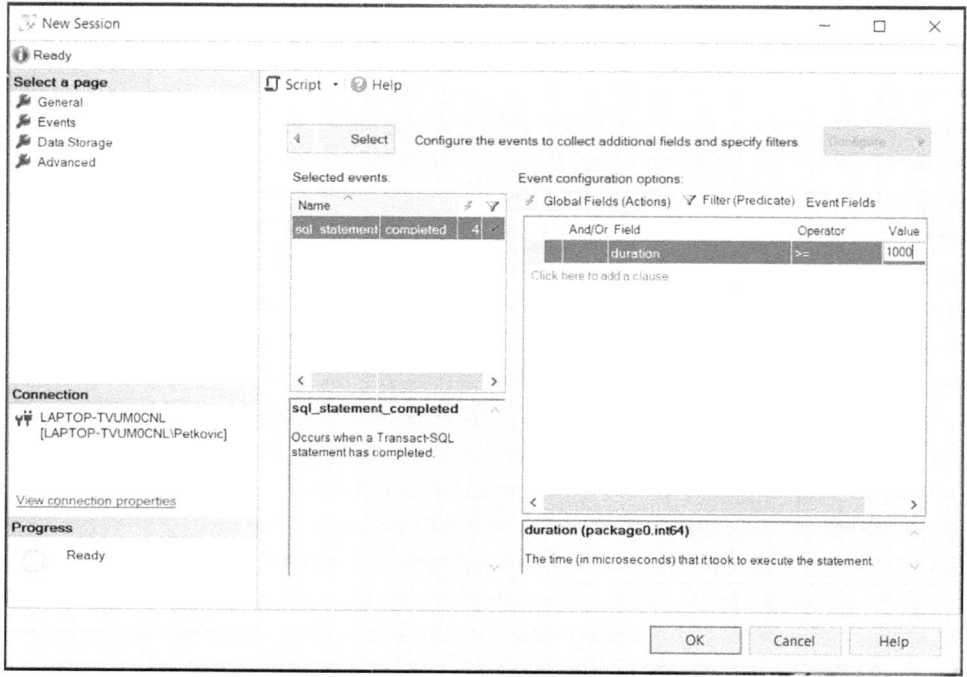

Figure 20-7 Extended Events: Events page

event_file option allows you to persistently save content of the event to a file.) Optionally, you can change the maximum file size in the field with the same name.

Now, select Advanced in the left pane. On the corresponding page, you can specify the event retention mode, the maximum dispatch latency, the maximum memory size, and the maximum event size.

You have completed the modification of all four pages so now you can click the OK button to create the session. If on the General page you chose Start Event Session Immediately After Session Creation, the session starts. If you chose Start the Event Session at Server Startup, the session is inactive. To start the session and track the corresponding event, select Management | Extended Events | Sessions and right-click the name of your session in Object Explorer. Select Start Session. This starts the tracking process. To stop capturing the data, choose Stop Session.

NOTE You can change properties of an already created session by right-clicking its name and selecting Properties. After that, you can modify properties by clicking the particular page (General, Events, Data Storage, and Advanced).

To display the collected data, double-click package0.event_file under the session name. A new query window will open with the list of tracked events and corresponding values. Select an event to see the details of the collected data in the result window (see Figure 20-8).

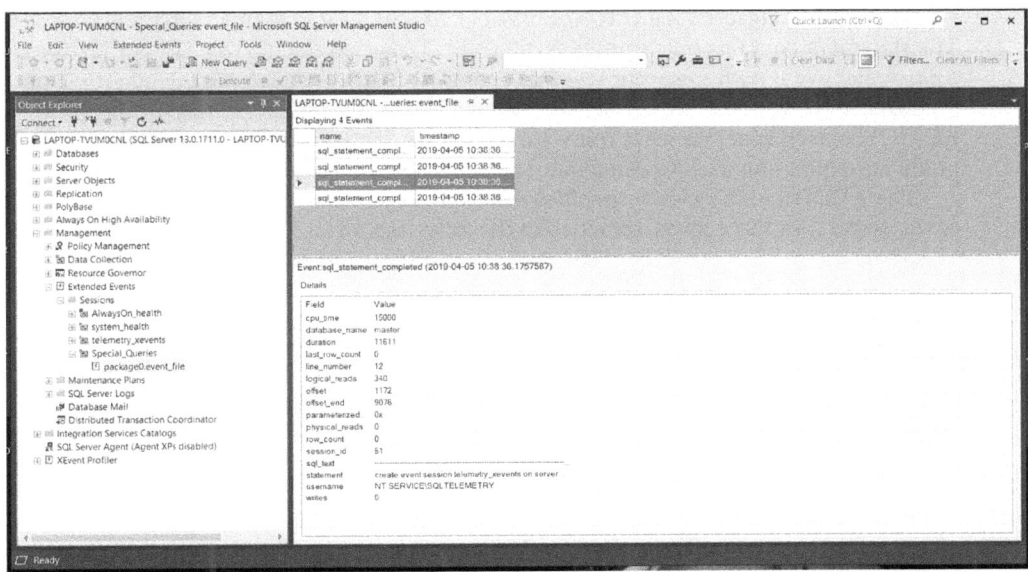

Figure 20-8   Details of collected data in the session

NOTE Another example in relation to Extended Events can be found in the "Memory Grant Feedback" section of Chapter 28.

Creating a Session Using Transact-SQL

You can use T-SQL statements to programmatically create a new event session. Example 20.14 creates a session called session1.

Example 20.14

```
CREATE EVENT SESSION session1 ON SERVER
ADD EVENT sqlserver.sql_batch_starting(ACTION(package0.event_
sequence,sqlserver.client_app_name,sqlserver.client_pid,sqlserver.
database_id,sqlserver.database_name,sqlserver.nt_username,sqlserver.query_
hash,sqlserver.server_principal_name,sqlserver.session_id)
    WHERE ([package0].[equal_boolean]([sqlserver].[is_system],(0))))
ADD TARGET package0.event_file(SET filename=N'Session1.xel',max_file_
size=(5),max_rollover_files=(4))
WITH (MAX_MEMORY=8192 KB,EVENT_RETENTION_MODE=ALLOW_SINGLE_EVENT_LOSS,
    MAX_DISPATCH_LATENCY=5 SECONDS,MAX_EVENT_SIZE=0 KB,
    MEMORY_PARTITION_MODE=PER_CPU,TRACK_CAUSALITY=ON,STARTUP_STATE=OFF)
```

NOTE The following discussion of the CREATE EVENT SESSION statement is superficial, because the easier way to create a session is to use SQL Server Management Studio. For the description of all options of this statement, see Microsoft Docs.

As its name suggests, the CREATE EVENT SESSION statement creates an event session. The statement contains one or more ADD EVENT clauses. Each ADD EVENT clause associates a particular event with the event session. The ADD EVENT clause can include three options: SET, ACTION, and WHERE. The SET option allows you to set attributes for the event. The ACTION option specifies what you want to capture, while the WHERE option specifies the predicate expression used to determine the condition for whether an event should be processed.

The ADD TARGET clause specifies the target to associate with the event session. The name of the target is in three parts, specified in the following form:

```
[event_module_guid].event_package_name.target_name
```

The first part of the name defines the global ID of the event module, the second part specifies the name of the event package, and the last part specifies the name of the target. The global ID is necessary only in the case that the target is not locally stored. Therefore, in Example 20.14 the global ID is omitted, while the event package is called **package0** and the target name is **event_file**. (You can use the optional SET clause to set different target parameters.)

As you already know, the session by default is in an inactive state. Examples 20.15 and 20.16 show you how to start and stop a session using Transact-SQL, respectively.

Example 20.15
```
USE master;
ALTER EVENT SESSION Session1 ON SERVER STATE=start
```

Example 20.16
```
USE master;
ALTER EVENT SESSION Session1 ON SERVER STATE=stop
```

Editing Information Concerning Extended Events

You can use several catalog views and/or dynamic management views to obtain information concerning events. This section discusses three catalog views:

- sys.server_event_sessions
- sys.server_event_session_events
- sys.server_event_session_actions

and two dynamic management views:

- sys.dm_xe_packages
- sys.dm_xe_objects

The **sys.server_event_sessions** catalog view lists all the event session definitions that exist in the Database Engine. The most important columns of this view are **event_session_id**, which specifies the unique ID of the event session, and **name**, which defines the unique name for identifying the event session.

The **sys.server_event_session_events** catalog view returns a row for each event in an event session. The most important columns of this view are **event_session_id**, **event_id**, and **name**. The first column specifies the unique ID of the event session, and the last two columns define the ID and the name of the event, respectively.

The **sys.server_event_session_actions** catalog view returns a row for each action on each event in an event session. The most important columns of this view are **event_session_id**, **event_id**, and **name**. The first two columns specify the unique ID of the event session and event, respectively, and the **name** column defines the name of the action.

Example 20.17 shows how you can display information concerning an event session, its events, and the corresponding actions.

Example 20.17

```
USE master;
SELECT sessions.name AS SessionName, event1.package as PackageName,
      event1.name AS EventName,action1.name AS ActionName
    FROM sys.server_event_sessions sessions
    INNER JOIN sys.server_event_session_events event1
      ON sessions.event_session_id = event1.event_session_id
    INNER JOIN sys.server_event_session_actions action1
      ON sessions.event_session_id = action1.event_session_id
    WHERE sessions.name = 'session1' ;
```

The result is

SessionName	PackageName	EventName	ActionName
session1	sqlserver	sql_batch_starting	event_sequence
session1	sqlserver	sql_batch_starting	client_app_name
session1	sqlserver	sql_batch_starting	client_pid
session1	sqlserver	sql_batch_starting	database_id
session1	sqlserver	sql_batch_starting	database_name
session1	sqlserver	sql_batch_starting	nt_username
session1	sqlserver	sql_batch_starting	query_hash
session1	sqlserver	sql_batch_starting	server_principal_name
session1	sqlserver	sql_batch_starting	session_id

Example 20.17 uses three catalog views described earlier to display the information concerning the session called **session1** (see Example 20.14). To display all events and actions contained in this session, you have to join the **sys.server_event_sessions** catalog view with two other views, **sys.server_event_session_events** and **sys.server_event_session_actions**, using the column **sessions.event_session_id**, which exists in all three views. (The **sessions.event_session_id** column specifies the unique ID of the corresponding session.)

NOTE Example 20.17 can easily be exetended to display where particular events are stored. The corresponding catalog view is called **sys.server_event_session_targets** (see Exercise 20.3).

As mentioned at the beginning of this section, you also can use two dynamic management views to obtain information about events. The **sys.dm_xe_packages** DMV lists all the packages registered with Extended Events of your system. The most important columns of this view are **name**, which specifies the name of package, and **guid**, which is the global identifier that identifies the package.

The **sys.dm_xe_objects** DMV returns a row for each object type that is exposed by an event package. (The object type can be an event, action, or target.) The most important columns of this view are **name, object_type**, and **package_guid. name** specifies the name of the object, while **object_type** specifies its type. **package_guid** is the global ID for the package that exposes the corresponding action.

Example 20.18 displays the names of all objects of the **event** type that belong to the system package called **filestream**.

Example 20.18

```
USE master;
SELECT pkg.name as PackageName, obj.name as EventName
    FROM sys.dm_xe_packages pkg
    INNER JOIN sys.dm_xe_objects obj on pkg.guid = obj.package_guid
    WHERE obj.object_type = 'event'  AND pkg.name = 'filestream'
    ORDER by 2;
```

The result is

```
PackageName    EventName
------------------------------------------------------------
filestream     filetable_application_error
filestream     filetable_file_io_request
filestream     filetable_file_io_response
filestream     filetable_nso_error
filestream     filetable_nso_kill
filestream     filetable_nso_operation
filestream     filetable_store_database_operation
filestream     filetable_store_enumerate_getitem
filestream     filetable_store_item_get
filestream     filetable_store_item_modify
filestream     filetable_store_item_moverename
filestream     filetable_store_operation
filestream     filetable_store_table_operation
```

Example 20.18 uses **sys.dm_xe_packages** and **sys.dm_xe_objects** to display the requested objects. Both DMVs are joined using the **guid** column of the former view and the **package_guid** of the latter view.

Advantages of Extended Events over SQL Server Profiler

The advantages of using extended events in relation to SQL Server Profiler are as follows:

- They are lightweight.
- They can trace and track more events.
- They provide better flexibility.

The most important advantage of Extended Events in relation to SQL Server Profiler is that the former does not have as much negative influence on the database server as the latter. SQL Server Profiler is a GUI that utilizes SQL Server Trace through the client side. This has impact on your system, and this impact can be severe. On the other hand, Extended Events is lightweight and uses a limited amount of resources, so it does not have any significant negative impact on the system performance.

Extended Events was introduced in SQL Server 2008. At that time, the system supported 253 events. Over time, the number of supported extended events significantly increased. For instance, SQL Server 2017 supports 564 different events, while in SQL Profiler there are altogether 235 events.

Extended events are more flexible than the events supported by SQL Server Profiler. For example, you can bind extended events to any target or you can specify any action with any event. Additionally, you can dynamically filter extended events using predicates. Finally, although extended events are fired synchronously in a host application, they can be processed asynchronously, too.

Besides SQL Server Profiler, a set of existing dynamic management views is another "tool" that provides similar functionality to extended events, because DMVs dynamically trace the state of the system as well as its performance. The main difference between them is that DMVs provide a snapshot of the system, while the trace made by extended events has a specific duration that is configurable.

Other Performance Tools of the Database Engine

The Database Engine supports four additional performance tools:

- Query Store
- Automatic tuning (as an "add-on" to Query Store)
- Performance Data Collector
- Resource Governor
- Live Query Statistics

The following sections describe these tools.

Query Store

As described in Chapter 19, the USE PLAN query hint advises the Database Engine to use the particular execution plan for the query execution. Query Store is a Database Engine component that has a similar task: to simplify performance troubleshooting by enabling you to find performance differences caused by changes in query plans. The main difference between the USE PLAN hint and Query Store is that the latter does not require any changes in user applications.

Query Store can be used for the following tasks:

- Store the history of all plans for your queries.
- Capture the performance of a query plan over time.
- Identify the top n queries (by execution time, memory usage, etc.).
- Identify queries with performance regression over time.
- Force a particular query plan for a query with performance regression.

Every time the Database Engine executes or compiles a query, the gathered information concerning each query is kept in memory and, after that, stored on disk. (You can specify the time interval for how long the data is kept in memory before being sent to disk.) Finally, the data is stored in the user database. This process describes roughly the functionality of the Query Store component. By default, Query Store is disabled. For this reason, I will first explain how you can enable it and modify its configuration parameters.

Enabling Query Store and Setting Configuration Options

As mentioned, Query Store must be enabled before you can use it. You can enable Query Store using Transact-SQL. Example 20.19 shows the use of Transact-SQL. (You have to change the name of the database if you do not use the **AdventureWorks** database as shown in the example.)

Example 20.19

```
USE master;
ALTER DATABASE AdventureWorks SET QUERY_STORE=ON;
```

Once you have enabled Query Store, you can configure its various options. Again, you can do so using SQL Server Management Studio or Transact-SQL. To configure options using Management Studio, right-click the database, choose Properties, and select Query Store in the left pane. The page in Figure 20-9 appears. (All configurable settings are in bold font and highlighted.) The following options, among others (which are self-explanatory), can be modified with SQL Server Management Studio:

- **Operation Mode**   Can be READ WRITE, READ ONLY, or OFF.
- **Data Flush Interval**   Determines the frequency at which data written to Query Store is persisted to disk.
- **Max Size**   Configures the maximum size of Query Store. If the data in Query Store hits the limit for maximum size, Query Store automatically changes the state from read-write to read-only and stops collecting new data.
- **Query Store Capture Mode**   Designates if Query Store captures all queries, captures only relevant queries based on execution count and resource consumption, or stops adding new queries and just tracks current queries.
- **Size Based Cleanup Mode**   Controls whether the cleanup process will be automatically activated when the total amount of data gets close to the maximum size.

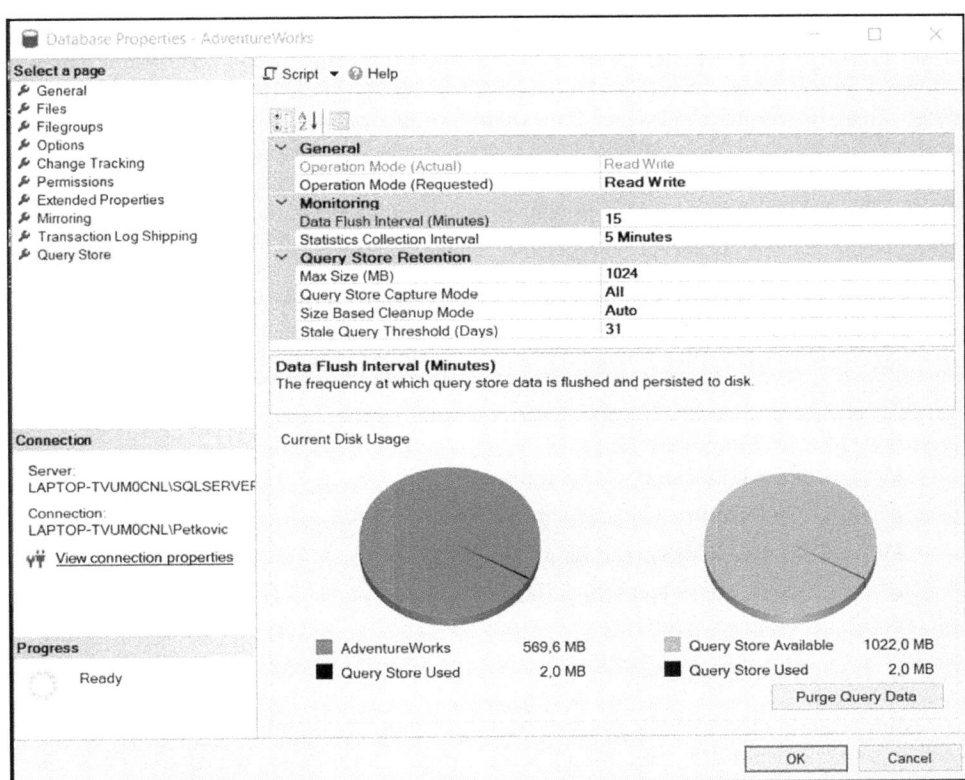

Figure 20-9   The Query Store page

The ALTER DATABASE statement can also be used to configure options of Query Store, as shown in Example 20.20.

Example 20.20

```
USE master;
GO
ALTER DATABASE AdventureWorks
SET QUERY_STORE ( OPERATION_MODE = READ_WRITE,
    CLEANUP_POLICY =  (STALE_QUERY_THRESHOLD_DAYS = 30),
    DATA_FLUSH_INTERVAL_SECONDS = 3000, MAX_STORAGE_SIZE_MB = 500,
    INTERVAL_LENGTH_MINUTES = 15, QUERY_CAPTURE_MODE = AUTO,
    MAX_PLANS_PER_QUERY = 1000);
```

The ALTER DATABASE statement and its QUERY_STORE clause are used to set configuration options of Query Store. The INTERVAL_LENGTH_MINUTES option determines the time interval at which run-time statistics data is aggregated into Query Store. The MAX_PLANS_PER_QUERY option represents the maximum number of plans maintained for each query. The default value for this option is 200.

The CLEANUP_POLICY option describes the data retention policy of Query Store, while STALE_QUERY_THRESHOLD_DAYS determines the number of days for which the information for a query is retained in Query Store. DATA_FLUSH_INTERVAL_SECONDS determines the frequency at which data written to Query Store is persisted to disk. (For optimization reasons, data collected by Query Store is asynchronously written to the disk.)

The QUERY_CAPTURE_MODE option designates the currently active query capture mode. The AUTO value of this option specifies that only relevant queries based on execution count and resource consumption are captured. (The default value is ALL.)

Analyze Query Plans with SQL Server Management Studio

Once Query Store is enabled and configured, you can start analyzing your query plans. To analyze plans, right-click the Query Store container in the folder of your database. The following options appear:

- View Regressed Queries
- View Overall Resource Consumption
- View Top Resource Consuming Queries
- View Queries with Forced Plans
- View Queries with High Variation
- Query Wait Statistics
- View Tracked Queries

After selecting the View Regressed Queries option, the Regressed Queries window shows queries and corresponding execution plans in Query Store. In this window you can select queries based on various criteria. Among other things, you can view the source query, force and unforce a particular query plan, and refresh the display. (The next section describes in detail how you can correct regressed queries.)

Selecting the View Overall Resource Consumption option opens a window that displays CPU Time, Duration, Execution Count, and Logical Reads for the specified time period and aggregated by an interval. (The default value for the time period is a week, and the default value for the aggregation interval is one hour.) Each metric appears in its own grid.

The View Top Resource Consuming Queries window displays all queries sorted by their use of a particular resource (metric). The possible metrics are CPU Time, Duration, Execution Count, Logical Reads and Writes, and Memory Consumption.

The View Queries with Forced Plans window enables you to see all queries where the stored execution plan will be forced when the corresponding query is executed. (To force an execution plans using Query Store, you can either click the Force Plan button in one of the Query Store reports or use the **sp_query_store_force_plan** system stored procedure.)

The View Queries with High Variation window shows all queries with high variation in query text. (If you have an ad hoc workload that has this property, you will have more individual queries stored, and thus more plans and more run-time and wait statistics.)

The View Query Wait Statistics window shows a bar chart containing the top wait categories in Query Store. Use the drop-down menu at the top to select an aggregate criteria for the wait time: avg, max, min, std dev, and total (default).

The View Tracked Queries window allows you to track a particular query. Additionally, you can track a query using the Track the Selected Query button in the Top Resource Consuming Queries window or Regressed Queries window.

All of these windows have a similar structure and functionality. For this reason, I will explain only one of them, the Top Resource Consuming Queries window.

Top Resource Consuming Queries

As you already know, the window that opens when you choose the View Top Resources Consuming Queries option displays all queries in relation to the particular metric in descending order. The chart in the top-left pane of the window displays total statistics based on selected metrics for the vertical and horizontal axes. It also has two drop-down menus: Metric and Statistic (see Figure 20-10). The Metric drop-down menu contains, among others, the following metrics: CPU Time, Duration, and Memory Consumption. The Statistic drop-down menu

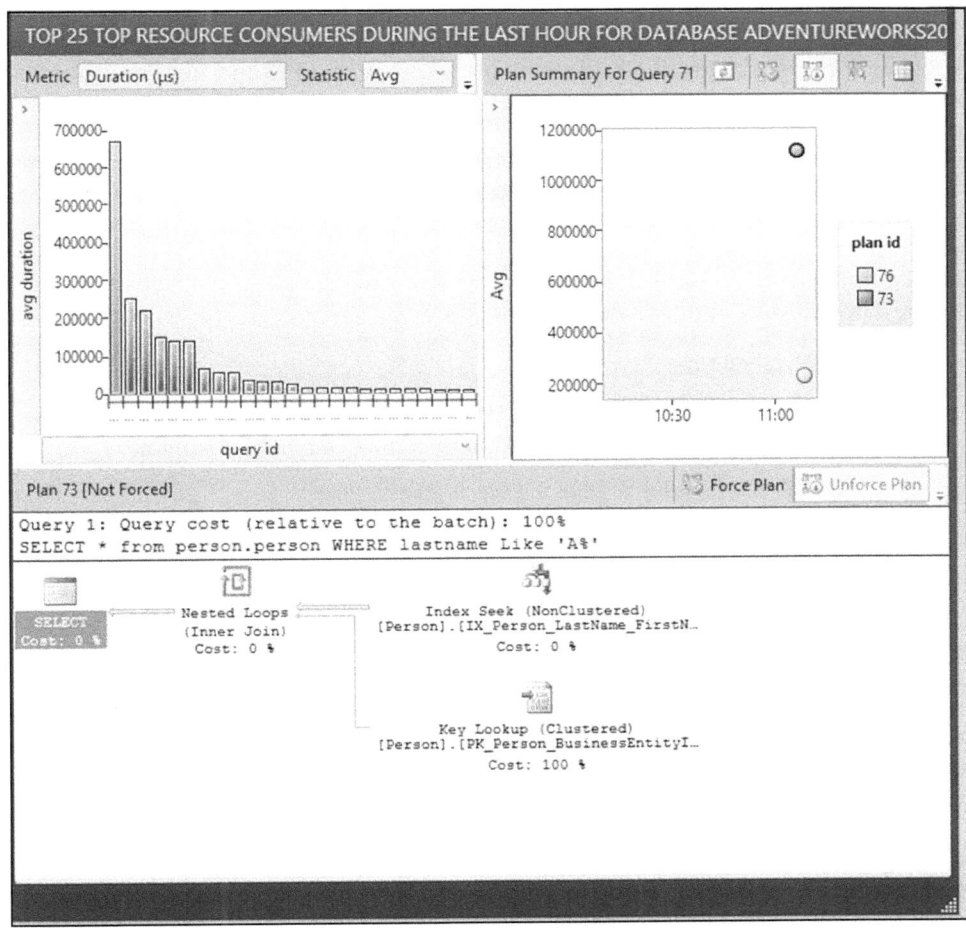

Figure 20-10   The Top Resource Consuming Queries window

enables you to select functions corresponding to the well-known aggregate functions (AVG, MAX, MIN, and COUNT).

Many metrics are available for the vertical axis of the left chart, so I will explain only three: execution count (**exec count**), number of execution plans (**num plans**), and the default option, which is dynamic. The default option depends on the values you choose in the Metric and Statistic drop-down menus. (Note, when the **exec count** metric is selected from the drop-down menu, only **exec count** and **num plans** are available for the vertical axis.)

There are three metrics available for the horizontal axis. The default value is **query id**, which contains values that uniquely identify each query. The second option, **plan_id**, contains values that uniquely identify execution plans. The third option is dynamic and, again, is based on the values you choose in the Metric and Statistic drop-down menus.

The chart in the top-right pane of the window, called Plan Summary, displays statistics of the selected/highlighted query (green bar on the left chart) based on the selected Metric and Statistic values (see Figure 20-10). The selected value of the Statistic drop-down menu becomes a vertical axis for the right chart. You can move the mouse cursor over one of the "bubbles" in the right part of the chart to see the detailed statistics for a specific query. Figure 20-10 has two bubbles in the right chart, which represent plans with **plan_id** 73 and 76. (The values 73 and 76 may be different when you create the same detailed statistics.) Also, the statistics details will be different depending on the metric selected. The size of these bubbles depends on total number of executions: the higher the number, the bigger the bubble. When you click a particular bubble, the bottom pane displays the corresponding execution plan.

Above the left pane, there are several additional buttons that are used to display queries and their plans in different forms. (If you have smaller screen resolution, these buttons will be hidden, as in Figure 20-10; click the drop-down arrow to the right of the Statistic field to expose them.) The following buttons belong to that group:

- Last refreshed time
- Track the selected query in a new Tracked Queries window
- View the query text of the selected query in the Query Editor window
- View top resource consuming queries in a grid format
- View top resource consuming queries in a grid format with additional details
- View top resource consuming queries in a chart form

When you move the mouse cursor over the Last Refreshed Time button, the last refreshed time will be shown. By clicking the second button, Track the Selected Query in a New Tracked Queries Window, a corresponding window will be opened, and you can select one of the queries represented by bubbles and track it. The third button, View the Query Text of the Selected Query in the Query Editor Window, shows the SELECT statement associated with the selected query. (The SELECT statement displays the state of the query from the corresponding catalog view.)

Each of the last three buttons orders all existing queries according to how they consume existing resources and displays them in descending order in a grid or chart in the left pane. The first two buttons display these queries in the grid format, while the last one displays them in the chart form. Figure 20-11 shows the left pane with the list of queries displayed in the grid format.

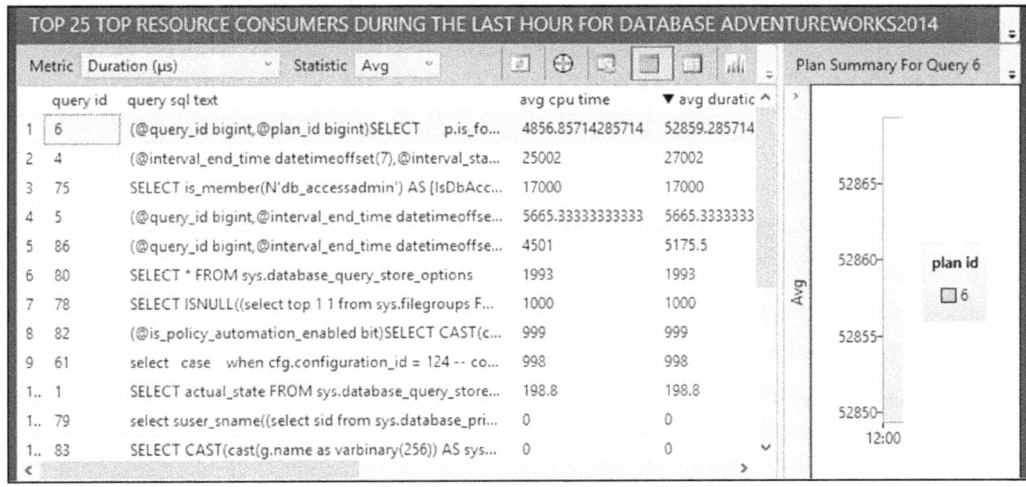

Figure 20-11   The left pane of the Top Resource Consuming Queries window

To force a plan, select a query and a corresponding plan, and then click Force Plan. (The Force Plan button is available below the Plan Summary pane.) You can only force plans that were saved and are still retained in the query plan cache. Next to the Force Plan button is the Unforce Plan button, which is used to unforce the corresponding plan.

Analyze Query Plans with Catalog Views

The Database Engine supports several catalog views in relation to Query Store. The following are the most important views:

- sys.query_store_query
- sys.query_store_query_text
- sys.query_store_plan
- sys.query_store_runtime_stats

The **sys.query_store_query** view contains the general information about the query. The most important columns are

- **query_id**   The unique ID number of the query
- **query_text_id**   The ID of the query text
- **object_id**   The ID of the database object to which the query belongs

The **sys.query_store_query_text** view contains the text of the corresponding SQL statement and the SQL handle of the query. The most important columns are

- **query_text_id**   The unique ID of the query text
- **query_sql_text**   SQL text of the query
- **statement_sql_handle**   The SQL handle of the particular query

The **sys.query_store_plan** view contains information about each execution plan associated with a query. Using this view, and joining it with other Query Store catalog views, you can track the history of all execution plans of a query. The most important columns are

- **plan_id**   The unique ID of the execution plan
- **query_id**   The ID of the query to which the plan belongs

The **sys.query_store_runtime_stats** view contains information about the run-time execution statistics information for the query. The most important columns are

- **runtime_stats_id**   The unique ID of the run-time statistics
- **plan_id**   The ID of the execution plan to which the statistics belong

With the help of the views described, you will see how some queries in relation to Query Store can be specified. For other queries that can be written using these and other views in relation to Query Store, please consult Microsoft Docs. Example 20.21 calculates the number of executions of a query. (The higher the number of query executions, the more important the design of such a query in relation to performance.)

Example 20.21
```
USE AdventureWorks;
SELECT q.query_id, qt.query_text_id, qt.query_sql_text,
    SUM(rs.count_executions) AS execution_count
FROM sys.query_store_query_text AS qt
JOIN sys.query_store_query AS q  ON qt.query_text_id = q.query_text_id
JOIN sys.query_store_plan AS p ON q.query_id = p.query_id
JOIN sys.query_store_runtime_stats AS rs ON p.plan_id = rs.plan_id
GROUP BY q.query_id, qt.query_text_id, qt.query_sql_text
ORDER BY execution_count DESC;
```

The four catalog views discussed previously are joined in Example 20.21 to calculate the number of executions. To display the calculation for each query, you must group the information using the values of the columns **query_id** and **query_text_id** as well as the text in the **query_sql_text** column. Finally, the **count_executions** column from the **sys.query_store_runtime_stats** catalog view gives you the number of executions of each query.

Example 20.22 shows how the number of queries (in this case, ten) with the longest average execution time can be calculated within a time period.

Example 20.22
```
USE AdventureWorks;
SELECT TOP 10 rs.avg_duration, qt.query_sql_text, q.query_id,
    qt.query_text_id, p.plan_id, GETUTCDATE() AS curr_time,
    rs.last_execution_time
FROM sys.query_store_query_text AS qt
JOIN sys.query_store_query AS q
    ON qt.query_text_id = q.query_text_id
```

```
JOIN sys.query_store_plan AS p
    ON q.query_id = p.query_id
JOIN sys.query_store_runtime_stats AS rs
    ON p.plan_id = rs.plan_id
WHERE rs.last_execution_time > DATEADD(HOUR, -1, GETUTCDATE())
ORDER BY rs.avg_duration DESC;
```

In Example 20.22, the four catalog views are again joined together. The TOP *n* clause in the SELECT list specifies that only top ten queries should be displayed. The DATEADD system function in the WHERE clause is used to specify the time period (in this case HOUR). The column **last_execution_time** of the **sys.query_store_runtime_stats** catalog view displays the time at which the query was last executed.

Automatic Tuning

As you already know, the query optimizer of the Database Engine is used to generate the query execution plan based on the existing statistics. The optimizer generates and evaluates multiple execution plans based on the calculation done for the least amount of resources. The most important factors negatively impacting selection of good execution plans are erroneous indexing and out-of-date statistics.

The previous section introduced Query Store, a feature that captures the compile-time and run-time statistics of statements being executed. With the data being collected, you can find out, among other things, whether the execution plans have regressed over time. With this information, you can identify a previous execution plan that performs better and *manually* force any future query executions to utilize the forced execution plan.

Starting with SQL Server 2017, the features of Query Store are taken one step further and the Database Engine can retrieve the information captured in Query Store and then make decisions *automatically*, forcing the most recent good execution plan. Therefore, the Database Engine now supports two kinds of correction:

- Manual plan tuning
- Automatic plan tuning

The following sections discuss both forms in detail so that you can see how automatic plan tuning is a big improvement over manual plan tuning.

Manual Plan Tuning

Without automatic plan tuning, administrators must periodically monitor the system and look for queries that have regressed. If any plan has regressed, the administrator should find some previous good plan and force it to be used instead of the current one using the **sp_query_store_force_plan** system stored procedure. The best practice would be to force the most recent known good plan because older plans might be invalid due to statistic or index changes.

The administrator who forces the last known good plan should monitor performance of the query that is executed using the forced plan and verify that the forced plan works as expected. Depending on the results of monitoring and analysis, the plan should be forced or the administrator should find some other way to optimize the query.

The following series of examples use Microsoft's new sample database called **WideWorldImporters** to show how you can manually correct execution plan regression.

Example 20.23 loads the **Order** and **OrderLine** tables of the **Sales** schema from the **WideWorldImporters** database into the **sample** database.

Example 20.23

```
-- Copy Sales.OrderLines and Sales.Orders in the sample DB
USE sample;
SELECT * INTO salesorderLines
    FROM WideWorldImporters.Sales.OrderLines;
SELECT * INTO salesorders
    FROM WideWorldImporters.Sales.Orders;
```

Next, Example 20.24 enables and configures Query Store.

Example 20.24

```
USE master;
GO
-- Enable Query Store
ALTER DATABASE sample SET QUERY_STORE = ON;
GO
-- Configure Query Store
ALTER DATABASE sample SET QUERY_STORE
  (OPERATION_MODE = READ_WRITE,
  DATA_FLUSH_INTERVAL_SECONDS = 600,
  MAX_STORAGE_SIZE_MB = 500,
  INTERVAL_LENGTH_MINUTES = 30);
```

The first ALTER DATABASE statement in Example 20.24 enables Query Store, while the second one sets several configuration options for this tool. (For the detailed description of the options in Example 20.24, see Example 20.20.)

In Example 20.25, the DBCC command clears the content of the procedure cache, while the ALTER DATABASE statement drops all persistently stored execution plans in Query Store. That way, all existing execution plans stored in the cache and in Query Store will be dropped.

Example 20.25

```
USE sample;
GO
-- Clear procedure cache
DBCC FREEPROCCACHE;
GO
-- Clear Query Store
ALTER DATABASE CURRENT SET QUERY_STORE CLEAR ALL;
```

All previous Transact-SQL statements in this subsection have been used to set the stage for execution of queries that will show plan change regression. Example 20.26 starts the SQL workload by executing the **sp_executesql** system procedure 600 times.

Example 20.26

```
EXEC sp_executesql  N'SELECT SUM(UnitPrice*Quantity)
    FROM SalesOrderLines sl
        JOIN salesorders so ON sl.OrderID=so.OrderID
    WHERE PackageTypeID = @typeID', N'@typeID int', @typeID = 7;
GO 600
```

The WHERE clause of the SELECT statement in Example 20.26 introduces a new variable called **@typeID** and specifies the data type of that variable (INT) as well as its value (7). This value then is assigned to the **PackageTypeID** column. Therefore, Example 20.26 uses the **sp_executesql** system stored procedure to execute 600 times the following statement:

```
SELECT SUM(UnitPrice*Quantity) FROM SalesOrderLines sl
  JOIN salesorders so ON sl.OrderID=so.OrderID
WHERE PackageTypeID = 7;
```

NOTE The reason to embed the query inside the **sp_executesql** procedure, instead of executing it directly, is that the particular plan created on first execution will be reused by subsequent executions of the same query.

Examples 20.27 and 20.28 again clear the procedure cache and start the **sp_executesql** system procedure twice, with different values for the **PackageTypeID** column.

Example 20.27

```
USE sample;
DBCC FREEPROCCACHE;
EXEC sp_executesql N'SELECT SUM(UnitPrice*Quantity)
FROM SalesOrderLines sl JOIN salesorders so ON sl.OrderID=SO.OrderID
    WHERE PackageTypeID = @typeID', N'@typeID INT', @typeID = 0;
```

Example 20.28

```
EXEC sp_executesql N'SELECT SUM(UnitPrice*Quantity)
FROM SalesOrderLines sl JOIN salesorders so ON sl.OrderID=so.OrderID
    where PackageTypeID = @typeID', N'@typeID INT', @typeID = 7;
GO 20
```

Now, Example 20.29 uses the **sys.dm_db_tuning_recommendations** DMV to see whether there is plan regression during the execution of the queries. (This DMV was introduced in SQL Server 2017.)

Example 20.29

```
USE sample;
SELECT reason, score,
    script=JSON_VALUE(details, '$.implementationDetails.script')
  FROM sys.dm_db_tuning_recommendations
  CROSS APPLY OPENJSON (Details, '$.planForceDetails')
    WITH ([query_id] int '$.queryId',
            regressedPlanId int '$.regressedPlanId',
            recommendedPlanId int '$.recommendedPlanId',
            regressedPlanErrorCount int,
            recommendedPlanErrorCount int,
            regressedPlanExecutionCount int,
            regressedPlanCpuTimeAverage float,
            recommendedPlanExecutionCount int,
            recommendedPlanCpuTimeAverage float
        ) AS planForceDetails;
```

The result is

```
reason                                                          score
Average query CPU time changed from 101.11ms to 627.7ms         84
script
-------------------------------------------------------------------------
exec sp_query_store_force_plan @query_id = 1, @plan_id = 1
```

As you can see, the three columns of the **sys.dm_db_tuning_recommendations** view are **reason**, **score**, and **script**. The **reason** column explains why this recommendation is being provided. The **score** column shows the estimated value for this recommendation on a scale from 0 (lowest) to 100 (highest). The **script** column displays the Transact-SQL script that should be executed to force the recommended plan. (Often, the execution of the DMV will not have any result, meaning that the system could not make any recommendations for the execution of the particular query.)

In Example 20.29, the system found that the execution time of the last query is approximately six times slower than the one stored in Query Store (the **reason** column). Therefore, we have the typical case of execution plan regression. Additionally, the probability that this is true is high (**score** = 84 percent). Finally, the system identifies the statement that should be executed to force the right execution plan (the **script** column).

Copying and executing the statement from the **script** column allows you to force the execution of the right plan.

NOTE Manually forced plans should not be forced forever. The database administrator should periodically test the performance of the forced plan. If the plan is not optimal enough, she can unforce the plan using the **sp_query_store_unforce_plan** system procedure and let the Database Engine find the optimal plan.

Automatic Plan Tuning

Automatic plan tuning is a database feature that identifies potential query performance problems, identifies recommend solutions, and *automatically* fixes identified problems. More precisely, it enables the Database Engine to identify and fix performance issues caused by query execution plan choice regressions. You can enable automatic tuning per database and specify that the most recent good plan should be forced whenever some plan change regression is detected. The ALTER DATABASE statement in Example 20.30 enables automatic tuning.

NOTE To automate the manual process, you have to repeat the execution of Examples 20.23 through 20.28, with one modification. Example 20.30 must be executed after Example 20.25.

Example 20.30

```
USE master;
ALTER DATABASE sample
     SET AUTOMATIC_TUNING (FORCE_LAST_GOOD_PLAN = ON );
```

Once you enable automatic tuning, the Database Engine automatically forces any recommendation where the estimated CPU gain is higher than 10 seconds, or the number of errors in the new plan is higher than the number of errors in the recommended plan. Additionally, the system will verify that the forced plan is better than the current one.

To show that the system automatically forces the obtained information, Example 20.31 again uses the **sys.dm_db_tuning_recommendations** DMV, this time with different columns in the SELECT list.

Example 20.31

```
USE sample;
SELECT name, state
FROM sys.dm_db_tuning_recommendations
CROSS APPLY OPENJSON (Details, '$.planForceDetails')
     WITH ([query_id] int '$.queryId',
             regressedPlanId int '$.regressedPlanId',
             recommendedPlanId int '$.recommendedPlanId',
             regressedPlanErrorCount int,
             recommendedPlanErrorCount int,
             regressedPlanExecutionCount int,
             regressedPlanCpuTimeAverage float,
             recommendedPlanExecutionCount int,
             recommendedPlanCpuTimeAverage float
          ) AS planForceDetails;
```

The result is

```
name    state
-----------------------------------------------------------------
PR_1    {"currentValue":"Success","reason":"PlanForcedByUser"}
```

The SELECT list in Example 20.31 uses only the columns that are related to the automatic tuning option: **name** and **state**. (The **reason**, **score**, and **script** columns have the same output as in Example 20.29.) The **name** column specifies the unique name of the recommendation (PR_1 in Example 20.31). The **state** column describes the state of the recommendation. The data type of this column is a JSON document. The document has two name-value pairs. The name of the first name-value pair, **currentValue**, describes the current state of the recommendation. The value "Success" means that the recommendation is successfully applied. The name of the second name-value pair, **reason**, is a string that describes why the recommendation is in the current state. (See Chapter 29 for the detailed description of JSON documents.)

Editing Information Concerning Automatic Tuning

There are two dynamic management views related to the automatic tuning option:

- sys.dm_db_tuning_recommendations
- sys.database_automatic_tuning_options

As you already know, the **sys.dm_db_tuning_recommendations** DMV returns detailed information about tuning recommendations. The discussion of Examples 20.29 and 20.31 already explained the most important columns of the view. While most of the columns of this view are JSON documents, you need the OPENJSON functions to retrieve their values. The OPENJSON function presents JSON documents as relational data. That way, the JSON values can be "joined" with values in relational columns, as has been done in Examples 20.29 and 20.31. (The OPENJSON function is described in detail in Chapter 29.)

The **sys.database_automatic_tuning_options** view returns the automatic tuning options for the current database. The most important columns are **name**, **actual_state**, **desired_state**, and **reason**. The **name** column specifies the name of the automatic tuning option (see the output of Example 20.32). The **actual_state** column indicates the operation mode of the automatic tuning. The **desired_state** column indicates the desired operation mode for the automatic tuning option, explicitly set by user. The possible values for the **actual_state** and **desired_state** columns are 0 (OFF) and 1 (ON). The **reason** column indicates why actual and desired states are different. Example 20.32 shows the use of this view.

Example 20.32

```
USE sample;
SELECT name, actual_state, desired_state, reason
   FROM sys.database_automatic_tuning_options;
```

If you execute this statement after the statement in Example 20.30, the result will be

```
name                    actual_state   desired_state   reason
---------------------------------------------------------------
FORCE_LAST_GOOD_PLAN    0              0               NULL
```

The result of Example 20.32 shows that the automatic tuning option of the system has forced the last good execution plan. Also, there is no difference between the operation mode of the system and the mode of the user. Therefore, the value of the **reason** column is NULL.

Performance Data Collector

Generally, it is very hard for DBAs to track down performance problems, primarily because DBAs usually are not there at the exact time a problem occurs, and thus they have to react to an existing problem by first tracking it down. Microsoft has included an entire infrastructure called Performance Data Collector to solve the problem.

Performance Data Collector is a component that is installed as a part of an instance of the Database Engine and can be configured to run either on a defined schedule or nonstop. The tool has three tasks:

- To collect different sets of data related to performance
- To store this data in the management data warehouse (MDW)
- To allow a user to view collected data using predefined reports

To use Performance Data Collector, you have to configure the MDW first. To do this using SQL Server Management Studio, expand the server, expand Management, right-click Data Collection, click Tasks, and select Configure Management Data Warehouse. The Configure Management Data Warehouse Wizard appears. The wizard has two tasks: to create the MDW and to set up data collection. After you complete these tasks, you can run Performance Data Collector and view the reports it generates.

Creating the MDW

After you click Next on the Welcome screen of the wizard, the Configure Data Warehouse Storage screen appears. Choose a server and database to host your management data warehouse and click Next. On the Map Login and Users screen, map existing logins and users to MDW roles. This activity has to be performed explicitly because no user is a member of an MDW role by default. Click Next when you are finished. On the Complete the Wizard screen, verify the configuration and click Finish.

Setting Up Data Collection

After setting up the MDW, you have to start data collection. Choose Tasks and then Configure Data Collection. Click Next. On the Setup Data Collection Sets screen (see Figure 20-12), specify the server name and the name of the data warehouse that you created in the prior section, and then specify where you want collected data to be cached locally before it is uploaded to the MDW. Click Next. On the Complete the Wizard screen, click Finish. The wizard finishes its work, giving you a summary of the executed tasks.

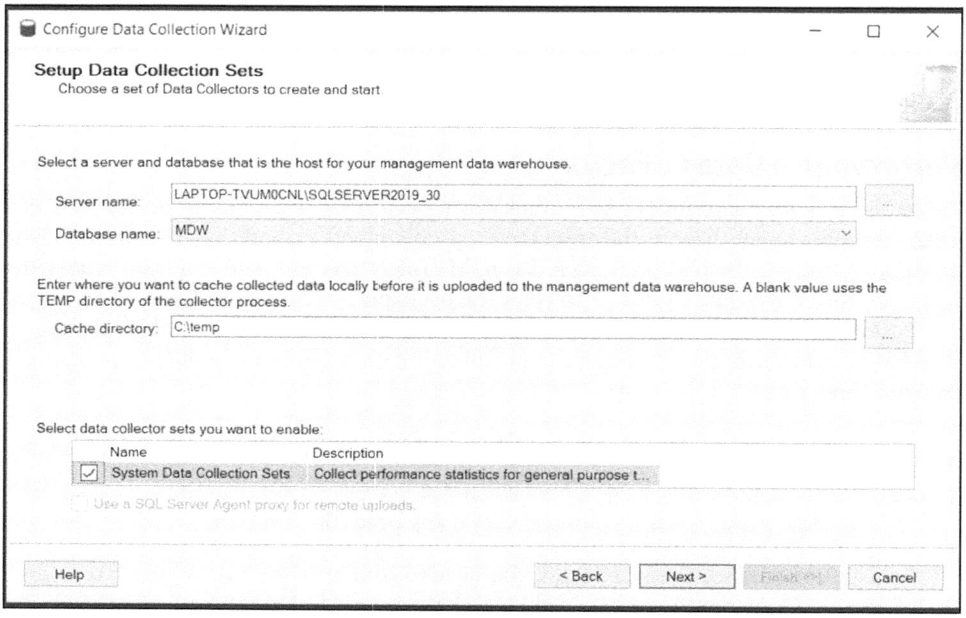

Figure 20-12   Setup Data Collection Sets wizard screen

Viewing Reports

Once Performance Data Collector is configured and active, the system will start collecting performance information and uploading the data to the MDW. Also, three new reports (Server Activity History, Disk Usage Summary, and Query Statistics History) will be created for viewing data collected by Performance Data Collector. To open the reports, right-click Data Collection, click Reports, and choose one of these reports under Management Data Warehouse.

The first report, Server Activity History, displays performance statistics for system resources described in this chapter. The second report, Disk Usage Summary, displays the starting size and average daily growth of data and log files. The last report, Query Statistics History, displays query execution statistics.

Resource Governor

One of the biggest problems in relation to performance tuning is trying to manage resources with the competing workloads on a shared database server. You can solve this problem using either server virtualization or several instances. In both cases, it is not possible for an instance to ascertain whether the other instances (or virtual machines) are using memory and the CPU. Resource Governor manages such a situation by enabling one instance to reserve a portion of a system resource for a particular process.

Generally, Resource Governor enables DBAs to define resource limits and priorities for different workloads. That way, consistent performance for processes can be achieved.

Resource Governor has two main components:

- Workload groups
- Resource pools

When a process connects to the Database Engine, it is classified and then assigned to a workload group based on that classification. (The classification is done using either a built-in classifier or a user-defined function.) One or more workload groups are then assigned to specific resource pools (see Figure 20-13).

As you can see in Figure 20-13, there are two different workload groups:

- Internal group
- Default group

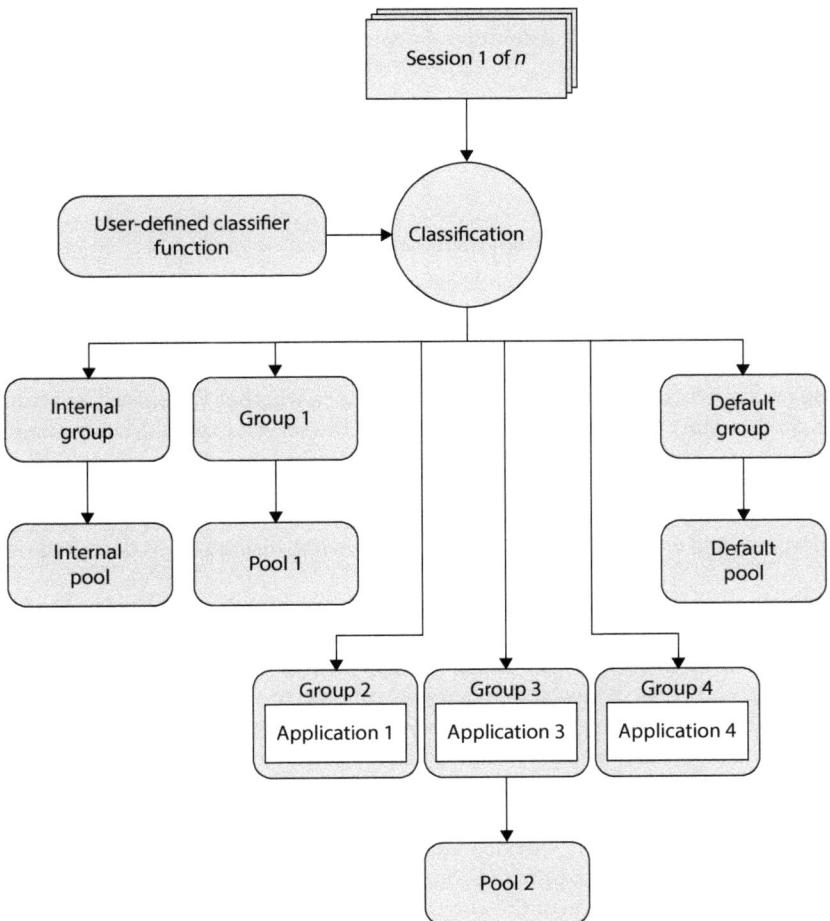

Figure 20-13   Architecture of Resource Governor

The internal group is used to execute certain system functions, while the default group is used when the process doesn't have a defined classification. (You cannot modify the classification for the internal group. However, monitoring of the workload of the internal group is possible.)

NOTE The internal and default groups are predefined workload groups. In addition to them, the tool allows the specification of 18 additional (user-defined) workload groups.

A resource pool represents the allocation of system resources of the Database Engine. Each resource pool has two different parts, which specify the minimum and maximum resource reservation. While minimum allocations of all pool resources cannot overlap, the sum of them cannot exceed 100 percent of all system resources. On the other hand, the maximum value of a resource pool can be set between its minimal value and 100 percent.

Analogous to workload groups, there are two predefined resource pools: the internal pool and the default pool. The internal pool contains the system resources, which are used by the internal processes of the system. The default pool contains both the default workload group and user-defined groups.

Creation of Workload and Resource Groups

The following steps are necessary to create workload and resource groups:

1. Create resource pools.
2. Create workload groups and assign them to pools.
3. For each workload group, define and register the corresponding classification function.

Resource Governor can be managed using SQL Server Management Studio or Transact-SQL. I will show you how you can manage it using Management Studio.

Before you create new resource pools, you have to check whether Resource Governor is enabled. To do this, start Management Studio, expand the server, expand Management, right-click Resource Governor, and click Enable. (Alternatively, you can use the ALTER RESOURCE GOVERNOR T-SQL statement.)

NOTE Resource pools and workload groups can be created in one dialog box, as described next.

To create a new resource pool, in Management Studio, expand the instance, expand Management, expand Resource Governor, right-click Resource Pools, and click New Resource Pool. The Resource Governor Properties dialog box appears (see Figure 20-14).

Generally, when you create a new resource pool, you have to specify its name and the minimum and maximum boundaries for CPU and memory. Therefore, in the Resource Pools table of the Resource Governor Properties dialog box, click the first column of the first empty row and type the name of the new pool. After that, specify the minimum and maximum values for CPU and memory.

You can specify a corresponding workload group in the same dialog box (see Figure 20-14). In the Workload Groups for Resource Pool table, double-click the empty cell in the Name column and type the name of the corresponding group. Optionally, you can specify several different properties for that workload group. Click OK to exit the dialog box.

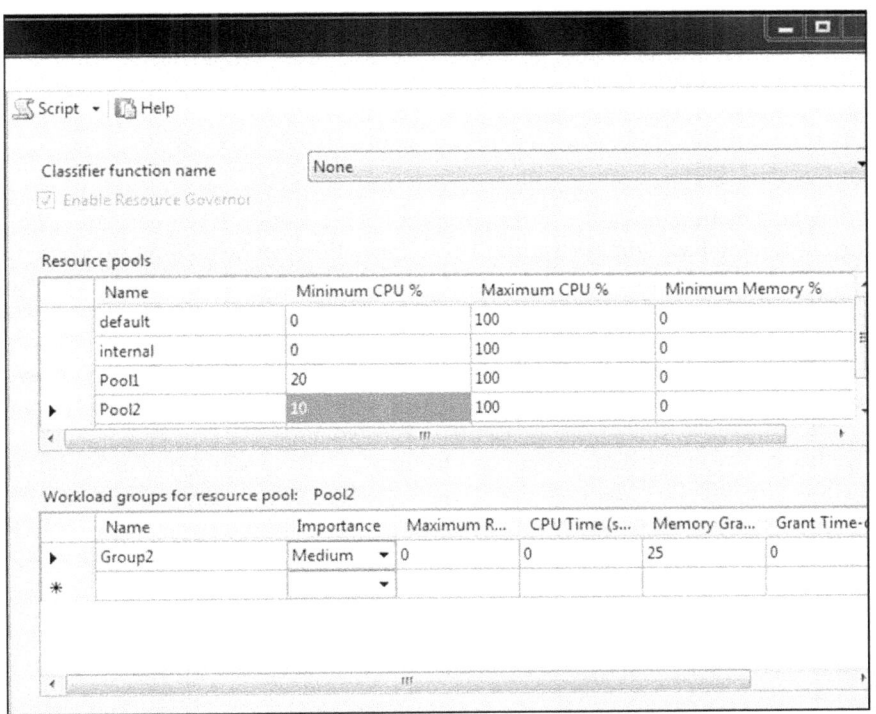

Figure 20-14   Resource Governor Properties dialog box

NOTE To create a new resource pool using T-SQL, use the CREATE RESOURCE POOL statement. To create a new workload group, use the CREATE WORKLOAD GROUP statement.

After you specify the new pool and its corresponding workload group, you have to create a classification function. This function is a user-defined function that is used to create any association between a workload group and users. (An example of such a function can be found in the description of the ALTER RESOURCE GOVERNOR statement in Microsoft Docs.)

Monitoring Configuration of Resource Governor
The following two DMVs can be used to monitor workload groups and resource pools:

- sys.dm_resource_governor_workload_groups
- sys.dm_resource_governor_resource_pools

The **sys.dm_resource_governor_workload_groups** view displays the information concerning workload groups. The **total_query_optimization_count** column of this view displays the cumulative count of query optimizations in this workload group; if the value is too high, it may indicate memory pressure.

The **sys.dm_resource_governor_resource_pools** view displays the information concerning resource pools. The **total_cpu_usage_ms** and **used_memory_kb** columns specify the total usage of CPU and the used memory, respectively, and indicate how your resource pools consume these two system resources.

Live Query Statistics

Accurate real-time estimation of the overall progress of a Transact-SQL query execution is very important for database administrators. This information can help DBAs to decide whether a long-running, resource-intensive query should be terminated or allowed to run to completion. Live Query Statistics allows the real-time monitoring of various execution metrics, including CPU memory usage, execution time, and progress of query execution.

Since SQL Server 2016, estimation of progress for *individual operators* in a query execution plan has been supported. This feature allows DBAs to rapidly identify issues with operators that require significantly more time or resources than expected.

This feature can be enabled in several ways using SQL Server Management Studio. First, you can click the Include Live Query Statistics icon, located in the toolbar of Management Studio. Second, you can launch it from Object Explorer to monitor queries executed by others by right-clicking the Server node, selecting Activity Monitor, and clicking the Active Expensive Queries option; all currently executing queries are listed. Figure 20-15 shows how you can monitor several parameters in relation to a given query.

NOTE The use of Live Query Statistics is demonstrated in Chapter 28 (see Example 28.3).

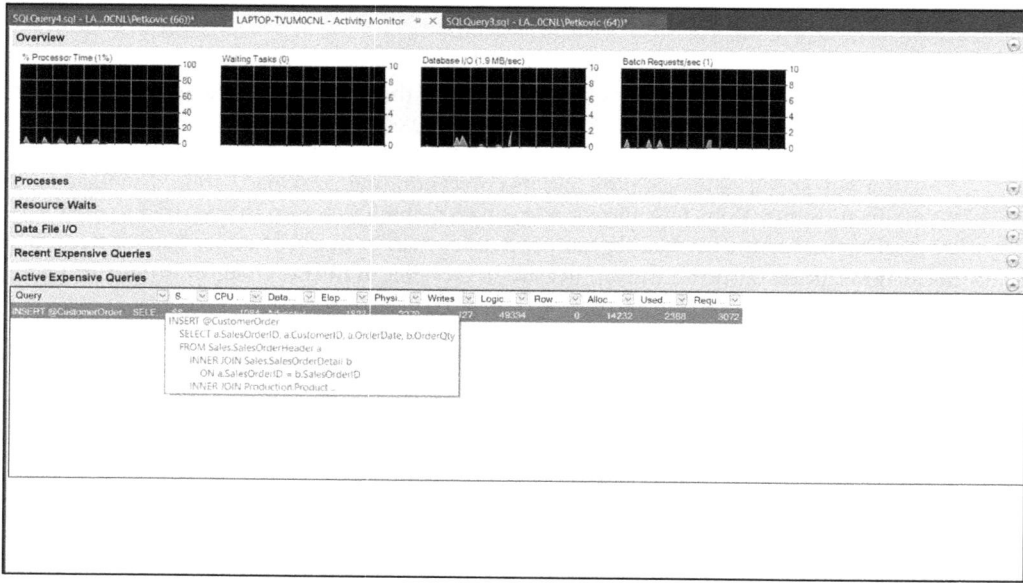

Figure 20-15   Activity Monitor

Summary

Performance issues can be divided into proactive and reactive response areas. Proactive issues concern all activities that affect performance of the overall system and that will affect future systems of an organization. Proper database design and proper choice of the form of Transact-SQL statements in application programs belong to the proactive response area. Reactive performance issues concern activities that are undertaken after the performance bottleneck occurs. The Database Engine offers a variety of tools (graphical components, Transact-SQL statements, and stored procedures) that can be used to view and trace performance problems of your system.

Of all components, Performance Monitor and dynamic management views are the best tools for monitoring because you can use them to track, display, report, and trace any performance bottlenecks.

The next chapter describes in-memory OLTP, a feature that enables you to achieve significant performance improvements.

Exercises

E.20.1   Discuss the differences between SQL Server Profiler and the Database Engine Tuning Advisor.

E.20.2   Discuss the differences between Performance Data Collector and Resource Governor.

E.20.3   Display information about an event session called **sessions1** with all its events, actions, and targets.

Part III

In-Memory OLTP

In This Chapter

- Memory-Optimized Tables
- Row and Index Storage
- In-Memory OLTP and Concurrency Control
- Accessing the Content of Memory-Optimized Tables
- Editing Information Concerning In-Memory Objects
- Tools for In-Memory OLTP

As previously discussed in Chapter 15, the Database Engine uses disk-oriented storage structures to store tables' rows. This means that all rows are stored on pages and are brought into memory as needed. (A page is a unit of storage that is moved from a disk to memory and back.) When a page is brought into memory, it must be locked before it can be accessed by an application that requests data on that page. Such locks may require thousands of instructions even when the required page is already in memory.

Another performance drawback of disk-oriented storage arises when disk reads and writes are performed by the system. These operations are two of the highest-cost operations that a database system performs. Therefore, if we succeed in eliminating disk reads and writes as well as page locking, significant performance gains can be achieved.

In-Memory OLTP (online transactional processing) is a relatively new component of the Database Engine that is optimized for OLTP queries. This means that a user can achieve significant performance improvements when database objects, such as tables and stored procedures, are stored in memory instead of on disk. These database objects are called memory-optimized tables and compiled stored procedures.

This chapter describes several features of In-Memory OLTP. First, it describes the creation of memory-optimized tables and their storage in main memory. Memory-optimized tables can be accessed in two different ways: either through interpreted T-SQL or through natively compiled stored procedures. The subsequent section explains both ways. After that, the way the content of memory-optimized tables is logged will be explained in detail. Finally, several tools in relation to In-Memory OLTP are covered.

Memory-Optimized Tables

SQL Server supports two forms of tables: disk-based tables and memory-optimized tables. Disk-based tables are stored on pages, and are read from and written to disk. These are traditional tables that the Database Engine has always supported.

Memory-optimized tables are stored completely in memory. This means that pages do not need to be read from disk when these tables are accessed, because the content of such tables is always stored in memory. (As you will see shortly, these tables do not even use pages as storage units.) Two other properties of these tables also improve performance significantly. First, delays caused by waiting for locks to be released are eliminated by the use of multiversion optimistic concurrency control. Second, the amount of data written to the transaction log is reduced to a minimum, using new algorithms.

Before we explore memory-optimized tables further, we'll review the similar feature supported in previous versions of SQL Server, *pinned tables*.

Pinned Tables as Predecessor of Memory-Optimized Tables

Earlier versions of SQL Server allow users to "pin" certain database objects into memory. This means that the system prevents such objects from being removed from the buffer cache, allowing fast access to the pages that must be read from the disk into memory. The objects that could be pinned are user-defined tables and indices.

The DBCC PINTABLE command is used to pin a database object. For instance, if a table is pinned, the DBCC command does not load pages from the table into memory. Instead, all tables' pages that are read into memory stay there and cannot be flushed by the Database Engine when it needs memory for new pages. However, if a pinned page is modified, the Database Engine logs the updated page and, if necessary, writes the modified page to disk.

As you can see from the description of pinned tables, these tables are just a special type of disk-based tables, and therefore require the same amount of locks and logging as any other table. As will be discussed later in the chapter, memory-optimized tables have a very different design and use different data storage structures. Also, no locking is used on memory-optimized tables.

Creating a Memory-Optimized Filegroup

To create memory-optimized tables, you must first create a memory-optimized filegroup. This filegroup stores, among other things, the checkpoint files that are used by the Database Engine in case of recovery, and these files have a different form than traditional checkpoint files. The memory-optimized filegroup holds one or more containers. Each container contains data files or delta files, or both. (Delta files will be explained in detail later in this chapter.) Example 21.1 shows creation of the new database called **sample2** with the **sample2_fg** memory-optimized filegroup.

Example 21.1
```
CREATE DATABASE sample2
ON PRIMARY
(NAME = sample2_data,
  FILENAME = 'C:\temp\sample2_data.mdf', size=500MB),
  FILEGROUP sample2_fg CONTAINS MEMORY_OPTIMIZED_DATA
(NAME = sample2_dir,
  FILENAME = 'C:\temp\sample2_dir')
LOG ON
(NAME = sample2_log,
  FILENAME='C:\temp\sample2_log.ldf', size=500MB);
```

The CREATE DATABASE statement in Example 21.1 first creates a new database called **sample2** and, as a part of that database, the **sample2_fg** filegroup, which is memory-optimized. Note that you can create only one memory-optimized filegroup per database. As can be seen in the example, such a filegroup must be specified with the CONTAINS MEMORY_OPTIMIZED_DATA clause.

It is possible to add a memory-optimized filegroup to an existing database, as shown in Example 21.2.

NOTE While attempting to add a memory-optimized filegroup to the **sample** database, you might get the following error message: "A database cannot be enabled for both Change Data Capture (CDC) and MEMORY_OPTIMIZED_DATA storage." In that case, disable CDC before you continue.

Example 21.2
```
ALTER DATABASE sample
ADD FILEGROUP sample_fg CONTAINS MEMORY_OPTIMIZED_DATA;
GO
ALTER DATABASE sample
ADD FILE (NAME = sample_dir, FILENAME = 'C:\temp\sample_file')
  TO FILEGROUP sample_fg;
```

Adding a new memory-optimized filegroup to an existing database in Example 21.2 is accomplished in two steps. In the first step, the ALTER DATABASE statement is used to add a filegroup to the database. In the second step, a new file is added to the memory-optimized filegroup. (If you do not have the directory C:\temp on your computer, please change the path to point to the directory where the file called **sample_file** will be stored.)

Creating Memory-Optimized Tables

Next, Example 21.3 creates a new memory-optimized table that will be stored in the **sample2_fg** filegroup.

Example 21.3
```
USE sample2;
CREATE TABLE employee_mem
( emp_no int not null, emp_lname CHAR(20) NOT NULL ,
  emp_fname CHAR(20) NOT NULL, dept_no CHAR(4) NOT NULL,
  PRIMARY KEY NONCLUSTERED HASH (emp_no)
    WITH (BUCKET_COUNT = 1000))
    WITH (MEMORY_OPTIMIZED = ON, DURABILITY = SCHEMA_AND_DATA);
```

The **employee_mem** table is similar to the **employee** table of the **sample** database that we have used up to this point in the book. The only difference is that the content of the new table will be stored in memory instead of on disk.

The CREATE TABLE statement has been extended with several clauses to allow the creation of memory-optimized tables. We will start with the final clause, DURABILITY. This clause specifies which part of a table is memory-optimized. There are two possible options for this clause:

- SCHEMA_AND_DATA
- SCHEMA_ONLY

The SCHEMA_AND_DATA option is the default value and indicates that the content of a memory-optimized table is persistently stored in memory. In other words, this option provides durability of both schema and data. If SCHEMA_ONLY is specified, the content of the table is volatile. This means that the table's schema is persistently stored, but any data updates are not durable upon a restart of the database with memory-optimized objects.

The MEMORY_OPTIMIZED clause indicates whether the table is memory-optimized (the value ON) or nonoptimized (the value OFF). The SCHEMA_ONLY option of the DURABILITY clause is not allowed with MEMORY_OPTIMIZED = OFF.

Memory-optimized tables have a few limitations. First, several data types, such as all LOB data types and XML and MAX data types, cannot be used to specify any column of a memory-optimized table. Also, the row length is limited to 8060 bytes.

All other options of the CREATE TABLE statement will be explained in "Index Storage" within the next section.

The ALTER TABLE statement can be used for the following tasks: changing the bucket count; adding and removing an index; changing, adding; and removing a column; and adding and removing a constraint.

Row and Index Storage

The content of a memory-optimized table is stored differently than the content of traditional tables. As you already know from Chapter 15, traditional tables are disk-based tables, meaning that their data and corresponding indices are stored on physical pages. On the other hand, memory-optimized tables are designed according to byte-addressable memory.

NOTE *Byte-addressable* means that, given an "address," this address refers to a single block of 8 bits. The most important consequence is that a byte becomes a working unit. In other words, a byte of memory will be read (or written) with a single read (write) call.

The following subsections give just a high-level overview of row storage of memory-optimized tables. For a detailed description, please refer to Microsoft Docs and white papers concerning these tables.

Row Storage

Rows of a memory-optimized table are connected using the tables' indices, and the index provides the structure for such tables. That is a reason why a memory-optimized table must have at least an index created on it.

Also, the structure of a row of a memory-optimized table is significantly different than the structure of a row of a disk-based table. Each row of a memory-optimized table consists of a header and the corresponding content. The header contains several fields, of which we will explore the two most important ones: Begin Timestamp (Begin-TS) and End Timestamp (End-TS). The value of Begin-TS is the timestamp of the transaction that inserted the row, and the End-TS value is the timestamp for the transaction that deleted the row. That way, there is always information for each row regarding when it was inserted and when it was deleted. If a row is not (yet) deleted, the special value called "infinity" is used for the End-TS value, so the system can differentiate between current and deleted rows. (All deleted rows have the End-TS value, which is different from infinity.)

Index Storage

As you already know, each memory-optimized table must have at least one index. There are two types of indices supported for memory-optimized tables:

- Hash index
- Nonclustered index

The following two subsections describe these index types.

Hash Index

A hash index consists of a collection of buckets organized in an array. Besides buckets, each hash index has a hash function, which determines the storage place for each index entry. In other words, a hash function is applied to each index key, and the result of the function determines the bucket where the particular row will be stored. All key values that hash to the same value are linked together in a chain, and can be accessed with the same pointer in the hash index. Also, when a row is inserted in an indexed table, the hash function will be applied to the index key value in that row and the value of the function again determines the corresponding bucket.

To explain hash indices, let's take another look at the code from Example 21.3, where such an index is created:

```
USE sample2;
CREATE TABLE employee_mem
( emp_no int not null, emp_lname CHAR(20) NOT NULL ,
  emp_fname CHAR(20) NOT NULL, dept_no CHAR(4) NOT NULL,
  PRIMARY KEY NONCLUSTERED HASH (emp_no)
    WITH (BUCKET_COUNT = 1000))
    WITH (MEMORY_OPTIMIZED = ON, DURABILITY = SCHEMA_AND_DATA);
```

First, you can create an index for a memory-optimized table only by using the CREATE TABLE statement. In other words, Transact-SQL does not support the CREATE INDEX statement for memory-optimized tables. The index created in Example 21.3 is a primary hash index, and it is created at table level for the **emp_no** column. (For a discussion about table-level and column-level constraints, see Examples 5.7 and 5.8 in Chapter 5.)

When a hash index is created, you must specify a number of buckets using the BUCKET_COUNT option. It is recommended to choose a number of buckets equal to or greater than the expected number of unique values of the index key column. That way you increase the probability that each bucket will have only rows with a single value in its chain. The number you supply is rounded up to the next power of two, so the value used in Example 21.3 (1000) will be rounded up to 1024.

NOTE A hash table uses a hash function to compute an index into an array of buckets or slots, from which the desired value can be found. This index is called hash index.

Memory-Optimized Nonclustered Index

The general structure of a memory-optimized nonclustered index is similar to the traditional B-tree data structure, except that index pages are not a fixed sized, and once they are built they are unchangeable. Like a regular B-tree page, each index page contains a set of ordered key values, and for each value there is a corresponding pointer. Example 21.4 shows the creation of a nonclustered index.

NOTE Hash indices and memory-optimized nonclustered indices exist only in memory. The index structure is created when the memory-optimized table is created in memory, both during CREATE TABLE and during database startup. All index operations are not logged in the transaction log.

Example 21.4

```
CREATE TABLE employee_mem2
( emp_no int not null, emp_lname CHAR(20) NOT NULL ,
  emp_fname CHAR(20) NOT NULL, dept_no CHAR(4) NOT NULL,
  PRIMARY KEY NONCLUSTERED  (emp_no)  )
    WITH (MEMORY_OPTIMIZED = ON, DURABILITY = SCHEMA_AND_DATA);
```

Again, a nonclustered index can be created only using the CREATE TABLE statement. The creation of a nonclustered index differs from the creation of a hash index in two ways. First, the NONCLUSTERED option is used instead of NONCLUSTERED HASH. Second, the number of buffers is not specified, because buffers exist only in relation to hash indices.

NOTE If most of your queries are range queries, the use of a nonclustered index is recommended.

Example 21.5 creates a table with two indices: one hash index and one nonclustered index.

Example 21.5

```
USE sample2;
CREATE TABLE employee_mem4
(emp_no int not null PRIMARY KEY NONCLUSTERED HASH WITH (BUCKET_COUNT=1000),
 emp_lname CHAR(20) NOT NULL ,
  emp_fname CHAR(20) NOT NULL, dept_no CHAR(4) NOT NULL,
  INDEX i_lname NONCLUSTERED  (emp_lname))
    WITH (MEMORY_OPTIMIZED = ON, DURABILITY = SCHEMA_AND_DATA);
```

The index on the **emp_no** column is a hash index, while the **i_lname** index is a nonclustered index on the **emp_lname** column. (The former is defined at the column level, while the latter is defined at the table level.)

There are two significant differences between a memory-optimized nonclustered index and the B+ tree structure of the Database Engine. First, a page pointer of the memory-optimized nonclustered index contains a logical page identifier (PID), while a pointer of the B+ tree is always a physical page number. A PID indicates a position in a corresponding mapping table, and it connects each index entry with a physical memory address. Index pages of a memory-optimized nonclustered index are never updated; instead, they are replaced with a new page and the mapping table is updated so that the same PID indicates a new physical memory address.

The second significant difference between memory-optimized nonclustered indices and B-trees is that at the leaf level, modifications caused by DML operations on memory-optimized tables with a nonclustered index are stored in a set of delta values, which together build a delta file. (*Delta value* means that only the modified part of a row is stored, while the unchanged part exists in the original row.) In other words, each update to a logical page, which can be an insert or delete of a key value on that page, produces a logical page containing a delta record indicating the change that was made. An update is represented by two new delta records, one for the delete and one for the insert of the new value.

In-Memory OLTP and Concurrency Control

As you already know from Chapter 13, transactions and concurrency are related topics because concurrency problems can be solved using transactions. (This is true only for so-called pessimistic concurrency models.) For this reason, in the context of In-Memory OLTP and concurrency control, we will handle both database concepts in one section, starting with logging.

Logging Memory-Optimized Objects

Transaction logs are used to ensure that effects of all committed transactions can be recovered after a failure. In-Memory OLTP achieves this by writing all logging processes to persistent storage. (The same is true for checkpoint processes, but checkpoint processing of memory-optimized tables is beyond the scope of this book and thus will not be described.)

Operations on memory-optimized tables use the same transaction log that is used for operations on disk-based tables, and as always, the transaction log is stored on disk. In case of a system failure, the rows of data in memory-optimized tables can be re-created from the log. The information written to disk consists of transaction log streams. The log contains required information about committed transactions to redo the transactions. The changes are recorded as inserts and deletes of row versions marked with the table identifier they belong to. Index operations on memory-optimized tables are not logged. In other words, all indices are completely rebuilt on recovery.

The important property of transaction logging of memory-optimized tables is that this process generates significantly less data than the same process on disk-based tables. There are several reasons for this. First, log records are generated only at commit time. Second, data generated by dirty reads (i.e., uncommitted changes) are never written to disk. Finally, the system tries to group multiple log records into one large log record of up to 24KB. That way, fewer log records are written to the log.

NOTE Logging is one of only a few operations on memory-optimized database objects where the system writes to disk. Therefore, it is important to minimize the number of such operations. Creating a single log record for the entire transaction and reducing the size of the logged information reduces significantly the number of write operations.

Optimistic Multiversion Concurrency Control

When accessing memory-optimized tables, the Database Engine implements an optimistic multiversion concurrency control. (Transaction conflicts are uncommon in most applications that access data in memory-optimized tables. For this reason, the optimistic approach, which has less overhead than the pessimistic one, is used.) This means that rows in memory-optimized tables can have different versions. Transactions that access memory-optimized tables use this row versioning to obtain a transactional consistent snapshot of rows in the tables. Data read by any statement in the transaction will be the consistent version of the data that existed at the time the transaction started. Therefore, any modifications made by concurrently running transactions are not visible to statements in the current transaction, and concurrent transactions access *potentially different* versions of the same row. (The system can obtain this information using the Begin-TS and End-TS values from the header of each row.)

Transactions on memory-optimized tables assume that there are no conflicts with concurrent transactions. (This is a general property of all optimistic concurrency models.) Transactions do not take locks on memory-optimized tables to guarantee isolation of transactions. Therefore, write operations do not block any read or other write operations. The Database Engine detects conflicts between concurrent transactions, as well as isolation violations. In that case, one of the conflicting transactions is aborted and must be restarted.

Accessing the Content of Memory-Optimized Tables

Memory-optimized tables can be accessed in two different ways: either through interpreted T-SQL or through natively compiled stored procedures. The following two subsections describe both access types.

Interpreted Transact-SQL

When you use interpreted T-SQL, you have access to all statements when working with your memory-optimized tables, but the performance of their execution will be worse than in the case of using compiled stored procedures. Interpreted T-SQL (also called "interop") should be used primarily when you need to access both memory-optimized and disk-based tables.

Two T-SQL statements that are not supported when accessing memory-optimized tables using interpreted SQL are TRUNCATE TABLE and MERGE (when a memory-optimized table is the target). There are also a few other limitations, which can be found in Microsoft Docs.

Compiled Stored Procedures

Generally, the Database Engine uses an interpreter to execute stored procedures. (An *interpreter* is a computer program that directly executes single instructions written in a programming or scripting language, without previously compiling them into a machine code.) On the other hand, compilation (or native compilation) generally means that programming instructions are converted to machine code. After that, such code can be executed directly by the CPU, without the need for further compilation or interpretation.

Particularly, the In-Memory OLTP compiler takes high-level constructs, such as SELECT or INSERT, and compiles them into machine code. The inputs of the In-Memory OLTP compiler are memory-optimized objects, which are first converted into C code, and after that the C compiler generates the machine code. The compilation of memory-optimized objects results in dynamic-link libraries (DLLs) that are loaded in memory and linked into the Database Engine process.

Example 21.6 shows the creation of the **project_mem** memory-optimized table. (It has the same structure as the **project** table from the **sample** database.)

Example 21.6

```
USE sample2;
CREATE TABLE project_mem (project_no INT NOT NULL PRIMARY KEY
                NONCLUSTERED HASH WITH (BUCKET_COUNT = 1024),
            project_name  CHAR (20), budget DEC(8,2))
            WITH (MEMORY_OPTIMIZED = ON, DURABILITY = SCHEMA_AND_DATA);
```

The **project_mem** memory-optimized table is referenced by the compiled stored procedure created in Example 21.7.

Example 21.7

```
USE sample2;
GO
CREATE PROCEDURE dbo.increase_budget_mem (@percent INT=10)
WITH NATIVE_COMPILATION, SCHEMABINDING, EXECUTE AS OWNER
AS  BEGIN ATOMIC WITH
    (TRANSACTION ISOLATION LEVEL = SNAPSHOT,LANGUAGE = N'us_english')
      UPDATE dbo.project_mem
                        SET budget = budget + budget * @percent/100
  END
```

The **increase_budget_mem** stored procedure has the same logic as the **increase_budget** procedure, which is used in Chapter 8 (see Example 8.6). The main difference is that the former will be natively compiled, while the latter is interpreted.

The NATIVE_COMPILATION option in the WITH clause specifies that the stored procedure will be compiled. Together with this option you have to specify two other options: SCHEMABINDING and EXECUTE AS. SCHEMABINDING declares that all tables referenced by the compiled stored procedure cannot be dropped without first dropping the stored procedure. The EXECUTE AS option is used to specify the security context under which to execute the stored procedure after it is accessed. By specifying the context in which the procedure is executed, you can control which user account the Database Engine uses to validate permissions on objects referenced by the procedure. In Example 21.7 we use the value OWNER to specify the owner context. (For more information about the EXECUTE AS option, see Chapter 8.)

Additionally, a natively compiled stored procedure needs to consist of one atomic block of code, which means that all statements in the code block are contained within a single transaction. In the ATOMIC WITH clause you need to specify the transaction isolation level and language.

Editing Information Concerning In-Memory Objects

Several interfaces that allow you to display metadata are enhanced in the new version of the Database Engine to provide information concerning memory-optimized tables. The following subsections describe different possibilities to edit metadata of memory-optimized objects.

Property Functions

The OBJECTPROPERTY function is extended with the **TableIsMemoryOptimized** property. Example 21.8 shows the use of this property.

Example 21.8

```
SELECT OBJECTPROPERTY(OBJECT_ID('sample2.dbo.employee_mem'),
                      'TableIsMemoryOptimized');
```

Example 21.8 uses the OBJECT_ID function to get the ID of the **employee_mem** table, because the first parameter of the OBJECTPROPERTY function must be the ID of the table. The result of Example 21.8 is 1, meaning that the **TableIsMemoryOptimized** property is true for the **employee_mem** table.

Catalog Views and System Stored Procedures

Several catalog views allow you to access information related to memory-optimized tables and their indices. I will discuss three of them: **sys.tables**, **sys.hash_indexes**, and **sys.sql_modules**.

The **sys.tables** view has three additional columns in relation to memory-optimized tables: **durability**, **durability_desc**, and **is_memory_optimized**. The **durability** column can contain either of two values, 1 or 0, depending on whether or not the DURABILITY clause is specified in the CREATE (ALTER) TABLE statement. The **durability_desc** column can contain either of two values, SCHEMA_AND_DATA or SCHEMA_ONLY, and these values correspond to the values of the DURABILITY clause in the CREATE (ALTER) TABLE statement. The **is_memory_optimized** column also has either of two values: 1 or 0. The value 1 is displayed if a table is memory-optimized.

Example 21.9 shows the use of the **sys.tables** view.

Example 21.9

```
SELECT object_id ,is_memory_optimized, durability, durability_desc
  FROM sys.tables WHERE name = 'employee_mem';
```

The result is

object_id	is_memory_optimized	durability	durability_desc
597577167	1	0	SCHEMA_AND_DATA

The **sys.hash_indexes** view shows the current hash indices and the hash index properties. This view contains the same columns as the **sys.indexes** view and an additional column named **bucket_count**. For more information concerning the **sys.indexes** view, see Chapter 10.

The most important catalog view in relation to compiled stored procedures is **sys.sql_modules**. This view is extended with the **uses_native_compilation** column, which has the BIT data type. The value true (1) tells you that the module is a compiled stored procedure.

Since SQL Server 2017, the **sp_spaceused** system stored procedure has been extended to display the disk space reserved and used for databases with at least one MEMORY_OPTIMIZED filegroup. Example 21.10 shows **sp_spaceused** in action.

Example 21.10

```
--Displaying space usage for a DB with MEMORY_OPTIMIZED filegroup
sp_spaceused
                @updateusage = 'FALSE',
                @mode = 'ALL',
                @oneresultset = '1',
                @include_total_xtp_storage = '1'
```

The result is

db_name	db_size	unallocated_space	reserved	data	index_size	unused	xtp_precreated	xtp_pending_trunc
sample2	1105.00MB	600.30MB	2760KB	992KB	1240KB	528KB	105472KB	2048KB

The two last parameters—**@oneresultset** and **@include_total_xtp_storage**—are relevant for the databases with memory-optimized filegroups and memory-optimized tables. The former indicates whether to return a single result set or not. When **@oneresultset** = 1, the parameter **@include_total_xtp_storage** determines whether the single result set includes columns for MEMORY_OPTIMIZED_DATA storage. These columns are **xtp_precreated** and **xtp_pending_trunc** (see the result of Example 21.10).

Dynamic Management Views

Generally, dynamic management views allow you to access metadata information that changes dynamically. While the content of memory changes steadily, DMVs are the right tool to get the necessary information about the storage of memory-optimized tables and compiled stored procedures.

This section describes several DMVs introduced by Microsoft to display information about memory-optimized database objects:

- sys.dm_db_xtp_table_memory_stats
- sys.dm_db_xtp_object_stats
- sys.dm_db_xtp_hash_index_stats

The **sys.dm_db_xtp_table_memory_stats** view returns memory usage statistics for each memory-optimized table (user and system) in the current database. The system tables have negative object IDs and are used to store run-time information for In-Memory OLTP. Unlike user objects, system tables are internal and only exist in memory; therefore, they are not visible through catalog views. System tables are used to store information such as metadata for all data files and delta files in storage, merge requests, dropped tables, and relevant information for recovery and backups. The In-Memory OLTP component can have up to 8192 data and delta file pairs; for large in-memory databases, the memory taken by system tables can be a few megabytes. Example 21.11 shows the use of this view.

Example 21.11
```
SELECT OBJECT_NAME(object_id) table_name,
       memory_allocated_for_table_kb, memory_allocated_for_indexes_kb
FROM sys.dm_db_xtp_table_memory_stats;
```

The result is

table_name	memory_allocated_for_table_kb	memory_allocated_for_indexes_kb
employee_mem	0	8
employee_mem2	0	128
employee_mem4	0	136

The **sys.dm_db_xtp_object_stats** view reports the number of rows affected by operations on each of the In-Memory OLTP objects since the last database restart. Statistics are updated when the operation executes. This view can help you to identify which memory-optimized tables are changing the most. You may decide to remove unused or rarely used indices on the table, as each index affects performance. If there are hash indices, you should periodically reevaluate the bucket count. You can do so with the **sys.dm_db_xtp_hash_index_stats** view.

Tools for In-Memory OLTP

There are two groups of tools that you can use in relation to In-Memory OLTP:

- Memory management tools
- Migration tools

After introducing memory management for In-Memory OLTP, we will discuss both groups of tools.

Memory Management for In-Memory OLTP: Overview

When running In-Memory OLTP, the Database Engine must be configured with sufficient memory for memory-optimized tables. Insufficient memory will cause transactions to fail at run time during any operations that require additional memory. This can happen when you execute any DML statement. The In-Memory OLTP memory manager is fully integrated with the Database Engine memory manager and can react to memory pressure by cleaning up old row versions.

A rule of thumb to predict the amount of memory you'll need for your memory-optimized tables is to have twice the amount of memory that your data will take up. Also, the total memory requirement depends on your workload; if there are a lot of data modifications due to OLTP operations, you'll need more memory for the row versions. If you're using mostly queries (i.e., read operations), less memory might be required.

For planning space requirements for indices, the calculation for the hash is easy. Each bucket requires 8 bytes, so you can just compute the number of buckets times 8 bytes. The calculation of the size of your memory-optimized nonclustered indices is difficult, because it depends on the size of the index key as well as the number of tables' rows.

Memory Management Tools

The following tools can be used to manage memory of memory-optimized objects:

- SQL Server Management Studio
- Resource Governor

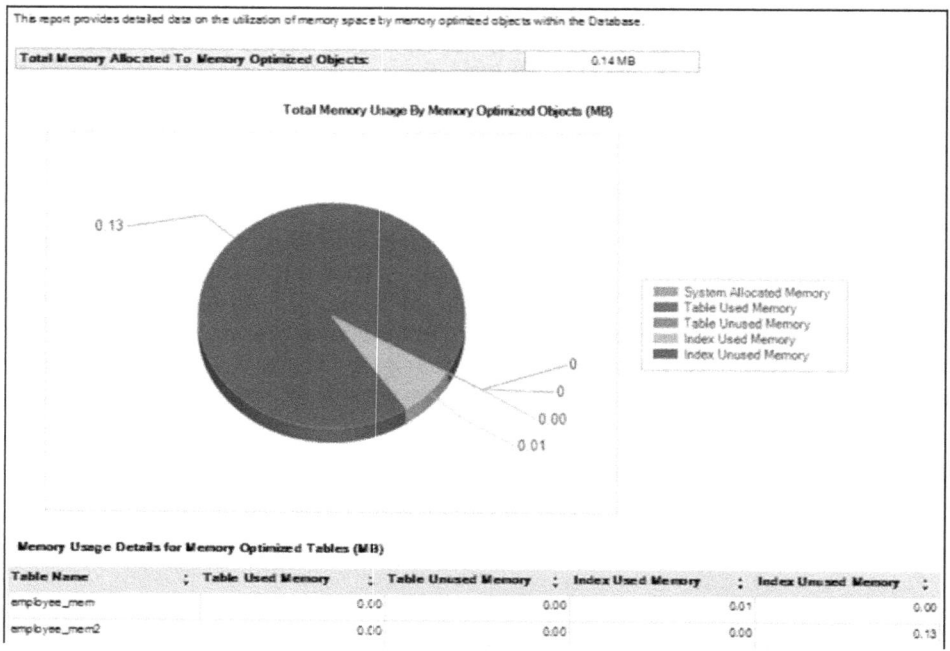

This report provides detailed data on the utilization of memory space by memory optimized objects within the Database.

Total Memory Allocated To Memory Optimized Objects:	0.14 MB

Total Memory Usage By Memory Optimized Objects (MB)

- System Allocated Memory
- Table Used Memory
- Table Unused Memory
- Index Used Memory
- Index Unused Memory

0.13   0   0   0.00   0.01

Memory Usage Details for Memory Optimized Tables (MB)

Table Name	Table Used Memory	Table Unused Memory	Index Used Memory	Index Unused Memory
employee_mem	0.00	0.00	0.01	0.00
employee_mem2	0.00	0.00	0.00	0.13

Figure 21-1   Memory usage report

Memory Usage Report Using SQL Server Management Studio

SQL Server Management Studio can be used to get a report of the current memory used by memory-optimized tables and their indices. To create such a report, right-click the name of the database containing memory-optimized tables and select Reports | Standard Reports | Memory Usage by Memory Optimized Objects. Management Studio displays a report similar to the one shown in Figure 21-1.

The report in Figure 21-1 shows the space used by the table rows and the indices, as well as the amount of space used by the system. Allocation of memory for hash and nonclustered indices is significantly different: hash indices will have memory allocated for the declared number of buckets as soon as they're created, and the corresponding report will show memory usage for those indices before any rows are inserted. For nonclustered indices, memory will not be allocated until rows are added, and the memory requirement will depend on the size of the index keys and the number of rows.

Managing Memory with Resource Governor

As you already know from Chapter 20, Resource Governor is a tool that allows database administrators to specify resource limits for different workloads. Therefore, you can use this tool to assign memory to a resource pool to which a database with memory-optimized tables is bound. After that, all memory-optimized tables in that database can maximally use that amount of memory. In other words, if the storage of all memory-optimized tables needs more memory than is available, some of them will be stored on the disk. The upper limit, which can be assigned to the pool, is 80 percent of the whole memory.

The whole process can be done in two steps, which are given in Examples 21.12 and 21.13.

Example 21.12

```
CREATE RESOURCE POOL Pool_mem WITH (MAX_MEMORY_PERCENT=50);
ALTER RESOURCE GOVERNOR RECONFIGURE;
```

Example 21.12 creates a memory pool for the **sample2** database using the MAX_ MEMORY_PERCENT clause. This specifies the percentage of the memory that may be allocated to memory-optimized tables in a database associated with the pool.

The second step, shown in Example 21.13, is to bind the databases that you want to manage to the respective pool. You can do this with the stored procedure **sp_xtp_bind_db_resource_ pool**. Note that one pool may contain many databases, but a database is only associated with one pool at any point in time.

Example 21.13

```
EXEC sp_xtp_bind_db_resource_pool 'sample2', 'Pool_mem';
```

Migration Tools for In-Memory OLTP

As you already know, the goal of In-Memory OLTP is to load in memory the tables and stored procedures that are crucial for performance. The Database Engine provides three tools that you can use to identify traditional tables and stored procedures that will provide the best performance gain after migration to In-Memory OLTP:

- Transaction Performance Analysis report
- Memory Optimization Advisor
- Native Compilation Advisor

The Transaction Performance Analysis report collects data concerning tables and stored procedures and analyzes their workloads. Reports based on this information give you recommendations for the best migration candidates among them.

After you identify a disk-based table to port to In-Memory OLTP, you can use the Memory Optimization Advisor to help you to migrate the table. In other words, the Advisor guides you through the process of migration.

The task of the Native Compilation Advisor is to help you to port a stored procedure to the corresponding natively compiled procedure.

NOTE This section discusses only the first component, Transaction Performance Analysis reports. After that, the use of the Memory Optimization Advisor and Native Compilation Advisor is straightforward.

Transaction Performance Analysis reports are installed as a part of SQL Server Management Studio. To generate such reports, right-click the database and select Reports | Standard Reports | Transaction Performance Analysis Overview. Choose Tables Analysis. (The database needs to have an active workload, or a recent run of a workload, in order to generate a meaningful analysis report.)

This report provides the details of your table's performance statistics over the period of time you monitored the instance with Transaction Performance Collection Set. This report includes the access characteristics of your queries on the table, and the detailed contention statistics including information on latches and locks.

Table Name	% of total waits	Page latch wait count	Latch Statistics Average wait time per latch wait (ms)	Page lock count	Lock Statistics Page lock wait count	Average wait time per lock wait (ms)
Person.Person	0.00	0	0	15441	0	0

Table Name	Index Name	% of total accesses	Total Singleton Lookup	Total Range Scans	Recommended In-Memory Index Type
Person.Person	PK_Person_BusinessEnti tyID	14.71	0	5	NONCLUSTERED
Person.Person	IX_Person_LastName_Fir stName_MiddleName	82.35	0	28	NONCLUSTERED
Person.Person	AK_Person_rowguid	2.94	0	1	NONCLUSTERED

Table Name	Number of Migration Blockers
Person.Person	20

Figure 21-2   The Recommended Tables Based on Usage report

Figure 21-2 shows the Recommended Tables Based on Usage report concerning the **Person .Person** table of the **AdventureWorks** database. (I executed several Transact-SQL statements to create an active workload for this table.) This report tells you generally which tables are the best candidates for migration to In-Memory OLTP based on their usage. On the left side, you can select how many tables you would like to choose from the given database. The chart shows the selected number of tables.

The horizontal axis shows the amount of work needed to transform a table to a memory-optimized one. The vertical axis shows the gains that can be achieved through the transformation. As you can see from Figure 21-3, migration of the **Person.Person** table gives

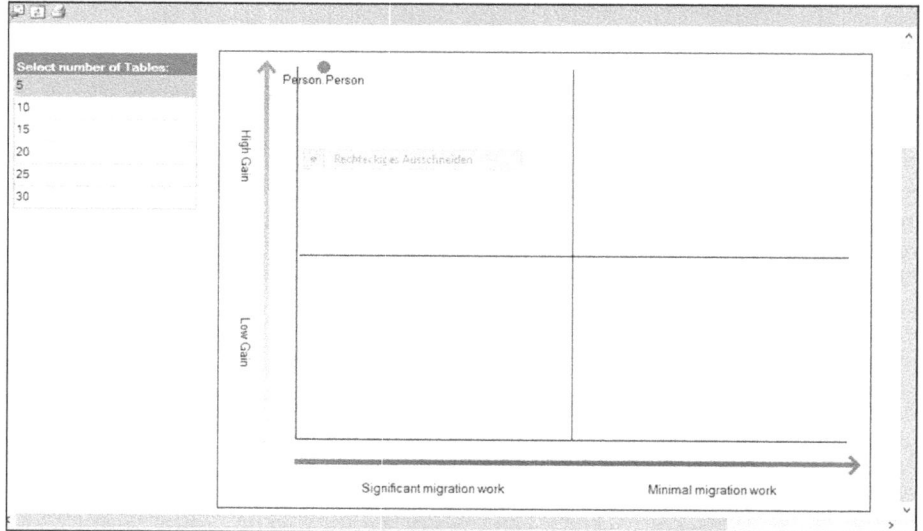

Figure 21-3 Performance statistics for the Person.Person table

very good performance gain, but requires significant work for its migration. Therefore, the best candidates for migration are those tables that appear in the top-right corner of the chart.

You can access a detailed report for a table by clicking its name in the chart. The report provides the overall access statistics for the table as well as contention statistics. (The contention statistics display information concerning locks and latches.)

Summary

In-Memory OLTP is a relatively new component of the Database Engine that allows you to achieve significant performance benefits for a group of OLTP queries storing a table's data in its entirety directly in memory. These database objects are called memory-optimized tables and compiled stored procedures.

The primary store for memory-optimized tables is main memory. Rows in the table are read from and written to memory. Natively compiled stored procedures are T-SQL stored procedures compiled to native code that access memory-optimized tables. Natively compiled stored procedures allow for efficient execution of the queries and business logic in the stored procedure.

Two groups of tools support In-Memory OLTP: memory management tools and migration tools. Memory management tools are usually used to configure memory for memory-optimized database objects. Migration tools are used to find out which traditional database tables and stored procedures will provide the best performance gain after migration to In-Memory OLTP. (The latter group of tools is usually used to find out which database objects from the previous versions are suitable to be stored entirely in main memory.)

Part III

PART

IV

SQL Server and Business Intelligence

22   Business Intelligence: An Introduction

In This Chapter

- Online Transaction Processing vs. Business Intelligence
- Data Warehouses and Data Marts
- Data Warehouse Design
- Cubes and Their Architectures
- Data Access

The goal of this chapter is to introduce you to an important area of database technology: business intelligence (BI). The first part of the chapter explains the difference between the online transaction processing world on one side and the BI world on the other side. A *data store* for a BI process can be either a data warehouse or a data mart. Both types of data store are discussed, and their differences are listed in the second part of the chapter. The design of data in BI and the need for creation of aggregate tables are explained at the end of the chapter.

Online Transaction Processing vs. Business Intelligence

From the beginning, relational database systems were used almost exclusively to capture primary business data, such as orders and invoices, using processing based on transactions. This focus on business data has its benefits and its disadvantages. One benefit is that the poor performance of early database systems improved dramatically, to the point that today many database systems can execute thousands of transactions per second (using appropriate hardware). On the other hand, the focus on transaction processing prevented people in the database business from seeing another natural application of database systems: using them to filter and analyze needed information out of all the existing data in an enterprise or department.

Online Transaction Processing

As already stated, performance is one of the main issues for systems that are based upon transaction processing. These systems are called online transaction processing (OLTP) systems. A typical example of an operation performed by an OLTP system is to process the withdrawal of money from a bank account using a teller machine. OLTP systems have some important properties, such as:

- Short transactions—that is, high throughput of data
- Many (possibly hundreds or thousands of) users
- Continuous read and write operations based on a small number of rows
- Data of medium size that is stored in a database

The performance of a database system will increase if transactions in the database application programs are short. The reason is that transactions use locks to prevent negative effects of concurrency issues. If transactions are long lasting, the number of locks and their duration for modification operations increases, decreasing the data availability for other transactions and thus their performance.

Large OLTP systems usually have many users working on the system simultaneously. A typical example is a reservation system for an airline company that must process thousands of requests for travel arrangements in a single country, or all over the world, almost immediately. In this type of system, most users expect that their response-time requirements will be fulfilled by the system and the system will be available during working hours (or 24 hours a day, seven days a week).

Users of an OLTP system execute their DML statements continuously—that is, they use both read and write operations at the same time and steadily. (Because data of an OLTP system is continuously modified, that data is highly dynamic.) All operations (or results of them) on a database usually include only a small amount of data, although it is possible that the database system must access many rows from one or more tables stored in the database.

In recent years, the amount of data stored in an *operational* database (that is, a database managed by an OLTP system) has increased steadily. Today, there are many databases that store several or even hundreds of gigabytes or petabytes of data. As you will see, this amount of data is still relatively small in relation to data warehouses.

Business Intelligence Systems

Business intelligence is the process of integrating enterprise-wide data into a single data store from which end users can run ad hoc queries and reports to analyze the existing data. In other words, the goal of BI is to keep data that can be accessed by users who make their business decisions on the basis of the analysis. These systems are often called *analytic* or *informative* systems because, by accessing data, users get the necessary information for making better business decisions.

The goals of BI systems are different from the goals of OLTP systems. The following is a query that is typical for a BI system: "What is the best-selling product category for each sales region in the third quarter of the year 2019?" Therefore, a BI system has very different properties from those listed for an OLTP system in the preceding section. The most important properties of a BI system are as follows:

- Periodic write operations (load) with queries based on a huge number of rows
- Small number of users
- Large size of data stored in a database

Other than loading data at regular intervals (usually daily), BI systems are mostly read-only systems. Therefore, the nature of the data in such a system is static. As will be explained in detail later in this chapter, data is gathered from different sources, cleaned (made consistent), and loaded into a database called a data warehouse (or data mart). The cleaned data is usually not modified—that is, users query data using SELECT statements to obtain the necessary information (and modification operations are very seldom).

Because BI systems are used to gain information, the number of users that simultaneously use such a system is relatively small in relation to the number of users that simultaneously use an OLTP system. Users of a BI system usually generate reports that display different factors concerning the finances of an enterprise, or they execute complex queries to compare data.

NOTE Another difference between OLTP and BI systems that actually affects the user's behavior is the daily schedule—that is, when those systems are available for use during a day. An OLTP system can be used nonstop (if it is designed for such a use), whereas a BI system can be used only as soon as data is made consistent and is loaded into the database.

In contrast to databases in OLTP systems that store only current data, BI systems also track historical data. (Remember that BI systems make comparisons between data gathered in different time periods.) For this reason, the amount of data stored in a data warehouse is large.

Data Warehouses and Data Marts

A *data warehouse* can be defined as a database that includes all corporate data and that can be uniformly accessed by users. That's the concise definition; explaining the notion of a data warehouse is much more involved. An enterprise usually has a large amount of data stored at different times and in different databases (or data files) that are managed by distinct DBMSs. These DBMSs need not be relational: some enterprises still have databases managed by hierarchical or network database systems. A special team of software specialists examines source databases (and data files) and converts them into a target store: the data warehouse. Additionally, the converted data in a data warehouse must be consolidated, because it holds the information that is the key to the corporation's operational processes. (*Consolidation* of data means that all equivalent queries executed upon a data warehouse at different times provide the same result.) The data consolidation in a data warehouse is provided in several steps:

- Data assembly from different sources (also called extraction)
- Data cleaning (in other words, transformation process)
- Quality assurance of data

Data must be carefully assembled from different sources. In this process, data is extracted from the sources, converted to an intermediate schema, and moved to a temporary work area. For data extraction, you need tools that extract exactly the data that must be stored in the data warehouse.

Data cleaning ensures the integrity of data that has to be stored in the target database. For example, data cleaning must be done on incorrect entries in data fields, such as addresses, or incompatible data types used to define the same date fields in different sources. For this process, the data cleaning team needs special software. An example will help explain the process of data cleaning more clearly. Suppose that there are two data sources that store personal data about employees and that both databases have the attribute **Gender**. In the first database, this attribute is defined as CHAR(6), and the data values are "female" and "male." The same attribute in the second database is declared as CHAR(1), with the values "f" and "m." The values of both data sources are correct, but for the target data source you must clean the data—that is, represent the values of the attribute in a uniform way.

The last part of data consolidation—quality assurance of data—involves a data validation process that specifies the data as the end user should view and access it. Because of this, end users should be closely involved in this process. When the process of data consolidation is finished, the data will be loaded in the data warehouse.

NOTE The whole process of data consolidation is called ETL (extraction, transformation, loading). Microsoft provides a component called SQL Server Integration Services (SSIS) to support users during the ETL process.

By their nature (as a store for the overall data of an enterprise), data warehouses contain huge amounts of data. (Some data warehouses contain dozens of terabytes or even petabytes of data.) Also, because they must encompass the enterprise, implementation usually takes a lot of time, which depends on the size of the enterprise. Because of these disadvantages, many companies start with a smaller solution called a data mart.

Data marts are data stores that include all data at the department level and therefore allow users to access data concerning only a single part of their organization. For example, the marketing department stores all data relevant to marketing in its own data mart, the research department puts the experimental data in the research data mart, and so on. Because of this, a data mart has several advantages over a data warehouse:

- Narrower application area
- Shorter development time and lower cost
- Easier data maintenance
- Bottom-up development

As already stated, a data mart includes only the information needed by one part of an organization, usually a department. Therefore, the data that is intended for use by such a small organizational unit can be more easily prepared for the end user's needs.

The average development time for a data warehouse is two years and the average cost is $5 million. On the other hand, costs for a data mart average $200,000, and such a project takes about three to five months. For these reasons, development of a data mart is preferred, especially if it is the first BI project in your organization.

The fact that a data mart contains significantly smaller amounts of data than a data warehouse helps you to reduce and simplify all tasks, such as data extraction, data cleaning, and quality assurance of data. It is also easier to design a solution for a department than to design one for the entire organization.

If you design and develop several data marts in your organization, it is possible to unite them all in one big data warehouse. This bottom-up process has several advantages over designing a data warehouse at once. First, each data mart may contain identical target tables that can be unified in a corresponding data warehouse. Second, some tasks are logically enterprise-wide, such as the gathering of financial information by the accounting department. If the existing data marts will be linked together to build a data warehouse for an enterprise, a global repository (that is, the data catalog that contains information about all data stored in sources and in the target database) is required.

NOTE Be aware that building a data warehouse by linking data marts can be very troublesome because of possible significant differences in the structure and design of existing data marts. Different parts of an enterprise may use different data models and have different instructions for data representation. For this reason, at the beginning of this bottom-up process, it is strongly recommended that you make a single view of all data that will be valid at the enterprise level; do not allow departments to design data separately.

Data Warehouse Design

Only a well-planned and well-designed database will allow you to achieve good performance. Relational databases and data warehouses have a lot of differences that require different design methods. Relational databases are designed using the well-known entity-relationship (ER) model, while the dimensional model is used for the design of data warehouses and data marts.

Using relational databases, data redundancy is removed using normal forms (see Chapter 1). Each step of the normalization process divides the particular table of a database that includes redundant data into two separate tables. The process of normalization should be finished when all tables of a database contain only nonredundant data.

The highly normalized tables are advantageous for OLTP because all transactions can be made as simple and short as possible. On the other hand, BI processes are based on queries that operate on a huge amount of data and are neither simple nor short. Therefore, the highly normalized tables do not suit the design of data warehouses, because the goal of BI systems is significantly different: there are few concurrent transactions, and each transaction accesses a very large number of records. (Imagine the huge amount of data belonging to a data warehouse that is stored in hundreds of tables. Most queries will join dozens of large tables to retrieve data. Such queries cannot be performed well, even if you use hardware with parallel processors and a database system with the best query optimizer.)

Data warehouses cannot use the ER model because this model is suited to design databases with nonredundant data. The logical model used to design data warehouses is called a *dimensional model*.

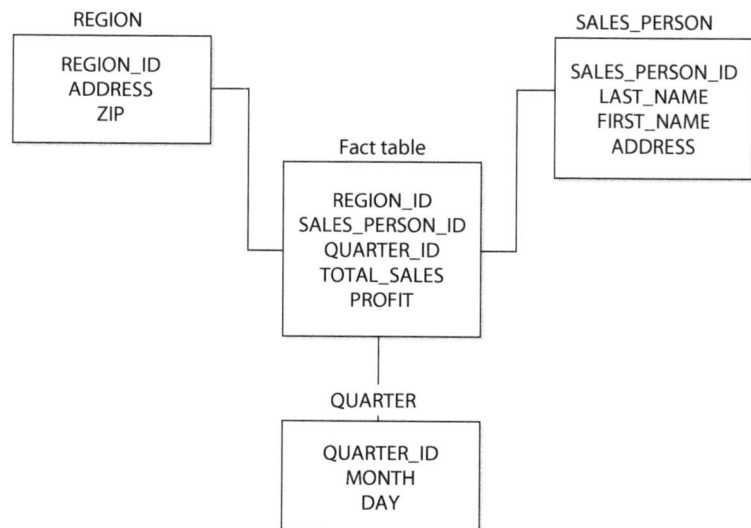

Figure 22-1 Example of the dimensional model: star schema

NOTE There is another important reason why the ER model is not suited to the design of data warehouses: the use of data in a data warehouse is unstructured. This means the queries are partly executed ad hoc, allowing a user to analyze data in totally different ways. (On the other hand, OLTP systems usually have database applications that are hard-coded and therefore contain queries that are not modified often.)

In dimensional modeling, every particular model is composed of one table that stores measures, called the *fact table*, and several other tables that describe dimensions, called *dimension tables*. Examples of data stored in a fact table include inventory sales and expenditures. Dimension tables usually include time, account, product, and employee data. Figure 22-1 shows an example of the dimensional model.

Each dimension table usually has a single-part primary key and several other attributes that describe this dimension closely. On the other hand, the primary key of the fact table is the combination of the primary keys of all dimension tables (see Figure 22-1). For this reason, the primary key of the fact table is made up of several foreign keys. (The number of dimensions also specifies the number of foreign keys in the fact table.) As you can see in Figure 22-1, the tables in a dimensional model build a star-like structure. Therefore, this model is often called *star schema*.

Another difference in the nature of data in a fact table and the corresponding dimension tables is that most nonkey columns in a fact table are numeric and additive, because such data can be used to execute necessary calculations. (Remember that a typical query on a data warehouse fetches thousands or even millions of rows at a time, and the only useful operation upon such a huge amount of rows is to apply an aggregate function, such as sum, maximum, or average.) For example, columns like **Units_of_product_sold**, **Total_sales**, **Dollars_cost**, or **Profit** are typical columns in the fact table. Such columns of the fact table are called *measures*.

Fact Table	Dimension Table
Usually one in a dimensional model.	Many (12–20).
Contains most rows of a data warehouse.	Contains relatively small amount of data.
Composite primary key (contains all primary keys of dimension tables).	One column of a table builds the primary key.
Nonkey columns are numeric and additive.	Columns are descriptive and therefore textual.

Table 22-1   Differences Between Fact Table and Dimension Tables

On the other hand, columns of dimension tables are strings that contain textual descriptions of the dimension. For instance, columns such as **Address**, **Location**, and **Name** often appear in dimension tables. (These columns are usually used as headers in reports.) Another consequence of the textual nature of columns of dimension tables and their use in queries is that each dimension table contains many more indices than the corresponding fact table. (A fact table usually has only one unique index composed of all columns belonging to the primary key of that table.) Table 22-1 summarizes the differences between the fact table and dimension tables.

NOTE   Sometimes it is necessary to have multiple fact tables in a data warehouse. If you have different sets of measures, each set has to be tied to a different fact table.

Columns of dimension tables are usually highly *denormalized*, which means that a lot of columns depend on each other. The denormalized structure of dimension tables has one important purpose: all columns of such a table are used as column headers in reports. If the denormalization of data in a dimension table is not desirable, a dimension table can be decomposed into several subtables. This is usually necessary when columns of a dimension table build hierarchies. (For example, the **product** dimension could have columns such as **Product_id**, **Category_id**, and **Subcategory_id** that build three hierarchies, with the primary key, **Product_id**, as the root.) This structure, in which each level of a base entity is represented by its own table, is called a *snowflake schema*. Figure 22-2 shows the snowflake schema of the **product** dimension.

The extension of a star schema into a corresponding snowflake schema has some benefits (reduction of used disk space, for example) and one main disadvantage: the snowflake schema requires more join operations to get information from lookup tables, which negatively impacts performance. For this reason, the performance of queries based on the snowflake schema is generally slow. Therefore, the design using the snowflake schema is recommended only in a few very specialized cases.

Cubes and Their Architectures

BI systems support different types of data storage. Some of these data storage types are based on a multidimensional database that is also called a cube. A *cube* is a subset of data from the data warehouse that can be organized into multidimensional structures. To define a cube, you

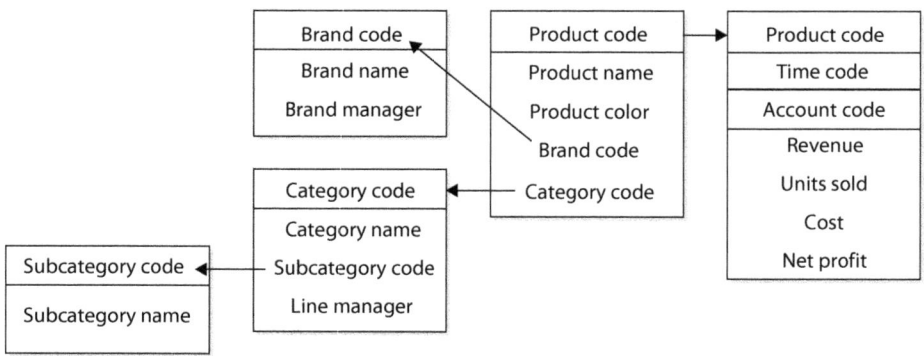

Figure 22-2   The snowflake schema of the product dimension

first select a fact table from the dimensional schema and identify numerical columns (measures) of interest within it. Then you select dimension tables that provide descriptions for the set of data to be analyzed. To demonstrate this, consider how the cube for car sales analysis might be defined. For example, the fact table may include the measures **Cars_sold**, **Total_sales**, and **Costs**, while the tables **Models**, **Quarters**, and **Regions** specify dimension tables. The cube in Figure 22-3 shows all three dimensions: **Models**, **Regions**, and **Quarters**.

In each dimension there are discrete values called *members*. For instance, the **Regions** dimension may contain the following members: ALL, North America, South America, and Europe. (The ALL member specifies the total of all members in a dimension.)

Additionally, each cube dimension can have a hierarchy of levels that allows users to ask questions at a more detailed level. For example, the **Regions** dimension can include the following level hierarchies: **Country**, **Province**, and **City**. Similarly, the **Quarters** dimension can include **Month**, **Week**, and **Day** as level hierarchies.

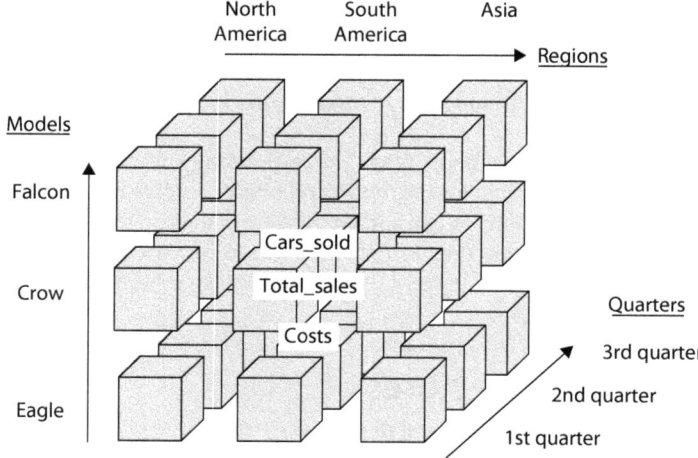

Figure 22-3   Cube with dimensions: Models, Quarters, and Regions

NOTE Cubes and multidimensional databases are managed by special systems called multidimensional database systems (MDBMSs). SQL Server's MDBMS is called Analysis Services, which is covered in Chapter 23.

The physical storage of a cube is described after the following discussion of aggregation.

Aggregation

Data is stored in the fact table in its most detailed form so that corresponding reports can make use of it. On the other hand (as stated earlier), a typical query on a fact table fetches thousands or even millions of rows at a time, and the only useful operation upon such a huge amount of rows is to apply an aggregate function (sum, maximum, or average). This different use of data can reduce performance of ad hoc queries if they are executed on low-level (atomic) data, because time- and resource-intensive calculations will be necessary to perform each aggregate function. For this reason, low-level data from the fact table should be summarized in advance and stored in intermediate tables. Because of their "aggregated" information, such tables are called *aggregate tables*, and the whole process is called *aggregation*.

NOTE An aggregate row from the fact table is always associated with one or more aggregate dimension table rows. For example, the dimensional model in Figure 22-1 could contain the following aggregate rows: monthly sales aggregates by salespersons by region and region-level aggregates by salespersons by day.

An example will show why low-level data should be aggregated. An end user may want to start an ad hoc query that displays the total sales of the organization for the last month. This would cause the server to sum all sales for each day in the last month. If an average of 500 sales transactions occur per day in each of 500 stores of the organization, and data is stored at the transaction level, this query would have to read 7,500,000 (500 × 500 × 30 days) rows and build the sum to return the result. Now consider what happens if the data is aggregated in a table that is created using monthly sales by store. In this case, the table will have only 500 rows (the monthly total for each of 500 stores), and the performance gain will be dramatic.

How Much to Aggregate?

Concerning aggregation, there are two extreme solutions: no aggregation at all, and exhaustive aggregation for every possible combination of queries that users will need. From the preceding discussion, it should be clear that no aggregation at all is out of the question, because of performance issues. (The data warehouse without any aggregation table probably cannot be used at all as a production data store.) The opposite solution is also not acceptable, for several reasons:

- Enormous amount of disk space needed to store additional data
- Overwhelming maintenance of aggregate tables
- Initial data load too long

Storing additional data that is aggregated at every possible level consumes an additional amount of disk space that increases the initial disk space by a factor of six or more (depending on the amount of the initial disk space and the number of queries that users will need). The creation of tables to hold the aggregates for all existing combinations is an overwhelming task for the system administrator. Finally, building aggregates at initial data load can have devastating results if this load already lasts for a long time and the additional time is not available.

From this discussion you can see that aggregate tables should be carefully planned and created. During the planning phase, keep these two main considerations in mind when determining what aggregates to create:

- Where is the data concentrated?
- Which aggregates would most improve performance?

The planning and creation of aggregate tables is dependent on the concentration of data in the columns of the base fact table. In a data warehouse, where there is no activity on a given day, the corresponding row is not stored at all. So if the system loads a large number of rows, as compared to the number of all rows that can be loaded, aggregating by that column of the base fact table improves performance enormously. In contrast, if the system loads few rows, as compared to the number of all rows that can be loaded, aggregating by that column is not efficient.

Here is another example to demonstrate the preceding discussion. For products in the grocery store, only a few of them (say, 15 percent) are actually sold on a given day. If we have a dimensional model with three dimensions, **Product**, **Store**, and **Time**, only 15 percent of the combination of the three corresponding primary keys for the particular day and for the particular store will be occupied. The daily product sales data will thus be *sparse*. In contrast, if all or many products in the grocery store are sold on a given day (because of a special promotion, for example), the daily product sales data will be *dense*.

To find out which dimensions are sparse and which are dense, you have to build rows from all possible combinations of tables and evaluate them. Usually, the **Time** dimension is dense, because there are always entries for each day. Given the dimensions **Product**, **Store**, and **Time**, the combination of the **Store** and **Time** dimensions is dense, because for each day there will certainly be data concerning selling in each store. On the other hand, the combination of the **Store** and **Product** dimensions is sparse (for the reasons previously discussed). In this case, the dimension **Product** is generally sparse, because its appearance in combination with other dimensions is sparse.

The choice of aggregates that would most improve performance depends on end users. Therefore, at the beginning of a BI project, you should interview end users to collect information on how data will be queried, how many rows will be retrieved by these queries, and other criteria.

Physical Storage of a Cube

Online analytical processing (OLAP) systems usually use one of the following three different architectures to store multidimensional data:

- Relational OLAP (ROLAP)
- Multidimensional OLAP (MOLAP)
- Hybrid OLAP (HOLAP)

Generally, these three architectures differ in the way in which they store leaf-level data and precomputed aggregates. (Leaf-level data is the finest grain of data that is defined in the cube's measure group. Therefore, the leaf-level data corresponds to the data of the cube's fact table.)

In ROLAP, the precomputed data isn't stored. Instead, queries access data from the relational database and its tables in order to bring back the data required to answer the question. MOLAP is a type of storage in which the leaf-level data and its aggregations are stored using a multidimensional cube.

Although the logical content of these two storage types is identical for the same data warehouse, and both ROLAP and MOLAP analytic tools are designed to allow analysis of data through the use of the dimensional data model, there are some significant differences between them. The advantages of the ROLAP storage type are as follows:

- Data is not duplicated.
- Materialized (that is, indexed) views can be used for aggregation.

If the data should also be stored in a multidimensional database, a certain amount of data must be duplicated. Therefore, the ROLAP storage type does not need additional storage to copy the leaf-level data. Also, the calculation of aggregation can be executed very quickly with ROLAP if the corresponding summary tables are generated using indexed views.

On the other hand, MOLAP also has several advantages over ROLAP:

- Aggregates are stored in a multidimensional form.
- Query response is generally faster.

Using MOLAP, many aggregates are precomputed and stored in a multidimensional cube. That way the system does not have to calculate the result of such an aggregate each time it is needed. In the case of MOLAP, the database engine and the database itself are usually optimized to work together, so the query response may be faster than in ROLAP.

HOLAP storage is a combination of the MOLAP and ROLAP storage types. Precomputed data is stored as in the case of the MOLAP storage, while the leaf-level data is left in the relational database. (Therefore, for queries using aggregation, HOLAP is identical to MOLAP.) The advantage of HOLAP storage is that the leaf-level data is not duplicated.

Data Access

Data in a data warehouse can be accessed using three general techniques:

- Reporting
- OLAP
- Data mining

Reporting is the simplest form of data access. A report is just a presentation of a query result in a tabular form. (Reporting is discussed in detail in Chapter 25.) With OLAP, you analyze data interactively; that is, it allows you to perform comparisons and calculations along any dimension in a data warehouse.

> **NOTE** Transact-SQL supports all standardized functions and constructs in relation to SQL/OLAP. This topic will be discussed in detail in Chapter 24.

Data mining is used to explore and analyze large quantities of data in order to discover significant patterns. This discovery is not the only task of data mining: using this technique, you must be able to turn the existing data into information and turn the information into action. In other words, it is not enough to analyze data; you have to apply the results of data mining meaningfully and take action upon the given results. (Data mining, as the most complex of the three techniques, will not be covered in this introductory book.)

Summary

At the beginning of a BI project, the main question is what to build: a data warehouse or a data mart. Probably the best answer is to start with one or more data marts that can later be united in a data warehouse. Most of the existing tools in the BI market support this alternative.

In contrast to operational databases that use ER models for their design, the design of data warehouses is best done using a dimensional model. These two models show significant differences. If you are already acquainted with the ER model, the best way to learn and use the dimensional model is to forget everything about the ER model and start modeling from scratch.

After this introductory discussion of general considerations about the BI process, the next chapter discusses the server part of Microsoft Analysis Services.

Exercises

E.22.1   Discuss the differences between OLTP and analytic systems.

E.22.2   Discuss the differences between the ER and dimensional models.

E.22.3   A data warehouse project starts with the ETL (extracting, transforming, loading) process. Explain the three subprocesses.

E.22.4   Discuss the differences between a fact table and corresponding dimension tables.

E.22.5   Discuss the benefits of the three storage types: MOLAP, ROLAP, and HOLAP.

E.22.6   Why is it necessary to aggregate data stored in a fact table?

SQL Server Analysis Services

In This Chapter

- Multidimensional Model
- Tabular Model
- Multidimensional Model vs. Tabular Model
- Analysis Services: Data Security

SQL Server Analysis Services (SSAS) is a group of services that is used to manage data stored in a data warehouse or data mart. The key features of SSAS are

- Ease of use
- Support of different architectures
- Support of different APIs

SSAS offer wizards for almost every task that is executed during the design and implementation of a data warehouse. For example, the Data Source Wizard allows you to specify one or more data sources, while the Cube Wizard is used to create a multidimensional cube in which to store aggregate data. In contrast to most other data warehouse systems, SSAS allows you to use the architecture that is most appropriate for your needs. You can choose between the three architectures (MOLAP, ROLAP, and HOLAP) discussed in detail in Chapter 22.

SSAS provides many different APIs that can be used to retrieve and deliver data. One of these is the OLE DB for OLAP interface that allows you to access SSAS cubes. Several APIs are described later in this chapter, in two sections: "Delivering Data from the Multidimensional Model" and "Delivering Data from the Tabular Project."

NOTE With Visual Studio 2019, the required functionality to enable Analysis Services projects has moved into the respective Visual Studio extensions. In other words, installation of stand-alone SQL Server Data Tools (SSDT) no longer is required, because SSDT has been integrated in Visual Studio.

Before I start to explain SSAS, you have to know what SQL Server Business Intelligence means. SQL Server Business Intelligence contains three services: SSAS, SQL Server Reporting Services (SSRS), and SQL Server Integration Services (SSIS). All three services are based on the BI Semantic Model (BISM), a single model that serves all levels of end users. The model can integrate data from a number of data sources, whether they are traditional data sources, such as databases or LOB applications, or nontraditional sources. This model contains the following submodels:

- Tabular model
- Multidimensional model

The idea behind these two models is to cover different forms of business intelligence: the Tabular model should be used for Personal BI and applications in relation to Team BI. Corporate BI is covered by the Multidimensional model.

The Tabular and Multidimensional models have totally different structure. For this reason, this chapter has two main parts. The first describes the Multidimensional model, while the second part discusses the Tabular model.

Before I describe the Multidimensional model, it's important to explain how both models can be installed. The installation of SSAS is a part of the installation process of the entire system, which is explained in detail in Chapter 2. The component called SQL Server Installation Center leads you through the whole process. To install SSAS, click Installation in SQL Server Installation Center and choose New SQL Server Stand-Alone Installation or Add Features to an Existing Installation. Proceed with the installation process as described in the "Installing SQL Server on Windows" section of Chapter 2 until you get to the Installation Type page (see Figure 2-4). Choose Add Features to an Existing Instance and select the instance to which you want to add SSAS. On the next page, Feature Selection (see Figure 2-5), check the Analysis Services check box in the Instance Features list, and click Next. (The actual beginning of the installation of SSAS will be started later, after the step called Database Engine Configuration.) On the Server Configuration page, click Next. The Analysis Services Configuration page shown in Figure 23-1 appears. As you can see, you can install either the Multidimensional model or the Tabular model, but not both of them. Therefore, you need *two instances* of SQL Server Analysis Services if you want to use both models; in the current installation process, choose Multidimensional and Data Mining model and click Next. (The installation of the Tabular model will be described in the second part of this chapter.) After that, the Instance Configuration page appears. Finish the installation process the same way as described in Chapter 2.

Multidimensional Model

This section shows you how to create and process a multidimensional cube. First, though, it is important to understand the terminology related to the Multidimensional model and to be familiar with the properties used to create and process a cube. These topics are addressed in turn next.

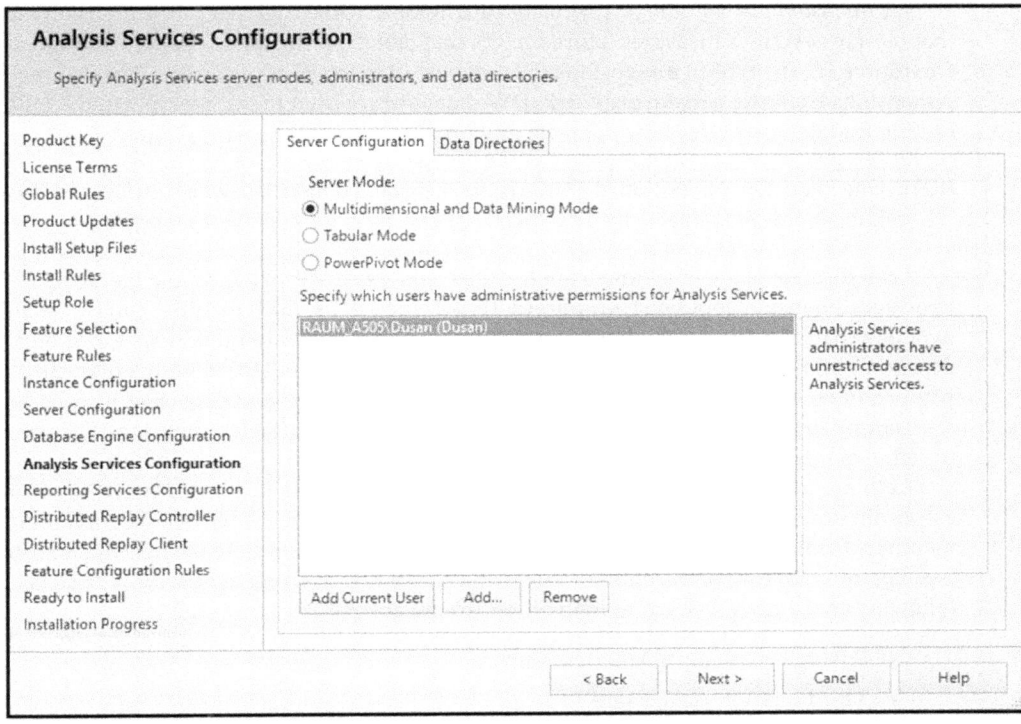

Figure 23-1   The installation process: Choosing the Multidimensional model

Multidimensional Model Terminology

The following are the most important terms in relation to the Multidimensional model:

- Cube
- Dimension
- Member
- Hierarchy
- Cell
- Level
- Measure group
- Partition

A *cube* is a multidimensional structure that contains all or part of the data from a data warehouse. Although the term "cube" implies three dimensions, a multidimensional cube generally can have many more dimensions. Each cube contains all other components in the preceding list.

A *dimension* is a set of logically related attributes (stored together in a dimension table) that closely describes measures (stored in the fact table). For instance, **Time**, **Product**, and **Customer** are the typical dimensions that are part of many BI applications. These three dimensions from the **AdventureWorksDW** database are used in the example in the following section that demonstrates how to create and process a multidimensional cube.

NOTE   One important dimension of a cube is the **Measures** dimension, which includes all measures defined in the fact table.

Each discrete value in a dimension is called a *member*. For instance, the members of a **Product** dimension could be **Computers**, **Disks**, and **CPUs**. Each member can be calculated, meaning that its value is calculated at run time using an expression that is specified during the definition of the member. (Because calculated members are not stored on the disk, they allow you to add new members without increasing the size of a corresponding cube.)

Hierarchies specify groupings of multiple members within each dimension. They are used to refine queries concerning data analysis.

Cells are parts of a multidimensional cube that are identified by coordinates (x-, y-, and z-coordinates, if the cube is three-dimensional). This means that a cell is a set containing members from each dimension. For instance, consider the three-dimensional cube in Chapter 22 (see Figure 22-3) that represents car sales for a single region within a quarter. The cells with the following coordinates belong, among others, to the cube:

- First quarter, South America, Falcon
- Third quarter, Asia, Eagle

When you define hierarchies, you define them in terms of their levels. In other words, *levels* describe the hierarchy from the highest (most summarized) level to the lowest (most detailed) level of data. The following list displays the possible hierarchy levels for the **Time** dimension:

- Quarter (Q1, Q2, Q3, Q4)
- Month (January, February, …)
- Day (Day1, Day2, …)

As you already know from Chapter 22, measures are numerical values, such as price or quantity, that appear in a fact table but do not build its primary key. A *measure group* is a set of measures that together build a logical unit for business purposes. Each measure group is built on the fly, using corresponding metadata information.

A cube can be divided into one or more partitions. *Partitions* are used by SSAS to manage and store data and aggregations for a measure group in a cube. Every measure group has at least one partition, which is created when the measure group is defined. Partitions are a powerful and flexible means of managing large cubes.

Creating a New Project Using Visual Studio 2019

As previously noted, with Visual Studio 2019 (VS 2019), the required functionality to enable Analysis Services projects has moved into the respective Visual Studio extensions. Therefore, you need Visual Studio to create and process multidimensional cubes.

If you have a license for Visual Studio 2019, just install the product. If you don't already have a license, you can install Visual Studio Community from https://visualstudio.microsoft.com/downloads. (Visual Studio Community is the most basic edition of Visual Studio and is available free of charge.)

After installation of Visual Studio 2019, you need to install SQL Server Analysis Services extensions for VS 2019. To do so, go to https://marketplace.visualstudio.com/items?itemName=ProBITools.MicrosoftAnalysisServicesModelingProjects.

Click Download to download the corresponding MS Visual Studio extension file (.vsix). Click the file to start the VSIX installer.

Now you have all necessary tools to create Analysis Services projects. Start Visual Studio 2019 and click Create a New Project (see Figure 23-2). On the Create a New Project page, choose Analysis Services Multidimensional and Data Mining Project (see Figure 23-3) and click Next.

The Configure Your New Project page appears. In the Name text box, type the name of the project. (I named my project **SSAS_Project1**.) In the Location drop-down list box, type or select the folder in which to store the files for the project, or click the Browse button (…) to select a folder. In the Solution Name text box, type the name of your solution (the default name of the solution is identical to the name of the created project).

NOTE The new project is always created in a solution. A *solution* is the largest management unit and comprises one or more projects.

Click Create, and the Visual Studio initial window for the new project opens. In the right pane, you can see the folders for the tasks, which are discussed in the next section.

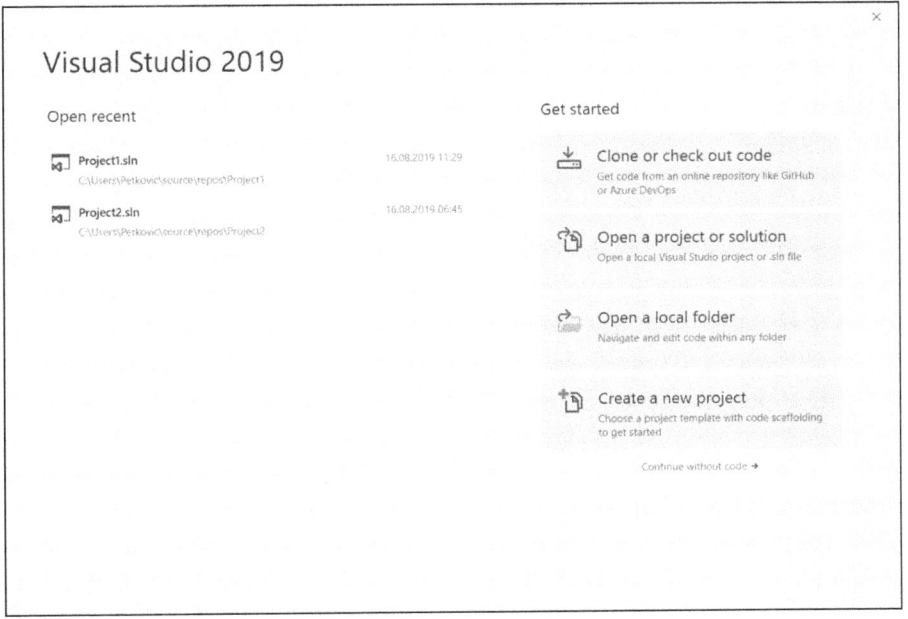

Figure 23-2   Creating a new project in Visual Studio 2019

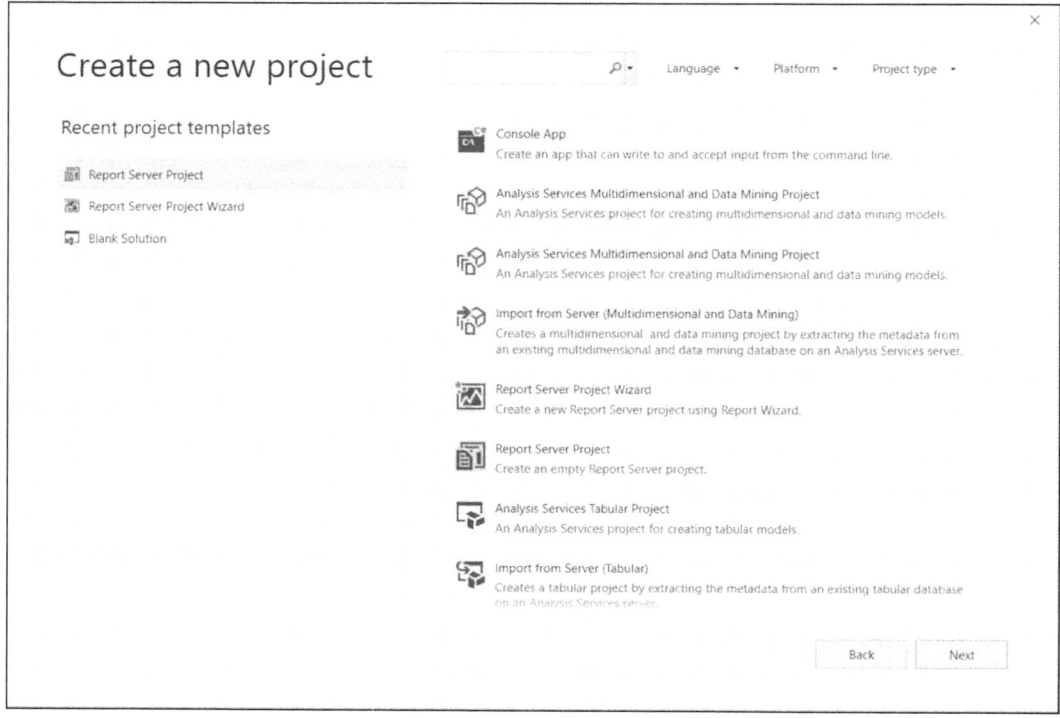

Figure 23-3   Choosing Analysis Services Multidimensional and Data Mining Project

Creating and Processing of a Multidimensional Cube

Now that you know how to create a BI project and configure it, I will show you how to perform the following steps that are necessary to create and process a multidimensional cube using Visual Studio:

1. Identify data sources.
2. Specify data source views.
3. Create a cube.
4. Design storage aggregation.
5. Deploy and process the cube.
6. Browse the cube.

The following subsections describe these steps.

Identifying Data Sources

To identify data sources in Visual Studio 2019, right-click the Data Sources folder in the right pane under your project and choose New Data Source. The Data Source Wizard appears, which guides you through the process of creating a data source. (This example uses the

AdventureWorksDW database as the data source.) Click Next on the Welcome to the Data Source Wizard page.

First, on the Select How to Define the Connection page, make sure that the Create a Data Source Based on an Existing or New Connection radio button is activated and click New. In the Connection Manager dialog box (see Figure 23-4), select Native OLE DB/SQL Server Native Client 11.0 in the Provider drop-down list, select the name of your database server in the Server Name drop-down list, and select Windows Authentication in the Authentication drop-down list. In the Connect to a Database section, with the Select or Enter a Database Name radio button enabled, choose the **AdventureWorksDW** database from the drop-down list. Click the Test Connection button to test the connection to the database. If the test is successful, click OK.

The next dialog box that appears could be either Select How to Define the Connection or Impersonation Information. The Select How to Define the Connection dialog box appears only when you connect to the data source for the first time. After selection, click Next.

The next step of the wizard is the Impersonation Information page. These settings determine which user account SSAS uses when connecting to the underlying source of data using Windows authentication. Which setting is appropriate depends on how this data source is

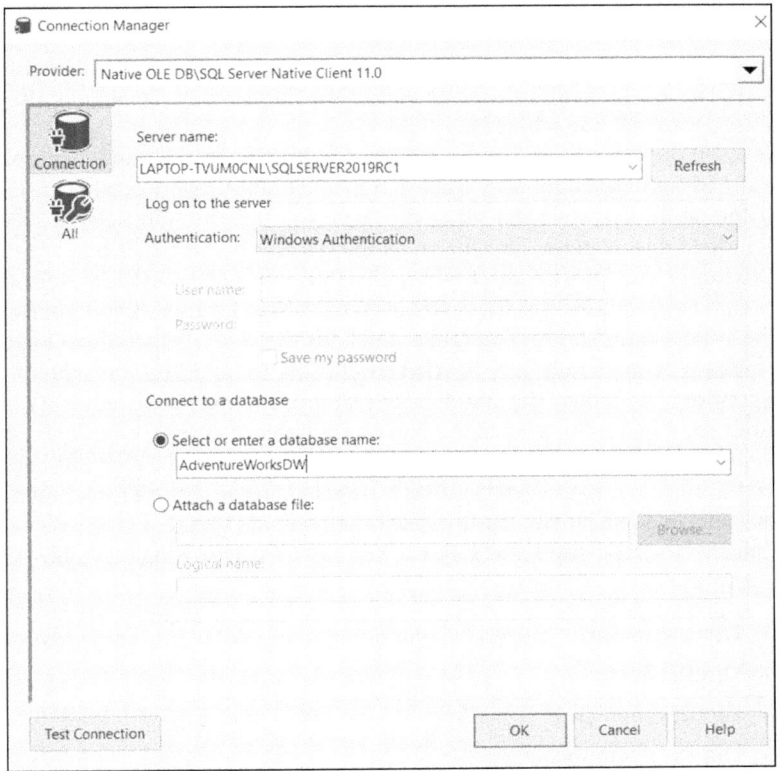

Figure 23-4   Connection Manager dialog box

being used. Click the Use a Specific Windows User Name and Password radio button and type your username and password in the corresponding fields. Click Next.

Finally, on the Completing the Wizard page, give the new data source a name (for this example, call it **BI_Source**) and click Finish. The new data source appears in the Solution Explorer pane in the Data Sources folder.

After identifying data sources in general, you have to determine which data in particular you want to select from the data source. In our example, it means that you have to select tables from the **AdventureWorksDW** database, which will be used to build a cube. This step involves specifying data source views, discussed next.

Specifying Data Source Views

To define data source views, right-click the Data Source Views folder under your project and choose New Data Source View. The Data Source View Wizard guides you through the steps that are necessary to create a data source view. (This example creates a view called **BI_View1**, which is based on the **Customer** and **Project** tables of the **AdventureWorksDW** database.) Click Next.

First, on the Select a Data Source page, select the **BI_Source** data source and click Next. (The name of this data source should appear in the left pane.) On the next wizard page, Select Tables and Views, choose tables that belong to your cube either as dimension tables or fact tables. To choose a table, select its name in the Available Objects pane and click the > button to move it to the Included Objects pane. For this example, choose the tables for customers and products (**DimCustomer** and **DimProduct**, respectively) in the **AdventureWorksDW** database. These tables will be used to build cube dimensions. They build the set of dimension tables used for your star schema.

Next, on the same wizard page, you need to specify one or more fact tables that correspond to the already selected dimension tables. (One fact table, together with the corresponding dimension tables, creates a star schema.) To do so, click the Add Related Tables button below the Included Objects pane. This instructs the system to find tables that are related to the **DimCustomer** and **DimProduct** tables. (To find related tables, the system searches all primary key/foreign key relationships that exist in the database for these two dimension tables.)

The system finds several fact tables and adds them to the Included Objects pane. Of these tables, you need only one, **FactInternetSales**, to build the star schema. Besides the corresponding fact tables, the system also searches for other tables that are created separately for a hierarchy level of the corresponding dimension. One such table to keep is **DimProductSubcategory**, which incarnates a hierarchy level called **Subcategory** of the **Product** dimension. Also keep the **DimDate** table, because the **Time** dimension is almost always a part of a cube. (The name of the schema is **BI_View1**.)

Thus, for the **BI_View1** star schema, you need the following five tables (as shown in Figure 23-5):

- DimCustomer
- DimProduct
- FactInternetSales
- DimProductSubcategory
- DimDate

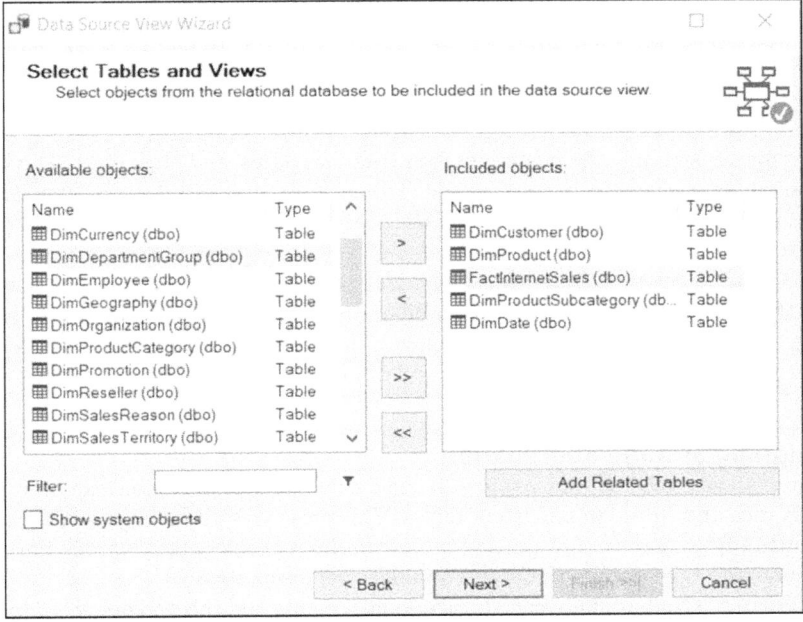

Figure 23-5   Select Tables and Views page of Data Source View Wizard

Exclude all other system-chosen tables that appear in the right pane by selecting them and clicking the < button. After restructuring the tables, click Next. On the Completing the Wizard page, choose the name for the new data source view and click Finish.

After you click Finish, double-click the **BI_View1** view in the right pane to display a graphical representation of the tables in the data schema you have defined, as shown in Figure 23-6. (The graphical representation is done with a tool called Data Source View Designer.)

NOTE Using drag and drop, I changed the design of the data source view in Figure 23-6 so that the tables have the convenient form of the star schema. Notice that the fact table is in the middle and the corresponding dimension tables build the circle around it. (Figure 23-6 actually has the form of a snowflake schema, because the **DimProductSubcategory** table presents the hierarchy level of the **Product** dimension.)

Data Source View Designer offers several useful functions. One of them is to view the data in individual tables. To do this, right-click a table and choose Explore Data. The content of the table appears in a separate window.

Creating a Cube

Before you create a cube, you must specify one or more data sources and create a data source view, as previously described. After that, you can use the Cube Wizard to create a cube.

To create a cube, right-click the Cubes folder under your project and choose New Cube. The welcome page of the Cube Wizard appears. Click Next. On the Select Creation Method page, choose Use Existing Tables, because the data source view exists and can be used to build a cube. Click Next.

On the Select Measure Group Tables page, you select measures from the fact table(s). Therefore, select only the fact table, **Fact Internet Sales**, and click Next. The wizard chooses all possible measures from the selected fact table and presents them on the Select Measures page. Check only the **Total Product Cost** column of the **Fact Internet Sales** table as the single measure (see Figure 23-7). Click Next.

On the Select New Dimensions page, select all three dimensions (**Dim Date**, **Dim Product**, and **Dim Customer**) to be created, based on the available tables. (If the **Fact Internet Sales** table was also selected, unselect that table.) The final page, Completing the Wizard, shows the summary of all selected measures and dimensions. Click Finish to finish creating the cube called **BI_Cube**.

Designing Storage Aggregation

As you already know from Chapter 22, basic data from the fact table can be summarized in advance and stored in persistent tables. This process is called *aggregation*, and it can significantly enhance the response time of queries, because scanning millions of rows to calculate the aggregation on the fly can take a very long time.

There is a tradeoff between storage requirements and the percentage of possible aggregations that are calculated and stored. Creating all possible aggregations in a cube and

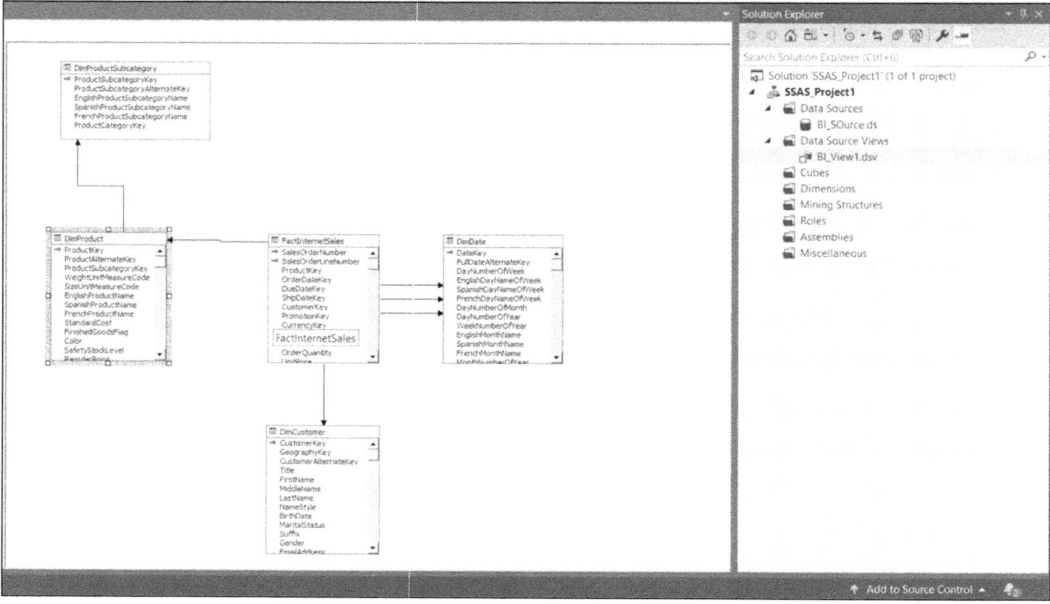

Figure 23-6 Graphical representation of the selected tables

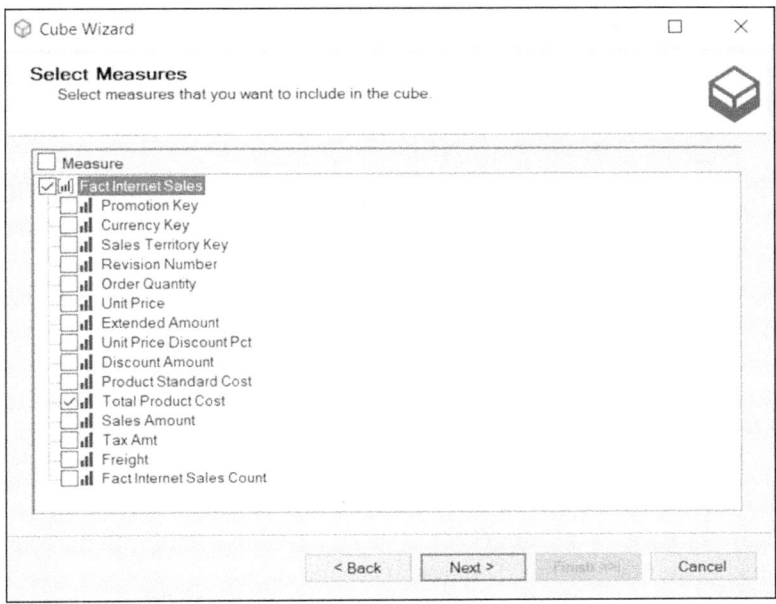

Figure 23-7   Select Measures page of Cube Wizard

storing all of them on the disk results in the fastest possible response time for all queries, because the response to each query is almost immediate. The disadvantage of this approach is that the storage and processing time required for the aggregations can be substantial.

On the other hand, if no aggregations are calculated and stored, you do not need any additional disk storage, but response time for queries concerning aggregate functions will be slow because each aggregate has to be calculated on the fly.

SSAS provides the Aggregation Design Wizard to help you design aggregations optimally. To start the wizard, you have first to start the Cube Designer. (The Cube Designer is used to edit various properties of an existing cube, including the measure groups and measures, cube dimensions, and dimension relationships.) To start it, right-click the cube in Solution Explorer and select View Designer from the context menu. Now, click the Aggregations tab in the main menu of the Cube Designer. In the table that appears in the Cube Designer (**Fact Internet Sales**), right-click the cell under the Aggregations column and choose Design Aggregations. That starts the Aggregation Design Wizard.

In the first step of the wizard, Review Aggregation Usage, you review aggregation usage settings. In this step, you can include or exclude the attributes that appear on the page. Leave the settings as they are and click Next.

The next step is to specify the number of members in each attribute. You do this on the Specify Object Counts page. For each selected cube object, you have to enter the estimated count value or partition count value, before the wizard starts to create and store the selected aggregations. If you click the Count button, the wizard automatically performs the object counts and displays the obtained counts. Click Next.

In the second-to-last step, the Set Aggregation Options page, choose one of the four options to specify up to what point (or not at all) aggregations should be designed:

- **Estimated storage reaches __ MB**   Specifies the maximum amount of disk storage that should be used for precomputed aggregations. The larger the amount, the more precomputed aggregations that will be created.

- **Performance gain reaches __ %**   Specifies the performance gain that you want to achieve. The higher the percentage of precomputed aggregations, the better the performance.

- **I click Stop**   Enables you to decide when to stop the design process.

- **Do not design aggregation (0%)**   Specifies that no precomputed aggregations should be created.

NOTE Generally, you should choose one of the first two alternatives. I prefer the second one, because it is very difficult to estimate the amount of storage for different star schemas and different sets of queries. A value between 65 percent and 90 percent is optimal in most cases.

Figure 23-8 shows the result of choosing the second option with the value set to 65 percent and clicking the Start button. The system created six aggregations and uses 315.2 KB for them.

Click Next to go to the Completing the Wizard page. On this page, you can choose whether to process aggregations immediately (Deploy and Process Now) or later (Save the Aggregations But Do Not Process Them). Choose the second option and click Finish.

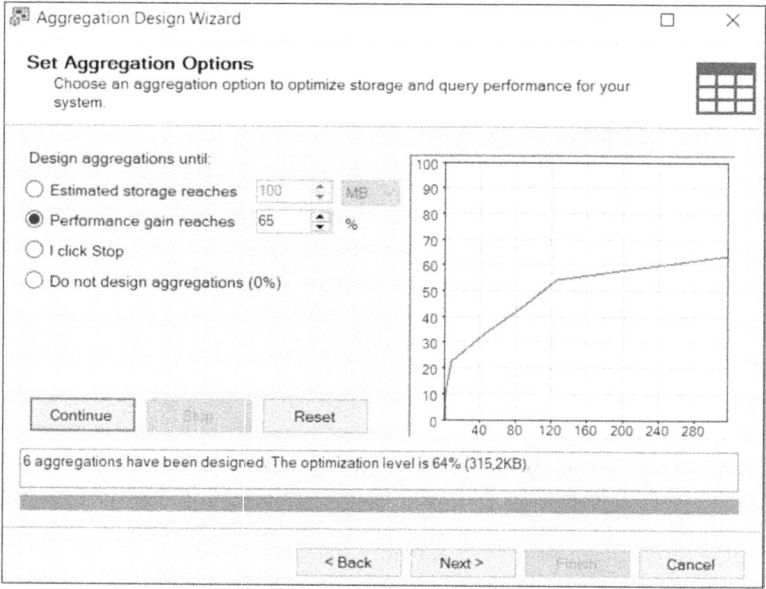

Figure 23-8   The Set Aggregation Options wizard page (after clicking Start)

Deploying and Processing the Cube

You have to complete two additional steps before you can use a cube that you have created. The first step is to deploy the cube to the Analysis Services database. The second step is to process already deployed structures in the database. The following two subsections give the detailed description of both steps.

Deploying the Cube   As you already know, you can create a cube either offline or online. Creating a cube offline means that all steps necessary for its creation are done without a connection to the Analysis Services server. (All steps described up to this point in the chapter can be executed offline.)

Deploying a cube means that all elements of the cube are moved from a "front-end" environment (in this case Visual Studio) to the server environment. The elements of the cube reside on the Analysis Services server as a part of a database. The name of the database is specified in the project definition. If the specified database does not exist, the Analysis Services server creates it.

Although the cube structures are copied to the server during the deployment process, the contents of dimensions and measures are not stored on the server side. To do this, you have to process the cube, which is explained next.

NOTE You don't have to start the deployment process explicitly. When you process your cube, the system will ask you whether the cube should be deployed.

Processing the Cube   If you chose the Save the Aggregations But Do Not Process Them option as the final step in the preceding section and deploy the cube to the SSAS database, you have still to process the cube.

NOTE A cube must be processed when you first create it and each time you modify it. If a cube has a lot of data and precomputed aggregations, processing the cube can be very time consuming.

To process the cube, right-click the name of your cube in the Cubes folder of Solution Explorer and select Process. The system starts processing the cube and displays the progress of this activity. To end the activity, just close the active dialog box(es).

Browsing the Cube

To browse a cube, right-click the cube name (in the Cubes folder of Solution Explorer) and choose Browse. The Browse view appears. You can add any dimension to the query by right-clicking the dimension name in the left pane and choosing Add to Query. You can also add a measure from that pane in the same way. (Adding measures first is recommended.) Figure 23-9 shows the tabular representation of the total product costs for Internet sales for different customers and products.

The approach is different if you want to calculate values of measures for particular dimensions and their hierarchies. For example, suppose that you want to deliver for customers with customer IDs 11008 and 11015 total product costs for all products they have ordered. In this case, you first drag and drop the measure (**Total Product Cost**) from the left pane into

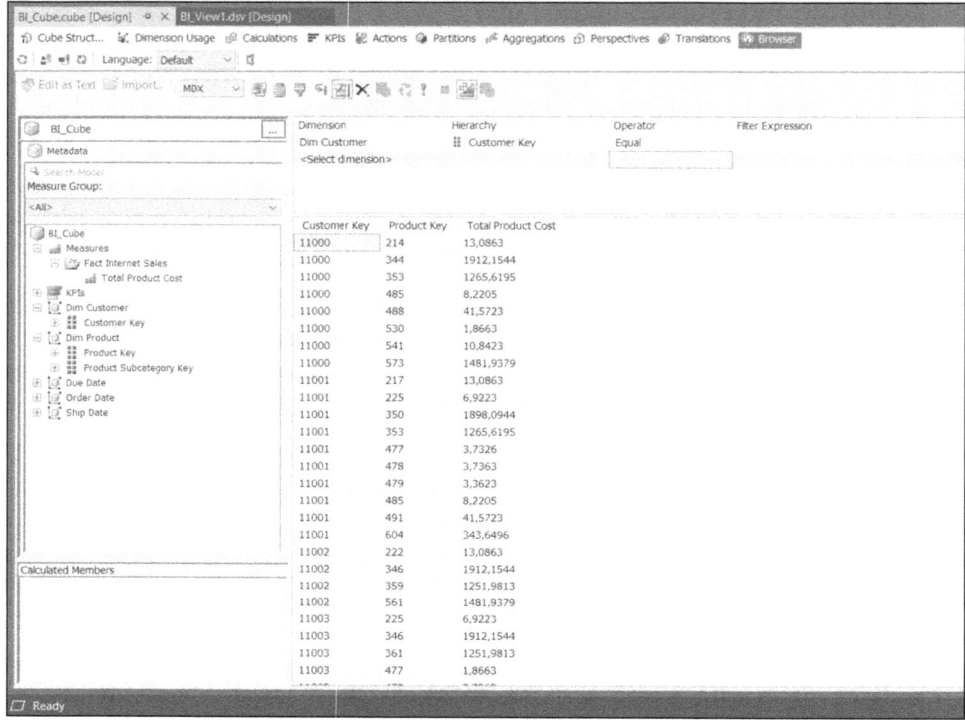

Figure 23-9 Total product costs for Internet sales for different customers and products

the editing pane, and then you choose values in the pane above it to restrict the conditions for each dimension (see Figure 23-10). First, in the Dimension column, choose the **Dim Customer** table, and in the Hierarchy column, choose the primary key of this table (**Customer Key**). In the Operator column, choose Equal, and in the Filter Expression column, choose both values 11008 and 11015, one after the other.

In the same way, choose the conditions for the **Dim Product** dimension table. The only difference is that all product values should be included. For this reason, in the Filter Expression column you should choose the root of the dimension. (The root of each dimension is specified by All.)

Delivering Data from the Multidimensional Model

Now that you have seen how to build and browse multidimensional cubes using Analysis Services, you are ready to learn how to retrieve data from a cube and deliver it to users using other interfaces. (The primary goal of Visual Studio is to develop BI projects, not to retrieve and deliver data to users.) For this task, there are many other APIs, such as:

- Power Pivot for Excel
- Multidimensional Expressions (MDX)

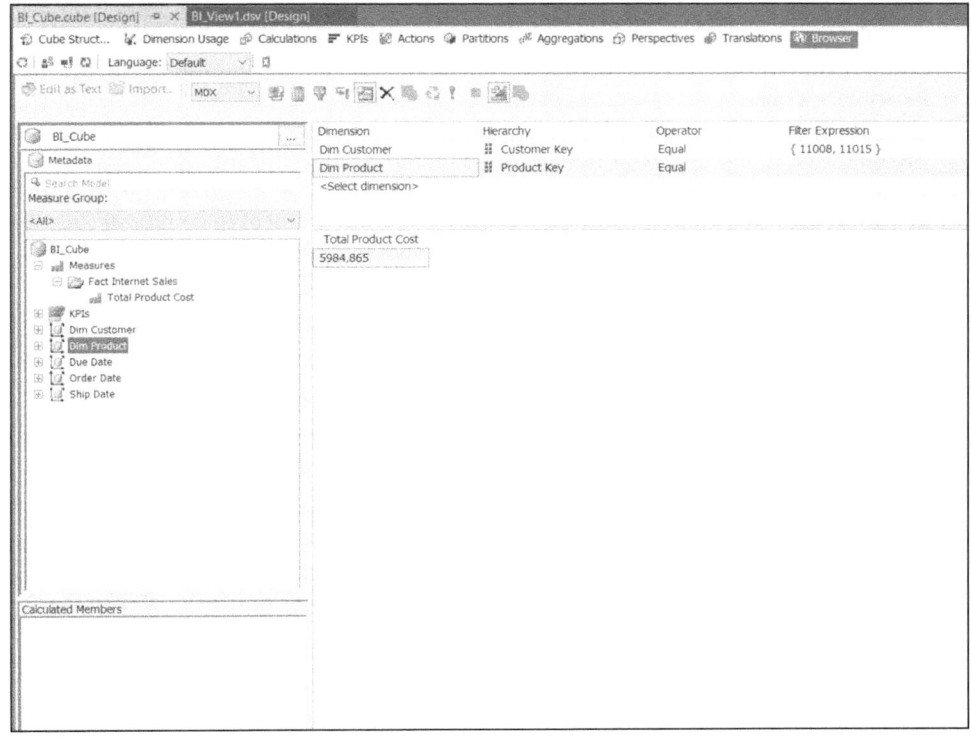

Figure 23-10   The total product costs as a crosstab

- OLE DB for OLAP
- ADOMD.NET
- Third-party tools

MDX is the only interface that I will discuss in detail, in the following subsection; the other interfaces in the list are briefly described here. (The reason I give more attention to MDX than the other interfaces is that MDX is a tool that is used by many third-party SSAS solutions.)

Power Pivot for Excel is a tool that allows you to analyze data using what is probably the most popular Microsoft tool for such purpose: Microsoft Excel. It is a user-friendly way to perform data analysis using features such as PivotTable, PivotChart views, and slices. The use of Power Pivot for Excel for the Multidimensional model is similar to the use of the same "front end" for the Tabular model, as you will learn in the section "Power Pivot for Excel" later in this chapter.

OLE DB for OLAP is an industry standard for multidimensional data processing, published by Microsoft. It is a set of entities and interfaces that extends the ability of OLE DB to provide access to multidimensional data stores. OLE DB for OLAP enables users to perform data analysis through interactive access to a variety of possible views of the underlying data. Many independent software vendors use the specification of OLE DB for OLAP to implement different interfaces that allow users to access cubes created by SSAS. Additionally, using OLE DB for OLAP, vendors can implement OLAP applications that can uniformly access both relational and nonrelational data stored in diverse information sources, regardless of location or type.

ADOMD (ActiveX Data Objects Multidimensional) is a Microsoft .NET Framework data provider that is designed to communicate with SSAS. With this interface, you can access and manipulate objects in a multidimensional cube, enabling web-based OLAP application development. This interface uses the XML for Analysis protocol to communicate with analytical data sources. Commands are usually sent in MDX. By using ADOMD.NET, you can also view and work with metadata.

After the description of MDX, I will describe Power BI Desktop, which comprises several Microsoft tools, such as Power Pivot for Excel.

Multidimensional Expressions (MDX)

NOTE The material in this section is complex. Therefore, you might want to skip it on the first reading of the book.

Multidimensional Expressions (MDX) is a language that you can use to query multidimensional data stored in OLAP cubes. (MDX can also be used to create cubes.) In MDX, the SELECT statement specifies a result set that contains a subset of multidimensional data that has been returned from a cube. To specify a result set, an MDX query must contain the following information:

- One or more axes that you use to specify the result set. (An axis in the SELECT statement corresponds to a dimension of the cube.) You can specify up to 128 axes in an MDX query. You use the ON COLUMNS clause to specify the first axis and the ON ROWS clause to specify the second axis. If you have more than two axes, the alternative syntax is to use numbers: ON AXIS(0) for the first axis, ON AXIS(1) for the second one, and so on.

- The set of members or tuples to include on each axis of the MDX query. This is written in the SELECT list.

- The name of the cube that sets the context of the MDX query, specified in the FROM clause of the query.

- The set of members or tuples to include on the "slicer axis," specified in the WHERE clause (see Examples 23.1 and 23.2).

NOTE The semantic meaning of the WHERE clause in SQL is different from its semantic meaning in MDX. In SQL it means *filtering* of rows by specified criteria. The WHERE clause in MDX means *slicing* the cube. While these concepts are somewhat similar, they are not equivalent.

Example 23.1 will be used to explain the syntax of the language. You can execute your MDX queries directly in SQL Server Management Studio.

To establish the connection to the server, click the database (in this case, the **BI_Cube** cube) under the corresponding SSAS server instance. Click the MDX icon in the menu of Management Studio. After that, the MDX Query Editor opens, which you use to design and execute statements and scripts written in the MDX language. First, type scripts in the Query Editor pane. Then, to execute the scripts, press F5 or click Execute on the toolbar.

Example 23.1
Display for each customer total product costs that are due on May 15, 2013:

```
SELECT [Measures].MEMBERS ON COLUMNS,
       [Dim Customer].[Customer Key].MEMBERS ON ROWS
FROM BI_Cube
WHERE ([Due Date].[Date Key].[20130515])
```

Example 23.1 queries data from the **BI_Cube** cube. The SELECT list of the first query axis displays all members of the **Measures** dimension. In other words, it displays the values of the **Total Product Costs** column, because the only existing measure in this cube is **Total Product Costs**. The second query axis displays all members of the **Customer Key** column of the **Dim Customer** dimension.

The FROM clause indicates that the data source in this case is the **BI_Cube** cube. The WHERE clause "slices" through the **Due Date** dimension according to the key values using the single date value 2013/05/15.

Example 23.2 shows another MDX query.

Example 23.2
Calculate for the customer with the customer key 11005 and for the product with the product key 562 the total product costs that are due in the period between 2013/05/13 and 2013/05/15:

```
SELECT [Measures].MEMBERS ON COLUMNS
                 FROM BI_Cube
WHERE ({[Due Date].[Date Key].[20130513]:[Due Date].[Date key].[20130515]},
                 [Dim Customer].[Customer Key].[11005],
                 [Dim Product].[Product Key].[562])
```

The SELECT list in the query in Example 23.2 contains only the members of the **Measures** dimension. For this reason, the query displays the value of the **Total Product Costs** column. The WHERE clause in Example 23.2 is more complex than in Example 23.1. First, there are three slices, which are separated using commas. Only one member of the **Customer** dimension and one member of the **Product** dimension are used for slicing, while from the **Due Date** dimension, the dates from 2013/05/13 through 2013/05/15 are sliced. (As you can see from the query, a colon is used to specify a range of dates.)

NOTE This section provides only a concise description of MDX. Refer to Microsoft Docs to learn more about this language.

Power BI Desktop
Power BI Desktop is an environment that comprises several Microsoft tools that enable you to build advanced queries, models, and reports that visualize data. With Power BI Desktop, you can build data models, create reports, and share your work by publishing to the Power BI service. Power BI Desktop is a free download.

NOTE This section gives you just a concise description of Power BI Desktop. See the section "Data Visualization in R" in Chapter 32 for a demonstration of how you can use this environment.

Microsoft developed Power BI Desktop by integrating the following components:

- **Power Pivot for Excel**   An environment whose main benefit is that it makes design easy.
- **Power View**   An add-in to Excel that enables you to visualize data. The main benefit of Power View is that you can create presentation-ready visualizations from the beginning.
- **Power Query**   Provides a way to retrieve data from a variety of data sources and cleanse the data before the load process.

Therefore, Power BI Desktop allows users to create a personal BI environment by gathering data from different sources, loading the data in the data model, and visualizing the data in the same workspace.

You can publish the content created in Power BI Desktop to PowerBI.com. First, you have to create an account on that site. After that, you can explore the content in many different ways. For instance, you can use existing visualizations to choose the one that best suits your purposes. Also, you can manipulate the model and create new visualizations.

NOTE Access to PowerBI.com is free, but there are certain limitations, published on the PowerBI.com site.

Tabular Model

The SSAS Tabular model uses relational constructs, such as tables and relationships between tables, to provide rapid access to BI data. The model uses the xVelocity analytic engine and different compression algorithms to achieve high performance of analytic queries. Data managed by this engine is always stored in memory. (For the detailed description of in-memory data, see Chapter 21.)

Working with the Tabular model is significantly different from working with the Multidimensional model. As you already know, the Multidimensional model allows you to work offline during the first phases of your project. In other words, you need the SSAS server only when you want to deploy and process your cube.

The Tabular model requires a connection to the SSAS database from the beginning, so that "front end" will be in constant communication with the server during the whole process.

NOTE The Tabular model requires a separate instance of SQL Server Analysis Services. In other words, you need to install *two instances* of the SSAS server if you want to work with both the Multidimensional model and the Tabular model.

To learn how to create and use a Tabular model project, you first need to understand the notion of the workspace database.

Workspace Database

The complete information concerning any Tabular model project is kept in a corresponding workspace database. A workspace database is created on the Analysis Services instance, specified in the **Workspace Server** property, when you create a new Business Intelligence project by using one of the Tabular model project templates in Visual Studio 2019. Each Tabular model project has its own workspace database. You can use SQL Server Management Studio to view the data stored in your workspace database.

The workspace database resides in memory while the Tabular model project is open. When you close the project, the location in which the workspace database is stored is determined by the **Workspace Retention** property. There are three different possibilities regarding what happens with your workspace database after the corresponding project is closed:

- Kept in memory
- Unloaded from memory and stored on disk
- Removed

NOTE The location and behavior of the workspace database is controlled by two properties of the Tabular model: **Workspace Server** and **Workspace Retention**. The default values for these properties come from the settings. The workspace server is where the particular workspace database is created. This must be a Tabular instance of SSAS. The workspace retention can have one of the three options that were described.

Creating a Tabular Model Solution

Similarly to how you create a Multidimensional model solution, you create a Tabular model solution by using Visual Studio 2019. Choose Create a New Project in the initial window of VS 2019, select Analysis Services Tabular Project (refer back to Figure 23-3), and click Next.

The Configure Your New Project page appears, shown in Figure 23-11. In the Name text box, type the name of the project. (I named my project **Tab_Project1**.) The name you enter will be used as the database name. In the Location drop-down list box, type or select the folder in which to store the files for the project, or click the Browse button (…) to select a folder. In the Solution Name text box, type the name of your solution. The default name of the solution is identical to the name of the created project. Additionally, you can choose the version of .NET Framework that suits your project.

Click Create, and the Tabular Model Designer dialog box appears (see Figure 23-12). Select Integrated Workspace. By choosing this option, Visual Studio 2019 uses a built-in instance, eliminating the need to install a separate Analysis Services server instance just for model authoring.

To create a new project, you must select a version of SQL Server from the Compatibility Level drop-down list. Choose SQL Server 2017 and click Next.

After your project is created, it opens in Visual Studio. On the right side, in Tabular Model Explorer, you see a tree view of the objects in your model. (All folders are empty because you have not yet imported your data.) You can right-click an object folder to perform actions, similar to the menu bar.

Part IV

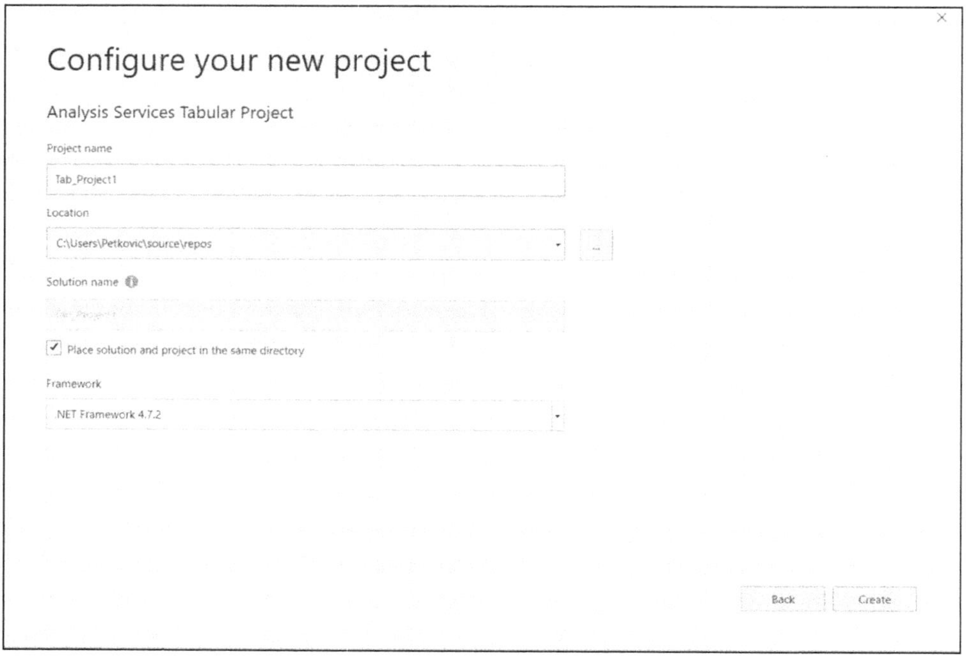

Figure 23-11 Tabular Model: Configure your new project

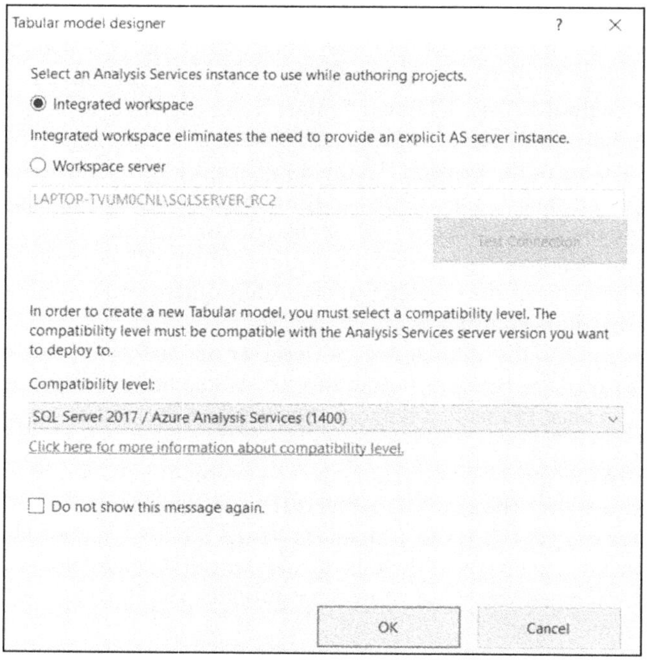

Figure 23-12 Tabular Model Designer dialog box

Click the Solution Explorer tab. Here, you see your Model.bim file. If you don't see the designer window to the left (the empty window with the Model.bim tab), in Solution Explorer, under your project, double-click the Model.bim file. The file contains the metadata for your model project.

Click Model.bim. In the Properties window, you see the model properties, most important of which is the DirectQuery Mode property. This property specifies if the model is deployed in In-Memory mode (Off) or DirectQuery mode (On). In our example we will deploy the model in In-Memory mode, so leave the DirectQuery Mode property set to the default, Off.

When you create a model project, certain model properties are set automatically according to the Data Modeling settings, which you can specify by choosing Tools | Options. Data Backup, Workspace Retention, and Workspace Server properties specify how and where the workspace database is backed up, retained in-memory, and built. You can change these settings later, if necessary.

After these introductory steps, you're ready to look at a Tabular model example.

A Tabular Model Example

The process of creating a Tabular model solution comprises, among others, the following steps:

1. Create a connection.
2. Import data.
3. Add measures.
4. Deploy the model to the server.

These steps are discussed in the following subsections.

NOTE You can find the detailed description of other possible steps, such as creation of relationships and calculated columns, in Microsoft Docs.

Creating a Connection

To begin, click the Tabular Model Explorer tab. The source database for this example is the **AdventureWorksDW** database. To create a connection to the database, in Tabular Model Explorer, right-click Data Sources and choose Import from Data Source. In the Get Data dialog box (see Figure 23-13), click Database in the left pane, choose SQL Server Database in the right pane, and then click Connect.

NOTE The component underlying the Get Data dialog box is Power Query, which provides a vast array of tools for connecting to and reshaping data for modeling and analysis.

In the SQL Server Database dialog box, in the Server field, type the name of the server where you created the **AdventureWorksDW** database, and then click Connect. When prompted to enter credentials, you need to specify the credentials Analysis Services uses to connect to the data source when importing and processing data. In the Impersonation Mode drop-down list, select Impersonate Account and enter your credentials. (Type your user account in the form computer_name\user_name.) Click Connect to start the next step, importing data.

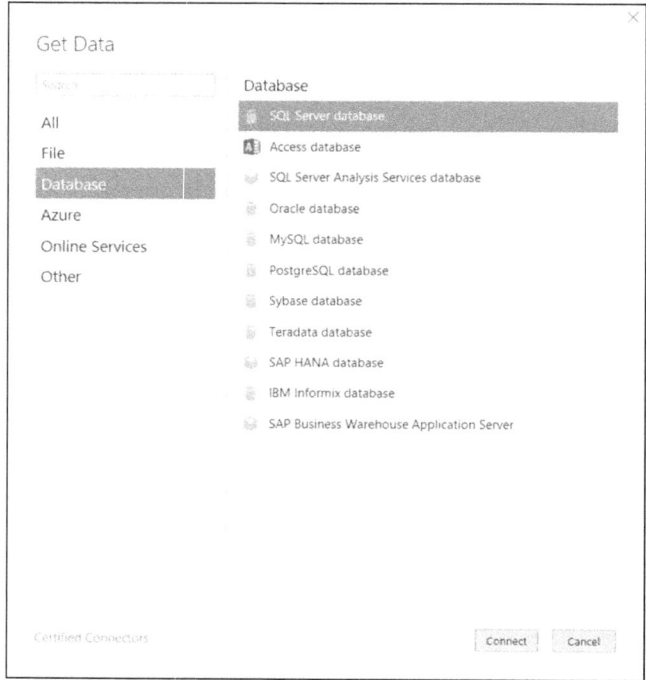

Figure 23-13 Connecting to a data source via the Get Data dialog box

NOTE The **Tab_Project1** project will be based on the same example that I used to create the Multidimensional cube. I intentionally chose to use the same example for both models so that you can compare the capabilities of both models.

Importing Data

After the successful connection, the Navigator dialog box appears (see Figure 23-14). First, select the **AdventureWorksDW** database and then click OK. This creates the connection to the database. (If the connection to the database already exists, the system omits this step.) After that, check the check boxes for the following tables:

- DimCustomer
- DimDate
- DimProduct
- DimProductSubcategory
- FactInternetSales

Next, click the Edit button. Data from the selected tables will be imported into the model. Figure 23-14 shows the part of this process, i.e., that the DimCustomer check box is checked,

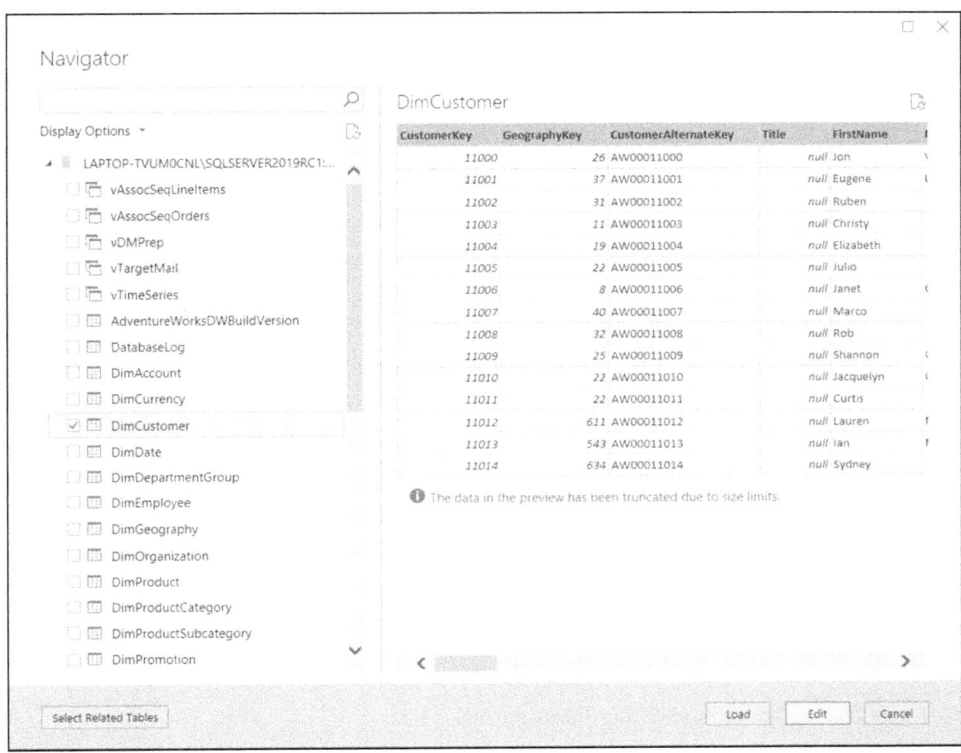

Figure 23-14   Tabular Model: Navigator dialog box

with the corresponding data shown in the right pane. Each imported table appears as a tab at the bottom of the design area.

Now, you can select all imported data or a part of it. To import the subset of data, you have to filter out unnecessary data. Suppose that you do not need import data for customers who have a Spanish or French occupation. To filter them out, in Query Editor, select the **DimCustomer** table. A view of the **DimCustomer** table at the data source appears. Right-click the **FrenchOccupation** column and then click Remove Columns. Repeat the process for the **SpanishOccupation** column.

After filtering out unnecessary data, you can import the rest of the data. To import your data for processing, click Import. The Data Processing dialog box shows the status of data being imported from your data source into your workspace database (see Figure 23-15). The wizard imports the table data and creates new tables in the model. To save your model project, choose File | Save All.

Adding Measures

Generally, to add a measure to a Tabular model project, you have to define a formula that combines a field with an aggregate function. In other words, measures are evaluated based on a filter. That way, you inform the model how to handle the quantity when groups are built.

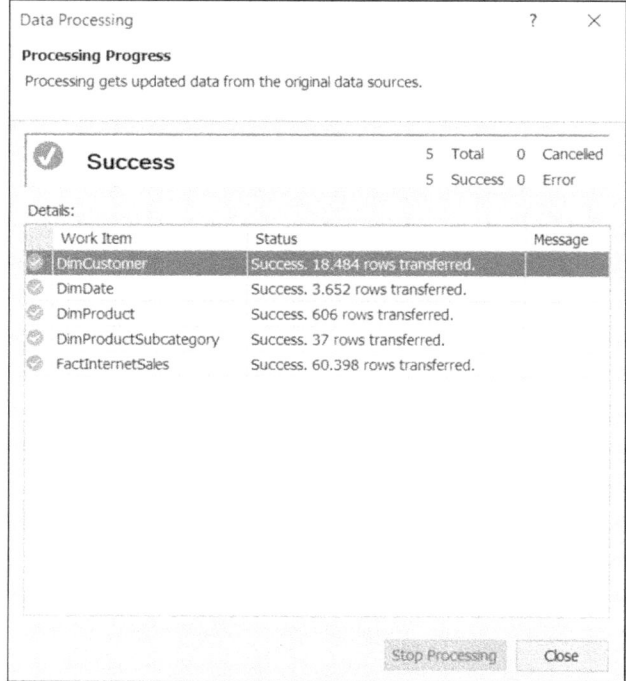

Figure 23-15 Tabular model: Data Processing dialog box

NOTE In contrast to the Multidimensional model, the Tabular model cannot calculate aggregations in advance.

To create measures, you use the *measure grid*. From the beginning, each table has an empty measure grid. The measure grid appears below a table in the model designer (see Figure 23-16). To hide or show the measure grid for a table, choose Table | Show Measure Grid. (The alternative way to find the Measure Grid menu is by expanding Tables in Tabular Model Explorer.)

You can create a measure by clicking an empty cell in the measure grid and then typing a formula in the formula bar. (The formula represents your filter for the corresponding table's column.) When you press ENTER to complete the formula, the measure then appears in the cell. You can also create measures using a standard aggregation function by clicking a column and then clicking the AutoSum button (Σ) on the toolbar. Measures created using the AutoSum feature appear in the measure grid cell directly beneath the column, but can be moved.

As an example, add to the **TotalProductCost** column of the **FactInternetSales** table a measure that calculates the sum of all values of the **TotalProductCost** column. To do this, click the FactInternetSalesTable tab to select the table with the same name. Click the **TotalProductCost** column title to select the entire column. Click the Σ button in the toolbar. That way you create a measure in the measure grid at the bottom of the design area, as shown in Figure 23-16. On the right side of the figure you can see the Properties pane, where the formula

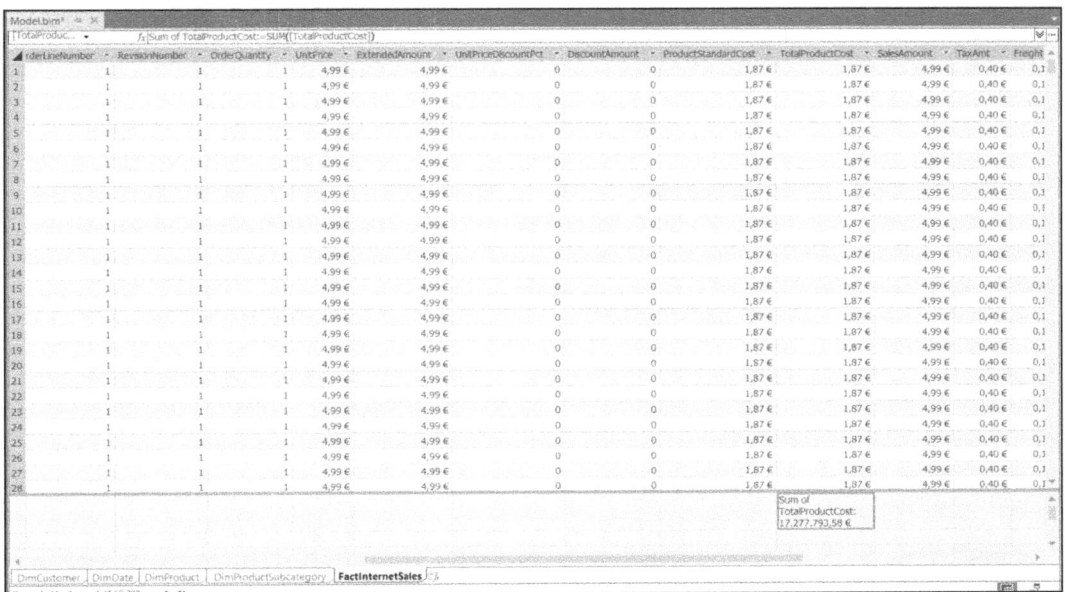

Figure 23-16   The TotalProductCost column with the sum measure

for the calculation, SUM([TotalProductCost]), and the name of the measure are presented. (You can rename a measure by clicking the name of the measure and then, in the Properties window, replacing the old name with the new one in the Measure Name field.)

Deploying the Model to the Server

Deploying a Tabular model project creates a model database in a test, staging, or production environment. Users can then connect to the deployed model by using a data connection directly from reporting client applications.

Deploying is a simple process. However, it requires completing a set of tasks to ensure that your model is deployed to the correct Analysis Services instance and with the correct configuration options.

Tabular models are defined with several deployment-specific properties. When you deploy, a connection to the Analysis Services instance specified in the **Server** property is established. A new model database with the name specified in the **Database** property is then created on that instance, if one does not already exist. Metadata from the model project's Model.bim file is used to configure objects in the model database on the deployment server. With the **Processing Option** property, you can specify to deploy just the model metadata to create the model database (the **Do Not Process** option), or you can specify **Default** or **Full** to have the impersonation credentials used to connect to data sources passed in memory from the model workspace database to the deployed model database. (The list of all configuration properties with their detailed description can be found in Microsoft Docs.)

When a model is deployed from Visual Studio, it is automatically processed, unless the **Processing Option** property is set to **Do Not Process**. To keep the data in the model up to date, you have to process the data several times. This can be done with the Analysis Services Processing Task editor.

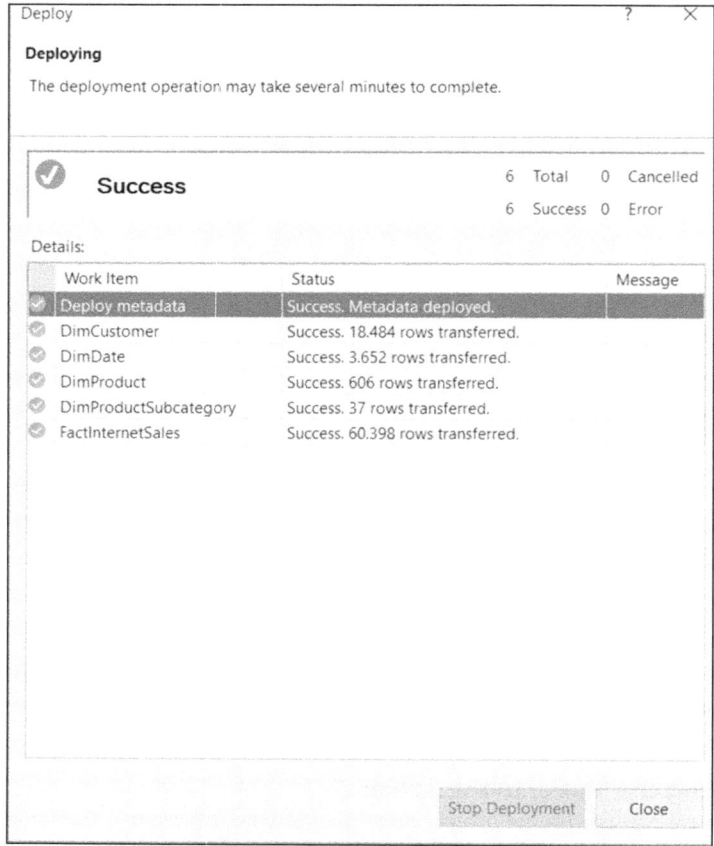

Figure 23-17 The Deployment process for the Tab_Project1 project

To deploy the project, right-click its name in Solution Explorer and select Deploy. Before the system starts the deployment process, you are prompted to enter your credentials in the Impersonation Information page. Figure 23-17 shows the deployment process for the **Tab_Project1** project.

Delivering Data from the Tabular Project

Now that you have seen how to create a Tabular model project and import data, let's take a look at how data can be retrieved and delivered. The list of interfaces that you can use to access data is almost the same as the corresponding list concerning the Multidimensional model. The following interfaces, among others, can be used:

- Power Pivot for Excel
- Data Analysis Expressions (DAX)
- OLE DB for OLAP
- ADOMD.NET

As you can see from the list, the only significant difference to the corresponding list of interfaces for the Multidimensional model is the existence of DAX instead of MDX. For this reason, this section discusses the Power Pivot for Excel interface as well as the DAX interface.

Power Pivot for Excel

As introduced earlier in the chapter, Power Pivot for Excel allows you to analyze data using Excel and is a user-friendly way to perform data analysis using features such as Pivot Table, PivotChart views, and slices.

Before you learn how to use this tool, take a look at the advantages of PowerPivot:

- Familiar Excel features for delivering data are available.
- Very large datasets can be loaded from virtually any source.

As you will see in a moment, you can use the same sources that you use for SSAS in almost the same way for PowerPivot.

The integration of PowerPivot with Excel is straightforward. In Visual Studio 2019, choose Extensions | Model | Analyze in Excel. You are prompted for your credentials. (For details, see the description in the previous section.) The Excel initial window appears. In the PivotTable Fields pane, add the created measure from the **FactInternetSales** table to Values, **CalendarYear** from the **DimDate** table to Columns, and **EnglishOccupation** to Rows. The PivotTable now gives an aggregated result from the measure by regions and year.

Data Analysis Expressions

Data Analysis Expressions (DAX) is a formula language used to create custom calculations in SSAS Tabular model projects. DAX formulas include functions, operators, and values to perform advanced calculations on data in tables and columns.

NOTE You can also use DAX with Power Pivot workbooks.

When using DAX to retrieve tabular data, your entire implementation is founded on the **evaluate** statement. Example 23.3 shows this statement in action.

NOTE All of the following examples can be implemented and executed using Management Studio similarly to the way in which Transact-SQL statements are implemented and executed. To execute these examples, open Management Studio and connect to the Tabular instance of Analysis Services. Expand the Database folder, right-click the database (the Tab_Project1 database in this example), and select New Query.

Example 23.3

```
evaluate ('DimCustomer')
    order by  'DimCustomer'[LastName]
```

The **evaluate** statement begins with the **evaluate** keyword, followed by a table expression enclosed in parentheses. As you can see from Example 23.3, the simplest table expression is one that specifies the name of the table enclosed in single quotes. When you specify only the table name, all rows and columns are returned. Therefore, Example 23.3 returns all rows of the **DimCustomer** table. (The **order by** clause is optional. It is used generally to sort rows.

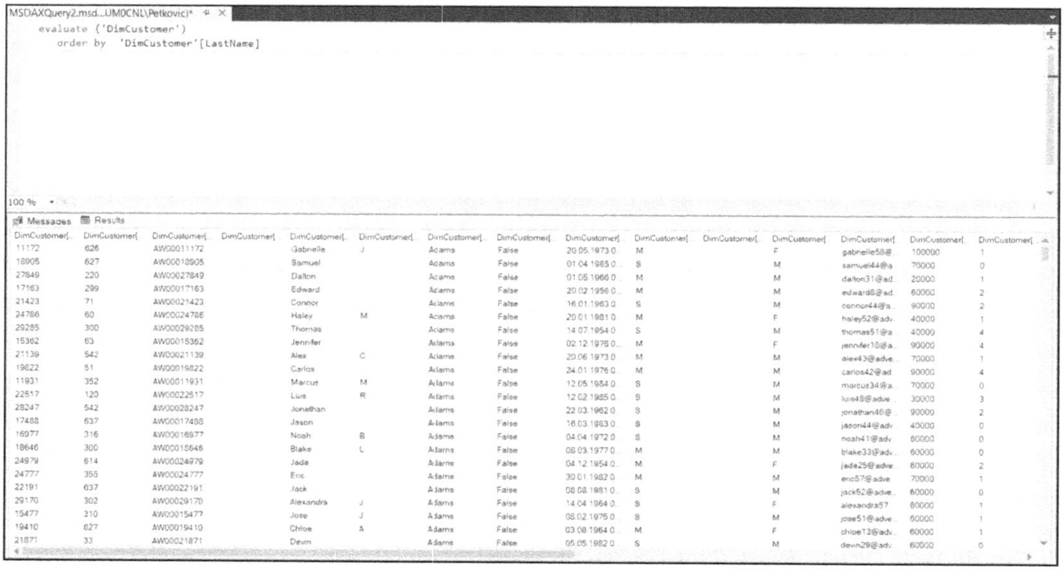

Figure 23-18   The result of Example 23.3

Therefore, in Example 23.3 this clause sorts all rows according to the values of the **LastName** column of the **DimCustomer** table.) The result of Example 23.3 is displayed in Figure 23-18.

NOTE The statement in Example 23.3 is analogous to the following SQL statement:

```
SELECT * FROM DimCustomer ORDER BY LastName;
```

As you already know from Chapter 22, a typical query on a fact table usually uses some form of grouping numerical values to apply one of the aggregate functions (sum, maximum, minimum, or average). DAX supports the **summarize** function, which groups data based on specified columns in order to aggregate data on some other column(s). Example 23.4 shows the use of this function.

Example 23.4

```
evaluate (
    summarize('FactInternetSales',
              'FactInternetSales'[TotalProductCost]))
```

The **summarize** function is part of the **evaluate** statement. When you use it, you specify the function name and the arguments passed to this function. The first argument is the name of the table. All subsequent arguments are the columns that you want to include in the result set. For all specified column names, the system returns a summary table for the requested totals over a set of groups. In other words, the combination of values of all specified columns is built, and the identical values are put together in a group. Therefore, in Example 23.4, the **summarize** function retrieves only the **TotalProductCost** column from the **FactInternetSales** table and the values in that column are grouped together.

NOTE The **summarize** function corresponds to the GROUP BY clause of the SELECT statement.

One of the most important operations on relational tables is to specify a condition that is used to restrict the result set to a subset of tables' rows. One of the ways to do this in DAX is to use the **filter** function. This function has two arguments: a table expression and a condition. The table expression can be the name of a table or an expression that returns a table. The condition is a Boolean expression that is evaluated for each row returned by the table expression. Any row for which the expression evaluates to TRUE is included in the result set. Example 23.5 shows how the **filter** function can be used.

Example 23.5

```
evaluate (filter
    ('FactInternetSales',
     'FactInternetSales'[TotalProductCost] > 1000))
        order by 'FactInternetSales'[ProductKey]
```

Example 23.5 selects all rows from the **FactInternetSales** table where the values of the **TotalProductCost** column are greater than 1000. The result of Example 23.5 is shown in Figure 23-19.

NOTE The **filter** function corresponds to the WHERE clause of the SELECT statement. All Boolean operators, as well as almost all other operators of the WHERE clause, are implemented for the **filter** function, too.

Another very important relational operator is projection—the operation you know as "SELECT list" in Transact-SQL. Generally, there is not a direct way to retrieve just a subset of table columns in DAX. So, we have to use a workaround to achieve this goal.

FactInternetSales[ProductKey]	FactInternetSales[OrderDateKey]	FactInternetSale...	FactInternetSale...	FactInternetSale...	FactInternetSale...	FactIn...
310	20110203	20110215	20110210	17051	1	6
310	20110124	20110205	20110131	16519	1	6
310	20110123	20110204	20110130	16350	1	6
310	20110120	20110201	20110127	16521	1	6
310	20110114	20110126	20110121	16612	1	6
310	20110113	20110125	20110120	16493	1	6
310	20110111	20110123	20110118	16482	1	6
310	20110107	20110119	20110114	16522	1	6
310	20101230	20110111	20110106	16624	1	6
310	20110209	20110221	20110216	28063	1	100
310	20110211	20110223	20110218	28070	1	100
310	20110212	20110224	20110219	28042	1	100
310	20110215	20110227	20110222	27920	1	100
310	20110215	20110227	20110222	28022	1	100
310	20110216	20110228	20110223	28061	1	100
310	20110227	20110311	20110306	28062	1	100
310	20110301	20110313	20110308	28156	1	100

Figure 23-19   The result of Example 23.5

One possible way is the use of the **summarize** function. As you already know, for all specified column names, this function builds the combination of values and displays each group separately. Therefore, we can specify the column(s) that build(s) the primary key of the corresponding table plus all other columns that should be displayed. That way, each combination of values belonging to the table's primary key builds a separate group, and the values of the rest of the columns will be displayed, too. Example 23.6 shows the use of the **summarize** function to implement this operator.

Example 23.6

Display the first and last names of all rows in the **DimCustomer** table:

```
evaluate (summarize
    ('DimCustomer',
        'DimCustomer'[CustomerKey],
        'DimCustomer'[FirstName],
        'DimCustomer'[LastName] ))
 ORDER BY 'DimCustomer'[CustomerKey]
```

The **CustomerKey** column is the primary key of the **DimCustomer** table. For this reason, **CustomerKey** must appear in the column list of the **summarize** function. The list is extended with the names of the columns (**FirstName** and **LastName**) whose values should be displayed. Figure 23-20 shows the output of Example 23.6.

NOTE This section describes only the most important DAX functions. Consult Microsoft Docs to learn more about this language.

DimCustomer[Cu...	DimCustomer[Fir...	DimCustomer[La...
11000	Jon	Yang
11001	Eugene	Huang
11002	Ruben	Torres
11003	Christy	Zhu
11004	Elizabeth	Johnson
11005	Julio	Ruiz
11006	Janet	Alvarez
11007	Marco	Mehta
11008	Rob	Verhoff
11009	Shannon	Carlson
11010	Jacquelyn	Suarez
11011	Curtis	Lu
11012	Lauren	Walker
11013	Ian	Jenkins

Figure 23-20 The result of Example 23.6

Multidimensional Model vs. Tabular Model

This section helps you to determine when to use the Multidimensional model and when to use the Tabular model. Several features can be used to evaluate both of these models. The following list shows the most important factors:

- The size of the dataset (a set of all data sources)
- Writeback support
- Ability to access many different data sources
- Need for complex modeling
- Easy to develop

NOTE One important factor is the use of the programming language. As you already know, you can use MDX exclusively with the Multidimensional model and DAX with the Tabular model. DAX is much easier to use and learn than MDX.

If the size of your dataset is extremely large, use the Multidimensional model. The reason is that the Multidimensional model is designed so that the huge volume of data is stored efficiently. (For this reason, the Multidimensional model is recommended for use of Corporate BI solutions, while the Tabular model should be used for Department and Team BI solutions.)

Among other things, Analysis Services is a tool that can modify the data you are analyzing. This capability is called writeback, and it is the enabling feature behind what-if analysis, forecasting, and financial planning in BI applications that use Analysis Services as the server. Processing writeback is possible only using the Multidimensional model.

NOTE The writeback process writes data back to Analysis Services rather than to the relational database system that provides the raw data. This feature is advantageous because when you write data back to a relational database, you have to wait until the cube is processed before the latest data becomes available. However, when you enable the writeback process, you can submit data into the cube in the current session, making it instantly visible to other users of the Analysis Services database.

If your solution has many external data sources that are different in their nature (tables, data files, etc.), the use of the Tabular model is recommended. The reason is that the Tabular model is based upon the relational data model, and the loading of external data for a relational database system is significantly faster than for multidimensional database systems.

If your solution requires complex modeling, choose the Multidimensional model. As you already know, BI solutions based on the Multidimensional model are designed using the dimensional model, while for the design of the Tabular model solutions, the ER model is used. The former model is easier to use in the case where many dimensions should be modeled. (See also the section "Data Warehouse Design" in Chapter 22.)

Development of BI solutions is easier if you use the Multidimensional model rather than a relational model, which is supported by the Tabular model.

Part IV

Analysis Services: Data Security

SQL Server Analysis Services data security issues correspond to the security issues of the Database Engine. This means that SSAS supports the same general features—authorization and authentication—that the Database Engine does, but in a restricted form.

NOTE The following description of data security is related to the Multidimensional model. Because data security is an issue of the server, security concerning the Tabular model is similar.

Authorization defines which user has legitimate access to SSAS. This issue is tightly connected to the operating system authorization. In other words, SSAS imposes user authorization based on the access rights granted to the user by the Windows operating system.

You can limit the number of users that can perform administrative functions for SSAS. You can also specify which end users can access data and delineate the types of operations they can perform. Additionally, you can control their access at different levels of data, such as the cube, dimension, and cube cell level. This is done using roles.

A *role* is a containing object for a group of users. Windows credentials mentioned previously are assigned to roles within the model. By specifying roles, you establish the membership of the users in Analysis Services. Because permissions are assigned by role, a user must be a member of a role before the user has access to any object.

There are two types of roles:

- Server role
- Database role

The server role defines administrative access of Windows users and groups to an instance of Analysis Services. Members of this role have access to all Analysis Services databases and objects on an instance of Analysis Services, and can perform all administrative tasks.

A database role defines user access to objects and data in an Analysis Services database. A database role is created as a separate object in an Analysis Services database, and applies only to the database in which that role is created. Windows users and groups are included in the role by an administrator, who also defines permissions within the role.

Summary

With SQL Server Analysis Services, Microsoft offers a BI server and a set of data warehousing components that can be used for entry- and intermediate-level data analysis. Visual Studio is the new front-end component of SSAS. In other words, SQL Server Data Tools has been integrated into Visual Studio.

SQL Server Business Intelligence is based on the BI Semantic Model (BISM), which supports two submodels: the Multidimensional model and the Tabular model. You should use the former for Corporate BI, while the latter supports entry-level data analysis and design.

SSAS is wizard oriented, meaning that for all tasks in relation to BI there is a corresponding wizard that you can use to create, deploy, and process a cube, in the case of the Multidimensional model, or to import tables' data and deploy a tabular solution for the Tabular model.

The next chapter describes SQL/OLAP extensions in Transact-SQL.

Exercises

E.23.1   Which model (Multidimensional or Tabular) is better, in relation to the data volume? Discuss this issue.

E.23.2   What does "writeback" mean? When is the writeback support important?

E.23.3   Discuss the differences between MDX and DAX.

24 Business Intelligence and Transact-SQL

In This Chapter

- Window Construct
- Extensions of GROUP BY
- OLAP Query Functions
- Standard and Nonstandard Analytic Functions

Have you ever tried to write a Transact-SQL query that computes the percentage change in values between the last two quarters? Or one that implements cumulative sums or sliding aggregations? If you have ever tried, you know how difficult these tasks are. Today, you don't have to implement them anymore. The SQL standard has adopted a set of online analytical processing (OLAP) functions that enable you to easily perform these calculations and many others that used to be very complex for implementation. This part of the SQL standard is called SQL/OLAP. Therefore, SQL/OLAP comprises all functions and operators that are used for data analysis.

Using OLAP functions has several advantages for users:

- Users with standard knowledge of SQL can easily specify the calculations they need.
- Database systems, such as the Database Engine, can perform these calculations much more efficiently.
- Because there is a standard specification of these functions, they're now much more economical for tool and application vendors to exploit.
- Almost all the analytic functions proposed by the SQL standard are implemented in enterprise database systems in the same way. For this reason, you can port queries in relation to SQL/OLAP from one system to another, without any code changes.

The Database Engine offers many extensions to the SELECT statement that can be used primarily for analytic operations. Some of these extensions are defined according to the SQL standard and some are not. The following sections describe both standard and nonstandard SQL/OLAP functions and operators.

The most important extension of Transact-SQL concerning data analysis is the window construct, which is described next.

NOTE The material in this chapter is complex. Therefore, you might want to skip it on the first reading of the book.

Window Construct

A window (in relation to SQL/OLAP) defines a partitioned set of rows to which a function is applied. The number of rows that belong to a window is dynamically determined in relation to the user's specifications. The window construct is specified using the OVER clause.

The standardized window construct has three main parts:

- Partitioning
- Ordering and framing
- Aggregation grouping

NOTE The Database Engine doesn't support aggregation grouping yet, so this feature is not discussed here.

Before you delve into the window construct and its parts, take a look at the table that will be used for the examples in this chapter. Example 24.1 creates the **project_dept** table, shown in Table 24-1, which is used in this chapter to demonstrate Transact-SQL extensions concerning SQL/OLAP.

Example 24.1

```
USE sample;
CREATE TABLE project_dept
    ( dept_name CHAR( 20 ) NOT NULL,
    emp_cnt INT,
    budget FLOAT,
    date_month DATE );
```

The **project_dept** table contains several departments and their employee counts as well as budgets of projects that are controlled by each department. Example 24.2 shows the INSERT statements that are used to insert the rows shown in Table 24-1.

dept_name	emp_cnt	budget	date_month
Research	5	50000	01.01.2017
Research	10	70000	02.01.2017
Research	5	65000	07.01.2017
Accounting	5	10000	07.01.2017
Accounting	10	40000	02.01.2017
Accounting	6	30000	01.01.2017
Accounting	6	40000	02.01.2018
Marketing	6	100000	01.01.2018
Marketing	10	180000	02.01.2018
Marketing	3	100000	07.01.2018
Marketing	NULL	120000	01.01.2018

Table 24-1   Content of the project_dept Table

Example 24.2

```
USE sample;
INSERT INTO project_dept VALUES
    ('Research', 5, 50000, '01.01.2017');
INSERT INTO project_dept VALUES
    ('Research', 10, 70000, '02.01.2017');
INSERT INTO project_dept VALUES
    ('Research', 5, 65000, '07.01.2017');
INSERT INTO project_dept VALUES
    ('Accounting', 5, 10000, '07.01.2017');
INSERT INTO project_dept VALUES
    ('Accounting', 10, 40000, '02.01.2017');
INSERT INTO project_dept VALUES
    ('Accounting', 6, 30000, '01.01.2017');
INSERT INTO project_dept VALUES
    ('Accounting', 6, 40000, '02.01.2018');
INSERT INTO project_dept VALUES
    ('Marketing', 6, 100000, '01.01.2018');
INSERT INTO project_dept VALUES
    ('Marketing', 10, 180000, '02.01.2018');
INSERT INTO project_dept VALUES
    ('Marketing', 3, 100000, '07.01.2018');
INSERT INTO project_dept VALUES
    ('Marketing', NULL, 120000, '01.01.2018');
```

Part IV

Partitioning

Partitioning allows you to divide the result set of a query into groups, so that each row from a partition will be displayed separately. If no partitioning is specified, the entire set of rows comprises a single partition. Although the partitioning looks like a grouping using the GROUP BY clause, it is not the same thing. The GROUP BY clause collapses the rows in a partition into a single row, whereas the partitioning within the window construct simply organizes the rows into groups without collapsing them.

The following two examples show the difference between partitioning using the window construct and grouping using the GROUP BY clause. Suppose that you want to calculate several different aggregates concerning employees in each department. Example 24.3 shows how the OVER clause with the PARTITION BY clause can be used to build partitions.

Example 24.3

Using the window construct, build partitions according to the values in the **dept_name** column and calculate the sum and the average for the Accounting and Research departments:

```
USE sample;
SELECT dept_name,    budget,
            SUM( emp_cnt ) OVER( PARTITION BY dept_name ) AS emp_cnt_sum,
            AVG( budget ) OVER( PARTITION BY dept_name )  AS budget_avg
               FROM project_dept
               WHERE dept_name IN ('Accounting', 'Research');
```

The result is

dept_name	budget	emp_cnt_sum	budget_avg
Accounting	10000	27	30000
Accounting	40000	27	30000
Accounting	30000	27	30000
Accounting	40000	27	30000
Research	50000	20	61666.6666666667
Research	70000	20	61666.6666666667
Research	65000	20	61666.6666666667

Example 24.3 uses the OVER clause to define the corresponding window construct. Inside it, the PARTITION BY option is used to specify partitions. (Both partitions in Example 24.3 are grouped using the values in the **dept_name** column.) Finally, an aggregate function is applied to the partitions. (Example 24.3 calculates two aggregates, the sum of the values in the **emp_cnt** column and the average value of budgets.) Again, as you can see from the result of the example, the partitioning organizes the rows into groups without collapsing them.

Example 24.4 shows a similar query that uses the GROUP BY clause.

Example 24.4

Group the values in the **dept_name** column for the Accounting and Research departments and calculate the sum and the average for these two groups:

```
USE sample;
SELECT dept_name, SUM(emp_cnt) AS cnt, AVG( budget ) AS budget_avg
    FROM project_dept
    WHERE dept_name IN ('Accounting', 'Research')
    GROUP BY dept_name;
```

The result is

dept_name	cnt	budget_avg
Accounting	27	30000
Research	20	61666.6666666667

As already stated, when you use the GROUP BY clause, each group collapses down to one row.

NOTE There is another significant difference between the OVER clause and the GROUP BY clause. As can be seen from Example 24.3, when you use the OVER clause, the corresponding SELECT list can contain any column name from the table. This is obvious, because partitioning organizes the rows into groups without collapsing them. (If you add the **budget** column in the SELECT list of Example 24.4, you will get an error.)

Ordering and Framing

The ordering within the window construct is like the ordering in a query. First, you use the ORDER BY clause to specify the particular order of the rows in the result set. Second, it includes a list of sort keys and indicates whether they should be sorted in ascending or descending order. The most important difference is that ordering inside a window is applied only *within* each partition.

The Database Engine also supports ordering inside a window construct for aggregate functions. In other words, the OVER clause for aggregate functions can contain the ORDER BY clause, too. Example 24.5 shows this.

Example 24.5

Using the window construct, partition the rows of the **project_dept** table using the values in the **dept_name** column, sort the rows in each partition using the values in the **budget** column, and additionally display the sum of budgets:

```
USE sample;
SELECT dept_name, budget, emp_cnt,
    SUM(budget) OVER(PARTITION BY dept_name ORDER BY budget) AS sum_dept
    FROM project_dept;
```

The query in Example 24.5, which is generally called "cumulative aggregations," or in this case "cumulative sums," uses the ORDER BY clause to specify ordering within the particular partition. This functionality can be extended using framing. *Framing* means that the result can be further narrowed using two boundary points that restrict the set of rows to a subset, as shown in Example 24.6.

Example 24.6

```
USE sample;
SELECT dept_name, budget, emp_cnt,
    SUM(budget) OVER(PARTITION BY dept_name ORDER BY budget
        ROWS BETWEEN UNBOUNDED PRECEDING AND CURRENT ROW)
        AS sum_dept
 FROM project_dept;
```

Example 24.6 uses two clauses, UNBOUNDED PRECEDING and CURRENT ROW, to specify the boundary points of the selected rows. For the query in the example, this means that based on the order of the budget values, the displayed subset of rows includes those prior to the current row that have no low boundary point. (The result set contains 11 rows.)

The frame bounds used in Example 24.6 are not the only ones you can use. The UNBOUNDED FOLLOWING clause means that the specified frame does not have an upper boundary point. Also, both boundary points can be specified using an offset from the current row. In other words, you can use the *n* PRECEDING or *n* FOLLOWING clauses to specify *n* rows before or *n* rows after the current one, respectively. Therefore, the following frame specifies all together three rows—the current row, the previous one, and the next one:

```
ROWS BETWEEN 1 PRECEDING and 1 FOLLOWING
```

The Database Engine also supports two functions related to framing: LEAD and LAG. LEAD has the ability to compute an expression on the next rows (rows that are going to come after the current row). In other words, the LEAD function returns the next *n*th row value in an order. The function has three parameters: The first one specifies the name of the column to compute the leading row, the second one is the index of the leading row relative to the current row, and the last one is the value to return if the offset points to a row outside of the partition range. (The semantics of the LAG function is similar: it returns the previous *n*th row value in an order.)

Example 24.6 uses the ROWS clause to limit the rows within a partition by physically specifying the number of rows preceding or following the current row. Alternatively, you can use the RANGE clause, which logically limits the rows within a partition. In other words, when you use the ROWS clause, the exact number of rows will be specified based on the defined frame. On the other hand, the RANGE clause does not define the exact number of rows, because the specified frame can contain duplicates, too.

You can use several columns from a table to build different partitioning schemas in a query, as shown in Example 24.7.

Example 24.7

Using the window construct, build two partitions for the Accounting and Research departments: one using the values of the **budget** column and the other using the values of the **dept_name** column. Calculate the sums for the former partition and the averages for the latter partition.

```
USE sample;
SELECT dept_name, CAST( budget AS INT ) AS budget,
 SUM( emp_cnt ) OVER( PARTITION BY budget ) AS emp_cnt_sum,
 AVG( budget ) OVER( PARTITION BY dept_name )  AS budget_avg
 FROM project_dept
WHERE dept_name IN ('Accounting', 'Research');
```

The result is

dept_name	budget	emp_cnt_sum	budget_avg
Accounting	10000	5	30000
Accounting	30000	6	30000
Accounting	40000	16	30000
Accounting	40000	16	30000
Research	50000	5	61666.6666666667
Research	65000	5	61666.6666666667
Research	70000	10	61666.6666666667

The query in Example 24.7 has two different partitioning schemas: one over the values of the **budget** column and one over the values of the **dept_name** column. The former is used to calculate the number of employees in relation to the departments with the same budget. The latter is used to calculate the average value of budgets of departments grouped by their names.

Example 24.8 shows how you can use the NEXT VALUE FOR expression of the CREATE SEQUENCE statement to control the order in which the values are generated using the OVER clause. (For the description of the CREATE SEQUENCE statement, see Chapter 6.)

Example 24.8

```
USE sample;
CREATE SEQUENCE Seq START WITH 1 INCREMENT BY 1;
GO
CREATE TABLE T1 (col1 CHAR(10), col2 CHAR(10));
GO
INSERT INTO dbo.T1(col1, col2)
  SELECT NEXT VALUE FOR Seq OVER(ORDER BY dept_name ASC), budget
  FROM (SELECT dept_name, budget
```

Part IV

```
FROM project_dept
ORDER BY budget, dept_name DESC
OFFSET 0 ROWS FETCH FIRST 5 ROWS ONLY) AS D;
```

The content of the T1 table is as follows:

col1	col2
1	10000
2	30000
3	40000
4	40000
5	50000

The first two statements create the **Seq** sequence and the auxiliary table **T1**. The following INSERT statement uses a subquery to filter the five departments with the highest budget, and generates sequence values for them. This is done using OFFSET/FETCH, which is described in Chapter 6. (You can find a few other examples using OFFSET/FETCH in the subsection with the same name later in this chapter.)

Extensions of GROUP BY

Transact-SQL extends the GROUP BY clause with the following operators and functions:

- CUBE
- ROLLUP
- Grouping functions
- Grouping sets

The following sections describe these operators and functions.

CUBE Operator

This section looks at the differences between grouping using the GROUP BY clause alone and grouping using GROUP BY in combination with the CUBE and ROLLUP operators. The main difference is that the GROUP BY clause defines one or more columns as a group such that all rows within any group have the same values for those columns. CUBE and ROLLUP provide additional summary rows for grouped data. These summary rows are also called *multidimensional summaries.*

The following two examples demonstrate these differences. Example 24.9 applies the GROUP BY clause to group the rows of the **project_dept** table using two criteria: **dept_name** and **emp_cnt**.

Example 24.9

Using GROUP BY, group the rows of the **project_dept** table that belong to the Accounting and Research departments using the **dept_name** and **emp_cnt** columns and additionally calculate the sum of the budgets:

```
USE sample;
SELECT dept_name, emp_cnt, SUM(budget) sum_of_budgets
    FROM project_dept
    WHERE dept_name IN ('Accounting', 'Research')
    GROUP BY dept_name, emp_cnt;
```

The result is

dept_name	emp_cnt	sum_of_budgets
Accounting	5	10000
Research	5	115000
Accounting	6	70000
Accounting	10	40000
Research	10	70000

Example 24.10 and its result set shows the difference when you additionally use the CUBE operator.

Example 24.10

Group the rows of the **project_dept** table that belong to the Accounting and Research departments using the **dept_name** and **emp_cnt** columns and additionally display all possible summary rows:

```
USE sample;
SELECT dept_name, emp_cnt, SUM(budget) sum_of_budgets
    FROM project_dept
      WHERE dept_name IN ('Accounting', 'Research')
      GROUP BY CUBE (dept_name, emp_cnt);
```

The result is

dept_name	emp_cnt	sum_of_budgets
Accounting	5	10000
Research	5	115000
NULL	5	125000
Accounting	6	70000
NULL	6	70000

Part IV

dept_name	emp_cnt	sum_of_budgets
Accounting	10	40000
Research	10	70000
NULL	10	110000
Accounting	NULL	120000
Research	NULL	185000
NULL	NULL	305000

The main difference between the previous two examples is that the result set of Example 24.9 displays only the values in relation to the grouping, while the result set of Example 24.10 contains, additionally, all possible summary rows. (Because the CUBE operator displays every possible combination of groups and summary rows, the number of rows is the same, regardless of the order of columns in the GROUP BY clause.) The placeholder for the values in the unneeded columns of summary rows is displayed as NULL. For example, the following row from the result set:

```
NULL                  NULL            305000
```

shows the grand total (that is, the sum of all budgets of all existing projects in the table), while the row:

```
NULL                  5                       125000
```

shows the sum of all budgets for all projects that employ exactly five employees.

NOTE The syntax of the CUBE operator in Example 24.10 corresponds to the standardized syntax of that operator. Because of its backward compatibility, the Database Engine also supports the old-style syntax:

```
USE sample;
SELECT dept_name, emp_cnt, SUM(budget) sum_of_budgets
    FROM project_dept
      WHERE dept_name IN ('Accounting', 'Research')
      GROUP BY dept_name, emp_cnt
      WITH CUBE;
```

ROLLUP Operator

In contrast to CUBE, which returns every possible combination of groups and summary rows, the group hierarchy using ROLLUP is determined by the order in which the grouping columns are specified. Example 24.11 shows the use of the ROLLUP operator.

Example 24.11
Group the rows of the **project_dept** table that belong to the Accounting and Research departments using the **dept_name** and **emp_cnt** columns and additionally display summary rows for the **dept_name** column:

```
USE sample;
SELECT dept_name, emp_cnt, SUM(budget) sum_of_budgets
            FROM project_dept
              WHERE dept_name IN ('Accounting', 'Research')
              GROUP BY ROLLUP (dept_name, emp_cnt);
```

The result is

dept_name	emp_cnt	sum_of_budgets
Accounting	5	10000
Accounting	6	70000
Accounting	10	40000
Accounting	NULL	120000
Research	5	115000
Research	10	70000
Research	NULL	185000
NULL	NULL	305000

As you can see from the result of Example 24.11, the number of retrieved rows in this example is smaller than the number of displayed rows in the example with the CUBE operator. The reason is that the summary rows are displayed only for the first column in the GROUP BY ROLLUP clause.

NOTE The syntax used in Example 24.11 is the standardized syntax. The old-style syntax for ROLLUP is similar to the syntax for CUBE, which is shown in the second part of Example 24.10.

Grouping Functions

As you already know, NULL is used in combination with CUBE and ROLLUP to specify the placeholder for the values in the unneeded columns. In such a case, it isn't possible to distinguish NULL in relation to CUBE and ROLLUP from the NULL value. Transact-SQL supports the following two standardized grouping functions that allow you to resolve the problem with the ambiguity of NULL:

- GROUPING
- GROUPING_ID

The following subsections describe in detail these two functions.

GROUPING Function

The GROUPING function returns 1 if the NULL in the result set is in relation to CUBE or ROLLUP, and 0 if it represents the group of NULL values.

Example 24.12 shows the use of the GROUPING function.

Example 24.12

Using the GROUPING function, clarify which NULL values in the result of the following SELECT statement display summary rows:

```
USE sample;
SELECT dept_name, emp_cnt, SUM(budget) sum_b, GROUPING(emp_cnt) gr
        FROM project_dept
            WHERE dept_name IN ('Accounting', 'Marketing')
            GROUP BY ROLLUP (dept_name, emp_cnt);
```

The result is

dept_name	emp_cnt	sum_b	gr
Accounting	5	10000	0
Accounting	6	70000	0
Accounting	10	40000	0
Accounting	NULL	120000	1
Marketing	NULL	120000	0
Marketing	3	100000	0
Marketing	6	100000	0
Marketing	10	180000	0
Marketing	NULL	500000	1
NULL	NULL	620000	1

If you take a look at the grouping column (**gr**), you will see that some values are 0 and some are 1. The value 1 indicates that the corresponding NULL in the **emp_cnt** column specifies a summary value, while the value 0 (in the fifth row) indicates that NULL stands for itself (i.e., it is the NULL value).

GROUPING_ID Function

The GROUPING_ID function computes the level of grouping. GROUPING_ID can be used only in the SELECT list, HAVING clause, or ORDER BY clause when GROUP BY is specified.

Example 24.13 shows the use of the GROUPING_ID function.

Example 24.13

```
USE sample;
SELECT dept_name, YEAR(date_month), SUM(budget),
GROUPING_ID (dept_name,  YEAR(date_month)) AS gr_dept
    FROM project_dept
    GROUP BY ROLLUP (dept_name,  YEAR(date_month));
```

The result is

dept_name	date_month	budget	gr_dept
Accounting	2017	80000	0
Accounting	2018	40000	0
Accounting	NULL	120000	1
Marketing	2018	500000	0
Marketing	NULL	500000	1
Research	2017	185000	0
Research	NULL	185000	1
NULL	NULL	805000	3

The GROUPING_ID function is similar to the GROUPING function, but becomes very useful for determining the summarization of multiple columns, as is the case in Example 24.13. The function returns an integer that, when converted to binary, is a concatenation of the 1s and 0s representing the summarization of each column passed as the parameter of the function. For example, the value 3 of the **gr_dept** column in the last row of the result means that summarization is done over both the **dept_name** and **date_month** columns. The binary value $(11)_2$ is equivalent to the value 3 in the decimal system.

Grouping Sets

Grouping sets are an extension to the GROUP BY clause that lets users define several groups in the same query. You use the GROUPING SETS operator to implement grouping sets. Example 24.14 shows the use of this operator.

Example 24.14

Calculate the sum of budgets for the Accounting and Research departments using the combination of values of the **dept_name** and **emp_cnt** columns first, and after that using the values of the single column **dept_name**:

```
USE sample;
SELECT dept_name, emp_cnt, SUM(budget) sum_budgets
        FROM project_dept
        WHERE dept_name IN ('Accounting', 'Research')
        GROUP BY GROUPING SETS ((dept_name, emp_cnt),(dept_name));
```

The result is

dept_name	emp_cnt	sum_budgets
Accounting	5	10000
Accounting	6	70000
Accounting	10	40000

dept_name	emp_cnt	sum_budgets
Accounting	NULL	120000
Research	5	115000
Research	10	70000
Research	NULL	185000

As you can see from the result set of Example 24.14, the query uses two different groupings to calculate the sum of budgets: first using the combination of values of the **dept_name** and **emp_cnt** columns, and second using the values of the single column **dept_name**. The first three rows of the result set display the sum of budgets for three different groupings of the first two columns (Accounting, 5; Accounting, 6; and Accounting, 10). The fourth row displays the sum of budgets for all Accounting departments. The last three rows display the similar results for the Research department.

You can use the series of grouping sets to replace the ROLLUP and CUBE operators. For instance, the following series of grouping sets:

```
GROUP BY GROUPING SETS ((dept_name, emp_cnt), (dept_name), ())
```

is equivalent, except for several duplicated rows, to the following ROLLUP clause:

```
GROUP BY ROLLUP (dept_name, emp_cnt)
```

Also,

```
GROUP BY GROUPING SETS ((dept_name, emp_cnt), (emp_cnt, dept_name),
(dept_name), (emp_cnt), ())
```

is equivalent to the following CUBE clause:

```
GROUP BY CUBE (dept_name, emp_cnt)
```

OLAP Query Functions

Transact-SQL supports two groups of functions that are categorized as OLAP query functions:

- Ranking functions
- Statistical aggregate functions

The following subsections describe these functions.

NOTE The GROUPING function, discussed previously, also belongs to the OLAP functions.

Ranking Functions

Ranking functions return a ranking value for each row in a partition group. Transact-SQL supports the following ranking functions:

- RANK
- DENSE_RANK
- ROW_NUMBER

Example 24.15 shows the use of the RANK function.

Example 24.15

Find all departments with a budget not greater than 30000, and display the result set in descending order:

```
USE sample;
SELECT RANK() OVER(ORDER BY budget DESC) AS rank_budget,
                  dept_name, emp_cnt, budget
    FROM project_dept
    WHERE budget <= 30000;
```

The result is

rank_budget	dept_name	emp_cnt	budget
1	Accounting	6	30000
2	Accounting	5	10000

Example 24.15 uses the RANK function to return a number (in the first column of the result set) that specifies the rank of the row among all rows. The example uses the OVER clause to sort the result set by the **budget** column in descending order. (In this example, the PARTITION BY clause is omitted. For this reason, the whole result set will belong to only one partition.)

NOTE The RANK function uses logical aggregation. In other words, if two or more rows in a result set are tied (have a same value in the ordering column), they will have the same rank. The row with the subsequent ordering will have a rank that is one plus the number of ranks that precede the row. For this reason, the RANK function displays "gaps" if two or more rows have the same ranking.

Example 24.16 shows the use of the two other ranking functions, DENSE_RANK and ROW_NUMBER.

Example 24.16
Find all departments with a budget not greater than 40000, and display the dense rank and the sequential number of each row in the result set:

```
USE sample;
SELECT DENSE_RANK() OVER( ORDER BY budget DESC ) AS dense_rank,
     ROW_NUMBER() OVER( ORDER BY budget DESC ) AS row_number,
     dept_name, emp_cnt, budget
  FROM project_dept
  WHERE budget <= 40000;
```

The result is

dense_rank	row_number	dept_name	emp_cnt	budget
1	1	Accounting	10	40000
1	2	Accounting	6	40000
2	3	Accounting	6	30000
3	4	Accounting	5	10000

The first two columns in the result set of Example 24.16 show the values for the DENSE_RANK and ROW_NUMBER functions, respectively. The output of the DENSE_RANK function is similar to the output of the RANK function (see Example 24.15). The only difference is that the DENSE_RANK function returns no "gaps" if two or more ranking values are equal and thus belong to the same ranking.

The use of the ROW_NUMBER function is obvious: it returns the sequential number of a row within a result set, starting at 1 for the first row.

In the previous two examples, the OVER clause is used to determine the ordering of the result set. As you already know, this clause can also be used to divide the result set produced by the FROM clause into groups (partitions), and then to apply an aggregate or ranking function to each partition separately.

Example 24.17 shows how the RANK function can be applied to partitions.

Example 24.17
Using the window construct, partition the rows of the **project_dept** table according to the values in the **date_month** column. Sort the rows in each partition and display them in ascending order.

```
USE sample;
SELECT date_month, dept_name, emp_cnt, budget,
     RANK() OVER( PARTITION BY date_month ORDER BY emp_cnt desc ) AS rank
  FROM project_dept;
```

The result is

date_month	dept_name	emp_cnt	budget	rank
2017-01-01	Accounting	6	30000	1
2017-01-01	Research	5	50000	2
2017-02-01	Research	10	70000	1
2017-02-01	Accounting	10	40000	1
2017-07-01	Research	5	65000	1
2017-07-01	Accounting	5	10000	1
2018-01-01	Marketing	6	100000	1
2018-01-01	Marketing	NULL	120000	2
2018-02-01	Marketing	10	180000	1
2018-02-01	Accounting	6	40000	2
2018-07-01	Marketing	3	100000	1

The result set of Example 24.17 is divided (partitioned) into eight groups according to the values in the **date_month** column. After that the RANK function is applied to each partition.

Statistical Aggregate Functions

Chapter 6 introduced statistical aggregate functions. There are four of them:

- **VAR**   Computes the variance of all the values listed in a column or expression.
- **VARP**   Computes the variance for the population of all the values listed in a column or expression.
- **STDEV**   Computes the standard deviation of all the values listed in a column or expression. (The standard deviation is computed as the square root of the corresponding variance.)
- **STDEVP**   Computes the standard deviation for the population of all the values listed in a column or expression.

You can use statistical aggregate functions with or without the window construct. Example 24.18 shows how the functions VAR and STDEV can be used with the window construct.

Example 24.18

Using the window construct, calculate the variance and standard deviation of budgets in relation to partitions formed using the values of the **dept_name** column:

```
USE sample;
SELECT dept_name,    budget,
              VAR(budget) OVER(PARTITION BY dept_name) AS budget_var,
              STDEV(budget) OVER(PARTITION BY dept_name)  AS budget_stdev
          FROM project_dept
          WHERE dept_name in ('Accounting', 'Research');
```

The result is

dept_name	budget	budget_var	budget_stdev
Accounting	10000	200000000	14142.135624731
Accounting	40000	200000000	14142.135624731
Accounting	30000	200000000	14142.135624731
Accounting	40000	200000000	14142.135624731
Research	50000	108333333,333333	10408.3299973306
Research	70000	108333333,333333	10408.3299973306
Research	65000	108333333,333333	10408.3299973306

Example 24.18 uses the statistical aggregate functions VAR and STDEV to calculate the variance and standard deviation of budgets in relation to partitions formed using the values of the **dept_name** column.

Standard and Nonstandard Analytic Functions

The Database Engine contains the following standard and nonstandard analytic functions:

- TOP
- OFFSET/FETCH
- NTILE
- PIVOT and UNPIVOT
- STRING_AGG
- APPROX_COUNT_DISTINCT

The following sections describe these analytic functions and operators.

TOP Clause

The TOP clause specifies the first *n* rows of the query result that are to be retrieved. This clause should always be used with the ORDER BY clause, because the result of such a query is always well defined and can be used in table expressions. (A table expression specifies a sample of a grouped result table.) A query with TOP but without the ORDER BY clause is *nondeterministic*, meaning that multiple executions of the query with the same data may not always display the same result set.

Example 24.19 shows the use of this clause.

Example 24.19
Retrieve the four projects with the highest budgets:

```
USE sample;
SELECT TOP (4) dept_name, budget
    FROM project_dept
    ORDER BY budget DESC;
```

The result is

dept_name	budget
Marketing	180000
Marketing	120000
Marketing	100000
Marketing	100000

As you can see from Example 24.19, the TOP clause is part of the SELECT list and is written in front of all column names in the list.

NOTE You should write the input value of TOP inside parentheses, because the system supports any self-contained expression as input.

The TOP clause is a nonstandard ANSI SQL implementation used to display the ranking of the top *n* rows from a table. A query equivalent to Example 24.19 that uses the window construct and the standardized RANK function is shown in Example 24.20.

Example 24.20
Retrieve the four projects with the highest budgets:

```
USE sample;
SELECT dept_name, budget
     FROM (SELECT dept_name, budget,
                  RANK() OVER (ORDER BY budget DESC) AS rank_budget
                     FROM project_dept) part_dept
     WHERE rank_budget <= 4;
```

The TOP clause can also be used with the additional PERCENT option. In that case, the first *n* percent of rows are retrieved from the result set. The additional option WITH TIES specifies that additional rows will be retrieved from the query result if they have the same value in the ORDER BY column(s) as the last row that belongs to the displayed set. Example 24.21 shows the use of the PERCENT and WITH TIES options.

Example 24.21
Retrieve the top 25 percent of rows with the smallest number of employees:

```
USE sample;
SELECT TOP (25) PERCENT WITH TIES  emp_cnt, budget
   FROM project_dept
ORDER BY emp_cnt ASC;
```

The result is

emp_cnt	budget
NULL	120000
3	100000
5	50000
5	65000
5	10000

The result of Example 24.21 contains five rows, because there are three projects with five employees.

You can also use the TOP clause with UPDATE, DELETE, and INSERT statements. Example 24.22 shows the use of this clause with the UPDATE statement.

Example 24.22
Find the three projects with the highest budget amounts and reduce them by 10 percent:

```
USE sample;
UPDATE TOP (3) project_dept
   SET budget = budget * 0.9
    WHERE budget in (SELECT TOP (3) budget
    FROM project_dept
    ORDER BY budget desc);
```

Example 24.23 shows the use of the TOP clause with the DELETE statement.

Example 24.23
Delete the four projects with the smallest budget amounts:

```
USE sample;
DELETE TOP (4)
  FROM project_dept
  WHERE budget IN
                (SELECT TOP (4) budget FROM project_dept
                    ORDER BY budget ASC);
```

In Example 24.23, the TOP clause is used first in the subquery, to find the four projects with the smallest budget amounts, and then in the DELETE statement, to delete these projects.

OFFSET/FETCH

Chapter 6 showed how OFFSET/FETCH can be used for server-side paging. This application of OFFSET/FETCH is only one of many. Generally, OFFSET/FETCH allows you to filter several rows according to the given order. Additionally, you can specify how many rows of the result set should be skipped and how many of them should be returned. For this reason, OFFSET/FETCH is similar to the TOP clause. However, there are certain differences:

- OFFSET/FETCH is a standardized way to filter data, while the TOP clause is an extension of Transact-SQL. For this reason, it is possible that OFFSET/FETCH will replace the TOP clause in the future.
- OFFSET/FETCH is more flexible than TOP insofar as it allows skipping of rows using the OFFSET clause. (The Database Engine doesn't allow you to use the FETCH clause without OFFSET. In other words, even when no rows are skipped, you have to set OFFSET to 0.)
- The TOP clause is more flexible than OFFSET/FETCH insofar as it can be used in DML statements INSERT, UPDATE, and DELETE (see Examples 24.22 and 24.23).

Examples 24.24 and 24.25 show how you can use OFFSET/FETCH with the ROW_NUMBER ranking function. (Before you execute the following examples, repopulate the **project_dept** table. First delete all the rows, and then execute the INSERT statements from Example 24.2.)

Example 24.24

```
USE sample;
 SELECT date_month, budget, ROW_NUMBER()
    OVER (ORDER BY date_month DESC, budget DESC) as row_no
    FROM project_dept
    ORDER BY date_month DESC, budget DESC
    OFFSET 5 ROWS FETCH NEXT 4 ROWS ONLY;
```

The result is

date_month	budget	row_no
2017-07-01	65000	6
2017-07-01	10000	7
2017-02-01	70000	8
2017-02-01	40000	9

Example 24.24 displays the rows of the **project_dept** table in relation to the **date_month** and **budget** columns. (The first five rows of the result set are skipped and the next four are displayed.) Additionally, the row number of these rows is returned. The row number of the first row in the result set starts with 6 because row numbers are assigned to the result set *before* the filtering. (OFFSET/FETCH is part of the ORDER BY clause and therefore is executed after the SELECT list, which includes the ROW_NUMBER function. In other words, the values of ROW_NUMBER are determined before OFFSET/FETCH is applied.)

If you want to get the row numbers starting with 1, you need to modify the SELECT statement. Example 24.25 shows the necessary modification.

Example 24.25

```
USE sample;
SELECT *, ROW_NUMBER()
 OVER (ORDER BY date_month DESC, budget DESC) as row_no
 FROM (SELECT date_month, budget
   FROM project_dept
   ORDER BY date_month DESC, budget DESC
   OFFSET 5 ROWS FETCH NEXT 4 ROWS ONLY) c;
```

The result of Example 24.25 is identical to the result of Example 24.24 except that row numbers start with 1. The reason is that in Example 24.25 the query with OFFSET/FETCH is written as a table expression inside the outer query with the ROW_NUMBER function in the SELECT list. That way, the values of ROW_NUMBER are determined before OFFSET/FETCH is executed.

NTILE Function

The NTILE function belongs to the ranking functions. It distributes the rows in a partition into a specified number of groups. For each row, the NTILE function returns the number of the group to which the row belongs. For this reason, this function is usually used to arrange rows into groups. Example 24.26 shows the use of the NTILE function.

NOTE The NTILE function breaks down the data based only on the count of values.

Example 24.26

```
USE sample;
SELECT dept_name, budget,
  CASE NTILE(3) OVER (ORDER BY budget ASC)
    WHEN 1 THEN 'Low'
    WHEN 2 THEN 'Medium'
    WHEN 3 THEN 'High'
  END AS groups
 FROM project_dept;
```

The result is

dept_name	budget	groups
Accounting	10000	Low
Accounting	30000	Low
Accounting	40000	Low
Accounting	40000	Low
Research	50000	Medium
Research	65000	Medium
Research	70000	Medium
Marketing	100000	Medium
Marketing	100000	High
Marketing	120000	High
Marketing	180000	High

Pivoting Data

Pivoting data is a method that is used to transform data from a state of rows to a state of columns. Additionally, some values from the source table can be aggregated before the target table is created.

There are two operators for pivoting data:

- PIVOT
- UNPIVOT

The following subsections describe these operators in detail.

PIVOT Operator

PIVOT is a nonstandard relational operator that is supported by Transact-SQL. You can use it to manipulate a table-valued expression into another table. PIVOT transforms such an expression by turning the unique values from one column in the expression into multiple columns in the output, and it performs aggregations on any remaining column values that are desired in the final output.

To demonstrate how the PIVOT operator works, let's use a table called **project_dept_pivot**, which is derived from the **project_dept** table specified at the beginning of this chapter. The new table contains the **budget** column from the source table and two additional columns: **month** and **year**. The **year** column of the **project_dept_pivot** table contains the years 2017 and 2018, which appear in the **date_month** column of the **project_dept** table. Also, the **month** columns of the **project_dept_pivot** table (**january**, **february**, and **july**) contain the summaries of budgets corresponding to these months in the **project_dept** table.

Example 24.27 creates the **project_dept_pivot** table.

Example 24.27
```
USE sample;
SELECT budget, month(date_month) as month, year(date_month) as year
 INTO project_dept_pivot
FROM project_dept;
```

The content of the new table is given in Table 24-2.

Suppose that you get a task to return a row for each year, a column for each month, and the budget value for each year and month intersection. Table 24-3 shows the desired result.

Example 24.28 demonstrates how you can solve this problem using the standard SQL language.

Example 24.28
```
USE sample;
SELECT year,
    SUM(CASE WHEN month = 1 THEN budget END ) AS January,
     SUM(CASE WHEN month = 2 THEN budget END ) AS February,
        SUM(CASE WHEN month = 7 THEN budget END ) AS July
        FROM project_dept_pivot
        GROUP BY year;
```

budget	month	year
50000	1	2017
70000	2	2017
65000	7	2017
10000	7	2017
40000	2	2017
30000	1	2017
40000	2	2018
100000	1	2018
180000	2	2018
100000	7	2018
120000	1	2018

Table 24-2   Content of the project_dept_pivot Table

Year	January	February	July
2017	80000	110000	75000
2018	220000	220000	100000

Table 24-3   Budgets for Each Year and Month

The process of pivoting data can be divided into three steps:

1. **Group the data.** Generate one row in the result set for each distinct "on rows" element. In Example 24.28, the "on rows" element is the **year** column and it appears in the GROUP BY clause of the SELECT statement.

2. **Manipulate the data.** Spread the values that will be aggregated to the columns of the target table. In Example 24.28, the columns of the target table are all distinct values of the **month** column. To implement this step, you have to apply a CASE expression for each of the different values of the **month** column: 1 (January), 2 (February), and 7 (July).

3. **Aggregate the data.** Aggregate the data values in each column of the target table. Example 24.28 uses the SUM function for this step.

Example 24.29 solves the same problem as Example 24.28 using the PIVOT operator.

Example 24.29

```
USE sample;
SELECT year, [1] as January, [2] as February, [7] July FROM
    (SELECT budget, year, month from project_dept_pivot) p2
    PIVOT (SUM(budget)  FOR month IN ([1],[2],[7])) AS P;
```

The SELECT statement in Example 24.29 contains an inner query, which is embedded in the FROM clause of the outer query. The PIVOT clause is part of the inner query. It starts with the specification of the aggregation function: SUM (of budgets). The second part specifies the pivot column (**month**) and the values from that column to be used as column headings—the first, second, and seventh months of the year. The value for a particular column in a row is calculated using the specified aggregate function over the rows that match the column heading.

The most important advantage of using the PIVOT operator in relation to the standard solution is its simplicity in the case in which the target table has many columns. In this case, the standard solution is verbose because you have to write one CASE expression for each column in the target table.

UNPIVOT Operator

The UNPIVOT operator performs the reverse operation of PIVOT, by rotating columns into rows. Example 24.30 shows the use of this operator.

Example 24.30

```
USE sample;
CREATE TABLE project_dept_pvt (year int, January float, February float, July float);
INSERT INTO project_dept_pvt VALUES (2017, 80000, 110000, 75000);
INSERT INTO project_dept_pvt VALUES (2018, 50000, 80000, 30000);
--UNPIVOT the table
SELECT year, month, budget
```

Part IV

```
FROM
    (SELECT year, January, February, July
        FROM project_dept_pvt) p
            UNPIVOT (budget FOR month IN (January, February, July)
        )AS unpvt;
```

The result is

year	month	budget
2017	January	80000
2017	February	110000
2017	July	75000
2018	January	50000
2018	February	80000
2018	July	30000

Example 24.30 uses the **project_dept_pvt** table to demonstrate the UNPIVOT relational operator. UNPIVOT's first input is the column name (**budget**), which holds the normalized values. After that, the FOR option is used to determine the target column name (**month**). Finally, as part of the IN option, the selected values of the target column name are specified.

NOTE UNPIVOT is not the exact reverse of PIVOT, because any NULL values in the table being transformed cannot be used as column values in the output.

STRING_AGG Function

The STRING_AGG function concatenates rows of alphanumerical values into a single string and places a separator between the values. The separator is not added at the end of the string. The syntax of the STRING_AGG function is as follows:

```
string_agg ( input_string, separator ) [ order_clause ]
```

input_string is of any type that can be converted to the VARCHAR or NVARCHAR data types. **separator** is an expression of the NVARCHAR (VARCHAR) data type that is used as a separator for concatenated strings. It can be a literal or a variable. **order_clause** is optional and specifies the sort order of the concatenated strings using the WITHIN GROUP clause in the following way:

```
WITHIN GROUP (ORDER BY expression [ ASC | DESC] )
```

The following two examples show the use of the STRING_AGG function.

Example 24.31 uses the STRING_AGG function to generate lists of e-mail addresses of persons from the **person.person** table of the **AdventureWorks** database with "Michael" as the first name, and grouped according to the last name. In other words, e-mail addresses of all

Michaels with the same last name are aggregated, separated by a semicolon (;), and displayed in one row.

Example 24.31

```
USE AdventureWorks;
SELECT p.LastName, STRING_AGG(e.EmailAddress,';') Email_list
FROM person.person p JOIN person.EmailAddress e
     ON p.BusinessEntityID = e.BusinessEntityID
     WHERE p.FirstName = 'Michael'
GROUP BY LastName;
```

The first six rows of the result set are as follows:

LastName	Email_list
Allen	michael10@adventure-works.com;michael31@adventure-works.com;michael32@adventure-works.com
Anderson	michael42@adventure-works.com
Blythe	michael9@adventure-works.com;michael11@adventure-works.com
Bohling	michael12@adventure-works.com
Brown	michael36@adventure-works.com
Brundage	michael13@adventure-works.com

Example 24.32 is the same as Example 24.31 but with the addition of the WITHIN GROUP clause.

Example 24.32

```
USE AdventureWorks;
SELECT p.LastName, STRING_AGG(e.EmailAddress,';')
             WITHIN GROUP (ORDER BY EmailAddress) Email_list
FROM person.person p JOIN person.EmailAddress e
     ON p.BusinessEntityID = e.BusinessEntityID
     WHERE p.FirstName = 'Michael'
GROUP BY LastName;
```

The only difference between Examples 24.31 and 24.32 is that each row with concatenated e-mail addresses in Example 24.32 will be additionally sorted in ascending sort order, which is the default sort order.

APPROX_COUNT_DISTINCT Function

As you already know, COUNT DISTINCT can be used to get the unique number of non-null values from a table. Using this function can be very time consuming, especially for larger tables with millions of rows. Starting with SQL Server 2019, Microsoft allows you to avoid this performance bottleneck by using the APPROX_COUNT_DISTINCT function. The function evaluates an expression for each row in a group, and returns the approximate number of unique non-null values in the group. This function is designed to provide aggregations across large data sets where minimizing the response time is important.

The APPROX_COUNT_DISTINCT function is one of the features of the topic called *intelligent query processing*. For this reason, this function and its performance advantages in relation to COUNT DISTINCT will be described in Chapter 28.

Summary

SQL/OLAP extensions in Transact-SQL support data analysis facilities. There are four main parts of SQL/OLAP that are supported by the Database Engine:

- Window construct
- Extensions of the GROUP BY clause
- OLAP query functions
- Standard and nonstandard analytic functions

The window construct is the most important extension. In combination with ranking and aggregate functions, it allows you to easily calculate analytic functions, such as cumulative and sliding aggregates, as well as rankings. There are several extensions to the GROUP BY clause that are described in the SQL standard and supported by the Database Engine: the CUBE, ROLLUP, and GROUPING SETS operators as well as the grouping functions GROUPING and GROUPING_ID.

The most important analytic query functions are ranking functions: RANK, DENSE_RANK, and ROW_NUMBER. Transact-SQL supports several nonstandard analytic functions and operators, TOP, NTILE, PIVOT, and UNPIVOT, as well as a standard one, OFFSET/FETCH.

There are also two new functions: STRING_AGG and APPROX_COUNT_DISTINCT. The former concatenates rows of alphanumerical values into a single string and places a separator between the values. The latter evaluates an expression for each row in a group and returns the approximate number of unique non-null values in the group.

The next chapter describes Reporting Services, a business intelligence component of SQL Server.

Exercises

E.24.1   Find the average number of the employees in the Accounting department. Solve this problem:

    a.   Using the window construct

    b.   Using the GROUP BY clause

E.24.2   Using the window construct, find the department with the highest budget for the years 2017 and 2018.

E.24.3   Find the sum of employees according to the combination of values in the departments (the **dept_name** column) and budget amounts (the **budget** column). Display all possible summary rows, too.

E.24.4   Solve E.24.3 using the ROLLUP operator. What is the difference between this result set and the result set of E.24.3?

E.24.5   Using the RANK function, find the three departments with the highest number of employees.

E.24.6   Solve E.24.5 using the TOP clause.

E.24.7   Calculate the ranking of all departments for the year 2018 according to the number of employees. Display the values for the DENSE_RANK and ROW_NUMBER functions, too.

25   SQL Server Reporting Services

In This Chapter

- Reports: An Introduction
- SQL Server Reporting Services Architecture
- Installation and Configuration of Reporting Services
- Creating Reports
- Managing and Tuning Reports
- Reporting Services Security

This chapter describes Microsoft's enterprise reporting tool called SQL Server Reporting Services (SSRS). After introducing the general structure of a report, the chapter explains the main components of Reporting Services and describes the installation and the configuration of an instance of Reporting Services. After that, the creation of reports is discussed using three examples. The first example shows the creation of a basic report, the second demonstrates how the functionality of reports can be extended using parameters, and the third example shows how SSRS supports visualization of reports through charts. After that, you'll see how reports can be managed and delivered using the Reporting Services web portal component. The last section discusses security and performance of reports.

Reports: An Introduction

A report is one of the interfaces through which users can interact with database systems. Using reports, the data is visualized and displayed to users. The data can be displayed in many different formats, which will be discussed later in this chapter.

Generally, data reports have the following properties:

- *A report can be used only to display data.* In contrast to forms-based interfaces, which can be used both to read and modify data, a report is a read-only user interface. You can use reports to view data statically or dynamically.

- *A report generator supports many different layouts and file formats.* A report presents data in a preformatted form. In other words, you use a report layout to compose items that correspond to result values of the report query. (You can generally design and visualize your reports either in graphical or tabular form.)

- *A report is always based on a corresponding SELECT statement.* Because reports can be used only to display data, each report is based on a SELECT statement, which retrieves data. The difference between the result of a retrieval operation and the corresponding report is that the latter uses various styles and formats to display data, while the result set of each SELECT statement has a static form, which is arranged by the system.

- *A report can use parameters, which are part of a particular query and whose values are set at run time.* You can use parameters, as a part of your query, to add flexibility to reports. The values of such parameters are passed from a user or an application program to the query.

- *One or more sources always provide input data for the report.* A query, which retrieves data, is stored in one or more sources that are usually relational databases, but can be other data sources, too.

Concerning its structure, each report has the following two instruction sets, which together specify the content of the report:

- **Data definition**   Specifies data sources and a dataset. Data sources are all databases and files that provide input data to report, as described. The content of the dataset is the output specified with the corresponding SELECT statement.

- **Report layout**   Enables you to present selected data to users. You can specify which column values correspond to which fields. You can also specify layout details such as the style and location of headings and numbers.

Now that you've had this general introduction to reports, you are ready to look at the architecture of SQL Server Reporting Services.

SQL Server Reporting Services Architecture

SQL Server Reporting Services is a Microsoft software system that you can use to create, manage, and deploy reports. SSRS includes three main components, which represent a server layer, a metadata layer, and an application layer, respectively:

- Reporting Services Windows service
- Report Catalog
- Reporting Services web portal

NOTE The first two components are described in the following sections, after a brief introduction to the Report Definition Language. The Reporting Services web portal is, since SQL Server 2016, a component that replaces Report Manager and significantly extends its functionality. The Reporting Services web portal will be discussed in detail later in this chapter.

When the information concerning data definition and report layout is gathered, SSRS stores it using the Report Definition Language (RDL). RDL is an XML-based language that is used exclusively for storing report definitions and layouts. RDL is an open schema language, meaning that developers can extend the language with additional XML attributes and elements. (For descriptions of XML in general and XML elements and attributes in particular, see Microsoft Docs.) RDL is usually generated via Visual Studio, although it may also be used programmatically.

A typical RDL file contains three main sections. The first concerns page style, the second section specifies field definitions, and the last section defines parameters.

NOTE It is not necessary for you to understand the details concerning RDL to develop reports. The language is important only when you want to create your own RDL files.

Reporting Services Windows Service

As its name implies, the Reporting Services Windows service is implemented as a Windows component that comprises two important services in relation to the report server. The first one, the Reporting Services Web service, is the application used for the implementation of the SSRS web site, while the second one, the Reporting Services Windows service, allows you to use it as a programmatic interface for reports.

NOTE Both services, the Reporting Services Windows service and the Reporting Services Web service, work together and constitute a single report server instance.

As you can see from Figure 25-1, Reporting Services Windows service includes the following components:

- Report processor
- Data providers
- Renderers
- Request handler

The report processor manages the execution of a report. It retrieves the definition of the report, which is done in RDL, and determines what is needed for the report. Also, it manages the work of other components that are used to produce a report. The report processor also retrieves data from data sources. After that, it selects a data provider that knows how to extract information from the data source.

SQL Server Reporting Services includes a set of renderers for exporting reports to different formats. Each renderer applies rules when rendering reports. When you export a report to a different file format, especially for renderers such as the Adobe Acrobat (PDF) renderer, you might need to change the layout of your report to have the exported report look and print correctly after the rendering rules are applied.

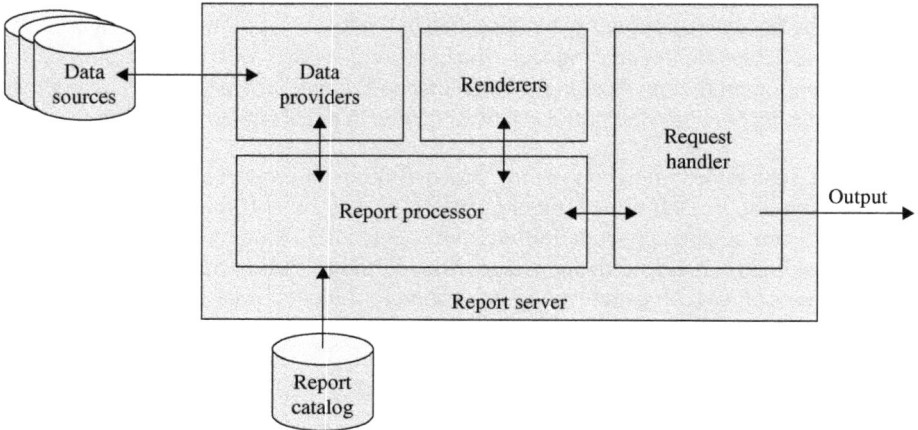

Figure 25-1 Components of Reporting Services Windows service

The task of the data providers is to connect to the data source, get the information for the report, and return it to the processor in the form of corresponding datasets. After that, the report processor turns this information into a dataset, to be used by the report. When data providers deliver the data for the report, the report processor can begin to process the report's layout. To render the report, the processor has to know its format (HTML or PDF, for instance).

The request handler receives requests for reports and sends them to the report processor. It also delivers the completed report. (The different forms of report delivery will be discussed later in the chapter.)

Report Catalog

The Report Catalog contains two databases that are used to store the definitions of all existing reports that belong to a particular service. The stored information includes report names, descriptions, data source connection information, credential information, parameters, and execution properties. The Report Catalog also stores security settings and information concerning scheduling and delivering data.

SSRS uses two databases, the Report Server database and the Report Server temporary database, to separate persistent data storage from temporary storage requirements. The databases are created together and bound by name. By default, the database names are **reportserver** and **reportservertempdb**, respectively. The former is used to store the Report Catalog, while the latter is used as temporary storage for cached reports and the work tables that they generate. If you do not want to use the default names for these databases, you can specify different names in the configuration process.

Installation and Configuration of Reporting Services

Before we delve into the installation and configuration of SSRS, you have to know what native mode is and how you can administer a Reporting Services installation.

In native mode, a report server is a stand-alone application server that provides all viewing, management, processing, and delivery of reports. A Reporting Services report server configured for native mode runs as an application server that provides all processing and management capability exclusively through Reporting Services components. You can use Reporting Services web portal to manage these reports. (The alternative to native mode, called SharePoint mode, is no longer available after SQL Server 2016.)

To fully administer a Reporting Services installation, you must have the following permissions:

- Membership in the local Administrator group on the report server computer
- Database administrator permissions for the SQL Server instance that hosts the database, which you use to create reports

SQL Server Reporting Services is no longer included in the SQL Server installation. Therefore, you need to download the setup file first. There are a few ways to download the corresponding file:

- As mentioned in Chapter 2, you can do it during the installation of SQL Server 2019 on the Feature Selection page. To the right of Looking for Reporting Services?, click the Download It from the Web link (see Figure 2-5 in Chapter 2).
- You can download it directly from the following page: https://www.microsoft.com/en-us/download/details.aspx?id=55252

After the download process is finished, launch the setup. Click Install Reporting Services. Enter the product key if you have one, or choose one of the free editions. For this installation, I selected the free edition (valid for 180 days). After this step, the installation process is straightforward and thus I won't discuss it further.

Once installation is complete, you can either finish the installation process or start to configure the report server. To perform configuration, click Configure Report Server. This opens the Report Server Connection dialog box with an auto-created instance name of SSRS. Click Connect.

Once connected, click Service Account in the left navigation pane to open the Service Account page (see Figure 25-2). Enter in the Account (Domain\user) field an existing service account you want to use for SSRS. As shown in Figure 25-2, I entered the name of my account, LAPTOP-TVUM0CNL\Petkovic, which I created previously. Click Apply and make sure the task is completed successfully.

Click Web Service URL in the left pane to open the corresponding page. Type in the Virtual Directory field a name of your choice. (The default name is ReportServer.) Click Apply and make sure the task is completed.

Click Database to open the Database page, and click Change Database to launch the Report Server Database Configuration Wizard, which takes you through the following steps to create a new report server database:

1. On the Action page of the wizard, click the Create a New Report Server Database radio button and click Next.

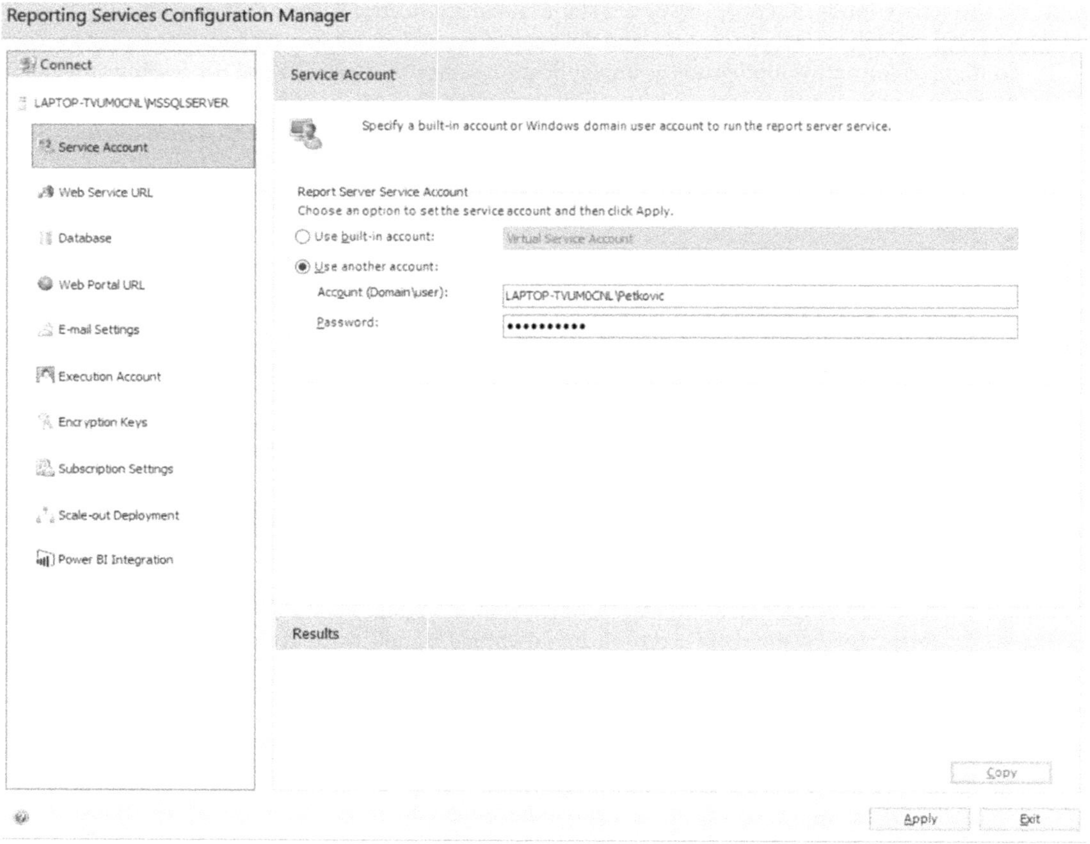

Figure 25-2 SSRS Configuration Manager: Service Account page

2. On the Database Server page, in the Server Name field (see Figure 25-3), specify a SQL Server name along with the instance name and click Test Connection to make sure that the connection to the specified database succeeds. Click Next.

3. On the Database page, in the Database Name field, provide a database name of your choice. (I kept the default name, ReportServer.) Click Next.

4. On the Credentials page, you should see the service account that you used previously. Click Next.

5. On the Summary page, verify that all the information you provided is correct and, if so, click Next. (If not, click Previous to return to the page where you need to make corrections.)

6. The Progress and Finish page shows a progress bar as the wizard configures the database, and then shows the configuration of Report Server Database Configuration with all parameters as Success (see Figure 25-4). Click Finish.

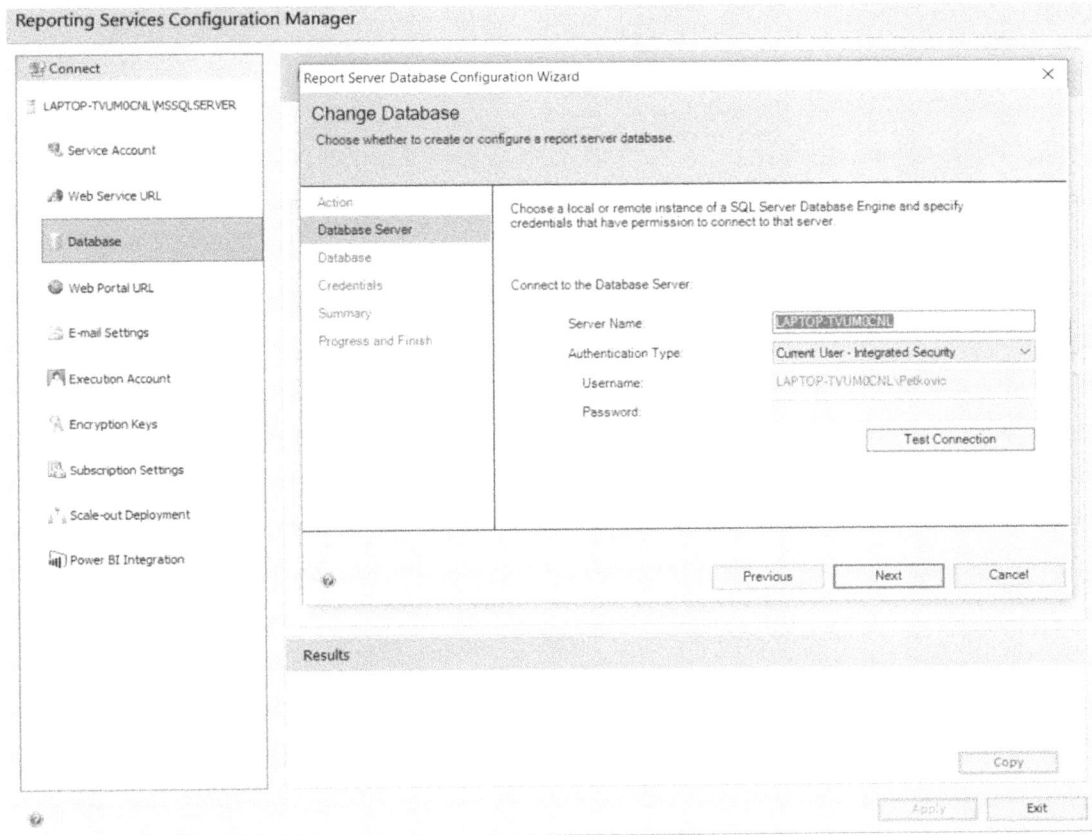

Figure 25-3   Report Server Database Configuration Wizard: Database Server page

Finally, back in the main SSRS Configuration Manager window, click web portal URL on the left to open the corresponding page and specify in the Virtual Directory field the name of the virtual directory to use as the website URL for reports. (I kept the default name, Reports.) Click Apply, and verify that the task completed successfully.

Now that you are familiar with the components of SQL Server Reporting Services as well as its installation and configuration, you will learn how to create, deploy, and deliver reports.

Creating Reports

SQL Server Reporting Services gives you two tools for creating reports:

- **SQL Server Data Tools (SSDT)**   A development tool that you use during the development phase. It is tightly integrated with Visual Studio and allows you to develop and test reports before their deployment.

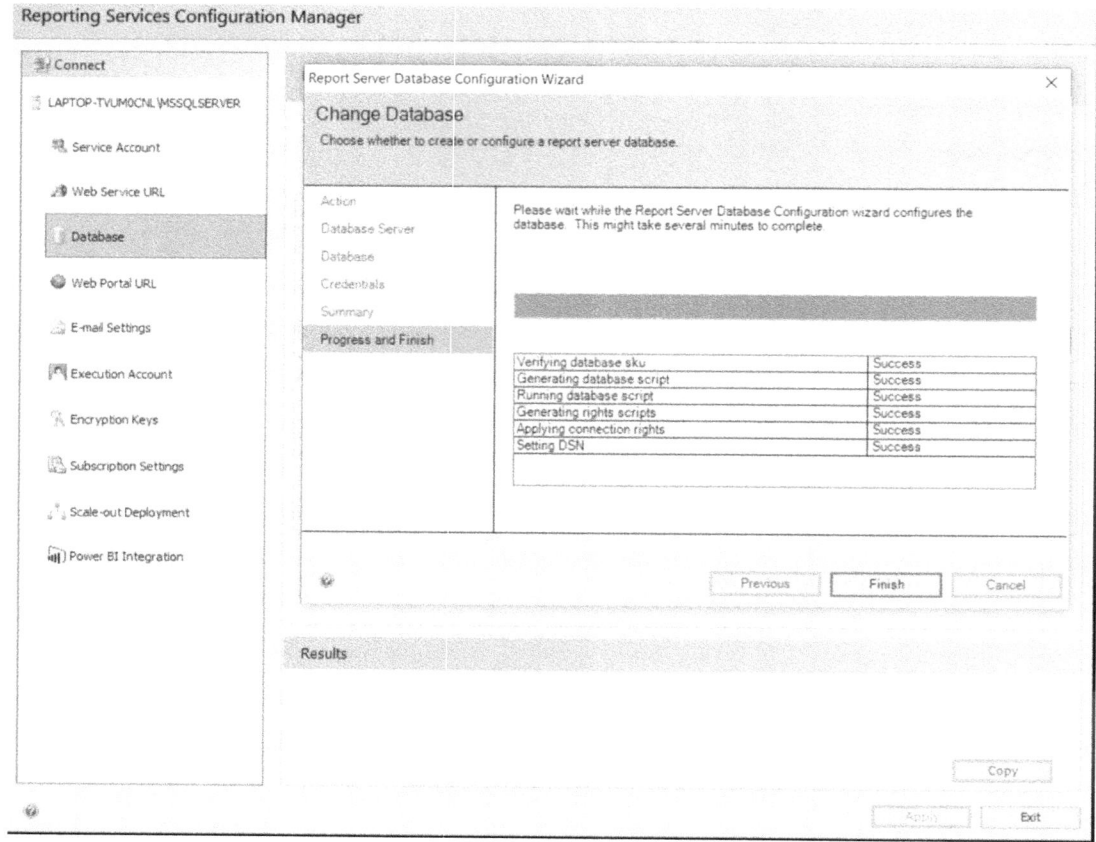

Figure 25-4   Report Server Database Configuration Wizard: Progress and Finish page

- **Report Builder**   A stand-alone tool that enables you to do ad hoc reporting without knowing anything about the structure of a particular database or how to create queries using SQL.

NOTE   This book discusses creating reports using SQL Server Data Tools only. The reason is that both tools deliver almost the same functionality, and therefore the description of one tool suffices to understand the basics of reporting with SSRS. For the description of Report Builder, see Microsoft Docs.

Before I show how you can create a report, I will describe the installation process of SSDT.

Installation of SQL Server Data Tools

As mentioned, SQL Server Data Tools is highly integrated with Visual Studio. In this book we use Visual Studio 2019 and SQL Server Data Tools 2017.

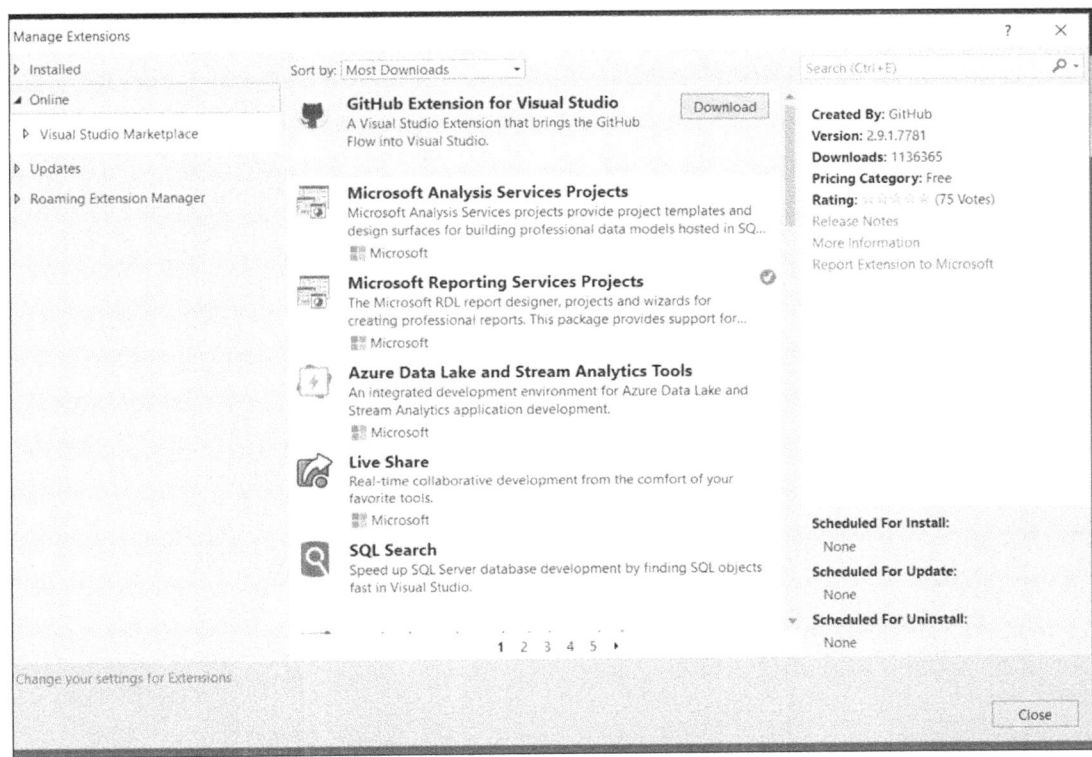

Figure 25-5   Installation of SSDT as an extension of Visual Studio

---

NOTE The user interface of Visual Studio 2019 is significantly different in many ways from the user interface of Visual Studio 2017. (The detailed description of the installation for Visual Studio 2019 is given in the section "Creating a New Project Using Visual Studio 2019" in Chapter 23.)

After installation of Visual Studio 2019, you need to install SSDT. To install SSDT, click Extensions in the main menu of Visual Studio 2019, click Manage Extensions, click Install, and select Online | Visual Studio Marketplace in the left navigation pane, as shown in Figure 25-5. The list of the components that you can install is shown in the middle of Figure 25-5.

Choose Microsoft Reporting Services Projects and click Install. The installation process starts when you close all active windows in Visual Studio.

Creating Your First Report

To create a report, start Visual Studio and click Create a New Project from the list of tasks. On the Create a New Project page (see Figure 25-6), you will see two project templates related to reports: Report Server Project Wizard and Report Server Project. The Report Server Project Wizard guides you during the creation phase of a new report. The Report Server Project creates an empty report and leaves you alone to do the rest of the work. This section shows you how to

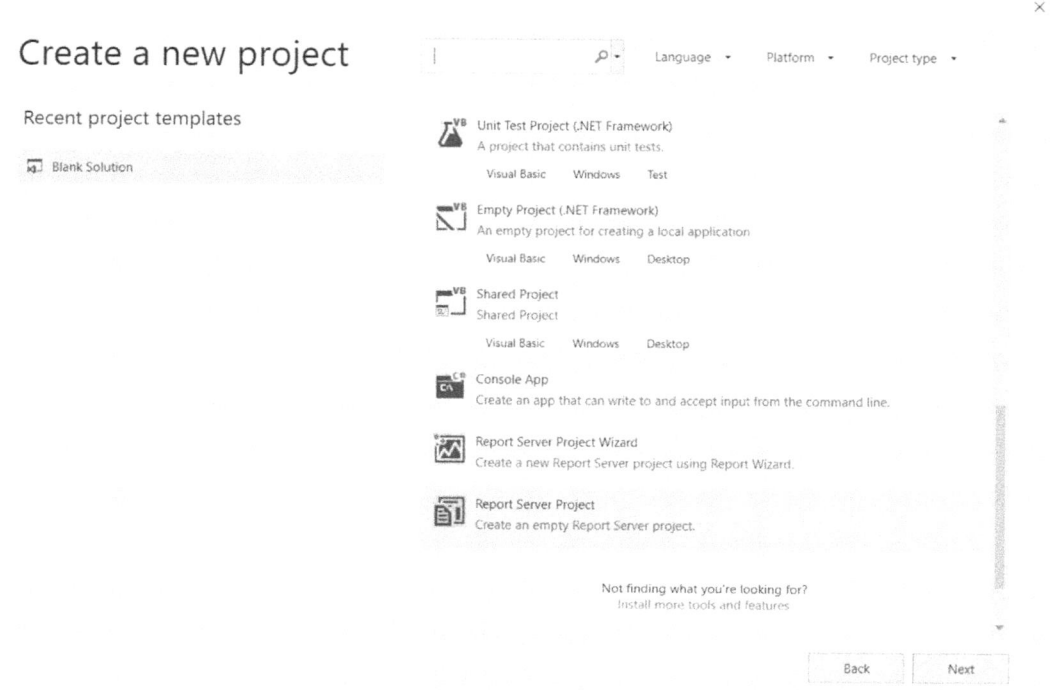

Figure 25-6 Create a New Project page

use the wizard to create reports, while the subsequent section shows you how to start with an empty report using the Report Server Project template.

NOTE Generally, SSRS reports are built using Report Designer, a collection of graphical query and design tools. It provides the Report Data pane, which you can use to organize data used in your report, and tabbed views for Design and Preview so that you can design a report interactively. Report Designer also includes Query Designer, to help you specify data for retrieval, and the Expression dialog box, to specify report data to use in the report layout. (All graphical and design tools will be explained later in this chapter.)

To create a new report with the wizard's guidance, select Report Server Project Wizard. Type the name for your project (**Project1**, in my case). Use the default filename in the Project Location dialog box. Click Create. This leads you to the welcome page of the wizard.

The Report Server Project Wizard welcome page introduces the major steps that it takes you through to create a report:

1. Select a data source from which to retrieve data.

2. Design a query to execute against the data source.

3. Choose the type of report you want to create.

4. Design the data in the table.

5. Choose the table layout.

These steps are described in the upcoming subsections, followed by a quick summary of how to preview the result set. First, though, you will learn how to plan your data sources and datasets.

Planning Data Sources and Datasets

Before you create a report, you should prepare your data sources for use. You use these sources to create the corresponding tables and/or views, which will be used to retrieve the particular result set. For this reason, the environment in which you prepare data sources is SQL Server Management Studio, with its Transact-SQL capabilities, rather than SSRS.

During the planning phase, you have to work with both data sources and datasets. The differences between these two concepts are described next, followed by a description of how you can use both of them.

Data Sources vs. Datasets   The most important difference between data sources and datasets is that data sources are not included in your report. A data source just delivers information for your report. This information can be stored in a database and/or a file. The task of SSRS is to generate datasets from the given data sources using the set of instructions. This set includes, among other things, the information concerning the type of the source, the name of the database (if the data source is stored in a database) or the file path, and optionally the connection information to the source.

During the execution of a report, SSRS uses this information to generate a new format, called a *dataset*. Therefore, a dataset is just an abstraction of underlying data sources and is used as a direct input for the corresponding report.

Using Data Sources   To include data in a report, you must first create data connections (a synonym for data sources). A data connection includes the data source type, connection information, and the login for connecting. (Creation of data connections is described in the next section, "Selecting a Data Source.")

There are two types of data sources: embedded and shared. An embedded data source is defined in the report and used only by that report. In other words, when you want to modify an embedded data source, you have to change the properties of that data source. A shared data source is defined independently from a report and can be used by multiple reports. Shared data sources are useful when you have data sources that are often used.

Using Datasets   As you already know, each dataset is an abstraction of corresponding data sources and therefore specifies the fields from the data sources that you plan to use in the report. All datasets that you create for a report definition appear in the Datasets window.

Similar to data sources, SSRS supports shared datasets, too. Simply put, a shared dataset is a dataset that allows several reports to share a query to provide a consistent set of data for multiple reports. Generally, you use shared data sources to generate shared datasets.

The query of a shared dataset can include parameters. You can configure a shared dataset to cache query results for specific parameter combinations on first use or by specifying a schedule.

Now we will begin to create a report with the wizard. The first step is selection of a data source, described next.

Selecting a Data Source

The data source contains information about the connection to the particular database. Click Next on the welcome page of the Report Server Project Wizard to select the data source. On the Select the Data Source page, type the name of the new data source in the Name field.

The Type drop-down list on the Select Data Source page allows you to choose one of the different data source types. SSRS can create reports from different relational databases (SQL Server, Teradata, and Oracle, among others) or multidimensional databases (SQL Server Analysis Services). OLE DB, ODBC, and Azure data sources can be used, too. After you choose a type (Microsoft SQL Server), click Edit. The Connection Properties dialog box appears (see Figure 25-7).

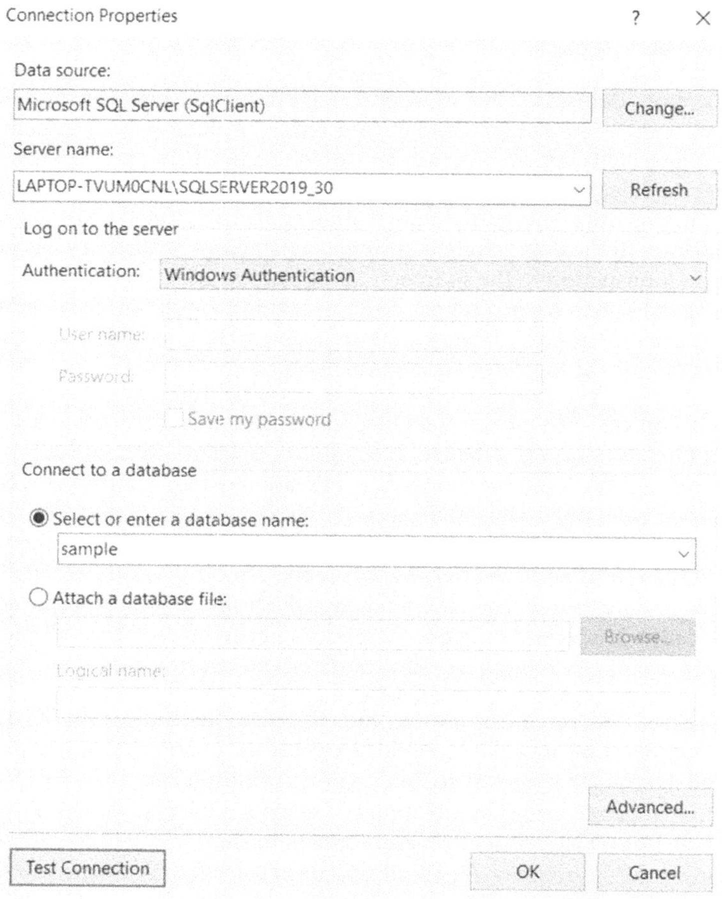

Figure 25-7 The Connection Properties dialog box

Type the name of your database server instance as the server name. Below that, choose Windows Authentication in the drop-down list. Click the Select or Enter a Database Name radio button and choose from the drop-down list one of the databases as the data source. (For the first report use the **sample** database.) Before you click OK, click the Test Connection button to test the connection to the database. Clicking OK takes you back to the Select the Data Source page. Click Next to start the next step of the wizard.

Designing a Query

The next step is to design a query to be executed against the selected data source. On the Design the Query page, you can either type (or paste) an existing query or use the Query Builder component to create a query from scratch.

NOTE Query Builder corresponds to the similar Access component that you can use to design queries even if you have no knowledge of the SQL language. This component is generally known as QBE (query by example).

For this first report, type the query given in Example 25.1 and click Next.

Example 25.1

```
SELECT dept_name, emp_lname, emp_fname, job, enter_date
   FROM department d JOIN employee e ON d.dept_no = e.dept_no
                    JOIN works_on w ON w.emp_no = e.emp_no
   WHERE YEAR(enter_date) = 2017
   ORDER BY dept_name;
```

The query in Example 25.1 is related to the **sample** database. It selects data for employees who entered their job in 2017. The result set of the query is then sorted by department names.

NOTE In the Designing a Query step, SSRS checks the names of the tables and columns listed in the query. If the system finds any syntax errors, it displays the corresponding message in the new window.

Choosing the Report Type

The next step in creating a report is to select the report type. You can choose between two report types:

- **Tabular**   Creates a report in tabular form. Columns of the table correspond to the columns from the SELECT list, while the number of rows in the table depends on the result set of the query.

- **Matrix**   Creates a report in matrix form, which is similar to table form but provides functionality of crosstabs. Unlike the tabular report type, which has a static set of columns, the matrix report type can be dynamic.

NOTE You should use the matrix report type whenever you want to create queries that contain aggregate functions, such as AVG or SUM.

The query in Example 25.1 does not contain any aggregate functions. Therefore, choose the tabular report type and click Next.

Designing the Data in the Table

The Design the Table page allows you to decide where selected columns will be placed in your report. The Design the Table page contains two groups of fields:

- Available fields
- Displayed fields

The page also has three views:

- Page
- Group
- Details

Available fields are the columns from the SELECT list of your query. Each column can be moved to one of the views. To move a field to the Page, Group, or Details view, select the field and then click the Page, Group, or Details button, respectively. A *displayed field* is an available field that is assigned to one of the existing views.

Page view lists all columns that appear at the page level, and Group view lists columns that are used to group the result set of the query. Details view is used to display columns that appear in the detail section of the table. In Example 25.1, the **dept_name** column will appear at the page level, while the **job** column will be used to group the selected rows. When you have chosen how to group the data in the table, click Next.

NOTE The order of the columns can be important, especially for the Group view. To change the order of the columns, select a column and click the up or down button to the right.

Choosing the Table Layout

The next step is to specify the layout of your report. The Choose the Table Layout page has these options:

- Stepped
- Block
- Include subtotals
- Enable drilldown

If you choose Stepped, the report will contain one column for each field, with grouping fields appearing in headers to the left of columns from the detail field. In this case, the group footer will not be created. If you include subtotals with this layout type, the subtotal is placed in the group header rows.

The Block option creates a report that contains one column for each field, with group fields appearing in the first detail row for each group. This layout type has group footers only if the Include Subtotals option is activated.

The Enable Drilldown option hides the inner groups of the report and enables a visibility toggle. (You can choose Enable Drilldown only if you select the Stepped option.)

To continue, choose Stepped and Include Subtotals and click Next. After that, you will be prompted to complete the wizard.

Finally, you complete the wizard's work by providing a name for the report. (This report has the name **Report1**.) Also, you can take a look at the report summary, where all your previous steps during the creation of the report are documented. Click Finish to finish the wizard.

Previewing the Result Set

When you finish the creation of your report using the wizard, the Report Designer pane automatically appears. (If the Report Designer pane is not visible, click View | Designer.) In the pane, there are two tabs, Design and Preview, which you can use to view the created report in different forms.

The Design tab allows you to view and modify the layout of your report. The Design mode consists of the following sections: body, page, header, and page footer. You can use the Toolbox and Properties windows to manipulate items in the report. To view these windows, select Toolbox or Properties Window in the View menu. Use the Toolbox window to select items to place them in one of the sections. Each item on the report design surface contains properties that can be managed using the Properties window.

To preview the report, click the Preview tab (see Figure 25-8). The report runs automatically, using already specified properties. To finish the report, select File | Save All. That way you save your project and all files in relation to the created report.

Creating a Parameterized Report

A *parameterized* report is one that uses input parameters to complete report processing. The parameters are then used to execute a query that selects specific data for the report. If you design or deploy a parameterized report, you need to understand how parameter selections affect the report.

Parameters in SQL Server Reporting Services are used to filter data. They are specified using the standard syntax for variables (**@year**, for instance). If a parameter is specified in a query, a value must be provided to complete the SELECT statement or stored procedure that retrieves data for a report.

You can define a default value for a parameter. If all parameters have default values, the report will immediately display data when the report is executed. If at least one parameter does not have a default value, the report will display data after the user enters all needed parameter values.

When the report is run in a browser, the parameter is displayed in a box at the top of the report. When the report is run in Preview mode, the value of the parameter is displayed in the corresponding box.

An example will be used to show you how to create a parameterized report. This example describes only those steps that are different from the steps already discussed in relation to Example 25.1.

To create a new report, go to Solution Explorer in Visual Studio, click Project1, and then right-click Reports and select Add New Report. The Report Wizard appears. Click Next.

Part IV

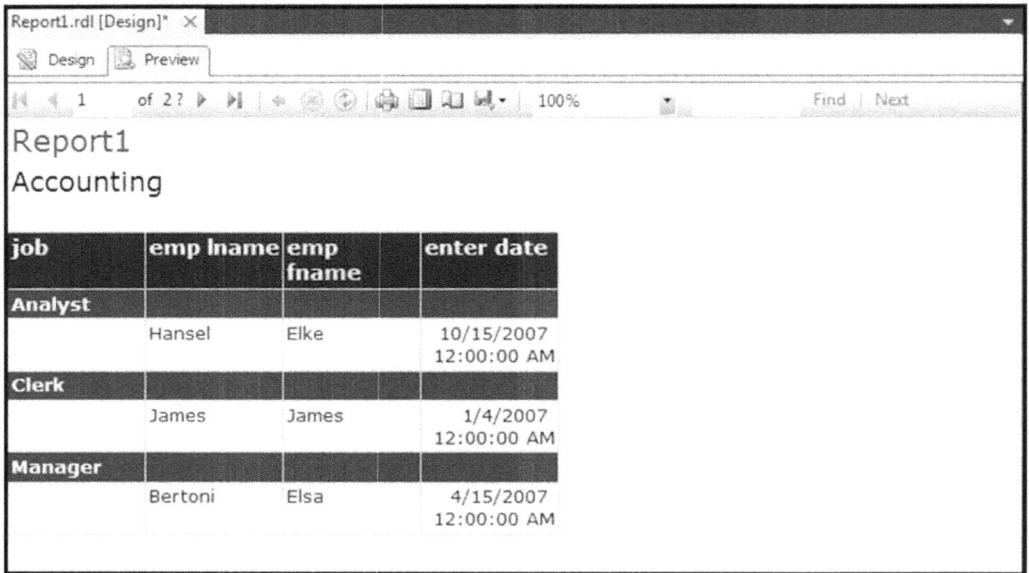

Figure 25-8 The result of the preview of the report

The selection of a new source is identical to the specification of the first report's source, except that you choose the **AdventureWorksDW** database instead of the **sample** database.

The query shown in Example 25.2 will be used to select data from the data source. It retrieves data from the **AdventureWorksDW** database. If you take a closer look at the SELECT statement, you will see that it calculates the number and the sum of unit product prices and groups the rows according to the list of column names in the GROUP BY clause, in that order. Again, type the query from Example 25.2 in the Design the Query page.

Example 25.2

```
SELECT  t.MonthNumberOfYear AS month,    t.CalendarYear   AS year,
        p.ProductKey AS product_id,    SUM(f.UnitPrice)  AS sum_of_sales,
        COUNT(f.UnitPrice)   AS total_sales
        FROM DimDate t, DimProduct p, FactInternetSales f
        WHERE t.DateKey    = f.OrderDateKey AND   p.ProductKey = f.ProductKey
            AND CalendarYear = @year
        GROUP BY t.CalendarYear, t.MonthNumberOfYear,  p.ProductKey
        ORDER BY 1;
```

The expression

```
CalendarYear = @year
```

in the WHERE clause of the example specifies that the input parameter **@year** in this query is related to the calendar year for which you want to query data. Therefore, this value will be used later to generate the report's preview.

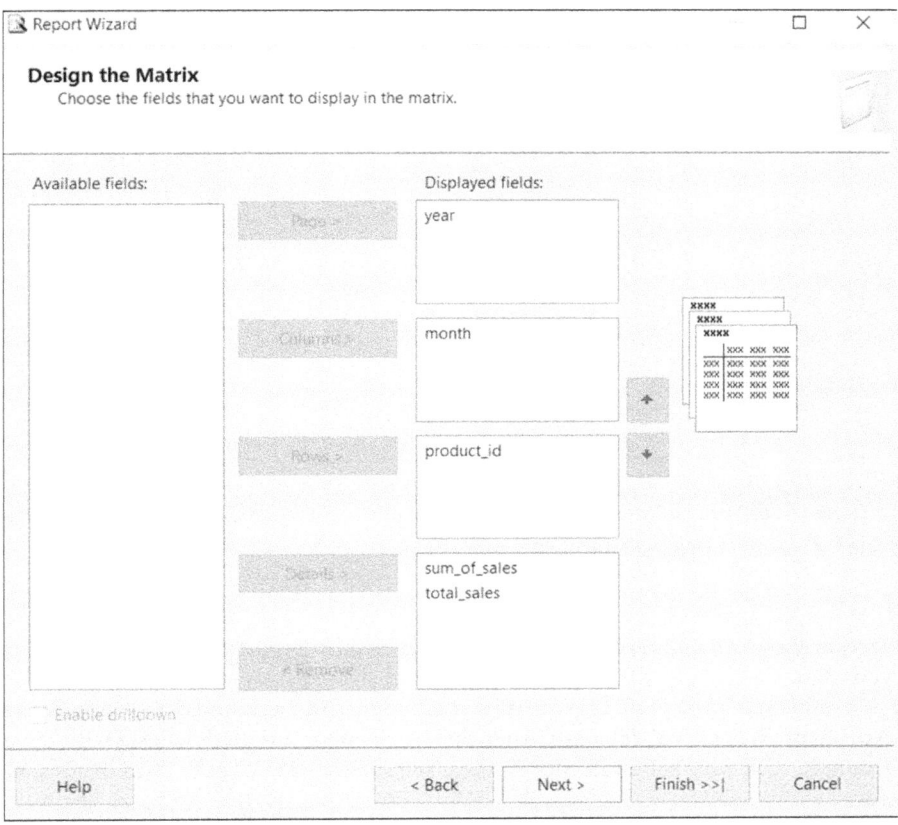

Figure 25-9   The Design the Matrix page

While the SELECT list of the query in Example 25.2 contains aggregate functions (COUNT and SUM), this report is a typical example of a matrix-type report. For this reason, choose Matrix on the Choose the Report Type page. On the Design the Matrix page, shown in Figure 25-9, values of the **year** column will be assigned to the Page view, values of the **month** column to the Columns view, and values of the **product_id** column to the Rows view. The Details view displays the aggregate values of **sum_of_sales** and **total_sales**. All remaining steps for the creation of this report are identical to the corresponding steps described in the section "Creating Your First Report."

To start the report in the Preview mode, type the value of the **CalendarYear** parameter and click the Preview tab. (I used the value 2013 to generate the report in Figure 25-10.)

Creating a Report Using a Chart

The output of both reports in the previous sections is given in tabular or matrix form. Today, almost all users prefer to see a report's output in graphical form. Charts are a popular method to present large quantities of data, because everyone can understand such an output. This

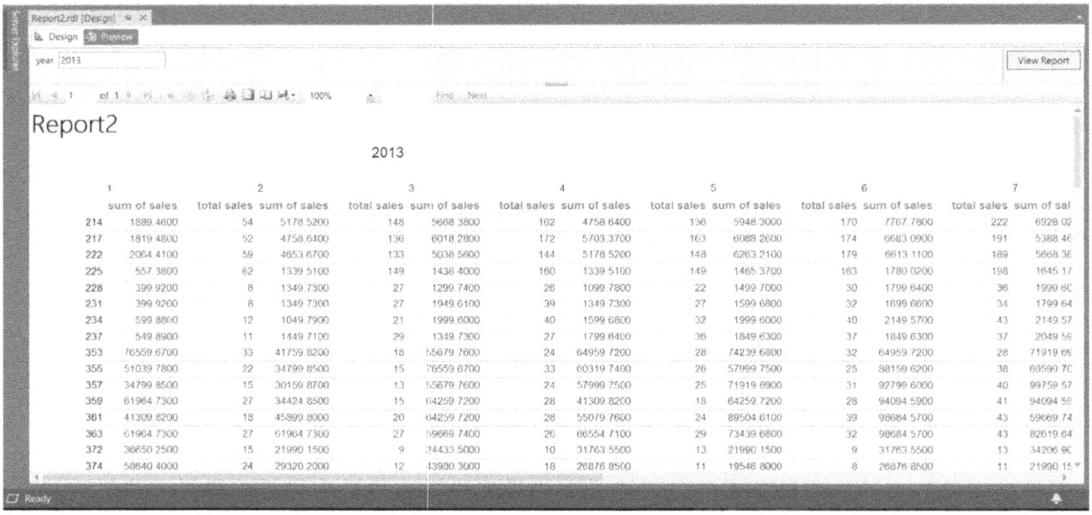

Figure 25-10 A preview of a part of the parameterized report

section shows you different tools that are provided by Reporting Services to graphically present your data.

In this section we will first create a blank report and a shared data source using the **AdventureWorksDW** database. After that, we'll create a shared data set with the query from Example 25.2, where the value for the year parameter will be set to 2013 in advance.

NOTE The main difference between the creation of reports in the previous section and the current section is that this section demonstrates how to create blank reports using the Report Service Project template instead of using the wizard.

To create a new report graphically, run Visual Studio. On the Start page, click Create a New Project to create a new project. Select Report Server Project (as shown earlier in Figure 25-6). Type the name of the project (**Project2**). The Location field should point to the default value. Click Create to create the new project.

In Solution Explorer, right-click the Shared Data Sources folder of the **Project2** project. Select Add New Data Source, and type the name of the data source (**Source3**). Click Build. The Connection Properties dialog box appears. In the Server Name text box, choose your database server instance that hosts the **AdventureWorksDW** database.

For the Authentication field, select Windows Authentication. In the Select or Enter a Database Name field, select or type the database name (**AdventureWorksDW**). Click Test Connection. If the connection succeeds, click OK in the Connection Test dialog box, and click OK again to exit the Connection Properties dialog box. Finally, click OK to exit the Shared Data Source Properties dialog box. A new data source, called **Source3.rds**, is stored in the Shared Data Sources folder of the **Project2** project.

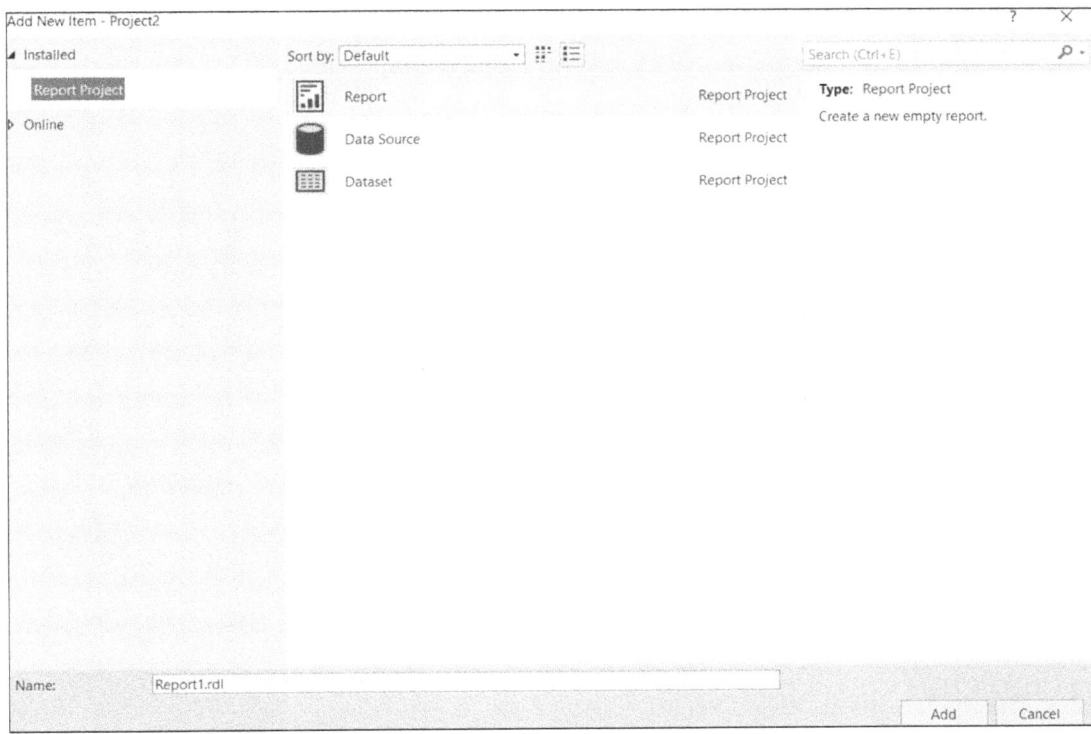

Figure 25-11 The Add New Item dialog box

Now, add the blank report in the following way: In Solution Explorer, under the **Project2** project, right-click the Reports folder, choose Add in the drop-down menu, and select the New Item submenu. The Add New Item – Project2 dialog box appears (see Figure 25-11). Select the Report item. Enter **Report3.rdl** in the Name field in the lower part of the window. Click Add. The blank report, Report3.rdl, is created in the **Project2** project. Now, we have to load the data and visualize it.

In the Report Data pane, click the New drop-down menu. Click Data Source from the submenu. The Data Source Properties dialog box appears. Enter **DataSource3** for the name and check the Use Shared Data Source Reference radio button. Select Source3 from the corresponding drop-down list. Click OK.

NOTE If you cannot open the Report Data pane, press CTRL-ALT-D to open it.

In the Report Data pane again, right-click the entry for the DataSource3 data source and select Add Dataset from the context menu. The Dataset Properties window appears with the Query page displayed. Enter **Sales_2013** in the Name Field. In the Query pane, type the content of the query from Example 25.3 (see Figure 25-12). Click OK.

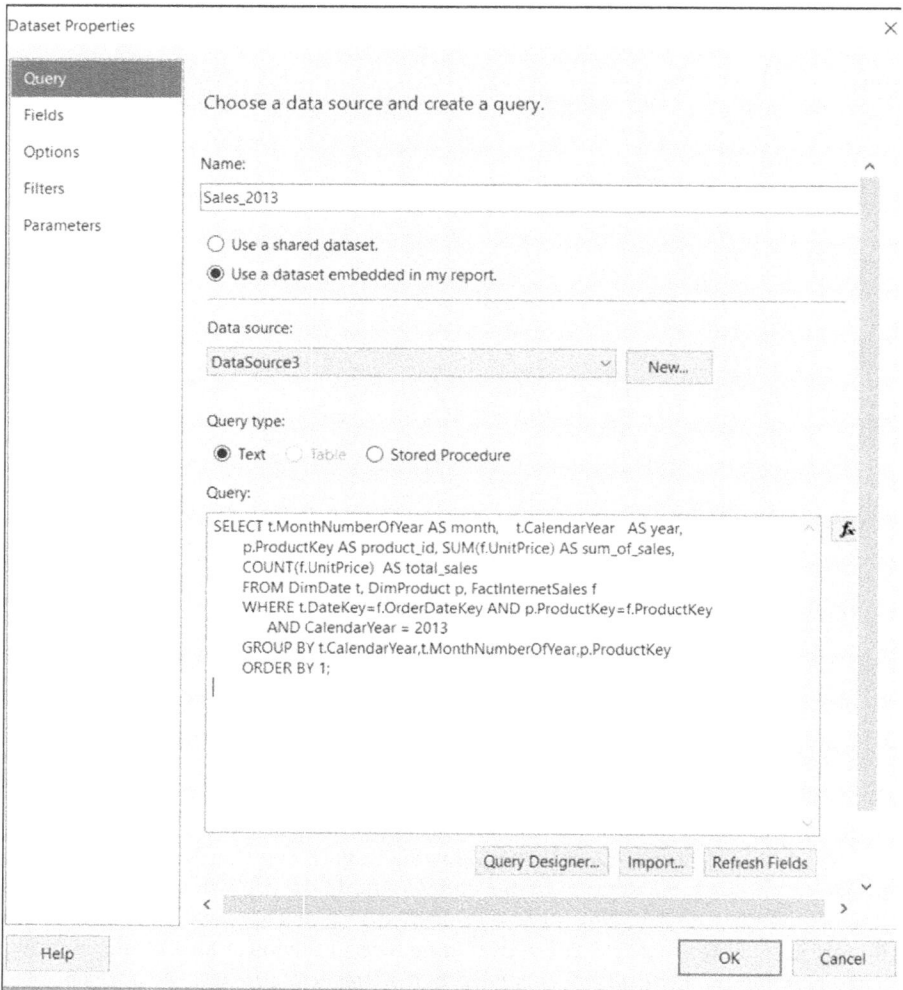

Figure 25-12 The Dataset Properties dialog box

Example 25.3

```
SELECT t.MonthNumberOfYear AS month,     t.CalendarYear    AS year,
      p.ProductKey AS product_id, SUM(f.UnitPrice) AS sum_of_sales,
      COUNT(f.UnitPrice)  AS total_sales
      FROM DimDate t, DimProduct p, FactInternetSales f
      WHERE t.DateKey=f.OrderDateKey AND p.ProductKey=f.ProductKey
          AND CalendarYear = 2013
      GROUP BY t.CalendarYear,t.MonthNumberOfYear,p.ProductKey
      ORDER BY 1;
```

Drag the edges of the design area so that this surface fills the available space on the screen. After that, click View, click Toolbox, and select Chart. The Select Chart Type dialog box appears. Choose the first chart report item on the left side under Column and click OK to exit the Select Chart Type dialog box. You will see a representation of that chart type in the design area.

Double-click anywhere on the chart. The Chart Data window with field areas appears to the right of the chart. The areas are Values, Category Groups, and Series Groups. (The Chart Data window disappears when you click somewhere on the Design tab that is not covered by the chart.) Now, click the Refresh Fields button on the Dataset Properties dialog to refresh all values.

Before you drag and drop particular fields in one of the three Chart Data window areas, check out the purpose of each:

- **Value**   The fields you select for this area provide the values for the data on the chart. In other words, when you add the field to the Values area, the text of the data field appears in the legend, and the data from this numeric field will be aggregated into one value. (The Values area in the chart is identical to the Data area in the matrix form.) Therefore, for our visualization, drag the **sum_of_sales** field from the folder and drop it in this field.

- **Category Groups**   The fields you select in this area provide the labels for the horizontal axis of the chart. These category fields also group the rows from the dataset into multiple categories. (The Groups area in the chart is identical to the Columns group in the matrix form.) Therefore, for our visualization, drag the **month** column.

- **Series Groups**   If you add a field to this area, the number of series depends on the data that is contained in that field. In other words, the number of values in the field will determine how many series will appear on the chart. For instance, if the field contains three values, the chart will display three series for every field in the Values area. For our visualization, we won't drop a field in this area because we don't have any series. (The Series Groups area in the chart is identical to the Rows group in the matrix form.)

Click Preview to run the report. Figure 25-13 shows what the bar chart looks like. Click Save All on the toolbar to save your report.

NOTE   As you can see from Figure 25-13, the name of the chart has not been changed. To change the chart title above the chart area of a bar chart, switch to the Design view, right-click the chart title at the top of the chart, and click Title Properties. Replace the Title text field with the new text and click anywhere outside the text. In the same way you can change the name of the vertical and horizontal axis titles.

Managing and Tuning Reports

The previous sections showed you how to create different types of reports. This section describes how you can move reports to a managed environment, where they can be used by different groups of users.

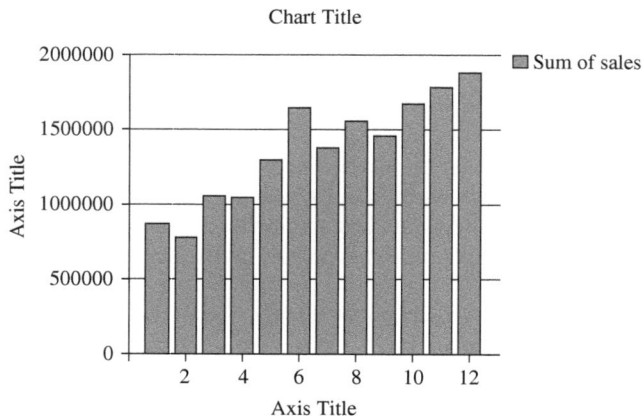

Figure 25-13 Bar chart for data retrieved with Example 25.3

NOTE Managing reports is an advanced task, performed by an administrator. For this reason, I will give you just an overview of the SSRS management tasks. For a detailed description of how reports can be managed, please see Microsoft Docs.

All created reports are available through the report server. The web application of SSRS allows you to create folders in the Report Catalog, which contains reports and all their supporting files. Each folder has a name and a description. Both the folder name and its description can be used by a user to find a report.

NOTE Report Catalog folders are different from usual file system folders because they are screen representations of records in the Report Catalog database, and therefore cannot be accessed in the usual way.

SSRS manages reports using a web application called Reporting Services web portal, which is described next.

Reporting Services Web Portal

Reporting Services web portal is a web application that allows you to create reports in the Report Catalog. During the installation process of SSRS, the home folder is created. The name of the default URL for accessing Reporting Services web portal is http://Computer_name/reports.

NOTE Reporting Services web portal replaces the Report Manager. This portal extends the functionality of the Report Manager by introducing several new features, which will be explained next.

The following features, among others, are supported by Reporting Services web portal:

- Publishing reports on mobile devices

- Exporting reports to several formats
- Features for subscriptions and delivery of reports

Besides these three features, which will be covered in the following subsections, SSRS allows you to access reports with different browser types. This feature will be described first.

Accessing Reports with Different Browser Types

When SSRS was initially added to SQL Server, it was optimized for Internet Explorer. Since then, several browsers that support newer web standards have emerged. The disadvantage of previous versions of SSRS is that they do not render reports consistently in these browsers. Since SQL Server 2016, SSRS is extended with a new renderer that supports HTML5 and has no dependency on features specific to Internet Explorer.

It is still possible that a report does not render correctly with the new rendering engine. In that case, the rendering process can be reverted to the previous rendering style by clicking the Switch to Compatibility Mode link on the right side of the report viewer toolbar.

Publishing Reports on Mobile Devices

In the earlier days, reports were delivered in the traditional form of paginated documents. The main property of these documents is that their layout is fixed for different computer types and different screen resolutions. In other words, paginated documents lay out their content on fixed-size pages.

In the meantime, users are doing more and more on their mobile devices and need to view reports on smartphones and tablets. Solutions based upon paginated documents often deliver a suboptimal solution for these devices. For this reason, SQL Server provides a solution that allows you to create mobile reports that are optimized for smartphones and tablets. You can create mobile reports using the SQL Server Mobile Report Publisher app.

NOTE Mobile Report Publisher is used to design and publish mobile reports in the same way as the Report Builder is used to design and publish paginated reports.

Exporting a Report to PowerPoint

One of the advantages of SSRS is the ability to export a report to a variety of different formats, such as Excel or Word. Since SQL Server 2016, PowerPoint has been added to the existing list of supported formats. To use this format, click the Export button in the report viewer toolbar and select PowerPoint.

When you select this export option, the .pptx file will be downloaded to your computer. If you have PowerPoint installed on your computer, you can open the file. In general, each page of your report becomes a separate slide in PowerPoint, although some report items might span multiple slides.

Subscriptions and Delivery of Reports

Before I discuss several features in relation to report subscriptions, let's take a look at how SSRS allows you to subscribe to and deliver reports.

SSRS supports standard and data-driven subscriptions. A standard subscription usually consists of specific parameters for parameterized reports as well as report presentation options and delivery options. You can use different tools to manage standard subscriptions.

Part IV

An important step in management of standard subscriptions is to configure a schedule. You can specify a schedule for a particular report only or use a shared schedule for several subscriptions. (If your report has parameters, you have to determine which values are assigned to parameters when the subscription is started.)

A data-driven subscription delivers reports to a list of recipients determined at run time. This type of subscription differs from a standard subscription in the way it gets subscription information: some settings from a data source are provided at run time, and other settings are static (that is, they are provided from the subscription definition). Static aspects of a data-driven subscription include the report that is delivered, the delivery extension, connection information to an external data source that contains subscriber data, and a query. Dynamic settings of the subscription are obtained from the row set produced by the query, including a subscriber list and user-specific delivery extension preferences or parameter values.

Subscriptions include a feature for enabling and disabling them. To enable or disable a subscription, browse to the subscription from the My Subscriptions page of an individual subscription. Select a subscription and then click either the Disable button or Enable button on the ribbon. (This task can be performed for several subscriptions if they are selected at once.) SSRS writes a row in the Reporting Services log when a subscription is disabled and another entry when the subscription is enabled again.

Another feature in relation to subscriptions is the option to include a file share delivery extension so that a report can be delivered to a folder. The file share delivery extension requires no additional configuration. In order for file delivery to succeed, you must set write access permissions on the shared folder. The account that requires write permissions is usually a file share account configured for the report server. To distribute a report to a file share, you define either a standard subscription or a data-driven subscription.

NOTE   Subscriptions are not available in every edition of SQL Server. See Microsoft Docs for information about which editions support standard and/or data-driven subscriptions.

Performance Issues

Each time a report is executed, SQL Server Reporting Services loads the data. That way, the user can always view the current data. This feature is advantageous, but has one drawback: the user has to wait for the data each time a record is executed. SSRS solves this performance problem through report caching, which is discussed next.

Cached Reports

Caching means that a report is generated only for the first user who opens it, and thereafter is retrieved from the cache for all subsequent users who work with the same report. A report server can cache a copy of a processed report and return that copy when a user opens the report. To a user, the only evidence available to indicate the report is a cached copy is the date and time that the report ran. If the date or time is not current and the report is not a snapshot, the report was retrieved from cache.

As you probably guessed, caching shortens the time to retrieve frequently accessed reports. Also, this technique is recommended for large reports. If the server is rebooted, all cached instances are reinstated when the server is restarted.

The contents of the cache are volatile and can change as new reports are added or existing ones dropped. If you require a more predictable caching strategy, you should create a report snapshot, which is described next.

Report Snapshots   The main disadvantage of cached reports is that the first user who wants to use the particular report has to wait until the system creates it. A more user-friendly method would be for the system to create the report automatically, so even the first user doesn't have to wait. This is supported by report snapshots.

A report snapshot is a way to create cached reports that contain data captured at a specific point in time. The benefit of a report snapshot is that no user has to wait, because the data has already been accessed from the report's data source(s) and stored in the Report Server temporary database. That way, the report will be rendered very quickly. The disadvantage of report snapshots is that data can become stale if the time difference between the creation of the report snapshot and access of the report is too long.

The main difference between report snapshots and cached reports is that cached reports are created as a result of a user action, while report snapshots are created automatically by the system.

Reporting Services Security

Reporting Services provides different authentication and authorization features for users to communicate with the report server. (For the definition of authentication and authorization, see Chapter 12.) The following subsections discuss these two features.

Authentication

Using the right authentication type for your report server enables your organization to achieve an appropriate level of security required by your organization. SSRS offers several options for authenticating users against the report server.

By default, the report server uses Windows Integrated authentication (see Chapter 12) and assumes that a trusted connection exists between clients and the server. (A *trusted connection* is a special form of connection between a client and the server, where the reporting server trusts that the operating system already validated the account and the corresponding password.)

Depending on your network topology and the needs of your organization, you can customize the authentication protocol that is used for Windows Integrated authentication or use a custom forms-based authentication extension. Each of the authentication types can be turned on or off individually. You can enable more than one authentication type if you want the report server to accept requests of multiple types.

Authorization

Authorization is based on roles that you assign to a principal. (For the definition of roles and principals, see Chapter 12.) Each role consists of a set of related tasks, which contain operations specific for each task. For example, the Manage Reports task grants access to the following report server operations: view reports, add report, update report, delete report, schedule report, and update report properties.

You can use SQL Server Management Studio to create, delete, or modify a role. To create a role, start Management Studio and connect to the particular report server instance. The corresponding report server node appears. Right-click the report server node and expand the Security folder. If you are creating an item-level role definition, right-click Roles and point to New Role. (The system roles can be created by right-clicking System Roles.) Type a unique name for the role. Optionally, type a description. Select the tasks that members of this role can perform and click OK. (The other operations can be performed in the similar way.)

Summary

SQL Server Reporting Services is the SQL Server–based enterprise reporting tool. To create a report, you can use the Report Server Project Wizard or Report Builder. The definition of a report, which comprises the corresponding query, layout information, and code, is stored using the XML-based Report Definition Language (RDL). SSRS processes the report definition into one of the standard formats, such as HTML, PDF, or PPTX.

Reports can be accessed on demand or delivered based on a subscription. When you execute a report on demand, a new instance of the report will usually be generated each time you run the report. Subscription-based reports can be either standard or data driven. Reports based on a standard subscription usually consist of specific parameters as well as report presentation options and delivery options. A data-driven subscription delivers reports to a list of recipients determined at run time.

To shorten the time of report execution, thereby increasing performance, you can either cache your reports or use report snapshots. The main difference between report snapshots and cached reports is that cached reports are created as a result of a user action, while report snapshots are created automatically by the system.

The next chapter describes optimization techniques for business intelligence.

Exercises

E.25.1   Get the employee numbers and names for all clerks. Create a report in the matrix report type using this query. Use Reporting Services web portal to view the report.

E.25.2   Use the **sample** database and get the budgets and project names of projects being worked on by employees in the Research department who have an employee number < 25000. Create a report in the tabular report type using this query. Use a browser to view the report.

26 Optimizing Techniques for Data Warehousing

In This Chapter

- Data Partitioning
- Star Join Optimization
- Indexed Views

This chapter describes three optimizing techniques pertaining to data warehousing. The first part of this chapter discusses when it is reasonable to store all entity instances in a single table and when it is better, for performance reasons, to partition the table's data. After a general introduction to data partitioning and the type of partitioning supported by the Database Engine, the steps that you have to follow to partition your table(s) are discussed in detail. You will then be given some partitioning techniques that can help increase system performance, followed by a list of important suggestions for how to partition your data and indices.

The second part of this chapter explains the technique called star join optimization. Two examples are presented to show you in which cases the query optimizer uses this technique instead of the usual join processing techniques. The role of bitmap filters will be explained, too.

The last part of this chapter describes an alternative form of a view, called an indexed view (or materialized view). This form of view materializes the corresponding query and allows you to achieve significant performance gains in relation to queries with aggregated data.

NOTE All the techniques described in this chapter can be applied only to relational OLAP (ROLAP).

Data Partitioning

The easiest and most natural way to design an entity is to use a single table. Also, if all instances of an entity belong to a table, you don't need to decide where to store its rows physically, because the database system does this for you. For this reason there is no need for you to do any administrative tasks concerning storage of table data, if you don't want to.

On the other hand, one of the most frequent causes of poor performance in relational database systems is contention for data that resides on a single I/O device. This is especially true if you have one or more very large tables with millions of rows. In that case, on a system with multiple CPUs, partitioning the table can lead to better performance through parallel operations.

By using data partitioning, you can divide very large tables (and indices too) into smaller parts that are easier to manage. This allows many operations to be performed in parallel, such as loading data and query processing.

Partitioning also improves the availability of the entire table. By placing each partition on its own disk, you can still access one part of the entire data even if one or more disks are unavailable. In that case, all data in the available partitions can be used for read and write operations. The same is true for maintenance operations.

If a table is partitioned, the query optimizer can recognize when the search condition in a query references only rows in certain partitions and therefore can limit its search to those partitions. That way, you can achieve significant performance gains, because the query optimizer has to analyze only a fraction of data from the partitioned table.

NOTE In this chapter, I discuss only horizontal table partitioning. Vertical partitioning is also an issue, but it does not have the same significance as horizontal partitioning.

How the Database Engine Partitions Data

A table can be partitioned using any column of the table. Such a column is called the *partition key*. (It is also possible to use a group of columns for the particular partition key.) The values of the partition key are used to partition table rows to different filegroups.

Two other important notions in relation to partitioning are the partition scheme and partition function. The *partition scheme* maps the table rows to one or more filegroups. The *partition function* defines how this mapping is done. In other words, the partition function defines the algorithm that is used to direct the rows to their physical location.

The Database Engine supports only one form of partitioning, called range partitioning. *Range partitioning* divides table rows into different partitions based on the value of the partition key. Hence, by applying range partitioning you will always know in which partition a particular row will be stored.

NOTE Besides range partitioning, there are several other forms of horizontal partitioning. One of them is called *hash partitioning*. In contrast to range partitioning, hash partitioning places rows one after another in partitions by applying a hashing function to the partition key. Hash partitioning is not supported by the Database Engine.

The steps for creating partitioned tables using range partitioning are described next.

Steps for Creating Partitioned Tables

Before you start to partition database tables, you have to complete the following steps:

1. Set partition goals.
2. Determine the partition key and number of partitions.

3. Create a filegroup for each partition.

4. Create the partition function and partition scheme.

5. Create partitioned indices (optionally).

All of these steps will be explained in the following sections.

Set Partition Goals

Partition goals depend on the type of applications that access the table that should be partitioned. There are several different partition goals, each of which could be a single reason to partition a table:

- Improved performance for individual queries
- Reduced contention
- Improved data availability

If your primary goal of table partitioning is to improve performance for individual queries, then distribute all table rows evenly. That way, the database system doesn't have to wait for data retrieval from a partition that has more rows than other partitions. Also, if such queries access data by performing a table scan against significant portions of a table, then you should partition the table rows only. (Partitioning the corresponding index will just add overhead in such a case.)

Data partitioning can reduce contention when many simultaneous queries perform an index scan to return just a few rows from a table. In this case, you should partition the table and index with a partition scheme that allows each query to eliminate unneeded partitions from its scan. To reach this goal, start by investigating which queries access which parts of the table. Then partition table rows so that different queries access different partitions.

Partitioning improves the availability of the database. By placing each partition on its own filegroup and locating each filegroup on its own disk, you can increase the data availability, because if one disk fails and is no longer accessible, only the data in that partition is unavailable. While the system administrator services the corrupted disk, limited access still exists to other partitions of the table.

Determine the Partition Key and Number of Partitions

A table can be partitioned using any table column. The values of the partition key are used to partition table rows to different filegroups. For the best performance, each partition should be stored in a separate filegroup, and each filegroup should be stored on a separate disk device. By spreading the data across several disk devices, you can balance the I/O and improve query performance, availability, and maintenance.

You should partition the data of a table using a column that does not frequently change. If you use a column that is often modified, any update operation of that column can force the system to move the modified rows from one partition to the other, and this could be time consuming.

Create a Filegroup for Each Partition

To achieve better performance, higher data availability, and easier maintenance, use different filegroups to separate table data. The number of filegroups to use depends mostly on the hardware you have. When you have multiple CPUs, partition your data so that each CPU can access data on one disk device. If the Database Engine can process multiple partitions in parallel, the processing time of your application will be significantly reduced.

Each data partition must map to a filegroup. To create a filegroup, you use either the CREATE DATABASE statement or ALTER DATABASE statement. Example 26.1 shows the creation of a database called **test_partitioned** with one primary filegroup and two other filegroups.

NOTE Before you create the **test_partitioned** database, you have to change the physical addresses of all .mdf and .ndf files in Example 26.1 according to the file system you have on your computer.

Example 26.1

```
USE master;
CREATE DATABASE test_partitioned
ON PRIMARY
  ( NAME='MyDB_Primary',
    FILENAME=
        'd:\mssql\PT_Test_Partitioned_Range_df.mdf',
    SIZE=2000,
    MAXSIZE=5000,
    FILEGROWTH=1 ),
FILEGROUP MyDB_FG1
  ( NAME = 'FirstFileGroup',
    FILENAME =
        'd:\mssql\MyDB_FG1.ndf',
    SIZE = 1000MB,
    MAXSIZE=2500,
    FILEGROWTH=1 ),
FILEGROUP MyDB_FG2
  ( NAME = 'SecondFileGroup',
    FILENAME =
        'f:\mssql\MyDB_FG2.ndf',
    SIZE = 1000MB,
    MAXSIZE=2500,
    FILEGROWTH=1 );
```

Example 26.1 creates a database called **test_partitioned**, which contains a primary filegroup, **MyDB_Primary**, and two other filegroups, **MyDB_FG1** and **MyDB_FG2**. The **MyDB_FG1** filegroup is stored on the D: drive, while the **MyDB_FG2** filegroup is stored on the F: drive.

If you want to add filegroups to an existing database, use the ALTER DATABASE statement. Example 26.2 shows how to create another filegroup for the **test_partitioned** database.

Example 26.2

```
USE master;
ALTER DATABASE test_partitioned
   ADD FILEGROUP MyDB_FG3
```

```
GO
ALTER DATABASE test_partitioned
ADD FILE
  ( NAME = 'ThirdFileGroup',
    FILENAME =
        'G:\mssql\MyDB_FG3.ndf',
    SIZE = 1000MB,
    MAXSIZE=2500,
    FILEGROWTH=1)
TO FILEGROUP MyDB_FG3;
```

Example 26.2 uses the ALTER DATABASE statement to create an additional filegroup called **MyDB_FG3** on the G: drive. The second ALTER DATABASE statement adds a new file to the created filegroup. Notice that the TO FILEGROUP option specifies the name of the filegroup to which the new file will be added.

Create the Partition Function and Partition Scheme

The next step after creating filegroups is to create the partition function, using the CREATE PARTITION FUNCTION statement. The syntax of the CREATE PARTITION FUNCTION is as follows:

```
CREATE PARTITION FUNCTION function_name(param_type)
    AS RANGE [ LEFT | RIGHT ]
      FOR VALUES ( [ boundary_value [ ,...n ] ] )
```

function_name defines the name of the partition function, while **param_type** specifies the data type of the partition key. **boundary_value** specifies one or more boundary values for each partition of a partitioned table or index that uses the partition function.

The CREATE PARTITION FUNCTION statement supports two forms of the RANGE option: RANGE LEFT and RANGE RIGHT. RANGE LEFT determines that the boundary condition is the upper boundary in the first partition. According to this, RANGE RIGHT specifies that the boundary condition is the lower boundary in the last partition (see Example 26.3). If not specified, RANGE LEFT is the default.

Example 26.3 shows the definition of the partition function.

Example 26.3

```
USE test_partitioned;
CREATE PARTITION FUNCTION myRangePF1 (int)
    AS RANGE LEFT FOR VALUES (500000);
```

The **myRangePF1** partition function specifies that there will be two partitions and that the boundary value is 500,000. This means that all values of the partition key that are less than or equal to 500,000 will be placed in the first partition, while all values greater than 500,000 will be stored in the second partition. (Note that the boundary value is related to the values in the partition key.)

The created partition function is useless if you don't associate it with specific filegroups. As mentioned earlier in the chapter, you make this association via a partition scheme, and you use the CREATE PARTITION SCHEME statement to specify the association between a partition function and the corresponding filegroups. Example 26.4 shows the creation of the partition scheme for the partition function in Example 26.3.

Example 26.4

```
USE test_partitioned;
CREATE PARTITION SCHEME myRangePS1
    AS PARTITION myRangePF1
    TO (MyDB_FG1, MyDB_FG2);
```

Example 26.4 creates the partition scheme called **myRangePS1**. According to this scheme, all values to the left of the boundary value (i.e., all values < 500,000) will be stored in the **MyDB_FG1** filegroup. Also, all values to the right of the boundary value will be stored in the **MyDB_FG2** filegroup.

NOTE When you define a partition scheme, you must be sure to specify a filegroup for each partition, even if multiple partitions will be stored on the same filegroup.

The creation of a partitioned table is slightly different from the creation of a nonpartitioned table. As you might guess, the CREATE TABLE statement must contain the name of the partition scheme and the name of the table column that will be used as the partition key. Example 26.5 shows the enhanced form of the CREATE TABLE statement that is used to define partitioning of the **orders** table.

Example 26.5

```
USE test_partitioned;
CREATE TABLE orders
    (orderid INTEGER NOT NULL,
     orderdate DATETIME,
     shippeddate DATETIME,
     freight money)
ON myRangePS1 (orderid);
```

The ON clause at the end of the CREATE TABLE statement is used to specify the already-defined partition scheme (see Example 26.4). The specified partition scheme ties together the column of the table (**orderid**) with the partition function where the data type (INT) of the partition key is specified (see Example 26.3).

The batch in Example 26.6 loads a million rows into the **orders** table. You can use the sys.partitions view to edit the information concerning partitions in the **orders** table.

Example 26.6

```
USE test_partitioned;
declare @i int , @order_id integer
        declare @orderdate datetime
        declare @shipped_date datetime
        declare @freight money
```

```
    set @i = 1
    set @orderdate = getdate()
    set @shipped_date = getdate()
    set @freight = 100.00
  while @i < 1000001
   begin
   insert into orders (orderid, orderdate, shippeddate, freight)
     values( @i, @orderdate, @shipped_date, @freight)
   set @i = @i+1
   end
```

Create Partitioned Indices

When you partition table data, you can partition the indices that are associated with that table, too. You can partition table indices using the existing partition scheme for that table or a different one. When both the indices and the table use the same partition function and the same partitioning columns (in the same order), the table and index are said to be aligned. When a table and its indices are aligned, the database system can move partitions in and out of partitioned tables very effectively, because the partitioning of both database objects is done with the same algorithm. For this reason, in the most practical cases it is recommended that you use aligned indices.

Example 26.7 shows the creation of a clustered index for the **orders** table. This index is aligned because it is partitioned using the partition scheme of the **orders** table.

Example 26.7

```
USE test_partitioned;
CREATE UNIQUE CLUSTERED INDEX CI_orders
 ON orders(orderid)
  ON myRangePS1(orderid);
```

As you can see from Example 26.7, the creation of the partitioned index for the **orders** table is done using the enhanced form of the CREATE INDEX statement. This form of the CREATE INDEX statement contains an additional ON clause that specifies the partition scheme. If you want to align the index with the table, specify the same partition scheme as for the corresponding table. (The first ON clause is part of the standard syntax of the CREATE INDEX statement and specifies the column for indexing.)

Partitioning Techniques for Increasing System Performance

The following partitioning techniques can significantly increase performance of your system:

- Table collocation
- Partition-aware seek operation
- Parallel execution of queries

Table Collocation

Besides partitioning a table together with the corresponding indices, the Database Engine also supports the partitioning of two tables using the same partition function. This partition form means that rows of both tables that have the same value for the partition key are stored together at a specific location on the disk. This concept of data partitioning is called *collocation*.

Suppose that, besides the **orders** table (see Example 26.3), there is also an **order_details** table, which contains zero, one, or more rows for each unique order ID in the **orders** table. If you partition both tables using the same partition function on the join columns **orders .orderid** and **order_details.orderid**, the rows of both tables with the same value for the **orderid** columns will be physically stored together. Suppose that there is a unique order with the identification number 49031 in the **orders** table and five corresponding rows in the **order_details** table. In the case of collocation, all six rows will be stored side by side on the disk. (The same procedure will be applied to all rows of these tables with the same value for the **orderid** columns.)

This technique has significant performance benefits when, accessing more than one table, the data to be joined is located at the same partition. In that case the system doesn't have to move data between different data partitions.

Partition-Aware Seek Operation

The internal representation of a partitioned table appears to the query processor as a composite (multicolumn) index with an internal column as the leading column. (This column, called **partitionedID**, is a hidden computed column used internally by the system to represent the ID of the partition containing a specific row.)

For example, suppose there is a **tab** table with three columns, **col1**, **col2**, and **col3**. (**col1** is used to partition the table, while **col2** has a clustered index.) The Database Engine treats internally such a table as a nonpartitioned table with the schema **tab (partitionID, col1, col2, col3)** and with a clustered index on the composite key (**partitionedID, col2**). This allows the query optimizer to perform seek operations based on the computed column **partitionID** on any partitioned table or index. That way, the performance of a significant number of queries on partitioned tables can be improved because the partition elimination is done earlier.

Parallel Execution of Queries

The Database Engine supports execution of parallel threads. In relation to this feature, the system provides two query execution strategies on partitioned objects:

- **Single-thread-per-partition strategy** The query optimizer assigns one thread per partition to execute a parallel query plan that accesses multiple partitions. One partition is not shared between multiple threads, but multiple partitions can be processed in parallel.

- **Multiple-threads-per-partition strategy** The query optimizer assigns multiple threads per partition regardless of the number of partitions to be accessed. In other words, all available threads start at the first partition to be accessed and scan forward. As each thread reaches the end of the partition, it moves to the next partition and begins scanning forward. The thread does not wait for the other threads to finish before moving to the next partition.

Which strategy the query optimizer chooses depends on your environment. It chooses the single-thread-per-partition strategy if queries are I/O-bound and include more partitions than the degree of parallelism. It chooses the multiple-threads-per-partition strategy in the following cases:

- Partitions are striped evenly across many disks.
- Your queries use fewer partitions than the number of available threads.
- Partition sizes differ significantly within a single table.

Editing Information Concerning Partitioning

You can use the following catalog views to display information concerning partitioning:

- sys.partitions
- sys.partition_schemes
- sys.partition_functions

The **sys.partitions** view contains a row for each partition of all the tables and some types of indices in the particular database. (All tables and indices in the Database Engine contain at least one partition, whether or not they are explicitly partitioned.)

The following are the most important columns of the **sys.partitions** catalog view:

- **partition_id**   Specifies the partition ID, which is unique within the current database.
- **object_id**   Defines the ID of the object to which this partition belongs.
- **index_id**   Indicates the ID of the index within that object.
- **hobt_id**   Indicates the ID of the data heap or B-tree that contains the rows for this partition.
- **partition_number**   Indicates a 1-based partition number within the owning index or heap. For nonpartitioned tables and indices, the value of the **partition_number** column is 1.

Example 26.8 displays a list of all partitioned tables in the **test_partitioned** database.

Example 26.8

```
USE test_partitioned;
SELECT DISTINCT t.name
    FROM sys.partitions p INNER JOIN sys.tables t
      ON p.object_id = t.object_id
    where p.partition_number <> 1;
```

The result is

```
name
------
orders
```

In Example 26.8 the **sys.partitions** catalog view is joined with **sys.tables** to get the list of all partitioned tables. Note that if you are looking specifically for partitioned tables, then you will have to filter your query with the condition

```
sys.partitions.partition_number <> 1
```

because for nonpartitioned tables, the value of the **partition_number** column is always 1.

The **sys.partition_schemes** catalog view contains a row for each data space with **type** = PS ("partition scheme"). (Generally, the data space can be a filegroup, partition scheme, or FILESTREAM data filegroup.) This view inherits the columns from the **sys.data_spaces** catalog view.

The **sys.partition_functions** catalog view contains a row for each partition function belonging to an instance of the Database Engine. The most important columns are **name** and **function_id**. The **name** column specifies the name of the partition function. (This name must be unique within the database.) The **function_id** column defines the ID of the corresponding partition function. This value is unique within the database.

Example 26.9 shows the use of the **sys.partition_functions** catalog view.

Example 26.9

For the **orders** table of the **test_partitioned** database, find the name of the corresponding partition scheme as well as the name of the partition function used by that scheme. Additionally, display names of all filegroups associated with that partition function.

```
SELECT ps.name PartScheme,pf.name PartFunc,fg.name FileGroupName
      FROM sys.indexes i
 JOIN sys.partitions p ON i.object_id=p.object_id
                    AND i.index_id=p.index_id
 JOIN sys.partition_schemes ps on ps.data_space_id=i.data_space_id
 JOIN sys.partition_functions pf on pf.function_id=ps.function_id
 JOIN sys.allocation_units au ON au.container_id=p.hobt_id
 JOIN sys.filegroups fg ON fg.data_space_id=au.data_space_id
     WHERE i.object_id = object_id('orders');
```

The result is

PartScheme	PartFunc	FileGroupName
myRangePS1	myRangePF1	MyDB_FG1
myRangePS1	myRangePF1	MyDB_FG2

Example 26.9 joins six catalog views to obtain the desired information: **sys.indexes**, **sys.partitions**, **sys.partition_schemes**, **sys.partition_functions**, **sys.allocation_units**, and **sys.filegroups**. To join **sys.indexes** with **sys.partitions**, we use the combination of the values of the **object_id** and **index_id** columns. Also, the join operation between **sys.indexes** and **sys.partition_schemes** is done using the **data_space_id** column in both tables. (**data_space_id** is an ID value that uniquely specifies the corresponding data space.)

sys.partition_functions and **sys.partition_schemes** are connected using the **function_id** column in both tables. **sys.filegroups** and **sys.allocation_units** are joined together using the already mentioned column **data_space_id**. Finally, to join the **sys.partitions** and **sys.allocation_units** views, we use the **hobt_id** column from the former and the **container_id** column from the latter. (For the description of the **hobt_id** column, see the definition of the **sys.partitions** view earlier in this section.)

Guidelines for Partitioning Tables and Indices

The following suggestions are guidelines for partitioning tables and indices:

- Do not partition every table. Partition only those tables that are accessed most frequently.
- Consider partitioning a table if it is a huge one, meaning it contains at least several hundred thousand rows.
- For best performance, use partitioned indices to reduce contention between sessions.
- Balance the number of partitions with the number of processors on your system. If it is not possible for you to establish a 1:1 relationship between the number of partitions and the number of processors, specify the number of partitions as a multiple factor of the number of processors. (For instance, if your computer has four processors, the number of partitions should be divisible by four.)
- Do not partition the data of a table using a column that changes frequently. If you use a column that changes often, any update operation of that column can force the system to move the modified rows from one partition to another, and this could be very time consuming.
- For optimal performance, partition the tables to increase parallelism, but do not partition their indices. Place the indices in a separate filegroup.

Star Join Optimization

As you already know from Chapter 22, the star schema is a general form for structuring data in a data warehouse. A star schema usually has one fact table, which is connected to several dimension tables. The fact table can have 100 million rows or more, while dimension tables are fairly small relative to the size of the corresponding fact table. Generally, in decision support queries, several dimension tables are joined with the corresponding fact table. The convenient way for the query optimizer to execute such queries is to join each of the dimension tables used in the query with the fact table, using the primary/foreign key relationship. Although this technique is the best one for numerous queries, significant performance gains can be achieved if the query optimizer uses special techniques for particular groups of queries. One such specific technique is called *star join optimization*.

Before you start to explore this technique, take a look at how the query optimizer executes a query in the traditional way, as shown in Example 26.10.

Example 26.10

```
USE AdventureWorksDW;
SELECT ProductAlternateKey
    FROM FactInternetSales f JOIN DimDate t ON f.OrderDateKey = t.DateKey
  JOIN DimProduct d ON d.ProductKey = f.ProductKey
```

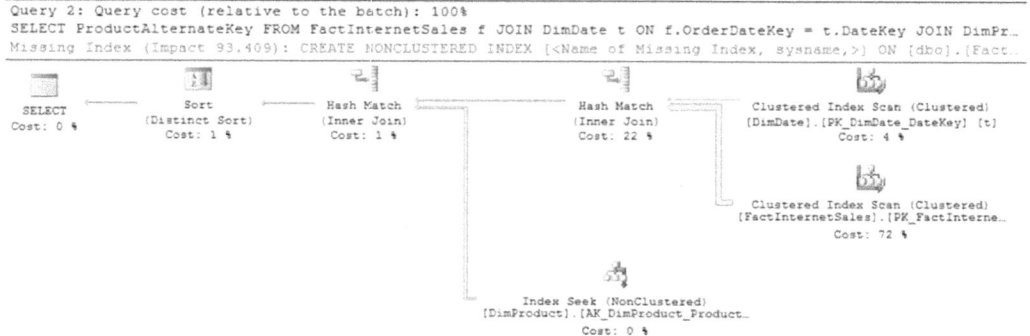

Figure 26-1 Execution plan of Example 26.10

```
WHERE CalendarYear BETWEEN 2013 AND 2014
AND ProductAlternateKey LIKE 'BK%'
GROUP BY ProductAlternateKey, CalendarYear;
```

The execution plan of Example 26.10 is shown in Figure 26-1.

As you can see from the execution plan in Figure 26-1, the query joins first the **FactInternetSales** fact table with the **DimDate** dimension table using the relationship between the primary key in the dimension table (**DateKey**) and the foreign key in the fact table (**DateKey**). After that, the second dimension table, **DimProduct**, is joined with the result of the previous join operation. Both join operations use the hash join method.

The use of the star join optimization technique will be explained using Example 26.11.

Example 26.11

```
USE AdventureWorksDW;
GO
SELECT F.ProductKey, F.CurrencyKey, D1.CurrencyName, D2.EndDate
FROM dbo.FactInternetSales AS F
JOIN dbo.DimCurrency AS D1 ON F.CurrencyKey = D1.CurrencyKey
JOIN dbo.DimProduct D2 ON F.ProductKey = D2.ProductKey
WHERE D1.CurrencyKey <= 12 AND D2.ListPrice > 50
OPTION (MAXDOP 32);
```

The query optimizer uses the star join optimization technique only when the fact table is very large in relation to the corresponding dimension tables. To ensure that the query optimizer would apply the star join optimization technique, I significantly enhanced the **FactInternetSales** fact table from the **AdventureWorksDW** database. The original size of this table is approximately 64,000 rows, but for this example I created an additional 500,000 rows by generating random values for the **ProductKey**, **SalesOrderNumber**, and **SalesOrderLineNo** columns. Figure 26-2 shows the execution plan of Example 26.11.

The query optimizer detects that the star join optimization technique can be applied and evaluates the use of bitmap filters. (A *bitmap filter* is a small to midsize set of values that is used to filter data. Bitmap filters are always stored in memory.)

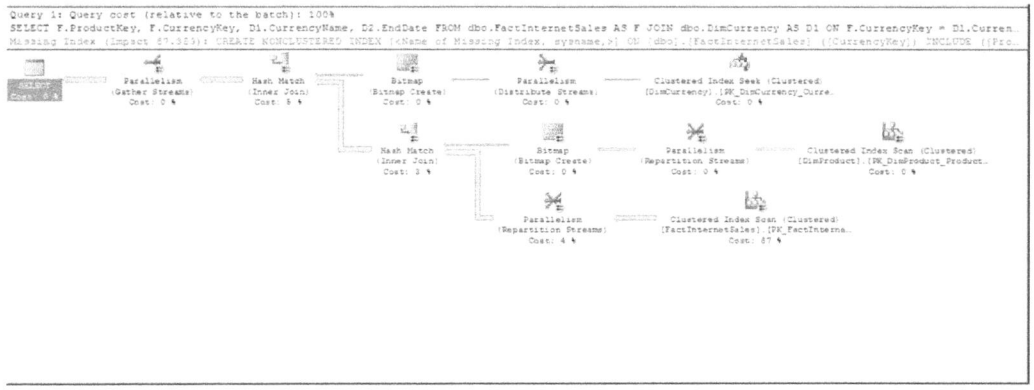

Figure 26-2   Execution plan of Example 26.11

As you can see from the execution plan in Figure 26-2, the fact table is first scanned using the corresponding clustered index. After that the bitmap filters for both dimension tables are applied. (Their task is to filter out the rows from the fact table.) This has been done using the hash join technique. At the end, the significantly reduced sets of rows from both streams are joined together.

NOTE Do not confuse bitmap filters with bitmap indices! Bitmap indices are persistent structures used in BI as an alternative to B^+-tree structures. The Database Engine doesn't support bitmap indices.

Indexed Views

As you already know from Chapter 10, there are several special index types. One of them is the indexed view, the topic of this section.

A view always contains a query that acts as a filter. Without indices created for a particular view, the Database Engine builds dynamically the result set from each query that references a view. ("Dynamically" means that if you modify the content of a table, the corresponding view will always show the new information.) Also, if the view contains computations based on one or more columns of the table, the computations are performed each time you access the view.

Building dynamically the result set of a query can decrease performance if the view with its SELECT statement processes many rows from one or more tables. If such a view is frequently used in queries, you could significantly increase performance by creating a clustered index on the view (demonstrated in the next section). Creating a clustered index means that the system materializes the dynamic data into the leaf pages on an index structure.

The Database Engine allows you to create indices on views. Such views are called indexed (or materialized) views. When a unique clustered index is created on a view, the view is executed and the result set is stored in the database in the same way a base table with a clustered index is stored. This means that the leaf nodes of the clustered index's B^+-tree contain data pages (see also the description of the clustered table in Chapter 10).

NOTE Indexed views are implemented through syntax extensions to the CREATE INDEX and CREATE VIEW statements. In the CREATE INDEX statement, you specify the name of a view instead of a table name. The syntax of the CREATE VIEW statement is extended with the SCHEMABINDING clause. For more information on extensions to this statement, see the description in Chapter 11.

Creating an Indexed View

Creating an indexed view is a two-step process:

1. Create the view using the CREATE VIEW statement with the SCHEMABINDING clause.
2. Create the corresponding unique clustered index.

Example 26.12 shows the first step, the creation of a typical view that can be indexed to gain performance. (This example assumes that **works_on** is a very large table.)

Example 26.12

```
USE sample;
GO
CREATE VIEW v_enter_month
  WITH SCHEMABINDING
  AS SELECT emp_no, DATEPART(MONTH, enter_date)  AS enter_month
      FROM dbo.works_on;
```

The **works_on** table in the **sample** database contains the **enter_date** column, which represents the starting date of an employee in the corresponding project. If you want to retrieve all employees that entered their projects in a specified month, you can use the view in Example 26.12. To retrieve such a result set using index access, the Database Engine cannot use a table index, because an index on the **enter_date** column would locate the values of that column by the date, and not by the month. In such a case, indexed views can help, as Example 26.13 shows.

Example 26.13

```
USE sample;
GO
CREATE  UNIQUE CLUSTERED INDEX
    c_workson_deptno ON v_enter_month (enter_month, emp_no);
```

To make a view indexed, you have to create a unique clustered index on the column(s) of the view. (As previously stated, a clustered index is the only index type that contains the data values in its leaf pages.) After you create that index, the database system allocates storage for the view, and then you can create any number of nonclustered indices because the view is treated as a (base) table.

An indexed view can be created only if it is deterministic—that is, the view always displays the same result set. In that case, the following options of the SET statement must be set to ON:

- QUOTED_IDENTIFIER
- CONCAT_NULL_YIELDS_NULL
- ANSI_NULLS
- ANSI_PADDING
- ANSI_WARNINGS

Also, the NUMERIC_ROUNDABORT option must be set to OFF.

There are several ways to check whether the options in the preceding list are appropriately set, as discussed in the upcoming section "Editing Information Concerning Indexed Views."

To create an indexed view, the view definition has to meet the following requirements:

- All referenced (system and user-defined) functions used by the view have to be deterministic—that is, they must always return the same result for the same arguments.
- The view must reference only base tables.
- The view and the referenced base table(s) must have the same owner and belong to the same database.
- The view must be created with the SCHEMABINDING option. SCHEMABINDING binds the view to the schema of the underlying base tables.
- The referenced user-defined functions must be created with the SCHEMABINDING option.
- The SELECT statement in the view cannot contain the following clauses and options: DISTINCT, UNION, TOP, ORDER BY, MIN, MAX, COUNT, SUM (on a nullable expression), subqueries, derived tables, or OUTER.

The Transact-SQL language allows you to verify all of these requirements by using the **IsIndexable** parameter of the **objectproperty** property function, as shown in Example 26.14. If the value of the function is 1, all requirements are met and you can create the clustered index.

Example 26.14

```
USE sample;
SELECT objectproperty(object_id('v_enter_month'), 'IsIndexable');
```

Modifying the Structure of an Indexed View

To drop the unique clustered index on an indexed view, you have to drop all nonclustered indices on the view, too. After you drop its clustered index, the view is treated by the system as a traditional one.

NOTE If you drop an indexed view, all indices on that view are dropped.

Part IV

If you want to change a standard view to an indexed one, you have to create a unique clustered index on it. To do so, you must first specify the SCHEMABINDING option for that view. You can drop the view and re-create it, specifying the SCHEMABINDING clause in the CREATE SCHEMA statement, or you can create another view that has the same text as the existing view but a different name.

NOTE If you create a new view with a different name, you must ensure that the new view meets all the requirements for an indexed view that are described in the preceding section.

Editing Information Concerning Indexed Views

You can use the **sessionproperty** property function to test whether one of the options of the SET statement is activated (see the earlier section "Creating an Indexed View" for a list of the options). If the function returns 1, the setting is ON. Example 26.15 shows the use of the function to check how the QUOTED_IDENTIFIER option is set.

Example 26.15

```
SELECT sessionproperty ('QUOTED_IDENTIFIER');
```

The easier way is to use the dynamic management view called **sys.dm_exec_sessions**, because you can retrieve values of all options discussed previously using only one query. (Again, if the value of a column is 1, the corresponding option is set to ON.) Example 26.16 returns the values for the first four SET statement options from the list in "Creating an Indexed View." (The global variable **@@spid** is described in Chapter 4.)

Example 26.16

```
USE sample;
SELECT quoted_identifier, concat_null_yields_null, ansi_nulls, ansi_padding
          FROM sys.dm_exec_sessions
          WHERE session_id = @@spid;
```

The **sp_spaceused** system procedure allows you to check whether the view is materialized—that is, whether or not it uses the storage space. The result of Example 26.17 shows that the **v_enter_month** view uses storage space for the data as well as for the defined index.

Example 26.17

```
USE sample;
EXEC sp_spaceused 'v_enter_month';
```

The result is

name	rows	reserved	data	index_size	unused
v_enter_month	11	16KB	8KB	8KB	0KB

Benefits of Indexed Views

Besides possible performance gains for complex views that are frequently referenced in queries, the use of indexed views has two other advantages:

- The index of an indexed view can be used even if the view is not explicitly referenced in the FROM clause.
- All modifications to data are reflected in the corresponding indexed view.

Probably the most important property of indexed views is that a query does not have to explicitly reference a view to use the index on that view. In other words, if the query contains references to columns in the base table(s) that also exist in the indexed views, and the optimizer estimates that using the indexed view is the best choice, it chooses the view indices in the same way it chooses table indices when they are not directly referenced in a query.

When you create an indexed view, the result set of the view (at the time the index is created) is stored on the disk. Therefore, all data that is modified in the base table(s) will also be modified in the corresponding result set of the indexed view.

Besides all the benefits that you can gain by using indexed views, there is also a (possible) disadvantage: indices on indexed views are usually more complex to maintain than indices on base tables, because the structure of a unique clustered index on an indexed view is more complex than the structure of the corresponding index on a base table.

The following types of queries can achieve significant performance benefits if a view that is referenced by the corresponding query is indexed:

- Queries that process many rows and contain join operations or aggregate functions
- Join operations and aggregate functions that are frequently performed by one or several queries

If a query references a standard view and the database system has to process many rows using the join operation, the optimizer will usually use a suboptimal join method. However, if you define a clustered index on that view, the performance of the query could be significantly enhanced, because the optimizer can use an appropriate method. (The same is true for aggregate functions.)

If a query that references a standard view does not process many rows, the use of an indexed view could still be beneficial if the query is used very frequently. (The same is true for groups of queries that join the same tables or use the same type of aggregates.)

Summary

The Database Engine supports range partitioning of data and indices that is entirely transparent to the application. Range partitioning partitions rows based on the value of the partition key. In other words, the data is divided using the values of the partition key.

If you want to partition your data, you must complete the following steps:

1. Set partition goals.
2. Determine the partition key and number of partitions.

Part IV

3. Create a filegroup for each partition.

4. Create the partition function and partition scheme.

5. Create partitioned indices (if necessary).

By using different filegroups to separate table data, you achieve better performance, higher data availability, and easier maintenance.

The partition function is used to map the rows of a table or index into partitions based on the values of a specified column. To create a partition function, use the CREATE PARTITION FUNCTION statement. To associate a partition function with specific filegroups, use a partition scheme. When you partition table data, you can partition the indices associated with that table, too. You can partition table indices using the existing partition schema for that table or a different one.

Star join optimization is an index-based optimization technique that supports the optimal use of indices on huge fact tables. The main advantages of this technique are the following:

- Significant performance improvements in case of moderately and highly selective star join queries.

- No additional storage cost. The system does not create any new indices, but uses bitmap filters instead.

Indexed views are used to increase performance of certain queries. When a unique clustered index is created on a view, the view becomes materialized—that is, its result set is physically stored in the same way a content of a base table is stored.

The next chapter describes columnstore indices, another technique for storing content of tables.

CHAPTER

27

Columnstore Indices

In This Chapter

- Benefits of Columnstore Indices
- Internal Storage of Columnstore Indices
- Types of Columnstore Indices
- Editing Information Concerning Columnstore Indices
- Columnstore Indices: Performance

As you already know from Chapter 10, *index access* means that indices are used to access entire rows that fulfill a condition of the given query. This is a general approach and doesn't depend on the number of columns being returned. In other words, the values of all columns of a particular row will be fetched, even if values of only one or two columns are needed. The reason for this is that the Database Engine and all other relational database systems store a table's rows on data pages, which are units of storage that are read into memory and written back to disk. This traditional approach of storing data is called *row store*.

Another approach to storing data promises to improve performance in cases where the values of only a few columns of a table need to be fetched. Such an approach is called *column store*, and Microsoft introduced this technique using the index called columnstore. In a column store, column values are grouped and stored *one column at a time*. The query processor of a database system that supports column store can take advantage of the new data layout and significantly improve query execution time of such queries that retrieve just a few of a table's columns, because the smaller the number of fetched columns, the smaller the number of I/O operations required.

Columnstore indices are designed for data warehouses where SELECT statements operate on many rows but usually use only a few columns. They can be created in fact tables of a data warehouse, but can also be applied in the case of extremely large dimension tables.

The first section of the chapter describes the benefits of columnstore indices over row store indices. After that, internal storage of these indices is discussed and the notion of segments is introduced. The third part of the chapter introduces clustered and nonclustered columnstore indices, gives the syntax for their creation, and describes how clustered columnstore indices can be modified.

The last section shows the performance benefits of column store in relation to row store, and introduces a new execution mode, called Batch Mode on Columnstore. The aim of this execution mode is to execute analytical queries more effectively.

Benefits of Columnstore Indices

As a performance-tuning technique, columnstore indices offer significant performance benefits for a certain group of queries. The following are the benefits of columnstore indices:

- *The system fetches only needed columns.* The smaller the number of fetched columns, the smaller the number of I/O operations required. For instance, if the values of a few columns of each row are retrieved, the use of columnstore indices can reduce I/O significantly because only a small part of the data has to be transferred from disk into memory. (This is especially true for data warehouses, where fact tables usually have millions of rows.)

- *No limitation exists on the number of key columns.* The concept of key columns exists only for row store. Therefore, the limitation on the number of key columns for an index does not apply to columnstore indices. Also, if a base table has a clustered index, all columns in the clustering key must be present in the nonclustered columnstore index. Otherwise, it will be added to the columnstore index automatically.

- *Columnstore indices work with table partitioning.* If you create a columnstore index, no changes to the table partitioning syntax are required. A columnstore index on a partitioned table must be partition-aligned with the base table (see Chapter 26). Therefore, a columnstore index can be created on a partitioned table only if the partitioning column is one of the columns in the columnstore index.

- *Buffer pool usage is improved.* Reading only the columns that are required improves buffer pool usage and therefore more data can be kept in memory.

Internal Storage of Columnstore Indices

Columnstore indices are based on Microsoft's xVelocity, an advanced storage and technology that Microsoft originally incorporated in PowerPivot and the Analysis Services Tabular model, and later adapted for SQL Server. Because xVelocity offers both storage and compression techniques, I describe these two features in the following subsections.

Index Storage

To explain index storage for columnstore indices, we'll start with a table that includes the PRIMARY KEY constraint, the **employee** table introduced in Chapter 1 and used throughout the book. Because the **emp_no** column of the **employee** table is specified as the primary key, the Database Engine will implicitly create a clustered index based on that column. As you already know, leaf nodes of the clustered index store data by rows, and each leaf node includes all the data associated with that row. Additionally, the data is spread across one or more data pages.

When the Database Engine processes a table scan on the **employee** table, it moves all data pages into memory, fetching the entire table even though most of the values of its columns are not needed. In doing so, the system wastes I/O and memory resources to retrieve unnecessary data. Now let's look at what happens when we create a columnstore index on the table.

The basic technique in relation to column index storage consists of dividing data into row groups, called *segments*. Each segment consists of approximately one million rows. Also, each segment contains values from one column only, which allows each column's values to be accessed independently. However, a column can span multiple segments, and each segment can be made up of multiple data pages. Data is transferred from disk to memory by segment, not by page. The data within each column's segment matches row-by-row so that the rows can always be assembled correctly, if needed. (This is an important performance issue when the system has to assemble data from a column store back to the corresponding row store.)

All data within a segment is encoded by using an internal algorithm. Additionally, for some of the columns that need dictionaries, an additional dictionary encoding conversion is used. The Database Engine uses two general forms of dictionaries: a global dictionary associated with the entire column, and a local dictionary associated with each segment.

NOTE The data in a column store is not sorted, not even within a column segment.

The metadata for columnstore indices is internally stored in *directories.* The metadata includes information about the allocation status of segments and dictionaries. It contains additional metadata about the number of rows, row size, and minimum (maximum) value inside of each segment. (You can retrieve information concerning directories using the **sys.column_ store_segments** DMV, described later in the chapter in the section "Editing Information Concerning Columnstore Indices.")

Compression

Before I discuss the compression technique of xVelocity, let me explain why compression is so important for column store. In the case of a row store, compressing data is generally suboptimal. The reason is that values of columns of a table have many different data types and forms: some of them are numeric and some are strings or dates. Most compression algorithms are based on similarities of a group of values. When data is stored by rows, the possibility to exploit similarity among values of different columns is thus limited. By contrast, data items from a single column have the same data type and are stored contiguously. Usually, there is repetition and similarity among values within a column. All these factors allow the system to apply compression very effectively on a column store.

As you already know, each set of rows is divided into segments of about one million rows each. Each segment is then compressed, independently and in parallel, using an internal supported technique. The result is one compressed column segment for each column included in the columnstore index.

Besides the form of compression just explained, the Database Engine supports an additional form called *archival compression.* This form of compression can be applied to already compressed data to further reduce the amount of storage space. In data warehouse systems, there is usually a group of data that is queried more often than other groups of data. For instance, data may be partitioned by date, and the most recent data (say, the data from the last two years) may be accessed much more frequently than older data. In such case, the older data can benefit from additional compression at the cost of slower query performance.

Archival compression should be applied to archival data that you keep for regulatory reasons. The advantage of archival compression is that you can partition a table with a columnstore index, and then change the compression mode for individual partitions.

NOTE The compression form can be specified using the DATA_COMPRESSION clause of the CREATE CLUSTERED COLUMNSTORE INDEX statement (see Example 27.1 in the next section).

After the system finishes the process of encoding and compression, the segments and the dictionaries are converted to large objects (LOBs) and are stored inside of the Database Engine. (If one of the LOBs spans more than 8KB, the regular storage mechanisms are used.)

Types of Columnstore Indices

The Database Engine supports two types of columnstore indices: clustered and nonclustered. The following subsections explain them.

Clustered Columnstore Index

This section describes how you can create a clustered columnstore index, load data into it, and modify already loaded data.

Creation of Clustered Columnstore Index

To create a clustered columnstore index, first create a row store table as a heap or clustered index, and then use the CREATE CLUSTERED COLUMNSTORE INDEX statement to convert the table to a clustered columnstore index.

Example 27.1 shows the creation of a clustered columnstore index. (The syntax of CREATE CLUSTERED COLUMNSTORE INDEX does not include the list of indexed columns, while each such index contains all columns of the corresponding table.)

Example 27.1

```
USE sample;
SELECT * INTO FactInternetSales
    FROM AdventureWorksDW.dbo.FactInternetSales;
GO
CREATE  CLUSTERED COLUMNSTORE INDEX
    cl_factinternetsales ON FactInternetSales
WITH ( DATA_COMPRESSION = COLUMNSTORE);
```

Example 27.1 uses the **FactInternetSales** table from the **AdventureWorksDW** database and copies its structure and data to the table with the same name in the **sample** database by using the SELECT ... INTO statement. After that, Example 27.1 creates the clustered columnstore index called **cl_factinternetsales**. As you can see from Example 27.1, the creation of a clustered columnstore index has the same syntax as the traditional CREATE INDEX statement except that it includes the additional CLUSTERED COLUMNSTORE clause.

The DATA_COMPRESSION clause specifies the data compression option for the specified table, partition number, or range of partitions. The options are COLUMNSTORE (the default value) and COLUMNSTORE_ARCHIVE.

NOTE An important property of a clustered columnstore index is that it can have one or more nonclustered *row store* indices.

Clustered Columnstore Index and Data Modification

Column storage significantly improves read operations, but data in that storage is very expensive to update directly, because it is compressed. For this reason, the Database Engine does not update data in compressed row groups during data modification operations. The modification operations are handled by two components in relation to clustered columnstore indices: delta stores and delete bitmaps.

NOTE As you will see in the next sections, both delta stores and delete bitmaps are *row store constructs*. Therefore, all update operations on them are row store operations, which are significantly faster than the same operations on highly compressed column structures.

Delta Stores   If you execute DML statements, new and updated rows are inserted into a delta store, which is a traditional B-tree structure. Delta stores are included in any scan of the clustered columnstore index.

A delta store contains the same columns as the corresponding clustered columnstore index. The B-tree key is a unique integer row ID generated by the system. (A clustered columnstore index does not have unique keys.) A clustered columnstore index can have zero, one, or more delta stores. New delta stores are created automatically as needed to accept inserted rows. A delta store is either open or closed. An open delta store can accept rows to be inserted. A delta store is closed when the number of rows it contains reaches a predefined limit. Every delta store has a **state** column value of either 1 (open) or 2 (closed).

The Database Engine automatically checks in the background for closed delta stores and converts them into segments. This background process is called Tuple Mover and by default is executed every 5 minutes. Tuple Mover is designed not to block any read operations on data. By contrast, concurrent delete and update operations have to wait until the compression process completes.

NOTE The implementation of delta stores for disk-based tables is done as a set of internal B-tree tables.

The large bulk insert operations are handled differently by the system. During a large bulk load, the rows are stored directly in the clustered columnstore index without passing through a delta store. The operation itself is very efficient, since it stores all inserted rows in memory and applies compression in-memory. After the process is terminated, it stores that data on the disk.

In case of DELETE, an additional structure called a delete bitmap is used, which will be explained next.

Delete Bitmap Each clustered columnstore index has an associated delete bitmap that is consulted during scans to disqualify rows that have been deleted. In other words, a *delete bitmap* is a storage component that contains information about the deleted rows inside segments. A delete bitmap has two different representations, depending on whether it is in memory or on disk. In memory, it is a bitmap, but on disk, it is stored in the following way: a record of a deleted row is inserted into the B-tree of the corresponding delete bitmap. (This is true only if a row is in a columnstore segment. If the row to be deleted is in a delta store, the row is simply deleted.)

> **NOTE** The implementation of delete bitmaps for disk-based tables is done as a set of internal B-tree tables.

Having discussed delta stores and delete bitmaps, we now can examine exactly what happens when an INSERT, UPDATE, or DELETE statement is executed: First, inserted data is simply added to one of the currently open delta stores. Second, if the deleted row is found inside of a segment, then the deleted bitmap information is updated with the row ID of the respective row. On the other hand, if the deleted row is actually inside of a delta store, then the direct process of removal is executed on the corresponding B-tree structure. Finally, data updates are basically represented as deletes and inserts. In other words, an update operation triggers the insertion of the old version of the row into the delete bitmap and insertion of a new version into the delta store.

Nonclustered Columnstore Index

Generally, a nonclustered columnstore index is stored in the same way as a clustered one. Only one columnstore index can be created in this situation. In other words, if a clustered (or nonclustered) columnstore index exists, it must be dropped before the new index can be created. The columnstore index requires extra storage since it contains a copy of the data in the regular ("row store") table.

Creation of a Nonclustered Columnstore Index

Generally, the CREATE NONCLUSTERED COLUMNSTORE INDEX statement is used to create a nonclustered columnstore index on a table. The underlying table can be a row store table with or without a clustered index. In all cases, creating a nonclustered columnstore index on a table stores a second copy of the data for the columns in the index.

> **NOTE** SQL Server does not allow you to create clustered and nonclustered columnstore indices at the same time. You have to drop the clustered index in Example 27.1 if you want to create **cs_index1** in Example 27.2.

Example 27.2

```
USE sample;
GO
DROP INDEX
    cl_factinternetsales ON factinternetsales;
```

```
GO
CREATE NONCLUSTERED COLUMNSTORE INDEX cs_index1
  ON FactInternetSales (OrderDateKey, ShipDateKey, UnitPrice);
```

The CREATE INDEX statement in Example 27.2 creates a columnstore index for three columns of the **FactInternetSales** table: **OrderDateKey**, **ShipDateKey**, and **UnitPrice**. This means that all values of each of the three columns will be grouped and stored separately.

Filtered Index

The nonclustered columnstore index definition supports the use of a filter. A *filtered index* is an index that is specified for a condition in the WHERE clause of the SELECT statement. (You can define a filtered index both for regular indices and for columnstore indices.) Example 27.3 shows creation of such an index. (Again, only one index can be specified for a columnstore. For this reason, drop the index from Example 27.2 before creating the index in Example 27.3.)

Example 27.3

```
USE sample;
CREATE NONCLUSTERED COLUMNSTORE INDEX i1 ON
        FactInternetSales (UnitPrice)
            WHERE OrderDateKey IS NULL;
```

Editing Information Concerning Columnstore Indices

As mentioned previously in the "Index Storage" section, principal storage for metadata concerning columnstore indices is called a *directory* The metadata includes information about the allocation status of segments and dictionaries. The directory also contains additional metadata about the number of rows, size, and minimum and maximum values inside each of the segments.

If you want to display this information, you can use the following dynamic management views:

- sys.column_store_segments
- sys.column_store_row_groups
- sys.column_store_dictionaries

The following subsections describe these DMVs.

sys.column_store_segments

The **sys.column_store_segments** view contains a row for each column in a columnstore index. Example 27.4 displays information about segments of existing columnstore indices.

Example 27.4

```
SELECT i.name, p.object_id, p.index_id, i.type_desc,
    COUNT(*) AS number_of_segments
```

```
FROM sys.column_store_segments AS s
INNER JOIN sys.partitions AS p
    ON s.hobt_id = p.hobt_id
INNER JOIN sys.indexes AS i
    ON p.object_id = i.object_id
WHERE i.type = 6
GROUP BY i.name, p.object_id, p.index_id, i.type_desc ;
```

The result of Example 27.4 is

name	object_id	index_id	type_desc	number_of_segments
Cs_index1	885578193	2	NONCLUSTERED COLUMNSTORE	4

The SELECT statement in Example 27.4 displays the name of the index, its ID, and the index type (NONCLUSTERED COLUMNSTORE). The last column displays the number of assigned segments.

sys.column_store_row_groups

The **sys.column_store_row_groups** view provides metadata information in relation to a clustered columnstore index on a per-segment basis. **sys.column_store_row_groups** has a column for the total number of rows physically stored (including those marked as deleted) and a column for the number of rows marked as deleted. Use **sys.column_store_row_groups** to determine which segments have a high percentage of deleted rows and should be rebuilt.

Example 27.5 shows the use of this DMV.

Example 27.5

```
USE sample;
SELECT i.object_id, o.name,
i.name AS IndexName, i.index_id, i.type_desc
FROM sys.indexes AS i
JOIN sys.column_store_row_groups AS row_group
    ON i.object_id = row_group.object_id
JOIN sys.objects o ON  i.object_id = o.object_id
AND i.index_id = row_group.index_id;
```

The result is

object_id	name	IndexName	index_id	type_desc
885578193	factinternetsales	cl_factinternetsales	1	CLUSTERED COLUMNSTORE

The query in Example 27.5 joins three tables, **sys.column_store_row_groups**, **sys.indexes**, and **sys.objects**, to display the name of the table and names and IDs of associated columnstore indices. The last column of the output, **type_desc**, displays the type of the associated index.

sys.column_store_dictionaries

The **sys.column_store_dictionaries** DMV contains a row for each dictionary used in columnstore indices. Dictionaries are used to encode some, but not all, data types; therefore, not all columns in a columnstore index have dictionaries.

The most important columns of the DMV are **hobt_id**, **column_id**, and **dictionary_id**. The **hobt_id** column value is the unique identifier of the B-tree index (hobt) for the table that has the corresponding columnstore index. The **column_id** column value is the ID of the columnstore column, starting with 1. The first column has ID = 1, the second column has ID = 2, and so on. The **dictionary_id** column value specifies the ID of the corresponding dictionary. The value 0 represents the global dictionary that is shared across all column segments (one for each row group) for that column. The value <> 0 specifies a local dictionary.

Columnstore Indices: Performance

As you already know from the first section of this chapter, the use of columnstore indices can reduce I/O significantly, because only a small part of the data has to be transferred from disk into memory. (The Microsoft website states that the columnstore index can achieve query performance gains up to ten times greater than traditional index, as well as ten times the data compression rate over the uncompressed data size.)

In the following two subsections we first take a closer look at a performance comparison of the columnstore index and a corresponding traditional index. After that, a *set-at-time execution*, called Batch Mode, will be introduced. The aim of this mode is to execute a query that processes millions and billions of rows more effectively.

Columnstore Indices vs. Rowstore Indices

Generally, columnstore indices should have significant performance benefits in relation to traditional (B-tree) indices. Example 27.6 uses the **AdventureworksDW2016_EXT** sample database to compare the CPU and execution time of two identical queries.

NOTE The **AdventureworksDW2016_EXT** database is another Microsoft sample database. This database is similar to the **AdventureworksDW** database, with significant extensions in relation to a number of tables as well as the cardinality of existing ones. You can download this database from the following site: https://github.com/Microsoft/sql-server-samples/releases/tag/adventureworks

---

Example 27.6

```
USE AdventureworksDW2016_EXT
GO
DBCC DROPCLEANBUFFERS
SET STATISTICS IO ON
SET STATISTICS TIME ON
SELECT s.SalesTerritoryRegion, SUM(f.SalesAmount) 'Total_Sales',
        COUNT(distinct f.Resellerkey) as 'Resellers'
FROM FactResellerSalesXL_PageCompressed f
```

```
INNER JOIN DimDate d ON f.OrderDateKey= d.Datekey
INNER JOIN DimSalesTerritory s on s.SalesTerritoryKey=f.SalesTerritoryKey
INNER JOIN DimEmployee e on e.EmployeeKey=f.EmployeeKey
WHERE FullDateAlternateKey between '1/1/2015' and '1/1/2017'
GROUP BY s.SalesTerritoryRegion
ORDER BY Total_Sales
SET STATISTICS IO OFF
SET STATISTICS TIME OFF
```

The result is

```
SalesTeritoryRegion     Total_Sales      Resellers
United Kingdom          127423931,5373   40
Australia               127665107,5174   40
Germany                 128176936,347    40
France                  130386374,0798   40
Northeast               179901556,0727   56
Central                 203630348,7865   63
Southeast               270641464,1219   85
Canada                  728011099,6868   114
Southwest               843191174,1162   131
Northwest               881346288,0295   92


SQL Server Execution Times:
   CPU time = 7782 ms,  elapsed time = 15772 ms.
```

We use the **AdventureworksDW_EXT** sample database in Examples 27.6 and 27.7 because it contains two tables: **FactResellerSalesXL_PageCompressed** (Example 27.6) and **FactResellerSalesXL_CCI** (Example 27.7). The only difference is that the second table has the clustered columnstore index. Therefore, Example 27.6 uses row store to retrieve results, while Example 27.7 displays the same result set using a columnstore index.

Example 27.7

```
USE AdventureworksDW2016_EXT
DBCC DROPCLEANBUFFERS
SET STATISTICS IO ON
SET STATISTICS TIME ON
SELECT s.SalesTerritoryRegion, SUM(f.SalesAmount) 'Total_Sales',
          COUNT(distinct f.Resellerkey) as 'Resellers'
FROM FactResellerSalesXL_CCI f
INNER JOIN dbo.DimDate d ON f.OrderDateKey= d.Datekey
INNER JOIN dbo.DimSalesTerritory s on s.SalesTerritoryKey=f.SalesTerritoryKey
INNER JOIN dbo.DimEmployee e on e.EmployeeKey=f.EmployeeKey
WHERE FullDateAlternateKey between '1/1/2015' and '1/1/2017'
```

```
GROUP BY s.SalesTerritoryRegion
ORDER BY Total_Sales
SET STATISTICS IO OFF
SET STATISTICS TIME OFF
```

The result set is the same as for Example 27.6. The execution time is

```
SQL Server Execution Times:
  CPU time = 250 ms, elapsed time = 1922 ms.
```

As you can see from the performance parameters of Examples 27.6 and 27.7, the execution time of the query using a columnstore index is eight times faster than the corresponding execution time of the same query using row store.

> **NOTE** You cannot expect that each query will have such huge performance advantages if you use the column store. Generally, using a columnstore index benefits a wide group of queries, but typically the execution time is not several times faster.

Batch Mode on Columnstore

When you run a query in the Database Engine, the Query Processor plans how it should actually get the optimal result. This plan is expressed as a tree with operators, where the operators build the nodes of the tree. Each operator with its properties represents a certain processing action applied to a portion of data.

The query optimizer generally uses *record-at-time execution*, meaning that each query plan operator processes one row at a time. Prior to SQL Server 2012, a unit of data transferred through a tree was a row. With the introduction of columnstore indices, a *set-at-time execution*, called Batch Mode, was designed. The aim of this mode is to execute a query that processes millions and billions of rows more effectively.

Therefore, the columnstore index allows the system to choose between two modes: Row Mode, which is the conventional execution mode, and Batch Mode, the new execution mode for analytical queries with columnstore indices.

A *batch* is a structure of 64KB, allocated for a bunch of rows, that contains column vectors and qualifying rows vector. Depending on the number of columns, it may contain up to 1000 rows.

The benefits of Batch Mode are

- Allows the use of algorithms that are optimized for multicore CPUs
- Increased memory throughput
- Significant reduction of database accesses

> **NOTE** The downside of Batch Mode is that not all operators of the execution plan can be executed in that mode. In other words, if there is just one unsupported operator in a particular query plan, the whole query will be executed in row-mode processing.

The following examples show you how a query is executed in Batch Mode. Example 27.8 creates the tables that are necessary for the query.

Example 27.8

```
USE sample;
SELECT * INTO FactInternetSales
    FROM AdventureWorksDW.dbo.FactInternetSales;
GO
INSERT INTO FactInternetSales
    SELECT * FROM AdventureWorksDW.dbo.FactInternetSales;
GO 6
```

The SELECT statement in Example 27.8 creates the **FactInternetSales** table in the **sample** database and loads all rows from the **FactInternetSales** table of the **AdventureWorksDW** database into it. The subsequent INSERT statement loads the same load six more times. That way, we created a new table in the **sample** database with 422,786 rows.

Example 27.9 repeats the process of creating and loading two other tables from the **AdventureWorksDW** database.

Example 27.9

```
USE sample;
SELECT * INTO DimCustomer
    FROM AdventureWorksDW.dbo.DimCustomer;
GO
SELECT * INTO DimDate
    FROM AdventureWorksDW.dbo.DimDate;
```

After the execution of the queries in Example 27.9, the **sample** database contains two new tables, **DimCustomer** and **DimDate**, with the same structure and content as the tables with the same names in the **AdventureWorksDW** database.

Example 27.10 creates a nonclustered columnstore index.

Example 27.10

```
USE sample;
CREATE NONCLUSTERED COLUMNSTORE INDEX CLI_CS_IFactInternetSales
  ON dbo.FactInternetSales(OrderDateKey, CustomerKey, SalesAmount)
```

The creation of a columnstore index is a condition for which the query optimizer can choose Batch Mode on Columnstore. (The creation of a columnstore index does not mean that the query optimizer will automatically choose to use Batch Mode for the execution of a query.)

As previously noted, Batch Mode processing uses algorithms that are optimized for multicore CPUs. This means the query optimizer will not choose Batch Mode if the number of processors is one, which is the default value. Therefore, Example 27.11 increases the number of processors to four, using the **max degree of parallelism** (MAXDOP) configuration option. This is an advanced configuration option that controls the number of processors that are used for the execution of a query in a parallel plan.

Example 27.11

```
EXEC sp_configure 'show advanced options', 1;
GO
RECONFIGURE WITH OVERRIDE;
GO
EXEC sp_configure 'max degree of parallelism', 4;
GO
RECONFIGURE WITH OVERRIDE;
GO
```

Example 27.12 shows the query that is processed in Batch Mode using a columnstore index.

Example 27.12

```
USE sample;
SELECT c.CommuteDistance, d.CalendarYear,
                SUM(f.SalesAmount) TotalSales
    FROM dbo.FactInternetSales as f
    INNER JOIN dbo.DimCustomer as c ON
          f.CustomerKey = c.CustomerKey
    INNER JOIN dbo.DimDate d ON
          d.DateKey = f.OrderDateKey
    GROUP BY c.CommuteDistance, d.CalendarYear;
```

Figure 27-1 shows the execution plan of the query in Example 27.12 in general and the properties of the **Columnstore Index Scan** operator in particular. Two properties are important in relation to Batch Mode on Columnstore: Estimated Execution Mode and Storage. The former has the value Batch and the latter the value ColumnStore. This means that the query is processed in Batch Mode using the columnstore index called **CLI_CS_IFactInternetSales**.

Summary

Column store is one of most important performance-tuning techniques supported by SQL Server for BI. This technique offers significant performance benefits for a group of queries with the following properties:

- When an aggregate needs to be computed over many rows but only for a notably smaller subset of all columns of data
- When new values of a column are supplied for all rows at once, because that column data can be written efficiently and replace old column data without touching any other columns

On the other hand, column store should not be used when the values of many columns but only a few rows are retrieved at the same time.

The next chapter describes a group of sophisticated performance features called Intelligent Query Processing.

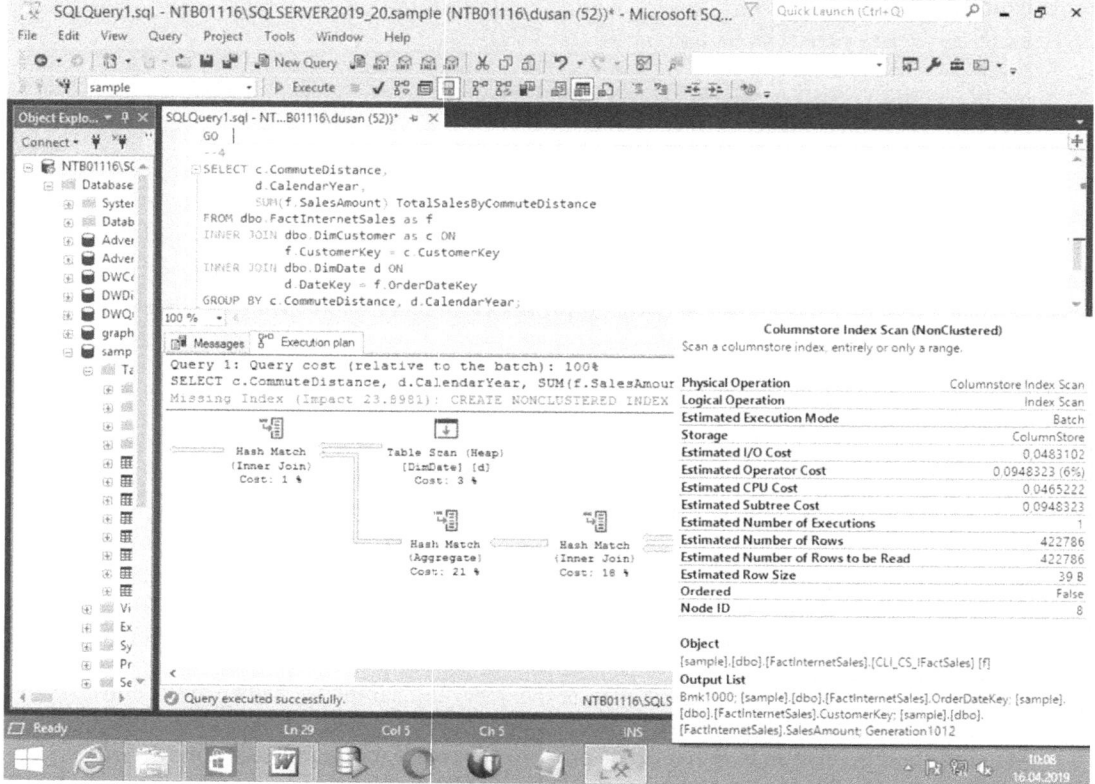

Figure 27-1   Execution plan of the query in Example 27.12

Exercises

E.27.1   Calculate the average unit price from the **FactInternetSales** table of the **AdventureWorksDW** database. (Use the **UnitPrice** column of the same table.) Do not use any indices. Using the STATISTICS TIME option of the SET statement, display the execution times and elapsed time for this query.

E.27.2   Create a nonclustered columnstore index for the following columns of the **FactInternetSales** table: **OrderDateKey**, **ShipDateKey**, **UnitPrice**. Again, calculate the average unit price from the **FactInternetSales** table of the **AdventureWorksDW** database. Using the STATISTICS TIME option of the SET statement, display the execution times and elapsed time for this query.

E.27.3   Calculate the performance gains of the second query in relation to the first one.

Intelligent Query Processing

In This Chapter

- Adaptive Query Processing
- Batch Mode on Rowstore
- Approximate Query Processing
- Scalar UDF Inlining
- Table Variable Deferred Compilation

This chapter describes a group of sophisticated and related features called Intelligent Query Processing (IQP). The main goal of these features is to improve performance of existing workloads while requiring only minimal implementation effort to adopt the features. The following features belong to the Intelligent Query Processing family:

- Adaptive Query Processing
- Batch Mode on Rowstore
- Approximate Query Processing
- Scalar UDF Inlining
- Table Variable Deferred Compilation

The following sections describe these features in detail.

NOTE You can make workloads automatically eligible for Intelligent Query Processing in all of the following examples by enabling compatibility level 150 using this statement:

```
ALTER DATABASE db_name SET COMPATIBILITY_LEVEL = 150;
```

Adaptive Query Processing

NOTE The term Adaptive Query Processing in SQL Server 2017 describes the group of techniques that you can use to improve performance of queries in the Database Engine. In SQL Server 2019, Adaptive Query Processing has the same meaning but now is considered a subset of the wider concept called Intelligent Query Processing.

As you already know from Chapter 19, the query optimizer generates a set of execution plans for a given query. During this process, the query optimizer estimates the cost of the various plan options and chooses the plan with the lowest estimated cost. The query execution process takes the plan chosen by the query optimizer and executes it. (All other plans are deleted and cannot be seen by the user.)

Sometimes the execution plan chosen by the query optimizer of the Database Engine is not the best one. Many factors can cause the optimizer to choose a plan that isn't the best one. For instance, the estimated number of input rows (called cardinality) may be incorrect. This factor is one of most important factors used to determine which execution plan is selected for execution of the particular query.

All the Adaptive Query Processing features allow the query optimizer to generate more accurate query plans with better cardinality. There are altogether three features for adapting to application workload characteristics:

- Memory Grant Feedback
- Adaptive Join
- Interleaved Execution

The following subsections describe these techniques.

Memory Grant Feedback

The ideal memory grant size and the minimum required memory needed for execution are two parameters that are used in an execution plan to calculate how all rows should fit in memory. If the required memory needed for execution is incorrectly sized, performance will suffer. Also, insufficient memory grants may cause memory overflow. The Memory Grant Feedback feature attempts to right-size memory grants for queries, correcting for both overestimation and underestimation. By addressing repeating workloads, Memory Grant Feedback recalculates the actual memory required for a query and then updates the grant value for the cached plan. That way, the query optimizer can use the modified values when an identical query statement is executed again.

The recalculation of the granted memory happens when the assigned memory is too large and there is not enough memory. First, if the granted memory is more than two times the size of the actual used memory, Memory Grant Feedback recalculates the memory grant and updates the cached plan. (Plans with memory grants under 1MB will not be recalculated.) Second, for an insufficiently sized memory grant condition, Memory Grant Feedback triggers a recalculation of the memory grant.

Memory Grant Feedback Caching

By default, the Memory Grant Feedback is stored in the cached plan and consecutive executions of that statement can benefit from updated values. This means that only the values in the cached plan are changed and that other components, such as Query Store, cannot capture these changes. (As you will see later in this chapter, this is a significant limitation that has been removed in SQL Server 2019.)

There are some special cases where the Memory Grant Feedback is not stored in the cached plan. For instance, a SELECT statement using OPTION (RECOMPILE) creates a new plan and does not cache it. Since it is not cached, no Memory Grant Feedback is produced and it is not stored for that compilation and execution. On the other hand, another statement without OPTION (RECOMPILE) using the same query hash will benefit, when re-executed, from the Memory Grant Feedback.

Tracking Memory Grant Feedback

Memory Grant Feedback can be tracked using several extended events. The two most important events in relation to Memory Grant Feedback are **memory_grant_updated by_feedback**, which occurs when the memory grant is updated by feedback, and **additional_ memory_grant**, which occurs when a query tries to get more memory during execution.

For these two extended events, we will use SQL Server Management Studio to create an extended events session called **Grant_Memory** and create an event file in which to store the messages from both events. (Refer to Chapter 20 for a detailed description of how to create an extended events session.)

On the General page in SQL Server Management Studio, specify the session name **Grant_ Memory**. Next, on the Events page, you have to select the two extended events mentioned previously, **memory_grant_updated by_feedback** and **additional_memory_grant**. To do so, type the phrase **memory_grant** in the Event Library field and move both events from the left pane to the right pane by selecting each event in turn and clicking the right-pointing arrow between the left and right panes. Move to the Data Storage page and enter the name of the file in which the captured data will be stored. Click OK.

After creation of the **Grant_Memory** session, you have to start capturing data. To do this, right-click the sessions name (under Management | Extended Events) and click Start Session. All subsequent statements will be tracked in relation to the **memory_grant_updated by_ feedback** and **additional_memory_grant** extended events. To end the capturing of data, click the Stop Session button.

Figure 28-1 shows the details of the session after executing two queries in relation to the **AdventureWorks** database.

As you can see from the session details, nine events are displayed with the corresponding timestamps and the additional memory required. All events are related to the **additional_ memory_grant** event. This means that the other event has not been fired in the time period between the start and end of the session.

Enabling/Disabling Memory Grant Feedback

By setting the compatibility level to 150, as shown in the Note at the beginning of this chapter, you enable all features belonging to Adaptive Query Processing. If you want to enable/disable

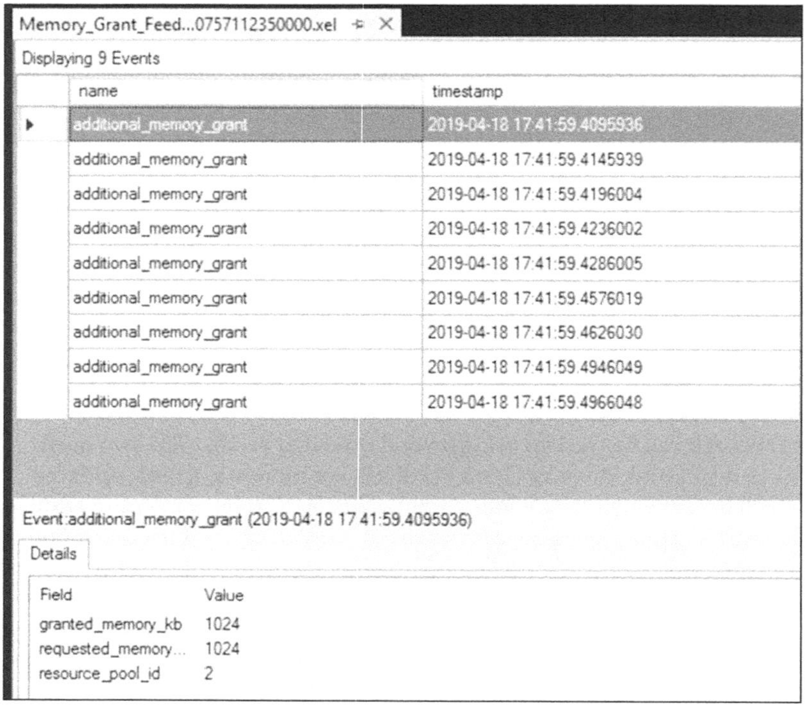

Figure 28-1 Displaying captured data in relation to Memory Grant Feedback

solely Memory Grant Feedback, you can use either the ALTER DATABASE statement or the query hint called DISABLE_BATCH_MODE_MEMORY_GRANT_FEEDBACK. (The value ON disables and the value OFF enables Memory Grant Feedback.)

Example 28.1 shows how you can use the ALTER DATABASE statement to disable only Memory Grant Feedback. (The statement modifies the mode of the current database.)

Example 28.1

```
ALTER DATABASE SCOPED CONFIGURATION SET
        BATCH_MODE_MEMORY_GRANT_FEEDBACK = ON;
```

Adaptive Join

The Adaptive Join feature enables the choice between the Hash Join operator and the Nested Loops operator to be deferred until after the first input has been scanned. In other words, the feature allows the query optimizer to choose dynamically between these two join processing techniques.

In a query execution plan, Adaptive Join is implemented as a new operator with the same name. This operator is an additional operator belonging to the group of join processing technique operators. The three already existing join operators are Nested Loops, usually used for joining a small data amount with a simple input; Merge Join, usually used for joining data streams that are already sorted by the clustered indices; and Hash Join, usually used for joining large, unsorted sets of rows.

The Adaptive Join operator has three inputs. The first input is an outer (build) input, the second is an input if the hash join technique is selected, and the third is an input if the nested loop method is picked. Additionally, the Adaptive Join operator defines a threshold that is used to decide when to switch to a query execution plan with nested loops. Your plan can therefore dynamically switch to a better join strategy during execution. Here is how it works:

- If the number of selected rows of the join input is small enough, the nested loop join technique would be more optimal than the hash join technique and your execution plan switches to the former technique.

- If the join input exceeds a specific row count threshold, no switch occurs and your plan continues with the hash join technique.

Example 28.2 provides a good starting point to explain the Adaptive Join feature and to introduce the Adaptive Join operator.

Example 28.2

```
USE AdventureWorks;
SELECT product.Name, COUNT(history.ProductID) AS Cnt,
          SUM(history.Quantity) AS Sum,
          AVG(history.ActualCost) AS Avg
   FROM Production.TransactionHistory AS history
   JOIN Production.Product AS product
          ON product.ProductID = history.ProductID
   GROUP BY history.ProductID, product.Name;
```

Example 28.2 joins two tables, **TransactionHistory** and **Product**, from the **Production** schema of the **AdventureWorks** database and calculates three different aggregate functions (SUM, COUNT, and AVG) for different columns of the **TransactionHistory** table. As you can see from the execution plan of the query given in Figure 28-2, the query optimizer uses the hash join technique to join both tables. (The SQL Server's Hash Match operator in Figure 28-2 corresponds to the hash join technique.)

To enable Adaptive Join, you have to

1. Create a corresponding nonclustered columnstore index (see Example 28.3).

2. Enable compatibility level 140 or higher. (Compatibility level 140 specifies SQL Server 2017, while 150 is the compatibility level of SQL Server 2019.)

3. Enable Live Query Statistics by clicking the icon with the same name in the toolbar of SSMS. (Live Query Statistics is described in Chapter 20.)

Example 28.3 creates the columnstore index, which is necessary to enable the Adaptive Join feature.

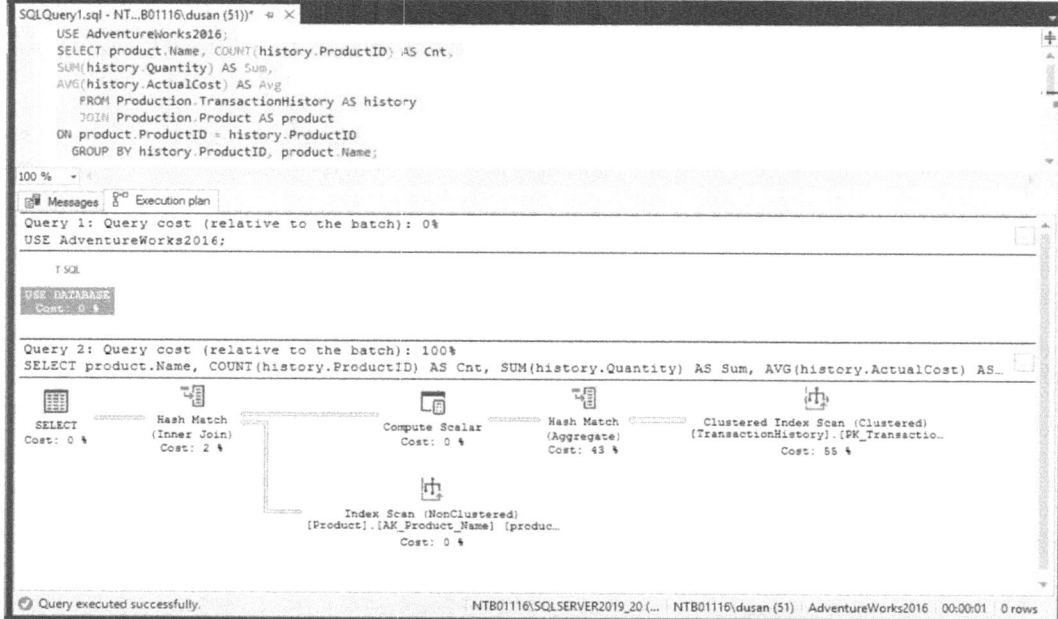

Figure 28-2 The execution plan of the query from Example 28.2 with the Hash Match operator

Example 28.3

```
USE AdventureWorks;
CREATE NONCLUSTERED COLUMNSTORE INDEX csi_history
  ON Production.TransactionHistory(ProductID,Quantity,ActualCost);
```

Now, if we execute the query in Example 28.2 again, we get the execution plan with the Adaptive Join operator, as shown in Figure 28-3.

Adaptive Join: Benefits and Disadvantages

Workloads with frequent oscillations between small and large join input scans will significantly benefit from the Adaptive Join feature. Additionally, if the Adaptive Join operator switches to Nested Loops, it uses the rows already read by Hash Join. In other words, the Adaptive Join operator does not reread the rows from the outer table.

On the other hand, the Adaptive Join operator introduces a higher memory requirement than an equivalent execution plan based on Nested Loops. The additional memory is requested as if the nested loops technique was Hash Join. There is also overhead for the build phase in relation to the corresponding Nested Loops operator.

NOTE You can disable Adaptive Join in a similar way as you disable Memory Grant Feedback, using the BATCH_MODE_ADAPTIVE_JOINS option. (For details, see Example 28.1.)

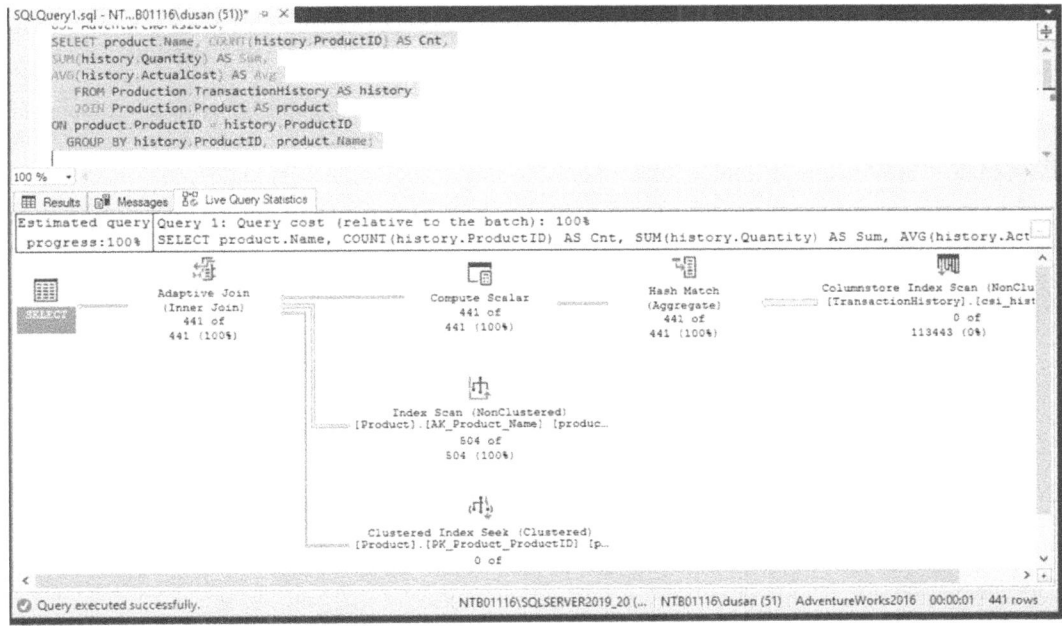

Figure 28-3   The modified execution plan with the Adaptive Join operator

Interleaved Execution

As you already know from Chapter 8, there are two types of user-defined functions (UDFs) in the Database Engine:

- Scalar functions
- Table-valued functions

Table-valued functions have two forms: inline table-valued functions and multistatement table-valued functions (MSTVFs). An inline table-valued function can contain only one root SELECT statement that is used to describe its result. This is good from an optimization perspective, because the query optimizer may inline a function's text into a query and can optimize a query as a whole.

MSTVFs can contain procedural language elements and may implement more complex logic. The benefit of MSTVFs is that they are more powerful for code reuse. On the other hand, they are more challenging in relation to the optimization process, because it is hard or impossible to estimate how many rows a function will return (i.e., estimate cardinality) before the query optimizer starts to execute the user-defined function.

Interleaved Execution is intended to correct cardinality estimation in MSTVFs. While selecting data from a table variable does not inhibit parallelism on its own, the low row estimates would often contribute to low query costs, where parallelism wouldn't be considered.

With Interleaved Execution, cardinality estimation is paused, the subtree for the MSTVF is executed, and optimization is resumed with a more accurate cardinality estimate.

NOTE In SQL Server 2012 and earlier, each MSTVF had, during its execution, a fixed estimate of one row. Starting from SQL Server 2016, the estimate is still fixed but is increased up to 100 rows.

To show how Interleaved Execution works, Example 28.4 introduces an MSTVF.

Example 28.4

```
USE AdventureWorks;
GO
CREATE FUNCTION GetLastShipped()
RETURNS @CustomerOrder TABLE
    (SaleOrderID    INT         NOT NULL,
     CustomerID     INT         NOT NULL,
     OrderDate      DATETIME    NOT NULL,
     OrderQty       INT         NOT NULL)
AS
BEGIN
    INSERT @CustomerOrder
    SELECT a.SalesOrderID, a.CustomerID, a.OrderDate, b.OrderQty
    FROM Sales.SalesOrderHeader a
        INNER JOIN Sales.SalesOrderDetail b
            ON a.SalesOrderID = b.SalesOrderID
        INNER JOIN Production.Product c
            ON b.ProductID = c.ProductID
          WHERE a.OrderDate = ( Select Max(SH1.OrderDate)
                        FROM Sales.SalesOrderHeader As SH1
                        WHERE SH1.CustomerID = A.CustomerId)
    RETURN
END
```

The **GetLastShipped** function created in Example 28.4 is an MSTVF because the body of the function is a complex INSERT ... SELECT statement. The function inserts several rows in the table variable called **@CustomerOrder**. All rows of the new table are generated from the **SalesOrderHeader** and **SalesOrderDetail** tables of the **Sales** schema as well as the **Product** table of the **Production** schema of the **AdventureWorks** database.

Example 28.5 executes the **GetLastShipped** function using a query and retrieves the number of rows from the result set.

Example 28.5

```
USE AdventureWorks;
SELECT C = COUNT_BIG(*)
    FROM GetLastShipped() C
```

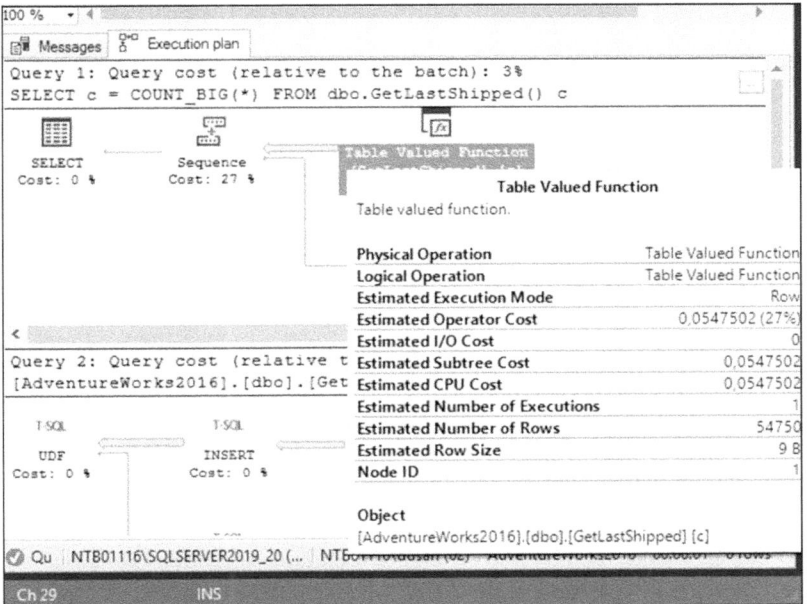

Figure 28-4   Properties of the Table Valued Function operator in Example 28.5

Figure 28-4 shows the corresponding execution plan of the SELECT statement in Example 28.5. Take a close look at the Table Valued Function operator and its properties in the execution plan and you will see that the value of Estimated Number of Rows is 54750. This means that Interleaved Execution has been performed and the correct cardinality of rows in the result set has been estimated.

Benefits and Disadvantages of Interleaved Execution

The benefits of the Interleaved Execution feature are twofold. First, estimating the correct cardinality is one of the most important parameters for the query optimizer to generate an optimal execution plan for the particular query. Second, in versions of SQL Server prior to SQL Server 2016, the query optimizer of the Database Engine used a fixed estimate of rows (one), which usually led to inefficient query plans. (See the Note before Example 28.4.)

Using the ALTER DATABASE statement, I set the compatibility level to 130, which is the level of SQL Server 2016, and executed the query in Example 28.5 again. Figure 28-5 shows that, in this case, Estimated Number of Rows is set to the fixed value (100).

On the other hand, there is only one (minor) disadvantage of using the Interleaved Execution feature: as with any plan affecting changes, some plans could change such that with better cardinality for the subtree of the corresponding execution plan tree, the result is a worse plan for the query overall.

Enabling/Disabling Interleaved Execution

If you want to enable/disable only the Interleaved Execution feature, you can use either the ALTER DATABASE statement or the DISABLE_INTERLEAVED_EXECUTION_TV query

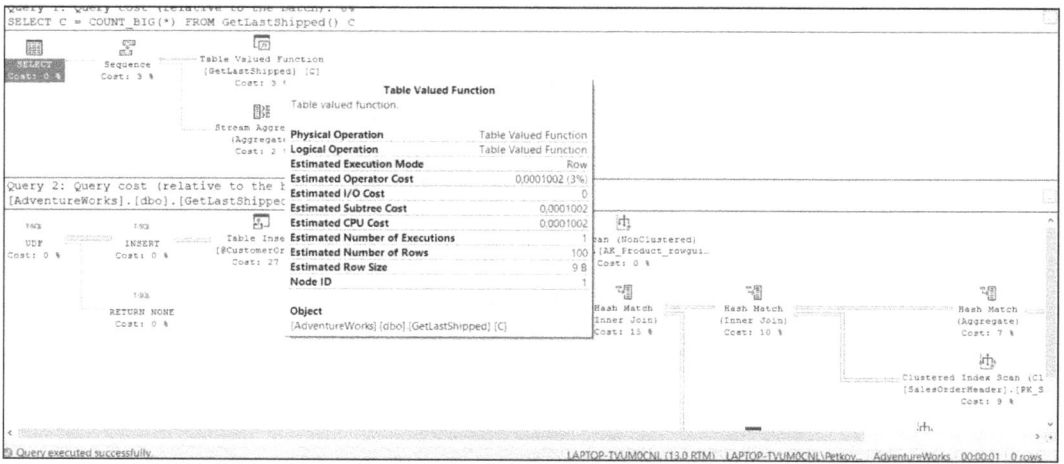

Figure 28-5   Properties of the Table Valued Function operator in SQL Server 2016

hint. Example 28.6 shows how you can use the ALTER DATABASE statement to disable solely Interleaved Execution. (The statement modifies the mode of the current database.)

Example 28.6

```
ALTER DATABASE SCOPED CONFIGURATION SET
          INTERLEAVED_EXECUTION_TV = ON;
```

Batch Mode on Rowstore

As you already know from Chapter 27, the Database Engine supports a feature called Batch Mode on Columnstore using columnstore indices. This feature can execute a query that processes millions or billions of rows more efficiently by including the rows in batches (up to 1000 rows in each batch) and executing batches rather than individual rows. That way, performance of queries can be significantly increased. (For a detailed discussion of this mode, see the "Batch Mode on Columnstore" section in Chapter 27.)

Even if a query does not involve any table with a columnstore index, the query optimizer of the Database Engine in SQL Server 2019 uses heuristics to decide whether to use Batch Mode or not. The heuristics consist of:

- An estimation of how many rows are in the input query, as well as which operators are used for the execution of that query

- Additional checkpoints to discover new and more optimal execution plans for the query

That way, the query optimizer can use batches of rows in certain cases even if the columnstore index is not specified.

We will use examples similar to ones presented in Chapter 27 to show a query in which the query optimizer uses Batch Mode on Rowstore. First, Example 28.7 (which corresponds to Example 27.9) increases the number of processors to four, using the **max degree of parallelism** (MAXDOP) advanced configuration option.

NOTE The query in Example 28.7 uses the same tables created in Examples 27.6 and 27.7. Therefore, if your **sample** database already contains the **FactInternetSales**, **DimCustomer**, and **DimDate** tables, start with Example 28.7. If not, first create these tables using Examples 27.6 and 27.7.

Example 28.7

```
EXEC sp_configure 'show advanced options', 1;
GO
RECONFIGURE WITH OVERRIDE;
GO
EXEC sp_configure 'max degree of parallelism', 4;
GO
RECONFIGURE WITH OVERRIDE;
GO
```

We increase the number of processors to four because Batch Mode processing uses algorithms that are optimized for multicore CPUs, and this processing mode is not a choice if the number of processors is one, which is the default value.

Example 28.8 creates a nonclustered non-columnstore index.

NOTE: If you have already created the columnstore index called **CLI_CS_IFactInternetSales** (see Example 27.10), you have to drop it to disable Batch Mode on Columnstore. To drop the index, use the following statement:

```
DROP INDEX FactInternetSales.CLI_CS_IFactInternetSales
```

Example 28.8

```
USE sample;
CREATE NONCLUSTERED INDEX nCLI_IFactSales
      ON  dbo.FactInternetSales
   (OrderDateKey, CustomerKey, SalesAmount) ;
```

The main difference between the sequence of statements in Chapter 27 (Examples 27.6 through 27.10) and the sequence of statements in this chapter is the type of index created in Example 28.8. Here, we create a *non-columnstore* index, whereas Example 27.10 created a *columnstore* index. That way, we disable Batch Mode on Columnstore, which requires creation of a columnstore index. In spite of this difference, the query optimizer uses Batch Mode for the query in Example 28.9, too. (The query in Example 28.9 is identical to the query in Example 27.10.)

Example 28.9

```
USE sample;
SELECT c.CommuteDistance,
       d.CalendarYear,
       SUM(f.SalesAmount) TotalSalesByCommuteDistance
FROM dbo.FactInternetSales as f
INNER JOIN dbo.DimCustomer as c ON
       f.CustomerKey = c.CustomerKey
INNER JOIN dbo.DimDate d ON
       d.DateKey = f.OrderDateKey
GROUP BY c.CommuteDistance, d.CalendarYear;
```

Figure 28-6 shows the execution plan of the query in Example 28.9 in general and the properties of the **Index Scan** operator in particular. Two properties are important in relation to Batch Mode on Rowstore: Estimated Execution Mode and Storage. The former has the value Batch and the latter the value RowStore. This means that the query is processed in Batch Mode, although the corresponding columnstore index is not created.

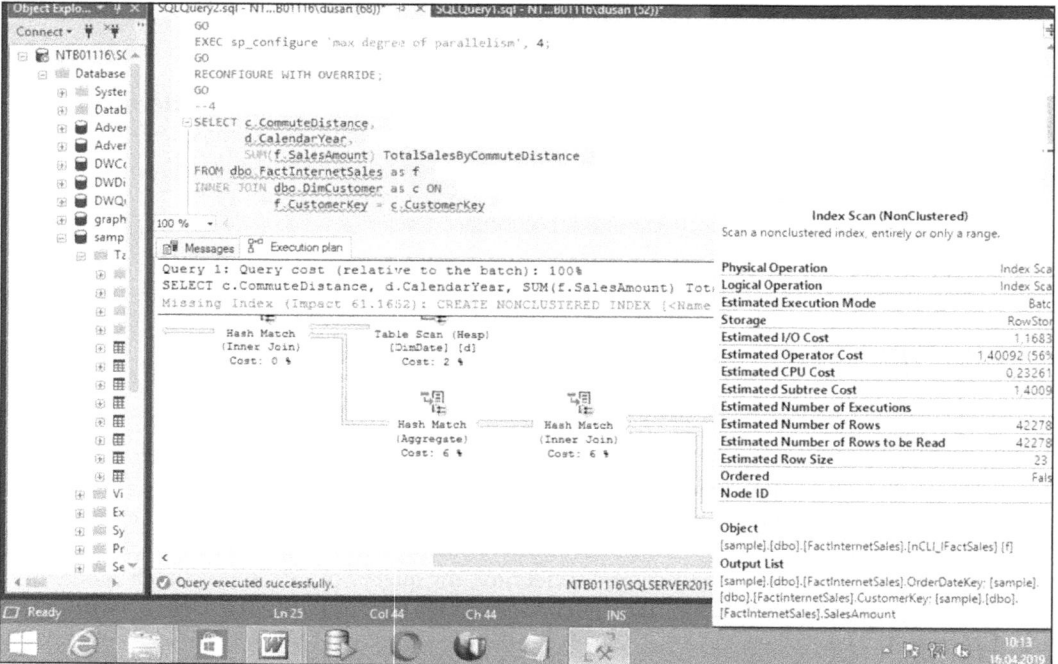

Figure 28-6 The execution plan of the query in Example 28.9

Approximate Query Processing

The APPROX_COUNT_DISTINCT function returns the approximate number of rows that contain distinct values of the given parameter, which is usually a column name. This function provides an alternative to the COUNT(DISTINCT) function, which returns the *exact* number of rows that contain distinct values for the column. The APPROX_COUNT_DISTINCT function processes large amounts of data significantly faster than COUNT(DISTINCT), with deviation from the exact result up to 3 percent.

Executing the query in Example 28.10 shows the differences between the execution of the APPROX_COUNT_DISTINCT function and the execution of COUNT(DISTINCT).

Example 28.10
```
USE AdventureWorksDW;
SET STATISTICS TIME ON;
SELECT count(DISTINCT(SalesOrderNumber))
               FROM FactInternetSales;
SELECT APPROX_COUNT_DISTINCT(SalesOrderNumber)
               FROM FactInternetSales;
```

Results for COUNT(DISTINCT):

```
27659 rows
   CPU time = 31 ms,   elapsed time = 196 ms.
```

Results for APPROX_COUNT_DISTINCT:

```
27990 rows
   CPU time = 63 ms, elapsed time = 60 ms.
```

As you can see from the elapsed times for executing the two functions, APPROX_COUNT_DISTINCT executes more than three times faster than COUNT(DISTINCT). On the other hand, the deviation from the exact result is about 1 percent.

NOTE The idea behind the APPROX_COUNT_DISTINCT function is to provide aggregations across large data sets where *response time is more critical than precision*. This idea can be extended to the other aggregate functions as well. Therefore, we can hope that Microsoft will implement similar aggregation functions (APPROX_SUM and APPROX_COUNT, for instance) in a future version of SQL Server.

Scalar UDF Inlining

As you already know from Chapter 8, a scalar user-defined function (UDF) is a function that returns an atomic (scalar) value. These functions have a significant performance drawback, because the query optimizer does not handle them optimally. The reason is that their execution is separated from the rest of the execution plans, meaning they get called once for every row and cannot be optimized based on the estimated number of rows.

SQL Server 2019 introduces a new feature, called Scalar UDF Inlining, which incorporates each scalar UDF into the overall plan. That way, the performance of queries that invoke scalar UDFs can be significantly improved.

In the following examples, we will create a function and execute it twice, first without using the new feature, and then with the Scalar UDF Inlining feature. Example 28.11 creates the function.

Example 28.11

```
ALTER DATABASE AdventureWorks SET COMPATIBILITY_LEVEL = 140;
GO
USE AdventureWorks;
GO
CREATE FUNCTION dbo.CustomerRate14 (@CustomerID INT)
   RETURNS CHAR(10) AS
BEGIN;
   DECLARE @sales DECIMAL (18,2);
   DECLARE @category CHAR(10);
   SET @sales = (SELECT SUM(Subtotal)
        FROM Sales.SalesOrderHeader WHERE CustomerID = @CustomerID);
IF @sales < 500000
            SET @category = 'REGULAR';
      ELSE IF @sales < 1000000
            SET @category = 'GOLD';
      ELSE
            SET @category = 'PLATINUM';
      RETURN @category;
END
```

The first statement in Example 28.11 sets the compatibility level of the **AdventureWorks** database to 140, which is the compatibility level of SQL Server 2017. That way, the Scalar UDF Inlining feature is indirectly disabled, because SQL Server 2017 does not support it.

After that, the **CustomerRate14** function is created. This function determines the sales category for a customer, which is identified using a unique customer key. The function first computes the total sales of all orders placed by that customer. Then, the function uses IF-ELSE logic to decide the category based on the customer's sales.

Example 28.12 defines a query that invokes the **CustomerRate14** function and measures the elapsed time of the query.

Example 28.12

```
USE AdventureWorks;
set STATISTICS TIME  ON;
SELECT c.CustomerID, dbo.CustomerRate14(c.CustomerID)
```

```
        FROM Sales.Customer AS c;
set STATISTICS TIME  OFF;
```

The elapsed time of this query on my computer is 1544 ms.

For every scalar UDF, you can use the **sys.sql_modules** catalog view to check whether or not a user-defined function can be inlined. This view includes a property called **is_inlineable**. A value of 1 indicates that the particular function can be inlined, and 0 indicates otherwise. Example 28.13 shows the use of the **sys.sql_modules** catalog view in relation to the function created in Example 28.12. (The result of this query is 1.)

Example 28.13

```
-- checking, whether a UDF can be inlined
SELECT is_inlineable
    FROM sys.sql_modules AS sm
    JOIN sys.objects AS o ON sm.object_id=o.object_id
        AND o.name = 'CustomerRate14'
```

NOTE If a scalar UDF is inlineable, it does not imply that it will always be inlined. The Database Engine will decide (on a per-query, per-UDF basis) whether to inline the particular UDF or not. For instance, if the UDF definition runs into thousands of lines of code, the Database Engine might choose not to inline it. This decision is made when the query referencing a scalar UDF is compiled.

Example 28.14 repeats the creation of the same function and the identical query, but with the compatibility level set to 150 for SQL Server 2019.

Example 28.14

```
ALTER DATABASE AdventureWorks SET COMPATIBILITY_LEVEL = 150;
GO
USE AdventureWorks;
GO
CREATE FUNCTION dbo.CustomerRate15 (@CustomerID INT)
   RETURNS CHAR(10) AS
BEGIN;
   DECLARE @sales DECIMAL (18,2);
   DECLARE @category CHAR(10);
   SET @sales = (SELECT SUM(Subtotal)
                   FROM Sales.SalesOrderHeader
                   WHERE CustomerID = @CustomerID);
IF @sales < 500000
            SET @category = 'REGULAR';
        ELSE IF @sales < 1000000
            SET @category = 'GOLD';
```

```
        ELSE
                SET @category = 'PLATINUM';
        RETURN @category;
END
GO
SET STATISTICS TIME  ON;
SELECT c.CustomerID, dbo.CustomerRate15(c.CustomerID)
       FROM Sales.Customer AS c;
set STATISTICS TIME  OFF;
```

The elapsed time of the query for Example 28.14 on my computer is 980 ms.

Enabling and Disabling Scalar UDF Inlining

This feature is enabled by default for SQL Server 2019, which means you don't have to make any changes to a particular UDF or a query using that UDF to take advantage of the feature.

To disable Scalar UDF Inlining at the database level, use the ALTER DATABASE statement for the current database, as follows:

```
ALTER DATABASE SCOPED CONFIGURATION
    SET TSQL_SCALAR_UDF_INLINING = OFF;
```

You can also disable Scalar UDF Inlining for a specific query by designating DISABLE_TSQL_SCALAR_UDF_INLINING as a USE HINT query hint. The USE HINT query hint takes precedence over the database-scoped configuration or compatibility level setting.

Table Variable Deferred Compilation

> **NOTE** This section gives you just a concise description of the Table Variable Deferred Compilation feature. The reason is twofold. First, the tests I have done with different forms of UDFs with a table variable show that only a small percentage of such functions benefit from the existence of the feature. Second, the Database Engine supports (and has for a long time) another feature with the same functionality: the OPTION(RECOMPILE) clause.

Generally, table variables do not have statistics. This means that the query optimizer of the Database Engine will assume that each table variable in a UDF has only a single row. Knowing that correctly estimating the number of input rows is one of the most important factors for the query optimizer to create an optimal execution plan, as discussed earlier in the chapter, you can appreciate that always using the fixed number without any calculation can cause all sorts of problems.

By delaying the compilation of the part of an execution plan that references a table variable, the Database Engine has a chance to capture the correct number of rows within the table variable, thus improving the execution plan for the following operators of that plan. Generally, if the table variable ends up containing a low number of rows, this assumption of a single row is not problematic. However, if the actual row count is higher, this can result in downstream inappropriate plan choices and significant performance disadvantages.

As stated at the beginning of this section, you can use the OPTION(RECOMPILE) clause to achieve the same functionality. (This option is the final part of the WHERE clause of a SELECT statement.) The clause forces the system to recompile the particular query every time it runs, but with correct row counts for table variables.

Generally, the functionality of OPTION(RECOMPILE) and the Table Variable Deferred Compilation feature is the same, but you need to remember that using OPTION(RECOMPILE) forces an execution query plan to be recompiled with every execution. The Table Variable Deferred Compilation feature does not have that problem, meaning that, once the code is compiled, the plan is cached away and can be reused like any other cached plan.

An additional benefit of this feature is that you do not have to change your code, as is the case with OPTION(RECOMPILE).

Summary

Intelligent Query Processing is Microsoft's name for a group of related SQL Server features whose main goal is to improve the performance of different forms of workload with only minimal changes to the existing code. Intelligent Query Processing comprises the following features:

- Adaptive Query Processing
- Batch Mode on Rowstore
- Approximate Query Processing
- Scalar UDF Inlining
- Table Variable Deferred Compilation

This is the final chapter of Part IV, "SQL Server and Business Intelligence." The next chapter starts the second-to-last part of the book, "Beyond Relational Data," and explains how the JSON data format is integrated in the Database Engine.

Part IV

PART

V

Beyond Relational Data

29 JSON Integration in the Database Engine

In This Chapter

- An Introduction to JSON
- Storing JSON Documents in the Database Engine
- Presenting and Querying JSON Documents
- Updating JSON Documents

This chapter describes the integration of JavaScript Object Notation (JSON) in the Database Engine. JSON is a data format used to store and transport data. In other words, you use JSON when you want to port data from one computer to another or from one operating system to another operating system.

The first major section introduces JSON and explains its basic concepts. The second section discusses relational storage of JSON documents. The third section discusses presentation and retrieval of data stored in the JSON format. The final section explains how to update JSON documents stored in relational tables.

An Introduction to JSON

JSON is a simple data format used for data interchange. The structure of JSON content follows the syntax structure for JavaScript. This data format is built on two structures:

- A collection of name/value pairs
- An ordered list of values

Example 29.1 shows an example of a JSON document.

Example 29.1

```
{"info": {"who": "Fred" ,"where": "Microsoft",
"friends":[{"name": "Lili", "rank":5},{"name": "Hank", "rank": 7}]}}
```

Example 29.1 shows a JSON string called **info**, which describes a single person, Fred, his affiliation, Microsoft, and his friends, Lili and Hank. Generally, a JSON string contains either an array of values or an object, which is an array of name/value pairs. An array is surrounded by a pair of square brackets and contains a comma-separated list of values. An object is surrounded by a pair of curly brackets and contains a comma-separated list of name/value pairs. A name/value pair consists of a field name (in double quotes), followed by a colon (:), followed by the field value (in double quotes).

From Example 29.1 we have the following strings:

- `"name": "Hank"`   This is a name/value pair.
- `{"name": "Hank", "rank": 7}`   This is an object.
- `[{"name":"Lili", "rank": 5} ,{"name": "Hank", "rank": 7}]`
  This is an array of objects.

A value in an array or object can be a number, a string, the NULL value, another array, or another string. A string must be written in double quotes, an array must be written in a pair of square brackets, and an object must be written in a pair of curly brackets.

NOTE To put actual double quotes inside strings, use the backslash character to escape the double quotes.

Why Support JSON in SQL Server?

The Database Engine has supported XML for a long time. Therefore, it is important to address why Microsoft added support for JSON as of SQL Server 2016. There are several reasons why JSON is supported in SQL Server:

- **Integration of semistructured data**   As you already know, relational tables contain structured data. The advantage of JSON is that it can contain both structured and semistructured data. By supporting the storage of JSON objects, the Database Engine extends its capabilities and integrates structured and semistructured data together.

- **Reduced administrative costs through use of data stores**   When JSON objects are stored individually and are used for separate programs, each program has to administer its own data. With JSON support through the Database Engine, the DBMS administers all data. The same is true for security and transaction management, because the DBMS takes over the management of security and transaction processing, meaning that this functionality does not need to be implemented in users' programs.

- **Increased developer productivity**   Using a DBMS to support JSON increases productivity because the DBMS takes over a lot of tasks that otherwise must be implemented in programs.

Storing JSON Documents in the Database Engine

Generally, there are three different ways in which data presented in a particular format can be stored in relational form:

- As "raw" documents
- Decomposed into relational columns
- Using native storage

If you store data presented in a particular format as a large object (LOB), an exact copy of the data is stored. In this case, data is stored "raw"—that is, in its character string form. The raw form allows you to insert documents very easily. The retrieval of such a document is very efficient if you retrieve the entire document. To retrieve parts of the documents, you need to create special types of indices.

To decompose data presented in a particular format into separate columns of one or more tables, you can use its schema. In this case, the hierarchical structure of the document is preserved, while order among elements is ignored.

NOTE The decomposition of JSON documents into separate columns is not possible, because JSON is (generally) a schema-less format.

Native storage means that documents are stored in their parsed form. In other words, the document is stored in an internal representation (Infoset, for instance) that preserves the content of the data.

Using native storage makes it easy to query information based on the structure of the document. On the other hand, reconstructing the original form of the document is difficult, because the created content may not be an exact copy of the document.

In the case of JSON, the Database Engine supports only the first form of storage, as raw documents. Therefore, JSON objects can be stored as values of the NVARCHAR data type.

NOTE The best way to store data is to use native storage. We can hope that the implementation of JSON objects as raw documents is just Microsoft's first step toward fully integrating JSON in the Database Engine.

Example 29.2 creates a table with a JSON document and inserts six rows in the table.

Example 29.2

```
USE sample;
CREATE TABLE json_table
(id INT PRIMARY KEY IDENTITY, person_and_friends NVARCHAR(2000));
-- Insert a couple of rows
INSERT INTO json_table (person_and_friends)  VALUES
    (N'{"info":{"who": "Fred" ,"where": "Microsoft" ,
 "friends":[{"name":"Lili","rank":5}, {"name":"Hank","rank":7}]}}');
```

```
INSERT INTO json_table (person_and_friends)  VALUES
   (N'{"info":{"who": "Tom", "where": "IBM", "friends": [ { "name":
 "Sharon", "rank": 2}, {"name": "Monty", "rank": 3} ] }}');
INSERT INTO json_table (person_and_friends)  VALUES
   (N'{"info":{"who":"Jack", "friends": [ { "name": "Connie" } ] }}');
INSERT INTO json_table (person_and_friends)  VALUES
   (N'{"info":{"who":"Joe","friends":[{"name":"Doris"},{"rank":1}]}}')
INSERT INTO json_table (person_and_friends)  VALUES
   (N' {"info": {"who":"Mabel",
 "where":"PostgresSQL","friends":[{"name":"Buck","rank": 6}]}}');
INSERT INTO json_table (person_and_friends)  VALUES
       (N' {"info":{"who": "Louise", "where": "Hanna" }}');
```

Example 29.2 creates a table called **jason_table** with a **person_and_friends** column, which contains JSON documents and is therefore specified using the NVARCHAR data type. The INSERT statements insert six JSON objects, the structure of which is identical to the JSON object presented in Example 29.1.

Because JSON documents are stored using the NVARCHAR data type, it is not necessary to support a customized JSON index. In other words, this data type is a standard data type, and columns specified using the NVARCHAR type can be indexed using the B-tree indices.

Presenting and Querying JSON Documents

Using the Database Engine, you can present data in either of the following ways:

- Present JSON documents as relational data
- Present relational data as JSON documents

The following sections describe these two methods. (Additionally, the last subsection describes how you can query JSON documents.)

Presenting JSON Documents as Relational Data

The Database Engine supports the OPENJSON function to present JSON documents as relational data. This function is a table-valued function that analyzes a given text to find an array of JSON objects. All objects found in the array are searched and, for each of them, the system generates a row in the output result. (For the description of table-valued functions, see Chapter 8).

There are two forms of the OPENJSON function:

- With a predefined result schema
- Without the schema

If a schema exists, it defines mapping rules that specify what properties will be mapped to the returned columns. Without such a schema, the result is a set of key-value pairs.

Example 29.3 shows the use of OPENJSON without the given schema.

Example 29.3

```
DECLARE @json NVARCHAR(MAX) =
N' {"info":{ "who": "Fred" ,"where": "Microsoft" ,
 "friends":[{"name":"Lili","rank":5}, {"name": "Hank", "rank": 7}]}'
SELECT [key], value FROM OPENJSON(@json, N' $.info.friends');
```

The result is

key	value
0	{"name": "Lili", "rank": 5}
1	{"name": "Hank", "rank": 7}

As you can see from the output of Example 29.3, the OPENJSON function searches in the array that is assigned to the @**json** variable. The variable is used as the first parameter of the OPENJSON function to return one row for each element in the array.

The second parameter of OPENJSON specifies the path, which is used to specify which part of the document will be displayed. (Examples with different path specifications will be given later in this chapter.)

The columns specified in the SELECT list of Example 29.3 define the output of key/value pairs discussed previously. By default, NULL is returned if the property is not found.

Presenting Relational Data as JSON Documents

If you want to convert the result set of a query to a JSON document, you can specify the FOR JSON clause at the end of your SELECT statement. In that case, the Database Engine takes the result, formats it as a JSON document, and returns it to the client. Every row is formatted as one object, with values generated as value objects and column names used as key names. The FOR JSON clause has two modes:

- AUTO
- PATH

The following subsections describe both modes.

AUTO Mode

With the AUTO option, the format of the JSON output is automatically determined based on the order of columns in the SELECT list and their source tables. Therefore, the user cannot change the output form.

AUTO mode returns the result set of a query as a simple, nested JSON tree. Each table in the FROM clause from which at least one column appears in the SELECT list is represented as an object. The columns in the SELECT list are mapped to the appropriate attributes of that object. Example 29.4 shows the use of AUTO mode.

Example 29.4

```
USE sample;
SELECT  dept_no, dept_name
    FROM department FOR JSON AUTO;
```

The result is

```
[{"dept_no":"d1","dept_name":"Research "},  {"dept_no":"d2 ","dept_name":
"Accounting "}, {"dept_no":"d3 ","dept_name":"Marketing "}]
```

As you can see from the output of Example 29.4, the result is displayed as a JSON array.

PATH Mode

In PATH mode, column names or column aliases are treated as expressions that indicate how the values are being mapped to JSON. (An expression consists of a sequence of nodes, separated by /. For each slash, the system creates another level of hierarchy in the resulting document.) Example 29.5 shows the use of PATH mode.

NOTE   In contrast to AUTO mode, using PATH mode allows you to maintain full control over the format of the JSON output.

Example 29.5

```
USE sample;
SELECT  dept_no AS [Department.Number], dept_name AS [Department.Name]
   FROM department FOR JSON PATH, ROOT ('Departments');
```

The result is

```
{"Departments":
[{"Department":{"Number":"d1 ","Name":"Research"}},{"Department":
{"Number":"d2 ","Name":"Accounting    "}},{"Department":{"Number":"d3 ",
"Name":"Marketing   "}}]}
```

To extend the user's control over the output for the PATH mode, the Database Engine supports two options:

- ROOT
- INCLUDE_NULL_VALUES

The ROOT option allows you to add a single, top-level element to the JSON output of the FOR JSON clause. Example 29.5 shows how the output of the corresponding query can be extended with the top-level element called "Departments."

Generally, the output does not include JSON properties for NULL values in the result of a query. To include NULLs, you have to specify the INCLUDE_NULL_VALUES option.

Querying JSON Documents

The Database Engine supports three functions that are used to query JSON documents:

- **isjson**
- **json_value**
- **json_query**

The following subsections describe these functions.

isjson

The **isjson** function tests whether a string contains a valid JSON document. This function is usually used to create a constraint that checks whether the document is well formed or not. The syntax of this function is

```
isjson(expression)
```

where **expression** is the name of a variable or a column that contains JSON text.

Example 29.6 shows the use of this function.

Example 29.6

```
USE sample;
SELECT id, person_and_friends
FROM json_table
WHERE isjson(person_and_friends) > 0;
```

The query in Example 29.6 tests whether the JSON documents stored in the table **json_table** are valid. (The function returns 1 if the string contains a valid JSON document; otherwise, it returns 0.) After that, it displays all valid documents. In this case, all values are valid and will be displayed.

json_value

The **json_value** function extracts a scalar value from a JSON string. The syntax of this function is

```
json_value(expression, path)
```

where **expression** is the name of a variable or a column that contains JSON text and **path** is a JSON path that specifies the property to extract.

Example 29.7 shows the use of this function.

Example 29.7

```
USE sample;
SELECT id,json_value(person_and_friends,'$.info.where') AS company
    FROM json_table
    WHERE isjson(person_and_friends) > 0
    AND json_value(person_and_friends, '$.info.who') = 'Fred';
```

Part V

The result is

id	company
1	Microsoft

Example 29.7 uses two **json_value** functions and an **isjson** function. The latter is used to test whether the JSON documents stored in the **person_and_friends** column are valid. The second **json_value** function searches for such objects where the name/value pair is equal to "who/Fred." After that, the first **json_value** function displays the value of the corresponding company, which is stored in the key name called **where**.

Path expressions, such as '$.info.who' and '$.info.where' in Example 29.7, are explained in the next subsection.

JSON Path Expressions   JSON path expressions are used to reference the properties of JSON objects. These expressions are a part of the OPENJSON, **json_value**, and **json_query** functions. Each path contains a set of path steps. Path steps can contain key names, array elements, or the dot operator, which indicates a member of an object.

The following list shows some examples of path expressions:

- **$**   The entire document
- **$.info[0].who**   Fred
- **$.info[1]**   "who": "Tom", "where": "IBM", "friends": [{ "name": "Sharon", "rank": 2}, {"name": "Monty", "rank": 3}] }}');

json_query

This function returns extracts of an object or an array from a JSON string. The result is of type NVARCHAR(MAX). The syntax is analogous to the syntax of the **json_value** function.

Example 29.8 shows the use of the **json_query** function.

Example 29.8

```
USE sample;
SELECT id, person_and_friends,
  json_query(person_and_friends, '$.info.where')
  FROM json_table;
```

The query in Example 29.8 returns the entire JSON document.

The key difference between **json_value** and **json_query** is that **json_value** returns a scalar value, while **json_query** returns an object or an array. For instance, if the path is specified as **$.info[0].who**, the result (**Fred**) will be displayed when the **json_value** function is applied. For the same path expression, **json_query** displays NULL or error. On the other hand, if the path is specified as **$**, the result (the entire document) will be displayed when **json_query** is applied. In the case of **json_value**, NULL or error will be displayed.

Updating JSON Documents

The Database Engine supports the **json_modify** function to update the value of a property in a JSON document and to return the updated document. (The **json_modify** function returns an error if the given JSON document does not contain valid JSON.)

The syntax of this function is

```
json_modify(expression, path, new_value)
```

expression specifies the name of a variable or a column that contains the JSON document. **path** is a JSON path expression that specifies the property to update. **new_value** is the new value for the property specified by **path**.

To explain the functionality of the **json_modify** function, Example 29.9 creates the **json_update_table** table and inserts a JSON document into it.

Example 29.9

```
CREATE TABLE json_update_table
(id INT PRIMARY KEY IDENTITY, person_and_friends NVARCHAR(2000));
-- Insert a  row
INSERT INTO json_update_table (person_and_friends)  VALUES
    (N'{"info":{"who": "Fred" , "friends":["Lili", "Hank"]}}');
```

The content of the **json_update_table** table after the execution of the INSERT statement is

id	person_and_friends
1	{"info":{"who": "Fred" , "friends":["Lili", "Hank"]}}

The following examples show the supported forms of updates that you can use with the **json_modify** function.

The UPDATE statement in Example 29.10 modifies the value of the object specified by the following path: '$.info.who'. The existing name/value pair ("who": "Fred") will be changed in ("who": "Peter").

Example 29.10

```
-- Update the value of the object
UPDATE json_update_table
  SET person_and_friends =  json_modify(person_and_friends,
            '$.info.who', 'Peter')
        WHERE id  = 1;
```

The UPDATE statement in Example 29.11 shows how you can append an element to the existing array.

Example 29.11
```
UPDATE json_update_table
SET person_and_friends= json_modify(person_and_friends,
     'append $.info.friends', 'Wendy')
  WHERE id = 1;
```

The modified JSON document is as follows:

```
{"info":{"who": "Peter" , "friends":["Lili", "Hank","Wendy"]}}
```

Example 29.12 inserts a new object (name/value pair) in the existing document. Note that the new object will be appended to the content of the existing document.

Example 29.12
```
-- INSERT a new object
UPDATE json_update_table
  SET person_and_friends = json_modify(person_and_friends,
                    '$.info.surname', 'Birch')
  WHERE id  = 1;
```

The result is

person_and_friends
{"info":{"who":"Peter","friends":["Lili","Hank","Wendy"],"surname":"Birch"}}

Example 29.13 shows how you can delete an object in the existing document.

Example 29.13
```
UPDATE json_update_table
  SET person_and_friends = json_modify(person_and_friends,
                    '$.info.surname', NULL)
  WHERE id = 1;
```

The UPDATE statement in Example 29.13 uses the NULL value to delete the existing object, the name/value pair "info.surname": "Birch". Note that this is valid only in the lax mode.

NOTE   JSON functions, such as **json_value** and **json_query**, can be in either of two modes: lax or strict. The lax mode generally means that an error will not be raised with an invalid operation, and a NULL value will be returned instead. If strict mode is specified, an error will be raised with an invalid operation. In this chapter, the lax mode, which is the default value, is implicitly used for all examples. (For particular differences between the lax mode and strict mode, see Microsoft Docs.)

Note that the **json_modify** function currently does not support the insertion of an array member before or after another array member. The only way you can do it is to delete the document first and then insert the modified one, as shown in Example 29.14.

Example 29.14

```
UPDATE json_update_table
  SET person_and_friends =  json_modify(person_and_friends,
          '$.info.friends', NULL)
  WHERE id = 1;
UPDATE json_update_table
  SET person_and_friends =  json_modify (person_and_friends,
          '$.info.friends', 'Lili, Wendy,Hank')
  WHERE id  = 1;
```

Summary

JSON is a simple way to store and present structured and semistructured data. This format is usually used in web applications to transfer data from the server to browsers. (The data transfer in the other direction is possible, but is not used much in practice.)

The Database Engine fully supports storing, presenting, and querying JSON objects. The only storage form the current Database Engine supports for JSON objects is as raw documents, meaning that each object is stored in a column of type NVARCHAR. There are two presenting modes: AUTO and PATH. AUTO mode returns the result set of a query as a simple, nested JSON tree. Each table in the FROM clause from which at least one column appears in the SELECT list is represented as an object. PATH mode treats column names or column aliases as expressions that indicate how the values are being mapped to JSON. Also, the built-in function **json_value** can be used to extract scalar values from a JSON object.

The modification of JSON documents is supported, too. The Database Engine supports the **json_modify** function to update the value of a property in a JSON document and to return the updated document.

The next chapter describes spatial and temporal data.

30

Spatial and Temporal Data

In This Chapter

- Spatial Data
- Working with Spatial Data Types
- Temporal Tables

This chapter describes two special data forms: spatial data and temporal data. The first part of the chapter covers spatial data and has two major sections. The introductory section describes the most important general issues about spatial data that you need to understand before you begin to work with spatial data. Besides different spatial models and formats, this section introduces both of the data types supported by the Database Engine, GEOMETRY and GEOGRAPHY.

The second major section presents several examples to show how spatial data can be used. In these examples, different methods of the GEOMETRY and GEOGRAPHY data types are applied. Additionally, this section describes the creation and use of spatial indices. The remainder of the section is dedicated to describing how spatial objects can be edited. You can display spatial data using SQL Server Management Studio. This section shows several examples of using SSMS to display spatial objects.

The second part of the chapter (the final major section) explains temporal data. First, it describes how to create temporal tables, which are used to store temporal data, and explains some special properties of temporal data. It then shows how the Database Engine can be used to modify temporal data using DML statements. Several examples using UPDATE and DELETE show how temporal data is stored in two different tables: a current table and a history table. After that, retrieval of data is explained. In contrast to INSERT, UPDATE, and DELETE, the SELECT statement contains extensions that allow you to query temporal data in various ways. The section wraps up by explaining how to convert nontemporal tables into temporal tables.

Spatial Data

In the past several years, the need of businesses to incorporate spatial data into their databases and to manage it using a database system has grown significantly. The most important factor leading to this growth is the proliferation of geographical services and devices, such as Microsoft Bing Maps, low-priced GPS devices, and mobile phones.

Generally, the support of spatial data by a database vendor helps users to make better decisions in several scenarios, such as:

- Real-estate analysis ("Find a suitable property within 500m of an elementary school.")
- Consumer-based information ("Find the nearest shopping malls to a given ZIP code.")
- Market analysis ("Define geographic sales regions and ascertain whether there is a necessity for a new branch office.")

As you already know from Chapter 5, you can use the CREATE TYPE statement to create user-defined data types. The implementation of such types is done using the Common Language Runtime (CLR). Developers use the CLR to implement two data types in relation to spatial data: GEOMETRY and GEOGRAPHY. These two data types are discussed after the following brief look at the different models for representing spatial data.

Models for Representing Spatial Data

Generally, there are two different groups of models for representing spatial data:

- Geodetic spatial models
- Flat spatial models

Planets are complex objects that can be represented using a flattened sphere (called a *spheroid*). A good approximation for the representation of Earth (and other planets) is a globe, where locations on the surface are described using latitude and longitude. (Latitude gives the location of a place on Earth north or south of the equator, while longitude specifies the location in relation to a chosen meridian.) Models that use these measures are called *geodetic models*. Because these models provide a good approximation of spheroids, they provide the most accurate way to represent spatial data.

Flat spatial models (or planar models) use two-dimensional maps to represent Earth. In this case, the spheroid is flattened and projected in a plane. The flattening process results in some deformation of shape and size of the projected (geographic) objects. Flat spatial models work best for small surface areas, because the larger the surface area being represented, the more deformation that occurs.

As you will see in the following two sections, the GEOMETRY data type is based on a flat spatial model, while the GEOGRAPHY data type is based on a geodetic spatial model.

GEOMETRY Data Type

The Open Geospatial Consortium (OGC) introduced the term "geometry" to represent spatial features, such as point locations and lines. Therefore, "geometry" represents data in

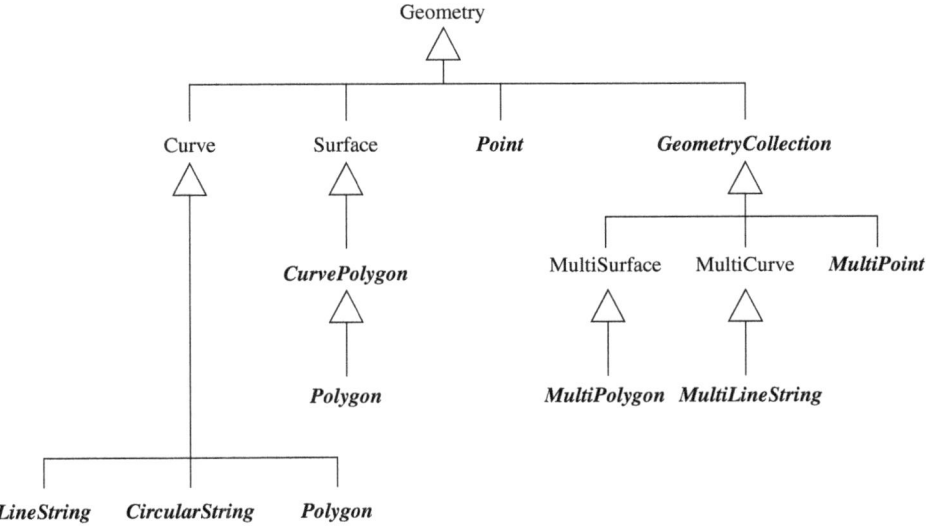

Figure 30-1   The type hierarchy with the GEOMETRY type as a root

a two-dimensional plane as points, lines, and polygons using one of the existing flat spatial models.

You can think of "geometry" as a data type with several subtypes, as shown in Figure 30-1. The subclasses are divided into two categories: the base geometry subclasses and the homogeneous collection subclasses. The base geometries include, among others, Point, LineString, and Polygon, while the homogeneous collections include MultiPoint, MultiLineString, and MultiPolygon. As the names imply, the homogeneous collections are collections of base geometries. In addition to sharing base geometry properties, homogeneous collections have their own properties.

The types in Figure 30-1 that appear in italic font are *instantiable*, which means they have instances. All instantiable types are implemented as user-defined data types in the Database Engine. The following are the instantiable types:

- **Point**   A point is a zero-dimensional geometry with single X and Y coordinate values. Therefore, it has a NULL boundary. Optionally, a point can have two additional coordinates: evaluation (Z coordinate) and measure (M coordinate). Points are usually used to build complex spatial types.

- **MultiPoint**   A multipoint is a collection of zero or more points. The points in a multipoint do not have to be distinct.

- **LineString**   A line string is a one-dimensional geometry object that has a length and is stored as a sequence of points defining a linear path. Therefore, a line string is defined by a set of points, which define the reference points of it. Linear interpolation between the reference points specifies the resulting line string. A line string is called *simple* if it does not intersect its interior. The endpoints (the boundary) of a *closed* line string occupy the same point in space. A line is called a *ring* if it is both closed and simple.

- **MultiLineString**   A multiline string is a collection of zero or more line strings.

- **Polygon**   A polygon is a two-dimensional geometry object with surface. It is stored as a sequence of points defining its exterior bounding ring and zero or more interior rings. The exterior and any interior rings specify the boundary of a polygon, and the space enclosed between the rings specifies the polygon's interior.

- **MultiPolygon**   A multipolygon is a collection of zero or more polygons.

- **GeometryCollection**   A geometry collection is a collection of zero or more geometry objects. In other words, this geometry object can contain instances of any subtype of the GEOMETRY data type.

- **CircularString**   A circular string is a collection of zero or more continuous circular arc segments. A circular arc segment is a curved segment defined by three points in a two-dimensional plane; the first point cannot be the same as the third point. If all three points of a circular arc segment are collinear, the arc segment is treated as a line segment.

- **CurvePolygon**   A curve polygon is a closed surface defined by an exterior bounding ring and zero or more interior rings

GEOGRAPHY Data Type

While the GEOMETRY data type stores data using X and Y coordinates, the GEOGRAPHY data type stores data as GPS latitude and longitude coordinates. (Longitude represents the horizontal angle and ranges from −180 degrees to +180 degrees, while latitude represents the vertical angle and ranges from −90 degrees to +90 degrees.)

The GEOGRAPHY data type, unlike the GEOMETRY data type, requires the specification of a Spatial Reference System. A Spatial Reference System is a system used to identify a particular coordinate system and is specified by an integer. Information on available integer values can be found in the **sys.spatial_reference_systems** catalog view. (This view will be discussed later in this chapter.)

NOTE   All instantiable types (see Figure 30-1) that are implemented for the GEOMETRY data type are implemented for the GEOGRAPHY data type, too.

GEOMETRY vs. GEOGRAPHY

As you already know, the GEOMETRY data type is used in flat spatial models, while the GEOGRAPHY data type is used in geodetic models. The main difference between these two groups of models is that with the GEOMETRY data type, distances and areas are given in the same unit of measurement as the coordinates of the instances. (Therefore, the distance between the points (0,0) and (3,4) will always be 5 units.) This is not the case with the GEOGRAPHY data type, which works with ellipsoidal coordinates that are expressed in degrees of latitude and longitude.

There are also some restrictions placed on the GEOGRAPHY data type. For example, each instance of the GEOGRAPHY data type must fit inside a single hemisphere.

External Data Formats

The Database Engine supports three external data formats that can be used to represent spatial data in an implementation-independent form:

- **Well-known text (WKT)**   A text markup language for representing spatial reference systems of spatial objects and transformations between spatial reference systems.
- **Well-known binary (WKB)**   The binary equivalent of WKT.
- **Geography Markup Language (GML)**   The XML grammar defined by OGC to express geographical features. GML is an open interchange format for geographic transactions on the Internet.

NOTE All examples shown in this chapter reference the WKT format, because this format is the easiest to read.

The following examples show the syntax of WKT for the selected types:

- **POINT(3,4)**   The values 3 and 4 specify the X coordinate and Y coordinate, respectively.
- **LINESTRING(0 0, 3 4)**   The first two values represent the X and Y coordinates of the starting point, while the last two values represent the X and Y coordinates of the end point of the line.
- **POLYGON(300 0, 150 0, 150 150, 300 150, 300 0)**   Each pair of numbers represents a point on the edge of the polygon. (The end point of the specified polygon is the same as the starting point.)
- **CIRCULARSTRING(0 –12.5, 0 0, 0 12.5)**   You need at least three points to define a circular string. The first point specifies the start, the second specifies the end of the circular string, and the third must be somewhere along the arc.
- **COMPOUNDCURVE(CIRCULARSTRING(0 –23.43778, 0 0, 0 23.43778),
  CIRCULARSTRING(0 23.43778, –45 23.43778, –90 23.43778),
  CIRCULARSTRING(–90 23.43778, –90 0, –90 –23.43778),
  CIRCULARSTRING(–90 –23.43778, –45 –23.43778, 0 –23.43778))**   The compound curve is a composition of either circular strings only or circular and linear strings.

Working with Spatial Data Types

As you already know, the Database Engine supports two different data types in relation to spatial data: GEOMETRY and GEOGRAPHY. Both data types have several subtypes, which are either instantiable or noninstantiable. For each instantiable subtype, you can create instances and work with them. These instances can be used as values of a table's columns, as well as variables or parameters. As you already guessed, noninstantiable types do not contain instances.

In a hierarchy of classes, the root class is usually noninstantiable, while the classes that build the leaves of the hierarchy tree are almost always instantiable.

The following two subsections describe how you can use these two data types to create and query spatial data. After that, the spatial indices will be introduced.

Working with the GEOMETRY Data Type

An example will help to explain the use of the GEOMETRY data type. Example 30.1 creates a table for nonalcoholic beverage markets in a given city (or state).

Example 30.1

```
            USE sample;
            CREATE TABLE beverage_markets
                 (id INTEGER IDENTITY(1,1),
                   name VARCHAR(25),
                   shape GEOMETRY);
INSERT INTO beverage_markets
        VALUES ('Coke', GEOMETRY::STGeomFromText
            ('POLYGON ((0 0, 150 0, 150 150, 0 150, 0 0))', 0));
INSERT INTO beverage_markets
        VALUES ('Pepsi', GEOMETRY::STGeomFromText
            ('POLYGON ((300 0, 150 0, 150 150, 300 150, 300 0))', 0));
INSERT INTO beverage_markets
        VALUES ('7UP', GEOMETRY::STGeomFromText
            ('POLYGON ((300 0, 150 0, 150 150, 300 150, 300 0))', 0));
INSERT INTO beverage_markets
    VALUES ('Almdudler', GEOMETRY::STGeomFromText('POINT (50 0)', 0));
```

The **beverage_markets** table has three columns. The first is the **id** column, the values of which are generated by the system because this column is specified with the IDENTITY property. The second column, **name**, contains the beverage name. The third column, **shape**, specifies the shape of the market area in which the particular beverage is the most preferred one. The first three INSERT statements create three areas in which a particular beverage is most preferred. All three areas happen to be a polygon. The fourth INSERT statement inserts a point because there is just one place where the particular beverage (Almdudler) can be bought.

NOTE If you take a look at the specification of the POINT (or POLYGON) data type in Example 30.1, you will see that this specification has an additional parameter as the last parameter, the spatial reference ID (SRID) parameter. This parameter is required, and for the GEOMETRY data type the default value is 0.

Example 30.1 introduces the first method in relation to the GEOMETRY data type: **STGeomFromText()**. This static method is used to insert the coordinates of geometric figures, such as polygons and points. In other words, it returns an instance of the GEOMETRY data type in the WKT format.

NOTE Generally, a type can have two different groups of methods: static methods and instance methods. Static methods are always applied on the whole type (i.e., class), while instance methods are applied on particular instances of the class. The invocation of methods from both groups is different. Static methods use the sign "::" between the type and the method (for instance, GEOMETRY::STGeomFromText; see Example 30.1), while instance methods use dot notation (for instance, @g.STContains; see Example 30.2).

Besides the **STGeomFromText()** method, the Database Engine supports three other similar static methods:

- **STPointFromText()**   Returns the WKT representation of an instance of the POINT data type
- **STLineFromText()**   Returns the WKT representation of an instance of the LINESTRING data type augmented with the corresponding elevation and measure values
- **STPolyFromText()**   Returns the WKT representation of an instance of the MULTIPOLYGON data type augmented with the corresponding elevation and measure values

Spatial data can be queried the same way as relational data. The following examples show a sample of the information that can be found from the content of the **shape** column of the **beverage_markets** table.

NOTE The Database Engine supports a lot of methods that can be applied to instances of the GEOMETRY data type. The following examples describe only some of the most important methods. For more information on other instance methods, refer to Microsoft Docs.

Example 30.2 shows the use of the **STContains()** method.

Example 30.2
Determine whether the shop that sells Almdudler lies within the area where Coke is the preferred beverage:

```
USE sample;
DECLARE @g geometry;
DECLARE @h geometry;
SELECT @h = shape FROM beverage_markets WHERE name ='Almdudler';
SELECT @g = shape FROM beverage_markets WHERE name = 'Coke';
SELECT @g.STContains(@h);
```

The result is 0.

The **STContains()** method returns 1 if an instance of the GEOMETRY data type completely contains another instance of the same type, which is specified as a parameter of the method. The result of Example 30.2 means that the shop that sells Almdudler does not lie within the area where the preferred beverage is Coke.

Example 30.3 shows the use of the **STLength()** method.

Example 30.3

Find the length and the WKT representation of the **shape** column for the Almdudler shop:

```
USE sample;
SELECT id, shape.ToString() AS wkt, shape.STLength() AS length
        FROM beverage_markets
        WHERE name = 'Almdudler' ;
```

The result is

id	wkt	length
4	POINT (50 0)	0

The **STLength()** method in Example 30.3 returns the total length of the elements of the GEOMETRY data type. (The result is 0 because the displayed value is a point.) The **ToString()** method returns a string with the logical representation of the current instance. As you can see from the result of Example 30.3, this method is used to load all properties of the given point and to display it using the WKT format.

Example 30.4 shows the use of the **STIntersects()** method.

Example 30.4

Determine whether the region that sells Coke intersects with the region that sells Pepsi:

```
USE sample;
DECLARE @g geometry;
DECLARE @h geometry;
SELECT @h = shape FROM beverage_markets WHERE name = 'Coke';
SELECT @g = shape FROM beverage_markets WHERE name = 'Pepsi';
SELECT @g.STIntersects(@h) ;
```

The result of Example 30.4 is 1 (TRUE), meaning that the two instances intersect.

In contrast to Example 30.3, where the column of a table is declared to be of the GEOMETRY data type, Example 30.4 declares the variables **@g** and **@h** using this data type. (As you already know, table columns, variables, and parameters of stored procedures can be declared to be of the GEOMETRY data type.) The **STIntersects()** method returns 1 if a geometry instance intersects another geometry instance. In Example 30.4, the method is applied to both regions, declared by variables, to find out whether the regions intersect.

Example 30.5 shows the use of the **STIntersection()** method.

Example 30.5

```
USE sample;
DECLARE @poly1 GEOMETRY = 'POLYGON ((1 1, 1 4, 4 4, 4 1, 1 1))';
DECLARE @poly2 GEOMETRY = 'POLYGON ((2 2, 2 6, 6 6, 6 2, 2 2))';
DECLARE @result GEOMETRY;
```

```
SELECT @result = @poly1.STIntersection(@poly2);
SELECT @result.STAsText();
```

The result is (the values are rounded):

```
POLYGON ((2 2, 4 2, 4 4, 2 4, 2 2))
```

The **STIntersection()** method returns an object representing the points where an instance of the GEOMETRY data type intersects another instance of the same type. Therefore, Example 30.5 returns the rectangle where the polygon declared by the **@poly1** variable and the polygon declared by the **@poly2** variable intersect. The **STAsText()** method returns the WKT representation of a GEOMETRY instance, which is the result of the example.

NOTE   The difference between the **STIntersects()** and **STIntersection()** methods is that the former method tests whether two geometry objects intersect, while the latter displays the intersection object.

Working with the GEOGRAPHY Data Type

The GEOGRAPHY data type is handled in the same way as the GEOMETRY data type. This means that the same (static and instance) methods that you can apply to the GEOMETRY data type are applicable to the GEOGRAPHY data type, too. For this reason, only Example 30.6 is used to describe this data type.

Example 30.6

```
USE AdventureWorks;
SELECT SpatialLocation, City
     FROM Person.Address
     WHERE City = 'Dallas';
```

The result is

SpatialLocation	City
0xE6100000010C4DD260393369404026C0A31BF73458C0	Dallas
0xE6100000010C10A810D1886240403A0F0653663158C0	Dallas
0xE6100000010C4346160AA26440406340F0E64F3B58C0	Dallas
0xE6100000010C107E16DAAD6540403DA892EAD52C58C0	Dallas
0xE6100000010C8044A1422D5F4040F66D784F983758C0	Dallas
0xE6100000010C8E345943826A4040839B00B8E03358C0	Dallas
0xE6100000010CAA5BBD5FAB69404087866D198D3C58C0	Dallas

The **Address** table of the **AdventureWorks** database contains a column called **SpatialLocation**, which is of the GEOGRAPHY data type. Example 30.6 displays the geographic location of all persons living in Dallas. As you can see from the result of this example, the value in the **SpatialLocation** column is the hexadecimal representation of the longitude and latitude of the location where each person lives. (Example 30.10, later in the chapter, uses Management Studio to display the result of this query.)

Spatial Indices

Spatial indices are necessary to speed up retrieval operations on spatial data. The Database Engine supports two different spatial index types:

- Spatial index
- Auto grid index

The following two subsections describe these indices. The subsequent two subsections show you how to edit information concerning spatial indices and how to display spatial objects.

Spatial Index

A spatial index is defined on a table column of the GEOMETRY or GEOGRAPHY data type. These indices are built using B-trees, which means that the indices represent two dimensions in the linear order of B-trees. Therefore, before reading data into a spatial index, the system implements a hierarchical uniform decomposition of space. The index-creation process decomposes the space into a four-level grid hierarchy.

The CREATE SPATIAL INDEX statement is used to create a spatial index. The general form of this statement is similar to the traditional CREATE INDEX statement, but contains additional options and clauses, some of which are introduced here:

- **GEOMETRY_GRID clause**   Specifies the geometry grid tessellation scheme that you are using. (*Tessellation* is a process that is performed after reading the data for a spatial object. During this process, the object is fitted into the grid hierarchy by associating it with a set of grid cells that it touches.) Note that GEOMETRY_GRID can be specified only on a column of the GEOMETRY data type.

- **BOUNDING_BOX option**   Specifies a numeric four-tuple that defines the four coordinates of the bounding box: the X-min and Y-min coordinates of the lower-left corner, and the X-max and Y-max coordinates of the upper-right corner. This option applies only within the GEOMETRY_GRID clause.

- **GEOGRAPHY_GRID clause**   Specifies the geography grid tessellation scheme. This clause can be specified only on a column of the GEOGRAPHY data type.

Example 30.7 shows the creation of a spatial index for the **shape** column of the **beverage_markets** table.

Example 30.7

```
USE sample;
GO
ALTER TABLE beverage_markets
 ADD CONSTRAINT prim_key PRIMARY KEY(id);
```

```
GO
CREATE SPATIAL INDEX i_spatial_shape
   ON beverage_markets(shape)
   USING GEOMETRY_GRID
   WITH (BOUNDING_BOX = ( xmin=0, ymin=0, xmax=500, ymax=200 ),
    GRIDS = (LOW, LOW, MEDIUM, HIGH),
    PAD_INDEX  = ON );
```

A spatial index can be created only if the primary key for the table with a spatial data column is explicitly defined. For this reason, the first statement in Example 30.7 is the ALTER TABLE statement, which defines this constraint.

The subsequent CREATE SPATIAL INDEX statement creates the index using the GEOMETRY_GRID clause. The BOUNDING_BOX option specifies the boundaries inside which the instance of the **shape** column will be placed. The GRIDS option specifies the density of the grid at each level of a tessellation scheme. (The PAD_INDEX option specifies that the FILLFACTOR setting should be applied to the index pages as well as to the data pages in the index.)

Auto Grid Index

The strategy that the auto grid index uses is to pick the right trade-off between performance and efficiency, which is different from the strategy used for a regular spatial index. The auto grid index uses eight levels of tessellation for better approximation of objects of various sizes. (The already described spatial index uses only four user-specified levels of tessellation.)

Example 30.8 shows the creation of a geometry auto grid index.

Example 30.8

```
USE sample;
CREATE SPATIAL INDEX auto_grid_index
    ON beverage_markets(shape)
    USING GEOMETRY_AUTO_GRID
        WITH (BOUNDING_BOX = (xmin=0, ymin=0, xmax=500, ymax=200 ),
    CELLS_PER_OBJECT = 32, DATA_COMPRESSION = page);
```

The GEOMETRY_AUTO_GRID clause describes the created index as a geometry auto grid index. The CELLS_PER_OBJECT clause specifies the maximum number of cells that tessellation can count per object. The DATA_COMPRESSION clause specifies whether the compression is enabled for the particular spatial index and, if so, which compression type is used. (In Example 30.8, page compression is used on data. Two other options are NONE and ROW.) All other clauses of the CREATE SPATIAL INDEX statement are described immediately following Example 30.7.

Editing Information Concerning Spatial Indices

The system supports, among others, three catalog views related to spatial data:

- **sys.spatial_indexes**
- **sys.spatial_index_tessellations**
- **sys.spatial_reference_systems**

The **sys.spatial_indexes** view represents the main index information of the spatial indices (see Example 30.9). Using the **sys.spatial_index_tessellations** view, you can display the information about the tessellation scheme and parameters of each of the existing spatial indices. The **sys.spatial_reference_systems** view lists all the spatial reference systems supported by the system. (Spatial reference systems are used to identify a particular coordinate system.) The main columns of the view are **spatial_reference_id** and **well_known_text**. The former is the unique identifier of the corresponding reference system, while the latter describes that system.

Example 30.9 shows the use of the **sys.spatial_indexes** catalog view.

Example 30.9

```
USE sample;
SELECT object_id, name, type_desc
    FROM sys.spatial_indexes;
```

The result is

object_id	name	type_desc
914102297	i_spatial_shape	SPATIAL
914102298	auto_grid_index	SPATIAL

The catalog view in Example 30.9 displays the information about the existing spatial indices, created in Examples 30.7 and 30.8.

Displaying Spatial Objects

Microsoft extended the functionality of SQL Server Management Studio to display spatial data in a graphical form. Two examples will be presented to show this functionality. Example 30.10 displays the union of two GEOMETRY data types (LineString and Polygon), while Example 30.11 displays a **MultiPolygon** instance, which is a collection of zero or more **Polygon** instances.

Example 30.10

```
USE sample;
DECLARE @rectangle1 GEOMETRY = 'POLYGON((1 1,  1 4,  4 4, 4 1, 1 1))';
DECLARE @line GEOMETRY = 'LINESTRING (0 2, 4 4)';
SELECT @rectangle1
UNION ALL
SELECT @line
```

To display the spatial object that is built as a result of Example 30.10, execute the batch in Example 30.10 and click the Spatial Results tab (which is next to the Results tab in the lower part of the editor in Management Studio). In the case of Example 30.10, Management Studio displays a rectangle showing the content of the **@rectangle1** variable and the line from the **@line** variable (see Figure 30-2).

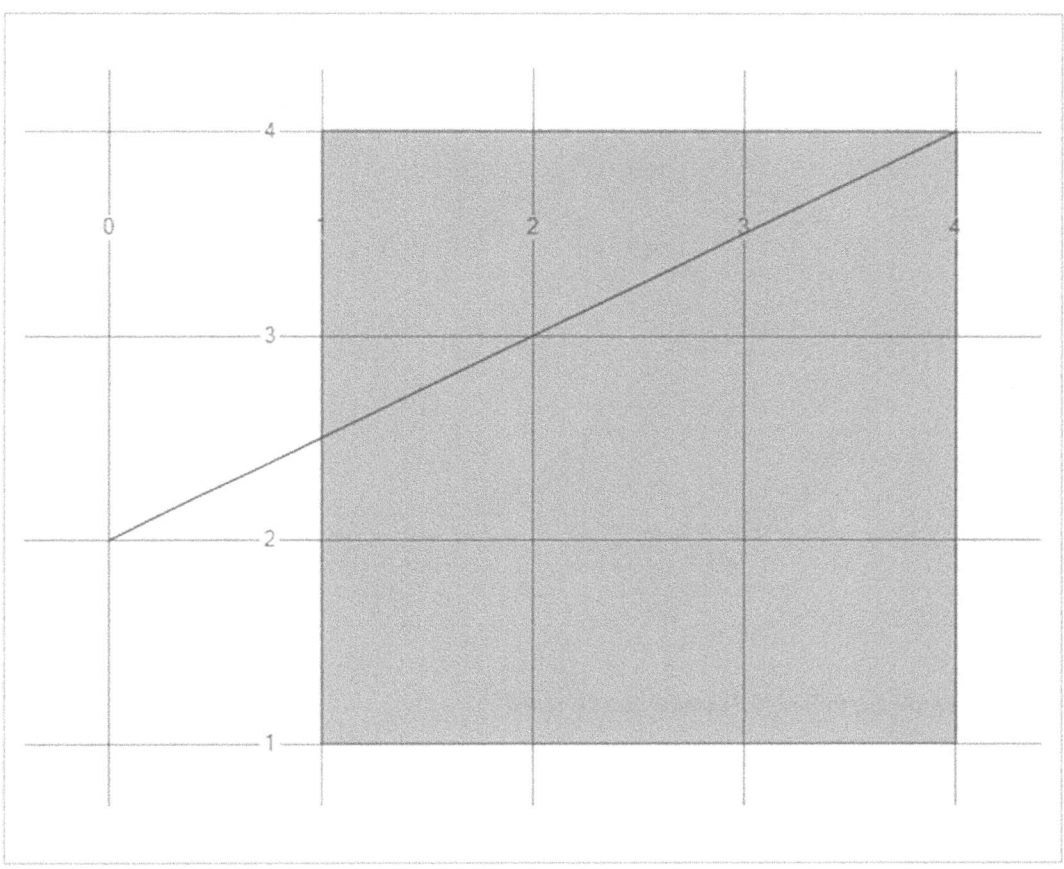

Figure 30-2   Displaying the result of Example 30.10 in SSMS

NOTE If you want to display multiple objects of the GEOMETRY data type using Management Studio, you have to return them as multiple rows in a single table. For this reason, Example 30.10 uses two SELECT statements combined into one using the UNION ALL clause. (Otherwise, only one point at a time will be displayed.)

Example 30.11 creates a **MultiPolygon** instance, which is shown in Figure 30-3.

Example 30.11

```
DECLARE @g1 geometry =
'MultiPolygon(((2 0, 3 1, 2 2, 1.5 1.5, 2 1, 1.5 0.5, 2 0)),
           ((1 0, 1.5 0.5, 1 1, 1.5 1.5, 1 2, 0 1, 1 0)))';
select @g1
```

Figure 30-3   Displaying the result of Example 30.11 in SSMS

Editing Information Concerning Spatial Data

You can use two system stored procedures to display information in relation to spatial data:

- **sp_help_spatial_geometry_histogram**
- **sp_help_spatial_geography_histogram**

NOTE The syntax and functionality of both system procedures is similar, so I will discuss only the first one.

The **sp_help_spatial_geometry_histogram** system stored procedure returns the names and values for a specified set of properties of a geometry spatial index. The result is returned in a table format. You can choose to return a core set of properties or all properties of the index. Example 30.12 shows the use of this system stored procedure.

Example 30.12

```
USE sample;
DECLARE @query geometry
  ='POLYGON((-90.0 -180.0, -90.0 180.0, 90.0 180.0, 90.0 -180.0, -90.0
  -180.0))';
```

```
EXEC sp_help_spatial_geometry_index
  'beverage_markets', 'auto_grid_index', 0, @query;
```

The **sp_help_spatial_geometry_index** system procedure in Example 30.12 displays the properties of the spatial index called **auto_grid_index**, created in Example 30.8.

Temporal Tables

Up to this point in the book, we've been dealing with databases that store facts concerning current time. To explain this, let's take a look at how a database system modifies a row. Logically, the row is first deleted and the modified row is inserted in the table. In other words, an UPDATE statement is a combination of a DELETE statement and an INSERT statement, which are executed in that order.

As of SQL Server 2016, a special form of table called a *temporal table* is supported. This is the first step in enabling this database system to handle temporal data. A temporal table is a table that contains two time-variant columns specifying the start and end time of the currentness of row data. The support for temporal tables enables you to track and manage multiple versions of your data. By defining a temporal table, you are instructing the database system to automatically capture changes made to the state of your table and to save old versions of a table's rows in a history table. (A *history table* is a separate table with the same structure as the table that contains current data.)

Temporal tables can be used in many application scenarios where time-variant attributes are involved. The following two cases belong to such scenarios:

- An insurance company needs to determine the terms of a client's policy in effect at the particular point in time when the accident occurred.

- A travel agency wants to detect inconsistent travel plans of a client, such as when, for instance, hotel bookings in two different cities overlap.

In case of database systems, which do not support temporal data, such scenarios are generally implemented using triggers. On the other hand, as you will see in this section, simple SQL statements allow users to automatically maintain a history of database changes or track time-variant data, eliminating the need for programming such logic.

The system uses a *closed-open approach* for modeling time periods. This means that the period's start time is included in the time period, but its end time is not.

Creation of Temporal Tables

Temporal tables are created using the CREATE TABLE statement, which is enhanced with several clauses concerning time-variant columns. Example 30.13 shows the creation of such a table.

Example 30.13

```
USE sample;
CREATE TABLE dept_temp
(dept_no CHAR(4) NOT NULL PRIMARY KEY CLUSTERED,
 dept_name CHAR(25) NOT NULL,
```

```
location CHAR(30) NULL,
  start_date DATETIME2 GENERATED ALWAYS AS ROW START HIDDEN,
  end_date   DATETIME2 GENERATED ALWAYS AS ROW END HIDDEN,
  PERIOD FOR SYSTEM_TIME (start_date, end_date))
WITH (SYSTEM_VERSIONING = ON (HISTORY_TABLE = dbo.Dept_History));
```

The **dept_temp** table is similar to the **department** table used in previous chapters but contains two additional columns: **start_date** and **end_date**. (All time-variant columns must be specified using the DATETIME2 data type.) The GENERATED ALWAYS AS ROW START option specifies the column with values identifying the start time, while GENERATED ALWAYS AS ROW END specifies the column with values identifying the end time of the corresponding time period. The additional PERIOD FOR SYSTEM_TIME clause contains the names of both time-variant columns, where the first name defines the beginning of a time period and the second the end of a time period. (This clause implicitly specifies that start_date <= end_date.)

The WITH SYSTEM_VERSIONING option must be enabled if you want to create a history table, where old versions of rows are stored. The name of the history table can be specified explicitly by the user, as in Example 30.13, or implicitly by the system, if the HISTORY_TABLE option is omitted. (I strongly recommend using the explicit definition of the history table.)

The most important property of temporal tables is that old versions of rows are preserved, but they are handled differently than rows with current values when modification operations (UPDATE and DELETE) are executed. In this case, only the rows stored in the current table are modified. Similarly, all specified constraints are valid only for the current table.

The HIDDEN option specifies the time-variant columns (the columns **start_date** and **end_date** in Example 30.13) as hidden columns. This means that values of these columns will not be displayed implicitly. In other words, the following SELECT statement,

```
SELECT * FROM dept_temp;
```

will display only nonhidden columns. To display values of all columns of the **dept_temp** table, you have to specify the hidden columns explicitly:

```
SELECT *, start_date, end_date FROM dept_temp;
```

Temporal Tables and DML Statements

As you will see in this section, the semantics of the DML statements UPDATE and DELETE in relation to temporal tables is different than the semantics of the same statements in relation to conventional tables. On the other hand, the syntax of these statements does not change.

Example 30.14 shows the insertion of three rows in the **dept_temp** table. (To make the examples easier to understand, all further results in this chapter show only the date, hour, and minute portions of time-variant columns.)

Example 30.14

```
USE sample;
INSERT INTO dept_temp (dept_no, dept_name, location)
        VALUES ('d1', 'Research', 'Dallas' );
INSERT INTO dept_temp (dept_no, dept_name, location)
        VALUES ('d2', 'Accounting', 'Seattle');
INSERT INTO dept_temp (dept_no, dept_name, location)
        VALUES ('d3', 'Marketing', 'Dallas');
```

The content of the **dept_temp** table after execution of three INSERT statements in Example 30.14 is

dept_no	dept_name	location	start_date	end_date
d1	Research	Dallas	2019-012-20 06:41	9999-12-31 23:59
d2	Accounting	Seattle	2019-12-20 06:41	9999-12-31 23:59
d3	Marketing	Dallas	2019-12-20 06:41	9999-12-31 23:59

Each INSERT statement in Example 30.14 contains the list of time-invariant columns. This is necessary, because the table's schema contains two additional columns, and the values of these columns are inserted by the system; that is, they cannot be inserted by the user. When inserting each row into the **dept_temp** table, the system generates appropriate DATETIME2 values for time-variant columns. Hence, the system assigns the current time to the **start_date** column and the value "forever" (9999-12-31 23:59) to the **end_date** column. (The required information will be automatically recorded by the system.)

Example 30.15 shows how temporal tables can be modified using the UPDATE statement.

Example 30.15

```
USE sample;
UPDATE dept_temp
   SET location = 'Houston'
   WHERE dept_no = 'd1';
```

The following result shows first the content of the current table and after that the content of the history table.

(current table)

dept_no	dept_name	location	start_date	end_date
d1	Research	Houston	2019-12-20 06:45	9999-12-31 23:59
d2	Accounting	Seattle	2019-12-20 06:41	9999-12-31 23:59
d3	Marketing	Dallas	2019-12-20 06:41	9999-12-31 23:59

Part V

(history table)

dept_no	dept_name	location	start_date	end_date
d1	Research	Dallas	2019-12-20 06:41	2019-12-20 06:45

Let's take a look at how the system executes the UPDATE statement. As shown in the result of Example 30.15, the system updates the value of the row in the current table. Also, it moves a copy of the "old" row to the history table. For both tables, the system determines the start and end times for these rows. In other words, the system sets the value for the end time of the row in the history table to the time of the transaction when UPDATE has been executed.

The deletion of rows is similar to their modification, as Example 30.16 demonstrates.

Example 30.16

```
USE sample;
DELETE FROM dept_temp
    WHERE dept_no = 'd2';
```

The content of the current table after deletion is

dept_no	dept_name	location	start_date	end_date
d1	Research	Houston	2019-12-20 06:45	9999-12-31 23
d3	Marketing	Dallas	2019-12-20 06:41	9999-12-31 23

and the content of the corresponding history table is

dept_no	dept_name	location	start_date	end_date
d1	Research	Dallas	2019-12-20 06:41	2019-12-20 06:45
d2	Accounting	Seattle	2019-12-20 06:41	2019-12-20 06:50

When you delete rows from the current table, the system removes data from that table and creates old versions of that data, which is stored in the corresponding history table. The value for the **end_time** column of each row stored in the history table is set by the system to the transaction start time of the DELETE statement. (As can be seen from the result of Example 30.16, the transaction start time of the DELETE statement has been set to 2019-09-20 06:50.)

NOTE   Rows in a history table cannot be modified or deleted. In other words, neither DELETE nor INSERT can be applied to a history table.

Querying Temporal Data

In contrast to INSERT, UPDATE, and DELETE, the SELECT statement contains extensions that allow you to query time-variant data in various ways. FOR SYSTEM_TIME is a clause, which has five subclauses:

- FOR SYSTEM TIME AS OF <date_time>
- FOR SYSTEM TIME FROM <start_date_time> TO <end_date_time>

- FOR SYSTEM TIME BETWEEN <start_date_time> AND <end_date_time>
- FOR SYSTEM TIME CONTAINED IN (<start_date_time> , <end_date_time>)
- FOR SYSTEM TIME ALL

These subclauses allow you to specify different forms of time periods in your query. The first subclause, AS OF <date_time>, enables you to query data as of a certain point in time. The FROM...TO subclause enables you to query data inside the specified time interval. In this case, the specified start time is included in the time interval but the specified end time is not.

The BETWEEN subclause is similar to FROM...TO. The only difference is that rows with the specified end time are also included in the result. The CONTAINED IN subclause returns a result with the values of all old versions that were opened and closed within the specified time range defined by the two parameters of that subclause. Records that became active exactly on the lower boundary or ceased being active exactly on the upper boundary are included. The ALL clause returns all the rows from both tables (the current table and the history table).

When we discuss temporal queries, the first issue to address is what happens if you do not use the SYSTEM_TIME clause at all in your query. In that case, to display current rows, you have to reference the current table. And to display historical rows, you have to reference the corresponding history table, separately. Now, we will see what happens when you use that clause. Examples 30.17 and 30.18 show the use of the first two subclauses of SYSTEM_TIME.

NOTE You need to update the date constants in the following example to produce any results.

Example 30.17

```
USE sample;
SELECT * FROM dept_temp
      FOR SYSTEM_TIME AS OF '2019-12-20 06:41:07.2902041';
```

The result is

dept_no	dept_name	location	start_date	end_date
d3	Marketing	Dallas	2019-12-20 06:41	9999-12-31 23:59
d1	Research	Dallas	2019-12-20 06:41	2019-12-20 06:45
d2	Accounting	Seattle	2019-12-20 06:41	2019-12-20 06:50

As you can see from the result, the SELECT statement in Example 30.17 returns all rows containing the specified point in time. Note that both the current table and history table are searched for the specified time. (The AS OF clause can be used to implement logic of the first application scenario listed at the beginning of the section "Temporal Tables.")

NOTE You need to update the date constants in the following example to produce any results.

Example 30.18 shows you how to use the FROM...TO subclause.

Example 30.18
```
USE sample;
SELECT * FROM dept_temp FOR SYSTEM_TIME
FROM '2019-06-19 06:41:07.2902041' TO '2019-06-20 06:41:07.2902041' ;
```

The result of this query is

dept_no	dept_name	location	start_date	end_date
d1	Research	Dallas	2019-06-20 06:41	2019-06-20 06:45
d2	Accounting	Seattle	2019-06-20 06:41	2019-06-20 06:50

If you compare the results of the last two queries, you will see that the row with dept_no=d3 is not selected in Example 30.18. The reason is that the FROM...TO subclause excludes the end time of the specified time period. If you want to include the third row to your result, you have to use the BETWEEN...AND subclause.

Editing Information Concerning Temporal Data

Metadata concerning temporal tables is displayed in three catalog views: **sys.tables**, **sys.columns**, and **sys.periods**. The first two views contain additional columns in relation to temporal tables , while **sys.periods** returns a row for each existing temporal table. The query in Example 30.19 displays the name and the type of all existing temporal tables in your database.

Example 30.19
```
USE sample;
SELECT name, type_desc FROM sys.tables
    WHERE object_id IN  (SELECT object_id FROM sys.periods);
```

If you want to determine whether the particular table is temporal, you can use the **temporal_type** column of the **sys.tables** catalog view. Example 30.20 shows this for the **dept_temp** table.

Example 30.20
```
SELECT temporal_type
FROM   sys.tables
WHERE  object_id = OBJECT_ID('dbo.dept_temp', 'U');
```

The value of the **temporal_type** column can be 0 (for nontemporal table), 1 (for history table), or 2 (for temporal table). So, the result of Example 30.20 is 2.

Converting Nontemporal Tables into Temporal Tables

Two additional statements that you can use when working with temporal tables are ALTER TABLE and DROP TABLE. The ALTER TABLE statement is used to convert an existing nontemporal table into a temporal one. (A scenario in which you might want to do such a conversion is if you have already implemented temporal logic using triggers with one of the previous versions of SQL Server and want to use features of temporal tables.)

Example 30.21 shows how a nontemporal table can be converted into a temporal one. (The **department** table is the well-known table of the **sample** database. The content of the table should be as it was when we started to use this table.)

Example 30.21

```
USE sample;
ALTER TABLE department ADD PRIMARY KEY (dept_no);
GO
ALTER TABLE department  ADD
SysStartTime datetime2 NOT NULL DEFAULT GETUTCDATE(),
SysEndTime datetime2 NOT NULL DEFAULT CONVERT(DATETIME2, '9999-12-31
23:59:59.99999999')
GO
ALTER TABLE department
ADD PERIOD FOR SYSTEM_TIME (SysStartTime, SysEndTime);
GO
ALTER TABLE department
alter column SysStartTime ADD HIDDEN;
GO
ALTER TABLE department
alter column SysEndTime ADD HIDDEN;
ALTER TABLE department
   SET (SYSTEM_VERSIONING = ON
   (HISTORY_TABLE = dbo.department_history, DATA_CONSISTENCY_CHECK = ON));
```

Example 30.21 contains six statements. The first ALTER TABLE statement specifies the PRIMARY KEY clause, because each temporal table must be defined with the corresponding primary key. The second statement adds two new columns of the DATETIME2 data type, while the third one declares them as the columns, which specify system start time and system end time, respectively. The fourth and the fifth statements specify both columns as hidden columns.

Finally, the last statement transforms the nontemporal table (**department**) into a temporal one. Also, the corresponding history table, **department_history**, is created. As we already know from Example 30.13, enabling the SYSTEM_VERSIONING clause is a requirement when creating a history table. (The DATA_CONSISTENCY_CHECK clause performs a data consistency check, which ensures that existing rows do not overlap and that temporal requirements are fulfilled for every individual row.)

To drop the current table and its corresponding history table, you have to disable system versioning first. After that both tables can be dropped using the DROP TABLE statement, as shown in Example 30.22.

Example 30.22

```
USE sample;
   ALTER TABLE department set (SYSTEM_VERSIONING = OFF);
   GO
   DROP TABLE dbo.department;
   GO
   DROP TABLE dbo.department_history;
```

Summary

The Database Engine supports two special data forms: spatial data and temporal data. Concerning spatial data, two standardized data types exist: GEOGRAPHY and GEOMETRY. The GEOGRAPHY data type is used to represent spatial data in geodetic models, while the GEOMETRY data type is used with flat spatial models. To work with these data types, you need a set of corresponding operations (methods). For both data types, Microsoft implemented methods specified by OGC that can be used to retrieve spatial data from a table's columns.

The Database Engine also supports two different types of spatial indices, the conventional spatial index and the auto grid index. The latter approximates spatial objects of various sizes significantly better.

The support of temporal data in the Database Engine is rudimentary. In other words, many important temporal concepts are not yet implemented. For instance, the main concept, the PERIOD data type, is not supported. The PERIOD data type in the temporal data model is a built-in data type, which specifies an anchored duration of contiguous time granules within the duration.

The next chapter describes a new type of database introduced in SQL Server 2017, the graph database.

SQL Server Graph Databases

In This Chapter

- Graph Databases: A General Introduction
- Creating Node Tables and Edge Tables
- Querying Graph Data
- Modifying and Editing Data in Graph Databases
- Querying Graph Data Using Relational Queries

Microsoft introduced graph capabilities in SQL Server 2017. This new component is very important because most social networks are based on graphs and Transact-SQL statements do not support operations on graphs optimally.

The chapter discusses all extensions in SQL Server to support Graph Databases. The first section provides a general introduction to graph databases and then lists all properties of this component in SQL Server. The second section explains in detail the creation of node tables and edge tables. The third section discusses how graphs stored in node tables and edge tables can be queried and provides several examples to show this. The fourth section explains how to modify data in node tables and edge tables. The final section explains how graph data can be queried using relational queries.

Graph Databases: A General Introduction

Generally, a *graph* is a collection of nodes and edges, or, in less formal language, a set of entities and the relationships that connect them. Therefore, *nodes* of a graph are entities, such as Employee, Product, and Customer, and *edges* are relationships. How the nodes relate to the world is specified by edges. For instance, employees sell products, and products are bought by customers. Using graphs, you can model a wide variety of different scenarios from the real world, anything from the design of social networks to the construction of space shuttles.

Graph Databases: Models

A graph database is a database management system that supports the usual properties of DBMSs (see Chapter 1) using a graph data model as its underlying model. Graph databases are generally built for use with online transaction processing (OLTP) systems. Two main properties of graph database technology are how the data is stored and how the data is processed.

The underlying storage of a graph database can be native or nonnative. Native storage is optimized and designed for storing and managing graphs. A graph database is classified as nonnative when the storage comes from a "non-graph" database model, such as a relational, columnar, or object-oriented model. (The nonnative approach can lead to performance problems, because the storage engine of non-graph database models is not optimized for graphs.)

The processing engine of a graph database doesn't necessarily need to support native graph processing. The key element of graph technology in relation to data processing is whether or not index-free adjacency is supported. (*Index-free adjacency* means that every node contains a direct pointer to its adjacent node. That way, creation of indices is not necessary, at least for queries that reference a node and its adjacent ones.)

Property Graph Model

The most popular model in relation to graphs is called the *property graph model*. Besides this model, there are two others: Resource Description Framework (RDF) and hypergraphs. The discussion of these two models is outside the scope of this book.

A property graph has the following characteristic: it contains nodes and edges. Each node and each edge has a unique identifier. Also, each node has a collection of properties (key-value pairs). Finally, a node has a set of outgoing and incoming edges.

Edges are named and always have a start and an end node. Each edge has a label that denotes its connection type between two nodes. Edges, like nodes, can also contain properties.

Advantages of Graph Databases

The power of graph databases becomes apparent in relation to two issues:

- Performance
- Flexibility

It is well known that SQL does not deal very well with connected data. The reason is that the number of join operations (inner joins and self-referencing joins) increases significantly as the dataset gets bigger and the traversal of a graph goes deeper and deeper. (This is discussed further in the section "Querying Graph Data Using Relational Queries" at the end of this chapter.) In contrast to relational database systems, the performance of a graph database system remains constant, even when the dataset significantly grows. This is because queries are localized to a particular subgraph of the entire graph. As a result, the execution time for each query is proportional only to the size of the subgraph traversed to satisfy that query.

One of the main advantages of graphs is that they are *additive*, meaning that you can add new substructures (nodes, relationships, and subgraphs) to an existing graph without disturbing existing user applications in their functioning. This flexibility has positive implications for developers, because they can significantly increase their productivity. Also, because of the graph model's flexibility, you don't have to model your domain in depth, making it significantly easier to change business requirements later, if necessary.

SQL Server Graph Databases: An Introduction

As previously introduced, a graph is a collection of nodes and relationships (edges) between nodes. A graph in SQL Server Graph Databases is a collection of node tables and edge tables, where nodes are stored in node tables and relationships between nodes are stored in edge tables. Using these two forms of tables, SQL Server Graph Databases supports several general features of graph databases:

- The nodes and edges of a graph are first-class citizens and can have attributes or properties associated with them. (An object is considered a "first-class citizen" of a system if it can be stored in variables and data structures and can be passed as a parameter to a subroutine.)

- Pattern matching can be expressed easily. (As you will see shortly, pattern matching in SQL Server Graph Databases is expressed using the MATCH function.)

- Edges can be specified as directed or undirected. (Directed edges are specified using edge constraints.)

- Polymorphic queries can be expressed easily. (Polymorphic queries in relation to graphs are queries that return instances of a particular node as well as instances of all subgraphs of that node.) Note that node instances are logically the same as instances of entities, as explained in Chapter 1.

Creating Node Tables and Edge Tables

As you will see in the following examples, the creation of node tables and edge tables is straightforward. In other words, if you know the syntax of the CREATE TABLE statement for regular relational tables, it will be very easy for you to create node tables and edge tables. All examples that follow use the model of a graph shown in Figure 31-1.

The graph in Figure 31-1 contains three nodes: **Employee, Company,** and **City**. The **Employee** entity has four attributes (properties): **ID, Name, Age,** and **Sex**. The **Company** entity has four attributes: **ID, Name, Sector,** and **City**, while **City** has three attributes: **ID, Cityname,** and **Statename**.

The graph contains three edges: **WorksIn, LocatedIn,** and **LivesIn. WorksIn** has a property called **Starts** that specifies the year in which the particular employee started to work for the specified company. Similarly, the **LivesIn** edge has the **Since** property that specifies the date on which an employee moved to a particular city. (The third edge does not have any additional properties.) All three relationships are nonrecursive, meaning that they all connect two *different* nodes.

Creating Node Tables

A node table can be created in any user-defined database. The creation of such a table is very similar to creating a regular relational table, with one extension. Example 31.1 shows the creation of three node tables, corresponding to the entities shown in Figure 31-1.

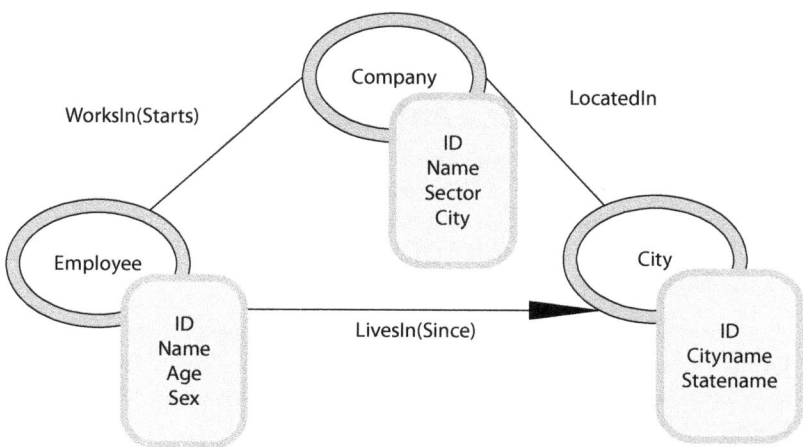

Figure 31-1   Graph used to demonstrate examples in this chapter

Example 31.1

```
CREATE DATABASE graph_db;
GO;
USE graph_db;
 CREATE TABLE   dbo.Company (
        ID    INT   NOT NULL PRIMARY KEY,
        name    VARCHAR (100) NULL,
        sector VARCHAR(25) NULL,
        city    VARCHAR (100) NULL) AS NODE;
 CREATE TABLE   dbo.Employee (
        ID    INT   NOT NULL PRIMARY KEY,
        name    VARCHAR (100) NULL,
        age    INT NULL,
        sex   char (10) NULL) AS NODE;
 CREATE TABLE   dbo.City (
        ID    INT   NOT NULL PRIMARY KEY,
        name    VARCHAR(100) NULL,
        stateName    VARCHAR(100) NULL) AS NODE;
```

The most important extension concerning node tables is the AS NODE clause, written at the end of the CREATE TABLE statement. This clause defines the corresponding table as a node table. When you specify this clause, the system adds two new columns of the BIT data type to the **sys.tables** catalog view: **is_node** and **is_edge**. For a node table, the value of **is_node** is set to 1, and the value of **is_edge** is set to 0. (A detailed description of metadata concerning graph databases is provided in the section "Editing Information Concerning Graph Databases" at the end of this chapter.)

Whenever you create a node table, along with the user-defined columns, an implicit **$node_id** column is created, which uniquely identifies each instance of the corresponding node table. The values in the **$node_id** column are automatically generated and are a combination of the **object_id** value of that node table and an internally generated value of the BIGINT data type. (When you display the values of the **$node_id** column, the corresponding computed values are displayed as JSON strings.) Also, **$node_id** is a pseudo column that maps to an internal name with a hex string appended to it. In other words, when you select **$node_id** from the table, the column name appears as **$node_id_\hex_string**.

After creation of node tables, you have to load data into them. As Example 31.2 shows, inserting rows into node tables works the same way as for any other regular table.

Example 31.2

```
USE graph_db;
INSERT INTO Employee (ID,Name,Sex)
     VALUES (1,'Matthew Smith','Male');
INSERT INTO Employee (ID,Name,Sex)
     VALUES (2,'Ann Jones','Female');
INSERT INTO Employee (ID,Name,Sex)
     VALUES (3,'John Barrimore','Male');
INSERT INTO Employee (ID,Name,Sex)
     VALUES (4,'James James','Male');
INSERT INTO Employee (ID,Name,Sex)
     VALUES (5,'Elsa Bertoni','Female');
INSERT INTO Employee (ID,Name,Sex)
     VALUES (6,'Elke Hansel','Female');

INSERT INTO Company VALUES (1,'Comp_A','Pharma','Kansas City');
INSERT INTO Company VALUES (2,'Comp_B','Manufacturing','Hoboken');
INSERT INTO Company VALUES (3,'Comp_C','Pharma','Indianopolis');
INSERT INTO Company VALUES (4,'Comp_D','IT','Lexington');
INSERT INTO Company VALUES (5,'Comp_E','IT','Madison');

INSERT INTO City VALUES (1,'Kansas City','Kansas');
INSERT INTO City VALUES (2,'Hoboken','New Jersey');
INSERT INTO City VALUES (3,'Indianopolis','Indiana');
INSERT INTO City VALUES (4,'Lexington','Kentucky');
INSERT INTO City VALUES (5,'Minneapolis','Wisconsin');
INSERT INTO City VALUES (6,'Madison','Wisconsin');
```

Creating Edge Tables

An edge table represents a relationship between two graph nodes. Therefore, each row of an edge table contains instances of the corresponding relationship. An edge table has, among others, three hidden columns: **$edge_id**, **$from_id**, and **$to_id**. The values of the **$edge_id** column specify the unique IDs and are stored as JSON documents. (For the description of

JSON, see Chapter 29.) The other two columns represent the references between the instances of the relationship—that is, the rows of the edge table to the instances of both connected node tables. The **$from_id** column stores the **$node_id** values of the nodes from which the edges originate, and the **$to_id** column stores the **$node_id** values of the nodes at which the edges terminate.

NOTE A *hidden column* is a column that exists in the table but cannot be selected. Besides the three hidden columns specified, there are several other hidden columns in relation to SQL Server Graph Databases. One of these columns is **graph_id**, which is used internally by the SQL Server system to manage graph data in the proper way (and is described later in the chapter, in the section "Graph Databases: System Functions").

Similar to node tables, for each edge table, the system adds two new columns to the **sys .tables** catalog view: **is_node** and **is_edge**. As expected, in contrast to node tables, the values of the **is_node** and **is_edge** columns are 0 and 1, respectively.

NOTE Generally, graphs can be undirected or directed. In an undirected graph, all the edges are bidirectional. In a directed graph, all the edges point in one direction. Single edges can be thought of in the same way: an undirected edge is bidirectional, while a directed edge points in a specified direction. Referring to Figure 31-1 earlier in the chapter, the edge **LivesIn** is directed, while the **WorksIn** and **LocatedIn** edges are undirected.

Example 31.3 shows the creation of the three edge tables: **WorksIn**, **LocatedIn**, and **LivesIn**.

Example 31.3

```
USE graph_db;
CREATE TABLE WorksIn (starts INT) AS EDGE;
CREATE TABLE LocatedIn AS EDGE;
CREATE TABLE LivesIn (Since DATE NULL
    CONSTRAINT Emp_to_City CONNECTION (Employee TO City)) AS EDGE;
```

The creation of the first two edge tables in Example 31.3 is straightforward. You just create a table in the regular way and append the AS EDGE clause. The creation of the **LivesIn** edge is different. The reason is that the first two edges (**WorksIn** and **LocatedIn**) are undirected edges, meaning that you can traverse them in both directions. For instance, in the case of the **LocatedIn** edge, you can traverse from the **Company** node to the **City** node and vice versa. In the case of the **LivesIn** edge, you can traverse only from the **Employee** node to the **City** node, because this edge is directed.

The creation of directed edges is possible in SQL Server 2019 by using *edge constraints.* With this feature, you can apply restrictions during the creation of an edge by using the CONNECTION clause. This clause is used in Example 31.3 to specify that you can traverse the **LivesIn** edge only from the **Employee** node to the **City** node. (This is explained further in the discussion after Example 31.6 in the next section.)

After creating graph objects, you can examine them using Object Explorer. You will see that there is a new subfolder called Graph in the Tables folder. All graph objects will be inside this subfolder. Note that names of auto-generated fields include a GUID, but you can reference these fields with their short names. (A short name is a pseudo column and you can use it in queries.)

Inserting Data into Edge Tables

In contrast to the creation of edge tables, which looks like the creation of regular relational tables (extended with the AS EDGE clause), loading data into edge tables is different than loading data into relational tables. Remember that an edge table represents a relationship between two nodes in a graph. For this reason, each INSERT statement, which loads an instance of such a relationship, must specify both the instance of the node where the relationship originates and the instance of the node where the relationship terminates. To do this, you use the already mentioned **$node_id** values from the **$from_id** and **$to_id** columns. (As you will see in Example 31.4, we use subqueries to solve this problem.)

Example 31.4 shows the insertion of rows into the **WorksIn** edge table.

Example 31.4

```
--To insert data into an edge table we need to provide the reference for
-- the $from_id and $to_id as a reference point to both nodes
USE graph_db;
INSERT INTO WorksIn VALUES ((SELECT $node_id FROM Employee WHERE id = 1),
       (SELECT $node_id FROM Company WHERE id = 1), 2015);
INSERT INTO WorksIn VALUES ((SELECT $node_id FROM Employee WHERE id = 2),
       (SELECT $node_id FROM Company WHERE id = 2), 2018);
INSERT INTO WorksIn VALUES ((SELECT $node_id FROM Employee WHERE id = 3),
       (SELECT $node_id FROM Company WHERE id = 3), 2015);
INSERT INTO WorksIn VALUES ((SELECT $node_id FROM Employee WHERE id = 4),
       (SELECT $node_id FROM Company WHERE id = 3), 2016);
INSERT INTO WorksIn VALUES ((SELECT $node_id FROM Employee WHERE id = 5),
       (SELECT $node_id FROM Company WHERE id = 3), 2017);
INSERT INTO WorksIn VALUES ((SELECT $node_id FROM Employee WHERE id = 6),
       (SELECT $node_id FROM Company WHERE id = 4), 2018);
```

The first INSERT statement in Example 31.4 defines a relationship in the **WorksIn** edge table between the instance of the **Employee** node (entity) with the ID value 1 and the instance of the **Company** node (entity) with the ID value 1. To add data to the **$from_id** column, you must specify the **$node_id** value associated with the **Employee** entity. One way to get this value is to include a subquery that targets the entity, using its primary key value. You can take the same approach for the **$to_id** column in relation to the **Company** entity.

NOTE Insertion of the data in this way demonstrates why it is useful to add primary keys to each node table (see Example 31.1), but not necessary for the edge tables. The primary keys on the node tables make it much easier to provide the **$node_id** value to the INSERT statements in Example 31.4.

Examples 31.5 and 31.6 show the insertion of rows in the other two edge tables.

Example 31.5

```
USE graph_db;
INSERT INTO LocatedIn VALUES ((SELECT $node_id FROM Company WHERE id = 1),
    (SELECT $node_id FROM City WHERE id=2))
INSERT INTO LocatedIn VALUES ((SELECT $node_id FROM Company WHERE id = 2),
    (SELECT $node_id FROM City WHERE id=1));
INSERT INTO LocatedIn VALUES ((SELECT $node_id FROM Company WHERE id = 3),
    (SELECT $node_id FROM City WHERE id=3));
INSERT INTO LocatedIn VALUES ((SELECT $node_id FROM Company WHERE id = 4),
    (SELECT $node_id FROM City WHERE id=2));
```

Example 31.6

```
USE graph_db;
INSERT INTO LivesIn VALUES ((SELECT $node_id FROM Employee WHERE id = 1),
    (SELECT $node_id FROM City WHERE id=6), '1.1.2018');
INSERT INTO LivesIn VALUES ((SELECT $node_id FROM Employee WHERE id = 2),
    (SELECT $node_id FROM City WHERE id=5), '2.1.2018');
INSERT INTO LivesIn VALUES ((select $node_id FROM Employee WHERE id = 3),
    (SELECT $node_id FROM City WHERE id=4), '3.1.2018');
INSERT INTO LivesIn VALUES ((SELECT $node_id FROM Employee WHERE id = 4),
    (SELECT $node_id FROM City WHERE id=2), '4.1.2018');
INSERT INTO LivesIn VALUES ((SELECT $node_id FROM Employee WHERE id = 5),
    (SELECT $node_id FROM City WHERE id=3), '5.1.2018')
INSERT INTO LivesIn VALUES ((SELECT $node_id FROM Employee WHERE id = 6),
    (SELECT $node_id FROM City WHERE id=1), '6.1.2018');
```

All the INSERT statements in Example 31.6 succeed because they all insert edge instances that connect *employees* to *cities.* However, if you try to insert an edge instance the other way around (from the **City** node to the **Employee** node), like the following example:

```
INSERT INTO LivesIn VALUES ((SELECT $node_id FROM City WHERE id = 1),
    (SELECT $node_id FROM Employee WHERE id=6), '6.1.2018');
```

the INSERT statement fails, because traversing from the **City** node to the **Employee** node is forbidden in the specification of the **LivesIn** table (see Example 31.3), and you get the following error message:

```
Msg 547, Level 16, State 0, Line 1
The INSERT statement conflicted with the EDGE constraint "Emp_to_City". The
conflict occurred in database "graph_db", table "dbo.LivesIn".
```

NOTE The CONNECTION constraint is similar to all other constraints that you can specify for a regular table. For this reason, you can drop such a constraint using the DROP CONSTRAINT clause of the ALTER TABLE statement (see Example 5.26 in Chapter 5).

Some Remarks to Directed and Undirected Relationships

As you already know, there are two different forms of relationships: undirected and directed. An undirected relationship specifies that the instances of two corresponding nodes can be connected in both directions, while in the case of directed edges the connection is possible only in one direction, which is specified with the corresponding edge constraint.

The implementation of edge constraint for SQL Server Graph Databases has been very important, because that way the main requirements in relation to directed relationships have been fulfilled.

This is not true for undirected relationships. Note that edges are actually binary relationships between *two particular* nodes. If you take a look at Example 31.3, you will see that the creation of the **WorksIn** edge table does not have any reference to the corresponding tables (**Employee** and **Company**), which are part of this relationship. Therefore, you can write any syntactically correct, but semantically wrong, INSERT statement and both of them will be accepted by the system. In other words, the first INSERT statement in Example 31.5

```
INSERT INTO WorksIn  VALUES ((SELECT $node_id FROM Employee WHERE id = 1),
   (SELECT $node_id FROM Company WHERE id = 1), 2015);
```

is semantically and syntactically correct. But, if you change "Company" to "City" such as shown here,

```
INSERT INTO WorksIn  VALUES ((SELECT $node_id FROM Employee WHERE id = 1),
   (SELECT $node_id FROM City WHERE id = 1), 2015);
```

the system will insert this statement even though it is semantically wrong. This makes a burden for the programmer, who has to worry about the correct names of the node tables during insertion of rows for a particular edge table.

For instance, the code to create an edge of a graph in the Cypher language, which is an alter ego of the query language of SQL Server Graph Databases, contains explicitly the names of the nodes:

```
CREATE (p:Person)-[:LIKES]->(t:Technology)
```

Therefore, in this case, the system takes care whether the logically correct names of nodes are specified and can use these names to check the semantic meaning of the INSERT statements used during insertion of rows. (Cypher is a declarative query language implemented to efficiently query graphs. It is a query language of another graph database system called neo4J.)

We can hope that the design of edge tables will be improved in one of the next versions of SQL Server to consider this issue.

Querying Graph Data

SQL Server Graph Databases supports the MATCH function to query graph data. The syntax and the semantics of this function are similar to those of the MATCH function from the Cypher query language of another graph database system, Neo4J. Cypher is a declarative query language that is implemented to efficiently query graphs. (Note that currently only part of the original language has been implemented in SQL Server Graph Databases.)

Part V

After the introduction of the MATCH function, we will discuss recursive relationships and how they can be implemented in SQL Server Graph Databases.

The MATCH Function

The MATCH function allows you to specify a search pattern based on relationships between two nodes. This function is a part of the WHERE clause of the SELECT statement that queries node and edge tables. Example 31.7 shows the use of this function.

Example 31.7

```
-- Get the names of the companies and the names of their employees
USE graph_db;
SELECT DISTINCT Cmp.Name CName, Emp.Name EName
            FROM Employee Emp, WorksIn, Company Cmp
            WHERE MATCH(Emp-(WorksIn)->Cmp);
```

The result is

CName	EName
Comp_A	Matthew Smith
Comp_B	Ann Jones
Comp_C	Elsa Bertoni
Comp_C	James James
Comp_C	John Barrimore
Comp_D	Elke Hansel

Each MATCH function must contain a search pattern. Such patterns represent one or more relationships. For each relationship, you must specify the originating node and the terminating node, as well as the edge table that connects the two nodes together. You must also specify the direction of the relationship, using dashes and arrows, with the edge table situated between the two node tables. In Example 31.7, the term

```
(Emp-(WorksIn)->Cmp)
```

specifies that the originating node is **Employee** (represented by its alias, **Emp**), the edge table is **WorksIn**, and the terminating node is **Company** (represented by its alias, **Cmp**). Generally, if you want to specify a single relationship called **Edge** with the originating node **Node1** and terminating node **Node2**, you would use the following syntax:

```
MATCH(Node1-(Edge)->Node2)
```

NOTE As shown in the syntax, the name of the edge table in the MATCH function is enclosed in parentheses, with a dash preceding its name and a dash and right arrow following its name. This specifies a relationship that moves from left to right. You can reverse this order and specify a relationship that moves from right to left by reversing the order of the node names (that is, MATCH(Node2-(Edge)->Node1). Generally, this relationship is semantically different from the left-to-right relationship, as demonstrated in Examples 31.15 and 31.16 later in the chapter.

Examples 31.8 and 31.9 show how to narrow the result of the query in Example 31.7 by using the AND operator.

Example 31.8

```
-- Get the name of the company where Matthew Smith works
USE graph_db;
SELECT Cmp.Name CName
        FROM Employee Emp, WorksIn, Company Cmp
          WHERE MATCH(Emp-(WorksIn)->Cmp)
                AND Emp.Name= 'Matthew Smith';
```

The result is

CName
Comp_A

Example 31.9

```
-- Get the list of the employees who live in Madison
USE graph_db;
SELECT Emp.Name EName
        FROM Employee Emp, LivesIn, City
          WHERE MATCH(Emp-(LivesIn)->City)
                AND City.Name='Madison';
```

The result is

EName
Matthew Smith

SQL Server Graph Databases allows you to use other clauses of the SELECT statement together with the MATCH function. Example 31.10 uses the ORDER BY clause to sort the result rows according to the first column in the SELECT list.

Example 31.10

```
-- Get the list of companies located in the city of Hoboken
-- Sort the companies according to their names
USE graph_db;
SELECT Cmp.Name CName
        FROM City C, LocatedIn, Company Cmp
          WHERE MATCH(Cmp-(LocatedIn)->C)
                AND C.Name='Hoboken'    ORDER BY 1;
```

The result is

CName
Comp_A
Comp_D

SQL Server Graph Databases allows you to traverse a graph as deep as you wish. In this case you have to specify multiple relationships using the MATCH function. One way to do it is to use the AND operator, as shown in Example 31.11.

NOTE The Boolean operators, OR and NOT, cannot be used with the MATCH function.

Example 31.11
```
USE graph_db;
SELECT Employee.name EName, Company.name CName
     FROM Employee, WorksIn, Company, LocatedIn, City
       WHERE MATCH(Employee-(WorksIn)->Company
             AND Company-(LocatedIn)->City )
             AND WorksIn.Starts='2017' AND Company.name='Comp_C';
```

The result is

EName	CName
Elsa Bertoni	Comp_C

Example 31.11 links the instances of the **Employee** node with the instances of the **Company** node using the **WorksIn** relationship. After that, it links the instances of the **Company** node with the instances of the **City** node using the **LocatedIn** relationship. By being able to link together multiple relationships, you can traverse a graph as deep as you wish.

Generally, you can link together multiple relationships without using the AND operator, as long as the particular search pattern specifies the same logic. In other words, the terminating node table of the previous pattern must be the originating table of the subsequent one. Example 31.12 solves the same problem as the previous one, linking together multiple relationships without using the AND operator.

Example 31.12
```
USE graph_db;
SELECT Employee.name, Company.name
    FROM Employee, WorksIn, Company, LocatedIn, City
      WHERE MATCH(Employee-(WorksIn)->Company-(LocatedIn)->City)
            AND WorksIn.Starts='2017' and Company.name='Comp_C';
```

Recursive Relationships

All relationships that have been used up to this point in this chapter are nonrecursive relationships, because each of them connects two *different* nodes. A recursive relationship is a special form of relationships in which a node is connected with itself.

In the following examples we will use a relationship called **Is_Liked**. This recursive relationship connects the **Employee** entity with itself to find employees who like other employees of the company. Example 31.13 creates the **Is_Liked** edge table and inserts several rows into this table.

Example 31.13

```
-- Create an edge table for the recursive relationship
USE graph_db;
 CREATE TABLE dbo.Is_Liked(start_date DATE) AS EDGE;
-- Insert several rows
INSERT INTO Is_Liked VALUES
  ((SELECT $node_id FROM Employee WHERE ID = 1),
   (SELECT $node_id FROM Employee WHERE ID = 2),'1.1.2017');
INSERT INTO Is_Liked VALUES
  ((SELECT $node_id FROM Employee WHERE ID = 1),
   (SELECT $node_id FROM Employee WHERE ID = 3),'2.1.2018');
INSERT INTO Is_Liked VALUES
  ((SELECT $node_id FROM Employee WHERE ID = 1),
   (SELECT $node_id FROM Employee WHERE ID = 4),'3.1.2019');
INSERT INTO Is_Liked VALUES
  ((SELECT $node_id FROM Employee WHERE ID = 2),
   (SELECT $node_id FROM Employee WHERE ID = 3),'4.1.2016');
INSERT INTO Is_Liked VALUES
  ((SELECT $node_id FROM Employee WHERE ID = 2),
   (SELECT $node_id FROM Employee WHERE ID = 5),'5.1.2017');
INSERT INTO Is_Liked VALUES
  ((SELECT $node_id FROM Employee WHERE ID = 2),
   (SELECT $node_id FROM Employee WHERE ID = 6),'6.1.2017');
INSERT INTO Is_Liked VALUES
  ((SELECT $node_id FROM Employee WHERE ID = 3),
   (SELECT $node_id FROM Employee WHERE ID = 4),'7.1.2018');
INSERT INTO Is_Liked VALUES
  ((SELECT $node_id FROM Employee WHERE ID = 4),
   (SELECT $node_id FROM Employee WHERE ID = 5),'8.1.2016');
INSERT INTO Is_Liked VALUES
  ((SELECT $node_id FROM Employee WHERE ID = 4),
   (SELECT $node_id FROM Employee WHERE ID = 6),'9.1.2017');
INSERT INTO Is_Liked VALUES
  ((SELECT $node_id FROM Employee WHERE ID = 5),
   (SELECT $node_id FROM Employee WHERE ID = 6),'10.1.2019');
```

Example 31.14 displays all employees who like other employees.

Example 31.14

```
USE graph_db;
SELECT E1.name AS SourceName, E2.name AS TargetName
```

```
   FROM Employee E1, Is_Liked, Employee E2
    WHERE MATCH(E1-(Is_Liked)->E2);
```

The result is

SourceName	TargetName
Matthew Smith	Ann Jones
Matthew Smith	John Barrimore
Matthew Smith	James James
Ann Jones	John Barrimore
Ann Jones	Elsa Bertoni
Ann Jones	Elke Hansel
John Barrimore	James James
James James	Elsa Bertoni
James James	Elke Hansel
Elsa Bertoni	Elke Hansel

The syntax of the SELECT statement in Example 31.14 is similar to the syntax of the same statement in previous examples. The only difference is that you have to use at least one alias for the **Employee** node table because that name appears twice in the query.

Example 31.15

```
-- Display all employees that like Matthew Smith
USE graph_db;
SELECT E2.name AS FriendName
    FROM Employee E1, Is_Liked, Employee E2
    WHERE MATCH(E1-(Is_Liked)->E2)
        AND E1.name = 'Matthew Smith';
```

The result is

FriendName
Ann Jones
John Barrimore
James James

The following example retrieves all employees who are liked by Matthew Smith.

Example 31.16

```
-- Display all employees who are liked by Matthew Smith
USE graph_db
SELECT E2.name AS FriendName
```

```
    FROM Employee E1, Is_Liked, Employee E2
    WHERE MATCH(E2-(Is_Liked)->E1)
        AND E1.name = 'Matthew Smith';
```

The result of Example 31.16 does not contain any rows. If you take a closer look at Examples 31.15 and 31.16, you will see that relationship is between the same nodes, one moving from left to right and the other from right to left. Semantically, the results of these two examples tell us that three employees in the company like Matthew Smith (Example 31.15), but Matthew does not like any other employee (Example 31.16).

Example 31.17 shows how you can specify "second-level" likes.

Example 31.17

```
-- Display all employees who like employees that like Ann Jones
USE graph_db;
SELECT Person3.name AS FriendName
    FROM Employee Person1, Employee Person2,
        Is_Liked, Is_Liked Is_Liked2, Employee Person3
    WHERE MATCH(Person1-(Is_Liked)->Person2-(Is_Liked2)->Person3)
        AND Person1.name = 'Ann Jones';
```

The result is

FriendName
James James
Elke Hansel

Example 31.17 also shows, generally, how you can navigate through a graph as deep as you want. Each time you reference the next sublevel of the graph, you have to add the name of the edge table, together with the name of the self-referencing node table, first in the FROM clause of the query and, after that, in the MATCH function, appending both names or their aliases to the single parameter of this function.

Modifying and Editing Data in Graph Databases

The first section of this chapter discussed how to use the INSERT statement to load data into node and edge tables. This section explains the other two modification operations, DELETE and UPDATE. Generally, deletion of rows in node and edge tables can be performed without any restrictions, while modification of values of columns is limited and depends on the type of the particular column.

Deleting Graph Data

Before we discuss how SQL Server Graph Databases deletes instances of node and edge tables, Example 31.18 provides one more look at how data is inserted into an edge table.

Example 31.18

```
USE graph_db;
INSERT into LivesIn VALUES ((SELECT $node_id FROM Employee WHERE id = 6),
      (SELECT $node_id FROM City WHERE id = 5), '2.1.2018');
```

The INSERT statement in Example 31.18 adds a new instance of the relationship **LivesIn**. This instance connects instance 6 of the **Employee** node table with instance 5 of the **City** node table. With the query in Example 31.19, you can check whether this instance is properly inserted.

Example 31.19

```
USE graph_db;
SELECT E.Name, C.Name
    FROM Employee E, LivesIn, City C
    WHERE MATCH(E -(LivesIn)->C) AND E.id = 6 AND C.id = 5;
```

The result is

Name	Name
Elke Hansel	Minneapolis

The query in Example 31.19 is based on two **$node_id** values. One approach to delete this instance is to use two nested SELECT statements, as shown in Example 31.20 (which is the same way they are used in Example 31.18). Note that the deletion is based on the values of the **$from_id** and **$to_id** columns of the **LivesIn** table and the corresponding values of **$node_id**.

Example 31.20

```
USE graph_db;
DELETE   FROM LivesIn
      WHERE $from_id =(SELECT $node_id FROM Employee WHERE id = 6)
         AND $to_id = (SELECT $node_id FROM City WHERE id = 5);
```

Another way to delete this instance of the relationship **LivesIn** is to use the MATCH function, as shown in Example 31.21. (Note that the syntax of the DELETE statement does not correspond to the standardized syntax of that statement. It is a proprietary extension in SQL Server.)

Example 31.21

```
USE graph_db;
DELETE   LivesIn
      FROM Employee E, LivesIn, City C
      WHERE MATCH(E-(LivesIn)->C) AND E.id = 6 AND C.id = 5;
```

The MATCH function in Example 31.21 searches for a pattern based on the specified relationship. In this case, MATCH selects all rows of the **LivesIn** relationship. The other two conditions in the WHERE clause of the query restrict the deletion to the particular instance of that relationship.

Updating Graph Data

In contrast to the DELETE statement, the UPDATE statement has some limitations. Generally, you can apply the UPDATE statement only to modify the values of the user-defined columns, as demonstrated in Example 31.22.

Example 31.22

```
USE graph_db;
UPDATE WorksIn  SET Starts = 2020
   WHERE  $from_id = (SELECT $node_id FROM Employee WHERE id = 6)
     AND $to_id = (SELECT $node_id FROM Company WHERE id = 1);
```

Example 31.22 modifies the value of the **Starts** column of the **WorksIn** edge table by using the values of the **$node_id** columns of the originating and terminating tables.

Editing Information Concerning SQL Server Graph Databases

To edit information concerning SQL Server Graph Databases, you can use either catalog views or system functions, both of which are described in the following subsections.

Graph Databases: Catalog Views

Two catalog views, **sys.tables** and **sys.columns**, have been extended to contain metadata concerning graph databases. Also, two new catalog views, **sys.edge_constraints** and **sys.edge_constraint_clauses**, are used to edit metadata concerning existing edge constraints.

As you already know, when you create a node table or an edge table, the system adds two new columns of the BIT data type, **is_node** and **is_edge**, to the **sys.tables** catalog view. For a node table, the value of **is_node** is set to 1, and the value of **is_edge** is set to 0. Conversely, the values of the **is_node** and **is_edge** columns of an edge table are 0 and 1, respectively.

Example 31.23 retrieves the names of all tables of the **graph_db** database that are either node tables or edge tables. (The corresponding values of the **is_node** and **is_edge** columns are displayed, too.)

Example 31.23

```
USE graph_db;
SELECT name, is_node, is_edge
FROM sys.tables
  WHERE is_node = 1 OR is_edge = 1;
```

The result is

Employee	1	0
City	1	0
Company	1	0

Part V

WorksIn	0	1
LocatedIn	0	1
LivesIn	0	1
Is_Liked	0	1

The **sys.columns** catalog view has been extended with two new columns: **graph_type** and **graph_type_desc**. They indicate the types of columns that the Database Engine generated. The type is indicated by a predefined numerical value and its related description. Microsoft does not provide a great deal of specifics about the columns, but you can find some details in the Microsoft documentation.

Example 31.24 uses the **sys.edge_constraints** and **sys.edge_constraint_clauses** catalog views to display metadata concerning existing edge constraints in the **LivesIn** edge table.

Example 31.24

```
USE graph_db;
SELECT
    EC.name AS Edge_constraint
    , OBJECT_NAME(EC.parent_object_id) AS Edge_table
    , OBJECT_NAME(ECC.from_object_id) AS From_node_table
    , OBJECT_NAME(ECC.to_object_id) AS To_node_table
FROM sys.edge_constraints EC
    INNER JOIN sys.edge_constraint_clauses ECC
        ON EC.object_id = ECC.object_id
WHERE EC.parent_object_id = object_id('LivesIn');
```

The result is

Edge_constraint	Edge_table	From_node_table	To_node_table
Emp_to_City	LivesIn	Employee	City

In Example 31.24, the **sys.edge_constraints** and **sys.edge_constraint_clauses** catalog views are joined together using the **object_id** column to display the information concerning the specified edge constraint (see Example 31.3). The constraint name (**Emp_to_City**) and the name of the corresponding edge table (**LivesIn**) are found in the **sys.edge_constraints** catalog view. Similarly, the names of the corresponding node tables are found in the **sys.edge_constraint_clauses** catalog view.

Graph Databases: System Functions

As you already know from Chapter 9, system functions are used to access catalog views. SQL Server contains six system functions related to Graph Databases. These functions are described in Table 31-1.

The next two examples demonstrate how these system functions can be used. Example 31.25 uses the NODE_ID_FROM_PARTS function.

System Function	Description
OBJECT_ID_FROM_NODE_ID	Extracts the object ID from a **$node_id** value
GRAPH_ID_FROM_NODE_ID	Extracts the graph ID from a **$node_id** value
NODE_ID_FROM_PARTS	Constructs a JSON node ID from an object ID and graph ID
OBJECT_ID_FROM_EDGE_ID	Extracts the object ID from an **$edge_id** value
EDGE_ID_FROM_PARTS	Constructs the edge ID from an object ID and identity
GRAPH_ID_FROM_EDGE_ID	Extracts the graph ID from an **$edge_id** value

Table 31-1   System Functions Related to SQL Server Graph Databases

Example 31.25

```
USE graph_db;
SELECT NODE_ID_FROM_PARTS(OBJECT_ID('dbo.Company'), 0);
```

The result is

```
{"type":"node","schema":"dbo","table":"Company","id":0}
```

Example 31.25 constructs a JSON document for a given **node_id** value from an **object_id** value, with the help of the NODE_ID_FROM_PARTS system function. This function has two parameters: the first is **object_id** value of the corresponding node table, and the second specifies the value of the **graph_id** column. As you can see from the result of this example, it returns the JSON document with four name/value pairs that correspond to the displayed value of the first row (ID = 0) in the **Company** node table.

NOTE The practical use of the NODE_ID_FROM_PARTS system function, as shown in Example 31.25, is in the case that you want to load data from another source and intend to assign the existing ID to each row as the graph ID.

Example 31.26 shows how the NODE_ID_FROM_PARTS function can be used to insert a row in an edge table.

Example 31.26

```
DECLARE @table1 INT = OBJECT_ID('dbo.Company');
DECLARE @table2 INT = OBJECT_ID('dbo.City');
INSERT INTO LocatedIn ($from_id, $to_id)
  VALUES (NODE_ID_FROM_PARTS(@table1, 1),
                   NODE_ID_FROM_PARTS(@table2, 2));
```

The batch in Example 31.26 obtains the object IDs from the relationship's originating and terminating node tables (**Company** and **City**) and saves them into the **@table1** and **@table2** variables, respectively. These variables are then used as the parameters of the NODE_ID_FROM_PARTS function to insert a new row into the **LocatedIn** edge table.

Part V

Querying Graph Data Using Relational Queries

This section demonstrates how Transact-SQL can be used to write queries over graphs. Graph query languages in general, and the query language of SQL Server Graph Databases in particular, are better suited to query *connected* data than Transact-SQL. In other words, query latency in a graph database is proportional to how much of the graph you choose to explore in a query, and is not proportional to the amount of data stored, which is the most important issue of Transact-SQL. The next five examples help to explain this.

First, Example 31.27 creates a relational table called **Employee1**, which has the same structure as the **Employee** node table (see Example 31.1).

Example 31.27

```
USE graph_db;
  CREATE TABLE  dbo.Employee1 (
  ID    INT  NOT NULL PRIMARY KEY,
  name    VARCHAR (100) NULL,
  sex    char (10) NULL);

INSERT INTO Employee1 VALUES (1,'Matthew Smith','Male');
INSERT INTO Employee1 VALUES (2,'Ann Jones','Female');
INSERT INTO Employee1 VALUES (3,'John Barrimore','Male');
INSERT INTO Employee1 VALUES (4,'James James','Male');
INSERT INTO Employee1 VALUES (5,'Elsa Bertoni','Female');
INSERT INTO Employee1 VALUES (6,'Elke Hansel','Female');
```

The only syntactical difference between the **Employee1** and the **Employee** table is that the former does not have the AS NODE option. (The six INSERT statements in Example 31.27 are identical to the corresponding statements for the **Employee** table in Example 31.2.)

To understand performance issues of performing queries over graphs in a relational database system such as the Database Engine, we will look at several examples concerning the **Is_Liked** relationship, introduced in the earlier section "Recursive Relationships." To implement such a recursive relationship using relational tables, you have to create a new table, as shown in Example 31.28.

Example 31.28

```
USE graph_db;
CREATE TABLE Employee1_Friend
  (EmployeeID INT NOT NULL,
   FriendID INT);
INSERT INTO  Employee1_Friend VALUES (1,2)
INSERT INTO  Employee1_Friend VALUES (1,3)
INSERT INTO  Employee1_Friend VALUES (1,4)
INSERT INTO  Employee1_Friend VALUES (2,3)
INSERT INTO  Employee1_Friend VALUES (2,5)
INSERT INTO  Employee1_Friend VALUES (2,6)
INSERT INTO  Employee1_Friend VALUES (3,4)
```

```
INSERT INTO  Employee1_Friend VALUES (4,5)
INSERT INTO  Employee1_Friend VALUES (4,6)
INSERT INTO  Employee1_Friend VALUES (5,6)
```

The **Employee1_Friend** table has two columns. The first, **EmployeeID**, represents the ID of an employee, while the second, **FriendID**, is the ID of another employee who likes the employee listed in the **EmployeeID** column. Therefore, the first INSERT statement in Example 31.28 can be interpreted as the employee with ID = 2 (Ann Jones) likes the employee with ID = 1 (Matthew Smith).

The next three examples show how you can implement the queries from Examples 31.15, 31.16, and 31.17 using T-SQL. Example 31.29 implements the query from Example 31.15.

Example 31.29

```
-- Corresponds to Example 31.15
  USE graph_db;
SELECT E1.name
   FROM Employee1 E1 JOIN Employee1_Friend
     ON Employee1_Friend.FriendID = E1.ID
                JOIN Employee1 E2
     ON Employee1_Friend.EmployeeID = E2.ID
   WHERE E2.name = 'Matthew Smith';
```

Concerning performance, the query in Example 31.29 is rather inexpensive because the number of qualified rows is restricted with the condition in the WHERE clause (E2.name = 'Matthew Smith'). If you create an index for the **name** column of the **Employee1** table, the query in Example 31.29 will be executed fairly quickly.

As you already know from Example 31.15, a recursive relationship such as **Is_Liked** must not be reflexive, meaning that an employee can be liked by another one, but the former must not share the same feelings. For this reason, Examples 31.15 and 31.16 display the different result sets.

The relational implementation of Example 31.16 is given in Example 31.30 and the performance of this query is similar to the performance of Example 31.29. (The result of Example 31.30 does not contain any rows.)

Example 31.30

```
-- Corresponds to Example 31.16
USE graph_db;
SELECT E1.name
   FROM Employee1 E1 JOIN Employee1_Friend
    ON Employee1_Friend.EmployeeID = E1.ID
                JOIN Employee1 E2
    ON Employee1_Friend.FriendID  =  E2.ID
  WHERE E2.name = 'Matthew Smith';
```

Example 31.31, which answers the query "Find employees who like the employees who like Ann Jones," is syntactically and computationally complex.

Example 31.31

```
-- Corresponds to Example 31.17
USE graph_db;
SELECT E1.name AS EMP_Name, E2.name AS FriendOfFriend
    FROM Employee1_Friend Ef1 JOIN Employee1 E1
    ON Ef1.EmployeeID = E1.ID
    JOIN Employee1_Friend Ef2 ON Ef2.EmployeeID = Ef1.FriendID
    JOIN Employee1 E2 ON Ef2.FriendID = E2.ID
    WHERE E1.name = 'Ann Jones' AND Ef2.FriendID <> E1.ID;
```

In Example 31.31, the number of JOIN operations is high and the WHERE clause contains the inequality (Ef2.FriendID <> E1.ID). As you already know from the section "Query Analysis" in Chapter 19, an expression with the NOT (<>) operator cannot be used by the optimizer as a search argument, and the only access the optimizer uses in this case is the table scan. Therefore, the execution of this query will be slow.

Things get more complex if you dig deeper into the graph with the **Is_Liked** relationship. Though it is possible to get an answer to the query given in Example 31.31 in a fairly short period of time, queries that extend to more degrees of liking will have very poor performance. Therefore, the higher the degree of liking, the poorer the performance.

Summary

SQL, as a common database language, does not deal very well with connected data because the number of join operations (inner joins and self-referencing joins) increases significantly as the dataset gets bigger and the traversal of a graph goes deeper. For this reason, Microsoft has implemented SQL Server Graph Databases as an important tool to store connected data efficiently.

A graph in SQL Server Graph Databases is a collection of node tables, which store nodes of the graph, and edge tables, which store relationships (edges of the graph) between nodes. As "first-class citizens" in Graph Databases, nodes and edges of a graph can have attributes (properties) associated with them. Graph extensions are fully integrated in the Database Engine. This means that the same storage engine used to store relational data is also used to store graph data. Additionally, you can retrieve graph data and relational data via a single query. SQL Server Graph Databases also supports all the security and compliance features available in SQL Server.

The next chapter describes SQL Server Machine Learning Services in general and support for R in particular.

Exercises

Using the following exercises, extend the graph presented in this chapter with a new node table called **Restaurant** and a new edge table called **Visits**.

E.31.1   Create a node table, **Restaurant**, with the following columns:

- ID (INT PK)
- name (VARCHAR(20))
- city (VARCHAR(20))

E.31.2   Insert three rows in the **Restaurant** table:

- (1, 'A', 'Hoboken')
- (2, 'B', 'Lexington')
- (3, 'C', 'Madison')

E.31.3   Create the edge table called **Visits**. Add a column called **rating**, for use by employees to rate restaurants where they eat. Note that the edge between the nodes **Employee** and **Restaurant** is directed, from **Employee** to **Restaurant**.

E.31.4   Insert the following six rows into the **Visits** table:

```
USE graph_db;
INSERT INTO Visits VALUES ((SELECT $node_id FROM Employee WHERE id = 1),
      (SELECT $node_id FROM Restaurant WHERE id = 1), 5);
INSERT INTO Visits VALUES ((SELECT $node_id FROM Employee WHERE id = 2),
      (SELECT $node_id FROM Restaurant WHERE id = 2), 3);
INSERT INTO Visits VALUES ((SELECT $node_id FROM Employee WHERE id = 3),
      (SELECT $node_id FROM Restaurant WHERE id = 3), 2);
INSERT INTO Visits VALUES ((SELECT $node_id FROM Employee WHERE id = 4),
      (SELECT $node_id FROM Restaurant WHERE id = 3), 3);
INSERT INTO Visits VALUES ((SELECT $node_id FROM Employee WHERE id = 5),
      (SELECT $node_id FROM Restaurant WHERE id = 3), 1);
INSERT INTO Visits VALUES ((SELECT $node_id FROM Employee WHERE id = 6),
      (SELECT $node_id FROM Restaurant WHERE id = 2), 4);
```

E.31.5   Find the names of restaurants that Ann Jones visits.

E.31.6   Find the names of restaurants visited by employees who Ann Jones likes.

VI

Machine Learning

SQL Server Machine Learning Services: R Support

SQL Server Machine Learning Services provides native support for R and Python on SQL Server. R and Python integration includes base open source distributions, plus Microsoft-specific libraries for high-performance analytics.

Machine learning is a subfield of artificial intelligence (AI), which is concerned with methods for improving the knowledge of an intelligent agent over time. Machine learning methods is a collection of features for extracting predictive models from data. The main application areas of R support in the Database Engine are data visualization, predictive analytics, and statistical modeling.

From an architecture standpoint, R scripts run as external processes. That way, the integrity of the database engine is preserved. However, all data access and security is through SQL Server database roles and permissions, meaning that any application with access to SQL Server can access your R and Python scripts when you deploy them as stored procedures, or serialize your scripts and save them as applications to a Database Engine database.

This chapter covers the integration of the R language into the Database Engine; the integration of the Python language is explained in detail in Chapter 33. The first major section introduces the R language, explains generally how the R language is embedded in the SQL Server system, and defines the concept of a data frame in relation to R. The second major section shows you how to visualize data using data frames. The chapter wraps up with an exploration of predictive modeling with R and presents an example of solving linear regression problems with R.

SQL Server R Services

SQL Server R Services is an add-on component of SQL Server Machine Learning Services and is used for executing R scripts on SQL Server. The compiled R scripts run isolated from Database Engine processes but can access relational data using either stored procedures or SQL scripts containing R statements. That way, the data stored persistently in SQL Server can be manipulated using R functions.

R Services includes a base distribution of R, overlaid with enterprise R packages from Microsoft so that you can load and process a large amount of data in parallel and aggregate it into a single consolidated output. The main application areas of Microsoft's R functions and algorithms are data visualization, predictive analytics, and statistical modeling.

R Language: An Introduction

R was created in the 1990s as an open source alternative to the proprietary language called S. Since then, the R language has been used by scientists, statisticians, and (more recently) data scientists as a convenient environment for exploratory data analysis. One of the most important properties of the R language is that it enables you to create and package entire scripts as libraries. That way, you can get more consistent and reliable results than when working with systems that require a lot of manual interaction with a graphical user interface. (Note that R packages can include datasets, too.)

R is best used to manipulate moderately sized datasets, to do statistical analysis, and to produce data-centric output that can be visualized using different forms of GUIs. Its popularity is based on its expressiveness and the huge collection of third-party libraries created for it.

On the other hand, R is not considered as a general-purpose programming language, because it does not have all the features typically offered in programming languages, such as control structures. R is usually compared to specialized statistical systems such as SPSS and SAS.

Getting Started with R in SQL Server

To run R scripts in the SQL Server environment, you have to use the **sp_execute_external_script** system stored procedure. This procedure allows you to execute any provided script at an external location. The syntax of this procedure is as follows:

```
sp_execute_external_script
    @language = N'language',
    @script = N'script'
    [ , @input_data_1 = N'SelectStatement' ]
    [ , @input_data_1_name = N'input_data_1_name' ]
    [ , @output_data_1_name = N'output_data_1_name' ]
    [ , @parallel = 0 | 1 ]
    [ , @params = N'@parameter_name data_type [ OUT | OUTPUT ] [ ,...n ]' ]
    [ , @parameter1 = 'value1' [ OUT | OUTPUT ] [ ,...n ] ]
```

Before we examine the most important parameters of the **sp_execute_external_script** system stored procedure, let's take a look at how a programming language can be integrated into SQL Server. Generally, when writing a script using a programming language other than Transact-SQL, three issues must be addressed:

- Specification of the particular programming language
- Implementation of the external language script as a string
- Implementation of the internal (Transact-SQL) language script

To address these issues, the **sp_execute_external_script** system stored procedure supports three parameters, **@language**, **@script**, and **@input_data_1**, respectively. In SQL Server 2019, the **@language** parameter can have the following values: R (for the R language) and Python. The **@script** parameter defines the external language script as a string. The **@input_data_1** parameter specifies the input dataset, which is generated using a Transact-SQL query. (The description of all other parameters can be found in Microsoft Docs.)

NOTE To be able to use the **sp_execute_external_script** system stored procedure, you need to enable it first by running the following batch:

```
EXEC sp_configure 'external scripts enabled', 1
GO
RECONFIGURE WITH OVERRIDE
GO
```

Example 32.1 is a very simple demonstration of how you can use the **sp_execute_external_ script** system stored procedure to display values of the **project** table.

NOTE In contrast to SQL, R is a case-sensitive language.

Example 32.1
```
USE sample;
GO
execute sp_execute_external_script
    @language = N'R'
  , @script = N'OutputDataSet <- InputDataSet;'
  , @input_data_1 = N'SELECT project_no, budget FROM project;'
    WITH RESULT SETS ((Name CHAR(20), Balanced_budget INT));
```

Example 32.1 uses the **project** table of the **sample** database to generate input data. The value **R** in the **@language** parameter of the **sp_execute_external_script** system stored procedure specifies to use the R language for the implementation of the script. As you already know, the **@input_data_1** parameter specifies the input dataset. Therefore,

```
    @input_data_1 = N'SELECT project_no, budget FROM project;'
```

specifies that the input data will be selected using the SELECT statement specified.

Two important parameters are related to the **@script** variable: **@input_data_1_name** and **@output_data_1_name**. The first parameter specifies the name of the variable used to represent the input data; that is, the result of the SELECT statement defined with the **@input_data_1** parameter. The default value is **InputDataSet**. As you can see from the query in Example 32.1, the input data contains all values of the columns **project_no** and **budget** of the **project** table.

The second parameter, **@output_data_1_name**, specifies the name of the variable in the external script that contains the data to be returned to the database system after the execution of the script. The default value is **OutputDataSet**. Thus, the expression

```
N'OutputDataSet <- InputDataSet;'
```

in Example 32.1 passes the input data contained in **InputDataSet** to R and then back to the result set *without any modifications*.

The last line of the code in Example 32.1 uses WITH RESULT SETS. By default, a result set returned by an R script is displayed as a table with unnamed columns (as shown in upcoming Example 32.2). In other words, names used within a script are local to the scripting environment and are not reflected in the corresponding result set. The WITH RESULT SETS clause allows you to name result set columns as you wish. Therefore,

```
WITH RESULT SETS ((Name CHAR(20), Balanced_budget INT))
```

assigns the names **Name** and **Balanced_budget** to the output of the columns **project_no** and **budget**, respectively.

The result of Example 32.1 is

Name	Balanced_budget
p1	120000
p2	95000
p3	186500

The script in Example 32.1 does not perform any manipulation of the input data. (I use this example just to demonstrate how to run the **sp_execute_external_script** system stored procedure and to explain the most important parameters of it.) The most important advantage of R Services is the capability to use data stored persistently in SQL Server, manipulate it using R functions, and return it to SQL Server. Example 32.2 shows this capability in action.

Example 32.2

```
USE sample;
DECLARE @rscript NVARCHAR(MAX);
SET @rscript = N'OutputDataSet <- SqlData;
OutputDataSet[,2] <- round(SqlData$budget/7, 2);';
EXEC sp_execute_external_script
@language = N'R',
@script = @rscript,
@input_data_1 = N'SELECT project_no, budget FROM project',
@input_data_1_name = N'SqlData';
```

Example 32.1 used the default name **InputDataSet** for the input dataset. In Example 32.2,

```
@input_data_1_name = N'SqlData';
```

assigns another name, **SqlData**, to the input dataset. (For the output data, we use the default name.) The following line of code uses the R function called **round** to round the obtained result:

```
OutputDataSet[,2] <- round(SqlData$budget/7, 2);';
```

Generally, to modify values of one of the selected columns of the corresponding SELECT statement, you have to reference the input dataset (**SqlData**) first, add a dollar sign (**$**), and name that column (in our case, **budget**). Each value of the selected column is then divided by 7 and rounded, up to two decimal digits. (The term **OutputDataSet[,2]** specifies that the obtained value will be assigned to the second column in the SELECT list.)

The result is

(NoColumnName)	(NoColumnName)
p1	17142.86
p2	13571.43
p3	26642.86

NOTE Because Example 32.2 does not use the WITH RESULT SETS statement, the columns in the result set will not have explicit names.

R Data Frames

A key concept when working with R is the *data frame*, because it allows you to pass data from a database system, modify the data using R, and send it back to the database system to be stored persistently. The structure of a data frame is similar to a relational table, consisting of rows and columns. Because you are already familiar with the relational model, you should not have any problems understanding and using data frames.

NOTE All subsequent examples use Microsoft's sample database **AdventureWorks**.

Example 32.3 shows the use of data frames in the R language.

Example 32.3

```
USE AdventureWorks;
GO
DECLARE @rscript NVARCHAR(MAX);
SET @rscript = N'
  purchase <- InputDataSet
```

```
   c1 <- levels(purchase$Units)
   print(c1)';
DECLARE @select NVARCHAR(MAX);
SET @select = N'
SELECT h.subtotal AS Total, v.UnitMeasureCode AS Units
   FROM Purchasing.PurchaseOrderHeader h
   INNER JOIN Purchasing.PurchaseOrderDetail d
                 ON h.PurchaseOrderID = d.PurchaseOrderID
  INNER JOIN Purchasing.ProductVendor v ON d.ProductID = v.ProductID';
EXEC sp_execute_external_script
@language = N'R',
@script = @rscript,
@input_data_1 = @select;
```

I will explain only the most important parts of the code in Example 32.3, starting with the following three lines of code:

```
purchase <- InputDataSet
c1 <- levels(purchase$Units)
print(c1)
```

The first line assigns **InputDataSet** to the **purchase** variable. That way, the input data is passed to the **purchase** data frame, which is a two-dimensional dataset. Once you have the data, you can start working with it. To understand which data is passed to the variable, let's look at the query assigned to the @**select** variable. The SQL statement refers to the **AdventureWorks** database to display the total purchase per unit measure code. To extract data, we join three tables: **PurchaseOrderHeader**, **PurchaseOrderDetail**, and **ProductVendor**. (All these tables belong to the **Purchasing** schema of the **AdventureWorks** database.) The first column in the SELECT list has the alias **Total** and specifies aggregated purchase amounts, while the second column is called **Units** and specifies the corresponding unit measure codes.

Moving on the second line, generally, the **levels** function of the R language provides access to the distinct values of a variable. In Example 32.3, this function retrieves a list of distinct values from the **UnitMeasureCode** column of the **ProductVendor** table. As you already know, when referencing a column, you must first specify the dataset (in this case, **purchase**), followed by a dollar sign (**$**), followed by the column name or its alias (**Units**). The second line also assigns the output of the **levels** function to the **c1** variable, which gives us what we need to retrieve the result set. The last statement, **print**(c1), displays the retrieved values.

To display the output, you have to click the column Messages in the lower part of the right pane. The output of Example 32.3 is

```
[1]  "CAN" "CS " "CTN" "DZ " "EA " "GAL" "PAK"
```

Although the SELECT statement in Example 32.3 returns aggregated purchase amounts for each unit measure code, the script in that example uses the values of only one "dimension" of the data frame—the values of the **UnitMeasureCode** column.

Example 32.4 goes a step further and uses the same data frame to get all data from the query in Example 32.3. In other words, Example 32.3 returns "row" values as well as "column" values of the data frame and displays them in the table form.

Example 32.4

```
USE AdventureWorks;
GO
DECLARE @rscript NVARCHAR(MAX);
SET @rscript = N'
  purchase <- InputDataSet
  c1 <- levels(purchase$Units)
  c2 <- round(tapply(purchase$total, purchase$Units, sum))
  purchase <- data.frame(c1, c2)
  print(c2)';
DECLARE @select NVARCHAR(MAX);
SET @select = N'
SELECT h.subtotal AS total, v.UnitMeasureCode AS Units
  FROM Purchasing.PurchaseOrderHeader h
        INNER JOIN Purchasing.PurchaseOrderDetail d
                 ON h.PurchaseOrderID = d.PurchaseOrderID
  INNER JOIN Purchasing.ProductVendor v ON d.ProductID = v.ProductID';
EXEC sp_execute_external_script
@language = N'R',
@script = @rscript,
@input_data_1 = @select;
```

The part of the program in Example 32.4 in relation to the **c1** variable is identical to the corresponding part in Example 32.3. For this reason, I will discuss only the following two lines of code:

```
c2 <- round(tapply(purchase$total, purchase$Units, sum))
        purchase <- data.frame(c1, c2)
```

In the first line of code, the **tapply** function is used. Generally, this function uses an operator to apply it on array elements, broken down by a given variable. In our case, the operator is **sum**, and it is specified as the third argument of the function. The array elements mentioned are the distinct values of the **UnitMeasureCode** column, and are specified in the second argument of the function. The first argument of the **tapply** function defines the subset (**subtotal**) to which the operator will be applied. Therefore, in Example 32.4, the aggregated sums of values in the **subtotal** column are calculated for each different value of the **UnitMeasureCode** column.

The result is

CAN	CS	CTN	DZ	EA	GAL	PAK
9609759	584963	4408405	958018	267092933	7526088	206097

Data Visualization

As previously mentioned, one of the most important advantages of the R language is that it enables you to visualize output data in many different (graphical) forms. In other words, the R language has powerful built-in functions, which are grouped together in different packages, to help you to create visualizations of your data in a form that best suits your needs.

There are two general ways to integrate R graphics with SQL Server:

- Output a dataset by using the **sp_execute_external_script** system stored procedure and apply one of many R packages that support data visualization. In other words, you do data visualization *inside* R.

- Integrate R into a SQL Server tool, such as Power BI, and use its functions to visualize data.

- The following sections describe data visualization in R and the integration of R in Power BI.

Data Visualization in R

The previous section discussed how data frames can be used to pass input values, store them in a data frame, manipulate them inside the script using different R functions, and finally return the output data in tabular form. In this section, Example 32.5 builds on Example 32.4 to demonstrate how to visualize the output data using a bar chart.

NOTE When executing Example 32.5, you might get an error similar to this:
Msg 39004, Level 16, State 20, Line 448 A 'R' script error occurred during execution of 'sp_execute_external_script' with HRESULT.
  This error is related to the SQL Server Launchpad service. This service is used to start Advanced Analytics Extensions processes, which are necessary for integration of the R system and Python with the Database Engine. The program in Example 32.5 writes the file in the C:\temp directory, which is by default not accessible by the Launchpad service. You can either grant access to this directory, or use another directory; for instance, C:\RScripts. To grant access to a directory, right-click that directory, select Grant Access To, and click Add Everyone.

Example 32.5

```
USE AdventureWorks;
GO
DECLARE @rscript NVARCHAR(MAX);
SET @rscript = N'
# Step 1: Import R packages
  library(scales)
  library(ggplot2)
# Step 2: Specify report file
  file <- "C:\\temp\\Figure32_1.tif"
  tiff(filename=file, width=1000, height=600)
```

```
# Step 3: Specify data frame
  purchase <- InputDataSet
  c1 <- levels(purchase$Units)
  c2 <- round(tapply(purchase$total, purchase$Units, sum))
  purchasedf <- data.frame(c1, c2)
  names(purchasedf) <- c("Total", "Units")
# Step 4: Generate bar chart
  barchart <- ggplot(purchasedf, aes(y=Units, x=Total)) +
    labs(title="Total Purchases per Unit Code", x="Unit Measure Codes",
y="Purchase Amounts") +
    geom_bar(stat="identity", color="blue", size=1, fill="lightblue")
  print(barchart)
  dev.off()';
 DECLARE @select NVARCHAR(MAX);
  SET @select = N'
  SELECT h.subtotal AS Total, v.UnitMeasureCode AS Units
  FROM Purchasing.PurchaseOrderHeader h
       INNER JOIN Purchasing.PurchaseOrderDetail d
                 ON h.PurchaseOrderID = d.PurchaseOrderID
    INNER JOIN Purchasing.ProductVendor v ON d.ProductID = v.ProductID';
 EXEC sp_execute_external_script
  @language = N'R',
  @script = @rscript,
  @input_data_1 = @select
  WITH RESULT SETS NONE;
```

As you can see in Example 32.5, the visualization process involves four steps:

1. Import R packages.
2. Specify the report file.
3. Specify the data frame.
4. Generate a bar chart.

These steps and their corresponding code in Example 32.5 are discussed in the following sections, but first, Example 32.6 shows how you can check whether or not a particular R package is installed on your system, if you are not sure.

Example 32.6

```
EXEC sp_execute_external_script
@language = N'R',
    @script = N'library(scales)'
```

Example 32.6 checks whether the **scales** package is installed on your system. If the package is already installed, you get the message, "Commands completed successfully." Otherwise, if some of the dependent packages fail, the system displays the name of the package you should install.

Step 1: Import R Packages

Generally, R packages are collections of functions in the R Services library. During the installation of R Services, SQL Server setup program adds a number of common R packages to your system, but you can install additional packages at any time. Once a package has been installed in the library, you can import it into your program using the **library** function.

There are several ways to install R packages to an instance of SQL Server. Which method you should choose depends on which version of SQL Server you have and whether or not your server has an Internet connection. The following three approaches are the most convenient ones:

- **Use conventional R package managers**   You can use standard R tools to install new packages on an instance of SQL Server. The only requirement is that you have administrator rights. The most popular standard tools are RGui, Rterm, and Rcmd.

- **Use RevoScaleR**   RevoScaleR is a machine learning package in R created by Microsoft. It is available as part of Machine Learning Services in SQL Server. The package contains functions for creating machine learning techniques, such as linear models, decision trees, and K-means clustering, in addition to functions for visualizing data.

- **Use the SQL statement CREATE EXTERNAL LIBRARY**   This statement makes it possible to add a package or set of packages to an instance or a specific database without running R code directly. In contrast to other methods, this method requires package preparation and additional database permissions.

This section describes the use of the R package manager called RGui. You need to have administrator rights to use the RGui package manager.

First, determine the location of the instance library and navigate to that location. The path for the SQL Server 2019 default instance library directory is as follows:

C:\Program Files\Microsoft SQL Server\MSSQL15.MSSQLSERVER\R_SERVICES\bin\x64

Right-click Rgui.exe and select Run As Administrator. In the new console window that opens, type the following at the command prompt:

```
install.packages("package-name")
```

As indicated in Example 32.5, you'll be installing the **scales** and **ggplot2** packages. You can get both packages by installing the **ggplot2** package. In other words, when you install this package, R Services downloads several additional packages, including the **scales** package. Therefore, replace *package-name* with **ggplot2** as the parameter of the **install.packages** command.

After successful installation of this package, import the **scales** and **ggplot2** packages into your script. As shown in Example 32.5, you do so with the **library** function:

```
library(scales)
library(ggplot2)
```

Step 2: Specify the Report File

For the bar chart that will be generated, you need to specify the image file that will hold it. As shown in Example 32.5, you do this in your program with the following code:

```
file <- "C:\\temp\\Figure32_1.tif"
tiff(filename=file, width=1000, height=600)
```

The first line specifies a string variable to hold the directory and filename of the image file and then uses the <- operator to assign the value to the **file** variable. (Note that you have to escape the backslashes in the path by doubling them.) The second line invokes the **tiff** device, which is used to create the .tif file for the corresponding bar chart. (The R language also allows you to create charts in other formats, such as .bmp and .png. For these formats, the corresponding devices are supported, too.) The first argument of the **tiff** device is **filename**, which contains the name of the file specified in the previous line of code. Additionally, you specify the width and height of the file, in pixels.

Step 3: Specify the Data Frame

Concerning the definition of the data frame, the only new line of code in Example 32.5 in relation to Example 32.4 is

```
names(purchasedf) <- c("Total", "Units")
```

This statement uses the **names** function to assign names to the parameters of the **purchasedf** data frame. These names will be then used in the next step.

Step 4: Generate a Bar Chart

Several functions from the **scales** and **ggplot2** packages are used to generate bar charts. As you will see in a moment, the main part of the code in this step is related to the bar chart definition. For this purpose, you use the **barchart** variable and assign different properties of the chart to it.

NOTE Microsoft R Services supports a lot of different possibilities to create graph types, such as histograms, box plots, and scatter plots. The description of all possibilities to plot results is out of the scope of this book and can be found in Microsoft Docs.

As shown in Example 32.5, the definition of all properties of the bar chart is made up of several elements, which are connected using the plus (+) sign. The first element uses the **ggplot** function to create the foundation for the bar chart:

```
ggplot(purchasedf, aes(y=Units, x=Total))
```

This function has two arguments. The first one, **purchasedf**, is the name of the dataset and delivers the data for the bar chart. The second argument uses the R **aes** function. This function specifies the aesthetic mappings. In other words, it describes how variables in the data are mapped to visual properties (aesthetics) of elements. The default values for aesthetic mappings are set in the **ggplot** function, but can be overwritten within a script. In our code, the **aes** function specifies the chart's coordinates that are identical to the columns in the **purchasedf** dataset.

The second element in the definition of the bar chart properties uses the **labs** function to provide labels for the title and each axis of the bar chart. The meaning of the three parameters of the **labs** function is straightforward. The first parameter, **title**, specifies the name of the chart ("Total Purchases per Unit Code"), which is positioned above the chart area. The other two parameters, **x** and **y**, specify the titles of the X axis ("Unit Measure Codes") and Y axis ("Purchase Amounts"). By using the **labs** function, you can override the default labels that are specified in the **ggplot** function.

The third element in the definition uses the **geom_bar** function to specify the intention to create a bar chart:

```
geom_bar(stat="identity", color="blue", size=1, fill="lightblue")
```

The function takes four arguments. The **stat** argument with the "identity" value ensures that data values map correctly to the chart points. The **color** argument sets the bar outlines to blue, the **size** argument sets the bar outlines to 1 point, and the **fill** argument sets the color of the bars to light blue.

NOTE Example 32.5 specifies the values of only a few arguments of the **ggplot** function. For all other arguments that are not specified, the system uses their default values.

The definition of the bar chart is now complete. The **print** function sends the bar chart to the .tif file and the **dev.off** function closes the **tiff** device. (Note that the last statement in Example 32.5, WITH RESULT SETS NONE, is necessary because there is no result set. In other words, the output is sent to a bar chart and not to a table.) Figure 32-1 shows the output of the R script in Example 32.5.

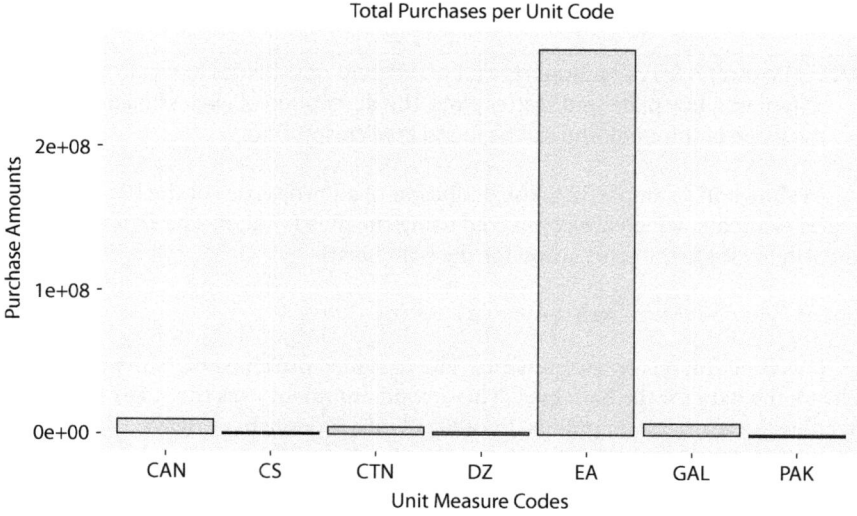

Figure 32-1   Bar chart of Example 32.5

Integrate R in Power BI Desktop

Power BI Desktop allows users to create a personal BI environment by gathering data from different sources, loading that data in the data model, and visualizing the data in the same workspace. If you have not installed Power BI Desktop, go to https://powerbi.microsoft.com and download the .exe file. The installation is simple and straightforward.

NOTE Power BI Desktop does not include, deploy, or install the R engine. To run R scripts in Power BI Desktop, you must separately install R on your local computer or use the R engine installed as a part of SQL Server ML Services. See the Chapter 23 section "Power BI" for more information about Power BI Desktop.

To enable R visuals, make sure your local R installation is specified. To do this, open Power BI Desktop and select File | Options and Settings | Options. In the Options window, select R Scripting under Global to open the R Script Options window (see Figure 32-2). Select Other in the Detected R Home Directories field. In the Set an R Home Directory field, specify the path where R_Services is stored. As you can see from Figure 32-2, I use the R installation that is part

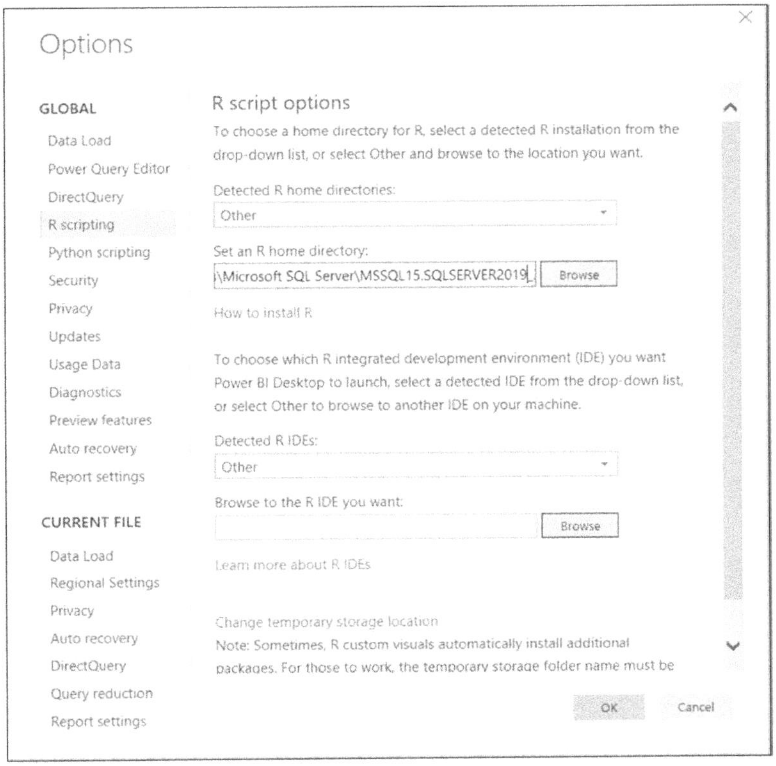

Figure 32-2   The R Script Options window

of SQL Server Machine Learning Services. On my computer, it is stored under C:\Program Files\Microsoft SQL Server\MSSQL15.SQLServer\R_Services. Click OK.

In the next step, you load the data. In the main menu of Power BI Desktop, click the Get Data button. From the list of all possible data sources, choose SQL Server Database. The Connection window for SQL Server appears. Type the name of your server instance and the database name. Click OK. In the Impersonation page, choose the Use My Current Credentials radio button and click Connect.

After that, click Advanced Options to provide the Transact-SQL query. To visualize the output data of the same query of the **AdventureWorks** database used in Example 32.5, type or copy the following statement in the SQL Statement field, as shown in Figure 32-3:

```
SELECT h.subtotal AS Total, v.UnitMeasureCode AS Units
FROM Purchasing.PurchaseOrderHeader h
    INNER JOIN Purchasing.PurchaseOrderDetail d
            ON h.PurchaseOrderID = d.PurchaseOrderID
  INNER JOIN Purchasing.ProductVendor v ON d.ProductID = v.ProductID;
```

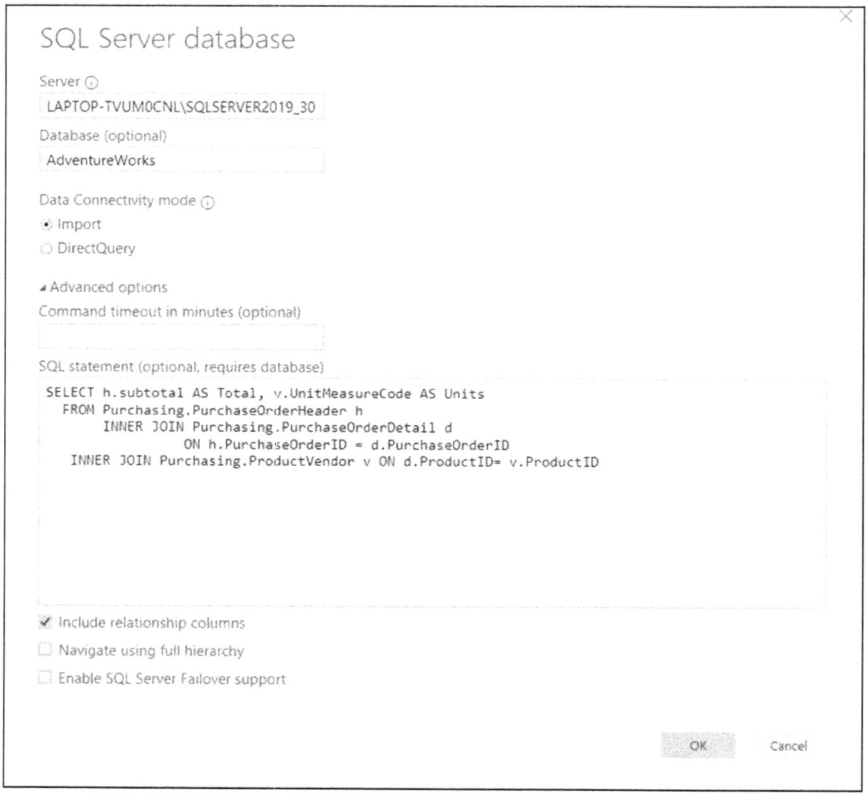

Figure 32-3   The SQL Server database credentials and the query

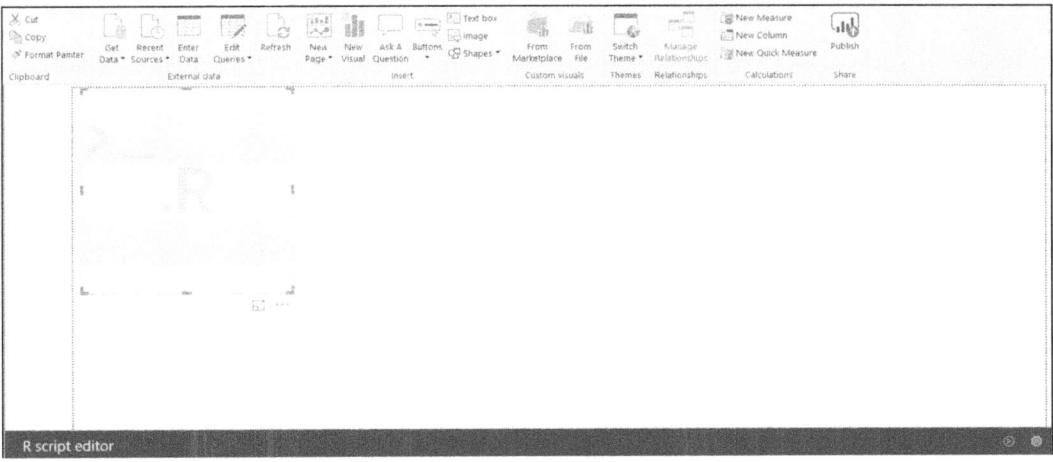

Figure 32-4   R Visualization box in Power BI Desktop

Click OK to see the preview of the query. Then click Load on the preview window. Now, the system creates the result set called Query1. (You can change the name, if you wish.) The table name and the names of selected columns, Total and Units, appear in the Fields pane in the right part the window.

From the Visualizations pane on the right, click the R Script icon. The R visualization box appears (see Figure 32-4). Drag and drop the Total and Units columns from the Fields pane into the Values box.

Then, scroll down to the R Script Editor located in the lower half of the Power BI screen (the title of which is shown at the bottom of Figure 32-4) and enter the following R code:

```
# Paste or type your script code here:
library(ggplot2);
a <- qplot(y = Total, x = Units, data = dataset,
color = Units, facets = ~Units,
        main = "Total Purchases");
a + scale_x_discrete("Units");
a + scale_y_continuous(label = scales::dollar);
```

Click the Run Script button (>) located on the right of the R Script Editor bar. Now, your Power BI report should look like Figure 32-5.

Predictive Modeling with R

Predictive analytics comprises a variety of statistical techniques from machine learning that analyze current and historical data to make predictions about future events. In the business sector, for example, predictive models analyze patterns found in historical and transactional data to identify possible risks and/or opportunities. To do this, a predictive model examines

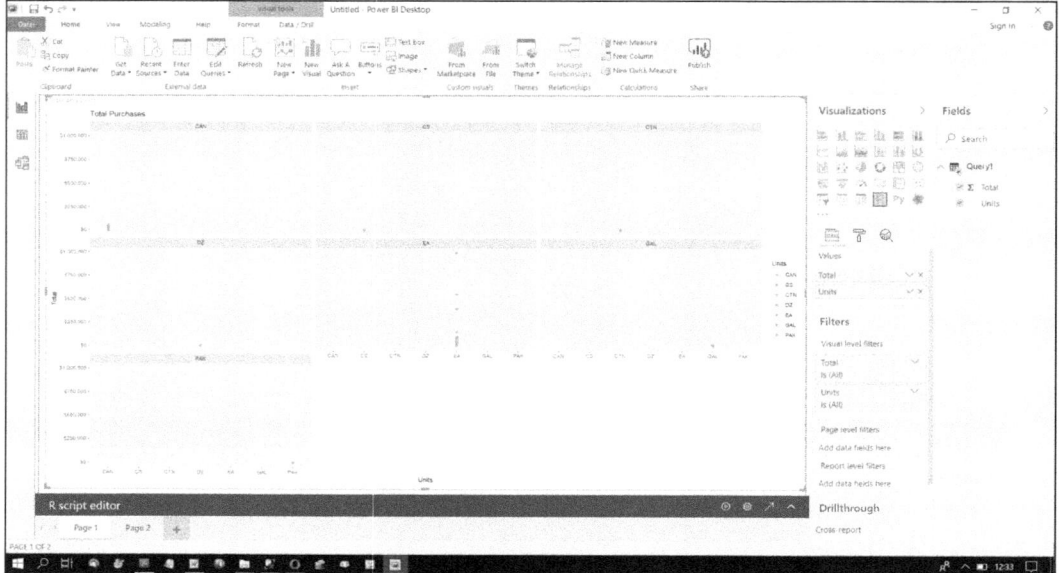

Figure 32-5   The visualization of output data of the Query1 query

relationships among several parameters. After the analysis, the predictive model presents the various scenarios for evaluation.

Solving Linear Regression Problems with R

Regression analysis is a statistical methodology that is most often used for numeric prediction. Generally, regression analysis can be used to model the relationship between one or more independent variables and a dependent variable. (An independent variable is also called a *predictor variable*, while a dependent variable is also called a *response variable*.) Therefore, the values of the independent variables are known and the goal of the regression analysis is to predict the dependent variable.

There are two forms of regression analysis:

- **Linear regression analysis**   A linear approach to modeling the relationship between a dependent variable and one or more independent variables. The simplest form of regression is straight-line regression analysis. In this case, the dependent variable is modeled as a linear function of independent variable(s).

- **Nonlinear regression analysis**   Existing data is modeled by a function that is a nonlinear combination of the model parameters and depends on one or more independent variables. In other words, the output of nonlinear regression cannot be represented with a straight line in the coordinate system. (Generally, problems based on nonlinear regression are usually converted to a linear approach.)

Linear regression is probably the simplest technique used in prediction analytics. I will show an R script that solves a linear regression problem.

As you already know, besides functions, R packages include datasets, too. A list of datasets in an R package can be displayed by calling the **data** function:

```
data(package='ggplot2')
```

One of the datasets available in the **ggplot2** package is named **mpg** and contains fuel economy data for several models of car. While the **ggplot2** package has already been loaded, you can use the **head** command to view the first few rows. Example 32.7 shows the corresponding R script.

Example 32.7
```
execute sp_execute_external_script
        @language = N'R' ,
               @script = N'
                        library (ggplot2)
                        c1 <- head(mpg)
                print(c1)'
```

The result is

	manufacturer	model	displ	year	cyl	trans	drv	cty	hwy	fl	class
1	audi	a4	1.8	1999	4	auto(l5)	f	18	29	p	compact
2	audi	a4	1.8	1999	4	manual(m5)	f	21	29	p	compact
3	audi	a4	2.0	2008	4	manual(m6)	f	20	31	p	compact
4	audi	a4	2.0	2008	4	auto(av)	f	21	30	p	compact
5	audi	a4	2.8	1999	6	auto(l5)	f	16	26	p	compact
6	audi	a4	2.8	1999	6	manual(m5)	f	18	26	p	compact

Suppose you want to perform a linear regression to determine the effect how a measurement of engine size (the **displ** columns) influences gas mileage in the city (the **cty** column). In this case, the **displ** column is the independent variable, and the **cty** column is the dependent variable. The R script in Example 32.8 calculates the effect.

NOTE When executing Example 32.8, you might get an error similar to this:
Msg 39004, Level 16, State 20, Line 448 A 'R' script error occurred during execution of 'sp_execute_external_script' with HRESULT.
   This error is related to the SQL Server Launchpad service. This service is used to start Advanced Analytics Extensions processes, which are necessary for integration of the R system and Python with the Database Engine. The program in Example 32.8 writes the file in the C:\temp directory, which is by default not accessible by the Launchpad service. You can either grant access to this directory, or use another directory; for instance, C:\RScripts. To grant access to a directory, right-click that directory, select Grant Access To, and click Add Everyone.

Example 32.8

```
DECLARE @rscript NVARCHAR(MAX);
SET @rscript = N'
# Import R packages
  library(scales)
  library(ggplot2)
# Specify report file
  file <- "C:\\Temp\\Figure32_6.tif"
  tiff(filename=file, width=1000, height=600)
# Generate plot
  plot(mpg$displ, mpg$cty);
  fit <- lm(cty ~ displ, mpg);
  chart <- abline(fit) ;
  print(chart);
  dev.off()';
EXEC sp_execute_external_script
  @language = N'R',
  @script = @rscript  ;
```

I will explain only the part of Example 32.8 that is added to the code in relation to Example 32.5. The following three lines of code define how to perform a linear regression and generate a regression line expressing the trend in the dataset:

```
plot(mpg$displ, mpg$cty);
fit <- lm(cty ~displ, mpg);
chart <- abline(fit) ;
```

The **plot** function generates graphs according to the type of the parameters specified. This function has many different forms. The form of the **plot** function given here is one of the simplest: the parameters specify the X and Y coordinates of the graphic, respectively. All points shown in the upcoming Figure 32-6 are generated after the **plot** function is executed.

The **lm** function is a generic function that can be used to display a large group of data mining techniques. The first parameter of this function specifies the particular technique. (The value lm means "linear model" and therefore specifies the linear regression model). The second parameter of the **lm** function specifies the independent variable (**displ**) on the right side and the dependent variable (**cty**) on the left side of the ~ operator. The last parameter, **mpg**, defines the dataset used to build the corresponding data mining model.

The **abline** function is used to add (vertical, horizontal, or regression) lines to a graph. It takes as the input argument the result of the **lm** function, produces the corresponding regression line according to the model, and assigns the result to the **chart** variable.

The last two lines of code print the graphics and close the specified plot explicitly:

```
print(chart);
dev.off()'
```

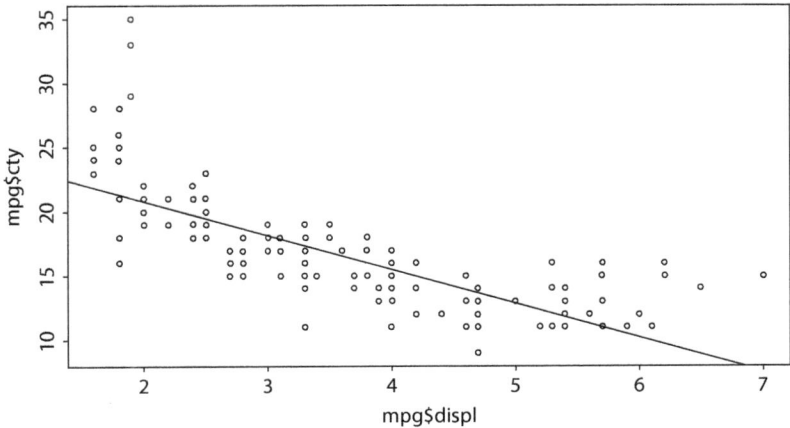

Figure 32-6   The output of Example 32.8

Figure 32-6 shows the points generated using the **plot** function, as well as the regression line generated by the **abline** function.

Summary

Microsoft R Services is an add-on component to SQL Server Machine Learning Services and is used for executing R scripts on SQL Server. The compiled R scripts run isolated from system processes but can access relational data using either stored procedures or SQL scripts containing R statements. The main application areas of Microsoft's R functions and algorithms are data visualization, predictive analytics, and statistical modeling.

The next chapter is the last chapter of this book and discusses SQL Server Machine Learning Services using the Python programming language.

33

SQL Server Machine Learning Services: Python Support

In This Chapter

- Python: An Introduction
- Data Visualization with Python
- Predictive Modeling with Python in SQL Server

Microsoft introduced support for Python with its Machine Learning Services in SQL Server 2017. The integration of Python in SQL Server allows developers to access the extensive Python libraries that are available in the open source community. Microsoft has added two SQL Server–specific features. First, the **sp_execute_external_script** stored procedure (introduced in Chapter 32) allows you to integrate Python scripts in SQL Server. Second, Python resources can be throttled using Resource Governor. That way, database administrators can make optimal decisions concerning Python processing workloads. (For the description of Resource Governor, see Chapter 20.)

This chapter covers the integration of the Python language into the Database Engine. The first major section introduces the Python language, explains generally how the language is embedded in the SQL Server system, and defines the concept of a data frame in relation to Python. The second major section shows you how to visualize data using data frames. The chapter wraps up with an exploration of predictive modeling with R and presents an example of solving linear regression problems with R.

Python: An Introduction

Python is a programming language that can be used for many purposes, including machine learning. The most important advantages of the language are

- Simple syntax
- High readability
- A large collection of libraries

Python aims to be simple in the design of its syntax, encapsulated in the slogan "There should be one *obvious* way to do it." Because of its simplicity, Python is known as a beginner's level programming language.

If you ask Python programmers what they like most about Python, they often cite its high readability. One reason for the high readability of Python code is its complete set of code style guidelines. Because of its simplicity, Python has attracted many developers to create new libraries for machine learning. Also, because of the existence of these libraries, Python is becoming very popular among machine learning experts.

NOTE   To be able to use the **sp_execute_external_script** system stored procedure, you need to enable it first by running the following batch:

```
EXEC sp_configure 'external scripts enabled', 1
GO
RECONFIGURE WITH OVERRIDE
GO
```

Getting Started with Python

If you are new to the Python programming language, it is important to know that Python code blocks are specified by their indentation. In other words, indentation of lines of code is a *requirement* and not a matter of style as in many other programming languages. Example 33.1 is a very simple example that shows how you can use the **sp_execute_external_script** system stored procedure to display values of the **project** table using Python. (For the syntax and description of this system procedure, see the section "Getting Started with R in SQL Server" toward the beginning of Chapter 32.)

Example 33.1
```
USE sample;
DECLARE @pscript NVARCHAR(MAX);
SET @pscript = N'
df = InputDataSet
OutputDataSet = df';
DECLARE @select NVARCHAR(MAX);
SET @select = N'
  SELECT project_no, budget
  FROM project';
EXEC sp_execute_external_script
  @language = N'Python',
  @script = @pscript,
  @input_data_1 = @select;
GO
```

As you can see from Example 33.1, the **sp_execute_external_script** system stored procedure has, among others, the following three parameters: **@language**, **@script**, and **@input_data_1**. The **@language** parameter specifies the name of the language integrated in

SQL Server. The value "Python" specifies that Python is used for implementation. The **@script** parameter defines the external language script as a string. You can specify it using a literal or a variable. The **@input_data_1** parameter defines the input data used by the external script in the form of a Transact-SQL query. In other words, the query generates a result set that is used as input data. The data type of this parameter is NVARCHAR(max), meaning that the query is passed as a string.

Two important parameters are related to the **@script** variable: **@input_data_1_name** and **@output_data_1_name**. The first parameter specifies the name of the variable used to represent the input dataset (i.e., the dataset defined by **@input_data_1**). You can omit the use of **@input_data_1_name**, in which case the default value **InputDataSet** is used. The second parameter, **@output_data_1_name**, specifies the name of the variable used to represent the output dataset (i.e., the result of the script). If you do not assign a value to **@output_data_1_name**, the default name **OutputDataSet** is used.

As you can see from the SELECT statement in Example 33.1, the input data contains all values of the columns **project_no** and **budget** of the **project** table. These values are assigned to the variable called **df**. After that, the input data is passed to the output data (**OutputDataSet** = **df**) without any modification. (As you can see from the following result, neither column has an explicit name. Example 33.2 shows how you can assign particular names to the columns in a result set.)

The result is

(No Column Name)	(No Column Name)
p1	120000
p2	95000
p3	186500

I present the script in Example 33.1 only to demonstrate how to run the **sp_execute_external_script** system stored procedure and to explain the meaning of some parameters of the procedure. In other words, Example 33.1 simply assigns query results to the Python script, which returns the same results as that of the SELECT statement. This is something you can do without using a Python script, but being able to assign your query results to the Python script means you can then use the analytical power built into Python to apply it to that data.

The most important advantage of the integration of Python in SQL Server is the capability to use data stored persistently in the database system, manipulate the data using Python functions, and return it to the Database Engine. Example 33.2 shows this.

Example 33.2

```
USE sample;
DECLARE @pscript NVARCHAR(MAX);
SET @pscript = N'
df1 = InputDataSet
OutputDataSet = round(df1/7, 2)';
DECLARE @select NVARCHAR(MAX);
```

```
SET @select = N'
  SELECT budget AS Balanced_budget
  FROM project';
 EXEC sp_execute_external_script
  @language = N'Python',
  @script = @pscript,
  @input_data_1 = @select
  WITH RESULT SETS ((Balanced_budget FLOAT));
GO
```

First, note the following two lines of code from Example 33.2:

```
df1 = InputDataSet
OutputDataSet = round(df1/7, 2)';
```

The SELECT statement assigned to **@input_data_1** using the **@select** variable generates the input dataset. The dataset contains values of the **budget** column of the **project** table. The second line applies a Python function to the **df1** variable, to which the input data is assigned. Precisely, each input value is then divided by 7 and rounded using the **round** function. After that, the modified values of the **budget** column are assigned to the result set, which is sent to the database system.

The last line of the code in Example 33.2 uses the WITH RESULT SETS clause to name a column of the result set. (By default, a result set returned by a Python script is displayed as a table with unnamed columns.)

The result is

Balanced_budget
17142.86
13571.43
26642.86

Python Data Frames

A data frame is used to store input data sent from the Database Engine, pass the data to Python, modify it using Python functions, and send it back to the database system. The structure of a data frame is "two-dimensional," meaning that each frame is made up of rows and columns.

In case of Python, input data is passed to a script and converted to a **DataFrame** object. Also, the data returned by a Python script is passed to the output variable as a **DataFrame** object. All **DataFrame** objects belong to the class with the same name, which is a part of the **pandas** library. (This library provides data structures designed to allow you to work with "relational" data—data provided in the table form.)

Example 33.3 shows the use of data frames.

Example 33.3

```
USE AdventureWorks;
DECLARE @pscript NVARCHAR(MAX);
SET @pscript = N'
df1 = InputDataSet
OutputDataSet = df1.groupby("Units", as_index=False).max()';
DECLARE @select NVARCHAR(MAX);
SET @select = N'
SELECT  v.UnitMeasureCode AS Units
  FROM Purchasing.PurchaseOrderHeader h
    INNER JOIN Purchasing.PurchaseOrderDetail d
        ON h.PurchaseOrderID = d.PurchaseOrderID
    INNER JOIN Purchasing.ProductVendor v ON d.ProductID=v.ProductID';
 EXEC sp_execute_external_script
    @language = N'Python',
    @script = @pscript,
    @input_data_1 = @select;
```

The input data of the script in Example 33.3 is generated from the tables of the **AdventureWorks** sample database. The SELECT statement displays the total purchase per unit measure code. To retrieve data, we join three tables: **PurchaseOrderHeader**, **PurchaseOrderDetail**, and **ProductVendor**. The single column in the SELECT list has the alias **Units**. The results of this query are all distinct unit measure codes.

The line of code following the assignment of the input data to the **df1** variable needs some explanation:

```
OutputDataSet = df1.groupby("Units", as_index=False).max()';
```

The **groupby** function is applied to the values stored in the **df1** variable. This function allows you to group records using distinct values. The first argument of the function specifies the object whose values will be grouped. In this case, the grouping is done on values of the **UnitMeasureCode** column of the **ProductVendor** table. (The alias of this column is **Units**.) To get one value from each group, the **max** function is applied to the result. (While all values in each group are identical, the function returns just one of them.)

The second argument of the **groupby** function is **as_index**. When using this function, the **as_index** parameter can be set to either True or False, depending on whether you want the grouping column to be the index of the output or not, respectively. When **as_index=False** is used, the key(s) you use in the **groupby** function is generated by the system and added as an additional column of the output data. (Taking a look at the following result, you can see that, besides the values of the **UnitMeasureCode** column, there is an additional column with index values: 1, 2, 3,...)

NOTE The use of **as_index=True** is discussed shortly in relation to Example 33.5.

The result is

	(NoColumnName)
1	CAN
2	CS
3	CTN
4	DZ
5	EA
6	GAL
7	PAK

As you already know, data frames provide data in table form. Example 33.3 uses only column data of the data frame for calculation. Example 33.4 uses the same input data as Example 33.3 but displays data in table form, with both row values and column values of the data frame.

Example 33.4

```
USE AdventureWorks;
DECLARE @pscript NVARCHAR(MAX);
SET @pscript = N'
df1 = InputDataSet
df2 = df1.groupby("Units", as_index=False).sum()
OutputDataSet = df2';
DECLARE @select NVARCHAR(MAX);
SET @select = N'
SELECT CAST(h.subtotal AS FLOAT)  AS Total, v.UnitMeasureCode AS Units
  FROM Purchasing.PurchaseOrderHeader h
    INNER JOIN Purchasing.PurchaseOrderDetail d
                ON h.PurchaseOrderID = d.PurchaseOrderID
    INNER JOIN Purchasing.ProductVendor v ON d.ProductID=v.ProductID';
 EXEC sp_execute_external_script
    @language = N'Python',
    @script = @pscript,
    @input_data_1 = @select
     WITH RESULT SETS((UnitCodes NVARCHAR(50), TotalSales MONEY));
```

First, the SELECT list of the query, assigned to the **@select** variable, contains two columns with the aliases **Total** and **Units**, respectively. In the following code,

```
df1 = InputDataSet
df2 = df1.groupby("Units", as_index=False).sum()
OutputDataSet = df2
```

the first line passes the input dataset to the **df1** data frame. In the second line, the **groupby** function is applied to the values of the unit measure codes and, for each distinct unit, the sum of all subtotals (the values of the **subtotal** column of the **PurchaseOrderHeader** table) is calculated. Finally, the content of the **df2** data frame is passed to the output data.

The result is

	UnitCodes	TotalSales
1	CAN	9609758.613
2	CS	584962.963
3	CTN	4408404.588
4	DZ	958018.1625
5	EA	267092933.073
6	GAL	7526088.36
7	PAK	206096.94

One more feature of the query in Example 33.4 requires further explanation. Note that the SELECT list of the query uses the CAST operator to select values from the **subtotal** column of the **PurchaseOrderHeader** table and to convert them into the values with the data type FLOAT. This is necessary, because if you do not use casting, the system displays an error with a message similar to the following: "Unsupported input data type in column 'total'. Supported types: bit, tinyint, smallint, int, bigint, uniqueidentifier, real, float, char, varchar, nchar, nvarchar, varbinary."

The reason for this error message is that the set of standard data types in Python is significantly smaller than the corresponding set for Transact-SQL. For this reason, the Python system cannot convert each Transact-SQL data type into an appropriate Python type. One example of such a data type is MONEY, which is the data type of the **subtotal** column. In such cases, you have to use the CAST operator and to explicitly convert the column's data type into a data type that is supported by Python.

Data Visualization with Python

As mentioned earlier in the chapter, one of the advantages of Python is that it supports a large number of open source libraries, which you can use for many different purposes. One of these libraries, **matplotlib**, provides extensive support to display data in different graphic forms, such as histograms, bar charts, and pie charts.

As with the R language, there are two general ways to integrate Python graphics with SQL Server:

- Output a dataset by using the **sp_execute_external_script** system stored procedure and apply one of many Python packages that support data visualization. In other words, you do data visualization *inside* Python.

- Integrate Python into a SQL Server tool, such as Power BI, and use its functions to visualize data.

NOTE Data visualization using Power BI in relation to Python is very similar to the same process using the R language. Please read the section "Integrate R in Power BI Desktop" in Chapter 32 for full details.

Example 33.5 uses the result set generated in Example 33.4 and displays it graphically using a bar chart.

NOTE When executing Example 32.5, you might get an error similar to this:
Msg 39004, Level 16, State 20, Line 448 A 'R' script error occurred during execution of 'sp_execute_external_script' with HRESULT.
This error is related to the SQL Server Launchpad service. This service is used to start Advanced Analytics Extensions processes, which are necessary for integration of the R system and Python with the Database Engine. The program in Example 33.5 writes the file in the C:\temp directory, which is by default not accessible by the Launchpad service. You can either grant access to this directory or use another directory; for instance, C:\PythonScripts. To grant access to a directory, right-click that directory, select Grant Access To, and click Add Everyone. This also applies later to Example 33.8.

Example 33.5

```
Use AdventureWorks;
DECLARE @pscript NVARCHAR(MAX);
SET @pscript = N'
import matplotlib
matplotlib.use("PDF")
import matplotlib.pyplot as plt
df1 = InputDataSet
df2 = df1.groupby("Units", as_index=True).sum()
pt = df2.plot.barh()
# Set title
pt.set_title (label="Total Purchases per Unit Code", y=1.1)
# Set labels for x and y axes
pt.set_xlabel("Purchase Amounts")
pt.set_ylabel("Unit Measure Codes")
# Set names for all items of Unit Code
pt.set_yticklabels (labels=df2.index, fontsize=8, color="green")
# Save bar chart to .pdf file
plt.savefig("c:\\temp\\Figure33_1.pdf",  bbox_inches="tight")';
DECLARE @sql NVARCHAR(MAX);
 SET  @sql = N'
SELECT CAST(h.subtotal AS FLOAT)  AS Total, v.UnitMeasureCode AS Units
   FROM Purchasing.PurchaseOrderHeader h
     INNER JOIN Purchasing.PurchaseOrderDetail d
                 ON h.PurchaseOrderID = d.PurchaseOrderID
     INNER JOIN Purchasing.ProductVendor v ON d.ProductID=v.ProductID';
```

```
EXEC sp_execute_external_script
  @language = N'Python',
  @script = @pscript,
  @input_data_1 = @sql;
GO
```

First, the following lines of code from Example 33.5 are related to the visualization of data:

```
import matplotlib
matplotlib.use("PDF")
import matplotlib.pyplot as plt
```

The first line imports the **matplotlib** library into the script. The second line applies the **use** function to the library to specify the format of the output file. This example uses the .pdf format to store the bar chart, but in practice you can choose among the many formats that are supported. The third line of the code imports the **pyplot** plotting framework that is included in the **matplotlib** library. You can use this framework when you intend to do simple plotting.

After executing the preceding three lines of code, all necessary information regarding the libraries is provided, and you can start setting up the data frame used in the script, as previously described in relation to Example 33.4.

After setup of the data frame, the following lines of code describe several properties of the bar chart:

```
pt = df2.plot.barh()
pt.set_title (label="Total Purchases per Unit Code", y=1.1)
pt.set_xlabel("Purchase Amounts")
pt.set_ylabel("Unit Measure Codes")
```

First, the **barh** function uses the modified input data and generates a graphic that contains a bar chart, which is then saved to the **pt** variable. After that, several properties of the bar chart are specified using subplots. One of the important features of the **matplotlib** library is that you can add multiple plots, called *subplots*, within a graphic. These subplots are helpful when you want to show different data presentations in a single view. So, the final three lines of code are regarded as the definition of subplots: the **set_title** function specifies the title of the bar chart, while **set_xlabel** and **set_ylabel** define the names of the X and Y coordinates, respectively. The second parameter of the **set_title** function specifies the font size.

After the properties of the bar chart are configured, the following line of code sets the tick labels in the bar chart:

```
pt.set_yticklabels (labels=df2.index, fontsize = 8, color="green")
```

The **set_yticklabels** function is used to adjust the tick labels in relation to the Y axis. The first argument of the function, **labels**, specifies which column will be used as the index of the result. Therefore, **labels = df2.index** specifies that the values of the **UnitMeasureCode** column will be used as the names of the tick labels on the Y axis in the bar chart.

NOTE The **labels** parameter is related to the **as_index** option of the **groupby** function (see Example 33.3). The use of the **labels** parameter implicitly specifies that an existing column will be the index of the output. Therefore, this is equivalent to the specification of **as_index=True**.

The next line of code in Example 33,

```
plt.savefig("c:\\temp\\Figure33_1.pdf",  bbox_inches="tight")';
```

uses the **savefig** function to store the graphic on a disk. (This function belongs to the **pyplot** framework.) The first parameter of the function specifies the file in which the corresponding bar chart will be stored. (Note that you have to escape the backslashes in the file path by doubling them and, as mentioned earlier, change the path name if your system does not have the **C:\temp** directory.)

The second argument of the **savefig** function, **bbox_inches**, is used to control how many whitespace characters are generated around the displayed graphics. The default value for **bbox_inches** is **None**, meaning that no action will take place. (The other meaningful value is **tight**, which removes unnecessary whitespace characters.) Figure 33-1 shows the output of Example 33.5.

NOTE If you want to create a visualization other than a bar chart, simply call another function by specifying the data frame and the name of the new function. For instance, if you want to display the data in Example 33.5 as a histogram, use the **hist** function instead of the **barh** function.

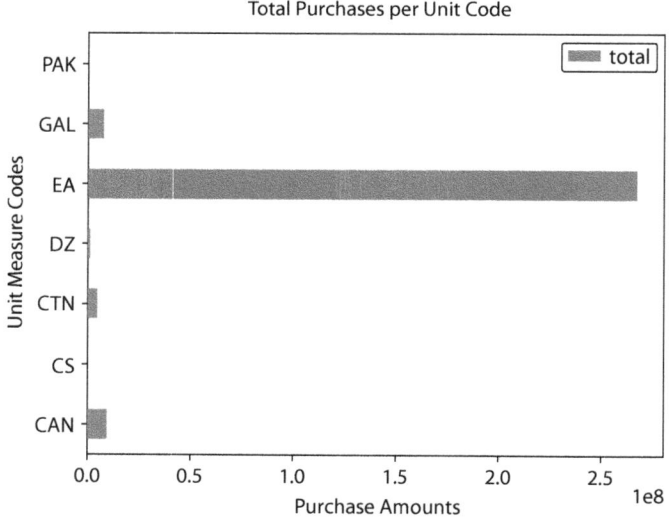

Figure 33-1 The output of Example 33.5

Predictive Modeling with Python in SQL Server

As briefly introduced in Chapter 32, predictive analytics comprises a variety of statistical techniques from machine learning that analyze current and historical data to make predictions about future events. In the business sector, for example, predictive models analyze patterns found in historical and transactional data to identify possible risks and/or opportunities. To do this, a predictive model examines relationships among several parameters. After the analysis, the predictive model presents the various scenarios for evaluation.

Solving Linear Regression Problems Using Python

Linear regression is one of the simplest techniques used in predictive modeling. For this reason, this section presents an example of using linear regression to explain predictive modeling with Python.

NOTE See the Chapter 32 section "Solving Linear Regression Problems with R" for an explanation of the two forms of regression analysis, linear and nonlinear.

For the following example, assume that for each given value, the corresponding value is measured. Measurements in real life, such as this one, usually do not have perfect linear relationships between the values of the independent variable and the values of the dependent variable. For this reason, the goal of linear regression is to find the straight line that best fits the given values (called the *best-fit line*). In other words, linear regression identifies the equation that produces the smallest difference between all the observed values and their fitted values.

The table created in Example 33.6 will be used to show how linear regression can be calculated using Python.

Example 33.6

```
USE sample;
CREATE TABLE Measures (x_value INT, y_value DEC (6,2));
INSERT INTO Measures VALUES (1,   33.5);
INSERT INTO Measures VALUES (2,   35.9);
INSERT INTO Measures VALUES (3,   37.9);
INSERT INTO Measures VALUES (4,   39.8);
INSERT INTO Measures VALUES (5,   41.6);
INSERT INTO Measures VALUES (6,   45.4);
INSERT INTO Measures VALUES (7,   44.6);
INSERT INTO Measures VALUES (8,   47.4);
INSERT INTO Measures VALUES (9,   48.2);
INSERT INTO Measures VALUES (10, 50.3);
```

The ten measured values are stored in the **Measures** table created at the beginning of Example 33.6. Our task is to find linear regression—the formula for the corresponding straight line to which all the given points are at a minimum distance. Example 33.7 calculates the line and displays its formula, together with some other parameters.

Example 33.7

```
USE sample;
EXEC sp_execute_external_script
       @language = N'Python' , @script = N'
     from revoscalepy import rx_lin_mod, rx_predict
linearmodel=rx_lin_mod(formula ="Y_Value~X_Value",data=InputDataSet);
print(linearmodel.summary())',
@input_data_1 = N'SELECT  x_value  AS X_Value,
        CAST (y_value AS FLOAT) AS Y_Value FROM Measures'
```

Let's take a look at the following three lines of code in Example 33.7:

```
from revoscalepy import rx_lin_mod, rx_predict
linearmodel=rx_lin_mod(formula="Y_Value~X_Value",data=InputDataSet);
print(linearmodel.summary())
```

The first line imports the **rx_lin_mod** and **rx_predict** functions from the **revoscalepy** module. (The **revoscalepy** module is a collection of Python functions that you can use for statistics-related tasks, such as linear models, regression, and classification.) The second line incorporates the **rx_lin_mod** function, which is used to fit linear models on small or large data sets. The two most important parameters of the function are **formula** and **data**. The former specifies which statistical model is used. The formula

```
"Y_Value ~ X_Value"
```

defines that the alphanumerical string on the right side of the ~ operator is the name of the independent variable, while the string on the left side is the name of the dependent variable. The **data** argument specifies the input data. The last line of the code prints the summary for the generated model.

After execution of this script, the result looks similar to the following (several lines of the result have been omitted):

```
Linear regression Results for: Y_Value ~ X_Value
Dependent variable(s): ['Y_Value']
Total independent variables: 2
Number of valid observations: 10
Number of missing observations: 0
(Intercept)   (Intercept)   32.360000
X_Value             X_Value   1.836364
Residual standard error: 0.8482 on 8.0 degrees of freedom
Multiple R-squared:0.9797
```

As you already know, when applying regression analysis, the program calculates the straight line that is generated so that all the data points are at a minimum distance from the line. Generally, the formula for the straight line is $y = a + bx$, where a is the y-intercept and b is the slope of the particular straight line. Therefore, by applying the algorithm for linear regression, we derive the values of the coefficients *a* and *b*. As you can see from the result of Example 33.7, the y-intercept of the straight line is 32.36 and the corresponding slope is 1.83.

Multiple R-squared (the last line of output) evaluates the scatter of the data points around the fitted regression line. It is also called the *coefficient of determination*. For the same data set, higher R-squared (aka R^2) values represent smaller differences between the observed data and the fitted values. R-squared is always between 0 percent and 100 percent. (The larger the value of multiple R-squared, the better the regression model fits your observations.) Therefore, our straight line is a very good choice, because the corresponding R^2 value is almost 98 percent.

Example 33.8 uses the measurements from the **Measures** table to show how data is plotted.

Example 33.8

```
USE sample;
EXEC sp_execute_external_script
      @language = N'Python'
   , @script = N'
#Importing Packages
import matplotlib
matplotlib.use("PDF")
from revoscalepy import rx_lin_mod, rx_predict
import matplotlib.pyplot as plt
import pandas as pd
linearmodel = rx_lin_mod(formula = "Y_Value ~ X_Value", data =
InputDataSet);
df = InputDataSet
plt.scatter(df.X_Value,df.Y_Value)
plt.xlabel("Values of Independent Variable ")
plt.ylabel("Values of Dependent Variable")
#plt.title("Graphical Output of Example 33.8")
plt.plot()
plt.savefig("C:\\temp\\Figure33_2.png") ',
@input_data_1 = N'SELECT x_value  AS X_Value,
        CAST (y_value AS FLOAT) AS Y_Value FROM dbo.Measures'
```

Three lines of code in Example 33.8 merit further explanation:

```
plt.scatter(df.X_Value,df.Y_Value)
plt.plot()
plt.savefig("C:\\temp\\Figure33_2.png") ',
```

The **scatter** function of the **matplotlib.pyplot** framework generates a scatter plot. This function has several parameters, but only the first two have to be specified. These two parameters specify the X and Y coordinates of data points, respectively. The **plot** function in the second line creates the corresponding scatter plot, shown in Figure 33-2.

Finally, the **savefig** function saves the figure. Note that the **use** function of the **matplotlib** library in Example 33.8 specifies the .pdf format as the format of the output file. This value can

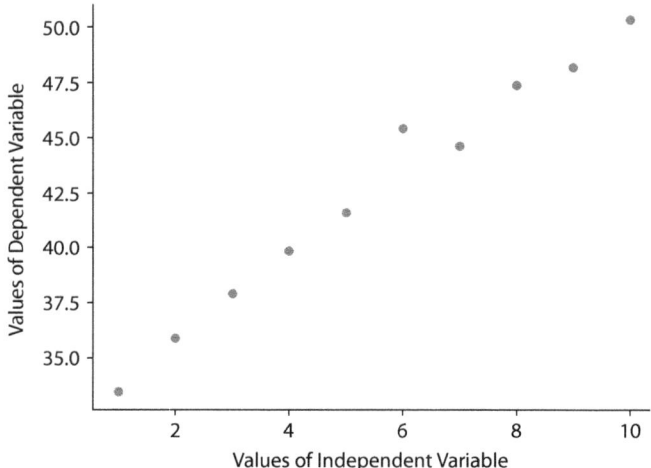

Figure 33-2   The graphical output of Example 33.8

be modified afterwards, using the **savefig** function. As you can see in the example, the format of the scatter plot is modified and stored as a .png file.

Summary

The Python programming language has the following general properties:

- Simple syntax
- High readability
- A large collection of libraries

Python aims to be simple in the design of its syntax, encapsulated in the slogan "There should be one *obvious* way to do it." (For instance, another related programming language, Perl, is known for its complexity in relation to syntax.) Python's design philosophy emphasizes code readability and the use of significant whitespace. With its extensive set of open source libraries, Python is very popular as a programming language for machine learning.

The integration between the Database Engine on one side and Python on the other is done using the **sp_execute_external_script** system stored procedure.

APPENDIX | Exercise Solutions

This appendix includes the solutions to the end-of-chapter exercises found in this guide. The solution number matches the exercise number found in the chapter. Not all chapters include exercises, as noted in the text.

Chapter 1   Relational Database Systems: An Introduction

S.1.1   *Data independence* means that database application programs are not dependent on the physical structure of the stored data in a database (physical data independence) and that database application programs are independent of the logical structure of the database (logical data independence).

S.1.2   The main concept of the relational model is a *relation*, commonly known as a table. The whole relational model consists of only tables with one or more columns and zero or more rows. At every position in the table there is only one data value.

S.1.3   The **employee** table represents an entity, while the data for Ann Jones specifies an instance of the entity.

S.1.4   The **works_on** table represents the relationship between employees and projects. The difference between the **works_on** table and the other tables of the **sample** database is that the **works_on** table shows the relationship between two entities, while each of the other tables represents an entity.

S.1.5

 a. Yes, **title** is a key of the table because it is unique.

 b. Yes, **isbn** functionally depends on **title** because each title uniquely determines the corresponding ISBN.

 c. Yes, the **book** table is in third normal form; there are no dependencies between the non-key attributes, because there is just one attribute.

S.1.6

    a. Yes, **order_no** is a key of the **order** table because the other columns, **customer_no** and **discount**, are dependent on it.

    b. No, **customer_no** is not a key of the table because of its functional dependency on the **order_no** column.

S.1.7   The **company** table is not in any normal form, because the **location** column is a multivalued attribute, which can't exist in 1NF.

S.1.8

    a. The **supplier** table is in 1NF, because the **city** column is functionally dependent on the **supplier_no** column, which is the partial key of the table.

    b. The existing functional dependencies can be resolved by using the following two tables:

       • supplier1 (supplier_no, article)

       • supplier2 (supplier_no, city)

S.1.9   The relation R is in 1NF, because the column **C** is dependent on the column **B**, which is the partial key of the table.

S.1.10   The relation R is in 3NF, but its design is not well done. (The primary key should be the combination of the columns **A** and **C**.)

Chapter 2   Planning the Installation and Installing SQL Server

There are no end-of-chapter exercises in this chapter.

Chapter 3   Front-End Tools for the Database Engine

S.3.1   To create the **test** database, in Object Explorer, expand the server, right-click Databases, and select New Database. In the New Database dialog box, type **test** in the Database Name field, click the ellipsis button in the Autogrowth/Maxsize column to open the Change Autogrowth dialog box, choose the In Megabytes radio button under File Growth and change the value to **2**. After that, choose the Limited To (MB) radio button under Maximum File Size and change the value to **20**, and then click OK to return to the New Database dialog box? In the Path column, change the path to **C:\tmp**. Change the value in the Initial Size column to **10MB**. Then click OK.

S.3.2   Select the **test** database in the Database folder of your instance. Right-click it and in the **test_log** row make the changes in the same way as you made the changes for the database properties in Exercise 3.1.

S.3.3   In the Properties dialog box of the **test** database, choose the Options page and set the Restrict Access option to SINGLE_USER. In this case, both users cannot use the database at the same time.

S.3.4   Right-click the Tables folder of the **sample** database. Select New Table and enter the name of the first table (**department**). Then enter the column names with the corresponding data types. Proceed in the same way with all four tables.

S.3.5   Click the **AdventureWorks** database in the Databases folder and select Tables. All user tables of the database are now shown in the detail pane. Double-click the **Person.Address** table to display its properties. (If you do not have this database, download it as described in the introduction of this book.)

S.3.6   Either type and execute the Transact-SQL statement **USE test** in the editor, or select **test** from the drop-down menu in the tool bar of SSMS.

S.3.7   To execute the statement, open Query Editor, select the **AdventureWorks** database from the drop-down menu, enter the statement, and press F5. To stop the current execution, click the red stop button.

S.3.8   Choose File | Save Query As. Then change the destination directory, enter the new filename (**createdb**), and click OK.

S.3.9   Choose Query in the main menu of SQL Server Management Studio, click Results To, and select Results to Text.

Chapter 4   SQL Components

S.4.1   The difference is in storage length and range of possible values. TINYINT is stored in 1 byte with a value range from 0 to 255. SMALLINT is stored in 2 bytes with a value range from −32768 to 32767. INT is stored in 4 bytes with a value range from −2147483648 to 2147483647.

S.4.2   CHAR is a string that can store up to 8000 characters. It stores the number of characters declared by the user, and the rest of the string is padded with blanks. VARCHAR also can store up to 8000 characters. However, the strings are stored with their actual length. Use CHAR when the data values in a column are expected to be approximately of the same size. Use VARCHAR when the data values in a column are expected to vary considerably in size.

S.4.3   To set the format of a column with the DATE data type so that its values can be entered in the form 'yyyy/mm/dd', use the following statement:

```
SET DATEFORMAT ymd
```

S.4.4   The following is used to find the ID number of the **test** database:

```
SELECT DB_ID('test');
```

S.4.5   The following system variables are used to display the current version of the database system software and the language that is used by this software:

```
SELECT @@VERSION, @@LANGUAGE
```

S.4.6
The result of (11100101) & (01010111) is (01000101).
The result of (10011011) | (11001001) is (11011111).
The result of (10110111) ^ (10110001) is (00000110).

S.4.7

A + NULL	The result is NULL, independent of A.
NULL = NULL	The result is NULL.
B OR NULL	The result is true if B is true, otherwise NULL.
B AND NULL	The result is false if B is false, otherwise NULL.

S.4.8 You can use both single and double quotation marks to define string and temporal constants by setting QUOTED_IDENTIFIER to OFF.

S.4.9 Delimited identifiers (also known as quoted identifiers) are a special kind of identifier that allow the use of reserved keywords as identifiers and allow spaces in the names of database objects, such as table names. Delimited identifiers can contain, or begin with, any character (other than the delimiters themselves), are enclosed in double quotation marks, and are case sensitive.

Chapter 5   Data Definition Language

S.5.1

```
-- Create first the C:\tmp directory, if necessary
USE master;
GO
CREATE DATABASE test_db
          ON (NAME = test_db_dat,
          FILENAME='C:\tmp\test_db.mdf',
          SIZE = 5, MAXSIZE = UNLIMITED, FILEGROWTH = 8%)
          LOG ON
            (NAME=test_db_log,
          FILENAME = 'C:\tmp\test_db_log.ldf',
          SIZE = 2, MAXSIZE = 10, FILEGROWTH = 500KB);
```

S.5.2

```
USE master;
ALTER DATABASE test_db
     ADD LOG FILE (NAME=emp_log1,
                      FILENAME='C:\tmp\test_db.ldf',
                      SIZE=2, MAXSIZE=UNLIMITED, FILEGROWTH=2)
```

S.5.3

```
USE master;
ALTER DATABASE test_db
          MODIFY FILE
            (NAME = test_db_dat, SIZE = 10MB)
```

S.5.4 The NOT NULL specification is necessary for all columns that are part of the primary key.

S.5.5 dept_no and **project_no** are defined as CHAR values because they may contain alphanumerical values.

S.5.6
```
USE test_db;
CREATE TABLE customers (customerid CHAR(5) NOT NULL,
                        companyName VARCHAR(40) NOT NULL,
                        contactName CHAR(30) NULL,
                        address VARCHAR(60) NULL,
                        city CHAR(15) NULL,
                        phone CHAR(24) NULL,
                        fax CHAR(24) NULL);
CREATE TABLE orders (orderid INTEGER NOT NULL,
                     customerid CHAR(5) NOT NULL,
                     orderdate DATE NULL,
                     shippeddate DATE NULL,
                     freight MONEY NULL,
                     shipname VARCHAR(40) NULL,
                     shipaddress VARCHAR(60) NULL,
                     quantity INTEGER NULL);
```

S.5.7
```
ALTER TABLE orders
        ADD shipregion INTEGER NULL;
```

S.5.8
```
ALTER TABLE orders
        ALTER COLUMN shipregion CHAR(8) NULL;
```

S.5.9
```
ALTER TABLE orders
        DROP COLUMN shipregion;
```

S.5.10   After deleting a table with the DROP TABLE statement, all data, indices, and triggers belonging to the table are also dropped. In contrast to that, all views that are defined using the table are not removed.

S.5.11
```
USE test_db;
DROP TABLE orders;
DROP TABLE customers;
CREATE TABLE customers (customerid CHAR(5) NOT NULL
        CONSTRAINT prim_cust PRIMARY KEY,
        companyName VARCHAR(40) NOT NULL,
        contactName CHAR(30) NULL,
        address VARCHAR(60) NULL,
```

```
                     city CHAR(15) NULL,
                     phone CHAR(24) NULL,
                     fax CHAR(24) NULL);
CREATE TABLE orders (orderid INTEGER NOT NULL,
                     customerid CHAR(5) NOT NULL,
                     orderdate DATE NULL,
                     shippeddate DATE NULL,
                     freight MONEY NULL,
                     shipname VARCHAR(40) NULL,
                     shipaddress VARCHAR(60) NULL,
                     quantity INTEGER NULL,
                     CONSTRAINT prim_ord PRIMARY KEY(orderid),
                     CONSTRAINT foreign_orders FOREIGN KEY(customerid) REFERENCES
customers(customerid));
```

S.5.12   It is not possible to insert the row, because of the referential constraint enforced in the CREATE TABLE statement (see S.5.11). The system prints the following message:

> The INSERT statement conflicted with the FOREIGN KEY constraint "foreign_orders".
> The conflict occurred in database "test_db", table "dbo.customers", column 'customerid'.

S.5.13
```
USE test_db;
ALTER TABLE orders
          ALTER COLUMN  orderdate DATE NULL;
ALTER TABLE orders
              ADD CONSTRAINT AddDateDflt DEFAULT getdate() FOR orderdate;
```

S.5.14
```
USE test_db;
ALTER TABLE orders
          ADD CONSTRAINT limit_qu
          CHECK (quantity BETWEEN 1 AND 30);
```

S.5.15   There are several ways to display integrity constraints of a table. The easiest way is to use the **sp_helpconstraint** system procedure:

```
sp_helpconstraint orders
```

S.5.16
```
USE test_db;
ALTER TABLE customers
              DROP CONSTRAINT prim_cust;
```

That statement will not work, because the primary key constraint **prim_cust** is referenced by the foreign key constraint defined in the **orders** table.

S.5.17
```
USE test_db;
ALTER TABLE employee
        DROP CONSTRAINT prim_empl;
```

S.5.18
```
EXEC sp_rename 'customers.city', town;
```

Chapter 6   Queries

S.6.1
```
USE sample;
SELECT *
      FROM works_on;
```

S.6.2
```
USE sample;
SELECT emp_no
        FROM works_on
        WHERE Job = 'Clerk';
```

S.6.3
```
USE sample;
SELECT emp_no
        FROM works_on
        WHERE project_no = 'p2'
          AND emp_no < 10000;
```

or:

```
USE sample;
SELECT emp_no
        FROM works_on
        WHERE project_no = 'p2'
          AND emp_no BETWEEN 1 AND 9999;
```

S.6.4
```
USE sample;
SELECT emp_no
    FROM works_on
  WHERE enter_date NOT BETWEEN
  '01.01.2017' AND '12.31.2017';
```

S.6.5
```
USE sample;
SELECT emp_no
```

```
        FROM works_on
        WHERE project_no = 'p1'
        AND (job = 'Manager' OR job = 'Analyst');
```

S.6.6
```
USE sample;
SELECT enter_date
        FROM works_on
         WHERE project_no = 'p2'
        AND Job IS NULL;
```

S.6.7
```
USE sample;
SELECT emp_no, emp_lname
        FROM employee
        WHERE emp_fname LIKE '%t%t%';
```

S.6.8
```
USE sample;
SELECT emp_no, emp_fname
        FROM employee
        WHERE emp_lname LIKE '_[ao]%es';
```

S.6.9
```
USE sample;
SELECT emp_no
        FROM employee
      WHERE dept_no =
              (SELECT dept_no FROM department
                        WHERE location = 'Seattle');
```

S.6.10
```
USE sample;
SELECT emp_lname, emp_fname
          FROM employee
           WHERE emp_no IN
                (SELECT emp_no
                        FROM works_on
                        WHERE enter_date = '04/01/2017');
```

S.6.11
```
USE sample;
SELECT location
        FROM department
        GROUP BY location;
```

S.6.12 There is no significant difference. If you use the GROUP BY clause without any additional specifications (aggregates, HAVING clause), it is exactly like DISTINCT. (It divides a table into groups and returns one row for each group.)

S.6.13 The GROUP BY clause puts all NULL values in one group. This is not exactly in accordance with the fact that each NULL value is a value per se, and it cannot be compared with other NULL values.

S.6.14 COUNT(expression) takes an argument (i.e., a column or expression) and displays all non-null occurrences of that argument. COUNT(*) counts all rows, whether or not any particular column contains a NULL value.

S.6.15
```
USE sample;
SELECT MAX(emp_no)
        FROM employee;
```

S.6.16
```
USE sample;
SELECT job
        FROM works_on
        GROUP BY job
        HAVING COUNT(*) > 2;
```

S.6.17
```
USE sample;
SELECT DISTINCT emp_no
        FROM works_on
        WHERE (Job = 'Clerk' OR  emp_no IN
                (SELECT emp_no
                    FROM employee
                    WHERE dept_no='d3'));
```

S.6.18 The inner SELECT statement can be used in conjunction with a comparison operator, such as =, if its result set has a maximum of one row (in this case = may be used). The result of the SELECT statement in E.6.18 has more than one row. Therefore, the comparison operator = has to be replaced with the set operator called IN.

The correct syntax form is

```
USE sample;
SELECT project_name
    FROM project
    WHERE project_no IN
        (SELECT project_no FROM works_on WHERE Job = 'Clerk');
```

S.6.19 Temporary tables can be used to store the intermediate result of a complex query.

S.6.20 Local temporary tables are removed at the end of the current session, while global temporary tables are removed at the end of the session that created the table.

S.6.21

a.   Natural join:

```
USE sample;
SELECT project.*, emp_no, job, enter_date
      FROM project JOIN works_on
      ON project.project_no = works_on.project_no;
```

b.   Cartesian product:

```
SELECT *
    FROM project CROSS JOIN works_on;
```

S.6.22   To join several tables in a query, you need at least $n - 1$ join conditions.

S.6.23

```
USE sample;
SELECT emp_no, job
        FROM works_on JOIN project
        ON works_on.project_no = project.project_no
        WHERE project_name = 'Gemini';
```

S.6.24

```
USE sample;
SELECT emp_fname, emp_lname
        FROM employee JOIN department
        ON employee.dept_no = department.dept_no
        WHERE (dept_name = 'Research' OR dept_name = 'Accounting');
```

S.6.25

```
USE sample;
SELECT enter_date
        FROM works_on JOIN employee
        ON works_on.emp_no = employee.emp_no
        WHERE job = 'Clerk'
        AND dept_no = 'd1';
```

S.6.26

```
USE sample;
SELECT project_name
        FROM project
        WHERE project_no IN
        (SELECT project_no
            FROM works_on
            WHERE Job = 'Clerk'
```

```
                    GROUP BY project_no
                    HAVING COUNT(*) > 1);
```

S.6.27

```
USE sample;
SELECT emp_fname, emp_lname
        FROM employee
        JOIN works_on ON employee.emp_no = works_on.emp_no
        JOIN project ON works_on.project_no = project.project_no
        WHERE project_name = 'Mercury'
        AND job = 'Manager';
```

S.6.28

```
USE sample;
SELECT emp_fname, emp_lname
        FROM employee
        WHERE emp_no IN
        (SELECT a.emp_no
                    FROM works_on a, works_on b
                    WHERE  b.enter_date=a.enter_date
                    AND a.emp_no != b.emp_no);
```

S.6.29

```
USE sample;
SELECT a.emp_no
          FROM employee_enh a, employee_enh b
        WHERE a.domicile = b.domicili
         AND a.dept_no = b.dept_no
        AND a.emp_no != b.emp_no;
```

S.6.30

a.   With the join operator:

```
-- Using subquery
USE sample;
SELECT emp_no
        FROM employee
        WHERE dept_no =
          (SELECT dept_no
                FROM department
                WHERE dept_name = 'Marketing');
```

b.   With the correlated subquery:

```
-- Using the JOIN operator
SELECT emp_no
        FROM employee
        JOIN department
```

```
              ON employee.dept_no = department.dept_no
              WHERE dept_name = 'Marketing';
```

Chapter 7   Modification of a Table's Contents

S.7.1
```
USE sample;
INSERT INTO employee values(11111,'Julia','Long',NULL);
```

S.7.2
```
USE sample;
CREATE TABLE emp_d1_d2   (emp_no INTEGER NOT NULL,
                          emp_fname CHAR(20) NOT NULL,
                          emp_lname CHAR(20) NOT NULL,
                          dept_no CHAR(4) NULL);
INSERT INTO emp_d1_d2
SELECT emp_no, emp_fname, emp_lname, dept_no
          FROM employee
          WHERE dept_no IN ('d1', 'd2');
```

or:

```
USE sample;
SELECT emp_no, emp_fname, emp_lname, dept_no
    INTO emp_d1_d2
    FROM employee
   WHERE dept_no IN ('d1', 'd2');
```

S.7.3
```
USE sample;
CREATE TABLE employee_three   (emp_no INTEGER NOT NULL,
                          emp_fname CHAR(20) NOT NULL,
                          emp_lname CHAR(20) NOT NULL,
                          dept_no CHAR(4) NULL);
INSERT INTO employee_three(emp_no, emp_fname, emp_lname, dept_no)
              SELECT emp_no, emp_fname, emp_lname, dept_no
                    FROM employee
                    WHERE emp_no IN
                    (SELECT emp_no FROM works_on
                              WHERE enter_date BETWEEN '01.01.2017' AND
'12.31.2017');
```

S.7.4
```
USE sample;
UPDATE works_on
        SET job = 'Clerk'
          WHERE job = 'Manager' AND project_no = 'p1';
```

S.7.5
```
USE sample;
UPDATE project
        SET budget = NULL;
```

S.7.6
```
USE sample;
UPDATE works_on
        SET job = 'Manager'
         WHERE emp_no = 28559;
```

S.7.7
```
USE sample;
UPDATE project
        SET budget = budget/10+budget
         WHERE project_no IN(
            SELECT project_no FROM works_on
            WHERE job='Manager' AND emp_no=10102);
```

S.7.8
```
USE sample;
UPDATE department
SET dept_name='Sales'
WHERE dept_no =
(SELECT dept_no FROM employee
        WHERE emp_lname='James');
```

S.7.9
```
USE sample;
UPDATE works_on
        SET enter_date='12.12.2017'
        WHERE project_no='p1' AND emp_no IN (
                    SELECT emp_no FROM employee
                JOIN department ON employee.dept_no = department.dept_no
                WHERE dept_name='Sales');
```

S.7.10
```
USE sample;
DELETE FROM department
    WHERE location = 'Seattle';
```

S.7.11
```
USE sample;
DELETE FROM works_on
                WHERE project_no = 'p3';
DELETE FROM project
                WHERE project_no = 'p3';
```

S.7.12

```
USE sample;
DELETE FROM works_on
WHERE emp_no IN (SELECT emp_no FROM employee
                     WHERE dept_no IN (SELECT dept_no
                        FROM department WHERE location ='Dallas'));
```

Chapter 8   Stored Procedures and User-Defined Functions

S.8.1

```
-- This procedure inserts 3000 rows in the employee table
USE sample;
declare @i integer
declare @first_name char(20)
declare @last_name char(20)
declare @department char(4)
set @i = 1
set @first_name = 'Jane'
set @last_name = 'Smith'
set @department = 'd1'
while @i < 3001
begin
insert into employee
   values (@i, @first_name, @last_name, @department)
set @i = @i+1
end
```

S.8.2

```
DECLARE @i INT
DECLARE @emp_no INT
SET @i = 0
SET @emp_no = (CONVERT(INT, (RAND() * 10000)))
WHILE @i < 1000
BEGIN
   WHILE (SELECT COUNT(*) FROM employee WHERE emp_no = @emp_no) > 0
   BEGIN
      SET @emp_no = (CONVERT(INT, (RAND() * 100000)))
   END
   INSERT INTO employee VALUES(@emp_no, 'Jane', 'Smith', 'd1')
   SET @i = @i + 1
END
```

Chapter 9   System Catalog

S.9.1

```
USE master;
SELECT filename from sys.sysdatabases
   WHERE name = 'sample';
```

S.9.2

```
USE sample;
SELECT COUNT(*)
  FROM sys.sysconstraints, sys.sysobjects
  WHERE sys.sysconstraints.id = sys.sysobjects.id
   AND sys.sysobjects.name = 'employee';
```

S.9.3

```
USE sample;
SELECT sysconstraints.status
  FROM syscolumns, sysobjects, sysconstraints
  WHERE syscolumns.id = sysconstraints.colid
  AND sysobjects.id = sysconstraints.id
  AND sysobjects.name = 'employee'
   AND syscolumns.name = 'dept_no';
```

S.9.4

```
USE AdventureWorks;
SELECT table_name from information_schema.tables
WHERE table_type = 'BASE TABLE';
```

S.9.5

```
USE sample;
SELECT column_name, data_type, ordinal_position
      FROM information_schema.columns
      WHERE table_name = 'employee';
```

Chapter 10   Indices

S.10.1

```
USE sample;
CREATE INDEX i_enterdate
     ON works_on(enter_date)
     WITH FILLFACTOR = 60;
```

S.10.2

```
USE sample;
CREATE UNIQUE INDEX i_lfname
     ON employee (emp_lname, emp_fname);
```

A composite index can be used for index access for the leading part of the index. Therefore, there is a significant difference if you change the order of the columns in a composite index.

S.10.3 An index that is implicitly created for the primary key of a table cannot be dropped using the DROP INDEX statement. It can be dropped only if you drop the constraint (using the ALTER TABLE statement with the DROP CONSTRAINT clause).

S.10.4 During index access, only the rows that satisfy the search criteria of the query are accessed. In most cases, this is an obvious advantage in relation to a table scan, where the system does not use an index. But besides this significant benefit, index access can have two disadvantages: In contrast to a table scan, the Database Engine uses smaller I/O units to read rows for index access; therefore, a number of read operations will be comparatively higher. The second disadvantage of the index access method (using a nonclustered index) is that data pages must be read repeatedly, because the rows to be selected are scattered on data pages.

S.10.5
```
USE sample;
CREATE INDEX i_employee_lname ON employee (emp_lname);
```

S.10.6
```
USE sample;
CREATE INDEX i_emp_name ON employee (emp_lname, emp_fname);
```

S.10.7
```
USE sample;
CREATE INDEX i_workson_empno ON works_on (emp_no);
CREATE INDEX i_employee_empno ON employee (emp_no);
```

S.10.8
```
USE sample;
CREATE INDEX i_department_deptno ON department (dept_no);
CREATE INDEX i_employee_deptno ON employee (dept_no);
CREATE INDEX i_department_deptname ON department (dept_name);
```

Chapter 11   Views

S.11.1
```
USE sample;
GO
CREATE VIEW v_10_1
  AS SELECT *
  FROM employee WHERE dept_no = 'd1';
```

S.11.2
```
USE sample;
GO
CREATE VIEW v_10_2
  AS SELECT project_no, project_name
    FROM project;
```

S.11.3
```
USE sample;
GO
CREATE VIEW v_10_3
  AS SELECT emp_lname, emp_fname
      FROM employee, works_on
      WHERE works_on.emp_no = employee.emp_no
      AND enter_date BETWEEN '06/01/2017' AND '12/31/2017';
```

S.11.4
```
USE sample;
GO
CREATE VIEW v_10_4 (first, last)
  AS SELECT emp_fname, emp_lname
      FROM v_10_3;
```

S.11.5
```
USE sample;
SELECT *
  FROM v_10_1 WHERE emp_lname LIKE 'M%;'
```

S.11.6
```
USE sample;
GO
CREATE VIEW v_10_6
   AS SELECT project.*
      FROM project, employee, works_on
      WHERE project.project_no = works_on.project_no
      AND employee.emp_no = works_on.emp_no
      AND emp_lname = 'Smith';
```

S.11.7
```
USE sample;
GO
ALTER VIEW v_10_1
  AS SELECT *
  FROM employee WHERE dept_no IN( 'd1', 'd2');
```

S.11.8
```
USE sample;
GO
DROP VIEW v_10_3;
```

The DROP VIEW statement also removes the view v_10_4.

S.11.9
```
USE sample;
INSERT INTO v_10_2 VALUES('p2, 'Moon');
```

S.11.10
```
USE sample;
GO
CREATE VIEW v_10_10
  AS SELECT emp_no, emp_fname, emp_lname, dept_no
        FROM employee
        WHERE emp_no  < 10000
        WITH CHECK OPTION;


INSERT INTO v_10_10 VALUES(22123, 'Michael',  'Kohn', 'd3')
-- doesn't work, because the employee number is greater than 10,000
```

S.11.11
```
USE sample;
GO
CREATE VIEW v_10_11
  AS SELECT emp_no, emp_fname, emp_lname, dept_no
        FROM employee
        WHERE emp_no  < 10000;


INSERT INTO v_10_11 VALUES(22123, , 'Michael',  'Kohn', 'd3');
-- works, because the value of the employee column won't be checked
```

S.11.12
```
USE sample;
GO
CREATE VIEW v_10_12
  AS SELECT emp_no, project_no, enter_date, job
      FROM works_on
      WHERE enter_date BETWEEN '01.01.2017' AND '12.31.2018';
      WITH check option
UPDATE v_10_12 SET enter_date = '06/01/2016'
   WHERE emp_no = 29346 AND project_no='p1';
-- doesn't work, because the date does not belong to the years 2017 or
2018.
```

S.11.13
```
USE sample;
GO
CREATE VIEW v_10_13
  AS SELECT emp_no, project_no, enter_date, job
```

```
        FROM works_on
        where enter_date between '01.01.2017' and '12.31.2018';
UPDATE v_10_12 SET enter_date = '06/01/2016'
    WHERE emp_no = 29346 AND project_no='p1';
--  this UPDATE statement works, because the value of the enter_date column
will not be checked
```

Chapter 12   Security System of the Database Engine

S.12.1   In Windows mode the Database Engine exclusively uses Windows user accounts, assuming that they already have been validated at the operating system level (trusted connection). In Mixed mode, there are two security options: SQL Server security and Windows security.

S.12.2   The login is used to allow a certain user to log in to the database system, whereas the user account is used to grant access to a particular database for a certain user or a role.

S.12.3
```
USE sample;
CREATE LOGIN ann WITH PASSWORD = 'a1b2c3d4e5!';
CREATE LOGIN burt WITH PASSWORD = 'd4e3f2g1h0!';
CREATE LOGIN chuck WITH PASSWORD = 'f102gh285!';
USE master;
SELECT name FROM sys.syslogins;
```

S.12.4
```
USE sample;
CREATE USER s_ann FOR LOGIN ann;
CREATE USER s_burt FOR LOGIN burt;
CREATE USER s_charles FOR LOGIN chuck;
```

S.12.5
```
USE sample;
GO
CREATE ROLE managers AUTHORIZATION s_ann;
GO
sp_addrolemember 'managers', 's_ann';
sp_addrolemember 'managers', 's_burt';
GO
sp_addrolemember 'managers', 's_charles';
GO
-- display the information using the sp_helpuser system procedure
EXEC sp_helpuser 'managers'
```

S.12.6
```
USE sample;
GRANT CREATE TABLE TO s_burt;
GRANT CREATE PROCEDURE  TO s_ann;
```

S.12.7
```
USE sample;
GRANT UPDATE ON employee(emp_lname,emp_fname)
            TO s_charles;
```

S.12.8
```
USE sample;
GO
CREATE VIEW readnames
        AS SELECT emp_lname,emp_fname FROM employee;
GO
GRANT SELECT ON readnames
        TO s_burt, s_ann;
```

S.12.9
```
USE sample;
GRANT INSERT ON project
        TO managers;
```

S.12.10
```
USE sample;
REVOKE SELECT ON readnames
            FROM s_burt;
```

S.12.11
```
USE sample;
DENY INSERT ON project
    TO s_ann;
```

S.12.12   The functionality of views in relation to the Transact-SQL statements GRANT, REVOKE, and DENY is limited, because with the former you can restrict only the access to one or more columns and one or more rows. (Using Transact-SQL statements, you can restrict read and write operations on data.)

S.12.13
```
USE sample;
GO
EXEC sp_helpuser s_ann;
```

Chapter 13   Concurrency Control

S.13.1   Transactions are used to keep the data consistent using their "all or nothing" property: either all statements of that particular transaction are (successfully) executed or no one of them is executed.

S.13.2   Unlike a local transaction, a distributed transaction needs a coordinator that coordinates the execution of all transaction parts on different servers. Also, you use the BEGIN TRANSACTION statement to start a local transaction and use the BEGIN DISTRIBUTED TRANSACTION statement to start a distributed transaction.

S.13.3   Each Transact-SQL statement always belongs either implicitly or explicitly to a transaction. When a session operates in implicit transaction mode, selected statements implicitly issue the BEGIN TRANSACTION statement. This means that you do nothing to start such a transaction. An explicit transaction is specified with the pair of statements BEGIN TRANSACTION and COMMIT TRANSACTION (or ROLLBACK TRANSACTION).

S.13.4   There are no locks that are compatible with an exclusive lock.

S.13.5   You can test the successful execution of each T-SQL statement by using the global variable **@@error** or the combination of the TRY and CATCH statements.

S.13.6   The SAVE TRANSACTION statement is used to execute parts of an entire transaction.

S.13.7   The advantage of row-level locking is that it maximizes concurrency because all other rows of the table that are stored on the same physical page can be used by other processes. On the other hand, it increases system overhead because each locked row requires one lock (and you need many more locks if you use row-level locking instead of page-level locking). The advantage of row-level locking is the disadvantage of page-level locking and vice versa.

S.13.8   Using the SET LOCK_TIMEOUT statement, a user can specify whether or not a transaction should wait for a lock to be released. Also, there are several options in the FROM clause of the SELECT statement that can be used by a user to influence locking behavior of the Database Engine, such as TABLOCK, ROWLOCK, and PAGLOCK.

S.13.9   Generally, an intent lock shows an intention to lock the next-lower resource in the hierarchy of the database objects. An intent lock is always placed at a level in the object hierarchy above the level the process intends to lock. An exclusive (X) lock or shared (S) lock is always used for the object that actually should be locked.

S.13.10   Lock escalation is the process of converting many page-, row-, or index-level locks into one table lock.

S.13.11   READ UNCOMMITTED is the simplest isolation level between transactions and therefore allows the maximum of data inconsistency of all isolation levels, which usually is very undesirable and should be used only when the accuracy of the data read is not important or the data is seldom modified. When a transaction retrieves a row at this isolation level, it acquires no locks and respects none of the existing locks. On the other hand, the advantage of this isolation level is that it allows the highest concurrency. SERIALIZABLE is the strongest isolation level, because it prevents all four concurrency problems (lost updates, dirty reads, nonrepeatable reads, and phantoms). It acquires a range lock on all data that is read by the corresponding transaction. The advantage of SERIALIZABLE is that there will be no data inconsistency at all when you apply this isolation level for a process. On the other hand, it decreases concurrency of the processes the most.

S.13.12   A deadlock is a special concurrency problem in which two transactions block the progress of each other. The first transaction has a lock on some database object that the other transaction wants to access, and vice versa. (It is possible that several transactions cause a deadlock, if the first transaction blocks the second, the second the third, and so on, and if the last transaction blocks the first one.)

S.13.13   The Database Engine always chooses the process that closed the loop in a deadlock and rolls it back. Users can use the SET DEADLOCK_PRIORITY statement to choose the "victim" process.

Chapter 14   Triggers

S.14.1

```
USE sample;
GO
CREATE TRIGGER tr_refint_dept
         ON department
         FOR DELETE, UPDATE
         AS
         IF UPDATE(dept_no)
                 BEGIN
                 IF (SELECT COUNT(*)
                         FROM employee, deleted
                         WHERE employee.dept_no = deleted.dept_no) >0
                         BEGIN
                         ROLLBACK TRANSACTION
                         PRINT 'Transaction failed!'
                         END
         ELSE PRINT 'Transaction succeeded'
         END;
GO
CREATE TRIGGER tr_refint_dept2
         ON employee
         FOR INSERT, UPDATE
         AS
         IF UPDATE(dept_no)
                 BEGIN
                 IF (SELECT department.dept_no
                         FROM department, inserted
                         WHERE department.dept_no = inserted.dept_no) IS NULL
                         BEGIN
                         ROLLBACK TRANSACTION
                         PRINT 'Transaction failed!'
                         END
                 ELSE PRINT  'Transaction succeeded'
                 END
```

S.14.2

```
USE sample;
GO
CREATE TRIGGER tr_refint_project
```

```
ON project
FOR DELETE, UPDATE
   AS
   IF UPDATE(project_no)
   BEGIN
       IF (SELECT COUNT(*)
               FROM works_on, deleted
               WHERE works_on.project_no = deleted.project_no) >0
       BEGIN
ROLLBACK TRANSACTION
PRINT 'Transaction failed!'
END
ELSE PRINT 'Transaction succeeded'
END
CREATE TRIGGER tr_ref_project2
ON works_on
FOR INSERT, UPDATE
 AS
    IF UPDATE(project_no)
 BEGIN
    IF (SELECT project.project_no
            FROM project, inserted
            WHERE project.project_no = inserted.project_no) IS NULL
BEGIN
ROLLBACK TRANSACTION
PRINT 'Transaction failed!'
END
ELSE PRINT 'Transaction succeeded'
END
```

Chapter 15   System Environment of the Database Engine

S.15.1   A temporary table is stored in the **tempdb** system database. To access the table, open SQL Server Management Studio, and follow the path: System Databases | tempdb | Temporary Tables.

S.15.2
```
USE master;
ALTER DATABASE model
   MODIFY FILE (NAME= modeldev,
           MAXSIZE = 4MB);
```

S.15.3   Policy-Based Management manages entities called *managed targets*, which may be server instances, databases, tables, or indices. All managed targets that belong to an instance form a hierarchy. A *target set* is the set of managed targets that results from applying filters to the target hierarchy. A *facet* is a set of logical properties that models the behavior or characteristics for certain types of managed targets. A *condition* is a Boolean expression that specifies a set of allowed states of a managed target with regard to a facet. A *policy* is a condition and its corresponding behavior. A policy belongs to one and only one *category*, which is a group of policies that is introduced to give a user more flexibility in cases where third-party software is hosted.

S.15.4   Conditions can be specified for server instances, databases, tables, and indices.

S.15.5   CLR cannot be disabled because Policy-Based Management requires CLR.

Chapter 16   Backup, Recovery, and System Availability

S.16.1   The benefit of differential backups is that you save time in the restore process, because to recover a database completely, you need a full database backup and only the *latest* differential backup. If you use transaction logs for the same scenario, you have to apply the full database backup and *all* existing transaction logs to bring the database to a consistent state.

A disadvantage of differential backups is that you cannot use them to recover data to a specific point in time, because they do not store intermediate changes to the database.

S.16.2   When you should back up your production database depends on several factors, such as the size of the database, the number of modification operations, and so on. However, you should back up any production database after creating it, after creating indices, after clearing the transaction log, and after performing nonlogged operations.

S.16.3   There is no way to make a differential backup of the **master** database. (You can make only the full database backup of this system database.)

S.16.4   One common technique is to configure database data files on a RAID 0 drive and place the transaction log and backups on a mirrored drive (RAID 1). If the data must be quickly recoverable, use RAID 5 for a database and RAID 1 for the corresponding transaction log(s).

S.16.5   Automatic recovery is done by the system, while manual recovery must be initiated by the system administrator. Also, a statement failure usually requires no manual action of DBA: the system automatically corrects it.

S.16.6   You can use the RESTORE VERIFYONLY statement to verify your backup without displaying information about your backup set. This statement checks the existence of all backup devices (tapes or files) and whether the existing information can be read.

S.16.7   Using the *full* recovery model, no work is lost due to a lost or damaged data file, and you can recover to any point in time. On the other hand, the corresponding transaction log may be very voluminous. Using the *bulk-logged* recovery model, you cannot recover to any point in time, but the corresponding transaction log is smaller, because minimal log space is used by bulk operations. The *simple* recovery model provides the simplest backup strategy, but all changes since the most recent database or differential backup must be manually redone.

S.16.8   Failover clustering provides server redundancy, but doesn't provide database and data file redundancy. Database mirroring provides both data file redundancy and database redundancy, but doesn't provide server redundancy. Log shipping provides database (and file) redundancy, but doesn't provide server redundancy.

Chapter 17   Automating System Administration Tasks

S.17.1   You can automate, among others, the following administrative tasks: data transfer, backing up the database and transaction log, maintaining indices, and checking data integrity.

S.17.2   To back up the transaction log of your database every hour during peak business hours and every four hours during nonpeak hours, you should create one job to back up your transaction log and specify two schedules.

S.17.3   To be notified automatically when the lock wait time is more than 30 seconds, you should create an alert on the **Lock Wait Time** counter of the **Locks** object of Performance Monitor.

S.17.4   A SQL Server error message contains the following parts: a unique error message number, a severity level number, a line number, which identifies the line where the error occurred, and the error text.

S.17.5   The three most important columns of the **sys.messages** catalog view are **message_id**, **severity**, and **text**.

Chapter 18   Data Replication

S.18.1   A primary key is required for data replication to uniquely identify the rows of the published table. All tables using transactional replication must explicitly contain a primary key. The primary key is required to uniquely identify the rows of the published table, because a row is the transfer unit in transactional replication.

S.18.2   You can limit network traffic and/or database size by partitioning tables and/or filtering data that you want to replicate.

S.18.3   You can minimize update conflicts by limiting subscriber update capabilities to an appropriate subset of data.

S.18.4   The Log Reader agent searches for marked transactions and copies them from the transaction log on the publisher to the distribution server. The transactions are stored in the **distribution** database. This agent is used for transactional replications. The synchronization job between all sites is done by the Merge agent. It is used for merge replications. Finally, the Snapshot agent generates the schema and data of the published tables during snapshot replication and stores them on the distribution server.

Chapter 19   Query Optimizer

S.19.1   The nested loop processing technique works by "brute force," meaning that for each row of the outer table, it retrieves and compares each row from the inner table. This technique is very slow if there is no index for the join column of the inner table, because without such an index, the Database Engine would have to scan the outer table once and the inner table n times, where n is the number of rows of the outer table.

The merge join processing technique provides a cost-effective alternative to the nested loop technique. The rows of the joined tables must be physically sorted using the values of the join column. Both tables are then scanned in order of the join columns, matching the rows with the same value for the join columns. This technique has a high overhead if the rows from both tables are unsorted.

The query optimizer usually chooses the nested loop technique if the join column of one of the tables is indexed, because that table is selected as the outer table and does not have to be scanned for each row in the outer table. The query optimizer prefers the hash join method when no indices exist for the join columns, or when the values of the join columns are sorted in advance (always the case when both join columns are primary keys of corresponding tables). Note that it is generally difficult to determine which join technique will be used by the query optimizer.

S.19.2   S.19.1 explains the nested loop technique. The hash join processing technique does not use indices on join columns. It considers both tables that have to be joined as two inputs, the build input and the probe input, and builds the result set by following a four-step process. Therefore, the query optimizer uses the former, when at least one of the join columns is indexed, and the latter technique, when there are no indices for join columns.

Again, it is generally difficult to determine which join technique the query optimizer will choose.

Chapter 20   Performance Tuning

S.20.1   SQL Server Profiler and the Database Engine Tuning Advisor are two complementary tools that are used together to automate tuning processes. The most important feature of SQL Server Profiler that is used by the Database Engine Tuning Advisor is its ability to capture activities in relation to queries. These activities are used as input for the Database Engine Tuning Advisor, which can then recommend several physical objects, such as indices, indexed views, and partitioning schema, that should be created.

S.20.2   Performance Data Collector helps DBAs to track down performance problems. It is used to collect different sets of data related to performance, store this data in the management data warehouse (MDW), and view the collected data using predefined reports. Resource Governor can be used by DBAs to define resource limits and priorities for different workloads, the goal being to achieve consistent performance for processes.

S.20.3

```
SELECT sessions.name AS SessionName, sevents.package as PackageName,
    sevents.name AS EventName,  sevents.predicate, sactions.name AS
ActionName,
    stargets.name AS TargetName
FROM sys.server_event_sessions sessions
      INNER JOIN sys.server_event_session_events sevents
            ON sessions.event_session_id = sevents.event_session_id
      INNER JOIN sys.server_event_session_actions sactions
            ON sessions.event_session_id = sactions.event_session_id
```

```
        INNER JOIN sys.server_event_session_targets stargets
             ON sessions.event_session_id = stargets.event_session_id
WHERE sessions.name = 'sessions1' ;
```

Chapter 21   In-Memory OLTP

There are no end-of-chapter exercises in this chapter.

Chapter 22   Business Intelligence: An Introduction

S.22.1   Online transaction processing (OLTP) systems have short transactions, many (possibly hundreds or thousands of) users, continuous read and write operations, and data of medium size that is stored in a database. Analytic (BI) systems have a small number of users, large size of data stored in a database, and, after the load process, only read operations.

S.22.2   The entity-relationship (ER) model is highly normalized and is suited to design databases with nonredundant data, while the dimensional model is usually denormalized because it uses nonredundant data. Also, the ER model produces very complex database design for large relational databases, while the dimensional model is used for the design of data warehouses and data marts.

S.22.3   Extracting specifies the process of loading source data from multiple, heterogeneous operational systems in a temporary staging area. Data transformation (or "cleansing") is the process of formatting and modifying data that is extracted from various sources to ensure the integrity of data that has to be stored in the target database. Finally, during the loading process, the cleaned data is loaded into the data warehouse.

S.22.4   A dimensional model usually has one fact table that stores measures and many dimension tables (12–20) that describe dimensions. Each dimension table usually has a single-part primary key, whereas the fact table has one composite primary key (combination of the primary keys of all dimension tables). A fact table contains a very large amount of the data stored in a data warehouse (about 70 percent), whereas a dimension table contains a relatively small amount of data. Also, most nonkey columns in a fact table are numeric and additive, whereas columns of dimension tables are strings that contain textual descriptions of the dimension.

S.22.5   The multidimensional OLAP (MOLAP) structure offers the best query performance for data analysis because aggregations and a copy of the base data are stored in a multidimensional structure allowing the high-speed query processor to retrieve data quickly. However, data in MOLAP is duplicated in the cube and consumes the most storage space.

The relational OLAP (ROLAP) structure allows you to use standard Transact-SQL statements to query against the relational tables. It also eliminates data duplication and does not require extra storage space. However, the query performance is not as fast as with the MOLAP and HOLAP structure.

The hybrid OLAP (HOLAP) structure retrieves data in the cube quickly and consumes less storage space than MOLAP.

S.22.6 Aggregation of data stored in a fact table is necessary because a very large amount of the data is stored in a data warehouse. If the low-level (atomic) data from the fact table is not summarized in advance and stored in intermediate tables (*aggregated*), performance of ad hoc queries will be reduced because time- and resource-intensive calculations will be necessary to perform each aggregate function.

Chapter 23   SQL Server Analysis Services

S.23.1   The use of cubes is recommended if the data warehouse contains a lot of data. Therefore, use the Multidimensional model in such a case. The Multidimensional model is recommended for use of Corporate BI solutions, while the Tabular model should be used for Department and Team BI solutions.

S.23.2   The writeback process writes data back to Analysis Services rather than to the relational database system that provides the raw data. This feature is advantageous because when you write data back to a relational database, you have to wait until the cube is processed before the latest data becomes available. However, when you enable the writeback process, you can submit data into the cube in the current session, making it instantly visible to other users of the Analysis Services database. It is important for what-if analysis, forecasting, and financial planning.

S.23.3   Multidimensional Expressions (MDX) and Data Analysis Expressions (DAX) are both analytical languages. MDX is used in the Multidimensional model to query multidimensional data stored in OLAP cubes, while DAX is used in the Tabular model to create custom calculations in SSAS projects. Data Analysis Expressions (DAX) is a formula language used to create custom calculations in SSAS Tabular model projects. DAX formulas include functions, operators, and values to perform advanced calculations on data in tables and columns. DAX is much easier to use and learn than MDX.

Chapter 24   Business Intelligence and Transact-SQL

S.24.1

a.   Using the window construct:

```
USE sample;
SELECT dept_name,    budget,
       AVG( emp_cnt ) OVER( PARTITION BY dept_name ) AS emp_cnt_avg
        FROM project_dept
       WHERE dept_name = 'Accounting';
```

b.   Using the GROUP BY clause:

```
USE sample;
SELECT dept_name,
       AVG( emp_cnt ) AS emp_cnt_avg
        FROM project_dept
       WHERE dept_name = 'Accounting'
       GROUP BY dept_name;
```

S.24.2

```
USE sample;
SELECT date_month, dept_name, budget
FROM (SELECT date_month, dept_name, MAX(budget)
 OVER (PARTITION BY date_month) max_budget_dept, budget
 FROM project_dept) part_deptname
  WHERE budget = max_budget_dept;
```

S.24.3

```
USE sample;
SELECT dept_name, budget, SUM(emp_cnt) sum_of_empcnt
     FROM project_dept
     GROUP BY CUBE (dept_name, budget);
```

S.24.4

```
USE sample;
SELECT dept_name, budget, SUM(emp_cnt) sum_of_empcnt
     FROM project_dept
     GROUP BY ROLLUP (dept_name, budget);
```

The result set of Exercise 23.4 is smaller than the result set of Exercise 23.3 (13 rows vs. 22 rows). The reason is that only one column is used to build summary rows.

S.24.5

```
USE sample;
SELECT dept_name, emp_cnt
   FROM (SELECT dept_name, emp_cnt,
   RANK() OVER( ORDER BY emp_cnt desc ) AS rank
   FROM project_dept) part_dept
        WHERE rank <=3;
```

S.24.6

```
USE sample;
SELECT TOP(3) dept_name, emp_cnt
    FROM project_dept
   ORDER BY emp_cnt DESC;
```

S.24.7

```
USE sample;
SELECT dept_name, emp_cnt, RANK() OVER (PARTITION BY year(date_month)
 ORDER BY emp_cnt desc) AS rnk,
       DENSE_RANK() OVER (PARTITION BY year(date_month) ORDER BY emp_cnt
 desc) AS dense_rnk,
     ROW_NUMBER() OVER (PARTITION BY year(date_month) ORDER BY
 emp_cnt desc ) AS row_no, date_month
     FROM project_dept
         WHERE year(date_month) = 2008;
```

Chapter 25   SQL Server Reporting Services

S.25.1   The query for this report is as follows:

```
SELECT employee.emp_no, emp_lname
    FROM employee JOIN works_on
    ON employee.emp_no = works_on. emp_no
      AND job = 'Clerk';
```

S.25.2   The query for this report is as follows:

```
SELECT budget, project_name
  FROM project JOIN works_on ON project.project_no = works_on. project_no
    JOIN employee ON employee.emp_no = works_on.emp_no
    JOIN department ON department.dept_no = employee.dept_no
  WHERE dept_name = 'Research
  AND emp_no < 25000;'
```

Again, create the report in Visual Studio by using the Report Server Project template or the Report Server Project Wizard and choosing Tabular as the report type.

Chapter 26   Optimizing Techniques for Data Warehousing

There are no end-of-chapter exercises in this chapter.

Chapter 27   Columnstore Indices

S.27.1

```
USE sample;
-- Copy all rows from the FactInternetSales table
-- in the new and corresponding table in the sample database.
-- All existing indices will be ignored.
SELECT * INTO FactInternetSales1
    FROM AdventureWorksDW.dbo.FactINternetSales;
-- Use STATISTICS TIME ON statement to measure
-- CPU time and elapsed time of the query
    SET STATISTICS TIME ON;
    GO
    SELECT AVG(UnitPrice)
    FROM FactInternetSales1;
    GO
    SET STATISTICS TIME OFF;
```

S.27.2

```
    USE sample;
SELECT * INTO FactInternetSales2
    FROM AdventureWorksDW.dbo.FactINternetSales;
    GO
    CREATE CLUSTERED COLUMNSTORE INDEX
```

```
    ci2_factinternetsales2 ON FactINternetSales2;

    SET STATISTICS TIME ON;
  GO
  SELECT AVG(UnitPrice)
  FROM FactInternetSales2;
  GO
  SET STATISTICS TIME OFF;
```

S.27.3   The differences between elapsed time and CPU time in S.27.1 and S.27.2 depend on the machine you use.

Chapter 28   Intelligent Query Processing

There are no end-of-chapter exercises in this chapter.

Chapter 29   JSON Integration in the Database Engine

There are no end-of-chapter exercises in this chapter.

Chapter 30   Spatial and Temporal Data

There are no end-of-chapter exercises in this chapter.

Chapter 31   SQL Server: Graph Database

S.31.1
```
-- create a node table called restaurant
USE graph_db;
CREATE TABLE restaurant
  (ID INT NOT NULL PRIMARY KEY,
  name VARCHAR(20), city VARCHAR(20)) AS NODE;
```

S.31.2
```
  -- Insert three rows in restaurant
  INSERT INTO restaurant VALUES (1, 'A', 'Hoboken');
  INSERT INTO restaurant VALUES (2, 'B', 'Lexington');
  INSERT INTO restaurant VALUES (3, 'C', 'Madison');
```

S 31.3
```
  -- create an edge table called visits
  CREATE TABLE visits (rating INT
      CONSTRAINT Emp_rates CONNECTION (Employee to Restaurant)) AS EDGE;
```

S.31.4
```
USE graph_db;
INSERT INTO Visits  VALUES ((SELECT $node_id FROM Employee WHERE id = 1),
      (SELECT $node_id FROM Restaurant WHERE id = 1), 5);
```

```
INSERT INTO Visits VALUES ((SELECT $node_id FROM Employee WHERE id = 2),
      (SELECT $node_id FROM Restaurant WHERE id = 2), 3);
INSERT INTO Visits VALUES ((SELECT $node_id FROM Employee WHERE id = 3),
      (SELECT $node_id FROM Restaurant WHERE id = 3), 2);
INSERT INTO Visits VALUES ((SELECT $node_id FROM Employee WHERE id = 4),
      (SELECT $node_id FROM Restaurant WHERE id = 3), 3);
INSERT INTO Visits VALUES ((SELECT $node_id FROM Employee WHERE id = 5),
      (SELECT $node_id FROM Restaurant WHERE id = 3), 1);
INSERT INTO Visits VALUES ((SELECT $node_id FROM Employee WHERE id = 6),
      (SELECT $node_id FROM Restaurant WHERE id = 2), 4);
```

S.31.5

```
USE graph_db;
SELECT restaurant.name
 FROM Employee, Visits, Restaurant
 WHERE MATCH (Employee - (Visits) -> Restaurant)
   AND employee.name = 'Ann Jones';
     (Result: 'B')
```

S.31.6

```
USE graph_db;
SELECT DISTINCT restaurant.name
 FROM Employee E1, Employee E2, Is_Liked, Visits, Restaurant
 WHERE MATCH (E1 - (Is_Liked) ->E2 - (Visits) -> Restaurant)
   AND E1.name = 'Ann Jones';
   (Result: 'C' , 'B')
```

Chapter 32   SQL Server Machine Learning Services: R Support

There are no end-of-chapter exercises in this chapter.

Chapter 33   SQL Server Machine Learning Services: Python Support

There are no end-of-chapter exercises in this chapter.

Index

9 781260 4588